PHYSICS OF MAGNETOSPHERIC SUBSTORMS

ASTROPHYSICS AND SPACE SCIENCE LIBRARY

A SERIES OF BOOKS ON THE RECENT DEVELOPMENTS OF SPACE SCIENCE AND OF GENERAL GEOPHYSICS AND ASTROPHYSICS PUBLISHED IN CONNECTION WITH THE JOURNAL SPACE SCIENCE REVIEWS

VOLUME 47

SYUN-ICHI AKASOFU
Geophysical Institute, University of Alaska, Fairbanks, Alaska 99701, U.S.A.

PHYSICS OF MAGNETOSPHERIC SUBSTORMS

D. REIDEL PUBLISHING COMPANY
DORDRECHT-HOLLAND/BOSTON-U.S.A.

Library of Congress Cataloging in Publication Data

Akasofu, Syun-Ichi.
Physics of magnetospheric substorms.

(Astrophysics and space science library; v. 47)
Includes bibliographies and index.
1. Magnetospheric substorms. 2. Auroral substorms.
3. Plasma (Ionized gases) I. Title. II. Series.
QC809.M35A37 538'.766 76–45158
ISBN 90–277–0748–0

Published by D. Reidel Publishing Company,
P.O. Box 17, Dordrecht, Holland

Sold and distributed in the U.S.A., Canada and Mexico
by D. Reidel Publishing Company, Inc.
Lincoln Building, 160 Old Derby Street, Hingham,
Mass. 02043, U.S.A.

Printed in The Netherlands

TABLE OF CONTENTS

PREFACE xiii

ACKNOWLEDGEMENTS xv

LIST OF FREQUENTLY USED SYMBOLS xvii

INTRODUCTION 1

CHAPTER 1 OPEN MAGNETOSPHERE AND THE AURORAL OVAL 13

1.1. Open Magnetosphere 13
1.2. Auroral Oval 15
1.3. Open Magnetosphere and the Auroral Oval 19
1.3.1. Solar Energetic Particles and the Auroral Oval 19
1.3.2. Field-Aligned Currents and the Auroral Oval 21
(a) Distribution of Field-Aligned Currents above the Polar Ionosphere 21
(b) Field-Aligned Currents and Auroral Arcs 24
1.3.3. Field-Aligned Currents and Ionospheric Currents 27
(a) Formulation 27
(b) Model 28
(c) Results 29
1.3.4. Electric Field Distribution and the Auroral Oval 31
(a) Observations 31
(b) Theoretical Studies 34
1.3.5. Plasma Convection in the Equatorial Plane and the Polar Cap 35
(a) Observations 37
(b) Theoretical Studies 41
1.3.6. Three-Dimensional Magnetospheric Current System 44
(a) Fejer-Swift-Vasyliunas-Wolf (FSVW) Model 44
(b) Boström-Rostoker (BR) Model 45
1.3.7. S_q^p Variation and the Auroral Oval 48
1.4. Solar Wind – Magnetosphere Dynamo 53
1.4.1. Electromotive Force and Power 53

1.4.2. Steady State Merging 55
(a) Sweet-Parker's Model 56
(b) Petschek's Model 58
(c) Sonnerup's Model 58
1.4.3. Production Rate of the Open Flux 61
References 62

CHAPTER 2 AURORAS AND AURORAL PARTICLES 71

2.1. Introduction 71
2.1.1. Montage Photographs of the Auroral Oval 71
2.1.2. Schematic Distribution Pattern of Auroras 75
2.2. Auroras in Different Local Time Sectors 76
2.2.1. Discrete Auroras and Diffuse Auroras in the Night Sector 76
2.2.2. Midday Auroras 83
2.2.3. Polar Cap Auroras 87
2.3. Auroral Electrons: The Statistical Precipitation Pattern 91
2.4. Auroral Electrons: Spectra of Auroral Electrons 93
2.4.1. Evening and Midnight Sectors 93
(a) Satellite Observations 93
(b) Rocket Observations 104
2.4.2. Morning Sector 107
2.4.3. Noon (Cusp) Sector 108
2.4.4. Polar Cap 109
2.5. Auroral Electrons and Field-Aligned Currents 111
2.5.1. Satellite Observations 111
2.5.2. Rocket Observations 111
2.6. Auroral Particles and Atmospheric Emissions 112
2.7. Auroral Protons 116
2.8. Auroral Helium Ions (He^{++}, He^{+}) and Oxygen Ions (O^{+}) 117
2.8.1. Auroral Helium Ions 117
2.8.2. Auroral Oxygen Ions 119
2.9. Auroral Oval and the Polar Ionosphere 120
2.9.1. F Region 120
2.9.2. E Region 126
2.10. Summary 127
References 128

CHAPTER 3 DISTRIBUTION OF PLASMAS IN THE MAGNETOSPHERE 137

3.1. Five Plasma Domains 137
3.2. Plasma Mantle 139
3.2.1. Frontal Region of the Magnetosphere 140
3.2.2. Magnetotail ($X > -30\,R_E$) 141

3.2.3. Magnetotail at the Lunar Distance ($X \simeq -60\ R_E$) 143
3.2.4. Distant Magnetotail ($X \sim -500 \sim -1000\ R_E$) 145
3.3. Polar Cusp 146
3.4. Plasma Sheet 146
3.4.1. Plasma Sheet During Very Quiet Periods 147
3.4.2. Average Energy Characteristics 151
(a) $-10\ R_E > X > -30\ R_E$ 151
(b) Plasma Sheet at the Lunar Distance ($X \simeq -60\ R_E$) 151
3.5. Origin and Dynamics of the Sheet Plasma 153
3.5.1. Meridional Convection of Magnetospheric Plasma 153
3.5.2. Electric Currents, Magnetic Field Structure and the Balance of Stresses in the Plasma Sheet 154
(a) Two-Dimensional Maxwell-Vlasov and Macroscopic Approaches 154
(b) Kinetic Approaches 155
3.6. Van Allen Belts 160
3.6.1. Ring Current Belt: the Quiet Time Belt 160
3.6.2. Electron Belt 162
3.7. Plasmasphere 163
3.8. Magnetospheric Plasmas and Auroral Particles 165
3.8.1. Oval Belt 165
3.8.2. Annular Belt 166
3.9. Acceleration Processes of Arc-Producing Auroral Electrons 166
3.9.1. Introduction 166
3.9.2. Auroral Arcs and Their Topological Relation with the Magnetospheric Structure 170
3.9.3. Arc Energization System 172
3.9.4. Possible Processes in the Wave-Particle Interaction Region 176
(a) Current Driven Instabilities 176
(b) Double Layer 178
References 180

CHAPTER 4 RESPONSES OF THE MAGNETOSPHERE TO INTERPLANETARY DISTURBANCES 190

4.1. Interplanetary Disturbances 190
4.2. Interplanetary Pressure Disturbances and Magnetospheric Responses 190
4.3. Changes of the IMF EW Component and Magnetospheric Responses 195
4.3.1. Merging of the Geomagnetic Field Lines with the IMF EW Component 195
4.3.2. Observations 199
4.4. Changes of the IMF NS Component and Magnetospheric Responses 206
4.4.1. Introduction 206
4.4.2. Merging with the IMF of an Arbitrary Angle 212
4.4.3. Erosion of the Dayside Magnetosphere 215

(a) Magnetopause Motion 215
(b) Cusp Motion 220
4.4.4. Magnetic Flux Transfer to the Magnetotail 222
(a) High Latitude Lobe Field B_T 222
(b) Radius of the Magnetotail R_T 229
(c) Plasma Sheet Thinning 230
4.4.5. Enhanced Convection in the Plasma Sheet 232
(a) Earthward Advance of the Plasma Sheet 232
(b) Tendency Toward the Tail-Like Field 235
(c) Slow Decrease of the H Component at the Synchronous Distance and in Low Latitudes 235
4.4.6. Polar Cap Phenomena 239
(a) Dawn-to-Dusk Electric Field in the Polar Region 239
(b) Expansion of the Auroral Oval 241
(c) DP-2 Variation 243
4.4.7. Correlation Between B_s ($= -B_z$) and the AE Index 247
(a) Arnoldy's Study of the Relationship between B_s and the AE Index 247
(b) The AE Index 249
4.4.8. Summary 256
References 257

CHAPTER 5 MAGNETOSPHERIC SUBSTORMS: INTRODUCTION 263

5.1. A New Classification of Magnetospheric Disturbances 264
5.1.1. Reversible or Quasi-Reversible Disturbances 264
5.1.2. Irreversible Disturbances 265
5.2. Substorm Energy ε_Σ and Substorm Function $\Sigma = \Gamma(\Phi_D)$ 265
5.2.1. Ground State of the Magnetosphere 265
5.2.2. Substorm Function $\Sigma = \Gamma(\Phi_D)$ 266
5.3. Substorm Intensity 273
5.3.1. Kinetic Energy of Auroral Particles 275
5.3.2. Joule Heat Energy of the Auroral Electrojet 276
5.4. Time-Dependent Merging 279
5.4.1. Stability of the Plasma Sheet 279
5.4.2. Numerical Simulation of Reconnection 280
5.4.3. Plasma Processes 281
5.4.4. Substorm Time Constants 288
5.4.5. Short Review of Theories of the Magnetospheric Substorm 289
5.5. Magnetospheric Substorms 291
5.5.1. Neutral Line Formation – Enhanced Reconnection or Plasma Sheet Deflation – Enhanced Reconnection 291
5.5.2. Enhancement of the Auroral Oval Circuit Current 292

5.5.3. Penetration of the Convection Electric Field into the Inner Magnetosphere and the Resulting Plasma Injection 293
5.6. Geomagnetic and Magnetospheric Storms 294
References 297

CHAPTER 6 MAGNETOTAIL PHENOMENA DURING MAGNETOSPHERIC SUBSTORMS 300

6.1. Introduction 300
6.2. **B** Vector Dipping 301
6.3. Plasma Sheet Thinning 315
6.3.1. Profile of Thinning 317
(a) Z-axis Dependence 317
(b) Profile in the Y-Z Plane 320
(c) Profile in the X-Y Plane 321
(d) Profile in the X-Z Plane 322
6.3.2. Timing of Thinning 323
6.3.3. Thinning and the B_z Component Reversal 327
6.3.4. Summary 330
6.4. Magnetotail Field B_T and Radius R_T 332
6.4.1. Decrease of the Magnetotail Lobe Field B_T 332
6.4.2. Decrease of the Radius of the Magnetotail R_T 333
6.5. Auroral Bulge 334
6.5.1. Auroral Bulge 335
6.5.2. Auroral Particles in the Bulge 340
6.5.3. Increase of the Magnetic Field Component B_z 343
6.5.4. Increase of Cosmic Ray Proton Cut-off 345
6.6. Plasma Sheet Expansion 345
6.7. Plasma Flow 348
6.7.1. Vela Satellite Observations 348
6.7.2. IMP-6 Satellite Observations 353
6.7.3. Flow Reversal 356
6.8. Other Important Magnetotail Phenomena 362
6.8.1. Brief Appearance of Energetic Electrons in the Plasma Sheet 362
6.8.2. Leakage of Plasma Sheet Particles into the Magnetosheath 364
6.8.3. 'Geomagnetic Storm Particles' 367
6.8.4. Dawn-Dusk Asymmetry of the Proton and Electron Distributions 369
(a) Production of Sub-Relativistic Protons and Electrons and Their Dawn-Dusk Asymmetry 369
(b) Dawn-Dusk Asymmetry of Precipitating Protons and Electrons 370
(c) Summary 371

6.8.5. Summary of the Magnetotail Observations 371
(a) Plasma Sheet Thinning 372
(b) Plasma Sheet Expansion 372
(c) Magnetic Field Variations 373
References 374

CHAPTER 7 MAGNETOSPHERIC CURRENTS DURING SUBSTORMS 381

7.1. Introduction 381
7.2. Field-Aligned Currents 382
7.2.1. Observations in the Magnetotail 382
7.2.2. Observations above the Ionosphere 383
7.2.3. Ground Observations 384
7.2.4. Positive Bays in Low Latitudes and Positive B_z Variations at the Synchronous Distance and in the Magnetotail 390
7.2.5. Modeling the Magnetospheric Substorm in Terms of the Substorm Current System 393
(a) Model 393
(b) Positive B_z Changes 394
(c) Poleward Shift of the Feet of the Geomagnetic Field Lines 394
(d) Equatorward Shift of the Cusp 394
(e) Summary 394
7.3. Field-Aligned Currents and the Auroral Electrojets 396
7.3.1. Introduction 396
7.3.2. Examples 396
(a) 1973, March 6 396
(b) 1973, March 9 398
7.3.3. Relative Location of the Auroral Electrojets with respect to Field-Aligned Currents in the Evening Sector: A Statistical Result and Summary 399
7.3.4. Model Calculation 402
7.4. Auroral Electrojets 405
7.4.1. Auroral Electrojet and Global Auroral Features 405
(a) Examples 405
(b) Summary 411
7.5. Cross-Section of the Electrojets 414
7.5.1. Latitudinal Profile of the Three-Component Changes 414
7.5.2. Development of the Electrojets 417
(a) A Substorm on 1970, June 15 417
(b) A Substorm on 1970, July 14 420
(c) A Substorm on 1971, December 23 421
7.5.3. Radar Auroras 424
7.6. Latitudinal Cross-Section of the Auroral Electrojet and its Relation to the Interplanetary Magnetic Field Polarity 428

7.7. Ionospheric Currents and Electric Fields 432
7.7.1. Model Study 433
7.7.2. Chatanika Radar Observations 435
(a) Electric Field, Current and Conductivity 435
(b) North-South Current and the *D* Component 441
(c) Electric Field and Auroral Activity 447
7.7.3. Balloon Observations 449
7.7.4. Barium Cloud Observations 452
(a) Auroral Oval 452
(b) Polar Cap 452
7.7.5. Rocket Observations 453
7.8. Thermospheric and Ionospheric Disturbances 453
References 462

CHAPTER 8 PENETRATING CONVECTION ELECTRIC FIELD, PLASMA INJECTION AND PLASMASPHERE DISTURBANCES 473

8.1. Introduction 473
8.2. Penetration of the Convection Electric Field into the Inner Magnetosphere and the Resulting Plasma Injection 474
8.2.1. Observations 474
(a) Plasma Injection at the Geosynchronous Distance 474
(b) Enhancement of the Westward Electric Field in the Plasmasphere 478
8.2.2. Theoretical Estimates 479
8.3. Relationship between Particles at the Geosynchronous Distance and Auroral Activity near the Geomagnetically Conjugate Point 486
8.4. 'The Fault Line' 498
8.4.1. Dawnside of the Fault Line 498
8.4.2. Duskside of the Fault Line 501
8.5. Drift Motions 501
8.5.1. Protons 501
(a) Satellite Observations of Drifting Protons 501
(b) Precipitating Protons 512
(c) IPDP Pulsations 514
8.5.2. Energetic Electrons 518
(a) Satellite Observations of Drifting Electrons 518
(b) Ground-Based Observations of Drifting Electrons 524
8.6. Deformation of the Plasmasphere and Associated Ionospheric Disturbances 533
References 537

CHAPTER 9 SOLAR-TERRESTRIAL RELATIONS AND MAGNETOSPHERIC SUBSTORMS 548

9.1. Interplanetary Disturbances 548
9.1.1. Basic Solar-Interplanetary Magnetic Field Structure 548
9.1.2. High Speed Solar Wind Streams and Geomagnetic Disturbances 551
9.1.3. Transient Solar Activities and Associated Interplanetary Disturbances 554
9.2. Morphological Model of Magnetospheric Substorms 558
9.2.1. Basic Requirements for Models 558
9.2.2. Description of a Model 561
(a) Deflation of the Plasma Sheet 565
(b) Magnetic Energy Conversion 575
9.2.3. Critical Tests and Unsolved Problems 579
9.3. Concluding Remarks 584
References 585

INDEX OF NAMES 588

INDEX OF SUBJECTS 596

PREFACE

Man, through intensive observations of natural phenomena, has learned about some of the basic principles which govern nature. The aurora is one of the most fascinating of these natural phenomena, and by studying it, man has just begun to comprehend auroral phenomena in terms of basic cosmic electrodynamic processes.

The systematic and extensive observation of the aurora during and after the great international enterprise, the International Geophysical Year (IGY), led to the concept of the auroral substorm. Like many other geophysical phenomena, auroral displays have a dual time (universal- and local-time) dependence when seen by a ground-based observer. Thus, it was a difficult task for single observers, rotating with the Earth once a day, to grasp a transient feature of a large-scale auroral display. Such a complexity is inevitable in studying many geophysical features, in particular the polar upper atmospheric phenomena. However, it was found that their complexity began to unfold when the concept of the auroral substorm was introduced. In a book entitled *Polar and Magnetospheric Substorms*, the predecessor to this book, I tried to describe the auroral phenomena as completely as possible in terms of the concept of the auroral substorm. At that time, the first satellite observations of particles and magnetic fields during substorms were just becoming available, and it was suggested that the auroral substorm is a manifestation of a magnetospheric phenomenon called the magnetospheric substorm.

Thanks are due to all my colleagues who have remarkably advanced the understanding of magnetospheric substorm processes during the last decade. There is now a challenge to synthesize the vast amount of satellite data by extending the concept of the auroral substorm and to unveil the cause of the substorm. This task has provided a serious test of the validity of the magnetospheric substorm concept. In this surge of progress during the last decade, there has also been some inevitable confusion. For these reasons, I have realized the need for a unified treatment of the subject and clarification by selecting those topics which I feel are most important to understanding the *cause* of magnetospheric substorms.

It is quite obvious that it is presumptuous for one person to attempt to cover the entire subject; however, a single author can, at least, present a unified treatment of a complex subject. This is especially true in this particular field where one has to deal with satellite data observed at different times and locations for different substorms, analyzed by different workers. In my treatment I frequently depart from what appear to be the more widely held views on certain topics. If in places there seems to be a bias unsupported by objective arguments, I can only apologize and promise to consider all suggestions for improvement.

It is worthwhile to mention that, on the basis of what has been learned about terrestrial magnetospheric substorms, a better understanding of certain phenomena in the solar system and beyond is emerging. Indeed, the magnetosphere of the Earth is the only region where man, in detail and by direct means, can study particular plasma processes in the natural environment. Therefore, *unless we can fully understand the terrestrial magnetospheric processes, interpretations of many plasma processes in planetary, solar and cosmic conditions would have to remain speculative and obscure.* For this reason, I believe that *a study of magnetospheric processes provides an important foundation for pursuing many cosmic electrodynamic processes.*

Throughout this book, I have tried to emphasize that magnetospheric phenomena are not the result of controlled experiments; rather they are natural occurrences having many uncontrollable parameters. Therefore, it is not necessarily a simple matter to establish a functional relation between two quantities, since quantities other than those chosen may be constantly changing. The complexity of dealing with the multi-functional relation must be overcome in studying many geophysical phenomena; this book presents, explicitly and implicitly, a methodology for doing it. In this respect, it is my hope that the general methodology developed in the preceding and present book will be useful in studying other geophysical phenomena as well.

September 1, 1976 SYUN-ICHI AKASOFU

ACKNOWLEDGMENTS

The chief inspiration for this work was the late Sydney Chapman, who not only pioneered the field of solar-terrestrial physics but led it for half a century. I would also like to thank Professor de Jager for his encouragement in writing this book.

I have a great many acknowledgments to make: This book would have been impossible without the help of a large number of colleagues who have shared the excitement triggered by many recent findings that have greatly increased our understanding of magnetospheric substorms.

Among the many who have stimulated and actively assisted me in writing this book, I gratefully thank my former students, Drs C.-I. Meng, K. Kawasaki, A. L. Snyder, F. Yasuhara, P. D. Perreault, T. Berkey and my own colleagues, past and present, at the Geophysical Institute. Most recently, Drs Y. Kamide, J. R. Kan, A. T. Y. Lui and J. Chao have worked closely with me in developing a new model of magnetospheric substorms. In particular, Chapter 9 was prepared after an intensive workshop was held at the Geophysical Institute on July 19–23, 1976, in which some of them participated.

It is also a great pleasure to thank my colleagues for their stimulating discussions during the preparation of this book; among them are Drs H. Alfvén, C. D. Anger, W. I. Axford, P. M. Banks, M. Baron, R. Boström, A. Brekke, J. L. Burch, D. Carpenter, F. V. Coroniti, T. N. Davis, J. W. Dungey, D. H. Fairfield, C. G. Fälthammar, Y. I. Feldstein, L. A. Frank, D. Gurnett, G. Haerendel, W. J. Heikkila, E. W. Hones, Jr., B. Hultqvist, C. F. Kennel, K. Lassen, C. E. McIlwain, R. L. McPherron, F. S. Mozer, T. Nagata, A. Nishida, C. G. Park, G. Parks, C. P. Pike, M. I. Pudovkin, D. Rossberg, G. Rostoker, K. Schindler, G. L. Siscoe, T. W. Speiser, V. A. Troitskaya, J. A. Van Allen, V. M. Vasyliunas, D. Venkatesan, J. R. Winckler and J. D. Winningham.

I would like to thank the colleagues who have provided me with many of the figures I used, and also the editors and publishers for their kind permission to reproduce these illustrations. Sources of figures are cited in the figure captions.

The greatest gratitude must go to the support given to my study of magnetospheric substorms by the Atmospheric Sciences Section of the National Science Foundation, the National Aeronautics and Space Administration, the Energy Research and Development Administration and the U.S. Air Force.

Special thanks are also due to Ms L. Rastall for her typing, Ms A. Swift, G. Brown and P. Woods for their careful editing of the manuscript and to Ms H. Horiuchi for drafting, and to the D. Reidel Publishing Company for its special care in the publication of this book. Finally, I would also like to thank my wife and children who have shared with me a long arctic life.

LIST OF FREQUENTLY USED SYMBOLS

AE	auroral electrojet index
$\boldsymbol{B}$	magnetic induction
B_P	magnetic field in the ionosphere
B_T	magnetic field in the high latitude lobe of the magnetotail
D	declination component of the Earth's magnetic field
dp. lat.	dipole latitude
$\boldsymbol{E}$	electric field
e	unit charge
H	horizontal component of the Earth's magnetic field
I	ionospheric current intensity
IMF	interplanetary magnetic field
inv. lat.	invariant latitude
j	electric current density
$j_{\parallel}$	field-aligned current density
J	electric current intensity
Kp	three-hourly planetary magnetic index
L	McIlwain's L parameter
L	length of the magnetotail
LT	local time
MLT	magnetic local time
m	mass of a particle
n	number density of particles
p	pressure
R_E	Earth's radius
R_T	radius of the magnetotail
S_q^p	solar magnetic daily variation in the polar region on a very quiet day
T	substorm time
T	temperature
t	time
UT	universal time
V_A	Alfvén wave speed
V_s	solar wind speed
v	velocity of a particle
V	$(E \times B)$ drift velocity
(X, Y, Z)	solar magnetospheric coordinates
(x, y, z)	Cartesian coordinates
Z	vertical component of the Earth's magnetic field

ε	energy of a particle
γ	10^{-5} G
μ	magnetic moment of a particle
Φ	electric potential
Φ_D	dayside merging rate of the interplanetary magnetic field lines with the geomagnetic field lines
Φ_N	nightside reconnection rate of the open field lines
Σ	substorm function
Σ	height integrated conductivity
Σ_P	Pedersen conductivity
Σ_H	Hall conductivity
Ω	gyro-(or cyclotron) frequency
λ	latitude
ψ	longitude

INTRODUCTION

The concept of magnetosphere has evolved considerably during the last decade. It was first introduced by Chapman and Ferraro (1931), who proposed that the Earth and its magnetic field were confined *temporarily* in a *cavity* which was formed during the passage of a hot gas from the Sun (Figure 1). However, when the presence of a continuous flow of solar plasma, the solar wind, was inferred by Biermann (1957), theorized by Parker (1958) and subsequently confirmed by space probes, it became apparent that the magnetosphere is a *permanent* feature of the Earth.

The shape of the magnetosphere had originally been thought to be a very long cylindrical cavity. In about 1960, however, it was suggested that the pressure associated with thermal motion of solar wind particles would close the cavity at a geocentric distance of about 20 R_E, forming a teardrop-shaped cavity. On the other hand, Piddington (1960) proposed that the magnetosphere has a long cylindrical tail, and its presence was confirmed by Ness (1965).

Meanwhile, the magnetic field topology of the magnetosphere has also evolved. In the Chapman-Ferraro theory, the solar plasma was assumed to be diamagnetic. Thus, the Earth's magnetic field was thought to be confined completely in the cavity. The assumption of diamagnetic solar plasma was criticized by Alfvén (1939, 1950), but it was Dungey (1961) who put forward the concept of the so-called 'open model' in which the geomagnetic field lines from the polar region are merged with the interplanetary magnetic field (IMF) lines; the merged field lines are called 'open field lines'.

The most important aspect of the open model is that it allows for the solar wind plasma to blow across the open field lines, providing the electromotive force and electrical power for most of the magnetospheric phenomena which we shall be concerned with in this book. That is, the magnetosphere is an MHD dynamo which converts the kinetic energy of solar wind plasma into electrical energy (Figure 2). For example, the aurora is an electrical discharge phenomenon which is powered by the solar wind-magnetosphere dynamo. Indeed, Zmuda and Armstrong (1974) showed that there is an inflow of electric current to the poleward half of the oval in the morning sector and an outflow from the poleward half of the oval in the evening sector. A significant part of the electrical energy is eventually converted into heat energy in the polar ionospheric plasma.

In the simplest situation (when the IMF vector is directed southward), a magnetic neutral line surrounds the magnetosphere (Figure 3). It is then possible to imagine a surface that consists of geomagnetic field lines which cross the neutral line. This surface separates the region of the open field lines from that of the closed field lines. Further, the intersection line between this surface and the ionosphere

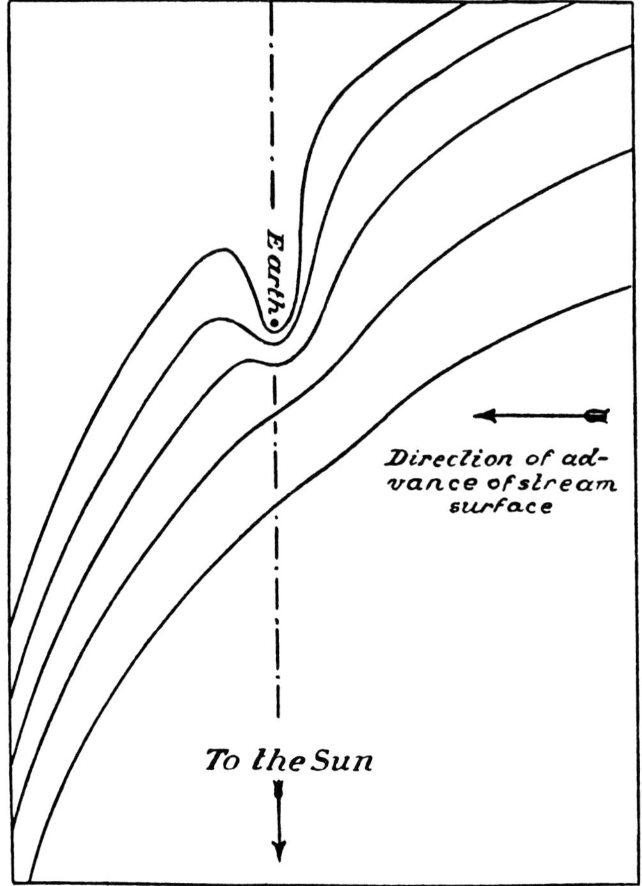

Fig. 1. Formation of a cavity around the Earth in a narrow beam of plasma (the M stream) from the Sun. (Chapman, S. and Ferraro, V. C. A.: *Terr. Mag. Atmosph. Elect.* **36**, 77, 1931.)

approximately coincides with the instantaneous belt of auroras, the auroral oval, which was discovered by Feldstein (1963). Thus the auroral oval is the projection, along the geomagnetic field lines, of the neutral line onto the polar ionosphere and delineates the boundary of the polar cap. It is this oval belt along which various coupling processes between the IMF, the magnetosphere and the ionosphere manifest themselves most conspicuously as the auroral phenomena. These subjects will be reviewed in Chapter 1.

The advent of scanning devices aboard polar orbiting satellites (ISIS-2 and DMSP) has considerably improved our knowledge of the large-scale distribution of auroras, and their morphological characteristics, during the last few years. Characteristics of auroral particles have also been extensively examined by ground-based, airborne, rocket-borne and satellite-borne instruments. It has become apparent that there is a broad oval-shaped region of precipitation. The

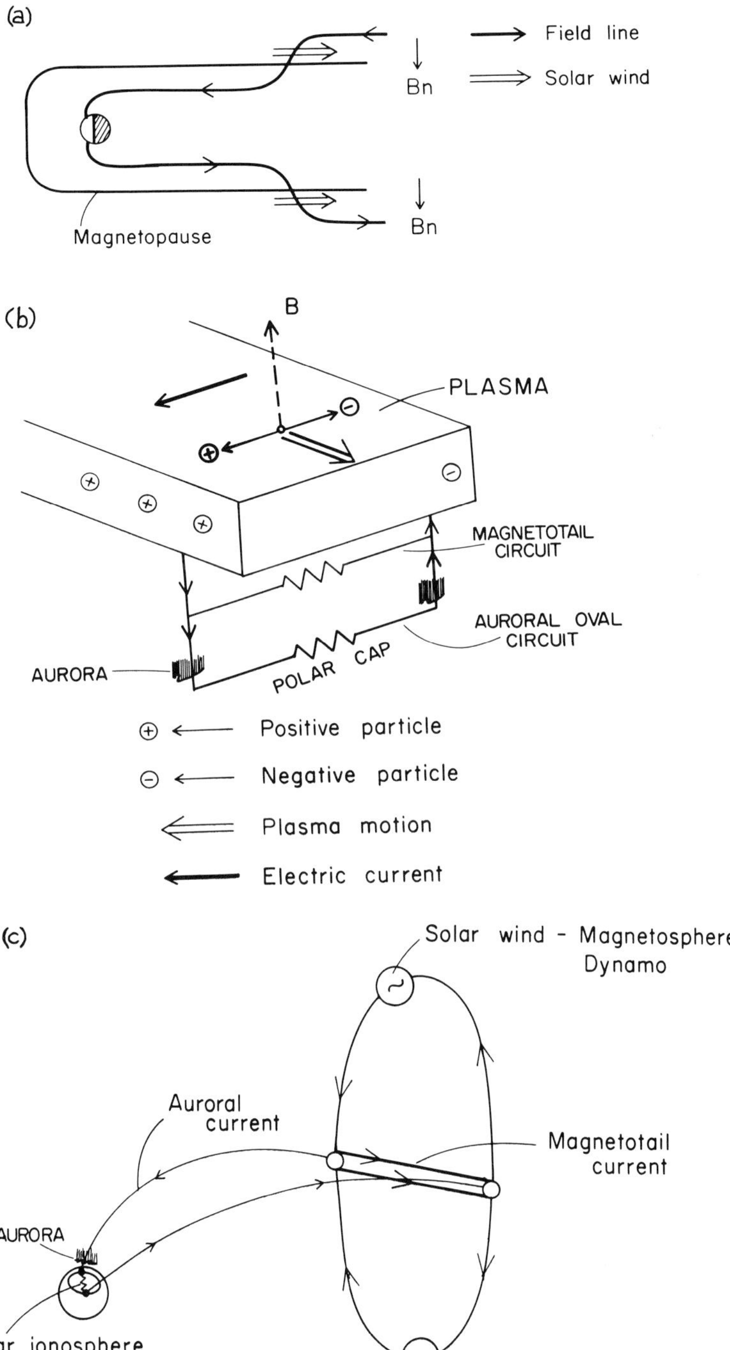

Fig. 2. Schematic diagram showing processes associated with the solar wind-magnetosphere dynamo. The location where the dynamo action takes place is schematically shown in (a). The basic processes associated with the dynamo and the connected circuits are shown in (b) and (c).

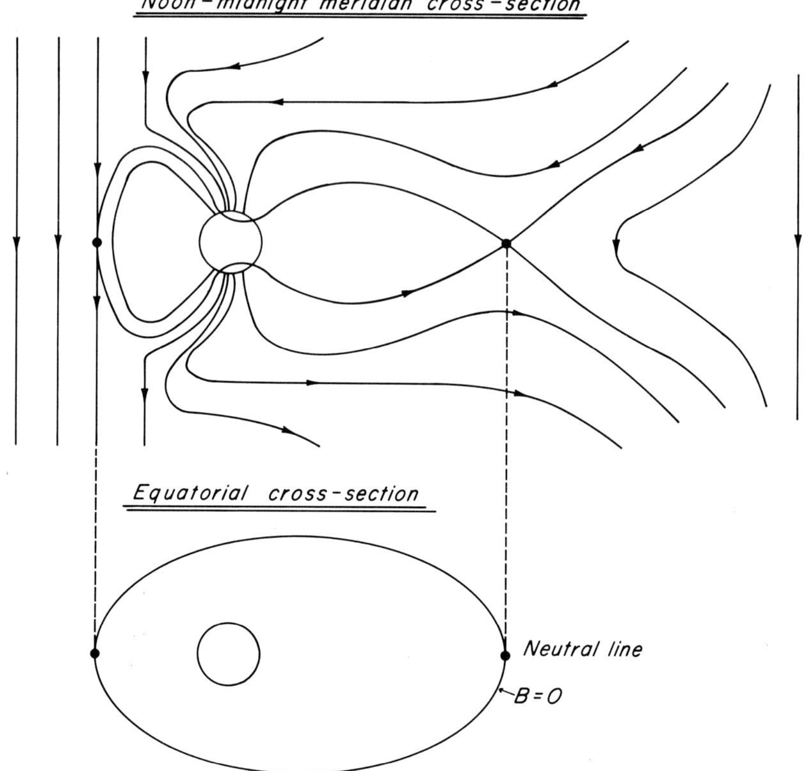

Fig. 3. Topology of the open magnetosphere proposed by J. W. Dungey (the noon-midnight cross-section). When the interplanetary magnetic field has only the southward component, the magnetosphere is bounded by the magnetic neutral line.

precipitating particles have energy (differential) spectra which consist of a Maxwellian and a power law spectrum. This precipitation along the broad oval band causes the diffuse aurora and arises from pitch-angle diffusion processes which result from wave-particle interactions. Within this precipitation region, a different type of precipitation and intense field-aligned currents develop along extremely narrow strips, resulting in discrete auroral arcs. The precipitating particles have a peaked component or the so-called 'monoenergetic' component, as well as a combination of a Maxwellian and a power law spectrum. Rocket observations indicate that those electrons carry a significant part of the intense field-aligned currents. This localized region appears to have a V-shaped potential structure and may be called the arc energization system, in which auroral electrons are accelerated towards the polar upper atmosphere. One of the most challenging problems in magnetospheric physics is to identify the processes associated with the arc energization system which should be an integral part of the magnetosphere-ionosphere coupling system. Chapter 2 deals with these subjects.

It has also been found that the magnetosphere is not an empty cavity carved in the solar wind, but is filled with tenuous plasmas. Five plasma domains of different energy characteristics have so far been identified: the plasma mantle, the plasma sheet, the cusp region, the Van Allen belts, and the plasmasphere. In particular, it has been revealed that the solar wind (or more accurately, the magnetosheath plasma) blows through the high latitude lobe region of the magnetotail, rather than blowing around a blunt-nosed obstacle, the magnetosphere.

The magnetospheric plasmas undergo a large-scale motion. The concept of the convective motion of magnetospheric plasmas was originally proposed by Axford and Hines (1961). It was then thought that a complete convective or circulatory motion of plasma took place within the teardrop-shaped magnetosphere. The driving force was inferred to be a 'viscous-like' interaction between the solar wind plasma and the magnetospheric plasma, forcing the latter to move towards the anti-solar direction near the magnetopause, generating the return flow along the Sun-Earth line and completing the circulatory motion.

This concept has now been incorporated in the open model. Magnetosheath plasma enters the magnetosphere through the merging process (Figure 4). During its passage through the high latitude lobe region, it is 'captured' by the plasma sheet and is then convected back towards the Sun, leaving the magnetosphere across the dayside magnetopause after a single circulatory motion. During its traverse through

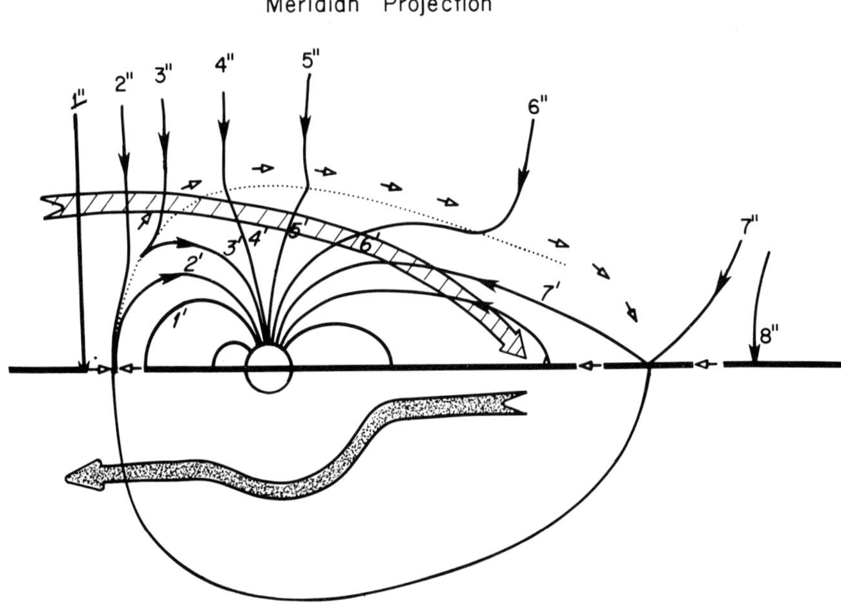

Fig. 4. Merging and reconnection processes, proposed by Dungey, together with the new convection pattern of plasma in the magnetosphere.

the plasma sheet region, the captured magnetosheath plasma is energized when magnetospheric substorms happen to occur. Thus, when the plasma particles leave the magnetosphere, they are often much more energetic (~ 10 keV) than before they entered the magnetosphere (~ 1 keV). Some of the energized plasma particles are also injected into the auroral ionosphere and the trapping region (the Van Allen belts) during magnetospheric substorms. Thanks are due to everyone who has contributed so much to the last decade of great progress in understanding magnetospheric plasma processes. Chapter 3 is concerned with these subjects.

The concept of magnetospheric disturbances has also evolved considerably during the last decade. As we shall see in this book, we have now successfully identified a specific mode of magnetospheric response to each specific type of interplanetary disturbance, such as shock waves, and changes of the east-west and north-south components of the IMF. IMF disturbances can often (~50%) be attributed to Alfvén waves which propagate away from the Sun, particularly in fast streams and their leading edges (Belcher and Davis, 1971). In their simplest form, IMF vector $\boldsymbol{B}$ of constant magnitude ($|\boldsymbol{B}| = \text{constant}$) moves on the surface of a cone, where $\boldsymbol{B} = \boldsymbol{B}_0 + \delta\boldsymbol{B}$ and $\delta\boldsymbol{B}$ is the wave vector, so that $\delta\boldsymbol{B}$ of constant magnitude ($|\delta\boldsymbol{B}| = \sqrt{\delta B_x^2 + \delta B_y^2 + \delta B_z^2} = \text{constant}$) rotates in a plane perpendicular to $\boldsymbol{B}_0$ (Figure 5).

The responses of the magnetosphere to a shock wave and the rotation of $\delta\boldsymbol{B}$ vector are classified as *reversible* or *quasi-reversible* disturbances. This is because there is no immediate explosive energy dissipation in the magnetosphere during such disturbances.

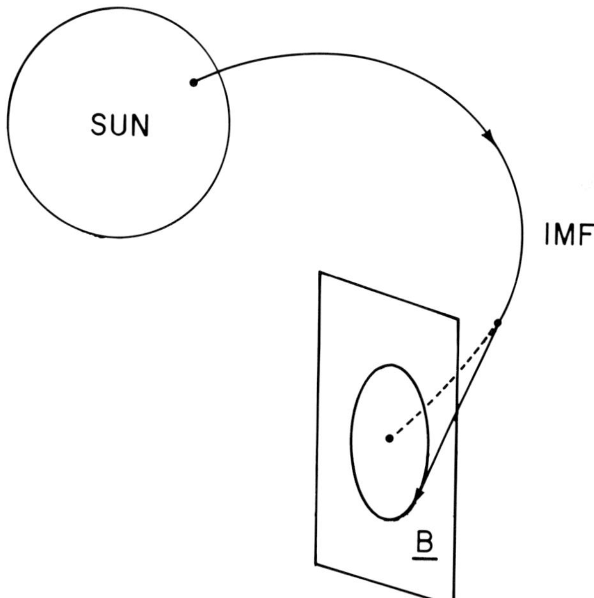

Fig. 5. Schematic diagram showing the motion of the interplanetary magnetic field vector as the Alfvén wave propagates along it.

The response of the magnetosphere to the north-south component of the IMF (the B_z component (δB_z) of the Alfvén wave) can be described in terms of changes of the efficiency of the solar wind-magnetosphere dynamo. When the IMF has a large ($\sim 5\gamma$) northward component for 6 h or more, the dynamo has the minimum efficiency, and the magnetosphere attains its 'ground state'; the magnetosphere has no energy to be dissipated in the polar ionosphere. The auroral oval contracts to its minimum size, and no auroral activity is seen along the minimum size oval.

When the magnitude of the IMF north-south component decreases from a large value ($\sim 5\ \gamma$), the efficiency of the solar wind-magnetosphere dynamo increases. Substorm energy is accumulated in the magnetotail so long as the rate of dayside merging Φ_D of the interplanetary and geomagnetic field lines at the front of the magnetosphere is greater than the rate of nightside reconnection Φ_N of the open field lines at the anti-sunward end of the magnetotail (that is, when there is a differential merging rate $(\Phi_D - \Phi_N) > 0$). As a result, the auroral oval expands equatorward until Φ_N catches up with Φ_D. This increased part of the magnetotail energy may be considered to be the excess energy. It is this excess energy which is available for magnetospheric substorms. The purpose of Chapter 4 is to examine these magnetospheric responses.

The magnetospheric substorm differs distinctly from the reversible disturbances in that there occurs a considerable energy dissipation, in a relatively short period, in the form of joule heat in the polar upper atmosphere and of kinetic energy of auroral particles, as well as in the form of ring current particles. Thus, it is an *irreversible* disturbance. After the magnetosphere is *once 'pulsed'* by the Alfvén wave, *a series of* magnetospheric substorms is often (but not always) observed.

Historically, one can find a hint of the concept of magnetospheric substorm in the term '*polar elementary storm*' introduced by Kristian Birkeland in as early as 1909. Birkeland (1913) believed that there were five elementary storms: the equatorial positive (which we denote ssc and the initial phase of a storm), the equatorial negative (which we call the main phase of a storm), polar positive and negative bays, together with the solar flare effect (which we denote S_q^a). The polar positive and negative elementary storms are a manifestation of the magnetospheric substorm.

Chapman established the present view of the typical geomagnetic storm, a storm which consists of the initial phase and the main phase, together with a characteristic onset, ssc. In his view, however, the polar positive and negative bays were considered to be minor disturbances. Indeed, in *Geomagnetism* (Chapman and Bartels, 1940, p. 338), they were discussed in Chapter X under the title 'Bays, Pulsations and Minor Disturbances'. It was only after the IGY that we found that a geomagnetic storm period can be defined as the period during which intense substorms occur very frequently; individual substorms cause a 'mini-ring current', and an intense ring current (and thus the main phase of a magnetic storm) results as a product of the frequent occurrence of intense substorms (Akasofu and Chapman, 1972).

The magnetospheric substorm is most dramatically manifested in the auroral substorm. The auroral substorm has two characteristic phases (Akasofu, 1964): the expansive phase and the recovery phase; see Figure 6. The first indication of a

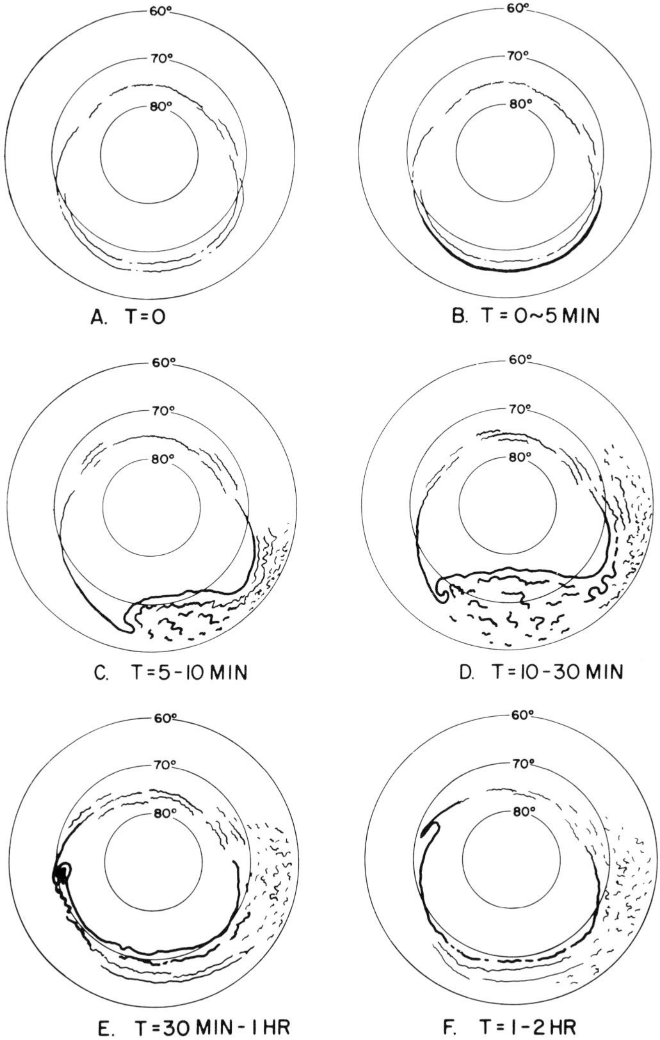

Fig. 6. Schematic diagram showing the growth and decay of the auroral substorm, viewed from above the Earth's dipole pole.

substorm is a sudden brightening of one of the quiet arcs lying in the midnight sector of the oval (or a sudden formation of an arc). This phenomenon is then followed by a rapid poleward motion of the brightened arc, resulting in an 'auroral bulge' in the midnight sector. As the auroral substorm progresses, the bulge expands in all directions. In the evening side of the expanding bulge, a large-scale fold appears and travels rapidly westward along the oval, namely the westward traveling surge. In the morning side of the bulge, arcs appear to disintegrate into 'patches' which drift eastward with a speed of order 300 m s^{-1}; the commonly used term 'break-up' refers to this 'disintegration process'.

When the expanding bulge attains its highest latitude, the recovery phase of the auroral substorm begins. The expanding bulge begins to contract, and at the end of the substorm the general situation will be similar to that just before its onset. Figure 7 shows eight DMSP satellite photographs chosen to show how the expansive phase of the auroral substorm might appear when the photographs could be taken several minutes apart.

This sudden energy dissipation process, the magnetospheric substorm, manifests itself in a variety of magnetospheric and polar upper atmospheric phenomena, as well as in the auroral substorm. Thus, different manifestations of the magnetospheric substorm reveal its different aspects. In the book entitled *Polar and*

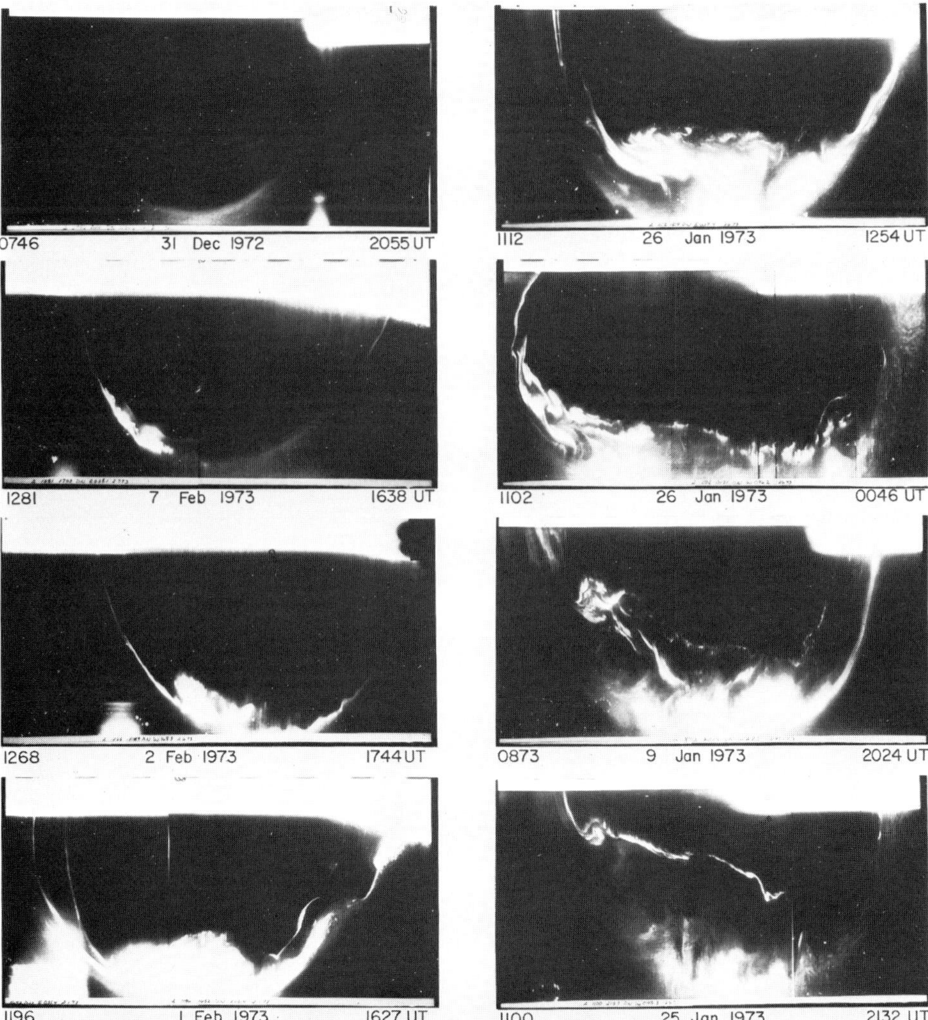

Fig. 7. Collection of DMSP-2 satellite photographs of the dark hemisphere which illustrate different epochs of the auroral substorm. They may be compared with Figure 6.

Magnetospheric Substorms, the predecessor to this one, the author attempted to identify and analyze each manifestation of the magnetospheric substorm (Figure 8) and to demonstrate that the concept of magnetospheric substorm can indeed synthesize complex auroral phenomena (Akasofu, 1968).

Thanks to all my colleagues who have advanced remarkably our understanding of magnetospheric substorm processes during the last decade, we can now attempt to synthesize systematically a vast quantity of both satellite and ground-based observations of magnetospheric substorms.

It has widely been accepted that the magnetosphere has an efficient mode of reducing the excess energy by forming a new magnetic neutral line in the near-Earth plasma sheet, where the reconnection rate Φ_N is greatly enhanced. The latest studies of magnetic field variations in the magnetotail do not support such a view.

It is more likely that the expansive phase of the magnetospheric substorm results from a 'deflation' of the plasma sheet, rather than from an enhanced reconnection. It is during the recovery phase when an enhanced reconnection may take place at a great distance, mostly beyond the lunar distance, perhaps at the anti-solar end of the magnetosphere. These are the subjects of Chapter 6.

In Chapter 7, we shall deal with magnetospheric current systems during the magnetospheric substorm and discuss in detail the auroral electrojets, ionospheric electric fields and related phenomena. Chapter 8 is concerned with a sudden penetration of the cross-tail electric field into the inner magnetosphere, the resulting plasma injection and associated phenomena.

In the first part of Chapter 9, the basic solar-interplanetary magnetic field structure, high speed solar wind streams, transient solar activities and associated interplanetary disturbances are briefly described. In particular, we shall be most interested in the nature of disturbances which give rise to changes of the north-south component of the IMF and also in knowing how and where such disturbances are generated.

In the second part of Chapter 9, an attempt will be made to construct a model of the magnetospheric substorm, which can describe the observed sequence of phenomena in the magnetosphere and also in the polar upper atmosphere.

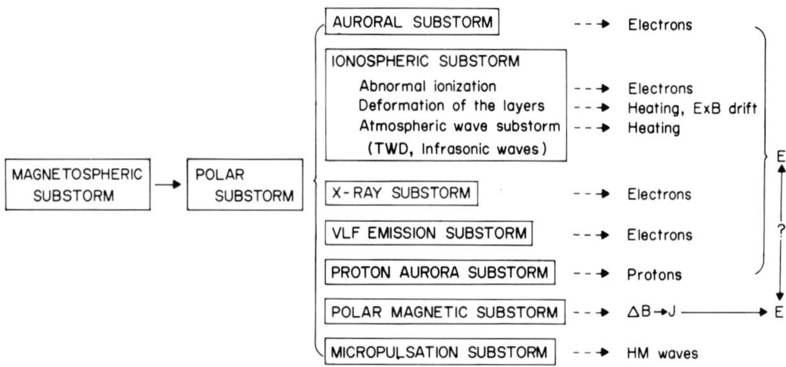

Fig. 8. Block diagram showing how the author's preceding book, *Polar and Magnetospheric Substorms*, was constituted.

Obviously, at this stage the model is not necessarily internally consistent and is also bound to be qualitative in many respects. It is for these reasons that deficiencies of the model are also critically examined. In constructing a model of the magnetospheric substorm, the following three points should be kept in mind.

First of all, we have been accustomed to using such terms as "erosion of the dayside magnetopause", "transfer of the merged field lines to the magnetotail", "pinching of the plasma sheet", "relaxation of the magnetosphere towards the dipolar configuration", "collapse of the magnetotail", etc. These terms tend to 'hypnotize' us to believe that basic physical processes associated with the interaction between the IMF and the magnetosphere are established and understood, and also that all observations fit 'theoretical' predictions. Actually, a painful effort is needed to establish the relationships between the enhanced merging rate Φ_D and individual magnetospheric quantities. This is because a magnetospheric quantity is, in general, a function of several parameters which we have no control of. Therefore, one must be cautious in establishing and substantiating a relationship between two magnetospheric quantities.

Secondly, in studying magnetospheric substorms, one has to determine how various magnetospheric quantities observed at different locations vary as a function of substorm time T, and then examine how they are related to each other. Ideally, such a study could be achieved if there were a large number of satellites, observing simultaneously all the desired quantities at a large number of locations. Since this is impossible, one has to compile satellite data observed at different locations and at different times for different substorms, and analyzed by different workers. Such a task will be possible only if there is the identical time frame of reference, namely the identical substorm time.

Thirdly, the lack of both theoretical and observational knowledge of the physical processes in the magnetotail makes it difficult to predict microscopic and macroscopic behaviors of plasma during substorms. In the field of magnetospheric physics, the highest degree of cooperation between theorists and experimenters has been achieved in the area of wave-particle interactions and the resulting pitch-angle diffusion and precipitation (Kennel and Petschek, 1966). A similar cooperating is urgently needed in studying the cause of magnetospheric substorms, one of the major unsolved problems in magnetospheric physics.

If we could finally succeed in constructing the framework within which magnetospheric data painstakingly analyzed can be fitted together without internal contradiction, it will take magnetospheric physics to the stage where a number of planetary, solar and stellar phenomena can be explained or predicted.

References

Akasofu, S.-I.: 1964, 'The Development of the Auroral Substorm', *Planetary Space Sci.* **12**, 273.
Akasofu, S.-I.: 1968, *Polar and Magnetospheric Substorms*, D. Reidel Publ. Co., Dordrecht-Holland.
Akasofu, S.-I. and Chapman, S.: 1972, *Solar-Terrestrial Physics*, Oxford Univ. Press.
Alfvén, H.: 1939, 'Theory of Magnetic Storms', *I. Kungl. Sv. Vet.-Akademiens Handl.* (3) **18**, No. 3.
Alfvén, H.: 1950, *Cosmical Electrodynamics*, Oxford University Press.
Axford, W. I. and Hines, C. O.: 1961, 'A Unifying Theory of High-Latitude Geophysical Phenomena and Geomagnetic Storms', *Can. J. Phys.* **39**, 1433.

Belcher, J. W. and Davis, L. Jr.: 1971, 'Large-Amplitude Alfvén Waves in the Interplanetary Medium, 2', *J. Geophys. Res.* **76**, 3534.

Biermann, L.: 1957, 'Solar Corpuscular Radiation and the Interplanetary Gas', *Observatory* **77**, 109.

Birkeland, K.: 1913, *The Norwegian Aurora Polaris Expedition 1902–1903*, Vol. I, H. Aschehoug, Christiania.

Chapman, S. and Bartels, J.: 1940, *Geomagnetism*, Oxford University Press.

Chapman, S. and Ferraro, V. C. A.: 1931, 'A New Theory of Magnetic Storms', *Terr. Mag. Atmosph. Elect.* **36**, 77.

Dungey, J. W.: 1961, 'Interplanetary Magnetic Field and the Auroral Zones', *Phys. Rev. Lett.* **6**, 47.

Feldstein, Y. I.: 1963, 'Some Problems Concerning the Morphology of Auroras and Magnetic Disturbances at High Latitudes', *Geomag. Aeronon.* **3**, 183.

Kennel, C. F. and Petschek, H. E.: 1966, 'Limit on Stably Trapped Particle Fluxes', *J. Geophys. Res.* **71**, 1.

Ness, N. F.: 1965, 'The Earth's Magnetic Tail', *J. Geophys. Res.* **70**, 2989.

Parker, E. N.: 1958, 'Dynamics of the Interplanetary Gas and Magnetic Fields', *Astrophys. J.* **128**, 664.

Piddington, J. H.: 1960, 'Geomagnetic Storm Theory', *J. Geophys. Res.* **65**, 93.

Zmuda, A. J. and Armstrong, J. C.: 1974, 'The Diurnal Flow Pattern of Field-Aligned Currents', *J. Geophys. Res.* **79**, 4611.

CHAPTER 1

OPEN MAGNETOSPHERE AND THE AURORAL OVAL

1.1. Open Magnetosphere

The impact of a unidirectionally streaming, unmagnetized plasma upon a dipole field was first studied by Chapman and Ferraro in a series of papers in 1931–33. The front of the stream was assumed to be an infinite plane or a cylindrical surface. Later, their study was extended by Zhigulev and Romishevskii (1960), Hurley (1961a, b) and Dungey (1961a) for a two-dimensional dipole and by Beard (1960, 1967), Midgley and Davis (1963) and Mead and Beard (1964) for a three-dimensional dipole.

There are several common features in all these extended studies, as well as in the original Chapman-Ferraro theory. The dipole field is completely confined in a cavity. Thus the magnetic field lines must lie parallel to the cavity boundary surface. In fact, the shape of this cavity is determined by knowing the configuration of the outermost field lines (cf. S.T.P., Sections 5.3 and 5.4).

This means that the cavity is a *closed system*, except that it has two singular points, one in each hemisphere, at the front of the magnetosphere. These points are called the neutral points. Consider the field line (the cusp line) which connects the neutral point in the northern hemisphere and the Earth, and also a very small bundle of field lines around it. The bundle of field lines forms a funnel-like surface near the neutral point and spreads over the entire northern half of the cavity surface. Figure 1.1(a) shows schematically the field line topology of the closed magnetosphere.

Consider now motions of solar electrons of relatively low energy ($\sim$10 keV) which reach the cavity surface after being ejected from an active region on the Sun. Their motions are restricted to movements more or less along the geomagnetic field lines, with helical paths. Some of them will first move towards the neutral point along the field lines which constitute the cavity surface, and will reach a very small area in the ionosphere along the cusp line. Thus the impact point on the Earth does not depend on the entry point of the electron on the cavity surface.

However, there are a number of phenomena which cannot be explained by assuming the closed cavity with the neutral point. We shall see in detail in the following several sections that the bundle of field lines which 'originate' in the polar cap are 'connected' to the interplanetary magnetic field (IMF) lines; here the polar cap is defined as the area bounded by the auroral oval, an instantaneous belt of auroras. For example, we shall see in Section 1.3.1 that solar electrons impinge over the entire polar cap region by following such connected field lines. In this situation, the field lines can be grouped into two. The field lines in the first group are connected with the IMF lines and are said to be 'open'. The field lines in the other group have conjugate points in the opposite hemisphere after crossing the equatorial plane.

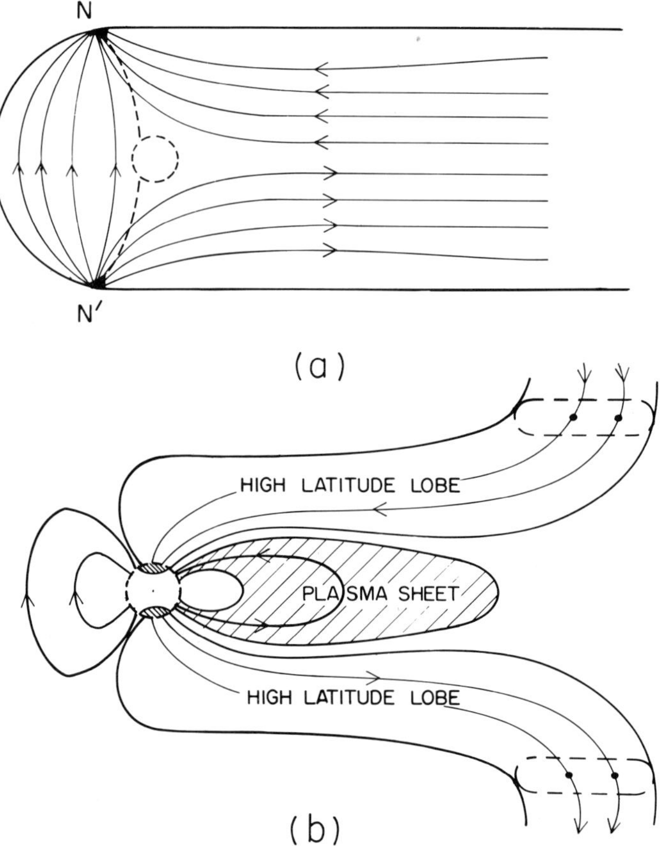

Fig. 1.1. Topological differences between the closed (a) and open (b) magnetosphere.

These field lines are said to be 'closed'. The magnetosphere with the open field lines is an open system, and the magnetosphere is said to be open. Figure 1.1(b) shows schematically the field line topology of the open magnetosphere. The open model was first proposed by Dungey (1961b). The topology of the open model has recently been discussed by a number of workers (Nishida, 1971; Nishida and Maezawa, 1971; Stern, 1973, 1975).

The Chapman-Ferraro theory was quite successful in predicting the shape of the dayside part of the magnetosphere. This is because the kinetic energy density of the solar wind is far greater than the magnetic energy density. Therefore, the flow pattern of the solar wind around the Earth's dipole field does not critically depend on the presence of the IMF. The Chapman-Ferraro theory is a reasonable first approximation in determining the boundary shape of the magnetosphere, but because of the assumption of unmagnetized solar wind plasma, it failed to predict that the magnetosphere is an open system.

It is generally believed that the magnetospheric cavity has a long cylindrical shape with a blunt nose at the sunward end. The cylindrical cavity extends mainly in the anti-solar direction, and the term 'magnetotail' is commonly used to define the anti-solar portion of the cavity beyond a geocentric distance of about 10 Earth radii (R_E) from the Earth. Its length is at least of the order 10^3 R_E (Ness *et al.*, 1967; Mariani and Ness, 1969; Fairfield, 1968; Intriligator *et al.*, 1969; Intriligator *et al.*, 1972; Walker *et al.*, 1975). A thin layer of plasma bisects the cylindrical magnetotail into northern and southern halves. This plasma layer is called the plasma sheet, and its density is of order 0.1 cm^{-3}. Characteristics of plasma sheet particles will be discussed in Section 3.4. It is generally believed that most of the magnetic field lines embedded in the plasma sheet have a closed configuration to contain the plasma sheet. The magnitude of the magnetic field in the plasma sheet is of order 2–5 γ.

In the high latitude lobe region of the magnetotail, the plasma density is much less ($\sim$0.01 cm^{-3}) than that in the plasma sheet, but the magnetic field intensity is of order of 10–30 γ. The magnetic field is approximately directed towards the Sun in the northern high latitude lobe, and towards the anti-solar direction in the southern high latitude lobe. There is also a slight flaring (out from the Earth) of the field lines in the lobes (Section 3.2.3).

1.2. Auroral Oval

The configuration of the auroral oval was first established by Feldstein (1963, 1973) on the basis of a statistical study of the IGY all-sky photographic data. His study indicated that auroral arcs tend to lie along an oval-shaped band, which is eccentric with respect to the dipole pole, rather than along the auroral zone; its

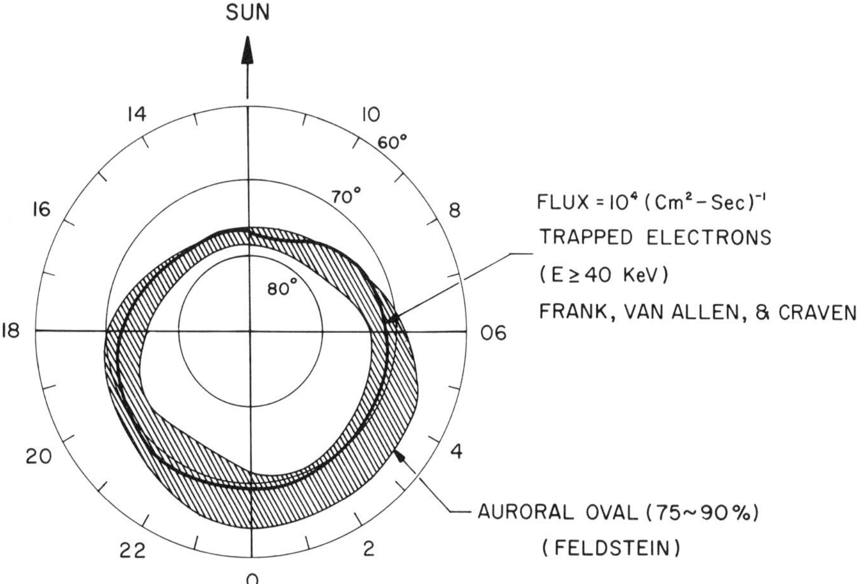

Fig. 1.2. Auroral oval determined by Feldstein and the outer boundary of the trapping region.

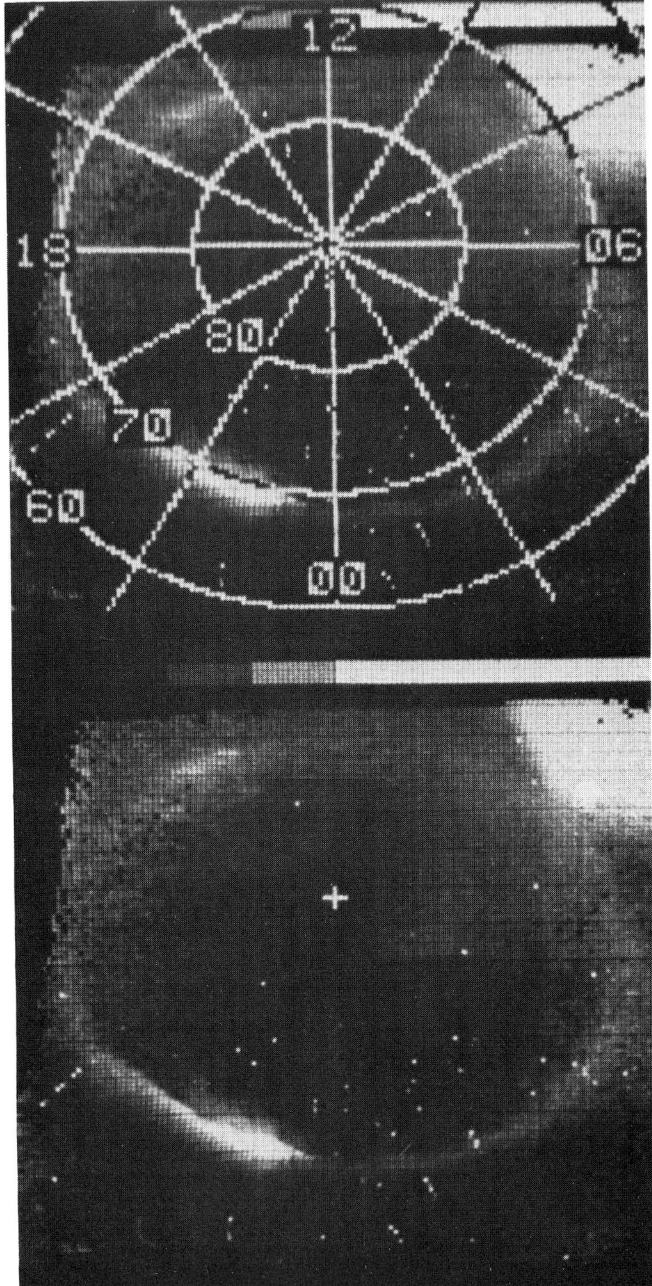

Fig. 1.3. Auroral oval (the 5577 Å emission) observed from the ISIS-2 satellite, 0247–0302 UT, December 16, 1971, in corrected geomagnetic coordinates (upper) and without the coordinate grid (lower). (Lui, A. T. Y., Anger, D., Venkatesan, D. and Sawchuk, W.: *J. Geophys. Res.* **80**, 1795, 1975.)

center is appreciably (~3°) displaced toward the dark hemisphere. Figure 1.2 shows the oval constructed by Feldstein (1963).

These results have recently been confirmed by airborne all-sky camera data (Buchau *et al.*, 1970) and by 'photographs' taken by an auroral scanner aboard polar orbiting satellites (Lui *et al.*, 1975). Figure 1.3 shows a photograph of the auroral oval taken by the scanning auroral photometer (5577 Å) aboard the ISIS-2 satellite, (Lui and Anger, 1973). It can be seen that there is a continuous belt of auroras at all local times. The eccentricity of the oval belt with respect to the pole is also evident. A discrete auroral arc is visible in the evening portion of the auroral oval and terminates sharply around midnight; the bright portion of the arc is a westward traveling surge.

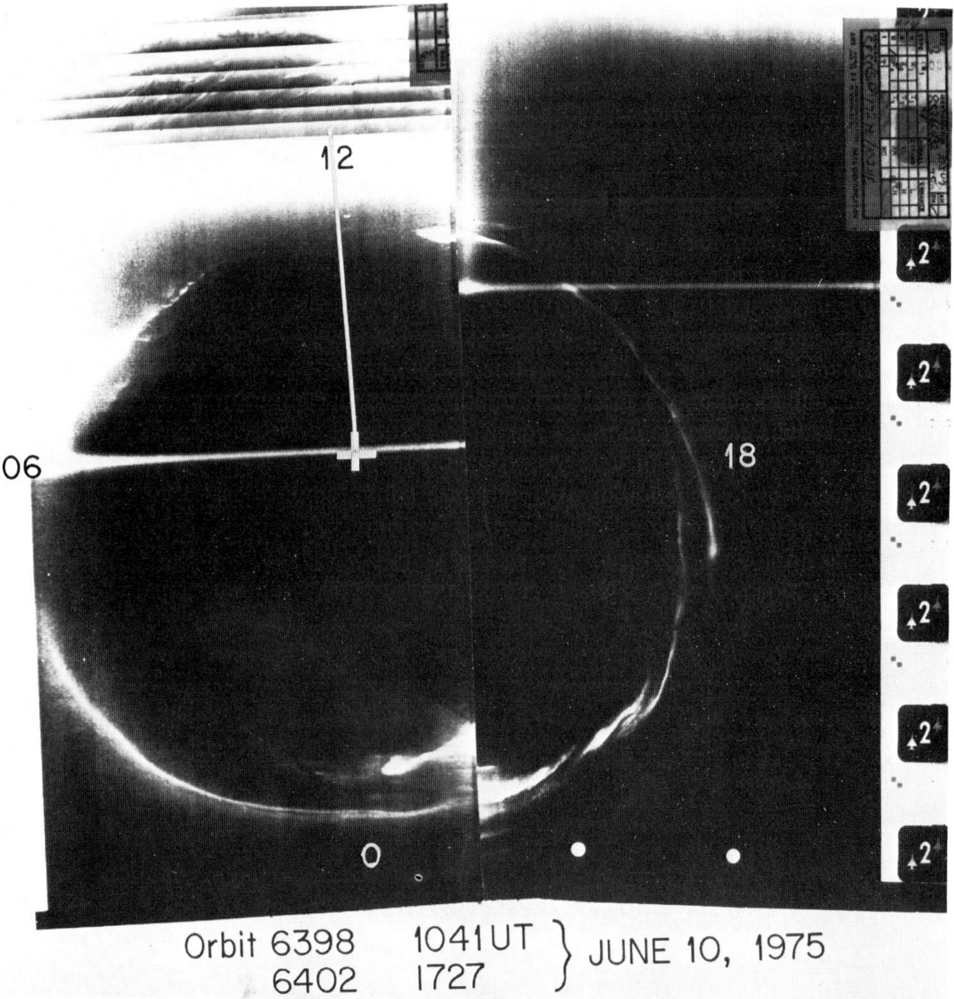

Fig. 1.4a. Montage photograph of the auroral oval taken from above the antarctic region by the DMSP-8531 satellite; the cross and the line indicate the invariant pole and the noon meridian, respectively. (Akasofu, S.-I.: *Space Sci. Rev.* **19**, 169, 1976.)

Photographs taken from DMSP (Defense Meteorological Satellite Program) satellites have a better spatial resolution but cover a slightly smaller field of view than ISIS-2 photographs. They can be used to construct montage photographs of the auroral oval, for days when the size of the oval did not change much during successive orbits.

Figure 1.4(a) is constructed by combining two DMSP-8531 photographs (taken from above the antarctic region) and shows one of the most structureless ovals. The cross indicates the location of the south invariant pole. The noon meridian extends from it towards the top of the figure. In this montage photograph, the morning part of the oval is delineated by a single band of luminosity, and the afternoon-evening sector of the oval by two faint arcs. The midday part of the oval appears to be lacking, but there are a few segments of arc-like structure.

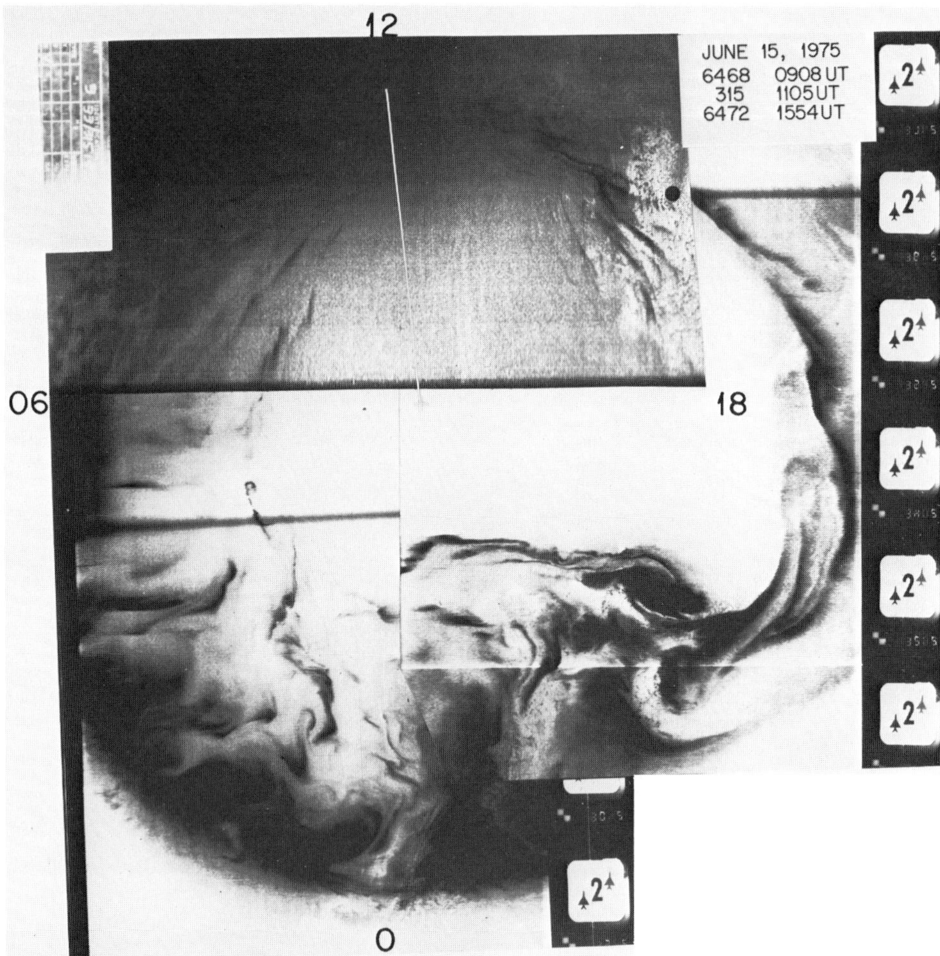

Fig. 1.4b. Montage photograph of the auroral oval taken from above the antarctic region by the DMSP-8531 and 10533 satellites; the cross and the line indicate the invariant pole and the noon meridian, respectively. (Akasofu, S.-I.: *Space Sci. Rev.* **19**, 169, 1976.)

The second montage photograph in Figure 1.4(b) consists of three DMSP photographs. These photographs were taken during one of the most active periods in June 1975. Several typical substorm features, such as a large poleward expanding bulge, an intense westward traveling surge and a well-developed Ω band, can clearly be identified.

In Chapter 2 we shall examine in detail differences of the characteristics of auroras at different local times and also of auroral particles which cause them. In the next section, we examine the significance of the open magnetosphere and its relation to the auroral oval.

1.3. Open Magnetosphere and the Auroral Oval

1.3.1. SOLAR ENERGETIC PARTICLES AND THE AURORAL OVAL

One of the most important indications that polar cap magnetic field lines are connected with interplanetary magnetic field lines is revealed by the fact that solar electrons and low energy solar protons (which are ejected from an active region of the Sun) impact almost uniformly over the entire polar cap region. Figure 1.5 shows the flux of solar electrons along an evening-morning meridian (the satellite orbit). It shows clearly that the flux was uniform across the polar cap, from about latitude 75° along an evening meridian to about 67° along a morning meridian. This feature was first shown by Stone (1964), Flindt (1970), Fanselow and Stone (1972) and Fennell (1973), who examined the impact region of low energy solar protons in high latitude regions.

Because of their low rigidity, solar electrons serve better than low energy protons in studying the topology of the magnetosphere. Vampola (1969, 1971 and 1973), McDiarmid and Burrows (1970), Paulikas *et al.* (1970), Evans and Stone

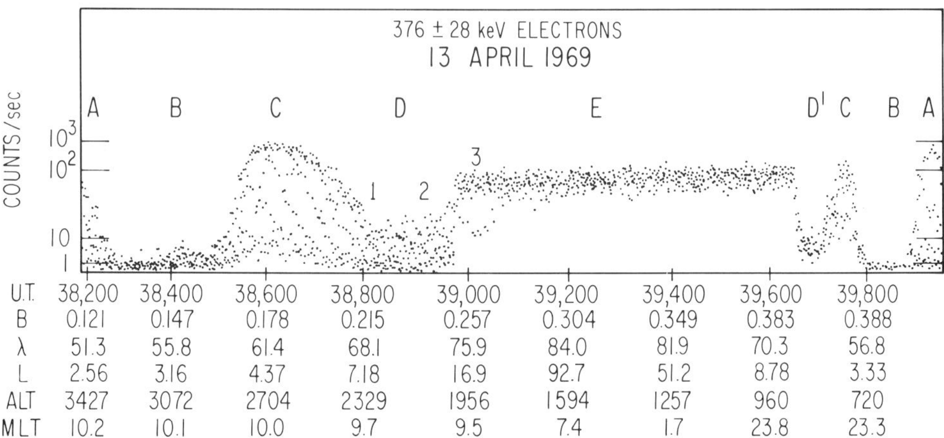

Fig. 1.5. Polar cap plateau, indicating a uniform bombardment of the polar cap by solar electrons, in an analog record of 376 ± 28 keV electron data plotted as counting rate versus time on a pseudo-logarithmic scale; UT, magnetic field intensity, invariant latitude, L, altitude and MLT are also shown along the time axis. (Vampola, A. L.: *J. Geophys. Res.* **76**, 36, 1971.)

(1972) and most recently Nielsen and Pomerantz (1975) obtained the latitude of the impact boundary as a function of MLT. The boundary is located at a much higher latitude in the noon sector than in the midnight sector. In Figure 1.6, an approximate mean location of this boundary is plotted in invariant latitude-MLT coordinates; it shows clearly that the impact area of solar electrons has an oval-like shape which is quite similar to the auroral oval.

Many simultaneous observations of solar electrons in interplanetary space and in the magnetotail indicate that their flux variations in time are quite similar, with a probable time delay of at most 100 s in the magnetotail (Anderson and Lin, 1969; Van Allen, 1970; Turtle *et al.*, 1972). Thus solar electrons have almost free access to the magnetotail. Van Allen (1970) inferred also that solar electrons enter the magnetotail at downstream distances between 64 and 900 R_E. Further, since the flux of solar electrons is uniform in the oval-shaped region, it is reasonable to conclude that the field lines from the oval-shaped region are connected with the IMF field lines.

Piddington (1965) and Akasofu (1966) noted that the boundary of the 45 keV

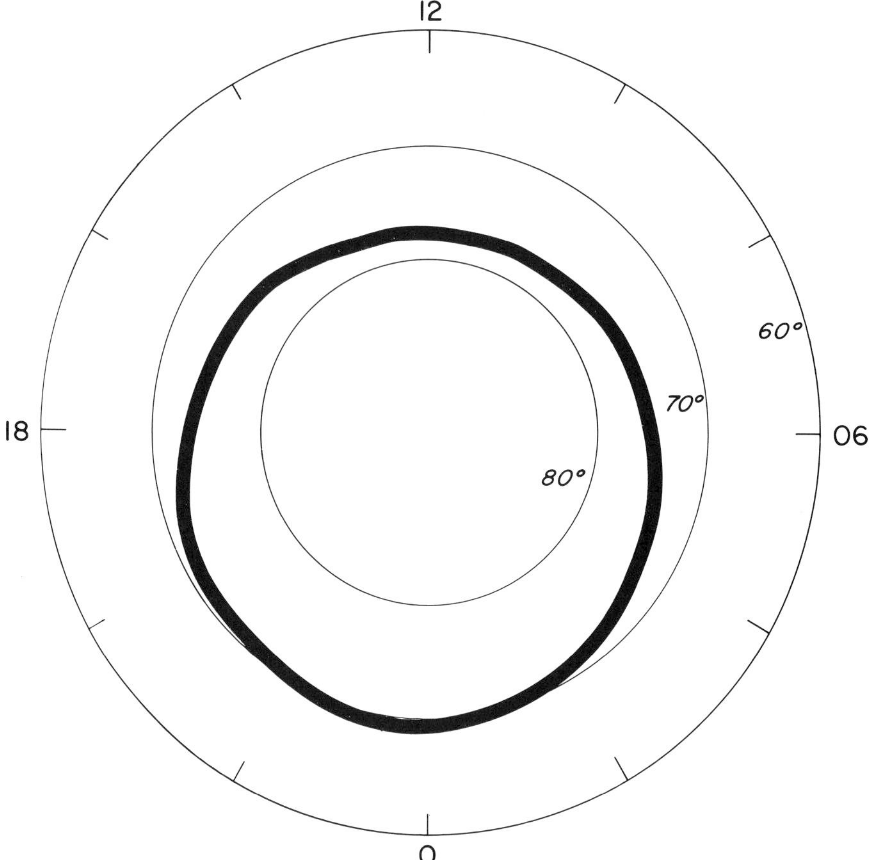

Fig. 1.6. Location of the boundary of the polar cap plateau, replotted from Figure 13 in Vampola (Aerospace Report No. TR-0074(4260-20)-5, Aug. 24, 1973), in polar invariant-MLT coordinates.

Van Allen belt at the ionospheric level coincides reasonably well with the auroral oval determined by Feldstein. This finding can be considered to be collateral to the above result, since the outer boundary surface of the Van Allen belt and the surface of the bundle of open field lines should lie closely together near the Earth.

Solar protons can penetrate deeper into the trapping region than energetic electrons because they have a higher rigidity. A detailed study of both solar protons and Van Allen belt electrons by McDiarmid and Burrows (1969) and McDiarmid *et al.* (1974) indicates that solar protons of energy as low as 0.6 MeV can penetrate deep into the outer part of the Van Allen belt, about 50° in latitude. Therefore, although proton trajectories in the magnetosphere have recently been studied extensively by a large number of workers (Gall, 1968; Gall *et al.*, 1968; Engelmann *et al.*, 1971; Gall *et al.*, 1972; Gall and Bravo, 1973; Morfill and Quenby, 1971; Quenby, 1972; Durney *et al.*, 1972; Durney and Morfill, 1972; Morfill and Scholer, 1972a, b, 1973a, b; Scholer, 1972, 1975; Page and Domingo, 1972; Scholer and Morfill, 1974; Morfill, 1973), solar electrons are much better suited for the particular purpose of determining the boundary of the polar cap. On the other hand, solar protons tend to retain their pitch-angle anisotropy in interplanetary space during their penetration into the magnetosphere, providing valuable information on the merging process between the geomagnetic and IMF lines (Cooper and Haskell, 1972; Aarsnes and Amundsen, 1972; Van Allen *et al.*, 1971; Scholer and Morfill, 1972; Scholer *et al.*, 1972; Innanen and Van Allen, 1973).

We have so far discussed only the average shape of the impact region of solar electrons and pointed out that it is quite similar to the average shape of the auroral oval. It is, at present, not certain how well the boundary of the impact region coincides with the instantaneous auroral oval.

1.3.2. FIELD-ALIGNED CURRENTS AND THE AURORAL OVAL

(a) *Distribution of Field-Aligned Currents above the Polar Ionosphere*

The location of intense field-aligned currents in the polar region was first inferred by Zmuda *et al.* (1966), Zmuda *et al.* (1970) and Armstrong and Zmuda (1970, 1973) on the basis of satellite-borne magnetometer data and also by Berko (1973) and Berko *et al.* (1975) on the basis of the distribution of field-aligned electron fluxes. They showed that there are circumpolar belts of field-aligned currents which approximately coincide with the auroral oval. They showed also that the field-aligned currents are a permanent feature, observable even on very quiet days.

Early studies by Zmuda and his colleagues were based on records obtained from a scalar magnetometer aboard a polar orbiting satellite. The TRIAD satellite was the first polar orbiting satellite which carried a tri-axial magnetometer. From TRIAD satellite records, Armstrong and Zmuda (1973) and Zmuda and Armstrong (1974a, b) showed that the largest magnetic perturbation appears in the east-west component, indicating that field-aligned currents have a form of sheet currents which are oriented in the east-west direction. Figure 1.7 shows the distribution of the eastward (solid line) and westward (dotted line) perturbations along orbits in

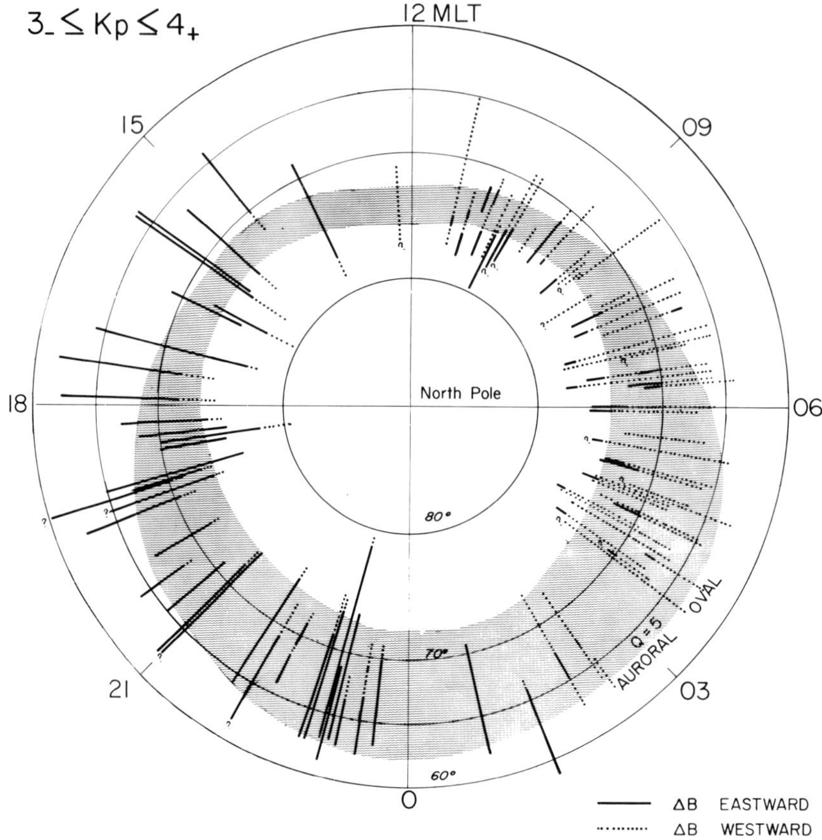

Fig. 1.7. Locations of the eastward (solid lines) and westward (dotted lines) magnetic perturbations approximately along the orbits of the TRIAD satellite for moderately disturbed conditions ($3- \leq \mathrm{Kp} \leq 4+$), together with the oval of Q = 5. (Yasuhara, F., Kamide, Y. and Akasofu, S.-I.: *Planet. Space Sci.* **23**, 1355, 1975.)

invariant latitude-MLT coordinates (Yasuhara *et al.*, 1975) for moderate magnetic conditions ($3- \leq \mathrm{Kp} \leq 4+$), together with the corresponding auroral oval. There is a gross agreement between the region of field-aligned currents and the statistical oval.

The eastward perturbation in the evening sector can be produced by a pair of field-aligned currents, one flowing away from the poleward half of the auroral oval and the other flowing into the equatorward half. In the morning sector, the direction of the current pair is reversed, namely one flowing into the poleward half of the oval and the other flowing out from the equatorward half of the oval. Figure 1.8 shows schematically the distribution of field-aligned currents inferred from the TRIAD magnetic field observation. Zmuda and Armstrong (1974a, b) concluded that in each pair the intensity of the inward flowing and outward flowing currents is the same. However, Yasuhara *et al.* (1975) have found that the intensity of the two currents differs significantly, the average ratio of the poleward current to the equatorward current being 2.0.

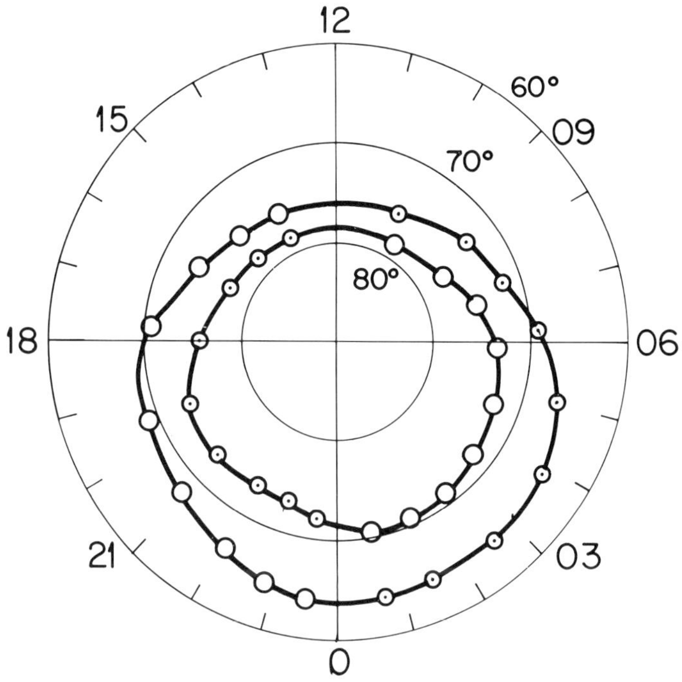

Fig. 1.8. Schematic diagram showing the locations of both inflow and outflow field-aligned currents in geomagnetic latitude-MLT coordinates.

It will be shown in Section 1.4 that the solar wind-magnetosphere dynamo generates the *primary* field-aligned current which flows into the poleward boundary of the oval in the morning sector and out from the poleward boundary of the oval in the evening sector. This primary current flows in a thin layer along the upper (in the northern hemisphere) and lower (in the southern hemisphere) surfaces of the plasma sheet, toward the Earth in the morning sector and away from the Earth in the evening sector. Sugiura (1975) showed that the declination of the magnetic field varies distinctly at the boundary between the plasma sheet and the high latitude lobe.

The primary field-aligned currents bend the field lines toward the midnight meridian in the high latitude lobe and away from the midnight meridian in the plasma sheet, namely toward the magnetopause. This feature is schematically shown in Figure 1.9. This distortion of the magnetic field in the magnetotail has been of fundamental importance in generating thick sheet currents (of order a few hundred kilometers in width) observed by the TRIAD satellite. The primary currents also generate secondary field-aligned currents which flow out from the equatorward half of the oval in the morning sector and into the equatorward half in the evening sector. These subjects will be discussed in Section 1.3.6. Both the

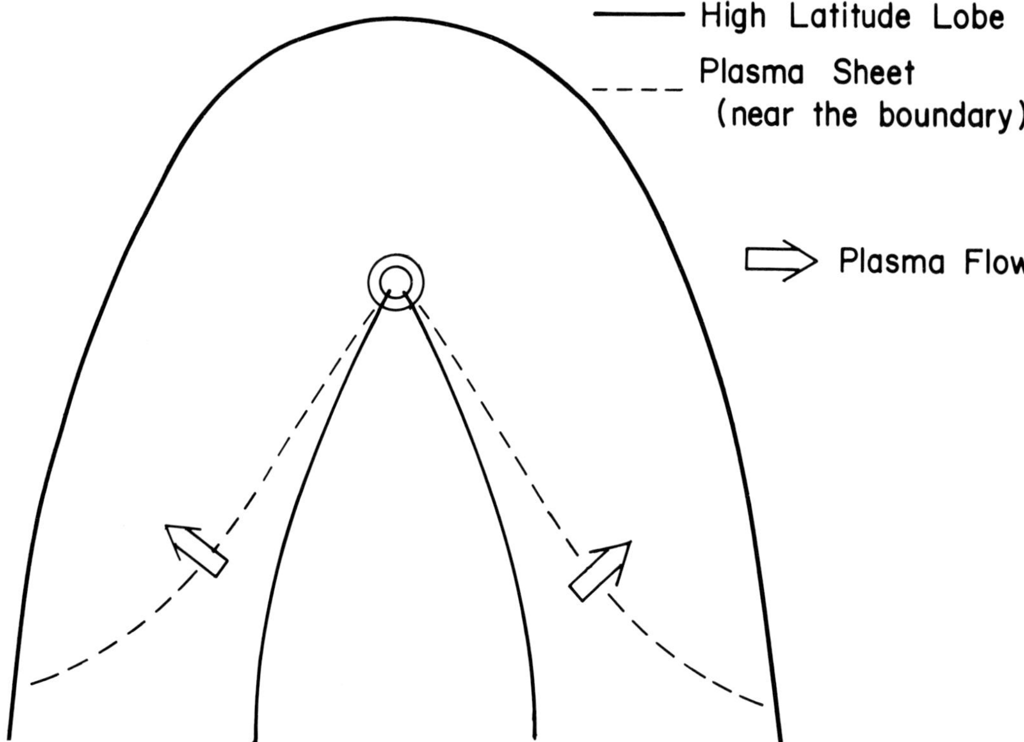

Fig. 1.9. Schematic diagram showing the distortion of two field lines, in the evening and morning sectors, which are located in the high latitude lobe (the solid line) and within the plasma sheet (the dashed line).

primary and secondary field-aligned currents generate ionospheric currents. Section 1.3.3 deals with this important subject of magnetosphere-ionosphere coupling.

(b) *Field-Aligned Currents and Auroral Arcs*

Armstrong *et al.* (1975) examined the spatial relationship between field-aligned currents and auroras for a single pass of the TRIAD satellite (together with two supplementary passes) and concluded that the poleward discrete arc marks the northernmost boundary of the field-aligned current region, and that all the visible arcs are confined within the latitudinal region occupied by the field-aligned current flows. Kamide and Akasofu (1976) examined a number of such simultaneous satellite and ground-based observations, with special reference to discrete and diffuse auroras. Figure 1.10(a) shows some TRIAD magnetometer (A sensor) data and the locations of the arcs which were crossed by the satellite. On the right-hand side, the superposed H component magnetic records from nine auroral zone stations are shown to indicate magnetic activity during a few hours before and after the passage time, which is marked by a vertical line. Most of the data were taken during evening hours in MLT, but the corresponding magnetic conditions

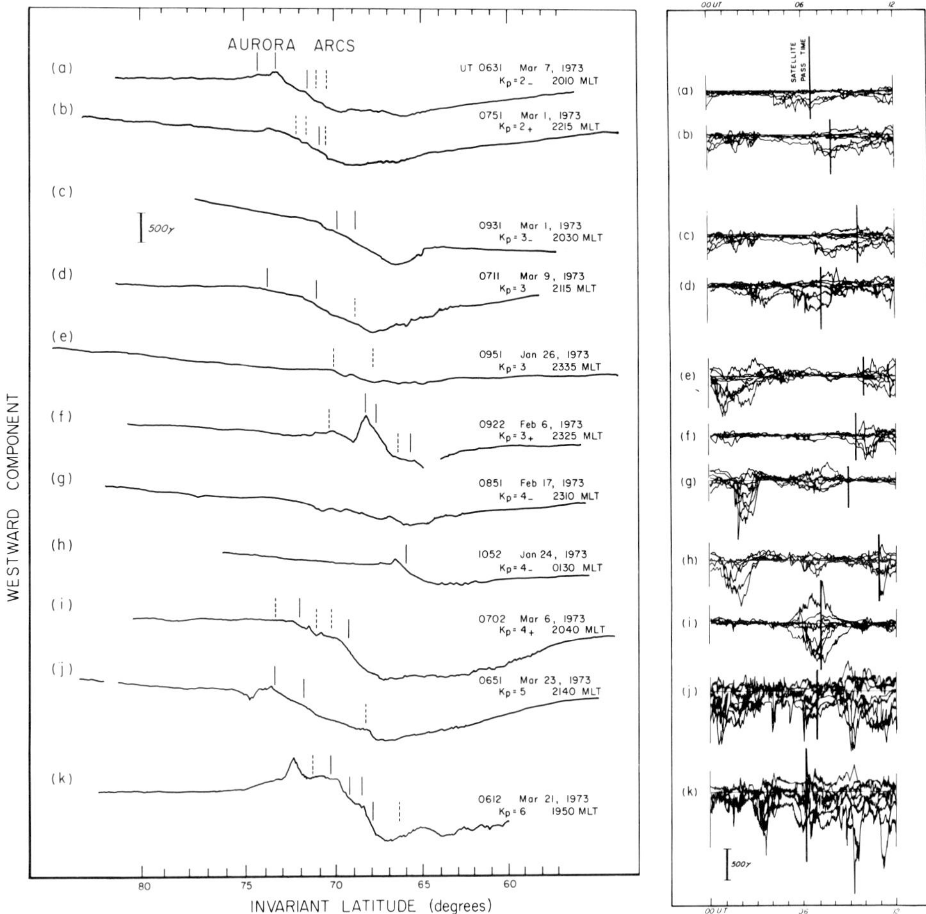

Fig. 1.10(a). Locations of auroral arcs relative to those of the field-aligned currents (TRIAD magnetometer data). (Kamide, Y. and Akasofu, S.-I.: *J. Geophys. Res.* **81**, 3999, 1976.)

(expressed by the Kp index) are quite different; the top example took place during a fairly quiet condition (Kp = 2) and the bottom example during a very disturbed condition (Kp = 6). There is a significant latitudinal expansion of the field-aligned current region as magnetic activity increases. The TRIAD data show, in general, a steep decrease (corresponding to the region of upward field-aligned currents) and the subsequent gradual increase or recovery (corresponding to the region of downward field-aligned currents).

The situation is, however, reversed in the morning sector. The downward field-aligned current occupies the poleward half of the auroral oval and is more intense than the upward field-aligned currents in the equatorward half of the oval (Figure 1.10(b)).

With this gross feature in mind, the figure shows that (i) discrete arcs are, in general, confined in the region of the upward field-aligned current; (ii) no discrete arcs are seen in the region of the downward field-aligned current, except for one

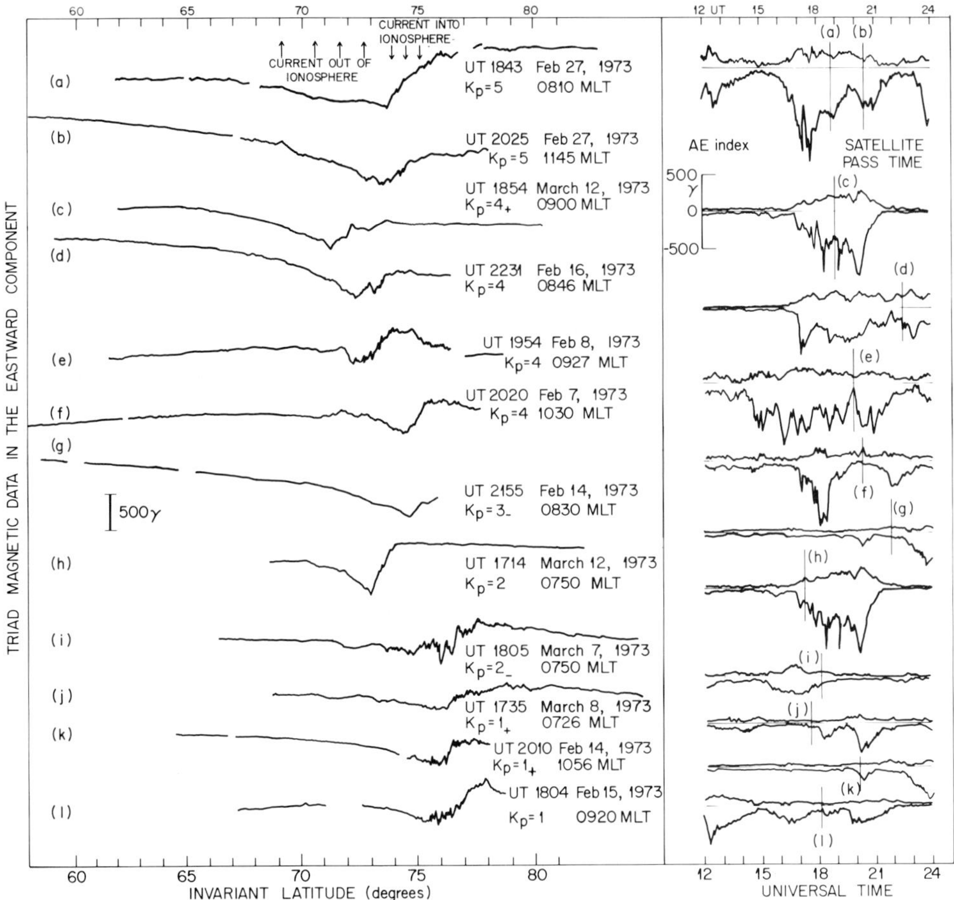

Fig. 1.10(b). Typical examples of TRIAD satellite data in the late morning sector. (Courtesy of Kamide, Y., Akasofu, S.-I. and Rostoker, G.)

faint arc in the last example; and (iii) the region of the downward field-aligned current appears to correspond to the region of the diffuse aurora. However, the spatial resolution of the data was not good enough to associate individual arcs with irregular features of the TRIAD magnetic perturbations, which presumably indicate concentrated upward or downward currents within the large-scale upward field-aligned current region. As we shall see in Section 2.5.2, sounding rocket magnetometer data indicate that a pair of intense small-scale field-aligned currents is associated with the auroral arc (see, review papers by Arnoldy, 1974, and Anderson and Vondrak, 1975); the upward current coincides with the arc, whereas the downward current is located on the equatorward side of the arc.

It may be noted that in one case (0851 UT on 1973, February 17), there was no visible arc recorded by the all-sky cameras (the minimum brightness for the camera is about 200 *R*), but the TRIAD data show the typical features. This particular pass occurred during a quiet period after a weak substorm, although the corresponding Kp (= 4−) was not low.

1.3.3. FIELD-ALIGNED CURRENTS AND IONOSPHERIC CURRENTS

The purpose of this subsection is to examine how the observed distribution of the field-aligned currents is related to ionospheric currents. More specifically, we examine theoretically how ionospheric currents are related to the observed distribution of field-aligned currents for a simple model of the ionosphere (Yasuhara, Kamide and Akasofu, 1975).

(a) *Formulation*

In a steady state, the current continuity equation for the ionosphere is given by

$$\nabla \cdot \boldsymbol{I} = j_{\parallel} \sin \chi \tag{1}$$

Here, $\boldsymbol{I}$ is the height-integrated current density (amp m^{-1}) in the ionosphere and $j_{\parallel}$ is the density of field-aligned current (amp m^{-2}) (positive for a downward current and negative for an upward current). The inclination angle χ of a geomagnetic field line with respect to the ionosphere is hereafter taken to be 90°; this is a reasonable approximation at latitudes higher than 60°.

The Ohm's law for ionospheric currents is given by

$$\boldsymbol{I} = -\begin{pmatrix} \Sigma_P & \Sigma_H \\ \Sigma_H & \Sigma_P \end{pmatrix} \cdot \nabla \Phi \tag{2}$$

where Σ_P and Σ_H denote the height-integrated ionospheric Pedersen and Hall conductivities, respectively, and Φ is the potential in the frame rotating with the Earth. Inserting (2) into (1), we have

$$-\nabla \cdot \begin{pmatrix} \Sigma_P & \Sigma_H \\ -\Sigma_H & \Sigma_P \end{pmatrix} \cdot \nabla \Phi = j_{\parallel} \tag{3}$$

Equation (3) can be solved for Φ for a given distribution of $j_{\parallel}$ when Σ_P and Σ_H are known as a function of longitude ψ and latitude λ (ψ is measured counter-clockwise from the midnight meridian (e.g., $\psi = 90°$ corresponds to the 06 LT meridian).

For simplicity, we assume that $j_{\parallel}$ has non-zero values only along two latitude circles. Thus, except for those latitude circles, we have

$$\nabla \cdot \boldsymbol{I} = 0 \tag{4}$$

Therefore, one may introduce the current function J defined as follows:

$$\boldsymbol{I} = \nabla J \times \boldsymbol{e}_r \tag{5}$$

where $\boldsymbol{e}_r$ denotes a unit vector in the radial direction from the center of the Earth. Since the gradient of the current function is perpendicular to the current vector $\boldsymbol{I}$, contours of J = constant give streamlines of $\boldsymbol{I}$.

The relationship between Φ and J is given by combining equations (2) and (4):

$$\begin{aligned} \frac{1}{\sin \theta} \frac{\partial J}{\partial \psi} &= -\Sigma_P \frac{\partial \phi}{\partial \theta} - \frac{\Sigma_H}{\sin \theta} \frac{\partial \theta}{\partial \phi} \\ \frac{\partial J}{\partial \theta} &= \Sigma_H \frac{\partial \phi}{\partial \theta} - \frac{\Sigma_P}{\sin} \frac{\partial \phi}{\partial \psi} \end{aligned} \tag{6}$$

where θ denotes the colatitude. The steps we have chosen to solve this problem are as follows: (i) solve Equation (2) for Φ for a given distribution of $j_{\parallel}$ and also for a given distribution of Σ_P and Σ_H; (ii) insert Φ thus obtained into (6) to obtain the current function J; (iii) examine relationships between $j_{\parallel}$ and the stream lines J = constant.

(b) *Model*

We consider an ionospheric model. The ionosphere is divided into three regions: the polar cap, the auroral oval, and the mid-low latitude belt. The polar cap is represented by a circular area which is bounded by the latitude circle of $\lambda_1 = 70°$. The auroral oval is represented by a circular belt which is bounded by two latitude circles of $\lambda_1 = 70°$ and $\lambda_2 = 65°$. In each region the conductivity is assumed to be uniform, and thus the model is azimuthally symmetric. The Pedersen and Hall conductivities in each region are given in Table 1.1.

TABLE 1.1
Conductivities (mho) used in the model study

Polar cap (Region I)		Auroral oval (Region II)		Middle- and low-latitude (Region III)	
Σ_P	Σ_H	Σ_P	Σ_H	Σ_P	Σ_H
1.0	2.0	5.0	10.0	1.0	2.0

The field-aligned currents are assumed to flow into or out only from the circles of λ_1 and λ_2; see Figure 1.11. Since the longitudinal dependence of the field-aligned current intensity is not available at the present time, we assume a simple distribution, illustrated in Figure 1.11.

Along the poleward boundary of the oval ($\lambda_1 = 70°$) the current flows into (namely, downward and positive) the oval in the sector bounded by $\psi = 0°$ and

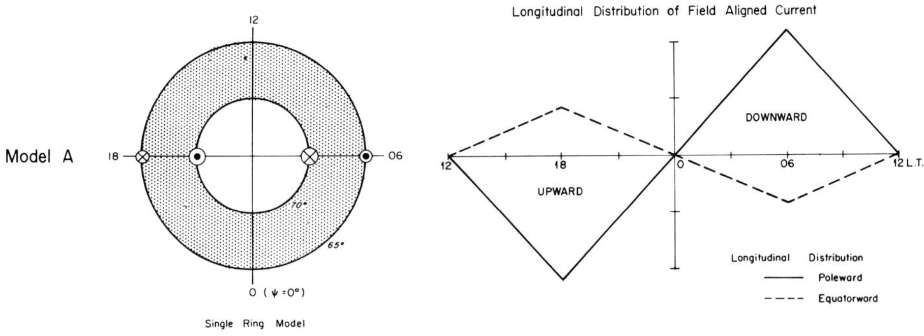

Fig. 1.11. Geometry of the high conductive belt with the peak locations of the field-aligned currents and the assumed longitudinal (or MLT) dependence of the field-aligned current. (Yasuhara, F., Kamide, Y., and Akasofu, S.-I.: *Planet. Space Sci.* **23**, 1355, 1975.)

$\psi = 180°$, while the current flows out (namely, upward and negative) from the oval in the sector bounded by $\psi = 180°$ and $\psi = 0°$. The current intensity has the maximum at $\psi = 90°$ (06 LT) and the minimum at $\psi = 270°$ (18 LT). Along the equatorward boundary of the oval ($\lambda = 65°$), the current flows out from the oval in the sector bounded by $\psi = 180°$ and $\psi = 0°$. The current intensity has the minimum at $\psi = 90°$ (06 LT) and the maximum at $\psi = 270°$ (18 LT). The ratio of the total equatorward current to the total poleward current is assumed to be 0.5 in both evening and morning sectors, as observed (Yasuhara *et al.*, 1975).

The conductivity in the oval (Region II) is assumed to be 5 times as large as that in Region I or III. The Hall conductivity in each region is assumed to be twice as large as the Pedersen conductivity in the same region (Section 7.7.2).

(c) *Results*

Figure 1.12(a) shows the stream lines of the ionospheric currents. The amount of current flowing between two adjacent flow lines is 20 kA, and the arrows indicate the direction of the currents. The discontinuity of the stream lines along the latitude circles of $\lambda = 70°$ and 65° indicates that there are field-aligned currents from or into those latitude circles. As a guide for Figure 1.12(a), Figure 1.12(b) is provided to illustrate the relationship between the field-aligned currents (shaded vertical arrows) and the ionospheric currents (thin and thick arrows). A number for each arrow indicates the normalized amount of current with respect to the total amount of field-aligned current which flows into the poleward boundary in the morning sector (taken to be 100). The reason for grouping the poleward incoming and outgoing field-aligned currents into three parts is self-explanatory. For the equatorward currents, the grouping depends on the direction of the connecting ionospheric currents, flowing approximately in the same meridian (the N-S direction), deflected westward or eastward. The criterion for the deflection is determined by the angle which the stream line makes with respect to the center line of the oval. If the angle is smaller than 45°, the current is considered to be deflected in either a westward or eastward direction. Note that three-dimensional divergence of the current is null with accuracy of ± 2. Note also that in order to see only the major features in Figures 1.12(a) and (b), ionospheric currents with the normalized intensity of more than 10 are indicated with thick arrows in Figure 1.12(b).

There is no doubt that details of the distribution of ionospheric currents vary for different models of the ionosphere. However, the basic features indicated by this model study will essentially remain unchanged. Some of the important results are:

(1) Of the total inflow (100) along the poleward boundary of the oval in the morning sector, about 50% flows across the oval and into the equatorward field-aligned currents. This will be called the N-S closure. It should be noted that a relatively large part of this component does not close in the same meridian, but has a large westward component in the early morning hours and a large eastward component in the late morning hours.

(2) Only about 10% of the total downward current flows westward along the nightside oval. This is because $(6 + 2\%)$ of the westward current (19%) flows

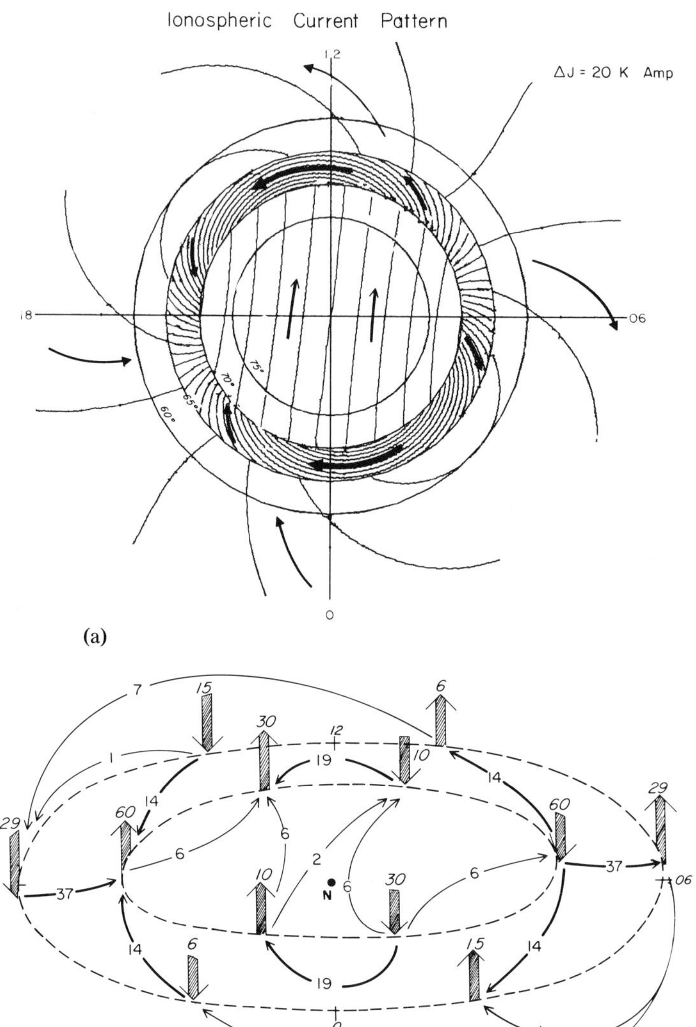

Fig. 1.12. (a) Computed ionospheric current distribution and (b) its schematic illustration, indicating the relationship with the field-aligned current, (Yasuhara, F., Kamide, Y., and Akasofu, S.-I.: *Planet. Space Sci.* **23**, 1355, 1975.)

across the polar cap. About 10% flows eastward along the dayside oval to form the E-W closure.

(3) Of the total inflow, about 20% ($= 6 + 6 + 2 + 6$) flows across the polar cap. However, this is not a direct connection between the field-aligned currents in the opposite sides; some parts of those circuits lie along the oval, contributing to westward currents in the oval.

(4) The total inflow along the equatorward boundary of the oval in the evening sector (50) is discharged from the poleward boundary in the evening sector. This is

the N-S closure, but again a large part of this does not close in the same meridian, as pointed out in (1).

(5) The amount of current in the mid-low latitude belt is much smaller than that which flows along the oval. The direction of this current is opposite to the expected direction of the low latitude return current from the auroral electroject (a pure two-dimensional current).

This study shows that the N-S segment of the ionospheric current which connects the pair of field-aligned currents has, in general, an appreciable east-west component. Another important feature is that the equatorward field-aligned currents in the evening and morning sectors do not connect each other. The entire amount of equatorward field-aligned currents and one half of the poleward field-aligned currents are connected. This does not seem to depend on models of the ionosphere.

The closure of the remaining half of the poleward field-aligned currents depends greatly on the distribution of the ionospheric conductivities. The direct E-W closure of this current increases when the conductivity is high in the oval compared with that in other regions. More currents flow in the polar cap and the mid-low latitude region when the auroral oval conductivity is reduced. The deflection of the N-S closure in the westward or eastward direction is also increased with an enhancement of the Hall conductivity in the oval.

The amount of ionospheric currents which flow in the middle-low latitude belt is surprisingly small. Perhaps the ground magnetic perturbations that have been ascribed to the two-dimensional currents might be mainly due to the magnetic fields of the field-aligned currents (Fukushima and Kamide, 1973; Lyatskiy *et al.*, 1974). Pudovkin (1974) made also an extensive study of the relationship between the field-aligned currents and ionospheric currents.

1.3.4. ELECTRIC FIELD DISTRIBUTION AND THE AURORAL OVAL

(a) *Observations*

During the last several years, several methods have been developed for measuring electric fields in the ionosphere and the magnetosphere. The methods can be grouped into two. One is to measure the potential directly with probes carried by balloons (Mozer and Serlin, 1969; Mozer and Manka, 1971; Mozer, 1972, 1973a, b, c; Mozer and Lucht, 1974; Mozer *et al.*, 1974), by rockets (Potter, 1970; Fahleson *et al.*, 1970; Kelley *et al.*, 1971; Kelley *et al.*, 1975) and by satellites (Cauffman and Gurnett, 1971; Heppner, 1972a, b, c; Gurnett, 1970, 1972a, b). The other is to observe either the drift velocity of thermal plasma particles or the motions of energetic particles in a known (or assumed) distribution of the magnetic field, and then to infer the electric field. Here we shall be concerned with the gross distribution of the electric field which is observed by balloon-, rocket- and satellite-borne probes. The electric field observations during substorms will be discussed in Sections 7.7.3–7.7.6.

Typical examples of the distribution of the electric field along the dawn-dusk meridian over the polar cap region are shown in Figure 1.13. In the upper part of the figure, the OGO-6 satellite moved toward the northern polar regions along the

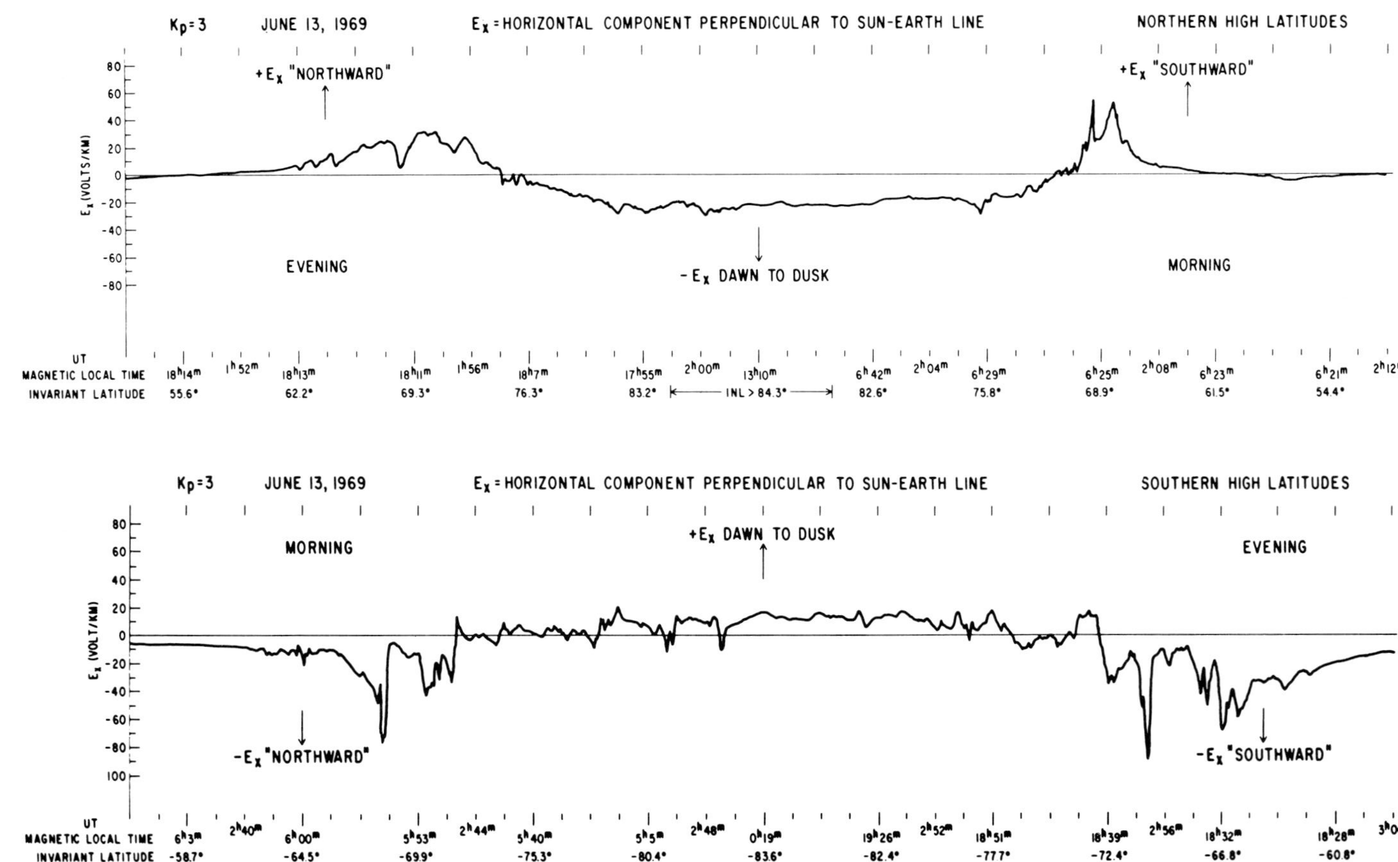

Fig. 1.13. Dawn-dusk component of the electric field along a dawn-dusk traverse across the northern and southern polar regions. (Heppner, J. P.: *Planet. Space Sci.* **20**, 1475, 1972.)

dusk meridian. The intensity of the poleward-directed electric field increased gradually, reaching a peak value of order $20 \sim 30$ mV m^{-1} at invariant latitude $\sim 70°$. Then it decreased steeply and reversed the sign at about inv. lat. $\sim 75°$. That is to say, beyond this point the satellite observed an electric field directed in the dawn-to-dusk direction, until it reached inv. lat. $\sim 70°$ along the dawn meridian, where the electric field reversed the direction (directed now equatorward) and increased sharply. The electric field reached a peak value of about 40 mV m^{-1} at inv. lat. $\sim 69°$ and then decreased. Subsequently, the satellite traversed the southern polar region, after crossing the equator; the profile is shown in the lower half of the figure. It can be seen that the electric field is directed from the dawn-to-dusk direction in the southern polar cap.

Figure 1.14 shows the observed location where the electric field reverses its direction, indicating that the reversal points are roughly distributed along the auroral oval (Gurnett, 1972b). Thus the gross distribution of the electric field over the polar cap region can be approximated by assuming that the charge distribution ρ along the oval is given by $\rho = \rho_0 \sin \psi$ where $\psi = 15° \times \text{MLT}$. The equipotential contours for such a potential distribution are shown in Figure 1.15.

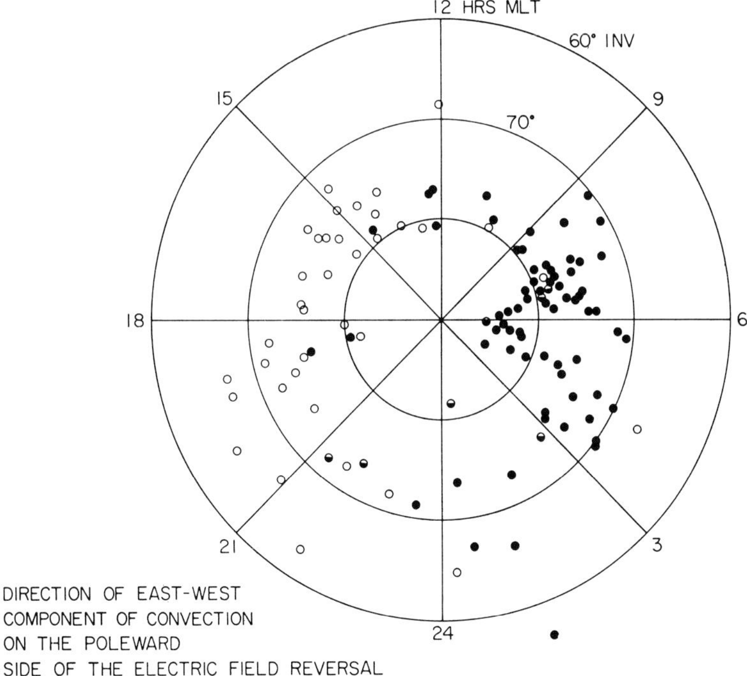

Fig. 1.14. Locations of the reversals of the electric field direction, observed by the Injun-5 satellite. The open circle indicates that the convection velocity is eastward on the poleward side of the reversal and westward on the equatorward side. The dots indicate the opposite. (Gurnett, D. A.: *Earth's Magnetospheric Processes*, B. M. McCormac (ed.), p. 233. D. Reidel Publ. Co., 1972.)

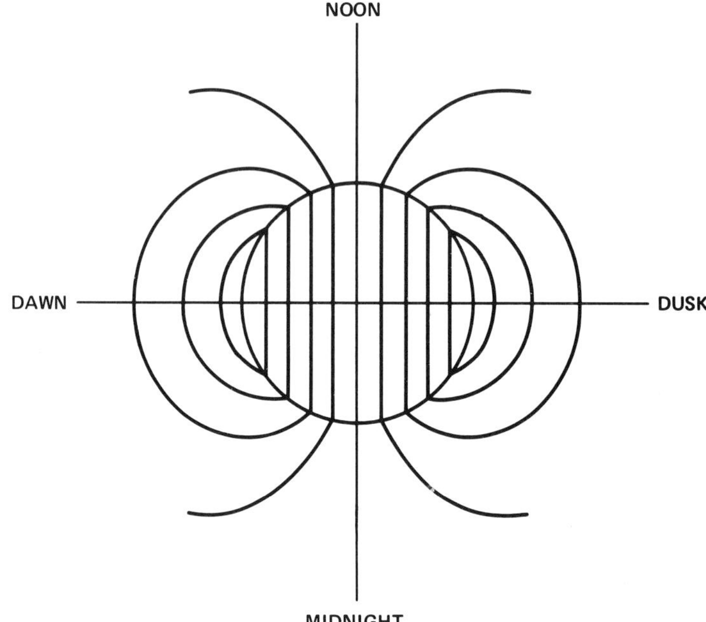

Fig. 1.15. Electric equipotentials for the charge distribution given by $\rho = \rho_0 \sin \psi$ where $\psi = 15° \times$ MLT. (Stern, D. P.: Goddard Space Flight Center Rep., X-602-75-17, 1975.)

Outside the auroral oval, the electric field distribution is complicated by the presence of the Harang discontinuity in the late evening sector. The north-south component of the electric field reverses the sign across the discontinuity. This feature is schematically shown in Figure 1.16. Maynard (1974) showed that the discontinuity is present even during quiet periods but becomes quite dynamic during substorms.

The dawn-to-dusk electric field in the polar cap has also been observed by balloon-borne probes (Mozer and Serlin, 1969; Mozer, 1973b; Mozer and Lucht, 1974). The balloon observations were based on the principle that the probe rotates (with the Earth) under the electric field pattern, which is steady and fixed with respect to the Sun-Earth line. Figure 1.17 shows a combined result of the electric field observations made by several balloons (Mozer and Lucht, 1974). Kelley and Mozer (1975) showed recently that the horizontal electric fields observed by the Chatanika incoherent scatter radar at the ionospheric level (~ 100 km) and by the balloon-borne probe at an altitude of 30 km are essentially equal.

(b) *Theoretical Studies*

The theoretical model presented in Section 1.3.3 provides the potential and electric field distribution, as well as the ionospheric current distribution. Figure 1.18(a) shows the electric field distribution along the dawn-dusk meridian, for both the dawn-to-dusk component and the noon-to-midnight component. The dawn-to-dusk component agrees well with the observed distribution in Figure 1.13. Figure 1.18(b) shows the potential distribution for the same model.

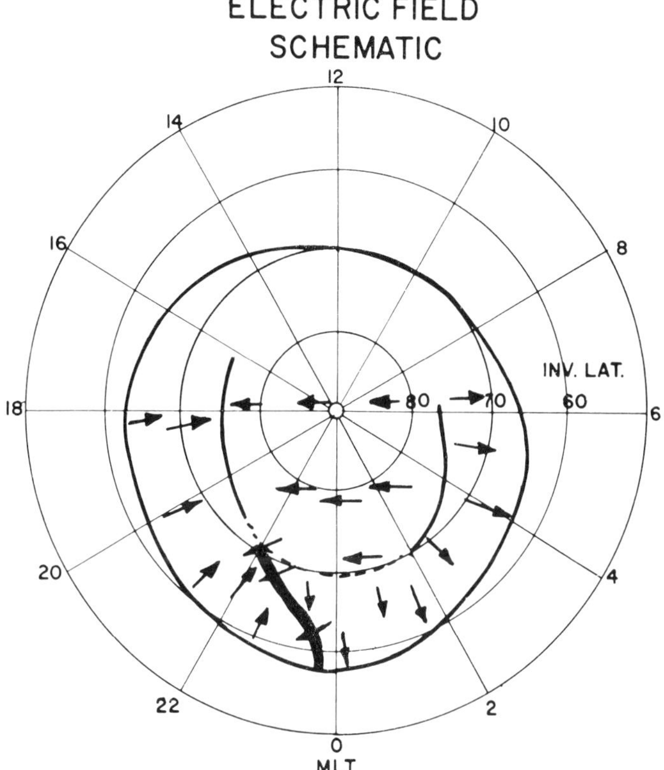

Fig. 1.16. Electric field distribution (schematic) in the polar region. (Maynard, N. C.: *J. Geophys. Res.* **79**, 4620, 1974.)

1.3.5. PLASMA CONVECTION IN THE EQUATORIAL PLANE AND THE POLAR CAP

The electric field drives a large-scale motion of plasma throughout the magnetosphere. By assuming that the geomagnetic field lines are equipotential, the electric field observed in the polar ionosphere can be projected onto the equatorial plane, provided that the field line geometry is known. In this subsection we deal with the noon-to-midnight convection of the ionospheric plasma and the corresponding sunward (or earthward) convective motion of plasma in the plasma sheet. It should be noted, however, that these are only one half of the entire convection of magnetospheric plasma, which will be discussed in Section 3.5.1.

When there is no electric field parallel to $\boldsymbol{B}$, the convection velocity $\boldsymbol{V}$ of magnetospheric plasma is given by

$$\boldsymbol{V} = \boldsymbol{E} \times \boldsymbol{B}/B^2$$

where $\boldsymbol{E}$ is given by the gradient of a scalar function

$$\boldsymbol{E} = -\nabla\Phi$$

so that

$$\boldsymbol{V} \times \boldsymbol{B} = \nabla\Phi$$

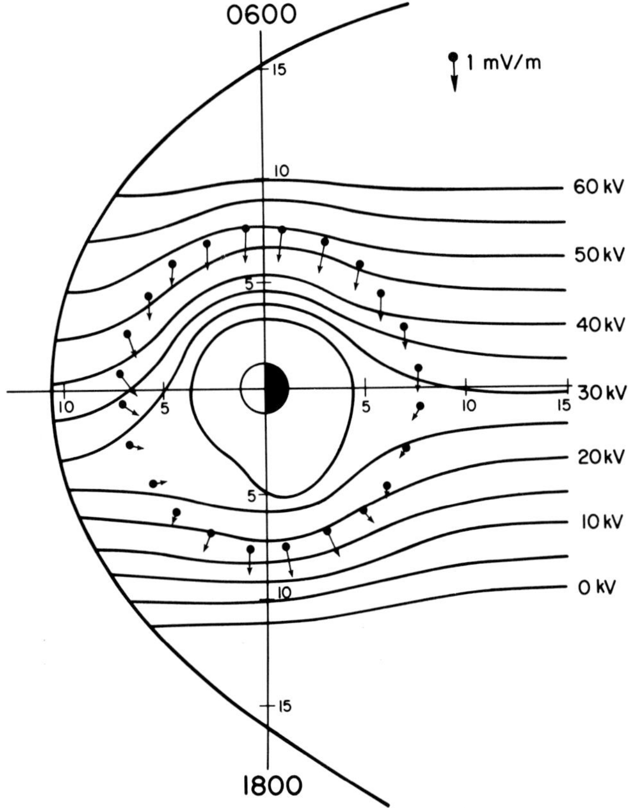

Fig. 1.17. Electric field (hourly averaged) vectors observed by balloon-borne instruments in the polar region. They are replotted in the equatorial plane of a nonrotating frame of reference, as viewed from above the north pole. The solid curves are electric equipotentials deduced from the observed data on the assumption that the electric field is uniform across the tail. (Mozer, F. S. and Lucht, P.: *J. Geophys. Res.* **79**, 1001, 1974.)

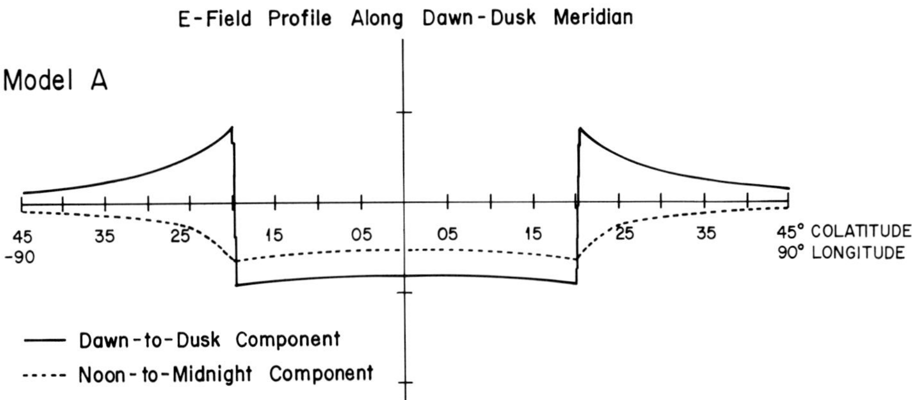

Fig. 1.18(a). Computed dawn-to-dusk component and noon-to-midnight component of the electric field along the dawn-dusk meridian. (Courtesy of Yasuhara, F.)

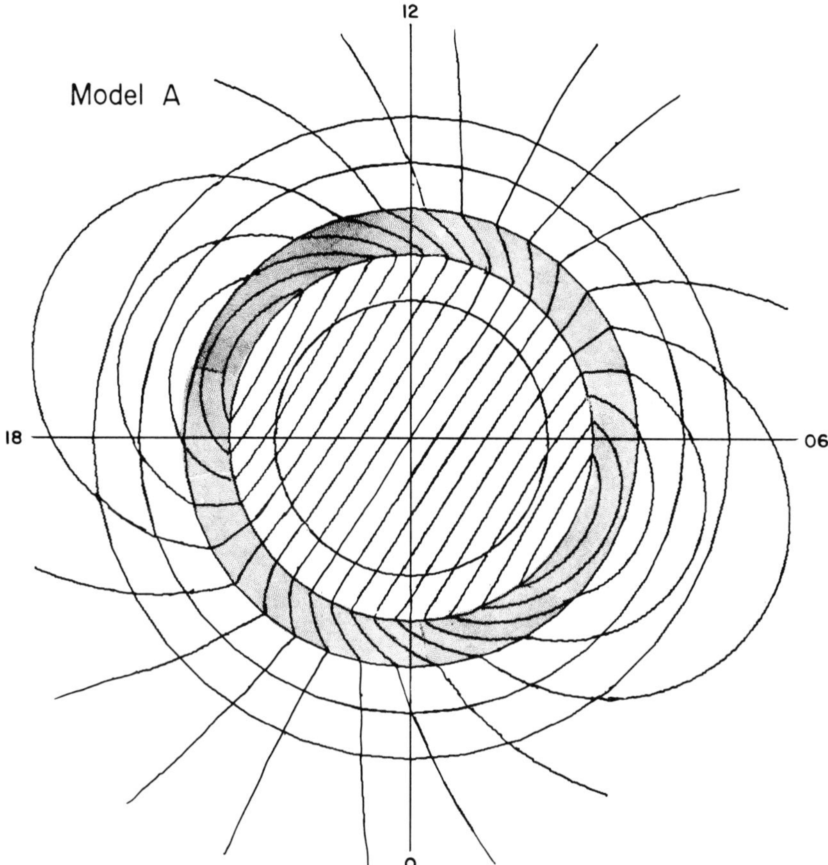

Fig. 1.18(b). Computed electric equipotentials in the ionosphere. (Courtesy of Yasuhara, F.)

Therefore, the stream lines of the convective motion are identical to the electric equipotential lines (Φ = constant). It is thus not difficult to infer the convection stream lines in the polar ionosphere. Such a large convective motion of magnetospheric plasma was discussed first by Axford and Hines (1961).

(a) *Observations*

The drift velocity of thermal plasma can be deduced by measuring directly (i) the drift motions of plasmas (Freeman, 1968; Galperin *et al.*, 1974; Gueth *et al.*, 1974; Whalen *et al.*, 1974; Whalen *et al.*, 1975; Heelis *et al.*, 1976): (ii) the drift motion of barium clouds released from rockets (Föppl *et al.*, 1968; Haerendel, 1972; Haerendel and Lüst, 1970; Wescott *et al.*, 1969, 1970; Jeffries *et al.*, 1975); (iii) the drift motion of ionospheric plasma by an incoherent scatter radar (Doupnik *et al.*, 1972; Banks, 1972); (iv) the drift motion of whistler ducts (Carpenter *et al.*, 1972);

(v) the drift motion of auroras (Davis, 1971); and (vi) the Doppler frequency drift of electrostatic electron cyclotron harmonic waves (Oya, 1975). Motions of non-thermal particles are also useful in inferring magnetospheric electric fields (Van Allen, 1970; Roederer and Hones, 1970; McIlwain, 1972, 1974; Palmer *et al.*, 1976; McCoy *et al.*, 1975).

(i) *Direct observations of drifting plasmas.* Galperin *et al.* (1974) measured directly the convective plasma flow over the polar cap region by a device aboard the Cosmos-184 satellite, which had a circular orbit at an altitude of 630 km. Orbits were approximately along the noon-midnight meridian. They showed that their results are consistent with the convective motion inferred from the electric field observation in the polar cap region.

A more recent observation of vector ion velocity by the AE-C satellite has shown that the convection stream lines tend to become parallel to both the afternoon and morning part of the oval, suggesting that a large portion of the dayside oval is nearly electrically equipotential (Heelis *et al.*, 1976). Jeffries *et al.* (1975) showed also that a barium cloud released in the midday cusp region drifted along the oval during a quiet time.

Whalen *et al.* (1974) and Whalen *et al.* (1975) made a detailed measurement of plasma drift motions in the vicinity of an active auroral arc. A little poleward of the arc the flow was directed poleward, and its speed was of order 3 km s^{-1} (equivalent electric field of 100–200 mV m^{-1}). Within the arc, however, the flow speed was of order 1 km s^{-1} (equivalent electric field of 10–30 mV m^{-1}).

(ii) *Barium cloud observation.* Figure 1.19 summarizes the results obtained by the barium plasma cloud technique (Haerendel, 1972). It can be seen that along the auroral zone the motions are predominantly westward (toward the dusk meridian) in the evening sector and eastward (toward the dawn meridian) in the morning sector, and that in the polar cap they are directed away from the Sun. In Section 7.7.4, we shall see that a barium cloud which was released along the field lines near the cusp (just inside the oval in the noon sector) during a disturbed period drifted rapidly in the anti-sunward direction with a speed of order of 1400 m s^{-1} (implying an electric field of 82 mV m^{-1} directed dawn-to-dusk).

The barium cloud release technique in the distant magnetosphere is considerably complicated by the fact that the ambient plasma is very tenuous there. As we shall discuss in detail in Section 7.7.4, Haerendel *et al.* (1971) inferred that the electric field of at least 0.08 mV m^{-1} was directed radially outward in the high latitude lobe at a geocentric distance of 12.5 R_E in the morning sector.

(iii) *Incoherent scatter radar observations.* Figure 1.20 shows an example of convection velocity vectors measured by the incoherent scatter radar located at Chatanika near Fairbanks during the course of 24 h, 1972, February 11–12. An intense westward flow in the late evening sector and an intense eastward flow in the early morning sector were associated with substorm activity. A similar flow pattern of much reduced speed is present even during a fairly quiet period. It is interesting to note also a poleward flow in the midday sector and an equatorward

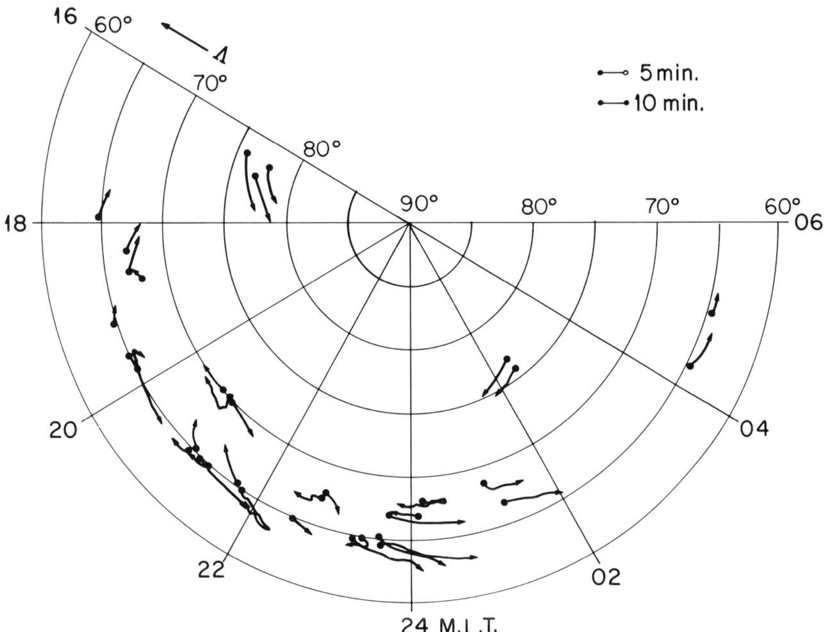

Fig. 1.19. Drift paths of barium clouds in invariant latitude-MLT coordinates. (Haerendel, G.: *Earth's Magnetospheric Processes*, B. M. McCormac (ed.), p. 246, D. Reidel Publ. Co., 1972.)

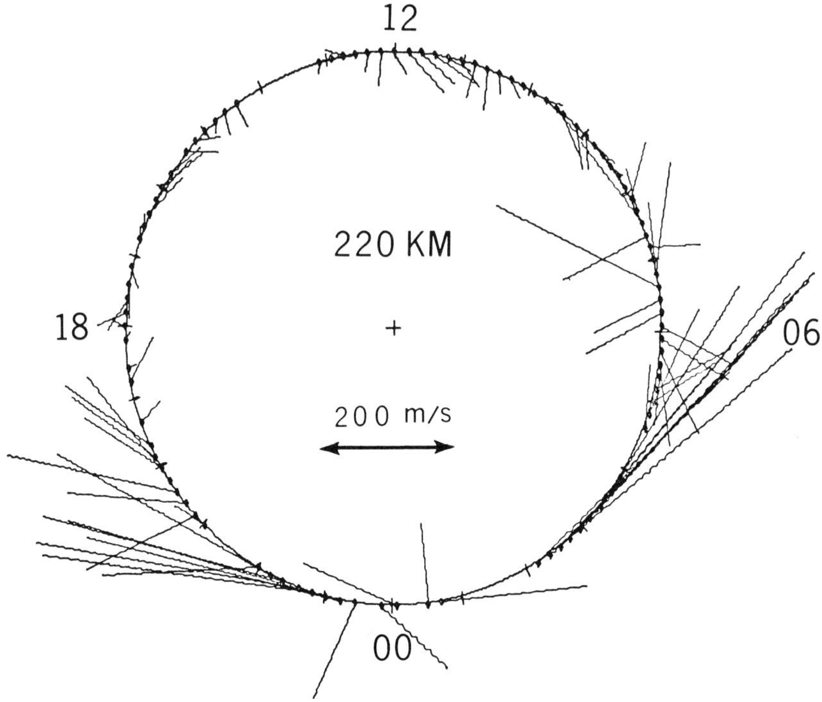

Fig. 1.20. Convection velocity vectors at 220 km altitude, observed by the Chatanika incoherent scatter radar, on 1972, February 11–12. (Doupnik, J. R., Banks, P. M., Baron, M. J., Rino, C. L., and Petriceks, J.: *J. Geophys. Res.* **77**, 4268, 1972.)

flow in the midnight sector. Note that the equatorward flow in the midnight sector is much more enhanced than the poleward flow in the midday sector; the substorm was in progress at that time.

(iv) *Particle motions.* On the basis of observations of both protons and electrons for a wide range of energies (0–20 keV) observed at the synchronous distance, McIlwain (1972, 1974) deduced the electric potential distribution for a given magnetic field distribution in the equatorial plane. His potential distribution for an observer in the rest frame is given in Figure 1.21.

Palmer *et al.* (1976) showed that solar protons enter into the magnetotail in the form of slab and that there is an appreciable anisotropy of the proton flux near the

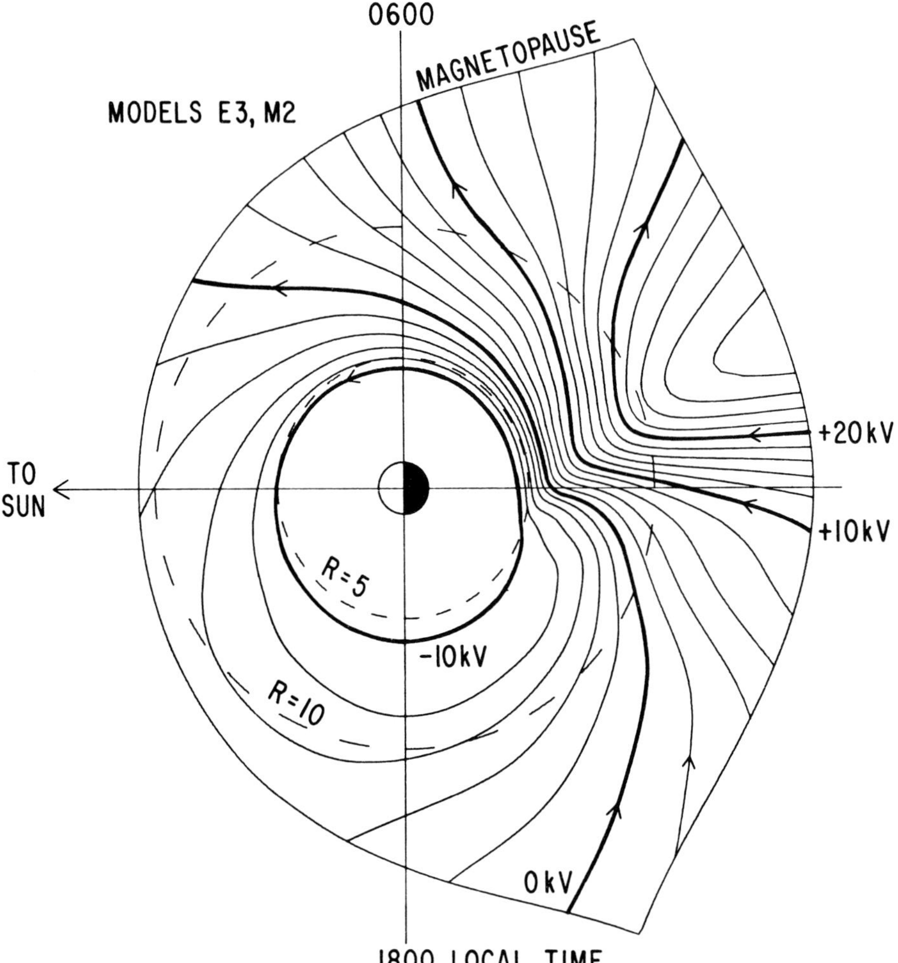

Fig. 1.21. Electric equipotentials (Model E3) in the magnetic equatorial plane, inferred from particle observations at the synchronous distance. (McIlwain, C. E.: *Earth's Magnetospheric Processes*, B. M. McCormac (ed.), p. 268, D. Reidel Publ. Co., 1972.)

surface of the slab. Assuming that the slab moves toward the neutral sheet with a speed $V = E/B$,

$$V = 2\eta_g/\tau_c$$

where η_g denotes the gyro-radius of solar protons and τ_c the observed duration of the anisotropy. Thus

$$E \approx \frac{250}{\tau_c}\ \text{mV m}^{-1} \qquad \text{for } 0.7\ \text{MeV protons}$$

The duration τ_c is of order a few hundred seconds. They obtained the average electric field of order 0.88 mV m^{-1}.

McCoy *et al.* (1975) devised an interesting method to infer electric fields in the magnetotail by detecting $(E \times B)$ drift displacement of low-energy (0.5–14 keV) solar electrons at the limb of the Moon. They found that the electric field is almost always directed from dawn to dusk, and its typical value is 0.15 mV m^{-1}.

(v) *Drifting whistler paths.* Carpenter and Stone (1967), Carpenter (1970), Carpenter *et al.* (1972) and Carpenter and Kirchoff (1975) showed that it is possible to infer the radial component of the drift velocity of plasma and thus of the east-west component of magnetospheric electric field on the basis of 'cross-L motions' of whistler ducts within the plasmasphere. This method provides important information on the depth of penetration of the large-scale electric field, particularly during substorms. We shall discuss the whistler results in Section 8.2.1(b).

(vi) *Auroral motions.* Davis (1971) inferred the convective motion from motions of the aurora over the polar region and projected the velocity vectors onto the equatorward plane by using a model of the magnetosphere (Figure 1.22).

(b) *Theoretical Studies*

In theoretical studies of the convection, the common practice has been to assume a particular potential distribution Φ along the auroral oval which is simulated by a latitude circle of λ_0 (Kavanagh *et al.*, 1968; Wolf, 1970, 1974, 1975; Swift, 1971; Vasyliunas, 1970, 1972). The potential Φ is similar to or the same as that shown in Figure 1.15. Such a potential distribution drives a sunward (or an earthward) convective motion of plasma in the plasma sheet. Vasyliunas (1970) showed that the field-aligned current from the convected plasma sheet is given by

$$(j_\parallel)_{eq} = \frac{B_p}{2B_e}\left(\nabla p \times \frac{\boldsymbol{B}_e}{B} \cdot \nabla \int \frac{dl}{B}\right)$$

where B_p and B_e denote the magnetic field at the ionospheric level and in the equatorial plane. The current continuity equation in the ionosphere is given (Section 1.3.3(a)) by

$$-\nabla \cdot \left\{\begin{pmatrix} \Sigma_P & \Sigma_H \\ \Sigma_H & \Sigma_P \end{pmatrix} \cdot \nabla\Phi\right\} = (1/2)\left(\frac{B_p}{B_e}\right)(j_\parallel)_{eq}$$

Jaggi and Wolf (1973) and Wolf (1975) examined a time-dependent solution of

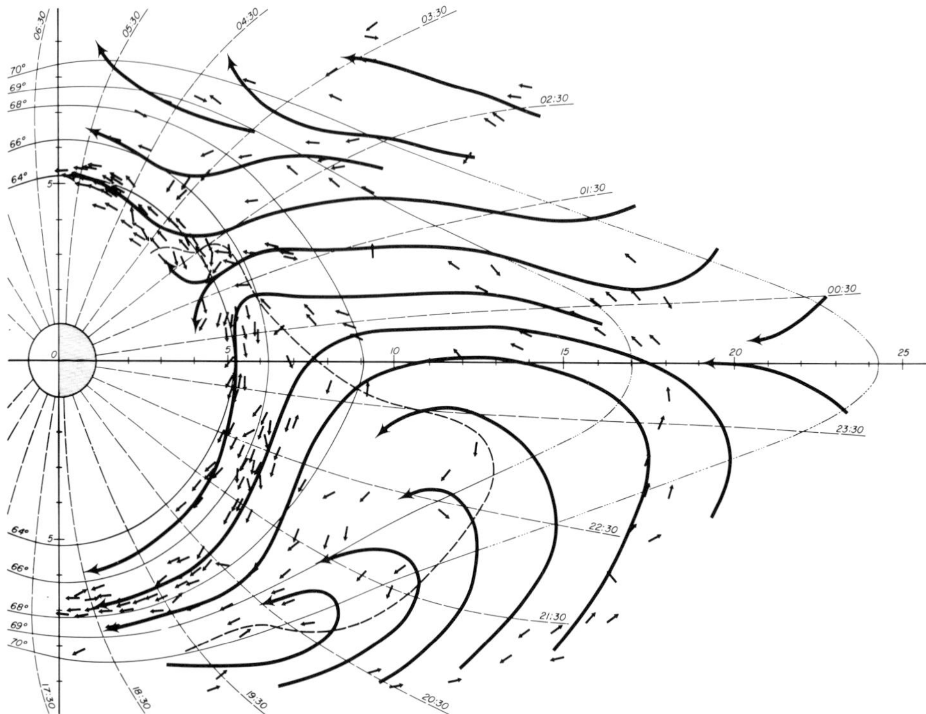

Fig. 1.22. Direction of motion of irregularities along auroral forms and also of segments of auroral forms, mapped from the ionosphere to the equatorial plane through a magnetospheric model. Heavy continuous lines are generalized flow lines. (Davis, T. N.: *J. Geophys. Res.* **76**, 5978, 1971.)

the above equation for an ion sheet which is injected toward the Earth from a great distance in the magnetotail. The field-aligned current $(j_{\|})_{eq}$ arises from the advancing front of the plasma sheet (Section 1.3.6(a)). Figure 1.23 is an example of such a calculation, showing the equi-potential contour lines (in kilovolts) which are mapped into the equatorial plane for a given model of the magnetosphere at $t = 2.5$ h after the injection.

The adopted parameters are:

Φ = the potential across the polar cap
= 33.4 kV

η = the number of particles per unit magnetic flux
= 6.3×10^{20} Wb^{-1} (corresponding to 22 cm^{-3} and 0.33 cm^{-3} at $L = 3.5$ and 10, respectively).

μ = the magnetic moment
= 200 eV/γ

$\langle \sigma_P \rangle$ = the average Pedersen conductivity in the afternoon sector
= 8 mho

The figure gives the potential distribution as measured in the frame rotating with the Earth. One of the important features depicted in this study is that the

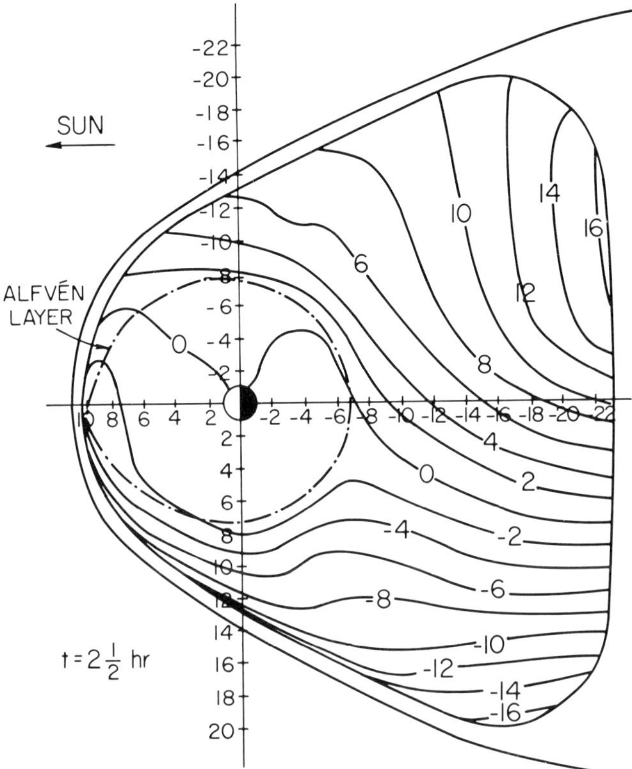

Fig. 1.23. Electric equipotentials (in kilovolts) in the equatorial plane, computed by assuming a near-circular inner boundary of the plasma sheet (the Alfvén layer) at $L \simeq 8$ at time 0, then allowing the plasma sheet and electric field to evolve self-consistently for 2.5 h. (Wolf, R. A.: *Space Sci. Rev.* **17**, 537, 1975.)

electric field becomes much weaker beyond the boundary called 'the Alfvén layer' by the authors. This shielding results from the electric polarization in the Alfvén layer, which is caused by a differential motion of protons and electrons in the non-uniform geomagnetic field.

For an observer in the rest frame, the potential distribution Φ_m is given by

$$\Phi_m = \Phi - \Omega R_E^2 B_e^2 \sin^2\left(\frac{\pi}{2} - \lambda\right)$$

where Ω and B_e denote the angular speed of the Earth and the magnetic field intensity on the Earth's surface at the equator, respectively.

Here we consider qualitatively effects of the ionosphere on the convection of magnetospheric plasma. When plasma particles in the plasma sheet are convected towards the Earth by the dawn-to-dusk electric field across the magnetotail, non-uniformity of the Earth's magnetic field causes their gradient drift motions. Protons drift towards the evening magnetopause, and electrons towards the morning magnetopause, creating space charges near the advancing front of the plasma sheet. Since the resulting electric field will be directed in the dusk-to-dawn direction, the primary (dawn-to-dusk) field tends to be cancelled by this newly induced electric field near the advancing front. Thus the minimum distance which the plasma can reach depends on the growth of this field. Since the ionosphere is a conducting layer (and the space charge region is connected to it by the geomagnetic field lines), it tends to discharge the space charges, allowing the plasma to advance deeper than without it. Jaggi and Wolf (1973) estimated that in a steady state the newly induced field reduces the primary field by an order of magnitude in the region between the advancing surface of the plasma sheet and the plasmasphere. Further, they showed that the minimum penetration depth of the plasma is approximately given by

$$L_G \simeq 7.93 \left(\frac{(31\,\gamma)\eta\mu}{\Phi\langle\sigma_p\rangle}\right)^{1/3} \simeq 5.7$$

Here the distance L_G is actually the geocentric distance of the plasma in the dusk sector (although the inner boundary of the plasma sheet is nearly circular) for the parameters $\Phi = 33.4$ kV, $\eta = 6.3 \times 10^{20}$ Wb^{-1} and $\mu = 50$ eV/γ and $\sigma_p = 8$ mho; for details of the derivation of this equation, see their paper (p. 2861).

The convection of magnetospheric plasma and its relation to the ionosphere has also been discussed by a number of workers (Coroniti and Kennel, 1973; Atkinson, 1975; Krylov and Shcherbakov, 1972; Mal'tsev, 1974). By noting that the dawn-dusk electric field is cancelled by the electric field due to the Earth's rotation at a point of geocentric distance Lpp along the 18 LT meridian and that this point may coincide with the plasmapause location (Nishida, 1966), Mendillo and Papagiannis (1971) obtained an empirical relation between the solar wind speed V_s (km s^{-1}) and the convection electric field E, given by $E = 3.63\,(V_s/533)^2$ kV/R_E and $L\text{pp} = 5.0\,(533/V)$. Freeman (1974) obtained also an empirical relationship between the Kp index, the convection electric field and the geocentric distance of the plasma sheet boundary in the midnight sector (L); they are given by

$$E \simeq 4.5 \times 10^{-4}\left(1 - \frac{\text{Kp}}{10}\right)^{-2} \text{(volt m}^{-1}\text{)}$$

$$L = \left\{3.48 \times 10^{-3} + \frac{9.7 \times 10^{-11}\cos(143 - 24\,\text{Kp})}{\mu\,[1 - (\text{Kp}/10)]^2}\right\}^{-1/3} \qquad (\mu \text{ in mks unit})$$

1.3.6. THREE-DIMENSIONAL MAGNETOSPHERIC CURRENT SYSTEM

In this subsection, we describe briefly two basic models of magnetospheric current systems which have been discussed in the past.

(a) *Fejer-Swift-Vasyliunas-Wolf (FSVW) Model*

Figure 1.24 shows the projection of the potential pattern in Figure 1.23 onto the northern polar ionosphere. Note that the electric field is very weak below the latitude of the projection line of the Alfvén layer. Wolf (1975) suggested that the model will be associated with the primary and secondary field-aligned currents. The primary field-aligned current flows downward along the morning half of the

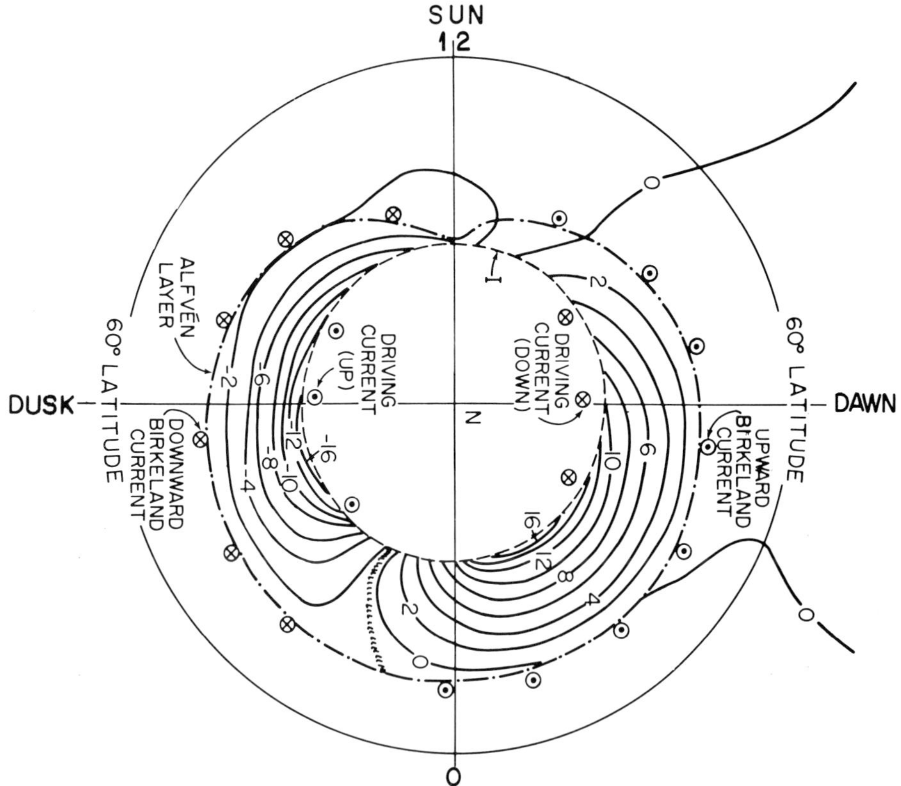

Fig. 1.24. Electric equipotentials (in kilovolts) in the ionosphere, together with the distribution of the field-aligned currents; see caption in Figure 1.23. (Wolf, R. A.: *Space Sci. Rev.* **17**, 537, 1975.)

oval and upward along the evening half of the oval; this is to maintain the impressed potential distribution along the oval. The secondary field-aligned currents are connected to the Alfvén layer (upward along the morning half of the Alfvén layer and downward along the evening half).

In this view, as well as those by Fejer (1963), Swift (1967, 1971), Schield *et al.* (1969) and Vasyliunas (1970), the primary field-aligned currents along the poleward boundary of the oval and those along the equatorward boundary of the oval are connected only through the ionosphere. They are not connected in the magnetosphere. An example of such three dimensional current systems is schematically illustrated in Figure 1.25.

(b) *Boström-Rostoker* (*BR*) *Model*

Boström (1964, 1968) and Rostoker and Boström (1974) proposed that field-aligned currents complete a single circuit in the plasma sheet in a meridian plane (Figure

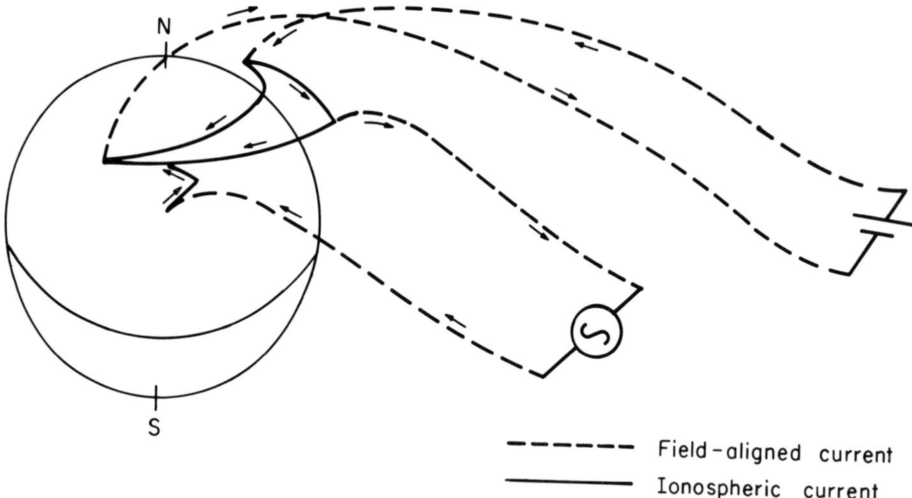

Fig. 1.25. Example of the FSVW type current system in which the poleward and equatorward field-aligned currents are not connected in the equatorial plane. (Yasuhara, F., Kamide, Y., and Akasofu, S.-I.: *Planet. Space Sci.* **23**, 1355, 1975.)

1.26). Thus, the current must flow perpendicular to the magnetic field ($j_\perp$) and is directed toward the Earth in the evening sector and away from the Earth in the morning sector. The most general equation for the current component perpendicular to a magnetic field is given (Parker, 1957) by

$$\boldsymbol{j}_\perp\left(1-\mu_0\frac{p_\parallel - p_\perp}{B^2}\right)=\frac{\boldsymbol{B}}{B^2}\times\left[\nabla p+\rho(\boldsymbol{V}\cdot\nabla)\boldsymbol{V}+(p_\parallel - p_\perp)\frac{\nabla B}{B}\right] \qquad (1)$$

Boström (1975) examined the importance of two terms in the above equation, the pressure gradient term (∇p) and the inertia term ($\rho(\boldsymbol{V}\cdot\nabla)\boldsymbol{V}$). Considering only the pressure gradient term and assuming that the pitch-angle distribution is isotropic, Boström obtained the vertical current j_v into the ionosphere. It is given by

$$j_v=\nabla_0 p_0\times\nabla_0 V_1+e_r \qquad (2)$$

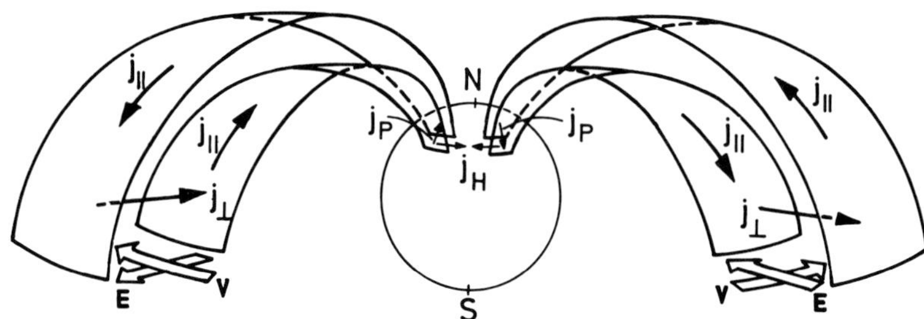

Fig. 1.26. Distribution of the field-aligned currents and the associated Pedersen (ionospheric) currents, plasma motions and electric fields in the equatorial plane. (Boström, R.: *Physics of the Hot Plasma in the Magnetosphere*, B. Hultqvist and L. Stenflo (eds.), p. 341, Plenum Press, 1975.)

where

$$V_1 = \int_0^L \frac{l}{B}\,\mathrm{d}l \tag{3}$$

where ∇_0 denotes the gradient evaluated at the foot (namely, the ionosphere) of a flux tube, $\boldsymbol{e}_r$ a vertical unit vector. For a dipole field of magnetic moment M, $\nabla_0 V_1$ can readily be evaluated to give

$$j_v = \frac{4R_E^2}{\mu M}\left(1 + 4\cos^2\Theta_0 - 2\cos^4\Theta_0 + \frac{4}{5}\cos^6\Theta_0 - \frac{1}{7}\cos^8\Theta_0\right)\sin^{-10}\Theta_0\,\frac{\partial p}{\partial \psi} \tag{4}$$

where Θ_0 denotes the co-latitude. The above equation shows that only the azimuthal gradient of pressure is important in driving the electric current which can close the pair of the field-aligned currents in each meridian plane.

Boström (1975) demonstrated also that the net field-aligned current to the ionosphere is zero in the region enclosed by contour defined by $V_1 = \text{constant}$. Thus, within any strip bounded by two contours of $V_1 = \text{constant}$, the net field-aligned current is also zero. Figure 1.27 shows the $V_1 = \text{constant}$ contour lines. Boström noted that the shape of a strip enclosed by two $V_1 = \text{constant}$ curves resembles the auroral oval. The fact that the net field-aligned current is zero in each strip is in agreement with the TRIAD observation (Section 1.3.2). The required pressure gradient for the observed field-aligned currents is shown in Figure 1.28; the total pressure drop over 3000 km is estimated to be about $1.8 \times 10^{-9} \sim 9 \times 10^{-8}$ dyne cm^{-2}.

Equation (1) indicates that the inertia term $(\rho(\boldsymbol{V} \cdot \nabla)\boldsymbol{V})$ is also capable of driving an electric current $\boldsymbol{j}_\perp$ perpendicular to the magnetic field. Therefore, Rostoker and Boström (1974) and Boström (1975) examined the types of flow pattern of plasma that can drive the two pairs of field-aligned currents. Suppose that the sunward convection flow of plasma in the plasma sheet has a component towards the magnetopause, towards the morning magnetopause in the morning half of the plasma sheet and towards the evening magnetopause in the evening half (as indicated in Figure 1.9). Since the convection arises from the $(E \times B)$ drift motion, such flows require that the electric field has a component towards the Earth in the morning sector and away from the Earth in the evening sector (see Figure 1.26). Projecting these electric fields along the geomagnetic field lines onto the ionosphere, one can see that there will be an equatorward electric field in the morning sector and a poleward electric field in the evening sector, driving the equatorward and poleward Pedersen currents, respectively. Thus, Rostoker and Boström considered that the assumed plasma flow constitutes a dynamo, one in the morning sector and the other in the evening sector, driving the Pedersen currents in the ionosphere. Within the dynamo (located in the equatorial plane in Figure 1.26), the current flows outward in the morning sector and inward in the evening sector. Since these currents flow across the magnetic field, there will be the Lorentz force $\boldsymbol{j} \times \boldsymbol{B}$ directed towards the midnight meridian, retarding the assumed plasma flow. It is this process by which the kinetic energy of the assumed

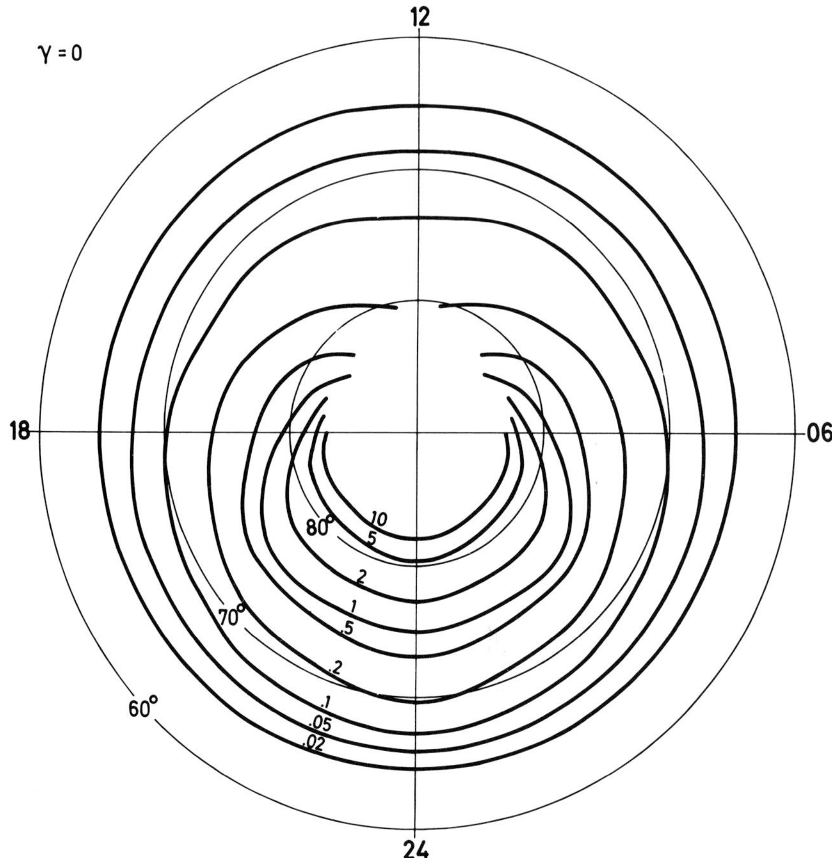

Fig. 1.27. Contours of constant V_1 in units of R_E/nT for the parameter $\gamma = 0\,(p_\parallel = p_\perp)$ by using a model magnetosphere. The shape of the contours resembles that of the auroral oval. (Boström, R.: *Physics of the Hot Plasma in the Magnetosphere*, B. Hultqvist and L. Stenflo (eds.), p. 341, Plenum Press, 1975.)

plasma flow is converted into the electrical energy, driving the current across the magnetic field. The dissipation takes place in the form of Ohmic heating mostly in the ionosphere where the Pedersen current flows.

Boström (1975) did not consider, however, how the assumed plasma flows towards the magnetopause can be driven. In Section 1.3.2, it was shown that the field lines in the plasma sheet should be 'bent' towards the magnetopause by the primary field-aligned current. Perhaps it is this bending which allows the convective $(E \times B)$ motion to have the component towards the magnetopause.

1.3.7. S_q^p VARIATION AND THE AURORAL OVAL

It has been known that there is a particular type of geomagnetic daily variation at polar stations, denoted by S_q^p. This was first recognized by Hasegawa (1940) and later by Nagata and Kokubun (1962). The exact nature of the S_q^p variation was further studied by Kawasaki and Akasofu (1967) and Feldstein and Zaitzev (1968).

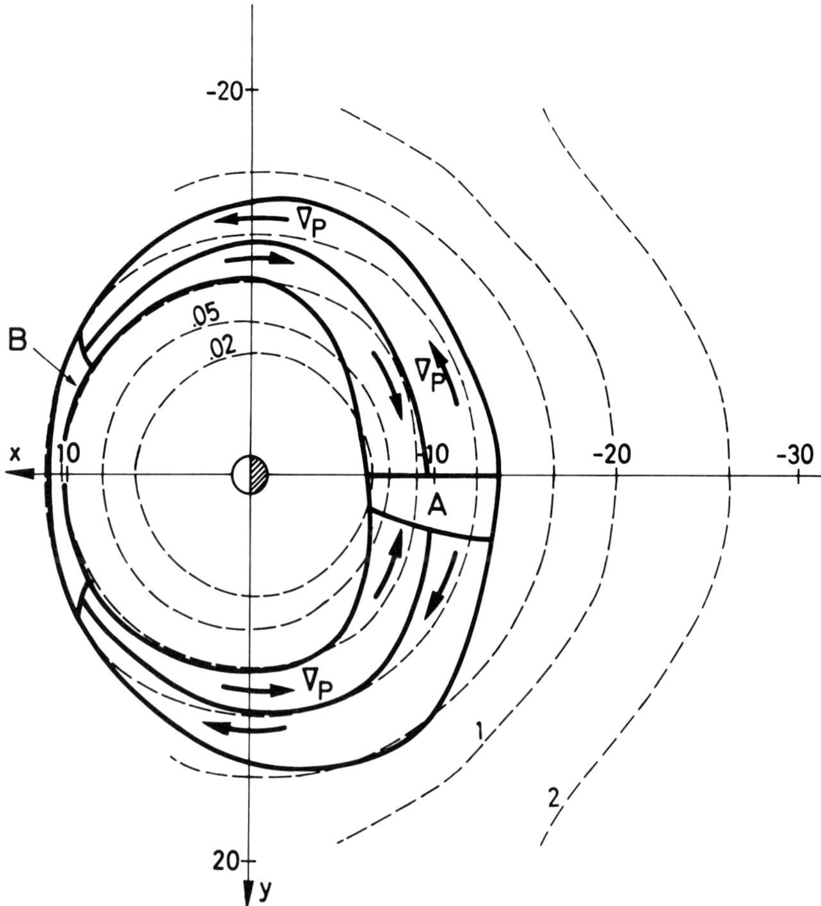

Fig. 1.28. Equatorial plane projection of the contours V_1 = constant in Figure 1.27 and the inferred direction of pressure gradients along the V_1 contours, which are needed in generating the observed distribution of the field-aligned currents. (Boström, R.: *Physics of the Hot Plasma in the Magnetosphere*, B. Hultqvist and L. Stenflo (eds.), p. 341, Plenum Press, 1975.)

The distribution of magnetic vectors associated with the S_q^p variation is shown in Figure 1.29.

Kawasaki and Akasofu (1973) suggested that the current system which is responsible for the S_q^p variation consists of an inward field-aligned current from the morning side of the magnetopause to the morning half of the auroral oval and an outward field-aligned current from the afternoon half of the oval to the afternoon side of the magnetopause, together with ionospheric currents across the polar cap. The presence of such field-aligned currents was discussed in Section 1.3.2. Figure 1.30 shows the computed distribution of magnetic vectors for the assumed S_q^p current system. The total current is assumed to be 10^6 A. It can be seen that there is a reasonable similarity between Figure 1.29 and Figure 1.30. Since we know now that the field-aligned currents are present as a pair of different

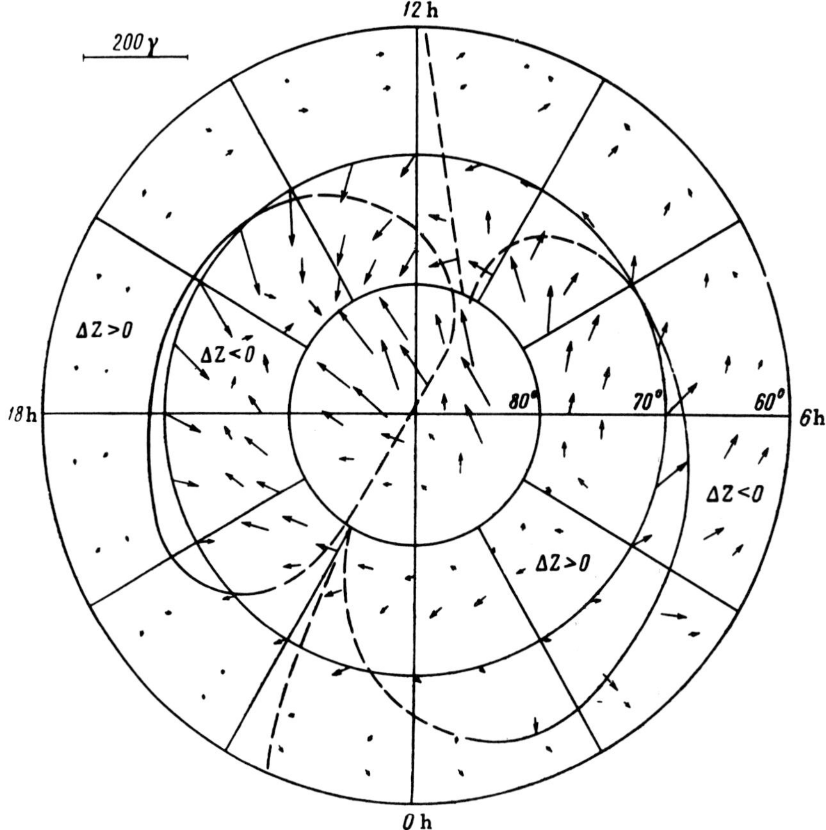

Fig. 1.29. Distribution of magnetic field vectors (S_q^p) at high latitudes on quiet days in summer during the IGY. (Feldstein, Y. I. and Zaytzev, A. N.: *Geomag. Aeronom.* 7, 160, 1967.)

intensities, the computed magnetic field should arise from the net current of order 5×10^5 A during an extremely quiet period. The net current intensity increases by an order of magnitude during disturbed periods (Section 7.2.2).

One of the important reasons for studying the S_q^p variation is that it allows us to infer time variations of this net field-aligned current intensity on a continuous basis. First of all, this magnetic variation is present even on almost absolutely quiet days ($\Sigma Kp = 1+$), as well as on disturbed days. The upper part of Figure 1.31(a) shows the daily record (solid curve) of the *X* component from Baker Lake on 1964, May 8, one of the quietest days during the IQSY. The figure also shows a shaded curve of which the upper and lower boundaries represent the upper and lower envelopes of the combined daily records of several moderately disturbed days for the *X* component from the same station. It can be seen that the May 8 record exhibits very similar variations to that of the shaded curve, but much less amplitude. In the lower part of Figure 1.31(a), the 1964, May 8, curve is reproduced with a 5-time magnification, so that the daily variation can more clearly be seen; the daily variation on 1964, May 12, another very quiet day, is also

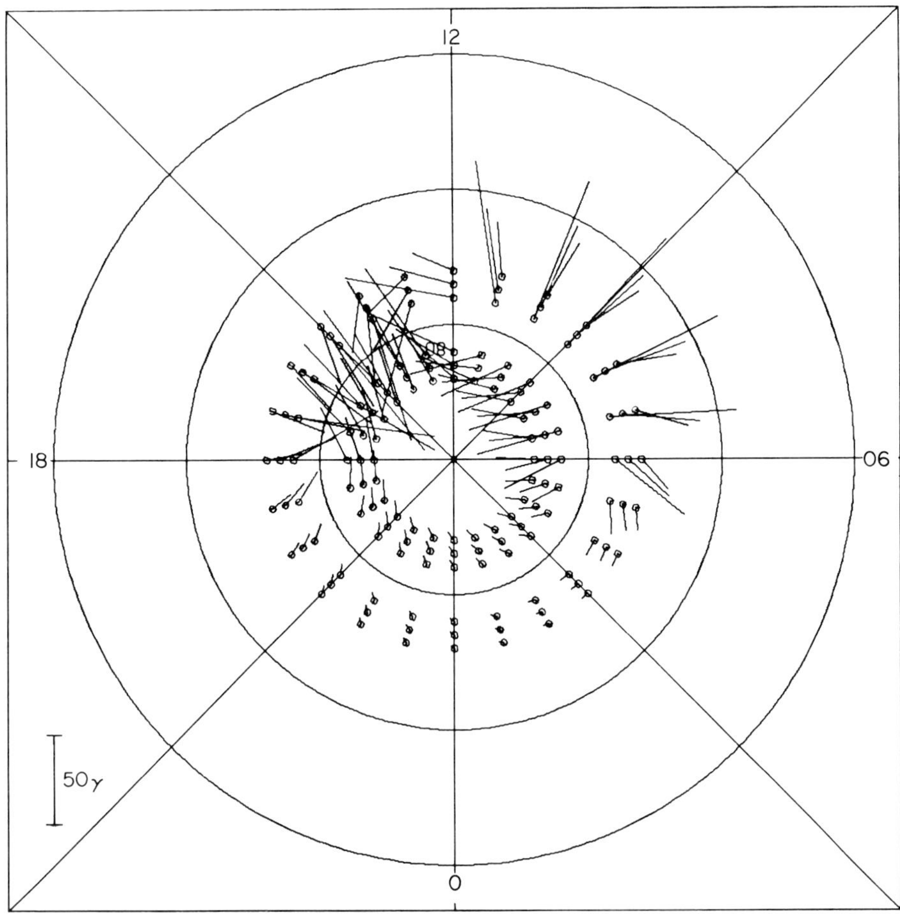

Fig. 1.30. Distribution of magnetic field vectors for the S_q^p variation, computed on the basis of the primary field-aligned currents and the associated ionospheric currents. (Kawasaki, K. and Akasofu, S.-I.: *Planet. Space Sci.* **21**, 329, 1973.)

shown, indicating that the major features of the daily variation on both days are very similar. Figure 1.31(b) shows the daily records of the *Y* component from Baker Lake on both 1964, May 8 and May 12. They show the same tendency as that in Figure 1.31(a). The intensification of the S_q^p current on disturbed days will be discussed in detail in Section 7.2. Langel (1974a, b) examined extensively magnetic field deviations in the total field observed by polar orbiting satellites (OGO-2, 4 and 6). It appears that a significant part of the deviations arise from the S_q^p variation (cf. Fukushima, 1975).

The permanent nature of the S_q^p variation suggests strongly that the S_q^p field-aligned current system is always present. Zmuda *et al.* (1970) showed, on the basis of TRIAD satellite records, that this is indeed the case. One can also infer that the solar wind-magnetosphere dynamo is also permanently operating, feeding

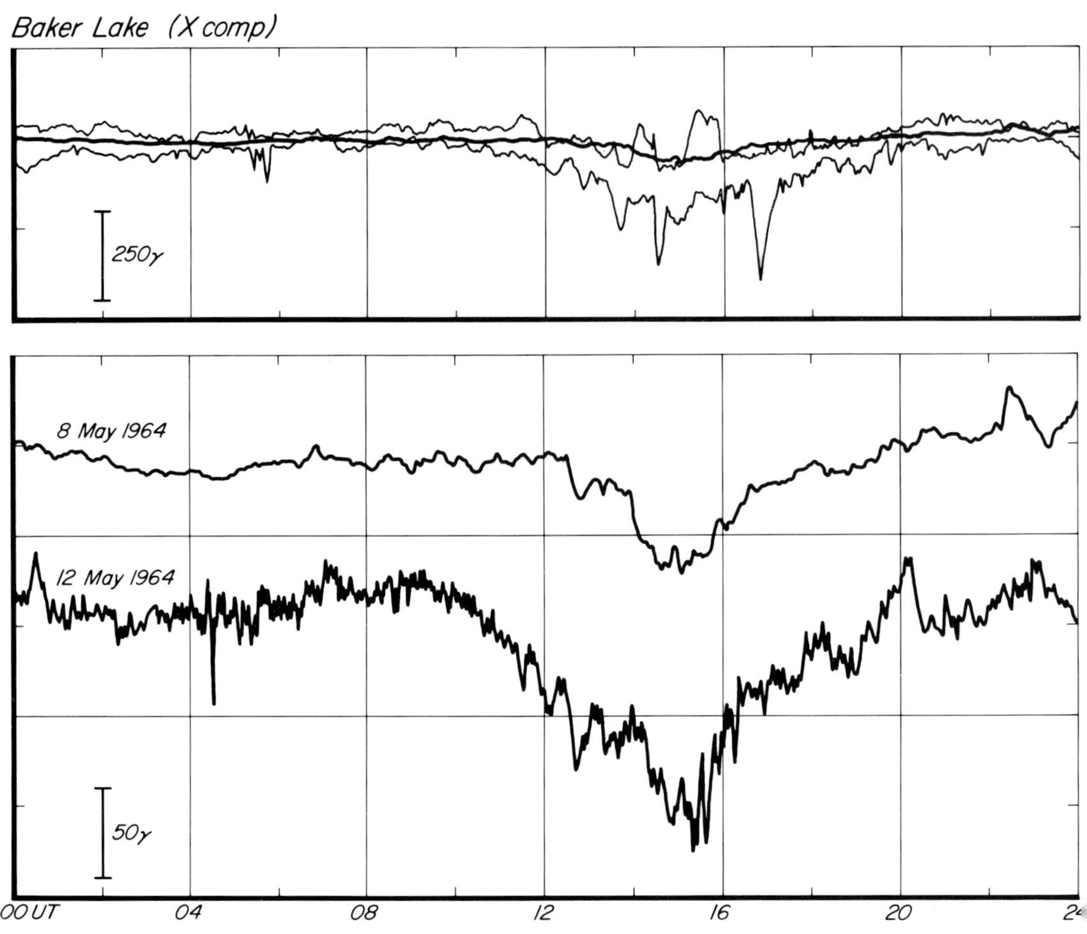

Fig. 1.31(a). S_q^p variation at a polar cap station. Upper and lower envelopes of the superimposed magnetic records (the X component) from Baker Lake on several moderately disturbed days, together with the same component on 1964, May 8, one of the most quiet days during the IQSY. In the lower part, the 1964, May 8 record is reproduced with a 5-time magnification, and the 1964, May 12 record is also shown for comparison. (Kawasaki, K. and Akasofu, S.-I.: *Planet. Space Sci.* **20**, 1163, 1972.)

the current to the auroral oval. We shall examine the dynamo process in the next section.

It is of interest to examine joule heat P_J generated in the polar ionosphere by the S_q^p current. Assuming the current intensity of $J \simeq 5 \times 10^5$ A,

$$P_J = 2 \times (E \cdot J) \times d = 6 \times 10^{17} \text{ erg s}^{-1}$$

where E denotes the electric field across the polar cap and is of order 20 mV m^{-1} and d denotes the diameter of the polar cap which is taken to be 3000 km. The value of power dissipation will be discussed in terms of the power of the solar wind-magnetosphere dynamo in the next section.

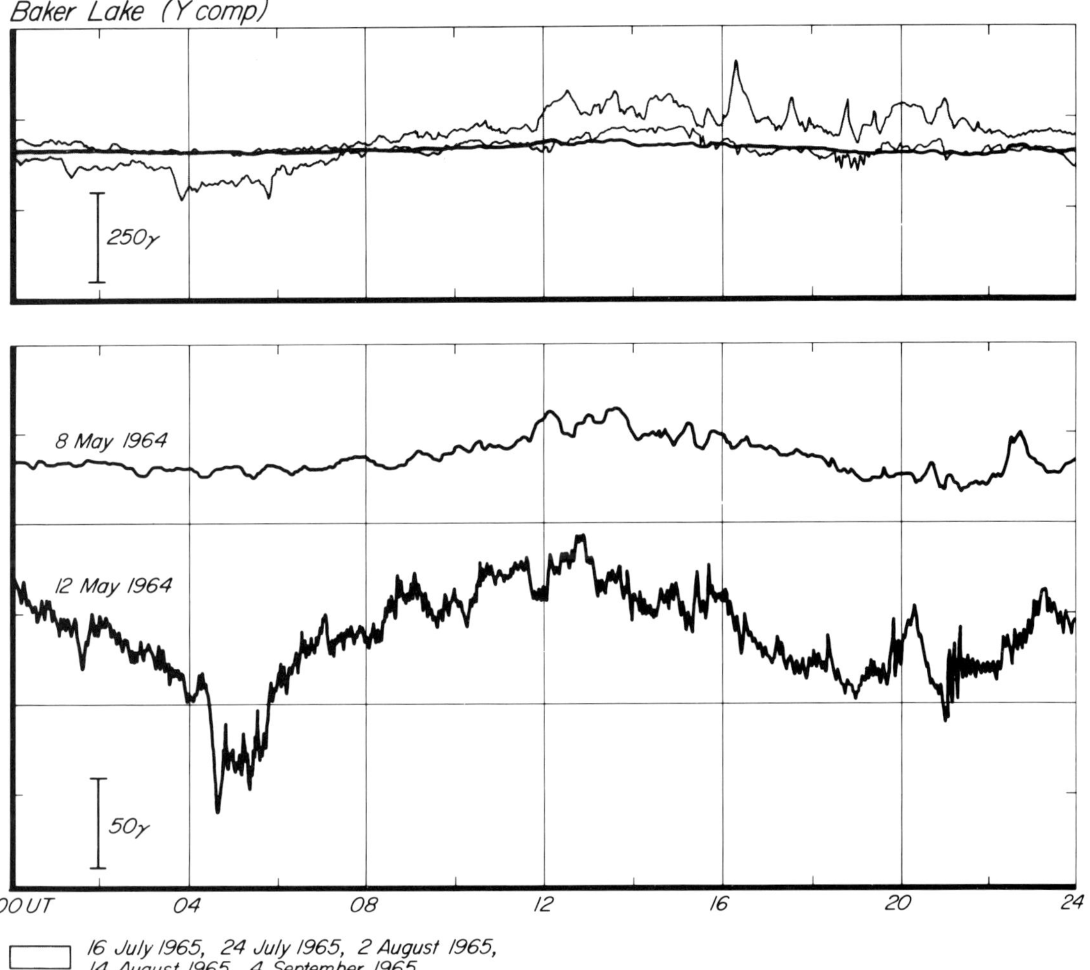

Fig. 1.31(b). Same as Figure 1.31(a); for the Y component.

1.4. Solar Wind – Magnetosphere Dynamo

1.4.1. ELECTROMOTIVE FORCE AND POWER

The purpose of this section is to examine the dynamo process by which the kinetic energy of the solar wind is converted into electrical energy (Siscoe, 1966; Heikkila, 1974; see also S.T.P. Section 5.6). The total electromotive force generated by the dynamo can be expressed by

$$\pi R_{\mathrm{T}} b_{\mathrm{norm}} V_{\mathrm{s}}$$

where R_{T} and V_{s} denote the radius of the cylindrical magnetotail and the speed of solar wind, respectively. Unfortunately, the magnitude of b_{norm}, the normal component of the magnetic field on the magnetopause, is not known, so that it is not possible to estimate the total electromotive force from the above equation.

Therefore, we examine first current circuits associated with the dynamo. Assuming the simplest situation in which the interplanetary magnetic field has only the southward component, the magnetic neutral line completely surrounds the magnetosphere; the morning half of the neutral line acts as the positive terminal of the dynamo and the evening half as the negative terminal. The major part of the current generated by the dynamo is discharged through the central part of the plasma sheet, so that the two bisected cylindrical parts constitute the two 'solenoids'. This part of the circuit may be called the magnetotail circuit (Figure 2 in Introduction). Thus, it is the solar wind-magnetosphere dynamo which feeds the current into the two solenoids constituting the magnetotail. That is to say, the magnetotail is a product of the solar wind-magnetosphere dynamo, which converts the kinetic energy of solar wind to electrical energy. This process is often expressed in terms of 'stretching' the merged field lines by the solar wind.

We note also that the neutral line is connected to the polar ionosphere by a group of geomagnetic field lines, constituting the boundary surface which separates the closed and open field lines (cf. Akasofu, 1974b, 1975). Therefore, a part of the dynamo-generated current can flow across the polar cap ionosphere. This part of the current may be called the auroral oval circuit (Figure 2 in Introduction). If the geomagnetic field lines are assumed to be highly conductive, the potential drop between the morning and evening halves of the neutral line should be approximately the same as the potential drop between the morning and evening halves of the auroral oval. As we saw in Section 1.3.3, the latter is about 50 kV. Thus, the total electromotive force generated by the solar wind-magnetosphere dynamo is of order 50 kV.

Following Siscoe and Cummings (1969), let us estimate the power generated by the dynamo. It is given by

$$\text{power} = F \cdot V_s = 10^{19}\ \text{erg s}^{-1} = 10^{12}\ \text{W}$$

where F is the Maxwell stress and is given by

$$F = \left(\frac{B_{\tan} B_{\text{norm}}}{4\pi}\right) S$$

$$= \frac{B_T^2}{8\pi}\, \pi R_T^2$$

where B_T denotes the magnetotail field ($\sim 15\ \gamma$), R_T the radius of the cylindrical tail ($\sim 20\ R_E$), and S surface area of the magnetotail. Noting that power (watt) = potential (volt) $\times$ current (amp) the total current generated by the dynamo is given by 10^{12} W/50 kV $= 2 \times 10^7$ A. This value may be compared with the current S_q^p of about 10^6 A. It appears that the major part of the power generated by the dynamo is fed into the magnetotail by deriving the magnetotail current and that only a very small portion of the current flows along the auroral circuit. Thus only a small portion of the energy is dissipated in the polar ionosphere as heat energy. If, however, the power of the solar wind-magnetosphere dynamo becomes much less than 10^{12} W, say 10^{11} W or less (namely 10^{18} erg s^{-1} or less) during a quiet time, it becomes comparable to Joule heat generated in the polar ionosphere by the S_q^p current. Thus the polar ionosphere can become the major sink of the electrical energy generated by the solar wind-magnetosphere dynamo.

1.4.2. STEADY STATE MERGING

Here, we examine a possible process by which the magnetosphere becomes open, the merging between the interplanetary and geomagnetic field lines. This process is schematically shown in Figure 1.32. An interplanetary magnetic field line is transported toward the Earth by the solar wind (a). After interacting with the geomagnetic field line (b), a pair of open field lines (that is, one end is connected to the Earth) is formed (c). The merging takes place where the interplanetary and geomagnetic field lines form an X-type configuration. In our simplest situation (that is, when the IMF has only the southward component), this process takes place along the dayside half of the neutral line.

Along the nightside half of the neutral line, the 'reverse' process takes place; a pair of open field lines 'reconnect' or merge to form a closed field line, leaving an interplanetary field line (d, e, f). However, the basic process associated with the reconnection process is the same as that of the merging.

The merging process has been studied by a number of workers (Sweet, 1969;

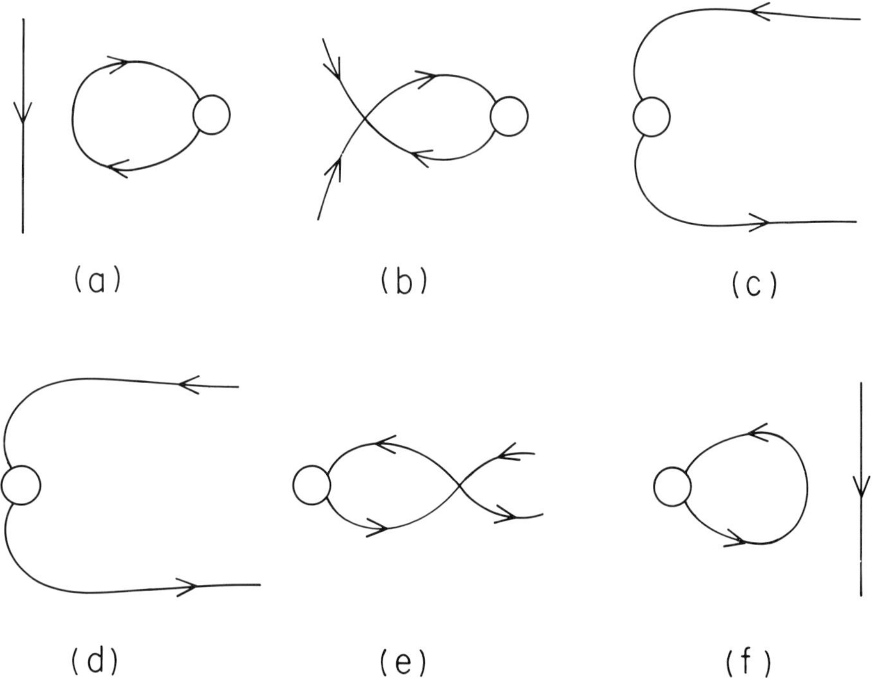

Fig. 1.32. Schematic presentation of the merging process (a → b → c) in the dayside magnetopause and of the reconnection process (d → e → f) in the nightside magnetopause.

Parker, 1957b, 1963; Petschek, 1964; Sonnerup, 1970, 1972, 1974a, b; Yeh and Axford, 1970; Speiser, 1965, 1967, 1970; Dessler, 1968, 1971; Alfvén, 1968; Cowley, 1971, 1973; Kropotkin, 1971; Gonzalez, 1973). Their studies are well summarized in a recent paper by Vasyliunas (1975). In this section we review briefly steady MHD models.

Most MHD models proposed so far assume a steady ($\partial/\partial t = 0$), two-dimensional configuration in an incompressible fluid; the magnetic field lines and the stream lines of plasma are assumed to be confined in the x-y plane (and $\partial/\partial z = 0$), and the neutral line coincides with the z-axis; see Figure 1.33(a). The domain of our interest can be divided into two regions. The first region is called the convective region, where the following equation is applicable:

$$\boldsymbol{E} + \boldsymbol{v} \times \boldsymbol{B} = 0 \tag{1}$$

The other region is a small region in the vicinity of the neutral line, namely the z-axis, where the above equation breaks down because $\boldsymbol{v} \times \boldsymbol{B} = 0$ along the neutral line, but $\boldsymbol{E}$ $(0, 0, E_z)$ is constant throughout the domain ($\partial/\partial t = 0$, $\partial/\partial z = 0$). There, Equation (1) must be replaced by

$$\boldsymbol{E} + (\boldsymbol{v} \times \boldsymbol{B}) = \frac{1}{\sigma} \boldsymbol{J} \tag{2}$$

where σ denotes the conductivity. This small region is called the diffusive region; the field lines 'diffuse' through it because of finite conductivity and lose their identity. Heikkila (1976) has objected to the presence of this electric field by stating that it violates Lenz's law.

(a) *Sweet-Parker's Model*

In Sweet-Parker's model, mass is carried into the merging (or reconnection) region at a rate of order $2\mathscr{L}nmv$ where n and m denote the number density and

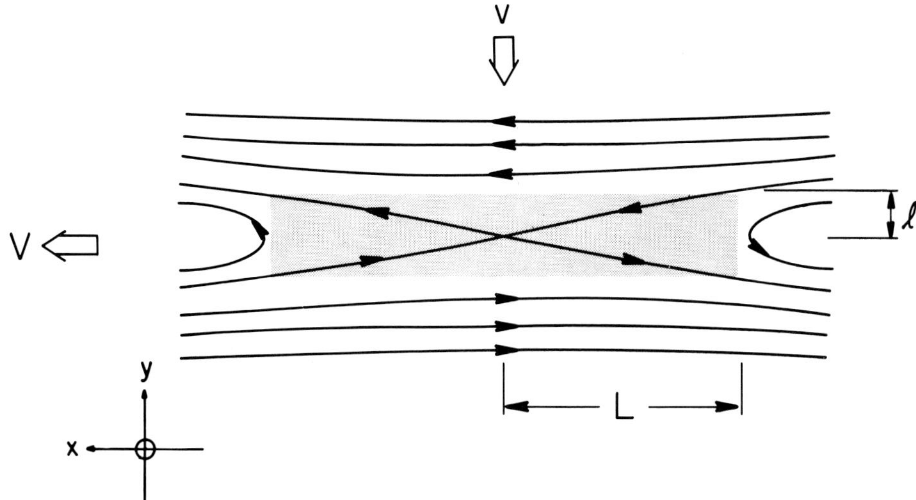

Fig. 1.33(a). Geometry of the merging region.

the mass of plasma particles, respectively; the velocity v with which the fields are diffusing into each other is given (Figure 1.33(a)) by

$$v = \frac{1}{\ell\sigma}$$

The resulting efflux is given by $2\mathscr{L}nmV$ where V denotes the velocity of the ejected plasma and is of order of the Alfvén wave velocity ($V_A = B_1/\sqrt{4\pi nm} = (10^{-4}\text{ G})/\sqrt{4\pi m(10^{-2}\text{ cm}^{-3})} \simeq 2\times10^8\text{ cm s}^{-1}$) in the influx region. The conservation of mass in the system can be expressed by $v\mathscr{L} = V\ell$. From these relations, we have

$$v = \sqrt{V/\mathscr{L}\sigma} = V/\sqrt{R_m}$$

where $R_m = \mathscr{L}\sigma V$ denotes the magnetic Reynolds number. The merging rate is conventionally defined by

$$M = \frac{v}{V} = 1/\sqrt{R_m}$$

The flow velocity V is related to the pressure p by

$$1/2mnV^2(x) = p(x) - p_1$$

and

$$p(x) + \frac{B^2(x)}{8\pi} = p_1 + \frac{B_1^2}{8\pi}$$

along the x-axis, where the suffix 1 denotes quantities in the influx region. At the center of the diffusion region ($x = 0$), we have $B = 0$, so that gas pressure is $p(0) = p_1 + B_1^2/8\pi$ and is higher than the ambient gas pressure by $B_1^2/8\pi$. It is this pressure differential between the neutral line and the surrounding spaces that causes plasma to flow rapidly out from between the two oppositely directed fields. The conductivity σ of plasma in the plasma sheet is of order of

$$\sigma = 2\times10^{-14}T^{3/2} \simeq 2\times10^{-5}\text{ emu}$$

where T denotes the temperature of electrons ($T \simeq 10^6$ K). Taking $\ell = 1.0\, R_E$, the characteristic time for this merging process is given by

$$t \simeq 4\pi\sigma\ell^2 \simeq 10^{14}\text{ s}$$

The magnetic Reynolds number R_m has a value of order

$$R_m = \mathscr{L}\sigma V \simeq (100\, R_E)(2\times10^{-5}\text{ emu})(2000\text{ km s}^{-1}) = 2.5\times10^{14}$$

Therefore, we have

$$v = (2000\text{ km s}^{-1})/(2.5\times10^{14})^{1/2} = 13\text{ cm s}^{-1}$$

$$M = \frac{v}{V} = \frac{13\text{ cm s}^{-1}}{2\times10^8\text{ cm s}^{-1}} = 6.5\times10^{-8}$$

(b) *Petschek's Model*

Petschek (1964) showed that Parker-Sweet's model overlooked the fact that the merging can be accomplished principally by means of waves (whose propagation speed is independent of σ in the medium) and that wave propagation characteristics can change Sweet-Parker's magnetic field configuration in such a way to increase significantly the merging rate $M \simeq 1/\ell n R_m$; for a detailed derivation of the merging rate for Petschek's model, see Vasyliunas (1975). For the magnetic Reynolds number given in the above, $M \simeq 1/\ell n (2.5 \times 10^{14}) \simeq 0.03$. It can be seen that this value of M is a great improvement from Sweet-Parker's merging rate $M \simeq 6.5 \times 10^{-8}$. Figure 1.33(b) shows both the magnetic field and flow configurations for the Petschek model.

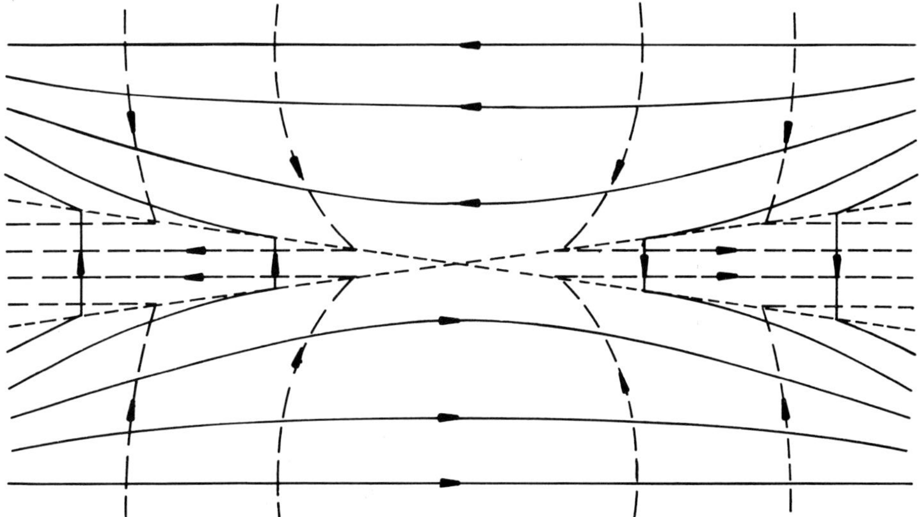

Fig. 1.33(b). Magnetic field configuration and plasma flow pattern of the merging region in Petschek's model. (Vasyliunas, V. M.: *Rev. Geophys. Space Phys.* **13**, 303, 1975.)

(c) *Sonnerup's Model*

Sonnerup (1970) showed that the convective region can be further subdivided into three regions by two planes. In Figure 1.33(c), the two planes are noted by OL and OT and represent large-amplitude standing Alfvén waves fronts. In a compressible fluid they may represent a slow expansion wave and a slow shock wave, respectively. The two planes divide the region of our interest into three: Region 1 (the convective region), Region 2 (the region between OL and OT) and Region 3 (the region between OT and the $-z$ axis).

In this situation, the angles $\theta_1(L, O, -x)$ and $\theta_3(T, O, -x)$ are given by

$$\tan \theta_1 = 1/M$$

$$\tan \theta_3 = (1 + \sqrt{2})^2/M$$

where $M = M_1$.

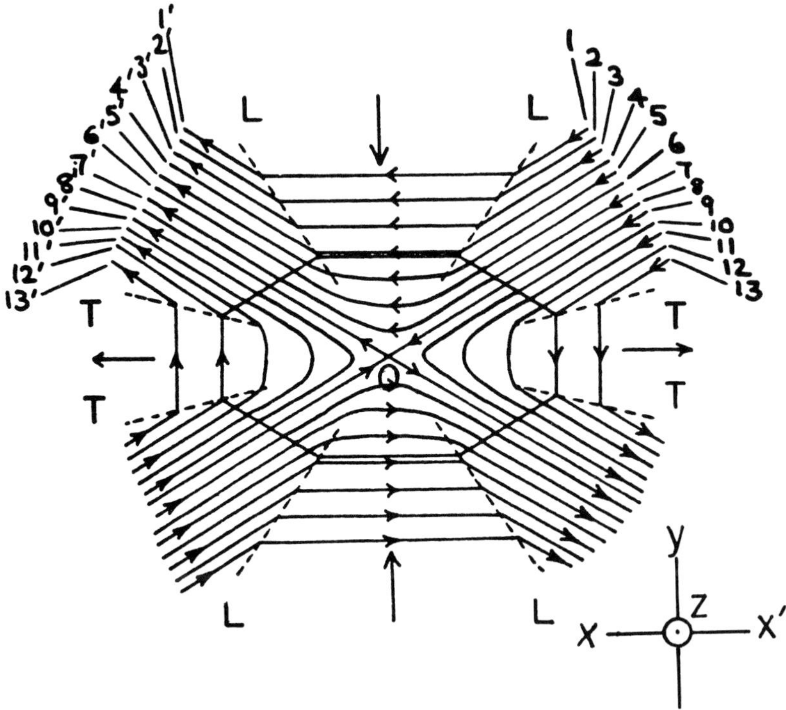

Fig. 1.33(c). Reconnection of field lines in Sonnerup's model. The diffusion region is indicated by the octagonal box surrounding the X-line. Both the leading (L) and trailing (T) waves are shown by dashed lines. (Cowley, S. W. H.: *Radio Sci.* **8**, 903, 1973.)

Thus, if the speed of plasma flow of v_1 is large, the merging rate is large, making also the angles θ_1 and θ_3 large. This tendency is illustrated in a set of examples which are constructed by Fukao and Tsuda (1973) in their models. In Figures 1.34(a) and (b), the values of M are taken to be 0.5 and 1.0, respectively. In each figure, both the magnetic field configurations (i) and the stream lines (ii) are shown. The flow speed of plasma V_3 and pressure p_3 in the outflow region (Region 3) are given respectively by

$$V_3 = V_A(1+\sqrt{2})$$

$$p_3 = p_1 + \frac{1}{2}\rho V_A^2\left[1 - \frac{M}{1+\sqrt{2}}\right]$$

The rate of magnetic energy conversion is given by the difference between the magnitude of the Poynting vectors S in Regions 1 and 3,

$$S_1 - S_3 = V_A B_1^2 M\left(1 - \frac{M}{1+\sqrt{2}}\right)$$

The above quantity has a maximum value at $M = (1+\sqrt{2})/2$ given by

$$(S_1 - S_3)_{\max} = V_A B_1^2 (1+\sqrt{2})/4$$

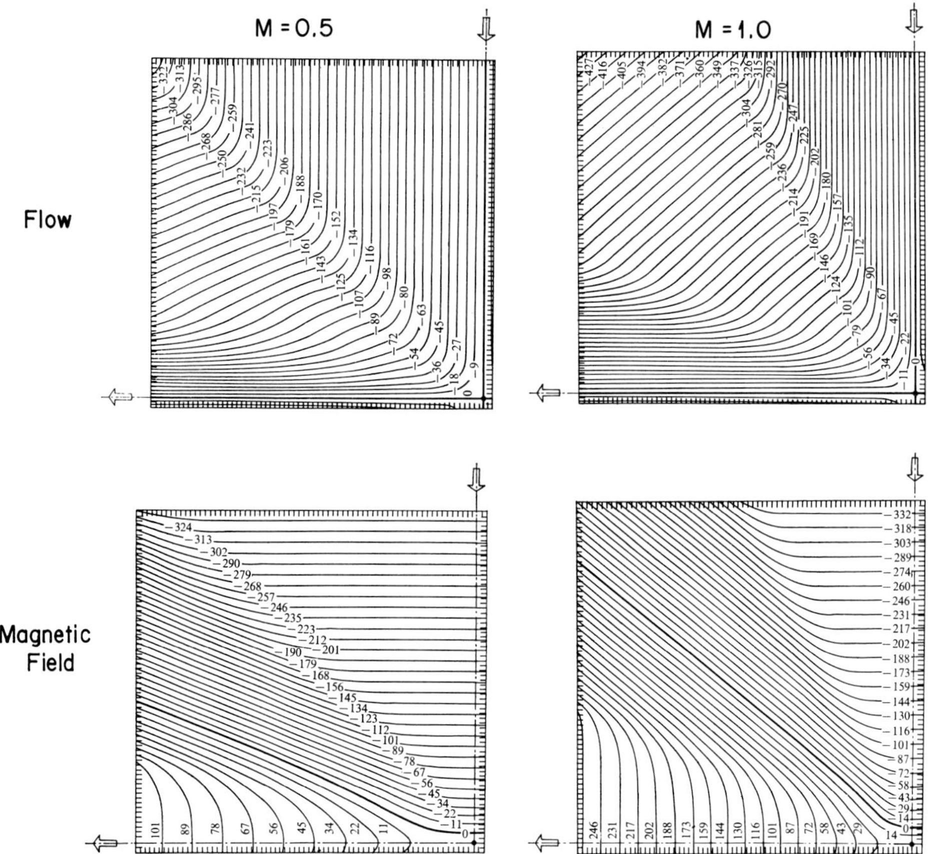

Fig. 1.34. Numerical solutions of the flow and magnetic field patterns of Sonnerup's model reconnection for $M_1 = 0.5$(a) and $M_1 = 1.0$(b), respectively. (Fukao, S. and Tsuda, T.: *J. Plasma Phys.* **9**, 409, 1973.)

Note also that the pressures in Regions 1 and 2 are related by

$$p_2 = p_1 - \tfrac{1}{4}\rho V_A^2[1 + 2\sqrt{2} + M^2]$$

and thus that for $p_2 > 0$

$$p_1 > \tfrac{1}{4}\rho V_A^2[1 + 2\sqrt{2} + M^2] = \tfrac{1}{2}p_m[1 + 2\sqrt{2} + M^2]$$

$$p_1/p_m = \beta = \tfrac{1}{2}[1 + 2\sqrt{2} + M^2] \gtrsim 2.0$$

Sonnerup showed that the presence of the diffusion region puts the upper limit to the value of M; it is given by $(1 + \sqrt{2})$. In general, M is a function of the magnetic Reynolds number $R_m = v_1\mathscr{L}/\nu$ but if the fluid is viscous, M is also a function of the Reynolds number $R = v_1\mathscr{L}/\nu$ where ν denotes the kinematic viscosity. In Fukao and Tsuda's study, they are $R_m = 200$ and $R = 200$.

1.4.3. PRODUCTION RATE OF THE OPEN FLUX

Gonzalez and Mozer (1974) showed that the power P transmitted to the magnetosphere can be expressed by the integral over the tail of the Poynting flux, namely

$$P = \frac{1}{4\pi}\int \boldsymbol{E}\times\boldsymbol{B}\cdot \boldsymbol{dS} = \frac{1}{4\pi}\int \mathrm{d}x\ \mathrm{d}y\ E_y B_x \tag{1}$$

where E_y denotes the cross-tail electric field and B_x the magnetotail field. Here, we assume that the integral $\int B_x\ \mathrm{d}x$ can be given by $B_{\mathrm{T}}L$, where B_{T} is the average tail field intensity and L is the length of the magnetotail. Thus, the above equation may be rewritten as

$$P = \frac{1}{4\pi}\,\Phi B_{\mathrm{T}}L \tag{2}$$

since $\int E_y\ \mathrm{d}y = \Phi$ is the potential drop across the tail.

Here, B_{T} may be estimated as the total amount of field line flux connected on the dayside in the time = L (the length of the magnetotail)/V_{s} (solar wind speed), divided by the half cross-sectional area of the magnetotail $\pi R_{\mathrm{T}}^2/2$. Thus,

$$B_{\mathrm{T}} = (2\Phi/\pi R_{\mathrm{T}}^2)(L/V_{\mathrm{s}}) \tag{3}$$

Thus,

$$P = \Phi^2 L^2/2\pi^2 V_{\mathrm{s}} R_{\mathrm{T}}^2 \tag{4}$$

or since

$$\Phi = \frac{B_{\mathrm{T}}}{2}\,\pi R_{\mathrm{T}}^2\,\frac{V_{\mathrm{s}}}{L} \tag{5}$$

$$P = \frac{B_{\mathrm{T}}^2}{8\pi}\,\pi R_{\mathrm{T}}^2 V_{\mathrm{s}} \tag{6}$$

The last equation is the same as (1) in Section 1.4.1 (Piddington, 1963). Further, the steady state magnetic energy ε in the magnetotail is given by

$$\varepsilon = P \times \frac{L}{V_{\mathrm{s}}} = \Phi^2 L^3/\pi^2 V_{\mathrm{s}}^2 R_{\mathrm{T}}^2 \tag{7}$$

The length of the magnetotail may be estimated by noting that the time required for a solar wind particle to reach the reconnection point from the dayside merging point should be approximately equal to the time required for the 'foot' of the newly merged field line to move across the polar cap (from the midday part of the oval to the midnight part of the oval). Denoting the noon-midnight dimension of the auroral oval to be d_1, we have

$$L/V_{\mathrm{s}} = d_1/V_{\mathrm{e}} \tag{8}$$

where

$$V_{\mathrm{e}} = E/B_{\mathrm{p}};\quad B_{\mathrm{p}} = 0.6\ \mathrm{G} \tag{9}$$

and

$$E = \Phi/d_2 \tag{10}$$

where

d_2 = the dawn-dusk dimension of the oval

Thus,

$$L = V_s d_1 B_p / E = d_1 d_2 V_s B_p / \Phi \tag{11}$$

In summary, we have obtained the following equations for Φ:

$$\Phi = \left(\frac{B_T}{2}\right) \pi R_T^2 \frac{V_s}{L} \tag{12}$$

$$\Phi = \frac{2\pi P}{B_T L} \tag{13}$$

$$\Phi = \frac{d_1 d_2 V_s B_p}{L} \tag{14}$$

$$\Phi = 5.12 \times 10^2 \left(\frac{31 \gamma \eta \mu}{\langle \sigma_p \rangle}\right) \frac{1}{L_G^3} \tag{15}$$

The last equation is added from Section 1.3.5. To these, it is useful to add the relationship between the total open flux ψ_0 and the geocentric distance D of the apex of the magnetopause (Kan and Akasofu, 1974) by

$$\psi_0 \simeq 4 B_e R_E^2 / D = 2 B_p R_E^2 / D \tag{16}$$

and the area A_p of the polar cap is then given by

$$A_p = 4(B_e / B_p) R_E^2 / D \simeq 2 R_E^2 / D \tag{17}$$

and D is given by

$$|B_i| D_0 = 2[(D_0/D)^3 - 1] \tag{18}$$

where

B_i = the interplanetary magnetic field intensity at the front of the magnetopause,

$D_0 = D$ for $B_i = 0$

For example, if the solar wind momentum is such that $D_0 = 12\ R_E$, then a 20% inward displacement, i.e., $D = 9.6\ R_E$, would require a southward field $|B_i| \simeq 33\ \gamma$.

References

Aarsnes, K. and Amundsen, R.: 1972, 'North/South Asymmetric Entry of Solar Protons during the November 18, 1968 Event', *Planet. Space Sci.* **20**, 1835.

Akasofu, S.-I.: 1960, 'Large-Scale Auroral Motions and Polar Magnetic Disturbance at about 1100 hours on 23 September 1957', *J. Atmospheric Terrestr. Phys.* **19**, 10.

Akasofu, S.-I.: 1966, 'The Auroral Oval, the Auroral Substorm and their Relations with the Internal Structure of the Magnetosphere', *Planet. Space Sci.* **14**, 587.

Akasofu, S.-I.: 1974a, 'A Study of Auroral Displays Photographed from the DMSP-2 Satellite and from the Alaska Meridian Chain of Stations', *Space Sci. Rev.* **16**, 617.

Akasofu, S.-I.: 1974b, 'The Aurora and the Magnetosphere: The Chapman Memorial Lecture', *Planet. Space Sci.* **22**, 885.

Akasofu, S.-I.: 1975, 'The Solar Wind-Magnetosphere Dynamo and the Magnetospheric Substorm', *Planet. Space Sci.* **23**, 817.

Alfvén, H.: 1968, 'Some Properties of Magnetospheric Neutral Surfaces', *J. Geophys. Res.* **73**, 4379.
Alksne, A. Y. and Webster, D. L.: 1970, 'Magnetic and Electric Fields in the Magnetosheath', *Planet. Space Sci.* **18**, 1203.
Anderson, K. A. and Lin, R. P.: 1969, 'Observation of Interplanetary Field Lines in the Magnetotail', *J. Geophys. Res.* **74**, 3953.
Anderson, H. R. and Vondrak, R. K.: 1975, 'Observations of Birkeland Currents at Auroral Latitudes', *Rev. Geophys. Space Phys.* **13**, 243.
Armstrong, J. C. and Zmuda, A. J.: 1970, 'Field-Aligned Current at 1100 km in the Auroral Region Measured by Satellite', *J. Geophys. Res.* **75**, 7122.
Armstrong, J. C. and Zmuda, A. J.: 1973, 'Triaxial Magnetic Measurements of Field-Aligned Currents at 800 Kilometers in the Auroral Region: Initial Results', *J. Geophys. Res.* **78**, 6802.
Armstrong, J. C., Akasofu, S.-I. and Rostoker, G.: 1975, 'A Comparison of Satellite Observations of Birkeland Currents with Ground Observations of Visible Aurora and Ionospheric Currents', *J. Geophys. Res.* **80**, 575.
Arnoldy, R. L.: 1974, 'Auroral Particle Precipitation and Birkeland Currents', *Rev. Geophys. Space Phys.* **12**, 217.
Atkinson, G.: 1975, 'Equations for Magnetospheric Convection and a Solution for Polar Cap Flows', *J. Geophys. Res.* **80**, 32.
Axford, W. I. and Hines, C. O.: 1961, 'A Unifying Theory of High-Latitude Geophysical Phenomena and Geomagnetic Storms', *Can. J. Phys.* **39**, 1433.
Banks, P. M.: 1972, 'Behavior of Thermal Plasma in the Magnetosphere and Top-Side Ionosphere', Presented at the COSPAR Symposium on "Critical Problems in Magnetospheric Physics", Madrid, Spain, May 10–12.
Banks, P. M. and Doupnik, J. R.: 1975, 'A Review of Auroral Zone Electrodynamics Deduced from Incoherent Scatter Radar Observations', *J. Atmosph. Terr. Phys.* **37**, 951.
Beard, D. B.: 1960, 'The Interaction of the Terrestrial Magnetic Field with the Solar Corpuscular Radiation', *J. Geophys. Res.* **65**, 3559.
Beard, D. B.: 1967, 'The solar wind', *Rep. Prog. Phys.* **30**, 409.
Berko, F. W.: 1973, 'Distributions and Characteristics of High-Latitude Field Aligned Electron Precipitation', *J. Geophys. Res.* **78**, 1615.
Berko, F. W., Hoffman, R. A., Burton, R. K. and Holzer, R. E.: 1975, 'Simultaneous Particle and Field Observations of Field-Aligned Currents', *J. Geophys. Res.* **80**, 37.
Block, L. P. and Carpenter, D. L.: 1974, 'Derivation of Magnetospheric Electric Fields from Whistler Data in a Dynamic Geomagnetic Field', *J. Geophys. Res.* **79**, 2783.
Boström, R. A.: 1964, 'A Model of the Auroral Electrojets', *J. Geophys. Res.* **69**, 4983.
Boström, R. A.: 1968, 'Currents in the Ionosphere and Magnetosphere', *Ann. Geophys.* **24**, 681.
Boström, R. A.: 1975, 'Mechanisms for Driving Birkeland Currents', *Physics of the Hot Plasma in the Magnetosphere*, p. 341, B. Hultqvist and L. Stenflo (eds.), Plenum Press, New York.
Buchau, J., Whalen, J. A. and Akasofu, S.-I.: 1970, 'On the Continuity of the Auroral Oval', *J. Geophys. Res.* **75**, 7147.
Cambou, F. and Galperin, Yu. I.: 1974, 'Resultats d'Ensemble Obtenus Grace a l'Experience ARCAD a Bord du Satellite AUREOLE', *Ann. Geophys.* **30**, 9.
Carpenter, D. L.: 1970, 'Whistler Evidence of the Dynamic Behavior of the Dusk Side Bulge in the Plasmasphere', *J. Geophys. Res.* **75**, 3837.
Carpenter, D. L. and Stone, K.: 1967, 'Direct Detection by a Whistler Method of the Magnetospheric Electric Field Associated with a Polar Substorm', *Planet. Space Sci.* **15**, 395.
Carpenter, D. L., Stone, K., Siren, J. C. and Crystal, T. L.: 1972, 'Magnetospheric Electric Fields Deduced from Drifting Whistler Paths', *J. Geophys. Res.* **77**, 2819.
Carpenter, L. A. and Kirchhoff, V. W. J. H.: 1975, 'Comparison of High-Latitude and Mid-Latitude Ionospheric Electric Fields', *J. Geophys. Res.* **80**, 1810.
Cauffman, D. P. and Gurnett, D. A.: 1971, 'Double-Probe Measurements of Convection Electric Fields with the INJUN-5 Satellite', *J. Geophys. Res.* **76**, 6014.
Cooper, W. A. and Haskell, G. P.: 1972, 'Anisotropy of Low-Energy Solar Protons at the Boundary of the Magnetotail', *J. Geophys. Res.* **77**, 6849.
Coroniti, F. V. and Kennel, C. F.: 1973, 'Can the Ionosphere Regulate Magnetospheric Convection?', *J. Geophys. Res.* **78**, 2837.
Cowley, S. W. H.: 1971, 'The Adiabatic Flow Model of a Neutral Sheet', *Cosmic Electrodynamics* **2**, 90.
Cowley, S. W. H.: 1973, 'A Self-Consistent Model of a Simple Magnetic Neutral Sheet System Surrounded by Cold, Collisionless Plasma', *Cosmic Electrodynamics* **3**, 448.
Davis, T. N.: 1971, 'Magnetospheric Convection Pattern Inferred from Magnetic Disturbance and Auroral Motions', *J. Geophys. Res.* **76**, 5978.

Dessler, A. J.: 1968, 'Magnetic Merging in the Magnetospheric Tail', *J. Geophys. Res.* **73**, 209.
Dessler, A. J.: 1971, 'Vacuum Merging: A Possible Source of the Magnetospheric Cross-Tail Electric Field', *J. Geophys. Res.* **76**, 3174.
Doupnik, J. R., Banks, P. M., Baron, M. J., Rino, C. L. and Petriceks, J.: 1972, 'Direct Measurements of Plasma Drift Velocities at High Magnetic Latitudes', *J. Geophys. Res.* **77**, 4268.
Dungey, J. W.: 1961a, 'The Steady State of the Chapman-Ferraro Problem in Two Dimensions', *J. Geophys. Res.* **66**, 1043.
Dungey, J. W.: 1961b, 'Interplanetary Magnetic Field and the Auroral Zones', *Phys. Rev. Lett.* **6**, 47.
Durney, A. C. and Morfill, G. E.: 1972, 'Entry of Energetic Solar Protons into the Tail', *Earth's Magnetospheric Processes*, B. M. McCormac (ed.), p. 101, D. Reidel Publ. Co., Dordrecht-Holland.
Durney, A. C., Morfill, G. E. and Quenby, J. J.: 1972, 'Entry of High-Energy Solar Protons into the Distant Geomagnetic Tail', *J. Geophys. Res.* **77**, 3345.
Engelmann, J.: 1972, 'Solar Particle Injection at Medium Energies ($25 < E < 250$ MeV)', *Earth's Magnetospheric Processes*, B. M. McCormac (ed.), p. 95, D. Reidel Publ. Co., Dordrecht-Holland.
Engelmann, J., Hynds, R. J., Morfill, G., Axisa, F., Bewick, A., Durney, A. C. and Koch, L.: 1971, 'Penetration of Solar Protons over the Polar Cap during the February 25, 1969 Event', *J. Geophys. Res.* **76**, 4245.
Evans, L. C. and Stone, E. C.: 1972: 'Electron Polar Cap and the Boundary of Open Geomagnetic Field Lines', *J. Geophys. Res.* **77**, 5580.
Fahleson, U. V., Kelley, M. C. and Mozer, F. S.: 1970, 'Investigation of the Operation of a D.C. Electric Field Detector', *Planet. Space Sci.* **18**, 1551.
Fairfield, D. H.: 1968, 'Simultaneous Measurements on Three Satellites and the Observation of the Geomagnetic Tail at 1000 R_E', *J. Geophys. Res.* **73**, 6179.
Fanselow, J. L. and Stone, E. C.: 1972, 'Geomagnetic Cutoffs for Cosmic-Ray Protons for Seven Energy Intervals between 1.2 and 39 MeV', *J. Geophys. Res.* **77**, 3999.
Fejer, J. A.: 1963, 'Theory of Auroral Electrojets'. *J. Geophys. Res.* **68**, 2147.
Feldstein, Y. I.: 1963, 'Some Problems Concerning the Morphology of Auroras and Magnetic Disturbances at High Latitudes', *Geomagn. Aeron.* **3**, 183.
Feldstein, Y. I.: 1973, 'Auroral Oval', *J. Geophys. Res.* **78**, 1210.
Feldstein, Y. I. and Starkov, G. V.: 1970, 'The Auroral Oval and the Boundary of Closed Field Lines of Geomagnetic Field', *Planet. Space Sci.* **18**, 501.
Feldstein, Y. I. and Zaitzev, A. N.: 1967, 'Magnetic Field Variations at High Latitudes on Quiet Days in Summer during the IGY', *Geomag. Aeronom.* **7**, 160
Feldstein, Y. I. and Zaitzev, A. N.: 1968, 'Quiet and Disturbed Solar-Daily Variations of Magnetic Field at High Latitudes during the IGY', *Tellus* **20**, 338.
Fennell, J. F.: 1973, 'Access of Solar Protons to the Earth's Polar Caps', *J. Geophys. Res.* **78**, 1036.
Flindt, H. R.: 1970, 'Local-Time Dependence of Geomagnetic Cutoffs for Solar Protons, $0.52 \leq E_p \leq 4$ MeV', *J. Geophys. Res.* **75**, 39.
Föppl, H., Haerendel, G., Haser, L., Lüst, R., Melzner, F., Meyer, B., Neuss, H., Rabben, H. H., Rieger, E., Stöcker, J. and Stoffregen, W.: 1968, 'Preliminary Results of Electric Field Measurements in the Auroral Zone', *J. Geophys. Res.* **73**, 21.
Freeman, J. W. Jr.: 1968, 'Observation of Flow of Low-Energy Ions at Synchronous Altitude and Implications for Magnetospheric Convection', *J. Geophys. Res.* **73**, 4151.
Freeman, J. W. Jr.: 1974, 'Kp Dependence of the Plasma Sheet Boundary', *J. Geophys. Res.* **79**, 4315.
Fukao, S. and Tsuda, T.: 1973, 'On the Reconnection of Magnetic Lines of Force', *J. Plasma Phys.* **9**, 409.
Fukushima, N.: 1975, 'Field-Aligned Current as a Possible Source for ΔB Observed by Low-Altitude Satellites', *Rep. Ionosph. and Space Res. Japan* **29**, 51.
Fukushima, N. and Kamide, Y.: 1973, 'Contribution of Magnetospheric Field-Aligned Current to Geomagnetic Bays and S_q Fields: A Comment on Partial Ring-Current Models', *Radio Sci.* **8**, 1013.
Fukushima, N. and Kawasaki, K.: 1974, 'A Simplified Model of Field-Aligned Current Sheet Pairs along the Auroral Oval in Connection with Geomagnetic S_q^p-Field', *Rep. Ionosph. Space Res. Japan* **28**, 83.
Gall, R.: 1968, 'Daily Variation of the Asymptotic Directions of Cosmic Rays', *J. Geophys. Res.* **73**, 4400.
Gall, R. and Bravo, S.: 1973, 'Role of the Neutral Sheet in the Illumination of Polar Caps by Solar Protons', *J. Geophys. Res.* **78**, 6773.
Gall, R., Bravo, S. and Orozco, A.: 1972, 'Model for the Uneven Illumination of Polar Caps by Solar Protons', *J. Geophys. Res.* **77**, 5360.
Gall, R., Jimenez, J. and Camacho, L.: 1968, 'Arrival of Low-Energy Cosmic Rays via the Magnetospheric Tail', *J. Geophys. Res.* **73**, 1593.
Galperin, Yu. I., Ponomarev, V. N. and Zosimova, A. G.: 1973a, 'Direct Measurements of Ion Drift

Velocity in the Upper Ionosphere during a Magnetic Storm, 1. Experiment Description and Some Results of Measurements during Magnetically Quiet Time', *Cosmicheski Issledovanija* **11**, 273.

Galperin, Yu. I., Ponomarev, V. N. and Zosimova, A. G.: 1973b, 'Direct Measurements of Ion Drift Velocity in the Upper Ionosphere during a Magnetic Storm, 2. Results of Measurements during the November 3, 1967 Magnetic Storm', *Cosmicheski Issledovanija* **11**, 284.

Galperin, Yu. I., Ponomarev, V. N. and Zosimova, A. G.: 1974, 'Plasma Convection in Polar Ionosphere', *Ann. de Geophys.* **30**, 1.

Gonzalez, W. D.: 1973, 'A Quantitative Three-Dimensional Model for Magnetopause Reconnection', Dissertation Doctor of Philosophy in Physics, U. of Calif., Berkeley.

Gonzalez, W. D. and Mozer, F. S.: 1974, 'A Quantitative Model for the Potential Resulting from Reconnection with an Arbitrary Interplanetary Magnetic Field', *J. Geophys. Res.* **79**, 4186.

Gueth, K., Hoerner, V. and Brommundt, G.: 1974, 'The Determination of Ionospheric Plasma Drifts by a Rocket-Borne Electrostatic Flux Meter', *Planet. Space Sci.* **22**, 1131.

Gurnett, D. A.: 1970, 'Satellite Measurements of D.C. Electric Fields in the Ionosphere', *Particles and Fields in the Magnetosphere*, B. M. McCormac (ed.), p. 239, D. Reidel Publ. Co., Dordrecht-Holland.

Gurnett, D. A.: 1972a, 'Injun 5 Observations of Magnetospheric Electric Fields and Plasma Convection', *Earth's Magnetospheric Process*, B. M. McCormac (ed.), p. 233. D. Reidel Publ. Co., Dordrecht-Holland.

Gurnett, D. A.: 1972b, 'Electric Field and Plasma Observations in the Magnetosphere', *Critical Problems of Magnetospheric Physics*, Proceedings of the Joint COSPAR/AGA/URSI Symposium, Madrid, Spain, 11–13 May 1972, 123.

Haerendel, G.: 1972, 'Plasma Drifts in the Auroral Ionosphere Derived from Barium Releases', *Earth's Magnetospheric Processes*, B. M. McCormac (ed.), p. 246, D. Reidel Publ. Co., Dordrecht-Holland.

Haerendel, G. and Lüst, R.: 1970, 'Electric Fields in the Ionosphere and Magnetosphere', *Particles and Fields in the Magnetosphere*, B. M. McCormac (ed.), p. 213, D. Reidel Publ. Co., Dordrecht-Holland.

Haerendel, G., Hedgecock, P. C. and Akasofu, S.-I.: 1971, 'Evidence of Magnetic Field Aligned Currents during the Substorms of March 18, 1969', *J. Geophys. Res.* **76**, 2382.

Hasegawa, M.: 1940, 'Provisional Report of the Statistical Study on the Diurnal Variations of Terrestrial Magnetism in the North Polar Regions', *Trans. Washington Meeting, I.U.G.G.-A.T.M.E., Bull.* No. 11, 311, A. H. R. Goldie (ed.), Edinburg.

Heelis, R. Burch, J. and Hanson, W.: 1976, 'Ion Convection Velocity Reversals in the Dayside Cleft', *J. Geophys. Res.* **81** (in press).

Heikkila, W. J.: 1974, 'Outline of a Magnetospheric Theory', *J. Geophys. Res.* **79**, 2496.

Heikkila, W. J.: 1976, 'Objection to the Reconnection Model for Magnetospheric Phenomena', *Space Sci. Rev.* (in press).

Heppner, J. P.: 1972a, 'Electric Fields in the Magnetosphere', Goddard Space Flight Center, *X-645-72-154*, May.

Heppner, J. P.: 1972b, 'Electric Field Variations during Substorms: OGO-6 Measurements', *Planet. Space Sci.* **20**, 1475.

Heppner, J. P.: 1972c, 'The Harang Discontinuity in Auroral Belt Ionospheric Currents', *Geofysiske Publikasjoner* **29**, 105.

Heppner, J. P., Stolarik, J. D. and Wescott, E. M.: 1971, 'Electric-Field Measurements and the Identification of Currents Causing Magnetic Disturbances in the Polar Cap', *J. Geophys. Res.* **76**, 6028.

Hurley, J.: 1961a, 'Interaction of a Streaming Plasma with the Magnetic Field of a Line Current', *Phys. Fluids* **4**, 109.

Hurley, J.: 1961b, 'Interaction of a Streaming Plasma with the Magnetic Field of a Two-Dimensional Dipole', *Phys. Fluids* **4**, 854.

Iijima, T. and Kokubun, S.: 1973, 'Geomagnetic S_q^p Variation on an Extremely Quiet Day', *Rep. Ionosph. Space Res. Japan* **27**, 195.

Innanen, W. G. and Van Allen, J. A.: 1973, 'Anisotropies in the Interplanetary Intensity of Solar Protons $E_p > 0.3$ MeV', *J. Geophys. Res.* **78**, 1019.

Intriligator, D. S., Wolfe, J. H., McKibbin, D. D. and Collard, H. R.: 1969, 'Preliminary Comparison of Solar Wind Plasma Observations in the Geomagnetospheric Wake at 1000 and 500 Earth Radii', *Planet. Space Sci.* **17**, 321.

Intriligator, D. S., Wolfe, J. H. and McKibbin, D. D.: 1972, 'Simultaneous Solar-Wind Plasma and Magnetic-Field Measurements in the Expected Region of the Extended Geomagnetic Tail', *J. Geophys. Res.* **77**, 4645.

Jaggi, R. K. and Wolf, R. A.: 1973, 'Self-Consistent Calculation of the Motion of a Sheet of Ions in the Magnetosphere'. *J. Geophys. Res.* **78**, 2852.

Jeffries, R. A., Roach, W. H., Hones, E. W. Jr., Wescott, E. M., Stenbaek-Nielsen, H. C., Davis, T. N.

and Winningham, J. D.: 1975, 'Two Barium Plasma Injections into the Northern Magnetospheric Cleft', *Geophys. Res. Letters* **2**, 285.
Kamide, F. and Akasofu, S.-I.: 1976, 'The Location of the Field-Aligned Currents with Respect to Discrete Auroral Arcs', *J. Geophys. Res.* **81**, 3999.
Kan, J. R. and Akasofu, S.-I.: 1974, 'A Model of the Open Magnetosphere', *J. Geophys. Res.* **79**, 1379.
Kavanagh, L. D. Jr., Freeman, J. W. Jr. and Chen, A. J.: 1968, 'Plasma Flow in the Magnetosphere', *J. Geophys. Res.* **73**, 5511.
Kawasaki, K. and Akasofu, S.-I.: 1967, 'Polar Solar Daily Geomagnetic Variations on Exceptionally Quiet Days', *J. Geophys. Res.* **72**, 5363.
Kawasaki, K. and Akasofu, S.-I.: 1972, 'Geomagnetic Disturbances in the Polar Cap: S_q^p and DP-2', *Planet. Space Sci.* **20**, 1163.
Kawasaki, K. and Akasofu, S.-I.: 1973, 'A Possible Current System Associated with the S_q^p Variation', *Planet. Space Sci.* **21**, 329.
Kawasaki, K. and Fukushima, N.: 1975, 'Derivation of Field-Aligned Current Distribution from Scalar B Observed by Low-Altitude Satellites', *Rep. Ionosph. and Space Res. Japan* **29**, 58.
Kelley, M. C. and Mozer, F. S.: 1975, 'Simultaneous Measurement of the Horizontal Components of the Earth's Electric Field in the Atmosphere and in the Ionosphere', *J. Geophys. Res.* **80**, 3275.
Kelley, M. C., Mozer, F. S. and Fahleson, U. V.: 1971, 'Electric Fields in the Nighttime and Daytime Auroral Zone', *J. Geophys. Res.* **76**, 6054.
Kelley, M. C., Haerendel, G., Kappler, H., Mozer, F. S. and Fahleson, U. V.: 1975, 'Electric Field Measurements in a Major Magnetospheric Substorm', *J. Geophys. Res.* **80**, 3181.
Knudsen, W. C.: 1974, 'Magnetospheric Convection and the High-Latitude F_2 Ionosphere', *J. Geophys. Res.* **79**, 1046.
Kropotkin, A. P.: 1971, 'Reconnection of the Lines of Force of the Interplanetary Magnetic Field and of the Geomagnetic Tail', *Geomag. Aeronom.* **11**, 902.
Krylov, A. L. and Shcherbakov, V. P.: 1972, 'Equations of Potential Electric Fields in the Earth's Magnetosphere and Ionosphere', *Geomag. Aeronom.* **12**, 189.
Langel, R. A.: 1974a, 'Near-Earth Magnetic Disturbance in Total Field at High Latitudes, I, Summary of Data from Ogo 2, 4, and 6', *J. Geophys. Res.* **79**, 2363.
Langel, R. A.: 1974b, 'Near-Earth Magnetic Disturbance in Total Field at High Latitudes, 2, Interpretation of Data from Ogo 2, 4, and 6', *J. Geophys. Res.* **79**, 2373.
Lassen, K.: 1972, 'On the Classification of High-Latitude Auroras', *Geofysiske Publikasjoner* **29**, 87.
Lloyd, K. H. and Haerendel, G.: 1973, 'Numerical Modeling of the Drift and Deformation of Ionospheric Plasma Clouds and of their Interaction with Other Layers of the Ionosphere', *J. Geophys. Res.* **78**, 7389.
Lui, A. T. Y. and Anger, C. D.: 1973, 'A Uniform Belt of Diffuse Auroral Emission Seen by the ISIS-2 Scanning Photometer', *Planet, Space Sci.* **21**, 799.
Lui, A. T. Y., Anger, C. D., Venkatesan, D., Sawchuk, W. and Akasofu, S.-I.: 1975, 'The Topology of the Auroral Oval as Seen by the ISIS-2 Scanning Auroral Photometer', *J. Geophys. Res.* **80**, 1795.
Lyatskiy, W. B., Mal'tsev, Yu. P. and Leont'yev, S. V.: 1974, 'Three-Dimensional Current System in Different Phases of a Substorm', *Planet. Space Sci.* **22**, 1231.
Mal'tsev, Yu. P.: 1974, 'The Effect of Ionospheric Conductivity on the Convection System in the Magnetosphere', *Geomag. Aeronom.* **14**, 128.
Mariani, F. and Ness, N. F.: 1969, 'Observations of the Geomagnetic Tail at 500 Earth Radii by Pioneer 8', *J. Geophys. Res.* **74**, 5633.
Maynard, N. C.: 1974, 'Electric Field Measurements across the Harang Discontinuity', *J. Geophys. Res.* **79**, 4620.
Maynard, N. C. and Heppner, J. P.: 1970, 'Variations in Electric Fields from Polar Orbiting Satellites', *Particles and Fields in the Magnetosphere*, B. M. McCormac (ed.), p. 247, D. Reidel Publ. Co., Dordrecht-Holland.
McCoy, J. E., Lin, R. P., McGuire, R. E., Chase, L. M. and Anderson, K. A.: 1975, 'Magnetotail Electric Fields Observed from Lunar Orbit', *J. Geophys. Res.* **80**, 3217.
McDiarmid, I. B. and Burrows, J. R.: 1969, 'Relation of Solar Proton Latitude Profiles to Outer Radiation Zone Electron Measurements', *J. Geophys. Res.* **74**, 6239.
McDiarmid, I. B. and Burrows, J. R.: 1970, 'Latitude Profiles of Low-Energy Solar Electrons', *J. Geophys. Res.* **75**, 3910.
McDiarmid, I. B., Burrows, J. R. and Wilson, M. D.: 1974, 'Solar Proton Flux Enhancements at Auroral Latitudes', *J. Geophys. Res.* **79**, 1099.
McIlwain, C. E.: 1972, 'Plasma Convection in the Vicinity of the Geosynchronous Orbit', *Earth's Magnetospheric Processes*, B. M. McCormac (ed.), p. 268, D. Reidel Publ. Co., Dordrecht-Holland.
McIlwain, C. E.: 1974, 'Substorm Injection Boundaries', *Magnetospheric Physics*, B. M. McCormac

(ed.), p. 143, D. Reidel Publ. Co., Dordrecht-Holland.
Mead, G. D. and Beard, D. B.: 1964, 'Shape of the Geomagnetic Field Solar Wind Boundary', *J. Geophys. Res.* **69**, 1169.
Mendillo, M. and Papagiannis, M. D.: 1971, 'Estimate of the Dependence of the Magnetospheric Electric Field on the Velocity of the Solar Wind', *J. Geophys. Res.* **76**, 6939.
Michel, F. C. and Dessler, A. J.: 1975, 'On the Interpretation of Low-Energy Particle Access to the Polar Caps', *J. Geophys. Res.* **80**, 2309.
Midgley, J. E. and Davis, L. Jr.: 1963, 'Calculation by a Moment Technique of the Perturbation of the Geomagnetic Field by the Solar Wind', *J. Geophys.* **68**, 5111.
Mikkelsen, I. S., Jørgensen, T. S. and Kelley, M. C.: 1975, 'Observation and Interpretation of Plasma Motions in the Polar Cap Ionosphere during Magnetic Substorms', *J. Geophys. Res.* **80**, 3197.
Morfill, G. E.: 1973, 'Non Adiabatic Particle Motion in the Magnetosphere', *J. Geophys. Res.* **78**, 588.
Morfill, G. E. and Quenby, J. J.: 1971, 'The Entry of Solar Protons over the Polar Caps', *Planet. Space Sci.* **19**, 1541.
Morfill, G. and Scholer, M.: 1972a, 'Reconnection of the Geomagnetic Tail Deduced from Solar-Particle Observations', *J. Geophys. Res.* **77**, 4021.
Morfill, G. and Scholer, M.: 1972b, 'Solar Proton Intensity Structures in the Magnetosphere during Interplanetary Anisotropies', *Planet. Space Sci.* **20**, 2113.
Morfill, G. and Scholer, M.: 1973a, 'Uneven Illumination of the Polar Caps by Solar Protons: Comparison of Different Particle Entry Models', *J. Geophys. Res.* **78**, 5449.
Morfill, G. and Scholer, M.: 1973b, 'Study of the Magnetosphere Using Energetic Solar Particles', *Space Sci. Rev.* **15**, 267.
Mozer, F. S.: 1972, 'Simultaneous Electric-Field Measurements on Nearby Balloons', *J. Geophys. Res.* **77**, 6129.
Mozer, F. S.: 1973a, 'Analyses of Techniques for Measuring D.C. and A.D. Electric Fields in the Magnetosphere', *Space Sci. Rev.* **14**, 272.
Mozer, F. S.: 1973b, 'On the Relationship between the Growth and Expansion Phases of Substorms and Magnetospheric Convection', *J. Geophys. Res.* **78**, 1719.
Mozer, F. S.: 1973c, 'Electric Fields and Plasma Convection in the Plasmasphere', *Rev. Geophys. Space Phys.* **11**, 755.
Mozer, F. S. and Gonzalez, W. D.: 1973, 'Response of Polar Cap Convection to the Interplanetary Magnetic Field', *J. Geophys. Res.* **78**, 6784.
Mozer, F. S. and Lucht, P.: 1974, 'The Average Auroral Zone Electric Field', *J. Geophys. Res.* **79**, 1001.
Mozer, F. S. and Manka, R. H.: 1971, 'Magnetospheric Electric Field Properties Deduced from Simultaneous Balloon Flights', *J. Geophys. Res.* **76**, 1697.
Mozer, F. S. and Serlin, R.: 1969, 'Magnetospheric Electric Field Measurements with Balloons', *J. Geophys. Res.* **74**, 4739.
Mozer, F. S., Serlin, R., Carpenter, D. L. and Siren, J.: 1974, 'Simultaneous Electric Field Measurements near $L = 4$ from Conjugate Balloons and Whistlers', *J. Geophys. Res.* **79**, 3215.
Nagata, T. and Kokubun, S.: 1962, 'An Additional Geomagnetic Daily Variation Field (S_q^p Field) in the Polar Region on Geomagnetically Quiet Day', *Rep. Ionosph. Res. Japan* **16**, 256.
Ness, N. F., Scearce, C. S. and Cantarano, S.: 1967, 'Probable Observations of the Geomagnetic Tail at 10^3 Earth Radii by Pioneer 7', *J. Geophys. Res.* **72**, 3769.
Nielsen, E. and Pomerantz, M. A.: 1975, 'Access of Solar Electrons to the Polar Regions', *Planet. Space Sci.* **23**, 945.
Nishida, A.: 1966, 'Formation of Plasmapause, or Magnetospheric Plasma Knee by the Combined Action of Magnetosphere Convection and Plasma Escape from the Tail', *J. Geophys. Res.* **71**, 5669.
Nishida, A.: 1971, 'Interplanetary Origin of Electric Fields in the Magnetosphere', *Cosmic Electrodynamics* **2**, 350.
Nishida, A. and Kokubun, S.: 1971, 'New Polar Magnetic Disturbances: S_q^p, SP, DPC, and DP2', *Rev. Geophys. Space Phys.* **9**, 417.
Nishida, A. and Maezawa, K.: 1971, 'Two Basic Modes of Interaction Between the Solar Wind and the Magnetosphere', *J. Geophys. Res.* **76**, 2254.
Ogawa, T., Tanaka, Y., Huzita, A. and Yasuhara, M.: 1975, 'Horizontal Electric Fields in the Middle Latitude', *Planet. Space Sci.* **23**, 825.
Osipov, N. K. and Pavlov, Ye. Ye.: 1971, 'Structure of the Geomagnetic Field and Currents in the Magnetosphere', *Geomag. Aeronom.* **11**, 483.
Oya, H.: 1975, 'Plasma Flow Hypothesis in the Magnetosphere Relating to Frequency Shift of Electrostatic Plasma Waves', *J. Geophys. Res.* **80**, 2783.
Page, D. E. and Domingo, V.: 1972, 'New Results on Particle Arrival at the Polar Caps', *Earth's Magnetospheric Processes*, B. M. McCormac (ed.), p. 107, D. Reidel Publ. Co., Dordrecht-Holland.

Palmer, I. D., Higbie, P. R. and Hones, E. W. Jr.: 1976, 'Gradients of Solar Protons in the High Latitude Magnetotail and the Magnetospheric Electric Field', *J. Geophys. Res.* **81**.

Parker, E. N.: 1957a, 'Newtonian Development of the Dynamical Properties of Ionized Gases of Low Density', *Phys. Rev.* **107**, 924.

Parker, E. N.: 1957b, 'Sweet's Mechanism for Merging Magnetic Fields in Conducting Fluids', *J. Geophys. Res.* **62**, 509.

Parker, E. N.: 1963, 'The Solar-Flare Phenomenon and the Theory of Reconnection and Annihilation of Magnetic Fields', *Astrophys. J.*, Suppl. 77, **8**, 177.

Paulikas, G. A., Blake, J. B. and Vampola, A. L.: 1970, 'Solar Particle Observations over the Polar Caps', *Particles and Fields in the Magnetosphere*, B. M. McCormic (ed.), p. 141, D. Reidel Publ. Co., Dordrecht-Holland.

Petschek, H. E.: 1964, 'Magnetic Field Annihilation, AAS-NASA Symposium on the Physics of Solar Flares', NASA Spec. Publ. SP-50, 425.

Piddington, J. H.: 1963, 'Theories of the Geomagnetic Storm Main Phase', *Planet. Space Sci.* **11**, 1277.

Piddington, J. H.: 1965, 'The Morphology of Auroral Precipitation', *Planet. Space Sci.* **13**, 565.

Potter, W. E.: 1970, 'Rocket Measurements of Auroral Electric and Magnetic Fields', *J. Geophys. Res.* **75**, 5415.

Potter, W. E. and Cahill, L. J. Jr.: 1969, 'Electric and Magnetic Field Measurements Near an Auroral Electrojet', *J. Geophys. Res.* **74**, 5159.

Pudovkin, M. I.: 1974, 'Electric Fields and Currents in the Ionosphere', *Space Sci. Rev.* **16**, 727.

Quenby, J. J.: 1972, 'Magnetospheric Field Fluctuations and the Penetration of Solar Protons to Low Geomagnetic Latitude', *Planet. Space Sci.* **20**, 1979.

Roederer, J. G. and Hones, E. W. Jr.: 1970, 'Electric Field in the Magnetosphere as Deduced from Asymmetries in the Trapped Particle Flux', *J. Geophys. Res.* **75**, 3923.

Rostoker, G. and Boström, R.: 1974, 'A Mechanism for Driving the Gross Birkeland Current Configuration in the Auroral Oval', Rep. TRITA-EPP-74-25, Dept. Plasma Phys. Royal Inst. Tech. Stockholm, Sweden.

Rostoker, G., Chen, A. J., Yasuhara, F., Akasofu, S.-I. and Kawasaki, K.: 1974, 'High Latitude Equivalent Current Systems During Extremely Quiet Times', *Planet. Space Sci.* **22**, 427.

Rostoker, G., Armstrong, J. C. and Zmuda, A. J.: 1975, 'Field-Aligned Current Flow Associated with Intrusion of the Substorm-Intensified Westward Electrojet into the Evening Sector', *J. Geophys. Res.* **80**, 3571.

Sato, T.: 1974, 'Possible Sources of Field-Aligned Currents', *Rep. Ionosph. and Space Res., Japan* **28**, 179.

Schield, M. A., Freeman, J. W. and Dessler, A. J.: 1969, 'A Source for Field-Aligned Currents at Auroral Latitudes', *J. Geophys. Res.* **74**, 247.

Scholer, M.: 1972, 'Polar-Cap Structures of Solar Protons Observed during the Passage of Interplanetary Discontinuities', *J. Geophys. Res.* **77**, 2762.

Scholer, M.: 1975, 'Solar Protons on Closed Magnetospheric Field Lines after an Interplanetary Flux Decrease', *Planet. Space Sci.* **23**, 1445.

Scholer, M. and Morfill, G.: 1972, 'Persistent Particle Anistropies and Magnetospheric Models', *Planet. Space Sci.* **20**, 1051.

Scholer, M. and Morfill, G.: 1974, 'On the Topology of the Geomagnetic Field', *Magnetospheric Physics*, B. M. McCormac (ed.), p. 61, D. Reidel Publ. Co., Dordrecht-Holland.

Scholer, M., Häusler, B. and Hovestadt, D.: 1972, 'Non-Uniform Entry of Solar Protons into the Polar Cap', *Planet. Space Sci.* **20**, 271.

Shabansky, V. P.: 1971, 'Some Processes in the Magnetosphere', *Space Sci. Rev.* **12**, 299.

Siscoe, G. L.: 1966, 'A Unified Treatment of Magnetospheric Dynamics with Applications to Magnetic Storms', *Planet. Space Sci.* **14**, 947.

Siscoe, G. L. and Cummings, W. D.: 1969, 'On the Cause of Geomagnetic Bays', *Planet. Space Sci.* **17**, 1795.

Sonnerup, B. U. O.: 1970, 'Magnetic-Field Re-Connexion in a Highly Conducting Incompressible Fluid', *J. Plasma Phys.* **4**, 161.

Sonnerup, B. U. O.: 1972, 'Magnetic Field Reconnection and Particle Acceleration', Invited paper, presented at the NASA Symposium on High-Energy Phenomena on the Sun, Sept. 28–30, Goddard Space Flight Center.

Sonnerup, B. U. O.: 1974a, 'Magnetopause Reconnection Rate', *J. Geophys. Res.* **79**, 1546.

Sonnerup, B. U. O.: 1974b, 'The Reconnecting Magnetosphere', *Magnetospheric Physics*, B. M. McCormac (ed.), p. 23, D. Reidel Publ. Co., Dordrecht-Holland.

Speiser, T. W.: 1965, 'Particle Trajectories in a Model Current Sheet, Based on the Open Model of the Magnetosphere, with Applications to Auroral Particles', *J. Geophys. Res.* **70**, 1717.

Speiser, T. W.: 1967, 'Particle Trajectories in Model Current Sheets, 2. Applications to Auroras Using a Geomagnetic Tail Model', *J. Geophys. Res.* **72**, 3919.
Speiser, T. W.: 1970, 'Conductivity without Collisions or Noise', *Planet. Space Sci.* **18**, 613.
Stern, D. P.: 1973, 'A Study of the Electric Field in an Open Magnetospheric Model', *J. Geophys. Res.* **78**, 7292.
Stern, D. P.: 1975, 'A Secondary Source of Electric Field in the Magnetosphere', Goddard Space Flight Center, *X-602-75-17*, January.
Stone, E. C.: 1964, 'Local Time Dependence of Non-Störmer Cutoff for 1.5 MeV Protons in Quiet Geomagnetic Field', *J. Geophys. Res.* **69**, 3577.
Sugiura, M.: 1975, 'Identifications of the Polar Cap Boundary and the Auroral Belt in the High-Altitude Magnetosphere: A Model for Field-Aligned Currents', *J. Geophys. Res.* **80**, 2057.
Sweet, P. A.: 1969, 'Mechanisms of Solar Flares', *Ann. Rev. Astron. Astrophys.* **7**, 149.
Swift, D. W.: 1967, 'Possible Consequences of the Asymmetric Development of the Ring Current Belt', *Planet. Space Sci.* **15**, 835.
Swift, D. W.: 1971, 'Possible Mechanisms for Formation of the Ring Current Belt', *J. Geophys. Res.* **76**, 2276.
Swift, D. W. and Gurnett, D. A.: 1973, 'Direct Comparison between Satellite Electric Field Measurements and the Visible Aurora', *J. Geophys. Res.* **78**, 7306.
Theile, B. and Praetorius, H. M.: 1973, Field-Aligned Currents between 400 and 3000 km in Auroral and Polar Latitudes, *Planet. Space Sci.* **21**, 179.
Turtle, J. P., Oelbermann, E. J. Jr., Blake, J. B., Lanzerotti, L. J., Vampola, A. L. and Yates, G. K.: 1972, 'Rapid Access of Solar Electrons to the Polar Caps', *J. Geophys. Res.* **77**, 730.
Vampola, A. L.: 1969, 'Energetic Electrons at Latitudes above the Outer-Zone Cutoff', *J. Geophys. Res.* **74**, 1254.
Vampola, A. L.: 1971, 'Access of Solar Electrons to Closed Field Lines', *J. Geophys. Res.* **76**, 36.
Vampola, A. L.: 1973, 'Solar Electron Access to the Magnetosphere', Aerospace Report No. TR-0074(4260-20)-5, August 24.
Van Allen, J. A.: 1970, 'On the Electric Field in the Earth's Distant Magnetotail', *J. Geophys. Res.* **75**, 29.
Van Allen, J. A., Fennel, J. F. and Ness, N. F.: 1971, 'Asymmetric Access of Energetic Solar Protons to the Earth's North and South Polar Caps', *J. Geophys. Res.* **76**, 4262.
Vasyliunas, V. M.: 1968, 'A Crude Estimate of the Relation between the Solar Wind Speed and the Magnetospheric Electric Field', *J. Geophys. Res.* **73**, 2529.
Vasyliunas, V. M.: 1970, 'Mathematical Models of Magnetospheric Convection and its Coupling to the Ionosphere', *Particles and Fields in the Magnetosphere*, B. M. McCormac (ed.), p. 60, D. Reidel Publ. Co., Dordrecht-Holland.
Vasyliunas, V. M.: 1972, 'The Interrelationship of Magnetospheric Processes', *Earth's Magnetospheric Processes*, B. M. McCormac (ed.), p. 29, D. Reidel Publ. Co., Dordrecht-Holland.
Vasyliunas, V. M.: 1975, 'Theoretical Models of Magnetic Field Line Merging', 1, *Rev. Geophys. Space Phys.* **13**, 303.
Walker, R. C., Villante, U. and Lazarus, A. J.: 1975, 'Pioneer 7 Observations of Plasma Flow and Field Reversal Regions in the Distant Geomagnetic Tail', *J. Geophys. Res.* **80**, 1238.
Wescott, E. M., Stolarik, J. D. and Heppner, J. P.: 1969, 'Electric Fields in the Vicinity of Auroral Forms from Motions of Barium Vapor Releases', *J. Geophys. Res.* **74**, 3469.
Wescott, E. M., Stolarik, J. D. and Heppner, J. P.: 1970, 'Auroral and Polar Cap Electric Fields from Barium Releases', *Particles and Fields in the Magnetosphere*, B. M. McCormac (ed.), p. 229, D. Reidel Publ. Co., Dordrecht-Holland.
Whalen, B. A., Green, D. W. and McDiarmid, I. B.: 1974, 'Observations of Ionospheric Ion Flow and Related Convective Electric Fields in and near the Auroral Arc', *J. Geophys. Res.* **79**, 2835.
Whalen, B. A., Verschell, H. J. and McDiarmid, I. B.: 1975, 'Correlations of Ionospheric Electric Fields and Energetic Particle Precipitation', *J. Geophys. Res.* **80**, 2137.
Wolf, R. A.: 1970, 'Effects of Ionospheric Conductivity on Convective Flow of Plasma in the Magnetosphere', *J. Geophys. Res.* **75**, 4677.
Wolf, R. A.: 1974, 'Calculations of Magnetospheric Electric Fields', *Magnetospheric Physics*, B. M. McCormac (ed.), p. 167, D. Reidel Publ. Co., Dordrecht-Holland.
Wolf, R. A.: 1975, 'Ionosphere-Magnetosphere Coupling', *Space Sci. Rev.* **17**, 537.
Yasuhara, F., Kamide, Y. and Akasofu, S.-I.: 1975, 'Field-Aligned and Ionospheric Currents', *Planet. Space Sci.* **23**, 1355.
Yeh, T. and Axford, W. I.: 1970, 'On the Re-Connexion of Magnetic Field Lines in Conducting Fluids', *J. Plasma Phys.* **4**, 207.
Zhigulev, V. N. and Romishevskii, E. A.: 1960, 'Concerning the Interaction of Currents Flowing in a

Conducting Medium with the Earth's Magnetic Field', *Soviet Phys. Dokl.* **4**, 859.
Zmuda, A. J. and Armstrong, J. C.: 1974a, 'The Diurnal Flow Pattern of Field-Aligned Currents', *J. Geophys. Res.* **79**, 4611.
Zmuda, A. J. and Armstrong, J. C.: 1974b, 'The Diurnal Variation of the Region with Vector Magnetic Field Changes Associated with Field-Aligned Currents', *J. Geophys. Res.* **79**, 2501.
Zmuda, A. J., Martin, J. H. and Heuring, F. T.: 1966, 'Transverse Magnetic Disturbances at 1100 km in the Auroral Region', *J. Geophys. Res.* **71**, 5033.
Zmuda, A. J., Armstrong, J. C. and Heuring, F. T.: 1970, 'Characteristics of Transverse Magnetic Disturbances Observed at 1100 Kilometers in the Auroral Oval', *J. Geophys. Res.* **75**, 4757.

CHAPTER 2

AURORAS AND AURORAL PARTICLES

2.1. Introduction

The purpose of this chapter is to review briefly recent morphological studies of the aurora: in particular, characteristics of auroras and of precipitating auroral particles as a function of local time and latitude.

2.1.1. MONTAGE PHOTOGRAPHS OF THE AURORAL OVAL

In this section we examine overall features of auroral characteristics in different local time sectors on the basis of several montage photographs (Akasofu, 1976). In each montage photograph, the invariant pole is indicated by a cross, and the magnetic noon meridian by a line extending from the invariant pole toward the top of each photograph. The midnight sector of the oval in each photograph is thus located toward the bottom. The afternoon-evening sector is located on the right-hand side and the forenoon-morning sector on the left-hand side, because all the photographs used in constructing the montage photographs were taken from above the antarctic region.

(i) *Example 1: Figure 2.1.* This montage photograph is constructed by using three DMSP-8531 photographs taken during orbits 6285, 6286 and 6289. During orbit 6285, auroral activity was rather weak in the midnight sector, although several arcs are clearly visible in the forenoon sector. A bright polar cap arc lies approximately along the midnight meridian and bends towards the evening sector. The appearance of such a polar cap arc is not very common in DMSP photographs from above the antarctic region.

During the second orbit, the typical substorm feature in the midnight sector (namely, the poleward expanding bulge) was observed. During the same orbit, there was a very bright arc in the early afternoon sector, but it was 'truncated' in the noon meridian. This is a very common feature and will be seen in more detail in the next example. In the afternoon-evening sector (orbit 6289), there were at least two arcs, one extending from the midday sector and the other from the midnight sector; the latter was particularly bright, lying poleward of the former. In the third photograph we can see the diffuse aurora, extending from the midnight sector. Note that the distance between discrete oval auroras and the diffuse aurora increases towards the noon sector.

(ii) *Example 2: Figure 2.2.* This montage photograph consists of four DMSP photographs, two of them (orbits 6424 and 6430) from the DMSP-8531, and the

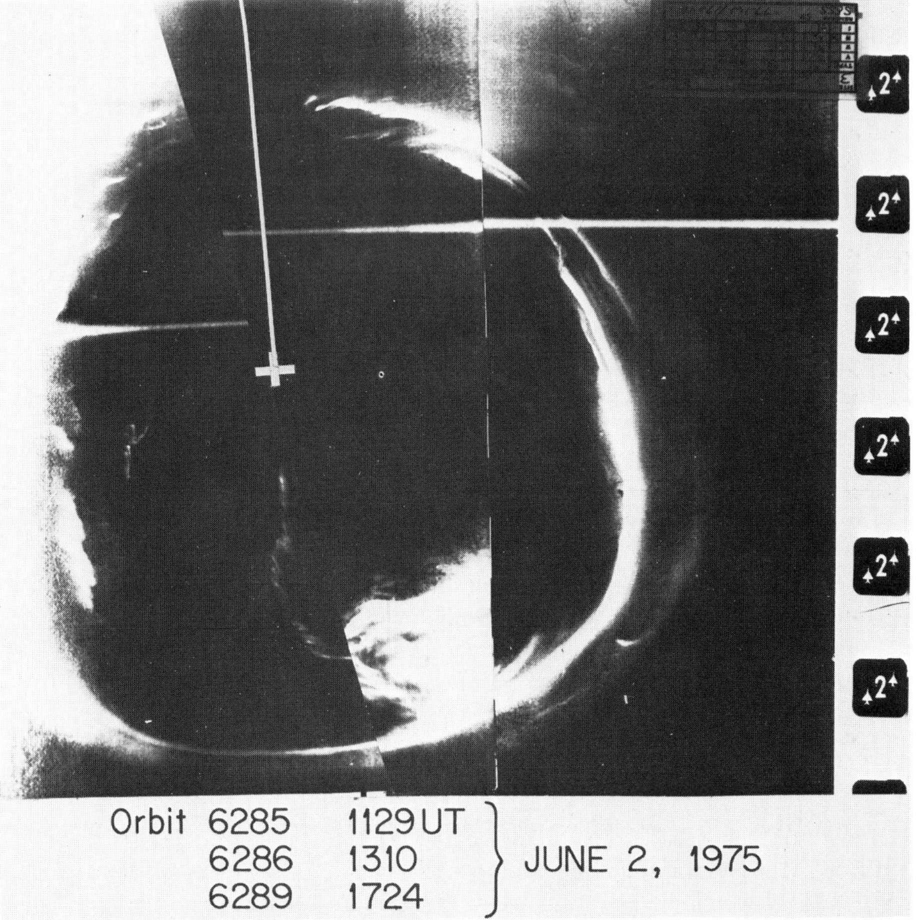

Fig. 2.1. Montage photograph of the auroral oval taken from above the antarctic region by the DMSP-8531 satellite; the cross and the line indicate the invariant pole and the noon meridian, respectively. (Akasofu, S.-I.: *Space Sci. Rev.* **19**, 169, 1976.)

other two from the DMSP-10533. It shows quite well complex auroral features in the morning sector. In particular, the whole morning half of the polar cap is filled with polar cap auroral arcs. Unfortunately, however, their relation to oval auroras has not been clearly established in this particular example, although such polar cap arcs are often 'rooted' in the morning sector of the oval, where Ω bands are a common feature (Akasofu, 1974).

Another important feature in the photograph is patchy auroras which are located a little equatorward of the forenoon sector of the oval auroras. Such patchy auroras were noted by Davis and DeWitt (1963) in all-sky photographs taken from Byrd during the IGY, and more recently by Snyder *et al.* (1972).

There was no significant auroral activity in the late evening sector, although a

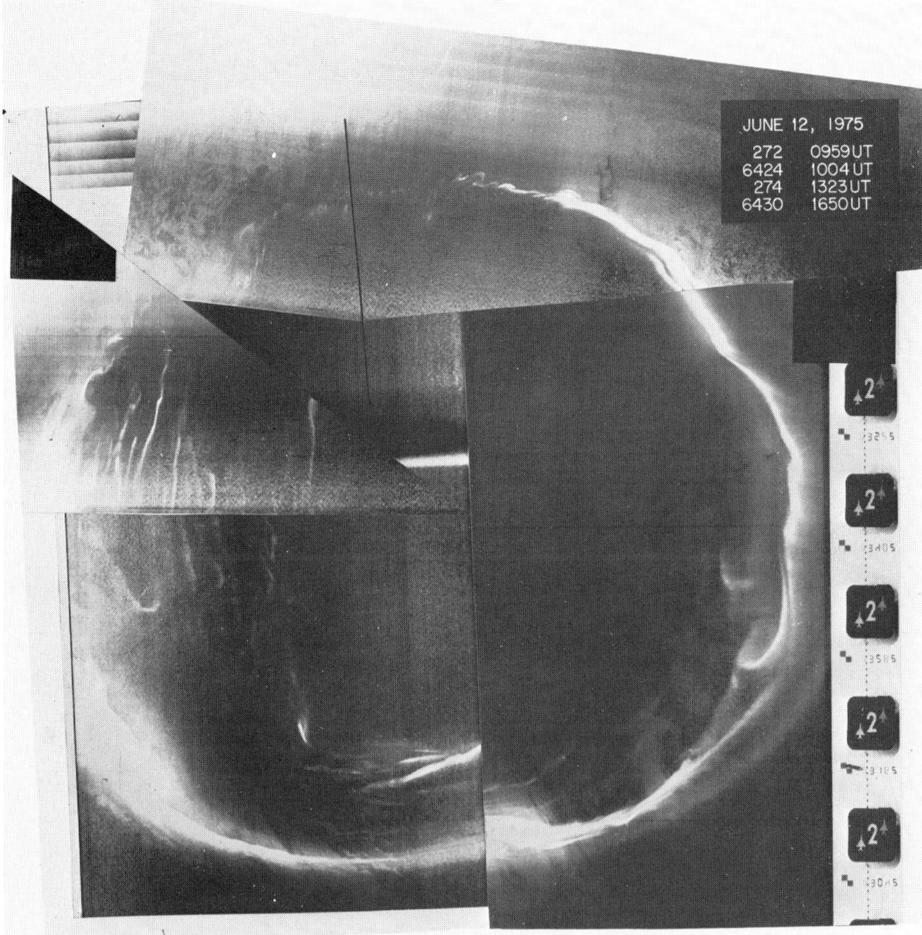

Fig. 2.2. Montage photograph of the auroral oval taken from above the antarctic region by the DMSP-8531 and -10533 satellites. (Akasofu, S.-I.: *Space Sci. Rev.* **19**, 169, 1976.)

very bright arc (or perhaps two arcs) was stretching along the afternoon sector of the oval. This afternoon arc was suddenly truncated in the midday sector.

(iii) *Example 3: Figure 2.3.* Another montage photograph is constructed by using three DMSP-8531 photographs on June 12 (orbits 6427, 6429 and 6432). In this example the morning part of the oval consisted of a few arcs. In the midnight sector there was a clear separation between discrete arcs and the diffuse aurora.

In this montage photograph, the midday part of the oval was not well photographed. On the other hand, the presence of the annular auroral belt, located a little equatorward of the oval and consisting of patchy auroras, is very clearly photographed in the day sector. There is also a horn-like luminosity which extends towards the midday sector from the afternoon sector. These faint auroras,

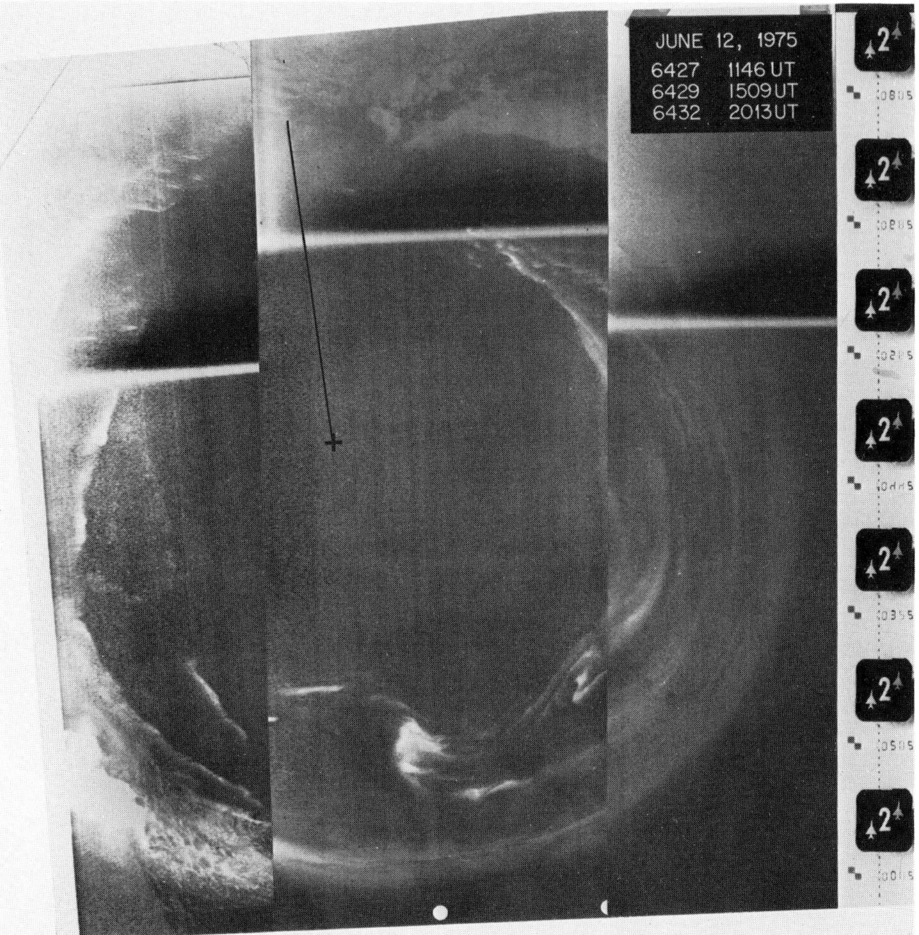

Fig. 2.3. Montage photograph of the auroral oval taken from above the antarctic region by the DMSP-8531 satellite. (Akasofu, S.-I.: *Space Sci. Rev.* **19**, 169, 1976.)

together with the diffuse aurora in the night sector, form an annular auroral belt. The third photograph shows a very broad diffuse aurora with some structures.

(iv) *Example 4: Figure 2.4.* This montage photograph consists of two DMSP-8531 photographs (orbits 6484 and 6488) and shows active auroras all along the oval. The most interesting feature in the morning half of the photograph is a number of arcs which point towards the midday sector of the oval. It is quite likely that they are polar cap auroras, but their orientation deviates considerably from the Sun-Earth line. It is not known, at present, how they grow and decay and how they might be related to the other types of polar cap auroras.

There are a few bright but short arcs in the early afternoon sector, which also align radially, pointing towards the poleward boundary of the midday part of the

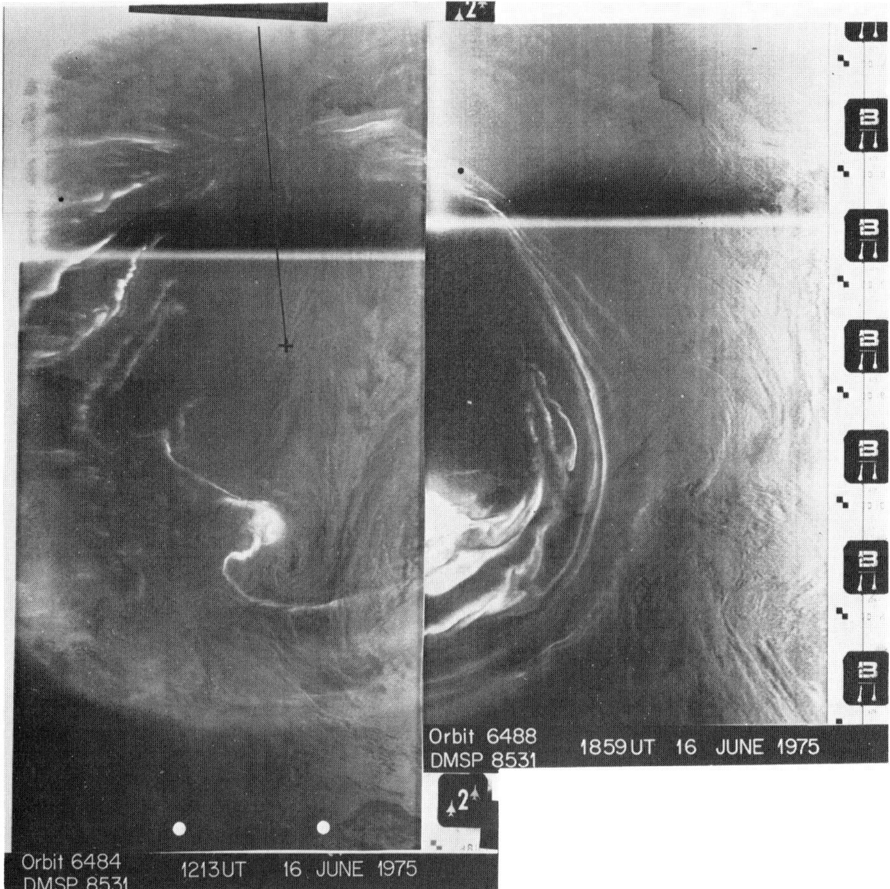

Fig. 2.4. Montage photograph of the auroral oval taken from above the antarctic region by the DMSP-8531 satellite. (Akasofu, S.-I.: *Space Sci. Rev.* **19**, 169, 1976.)

oval. In fact, it can be seen that all arcs and the tips of patchy auroras in the midday sector are radially aligned.

The second photograph shows the front of an intense surge in the late evening sector. Note that arcs extending from the surge appear to end rather abruptly in the mid-afternoon sector. Further, at least two arcs extend from the midday sector to the late evening sector; there are also several other arcs which do not extend that far. A structured diffuse aurora is seen a little equatorward of these arcs. The horn-like structure is visible in the early afternoon sector.

2.1.2. SCHEMATIC DISTRIBUTION PATTERN OF AURORAS

Here we combine all the major auroral features we have learned about, to improve the schematic pattern proposed earlier by Akasofu (1974). Figure 2.5 is the product of such an attempt. Note that the figure is for the northern

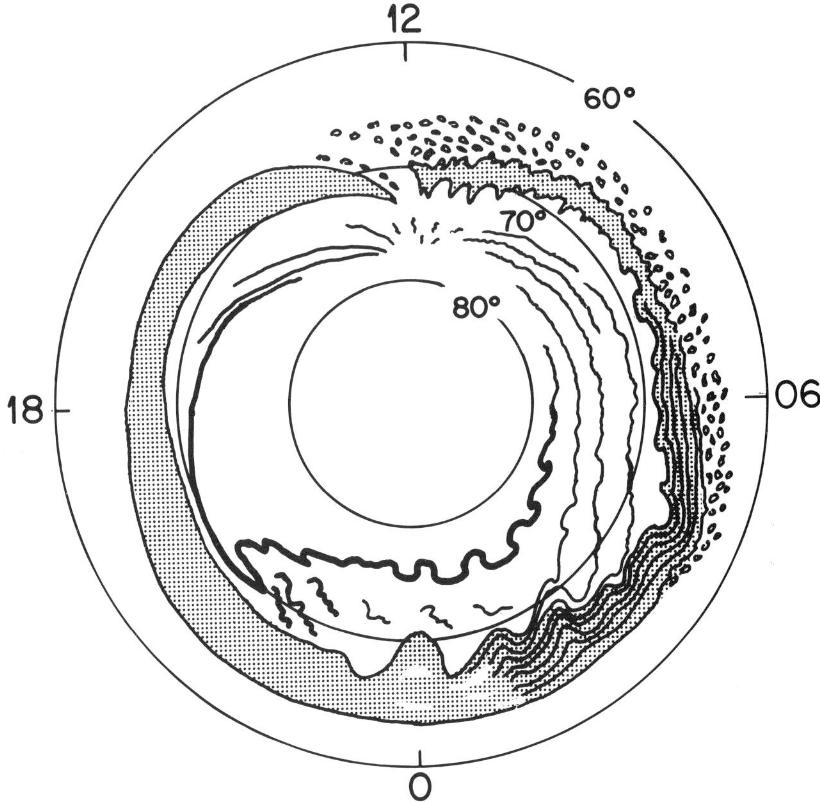

Fig. 2.5. Schematic diagram showing the main characteristics of auroras during an auroral substorm in dipole-MLT coordinates. Discrete arcs are indicated by lines and the diffuse auroral regions are shaded. (Akasofu, S.-I.: *Space Sci. Rev.* **19**, 169, 1976.)

hemisphere, so that the morning sector lies on the right-hand side. Perhaps the only serious uncertainty at the present time is the 'origin' of polar cap arcs in the morning sector, although the proposed pattern in the morning sector appears to occur quite frequently.

2.2. Auroras in Different Local Time Sectors

2.2.1. DISCRETE AURORAS AND DIFFUSE AURORAS IN THE NIGHT SECTOR

Recent airborne and satellite observations of auroras have identified two diffuse broad bands of luminosity, in addition to the well-established oval band of discrete auroras. One of these is an oval-shaped belt which is partially overlapped with and located slightly equatorward of the oval of discrete auroras. The other is the annular auroral belt which we examined in the previous section (Lui and Anger, 1973; Shepherd *et al.*, 1973; Lui *et al.*, 1973; Whalen *et al.*, 1971; Pike and Whalen, 1974; Snyder *et al.*, 1974; Snyder and Akasofu, 1974; see also a review paper by Akasofu, 1974). Undoubtedly, the mantle aurora (Sandford, 1964, 1967), the so-called 'hydrogen aurora' (cf. Eather, 1967; Reid and Rees, 1961) and the continuous aurora (Whalen *et al.*, 1971; Buchau *et al.*, 1972) constitute parts of these two diffuse bands of luminosity. Here we tentatively define the discrete and diffuse auroras, as follows (Snyder and Akasofu, 1974):

Discrete aurora. A discrete aurora appears as a single, bright strand, separated from others by a dark space of order of a few tens of kilometers in width. When it is seen from the ground it has a curtain-like structure.

Diffuse aurora. A diffuse aurora appears as a broad band of auroral luminosity with a width of at least several tens of kilometers. It may not be easily visible from the ground, but can cover one half of the sky. Its equatorward boundary may be identified by experienced eyes but could be misidentified as a homogeneous arc.

However, such a distinction between these two main auroral types cannot always be made. This problem will be pointed out later in a number of auroral photographs taken from satellites.

In the following, we examine in detail a typical example of photographs taken from the DMSP-2 satellite. In the upper part of Figure 2.6, the subtrack of the satellite runs horizontally, approximately along the dawn-dusk meridian, across the middle of the photograph. A photometric device scans rapidly a narrow strip of ground which lies perpendicular to the subtrack, so that the satellite 'photograph' results from a number of scannings as the satellite traverses across the polar region. The lower part of the figure shows the approximate geographic area covered by the photograph; it shows also the invariant latitude lines at 100 km altitude (thick lines) and geographic latitude lines (thin lines), together with approximate isochrons (dot-dashed lines), namely loci on the Earth's surface where magnetic noon occurs at the same UT (Lebau, 1970). This particular photograph was taken during orbit 873. The satellite passed nearest to the geographic north pole at 2024 UT on 1973, January 9; the magnetic midnight meridian is thus approximately along the isochron 0824 UT (= 2024 UT − 1200).

On the basis of the definition above, it can be seen that the poleward boundary of the belt of auroras is marked by a bright discrete aurora. Its curtain-like vertical structure is scanned, as is the lower border, and is seen as a horizontally projected image. Note that because the lower border is located at an altitude of 100 km (instead of ground level), the images are projected, along the line of sight of the satellite, slightly towards the top of the photograph with respect to ground images such as city lights when they are located in the upper half of the photograph. A large mass of luminosity over the Norwegian Sea is a westward traveling surge.

The equatorward boundary of the auroral belt can easily be recognized by a smooth boundary of the diffuse aurora. A relatively uniform band of luminosity, which extends from Iceland to the northern Scandanavian Peninsula, is a typical diffuse aurora in the evening sector. The reason for faintness over the Norwegian Sea is partly due to the small albedo of an open sea compared with the large albedo of a snow-covered land or ice-covered sea.

In the midnight and early morning sectors, the diffuse aurora covers an

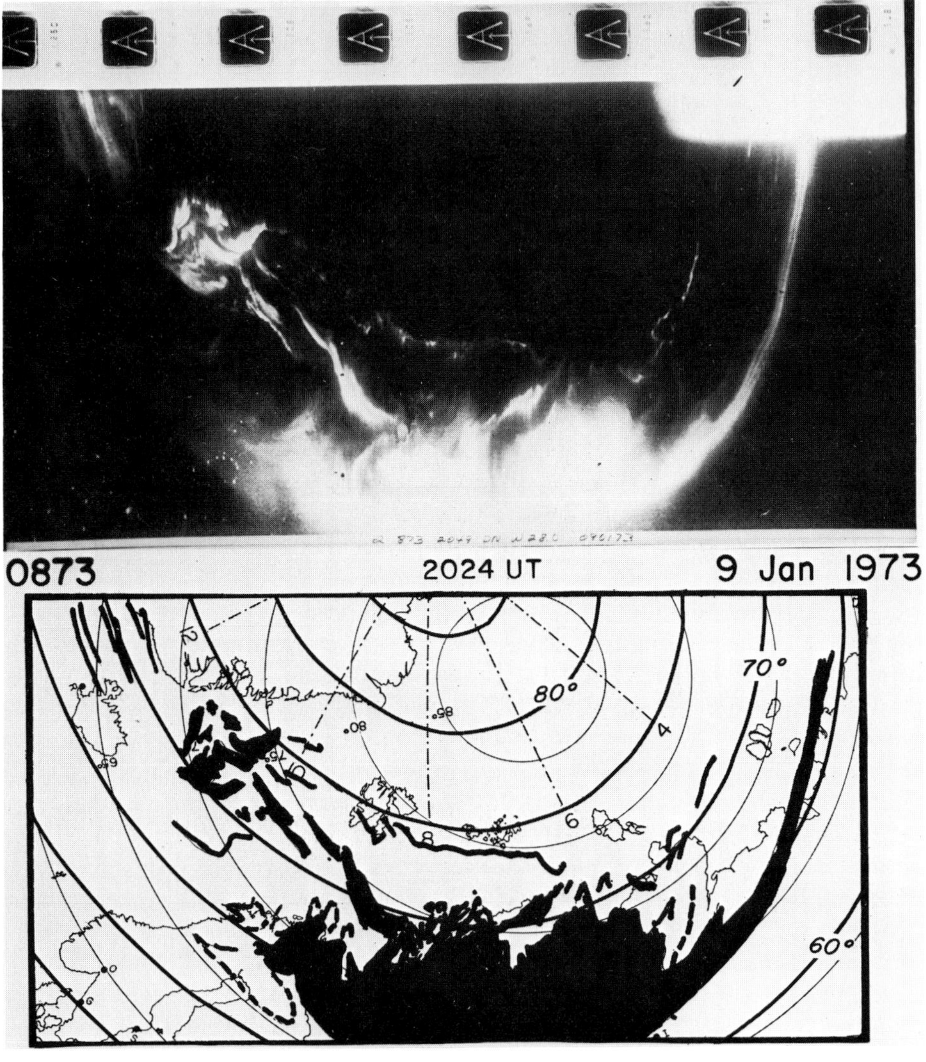

Fig. 2.6. DMSP-2 satellite photographs of the dark hemisphere showing the main features of auroral substorms. (Akasofu, S.-I.: *Space Sci. Rev.* **16**, 617, 1974.)

extensive area over the northern part of Siberia. Unlike the diffuse aurora in the evening sector, it is not a uniformly bright region and consists of a number of discrete arcs which are closely packed along a narrow east-west belt.

One of the striking features in many photographs is that there are distinct differences in the characteristics of auroras between the evening and morning sectors. Roughly speaking, discrete auroras are predominantly an evening feature and are located in the poleward half of the oval. On the other hand, the diffuse aurora occupies at least the equatorward half of the oval over the entire dark sector. Figure 2.7 shows a striking example of this asymmetry. This feature was

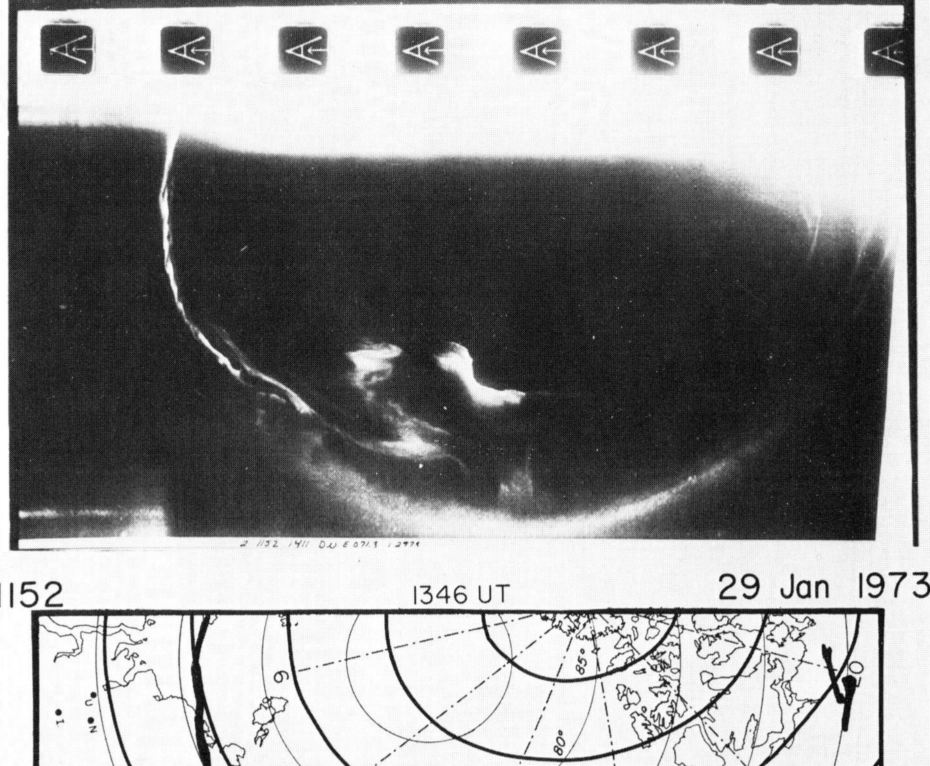

Fig. 2.7. A single bright discrete arc in the evening sector (DMSP-2 photograph). It is one of the characteristic features of auroras in the evening sector. (Akasofu, S.-I.: *Space Sci. Rev.* **16**, 617, 1974.)

not necessarily obvious in either past visual or all-sky observations; often the diffuse aurora covers a significant part of the field of view of an all-sky camera, making it difficult to even recognize its presence.

The regions of discrete and diffuse auroras cover a large latitudinal range, often well beyond the field of view of a single all-sky camera. Thus an airborne all-sky camera is an ideal tool in studying in detail the latitudinal structure of both regions. Figure 2.8 shows an example of such a study (Stenbaek-Nielsen *et al.*, 1973). It shows all-sky photographs taken from two aircraft at geomagnetically conjugate points, one in the northern hemisphere (over Alaska) and the other in the southern hemisphere (over the Antarctic Sea); the centers of the northern and

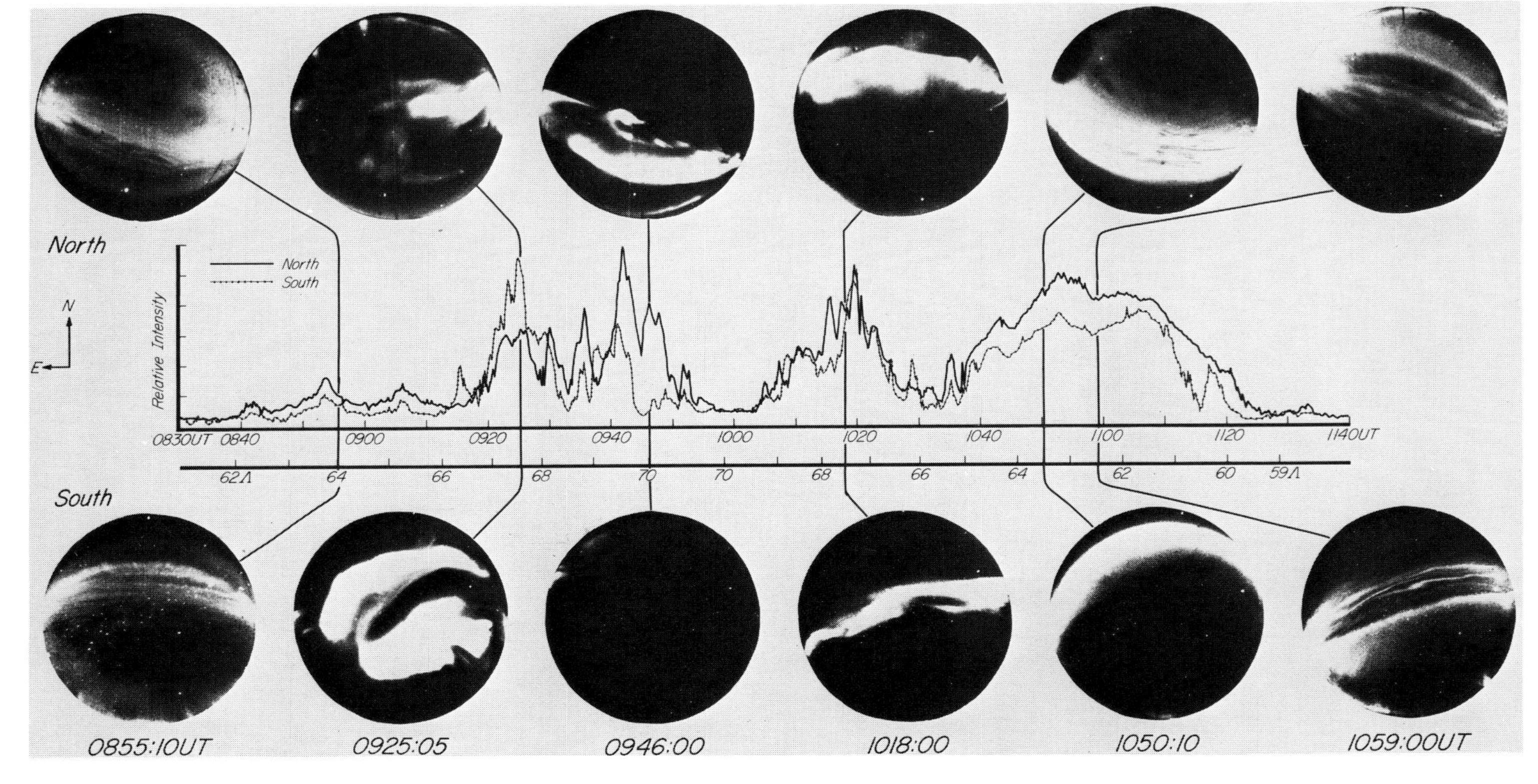

Fig. 2.8. Simultaneous airborne all-sky photographs at conjugate areas along the Alaska-New Zealand meridian (1968, March 19), together with auroral intensities during the flight. Intensity variations are well-correlated in the diffuse auroral region, but non-correlated variations occur in the discrete auroral region. (Stenbaek-Nielsen, H. C., Westcott, E. M., and Davis, T. N.: *J. Geophys.* **78**, 659, 1973.)

southern photographs are connected by a single field line. The first pair of photographs taken at dp. lat. 64° indicates clearly the diffuse aurora; note complicated fine structures in the aurora, particularly the dark filamentary structure in the diffuse aurora. One important aspect of the diffuse aurora is that it is nearly conjugate, and even fine structures in one hemisphere can often be identified in the other hemisphere. Beyond the latitude range of the diffuse aurora, there is the band of discrete auroras. The conjugacy of the northern and southern aurora is excellent during quiet periods (Belon *et al.*, 1969), but breaks down at times during disturbed periods (Stenbaek-Nielsen *et al.*, 1973). The second pair

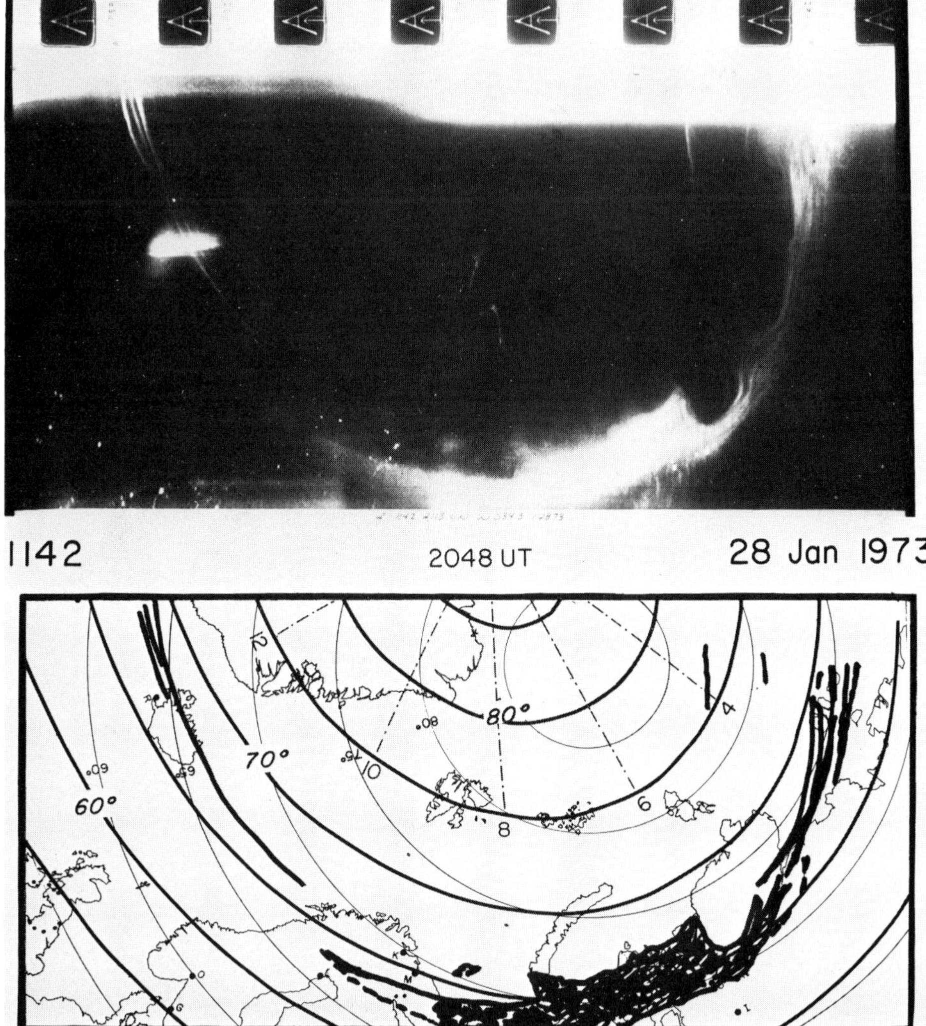

Fig. 2.9. Example of omega (Ω) bands in the morning sector (DMSP-2 photograph). (Akasofu, S.-I.: *Space Sci. Rev.* **16**, 617, 1974.)

(0925:05 UT) of all-sky photographs shows that the bright display (a fold) in the southern photograph is seen in the eastern sky in the northern photograph. The third pair of photographs (0946:00 UT) shows a complete breakdown of conjugacy. In the next pair, there is a meridional dislocation. The last two pairs show the diffuse aurora with a considerable internal structure; they were taken during the equatorward return flights in both hemispheres.

The poleward boundary of the diffuse aurora develops various wavy features. Some of the wavy structures have a particular shape like an inverted letter omega (Ω) and are often called 'omega bands'. Figure 2.9 shows a DMSP photograph with a series of omega bands. Omega bands 'drift' eastward. Another interesting structure of the poleward boundary of the diffuse aurora is a torch-like structure.

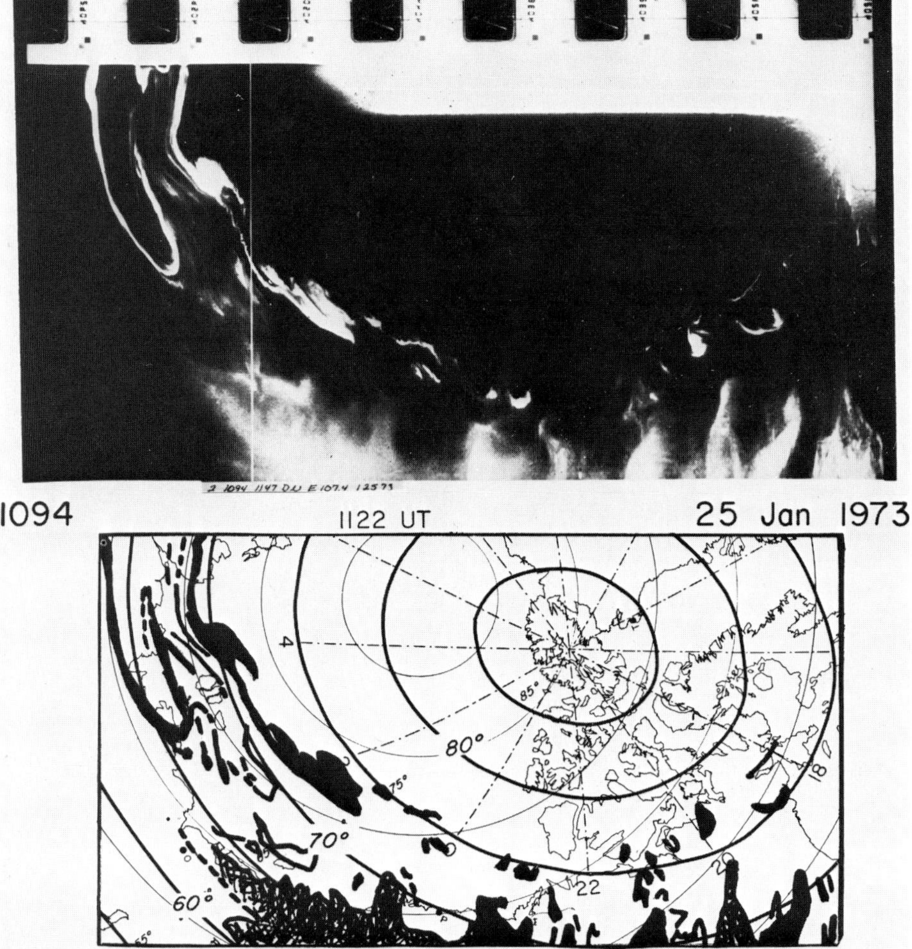

Fig. 2.10(a). DMSP-2 satellite photograph of the dark hemisphere showing the main features of auroral substorms. Note in particular the torch-like structure. (Akasofu, S.-I.: *Space Sci. Rev.* **16**, 617, 1974.)

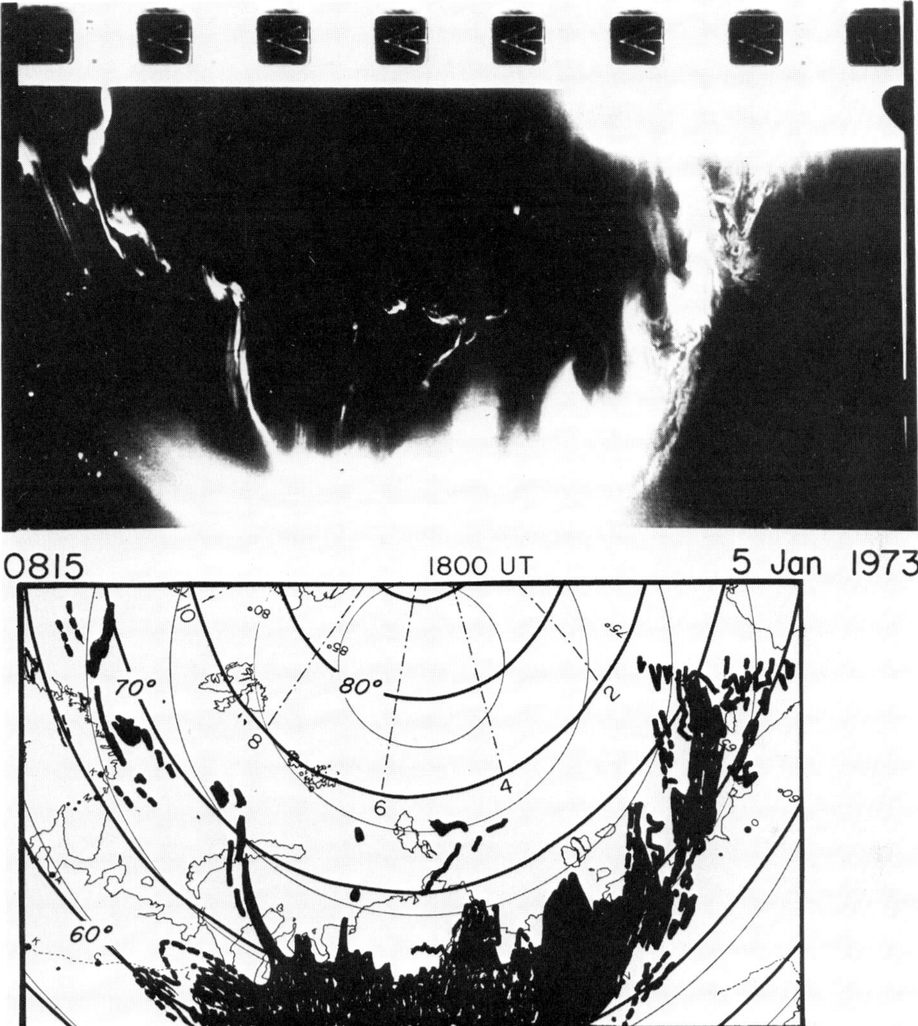

Fig. 2.10(b). DMSP-2 satellite photograph of the dark hemisphere showing the well-developed torch-like structure near the poleward boundary of the diffuse aurora. (Akasofu, S.-I.: *Space Sci. Rev.* **16**, 617, 1974.)

Figures 2.10(a) and 2.10(b) show a well-developed series of torch-like structures during intense substorms. They drift westward (Royrvik, 1976).

2.2.2. MIDDAY AURORAS

The midday part of the auroral oval was first studied with airborne instruments by Buchau *et al.* (1969). Whalen *et al.* (1971), Eather and Mende (1971a, b, 1972), Romick and Brown (1971), Heikkila *et al.* (1972) and Whalen and Pike (1973) showed that the midday part of the oval is a band 2° to 5° wide, in which the O I

6300 Å emission is greatly enhanced. Discrete auroras grow and decay in this red band.

The presence of the red band was confirmed by a scanning photometer aboard the ISIS-2 satellite. Shepherd and Thirkettle (1973) and Shepherd *et al.* (1973) showed that the maximum intensity of the O I 6300 Å emission was a little more than 2 kR and that its 1 kR contour line extended from the 6 to 18 LT meridian. Figure 2.11 shows the intensity distribution of the 6300 Å emission over the entire polar region. Recently, Derblom (1975) found that the 6300 Å emission in the cusp region is associated with enhanced hydrogen emissions.

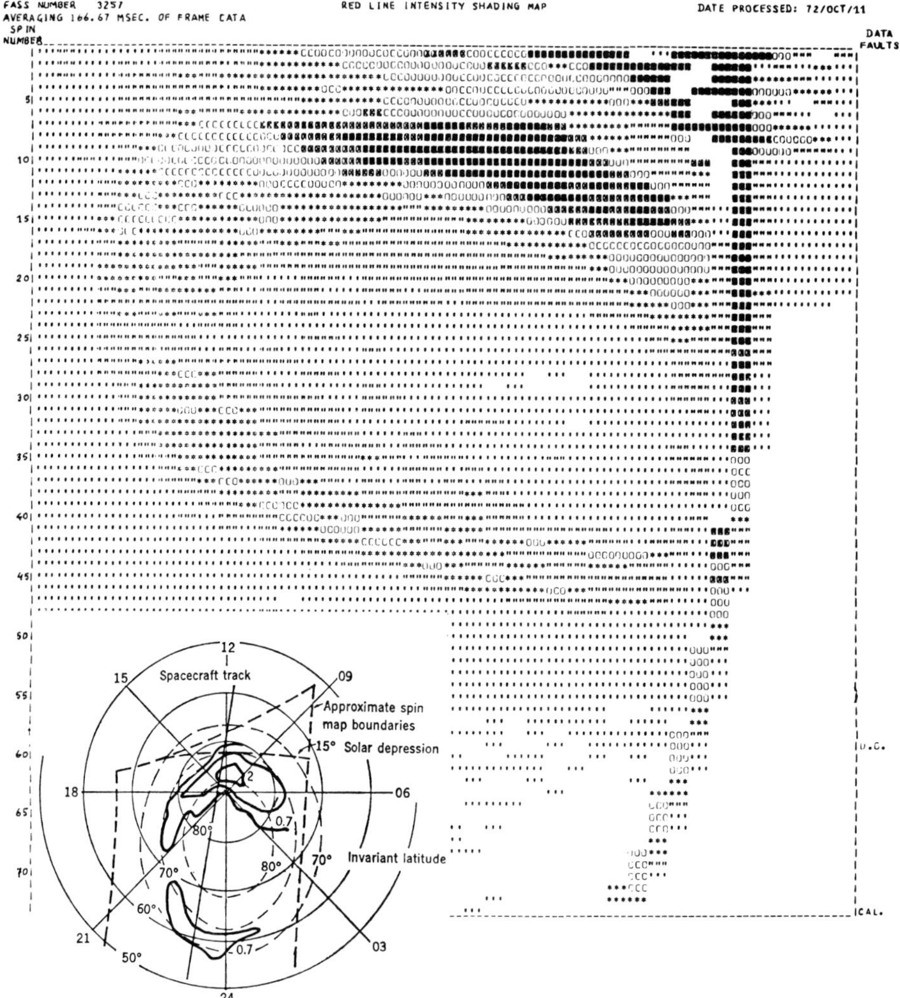

Fig. 2.11. Map of the oxygen 6300 Å emission obtained from the ISIS-2 satellite on orbit 3257, 0518–0540 UT, 1971, December 11. Each line of print corresponds to one scan across the Earth, and the intensities are indicated by the blackness of the symbols. The insert shows the iso-intensity contours of the emission in geographic latitude-LT coordinates. (Shepherd, G. G. and Thirkettle, F. W.: *Science* **180**, 737, 1973.)

Whalen *et al.* (1971) showed that the midday part of the auroral oval consists of three regions (Figure 2.12). The first is the red band, which extends to a very high altitude. The second region is a band of a uniform glow located a little equatorward of the red band (although its poleward boundary is located in the red band). They called this glow the continuous aurora. The third region is located equatorward of the continuous auroral region, although its poleward boundary is embedded in it. This region is characterized by ionization in the D region, which is caused by electrons with energies much higher than those in the red band and the continuous auroral region. When this belt is observed from above, it appears as a quasi-circular (annular) belt of luminosity. Figures 2.13(a) and (b) show multi-photometric observations across the midday part of the oval, both from above by the ESRO IA satellite and from below by the NASA 711 aircraft. In this particular case, the quasi-circular belt was located between 68° and 70°. The continuous (diffuse) aurora was located between 72° and 77°. Beyond dp. lat. 77°, the relative intensity of the 6300 Å with respect to the N_2^+ (4278 Å) is considerably enhanced. Thus the red band in Figure 2.13 extended from about 77° to 82°. An intense emission, narrowly confined around 80°, was caused by a discrete aurora.

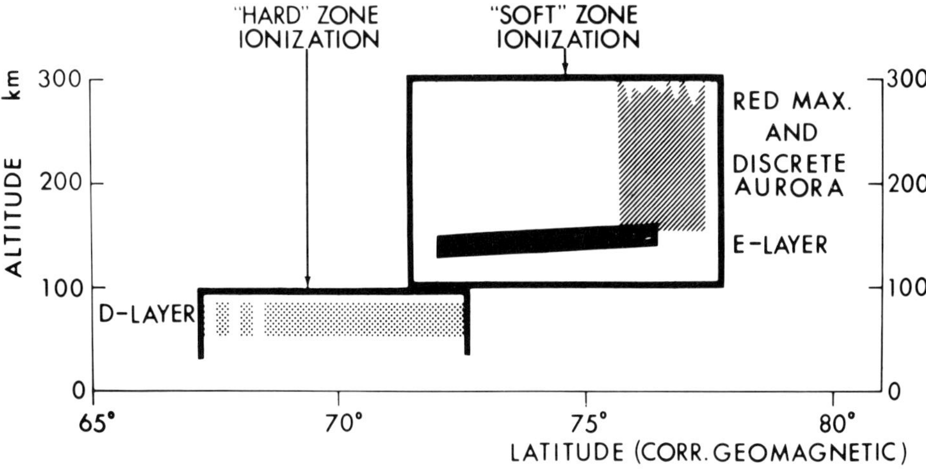

Fig. 2.12. Auroral structure in the noon meridian (altitude versus latitude cross-section). The E layer indicates the zone of diffuse (or continuous) aurora. 'Red Maximum' indicates the enhanced band of the 6300 Å emission in which discrete auroras are embedded. The D layer is present in the annular diffuse auroral region (the 'hard' zone). (Whalen, J. A., Buchau, J., and Wagner, R. A.: *J. Atmosph. Terr. Phys.*, **33**, 661, 1971.)

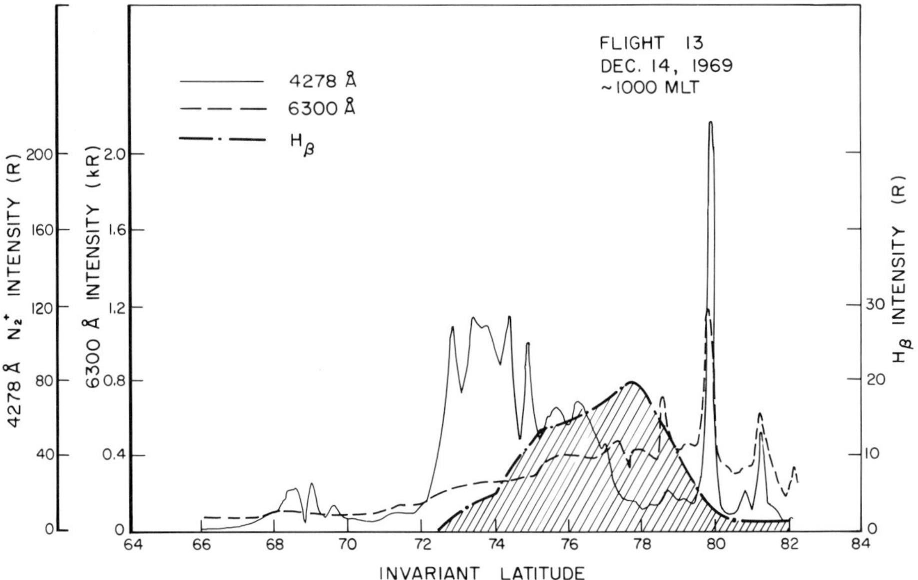

Fig. 2.13(a). Airborne observation of the midday part of the auroral oval by the NASA 711 aircraft on 1969, December 14. (Eather, R. H. and Mende, S. B.: *J. Geophys. Res.* **76**, 1746, 1971.)

We examine here in some detail the structure of discrete auroras in the midday sector. One of their important features is a tendency to fragment and to align radially toward the midday point of the poleward boundary of the oval. Figure 2.14 shows an example of the midday part of auroral photographs taken from the DMSP-10533 satellite. The 'fragmentation' of auroral arcs in the midday sector is quite clear in the photograph. The photograph also shows clearly a belt of patchy auroras, the annular auroral belt, which is located a little equatorward of the oval belt.

Figure 2.15 shows more clearly the remarkable radial alignment and convergence of the midday auroral arcs, photographed by the DMSP-8531 satellite. There is a definite tendency for the arcs to converge toward the poleward boundary of the midday part of the auroral oval. These arcs in the afternoon sector do not extend to the evening sector. Similarly, evening arcs do not extend to the early afternoon sector, and lie poleward of the afternoon arcs. Figure 2.16 shows an example of DMSP photographs which show these features. It appears that auroral arcs in the day and night sectors arise from different source *regions* despite the fact that they form a well-defined singly connected oval belt.

Lassen (1970, 1972, 1973) discussed the auroral distribution in terms of two auroral groups (the daytime and nighttime groups) which combine to form the auroral oval. He was particularly concerned about the fact that during quiet periods, the occurrence frequency of auroras in the midnight sector is very low, in contrast with a very high frequency in the noon sector. Mishin *et al.* (1970) suggested that the auroral distribution consists of two horseshoe patterns, one on the dayside of the polar region and the other slightly larger one on the darkside, and that there is a distinct gap between the two horseshoe patterns.

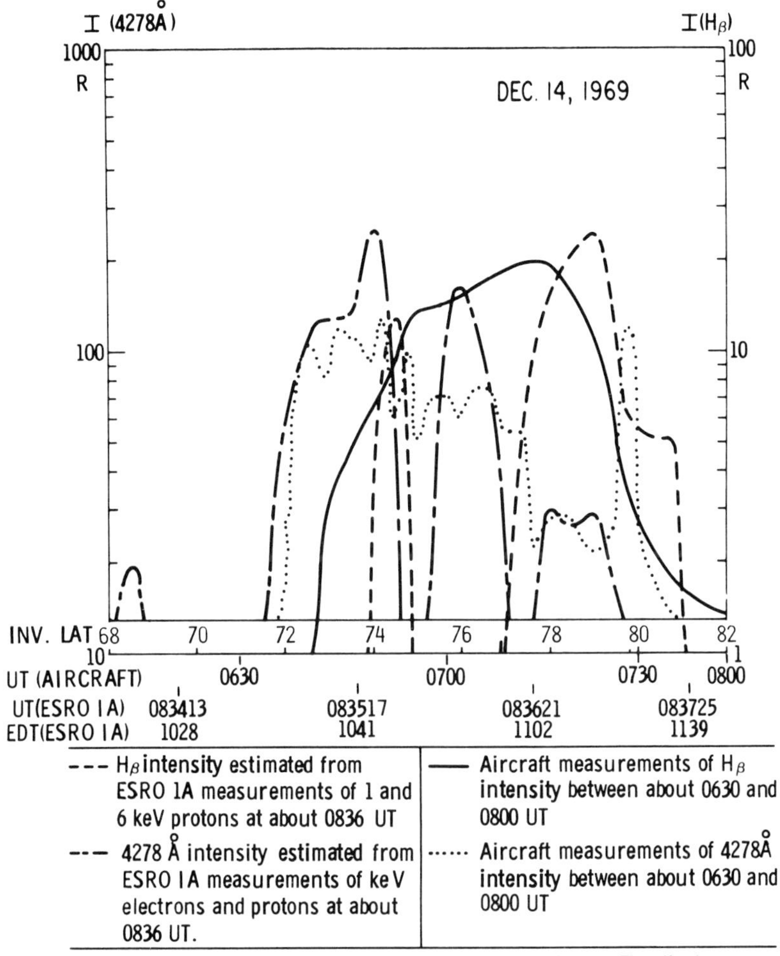

--- Hβ intensity estimated from ESRO 1A measurements of 1 and 6 keV protons at about 0836 UT	—— Aircraft measurements of Hβ intensity between about 0630 and 0800 UT
--- 4278 Å intensity estimated from ESRO IA measurements of keV electrons and protons at about 0836 UT.	······ Aircraft measurements of 4278Å intensity between about 0630 and 0800 UT

The satellite height varied between 580 and 470 km. The dipole magnetic time for the aircraft was between about 0930 and 1100 during the period shown.

Fig. 2.13(b). Approximately simultaneous optical observation of the midday part of the auroral oval from above (the ESRO IA satellite) and from below (the NASA 711 aircraft) on 1969, December 14 over the Greenland Sea. (Hultqvist, B.: *Ann. Geophys.* **30**, 223, 1974.)

2.2.3. POLAR CAP AURORAS

The polar cap is defined as the region bounded by the auroral oval, but it is not void of auroras and auroral particles. It has been known that there are discrete auroras, called the polar cap auroras. One of their characteristic features is that they are oriented approximately along the Sun-Earth line (Denholm, 1961; Davis, 1962). Thus, when the orientation of polar cap auroras is observed at a polar cap station, it varies systematically during the course of a day.

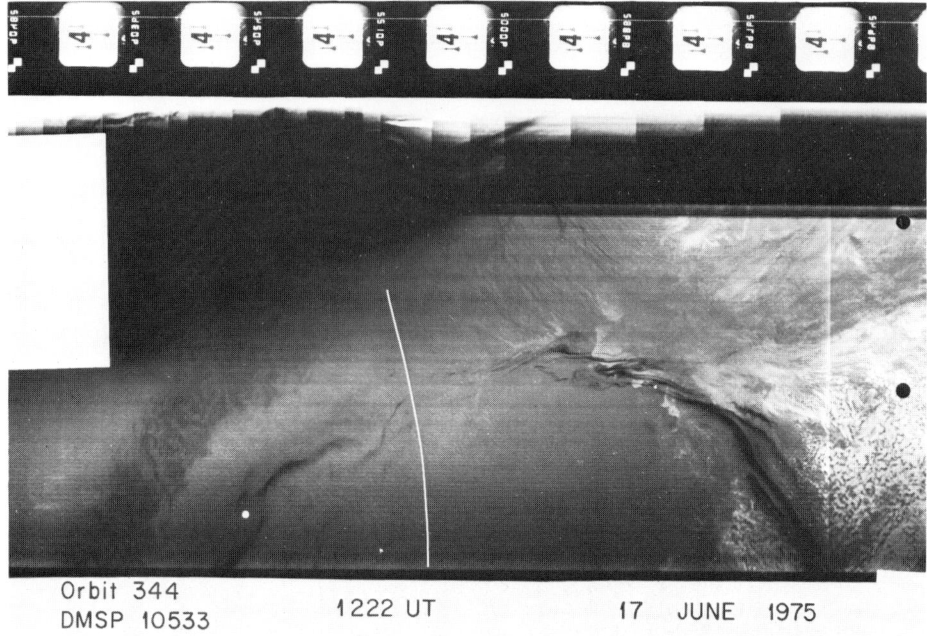

Fig. 2.14. Midday part of the auroral oval (DMSP-10533 photograph) over the antarctic region. The dot and line indicate the location of the South Pole and the noon meridian. (Akasofu, S.-I.: *Space Sci. Rev.* **19**, 169, 1976.)

An extensive study of DMSP photographs suggests that there are two types of polar cap auroras. The first type tends to appear during a fairly quiet period as a single or double arc which runs approximately along the noon-midnight meridian and bends itself toward the evening sector, lying parallel to arcs in the evening part of the oval. Figure 2.17 shows an example of such an arc. A variety of polar cap auroras has also been studied by Anger *et al.* (1974).

The second type of polar cap aurora tends to appear during the recovery phase of substorms and occupies the morning half of the polar cap. Further, it appears to originate from the oval in the morning sector. Figure 2.18 shows such polar cap arcs filling the entire morning half of the polar cap in the northern hemisphere. Eather and Akasofu (1969) and Akasofu and Yasuhara (1973) showed that the 6300 Å radiation is richer in polar cap auroras than in oval auroras.

The polar cap is often completely void of such polar cap auroras during an intense substorm, as noted first by Davis (1962). Figure 2.19 shows an example of a very dark polar cap which was taken during an intense substorm; an intense westward traveling surge is seen in the evening sky, while patchy auroras occupy the entire oval in the morning sector. It may well be that polar cap auroras in the morning sector shift equatorward during the early epoch of a substorm, but they shift back poleward or are formed during the recovery phase when the torch-like structure develops in a diffuse aurora.

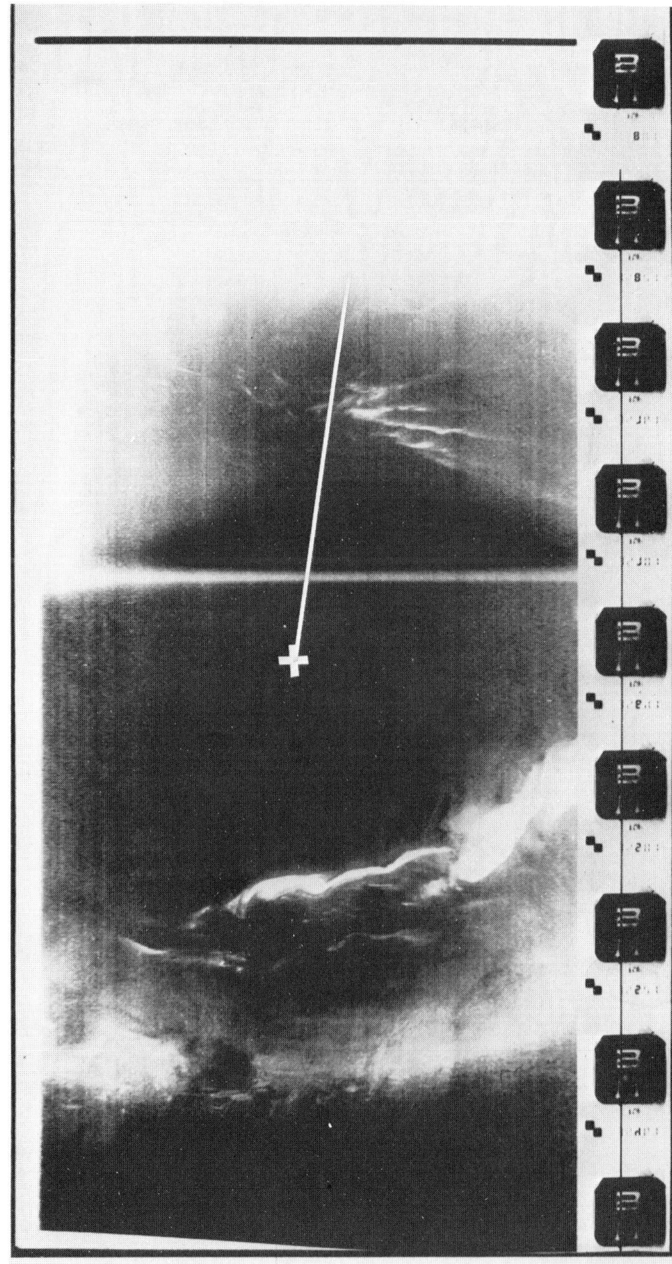

Fig. 2.15. Noon-midnight part of the auroral oval (DMSP-8531 photograph). Note in particular the radial structure of auroral arcs in the midday sector. (Snyder, A. L. and Akasofu, S.-I.: *J. Geophys. Res.* **81**, 1799, 1976.)

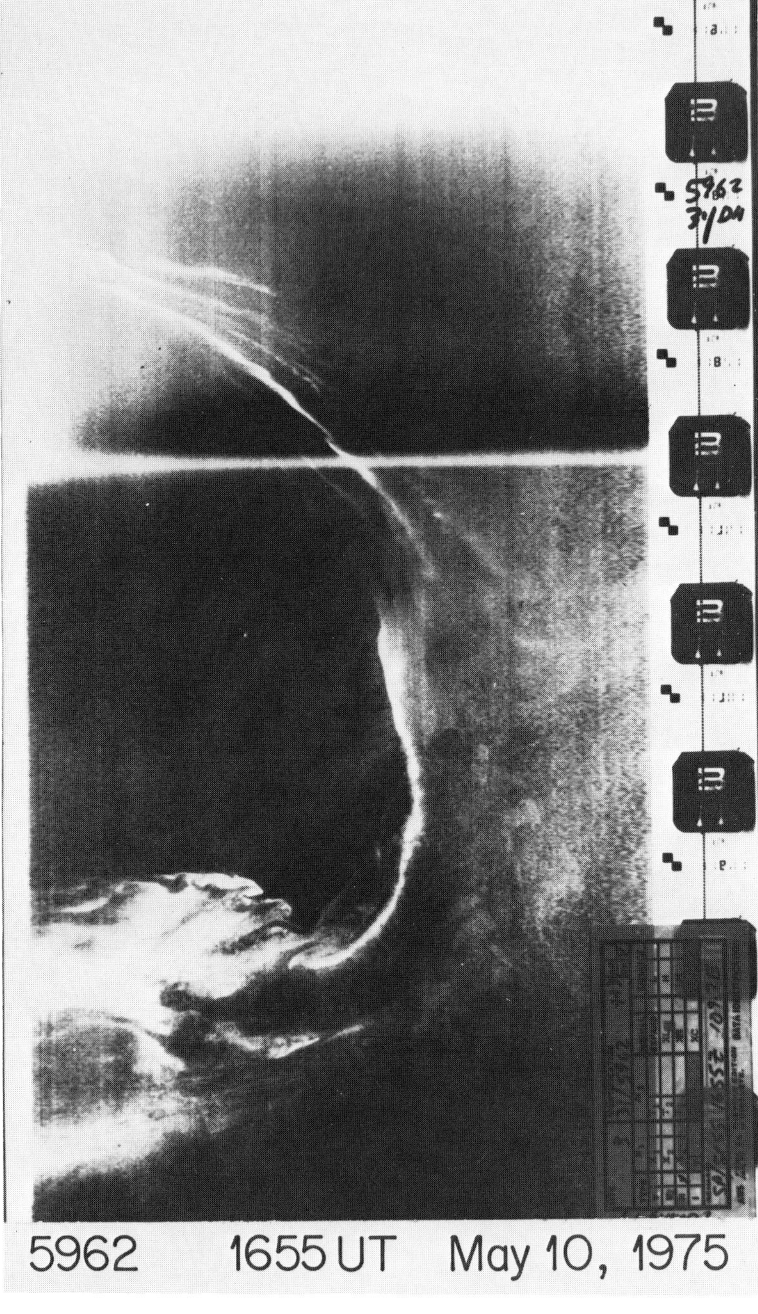

Fig. 2.16. Auroral distribution in the afternoon sector of the auroral oval (DMSP-8531 photograph).

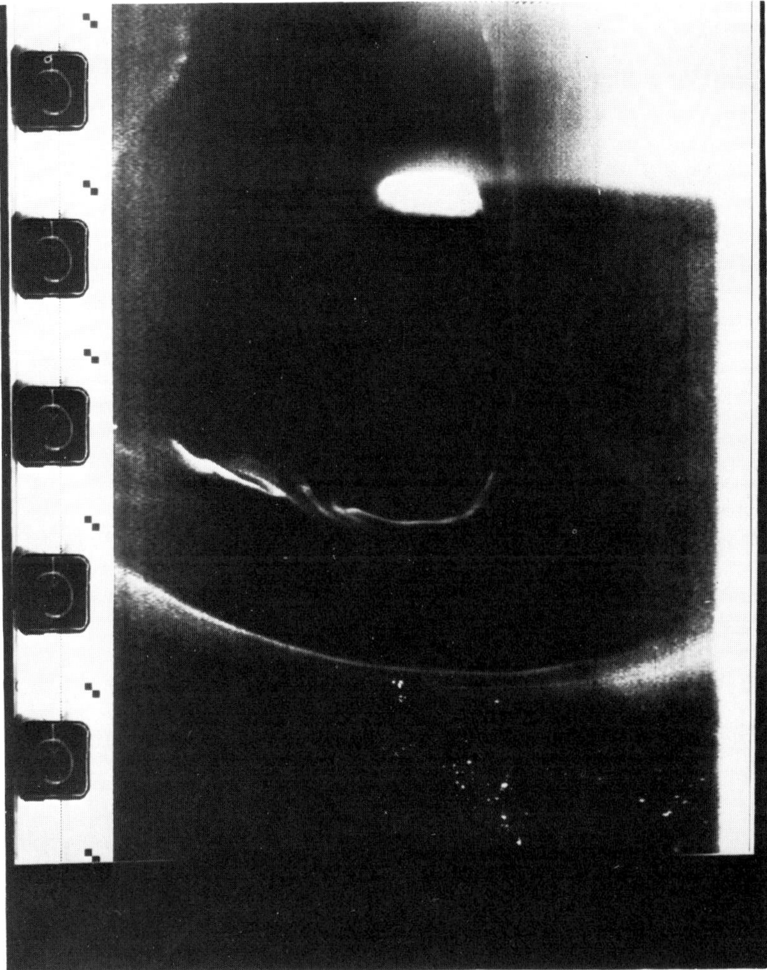

Fig. 2.17. A single polar cap auroral arc which runs approximately along the noon-midnight meridian and bends toward the evening sector. (Meng, C.-I. and Akasofu, S.-I.: *J. Geophys. Res.* **81**, 4004, 1976.)

2.3. Auroral Electrons: The Statistical Precipitation Pattern

The distribution of precipitating electrons over the entire polar region has been investigated by a number of workers. Figure 2.20 shows the distribution of high fluxes of 0.7 keV electrons with 0° pitch-angle during relatively quiet periods (Kp = 0–2) (Hoffman and Berko, 1971). It can be seen that the region of high fluxes forms an oval-shaped belt. There seems to be a general agreement of the shapes between the oval of visible auroras and the oval of high fluxes of low energy electrons. However, Hultqvist (1972) and Riedler (1972) pointed out that such an agreement is not necessarily obvious for electrons of energies greater than 1 keV. Indeed, Craven (1970) showed also that the precipitating pattern of 5 keV electrons differs considerably from Feldstein's oval. In general, higher

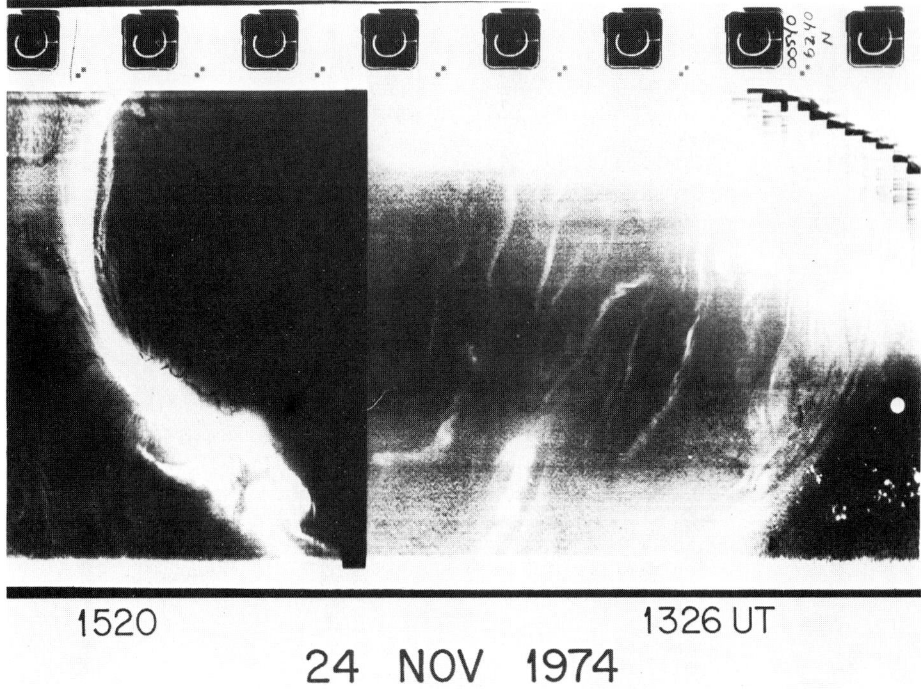

Fig. 2.18. Polar cap auroral arcs which occupy the entire morning half of the polar cap (DMSP-8531 photograph).

energy electrons tend to precipitate along a constant L curve, rather than along the oval as lower energy electrons do.

These features can most clearly be seen in the precipitation patterns constructed by McDiarmid *et al.* (1975). They showed first of all that the average energy of electrons in the range 150 eV to 9.6 keV has a minimum approximately along the trapping boundary (namely, the outer boundary of the Van Allen belt). This particular feature was noted by Burch (1968). Figure 2.21(a) shows McDiarmid *et al.*'s iso-energy contours of precipitative electrons in invariant latitude-MLT coordinates. The dashed curve indicates the trapping boundary where their 35 keV counting rate drops to the background level.

Figures 2.21(b), (e), (d), (e) and (f) show the average flux (per cm^2 s ster keV) contours for 150 eV, 1.3 keV, 9.6 keV, 22 keV, 35 keV and 210 keV electrons. Comparing the five contour plots, one interesting feature is that there is a maximum flux in the afternoon sector for lower energy electrons (150 eV and 1.3 keV), while there is a minimum of flux in the same sector for higher energy electrons (9.6 keV and 22 keV); further, for very high energy electrons (>210 keV), there is neither a maximum nor a minimum, and their iso-intensity contour lines are nearly circles.

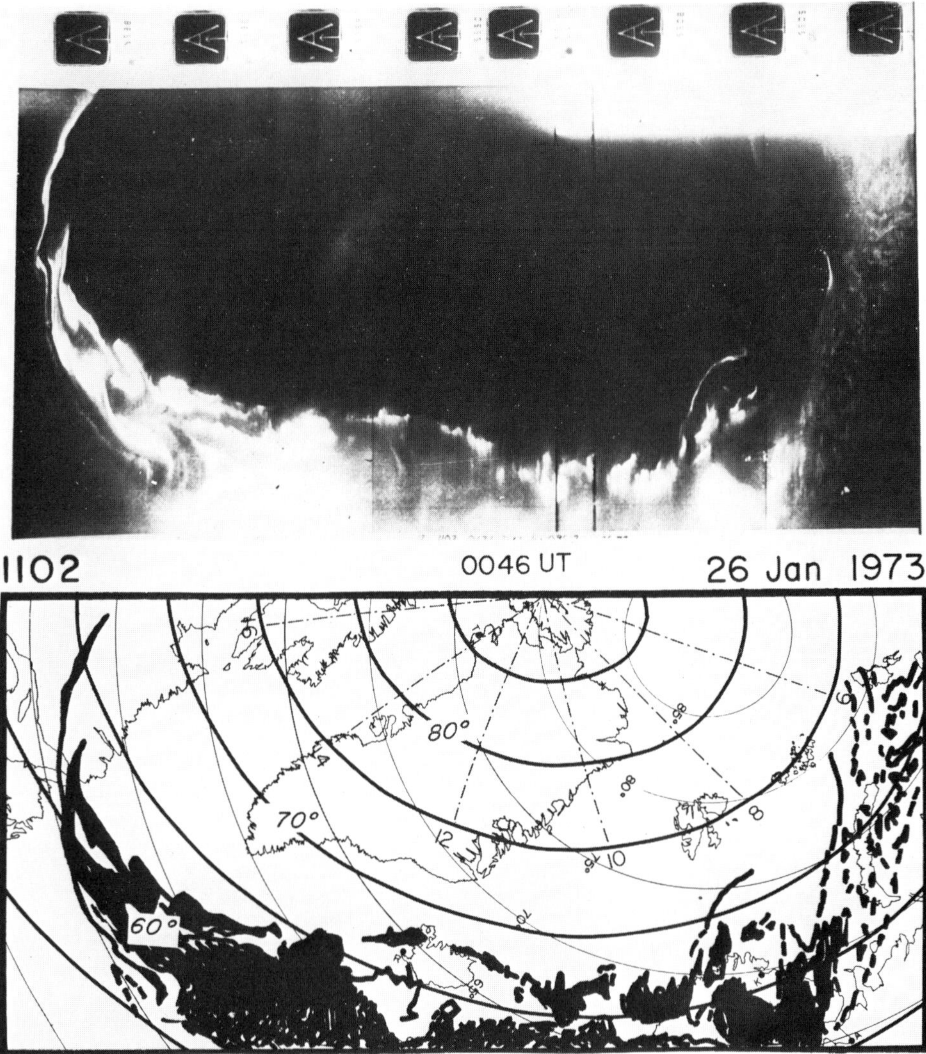

Fig. 2.19. DMSP-2 photograph showing the major auroral substorm features in the dark hemisphere. Note in particular the absence of auroras in the polar cap. (Akasofu, S.-I.: *Space Sci. Rev.* **16**, 617, 1974.)

2.4. Auroral Electrons: Spectra of Auroral Electrons

2.4.1. EVENING AND MIDNIGHT SECTORS

(a) *Satellite Observations*

In Section 2.1 we noted that in the evening and midnight sectors the aurora consists of two groups, discrete and diffuse auroras. In this section we examine characteristics of auroral particles which impinge into these two auroral regions (Frank and Ackerson, 1971; Heikkila, 1972; Hultqvist, 1974; Lui *et al.*, 1976).

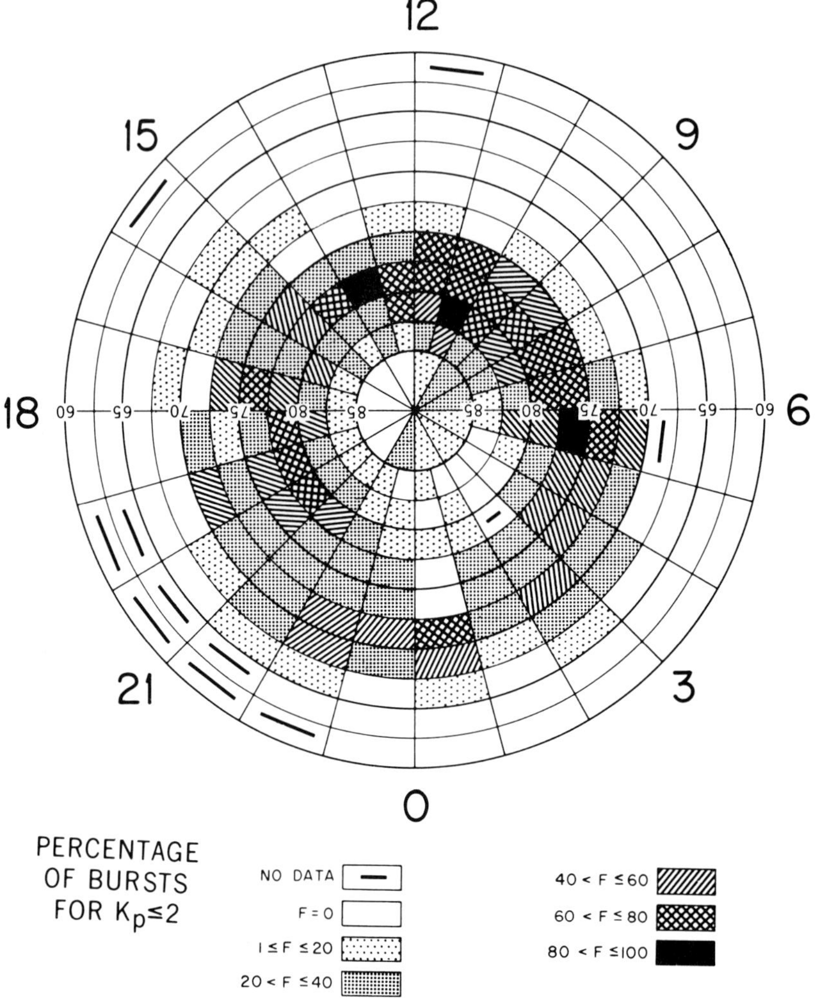

Fig. 2.20. Probability of occurrence of precipitating low energy electrons (0.7 keV) in invariant latitude-MLT coordinates for Kp < 2. (Hoffman, R. A. and Berko, F. W.: *J. Geophys. Res.* **76**, 2967, 1971.)

Figure 2.22(a) shows a photograph of the auroral oval taken by the auroral scanner aboard the ISIS-2 satellite. It shows also a segment of the satellite trajectory, indicated by a narrow strip, projected onto the ionosphere along the geomagnetic field lines. Dots in the strip are time marks. Note that a medium intensity substorm was in progress at that time, since discrete auroras advanced well poleward of the diffuse auroras with a dark space between them.

Figure 2.22(b) shows both the 5577 Å and 3915 Å photometer records which were taken from the same satellite. It can be seen that both photometers show a gradual increase of intensity well before the satellite 'crossed' the bright discrete arc. The peak of luminosity coincided well with the 'crossing time' of the bright

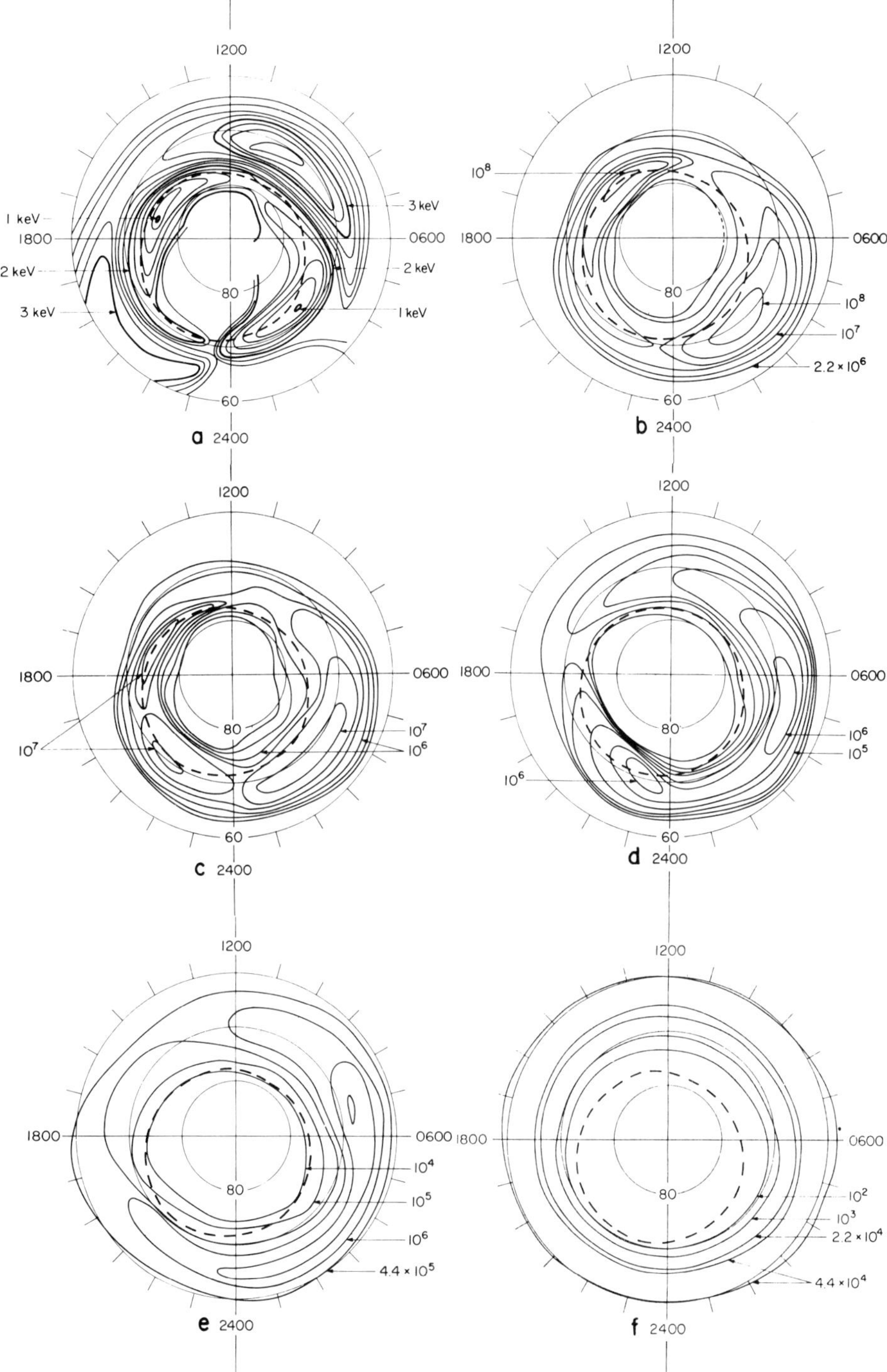

Fig. 2.21. (a) Iso-energy contours (0.2 keV interval) for precipitated electrons in the energy range 150 eV to 9.6 keV in invariant latitude-MLT coordinates for $Kp \leq 3$. The dashed line indicates the 35 keV background boundary.

(b) Average intensity contours for 150 eV electrons for $Kp \leq 3$. Intensity units are $cm^{-2}\ s^{-1}\ ster^{-1}\ keV^{-1}$, and the adjacent contours differ in intensity by a factor of about 2.2.

(c) Average intensity contours for 1.3 keV electrons.

(d) Average intensity contours for 9.6 keV electrons.

(e) Average intensity contours for electrons with energies above 22 keV.

(f) Average intensity contours for electrons with energies greater than 210 keV. (McDiarmid, I. B., Burrows, J. R., and Budzinski, E. E.: *J. Geophys. Res.* **80**, 73, 1975.)

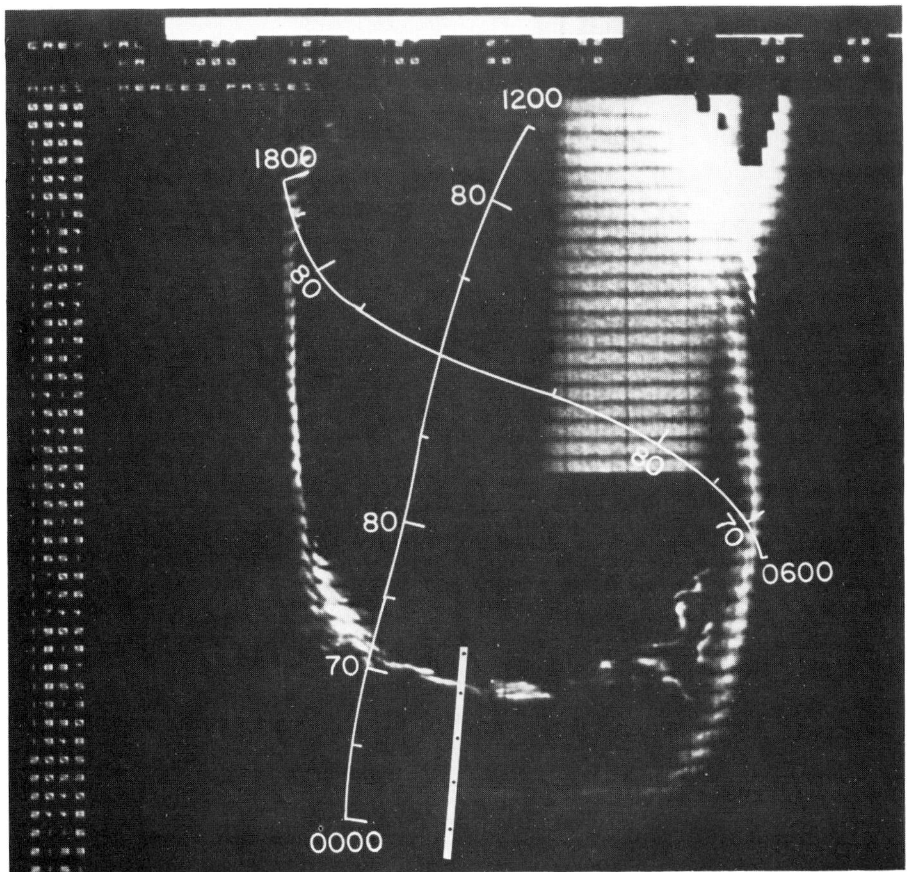

Fig. 2.22(a). ISIS-2 scanner photograph of the auroral oval, taken at 0209–0221 UT, 1971, December 15. Both the noon-midnight and dawn-dusk meridians, together with invariant latitudes and the satellite trajectory across the midnight sector of the oval, are indicated. (Courtesy of C. D. Anger.)

arc which is seen in Figure 2.22(a). After crossing the arc, the intensities remained fairly high until the satellite reached the equatorward boundary of the diffuse aurora.

Figure 2.22(b) shows also the spectrogram (SPS) obtained by the satellite. One 22-sample energy spectrum is obtained during 333 ms every 2 s. The number of electrons counted during each 1/90-s accumulation period is coded as 1 of 20 shades of grey, increasing greyness indicating a larger count. Each vertical line (composed on 22 segments) in the upper panel of the spectrogram represents a log-log plot of an electron spectrum with the energy indicated on the left-hand side. More quantitative information in the form of directional number and energy flux of auroral electrons from 10 eV to 10 keV is given in the middle and lower panels, respectively.

From the above spectrogram, the precipitation region can be divided into two regions. From the highest latitude end, there is a precipitation region in which both

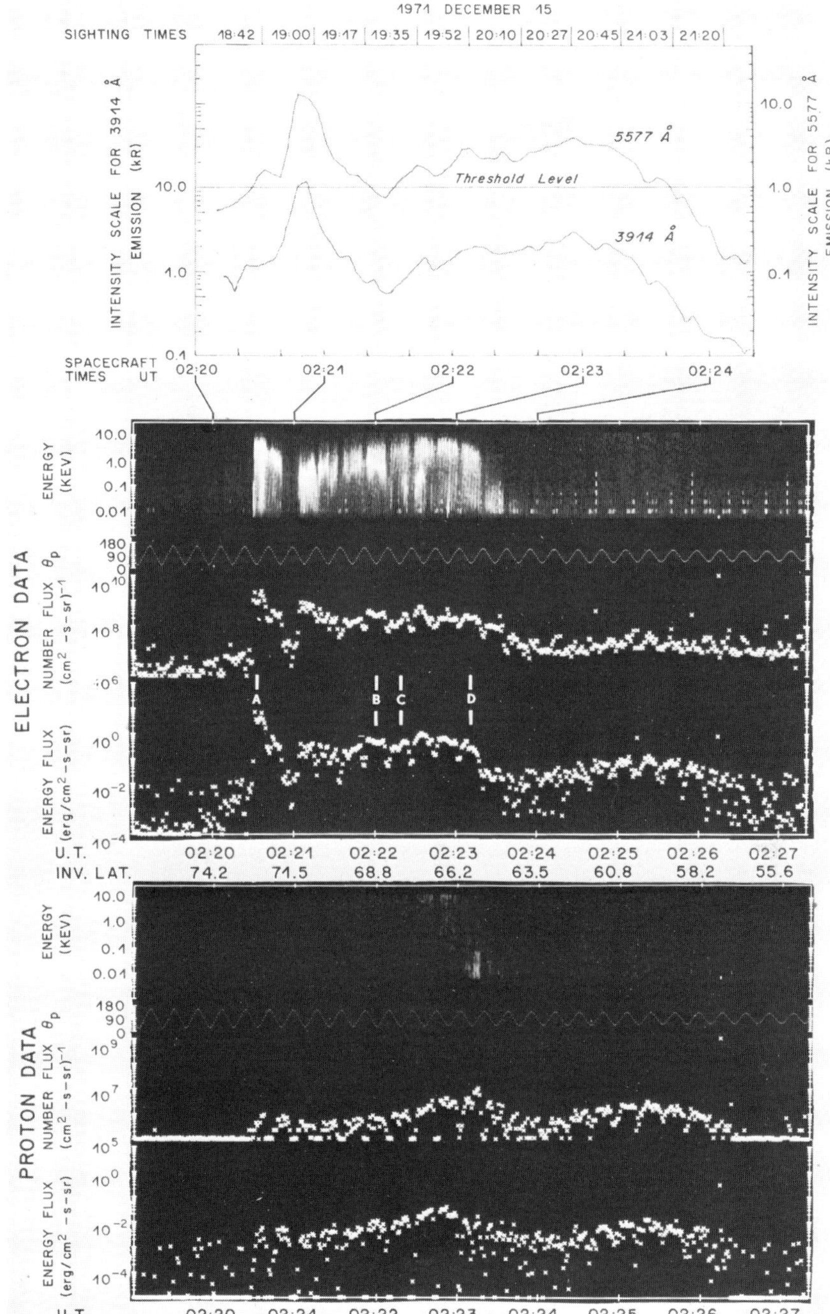

Fig. 2.22(b). Profiles of auroral intensity (the 5577 Å and 3914 Å) along the satellite trajectory indicated in (a), SPS electron and proton data (the energy, number flux and energy flux).

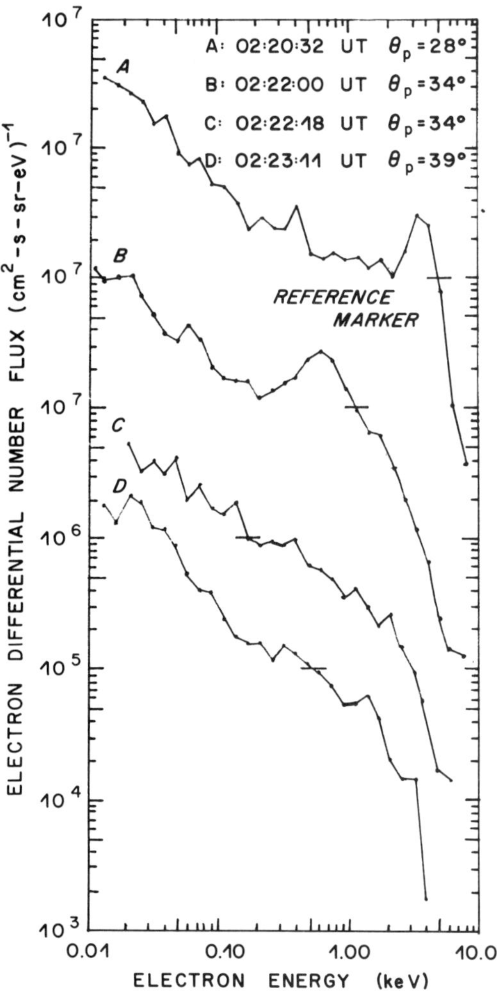

Fig. 2.22(c). Electron energy spectra at the four instants A, B, C and D indicated in the electron data (b). (Courtesy of Lui, A. T. Y., Venkatesan, D., Anger, C. D., Akasofu, S.-I., Heikkila, W. J., Winningham, J. D. and Burrows, J. R.)

the intensity and energy of electrons are highly variable. Its latitudinal range is between ~73.5° and 70°. This region is called the BPS region and corresponds to the region where discrete auroral arcs are present.

Equatorward of the BPS region, there is an extensive precipitation region with fairly uniform intensity. This region is called the CPS region. The vertical blank strips embedded in the spectrogram indicate that there are fewer electrons of pitch-angle 180° (namely, back-scattered electrons) than those with other pitch-angles. Another interesting feature of this region is that there is a very characteristic decrease of electron energies toward the equatorward boundary. Note that the CPS region corresponds to the region of the uniform glow, the diffuse aurora.

The CPS region overlaps with the most equatorward region of precipitation, the Van Allen belt, although its presence is not obvious in the spectrogram.

Figure 2.22(c) shows four energy spectra of auroral electrons which were taken at the times A, B, C and D, indicated in Figure 2.22(b). Comparing spectra A and D, which were taken in the discrete auroral region and the diffuse auroral region, respectively, the most important difference between them is the presence of a peak of the flux at about 3.5 keV in spectrum A, while it is absent in spectrum D. Spectrum B is an intermediate one between A and D, and spectrum C is similar to D.

Ackerson and Frank (1972) showed that an auroral arc was observed at the projected location (along geomagnetic field lines) of an 'inverted V' precipitation. Figure 2.23 shows an example of their spectrograms. The ordinate scale is electron energy in units of electron volts, and the abscissa is universal time. This particular example shows two inverted V precipitation regions, in which electron average energies increase to a maximum and subsequently decrease as the satellite passes through these regions. A little equatorward of these two inverted V precipitation regions lies the diffuse, structureless precipitation; it is likely that this region corresponds to the CPS region.

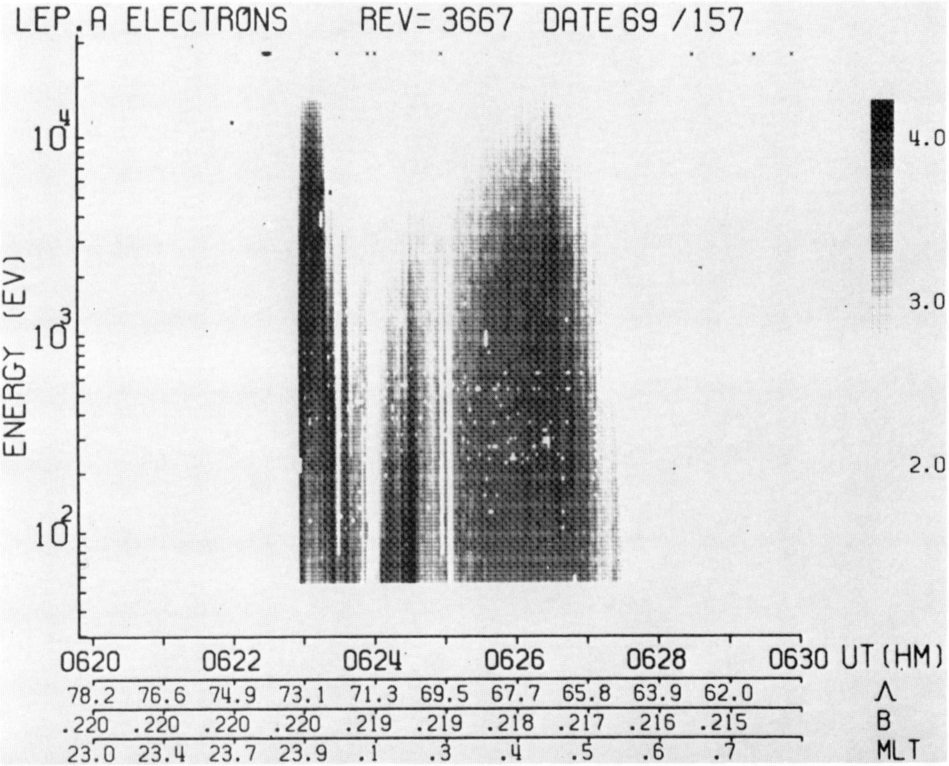

Fig. 2.23. Energy-time (ε-f) spectrogram, obtained by LEPEDEA aboard the INJUN-5 satellite, observed during a northern hemisphere pass across the auroral oval approximately along the midnight sector. (Gurnett, D. A. and Frank, L. A.: *J. Geophys. Res.* **78**, 145, 1973.)

Frank and Ackerson (1971) showed that an auroral electron spectrum in the 'inverted V' region can be approximated by a combination of two types of spectra, the power law type $dJ/d\varepsilon \propto \varepsilon^{-\delta}$ with δ ranging from approximately 1.5 to 2.5 and the Maxwellian type of $\varepsilon_0 \simeq 1$–10 keV. Frank and Ackerson (1972) showed also that 'inverted V' precipitation bands in the evening and midnight sectors are typically more energetic and have greater latitudinal width than those in the noon and morning sectors. Venkatarangan *et al.* (1975) found that the fluxes of electrons of energies 0.15–9.7 keV in the inverted V structures peak at 90° pitch-angle. Paschmann *et al.* (1972) showed that auroral electrons are, in general, isotropic in pitch-angle distribution.

Figure 2.24 shows an interesting correlative study between the satellite and all-sky data (Caverly, 1975). Figure 2.24(a) shows the total number and energy fluxes; (b) shows seven-point electron energy spectra associated with each of the one second data points (time progresses horizontally from left to right in columns of 10-s intervals and for the numbers 1–6, see (a)); and (c) the concurrent all-sky photograph from Fort Yukon. There are several peaks in both the electron number flux and energy flux; four major ones are denoted by A, B, C and D,

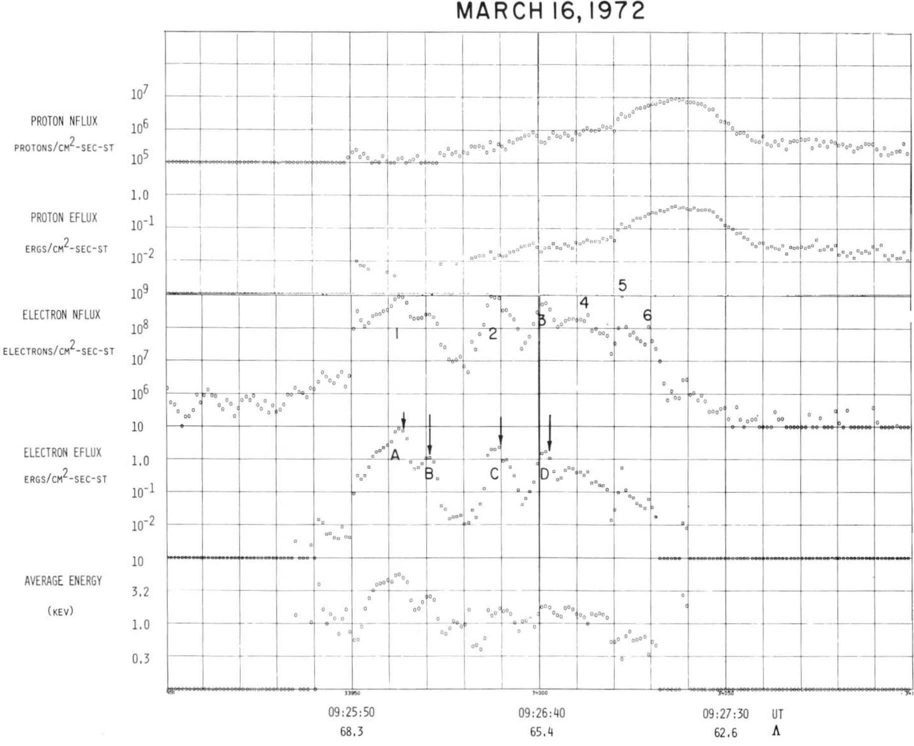

Fig. 2.24(a). Particle precipitation profiles (the number- and energy-flux for both protons and electrons and the average energy of electrons) approximately along the midnight meridian, observed by the Lockheed instrumented USAF satellite 1971a on March 16, 1972. The letters A, B, C and D refer to the auroral arcs in Figure 2.24(c).

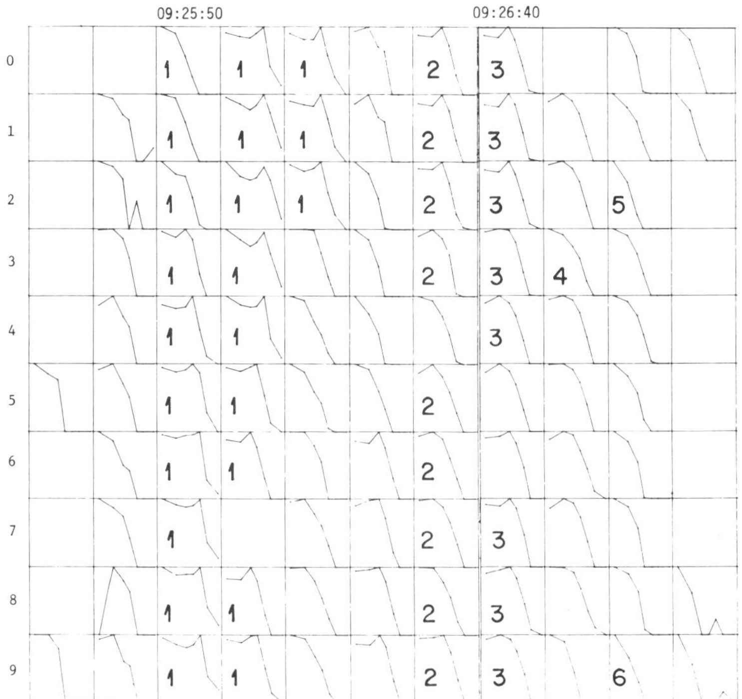

Fig. 2.24(b). Seven-point electron energy spectra associated with each of the one-second data points in Figure 2.24(a). Time progresses horizontally from left to right in columns of ten-second intervals; each column progresses in time vertically from top to bottom in one-second intervals. Spectra 1, 2, 3, 4, 5 and 6 correspond to the numbered peaks in the flux profiles in Figure 2.24(a).

respectively, and among them Peak A was most intense in the energy flux; the corresponding auroral arcs are also labelled in the same way in Figure 2.24(c). In Figure 2.24(b), it can be seen that a peak spectrum grows from a combination of the power law and Maxwellian spectra in the peak regions, so that the power law-Maxwellian spectrum can be considered to be the 'background spectrum'.

Figure 2.25 shows another example of the correlative study, a simultaneous observation of the two auroral regions, from above by the ISIS-2 satellite and from below by the NASA 711 jet aircraft, which was conducted over the Arctic Ocean on 1972, October 10 (Deehr *et al.*, 1976). The satellite passed the nearest point to the aircraft between 1104 and 1105 UT, when the aircraft was located at about the boundary of the two auroral regions, after an intense poleward expansive motion of active auroras. This can clearly be seen in the all-sky photograph which was taken at 1104 UT.

The satellite traversed over the auroral region in the period between 1103 UT and 1107 UT. The satellite path was projected (along the geomagnetic field lines) down to 110 km altitude and is plotted every ten seconds from 1103:20 to

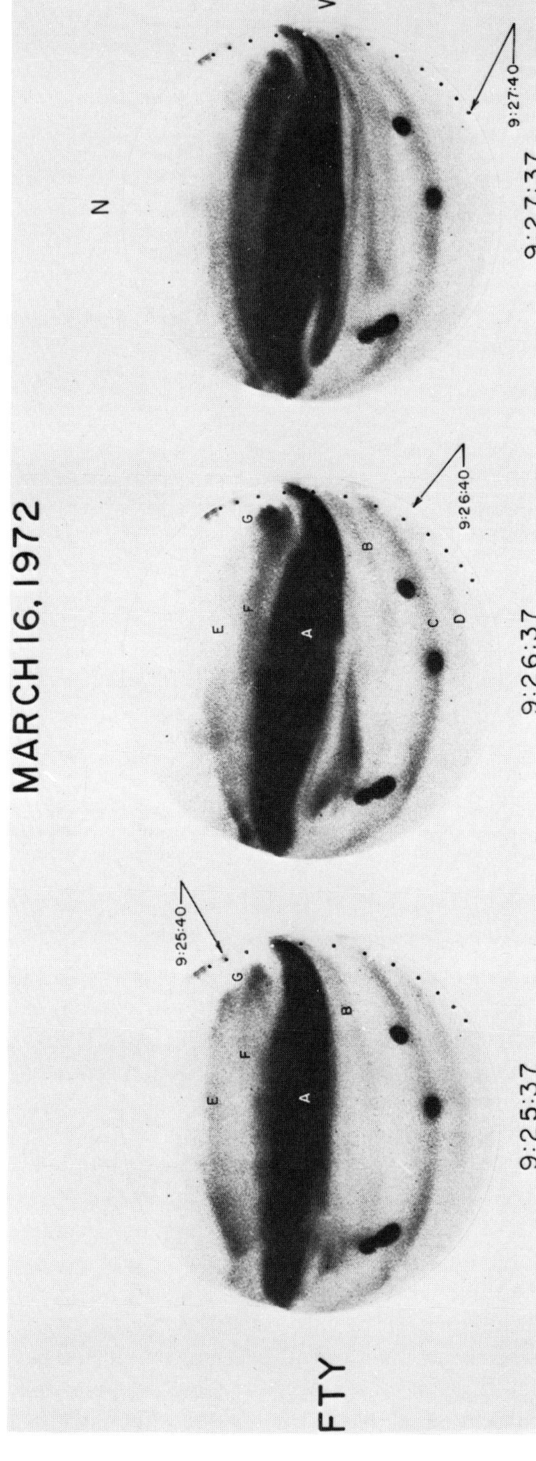

Fig. 2.24(c). All-sky photographs taken from Fort Yukon at the time of satellite passage in Figures 2.24(a) and (b). Auroral arcs A, B, C and D were traversed by the satellite whose 110-km altitude trajectory is represented by the sequence of dots on the western edge of each photograph. (Caverly, R. S.: M.S. Thesis, Univ. of Alaska, 1975.)

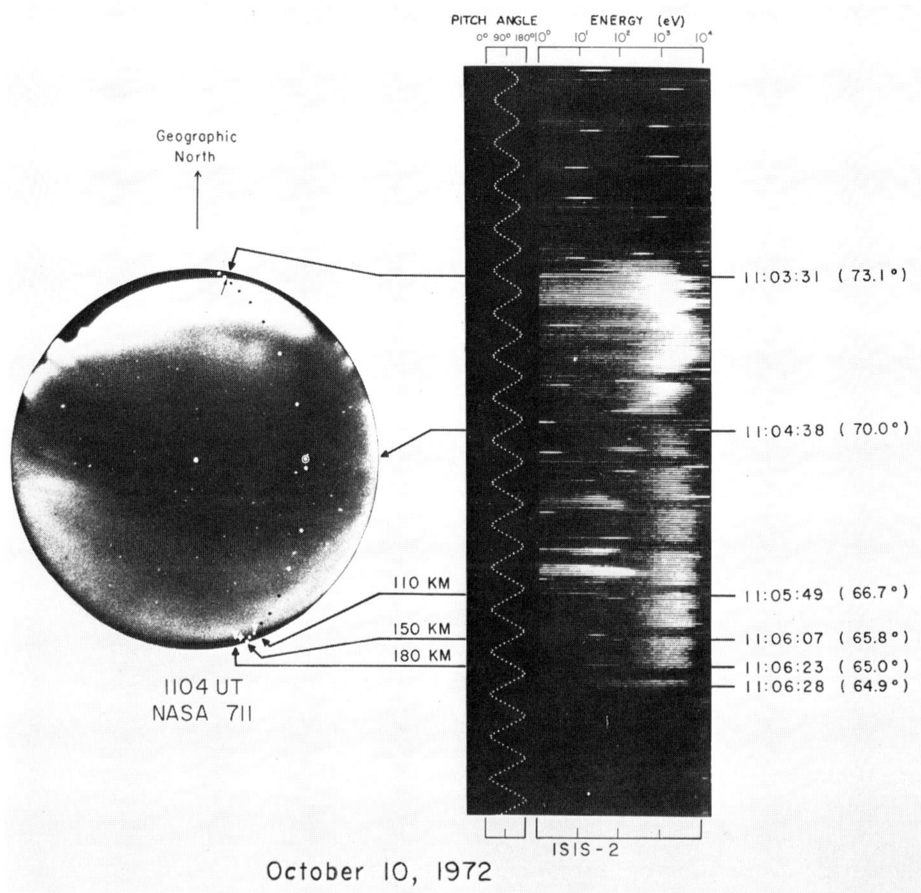

Fig. 2.25. All-sky photograph corresponding to the closest time of passage of the ISIS-2 satellite whose SPS data are displayed on the right-hand side. The path of the satellite is projected downward along magnetic field lines to a 110 km latitude and shown as dots on the photographs. (Courtesy of Deehr, C. S., Winningham, J. D., Yasuhara, F., and Akasofu, S.-I.)

1106:00 UT. The circle with a dot indicates the point where the projected latitude of the satellite is 70.0° N, which is also the latitude of the zenith of the all-sky photograph. The longitudes of the zenith and the dotted circle are different by approximately 2°.

The intersection between the poleward edge of the auroral luminosity and the projected (110 km) satellite trajectory (hereafter, the satellite location refers to the projected location) is indicated with a black arrow. The intersection points between the equatorward edge of the auroral luminosity and the satellite path projected down to 110, 150 and 180 km heights are shown with the white arrow head, the star mark and the triangle, respectively.

The SPS spectrogram for this pass is shown on the right-hand side of Figure 2.25. The satellite crossed the poleward edge of the auroral luminosity at

1103:31 UT; this time is marked on the spectrogram. It passed the aircraft latitude at 1104:38 UT.

In the all-sky photograph, we can easily distinguish two types of aurora: an active discrete aurora located poleward on the aircraft and the diffuse aurora, covering the equatorward half of the sky.

In the spectrogram one can also recognize two types of particle precipitations, poleward and equatorward of about 68° invariant latitude (~70° geographic). The poleward precipitation shows a considerable structure and has an intense flux and higher average energy, while the equatorward flux region shows almost no structure (except for very low energy bursts around 1105 to 1106 UT). In addition the diffuse region has a weaker flux, a lower average energy and a well-developed loss cone. Similar correlative studies have recently been made by Mizera *et al.* (1975) and Meng (1976).

(b) *Rocket Observations*

Peaks around a few kilovolts in differential energy spectra of auroral electrons have been observed by a number of rocket-borne detectors (Ogilvie, 1968; Evans, 1968, 1971; Chase, 1970; Rearwin, 1971; Whalen *et al.*, 1971a, b; Whalen and McDiarmid, 1972a, b; McEwen and Sivjee, 1972; Bryant *et al.*, 1972, 1973; Reasoner and Chappell, 1973; Arnoldy and Choy, 1973; Tulinov *et al.*, 1973; Whalen *et al.*, 1974; Vij *et al.*, 1975). Such peaks have often been referred to as 'monoenergetic' peaks. Westerlund (1969) was the first to point out that there is a 'continuum' which extends below 1 keV in addition to the monoenergetic peak. The 'continuum' corresponds to the power law or the exponential spectrum. The lowest range of energies of auroral electrons was studied by Feldman *et al.* (1971), Reasoner and Chappell (1973), and by Sharp and Hays (1974).

Rocket observations have been particularly useful in examining in detail the pitch-angle distribution of auroral electrons. Rearwin (1971) found that incident electron fluxes into a diffuse aurora were isotropic over the upper atmosphere. Whalen and McDiarmid (1972a, b) observed intense fluxes of field-aligned electrons (i.e., peaked near zero degrees of the pitch-angle) near the poleward boundary of a band of auroral precipitation. They estimated the thickness of this field-aligned precipitation region to be about 10 km and the corresponding electric current density to be 2×10^4 A m^{-2}.

Figure 2.26(a) shows the pitch-angle distribution for 1, 2 and 3 keV electrons which precipitate into a diffuse post-breakup aurora. It shows that 1 and 3 keV electrons were isotropic in pitch-angle, while 2 keV electrons were well collimated along geomagnetic field lines, indicating that an intense field-aligned electron flux is confined in a very narrow spectral range (Arnoldy, 1974). This feature can also be seen in the differential energy spectrum as shown in Figure 2.26(b); here the field-aligned flux was confined in a narrow spectral range. Bosqued *et al.* (1974) showed also that auroral electrons of energies 300 eV–5 keV showed a large anisotropy (the ratio of the fluxes measured at 0° and 30° pitch-angle being 40), while electrons of energies greater than 5 keV were almost isotropic.

Another interesting observation of auroral electrons by Bryant *et al.* (1972) indicates a systematic transition of the electron spectra, a Maxwellian-like

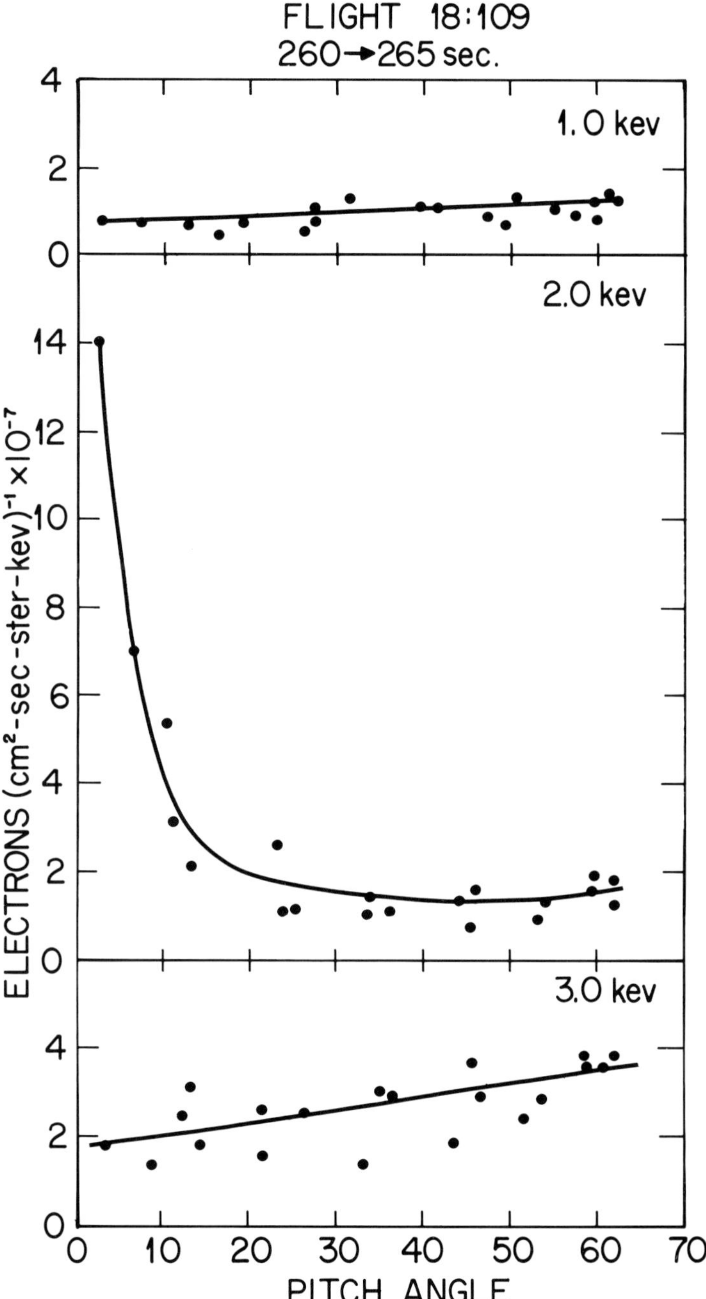

Fig. 2.26(a). Pitch-angle distribution of auroral electron at three energies, 1.0, 2.0 and 3.0 keV.

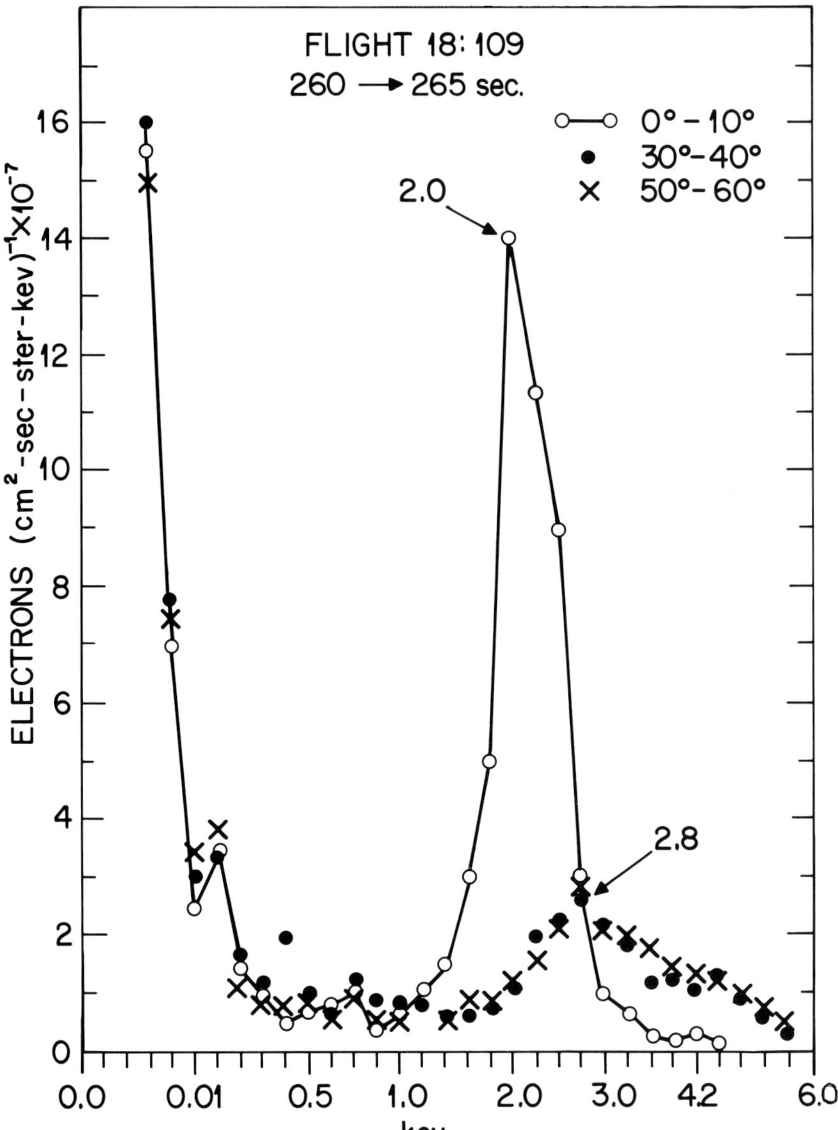

Fig. 2.26(b). Pitch-angle sorted spectra of auroral electrons. (Arnoldy, R. L., Lewis, P. B. and Isaacson, P.O.: *J. Geophys. Res.* **79**, 4208, 1974.)

distribution a little equatorward of an auroral arc to a sharply peaked one in the arc. This change of the electron spectra is quite similar to that described in the previous subsection. We shall discuss their result in more detail in Section 3.9.

It may also be noted that McEwen and Sivjee (1972) observed unusual electron spectra, a high electron flux of average energy of about 300 eV, in the type A aurora which is characterized by a great enhancement of the O I 6300 Å emission; in fact, the red brightness was about 14 kR.

2.4.2. MORNING SECTOR

We noted in auroral photographs taken from satellites that discrete auroras are either absent or not common in the morning sector. This feature is reflected in the precipitation pattern of auroral electrons in terms of the absence (or only rare occurrence) of an intense precipitation or of a clear 'inverted V' structure. Figure 2.27 shows an interesting example of simultaneous observation of auroral luminosity and auroral electrons (6.5 keV). The satellite traversed first the morning sector of the oval, then the polar cap in the midnight sector and finally the

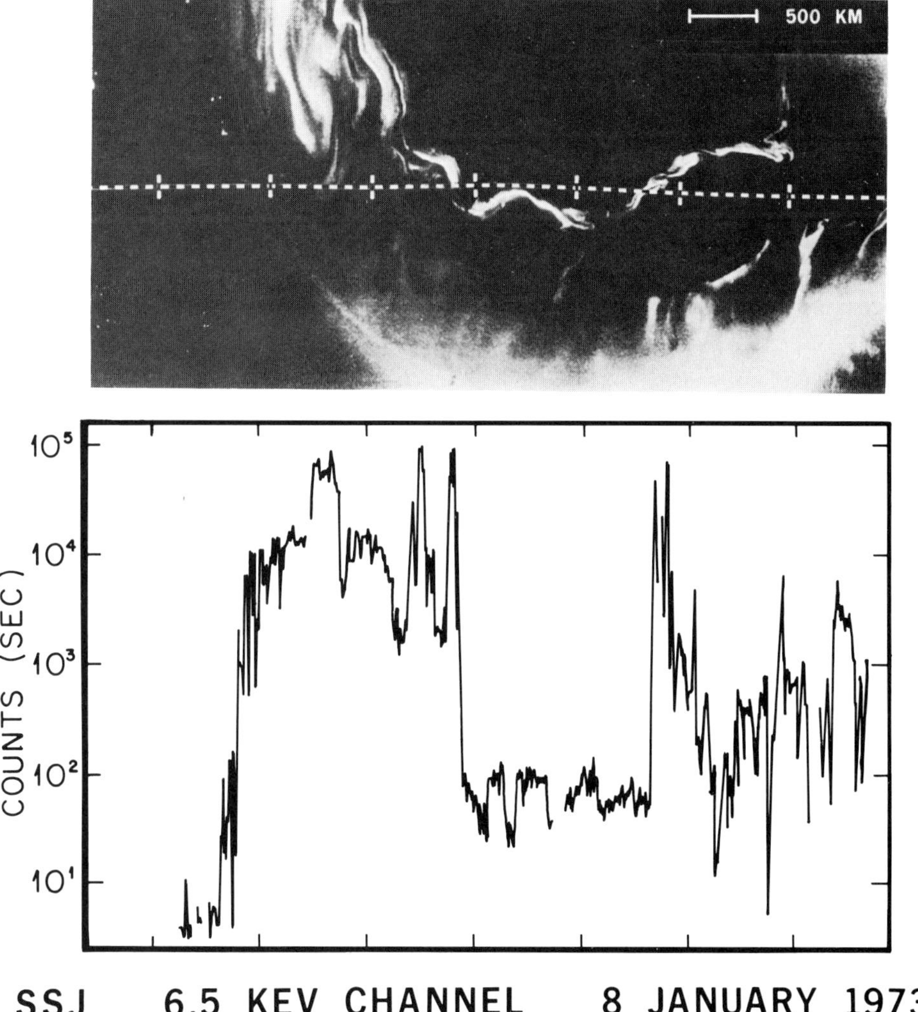

Fig. 2.27. DMSP-2 photograph of the dark hemisphere and the simultaneous measurement of 6.5 keV electrons. The subtrack (dashed-line) of the satellite runs horizontally across the middle of the photograph, taken on January 8, 1973. (Arnoldy, R. L.: *Rev. Geophys. Space Phys.* **12**, 217, 1974.)

oval in the evening sector. The precipitation is much more structured in the morning sector than in the evening sector, reflecting the complexity of luminous structure of the aurora. On the basis of what we learned in Section 2.2.1, it is not difficult to infer that a rather smooth precipitation region in the evening sector corresponds to the CPS precipitation region. An intense precipitation region just poleward of the CPS region can be identified as the BPS region; a bright discrete aurora is located there. The satellite traversed also the BPS region in the midnight sector.

2.4.3. NOON (CUSP) SECTOR

Hoffman and Berko (1971) showed that bursts of auroral electrons are observed with a high occurrence frequency in the region over the midday part of the auroral oval. Heikkila and Winningham (1971), and Winningham (1972) examined extensively characteristics of auroral particles, using detectors carried by the ISIS-1 satellite, over the midday part of the auroral oval. They showed that the particles appear sometimes a little poleward of the boundary of trapped electrons (namely, of the Van Allen belt). Figure 2.28 shows an example of the cusp observation. It can be seen that the cusp electrons are present beyond the latitude (~75°) where the energetic electrons in the Van Allen belt are 'cut-off'. The electron energy spectrum peaks in the cusp region at about 100–200 eV, with a total flux of order

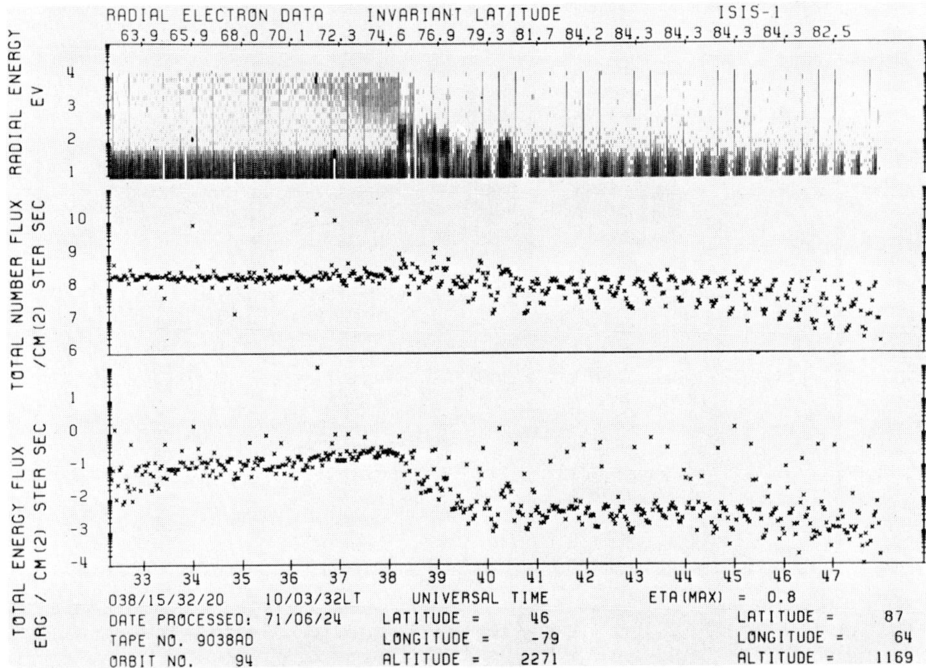

Fig. 2.28. ISIS-1 energy-time (ε-t) electron spectrogram across the midday part of the auroral oval. The invariant latitude is given at the top. (Courtesy of W. J. Heikkila.)

10^9 electrons cm^{-2} $ster^{-1}$ s^{-1}, carrying an energy of a few tenths of an erg cm^{-2} $ster^{-1}$ s^{-1}. The electron spectrum in the cusp is similar to that found in the magnetosheath (Section 3.3). The proton energy spectrum peaks at about 300 eV, and the flux is greater than 10^7 cm^{-2} $ster^{-1}$ s^{-1}. The spectrum of cusp protons is also quite similar to that of magnetosheath protons. Such similarity led Heikkila and Winningham (1971) and Frank (1971) to infer that the cusp plasma is of magnetosheath origin. Maynard and Johnstone (1974) made also a detailed study of the precipitating electrons and protons in the cusp region.

An accurate determination of the location of an auroral arc with respect to the cusp precipitation region was made by Winningham *et al.* (1973) on the basis of all-sky photographs of the auroras taken from the South Pole (Amundsen-Scott) station when the ISIS-1 satellite passed over the station. It was found that arcs appear within the region which is projected, along the field lines, onto the height where the cusp particles are thought to produce maximum luminosity.

As mentioned earlier (Section 2.2.1), spectral characteristics of midday discrete auroras were studied by Eather and Mende (1971a, b, 1972) and Heikkila *et al.* (1972). They are essentially red (O I 6300 Å) arcs with a slight enhancement of the O I 5577 Å line near the lower border. Kennel and Rees (1972) estimated the downward heat fluxes into the cusp region to be of order 1–10 erg cm^{-2} s^{-1} during a quiet condition.

2.4.4. POLAR CAP

Winningham and Heikkila (1974) showed that there are three types of auroral particle precipitation in the polar cap region. The first type is a fairly uniform precipitation of soft (~100 eV) electrons, which are present nearly all the time. They suggested that these electrons are of magnetosheath origin, but their flux is about two orders of magnitude less than that of cusp electrons. Winningham and Heikkila called this type of precipitation 'polar rain'.

The second type, called 'polar shower', has a structured pattern when it is observed by a polar orbiting satellite. The electrons have energies of about 1 keV. Winningham and Heikkila suggested that the shower precipitation is responsible for polar cap (discrete) auroras. Figure 2.29 shows an example of SPS spectrograms which show 'polar shower'. The shower precipitation is greatly enhanced during great magnetic substorms; such an intensified shower precipitation is called 'polar squall'.

Andrews and Strömman (1973) examined the excitation of the O I 6300 Å red emission within the polar cap on the basis of simultaneous ground-based observations and satellite (ESRO-IA) observations of low energy electrons. They showed that thermal excitation by ambient electrons (heated by secondary electrons) is most likely to be responsible for the red emission.

Whalen *et al.* (1971b) observed (with a rocket) the energy spectra and pitch-angle distribution of a polar cap aurora of a relatively stable 1-kR brightness. They showed that the differential energy spectrum has a peak in the 1.5 and 2.0 keV energy range with a peak flux of ~10^7 cm^{-2} s^{-1} $ster^{-1}$ keV^{-1} (see also Meng and Akasofu, 1976). The pitch-angle distribution was isotropic.

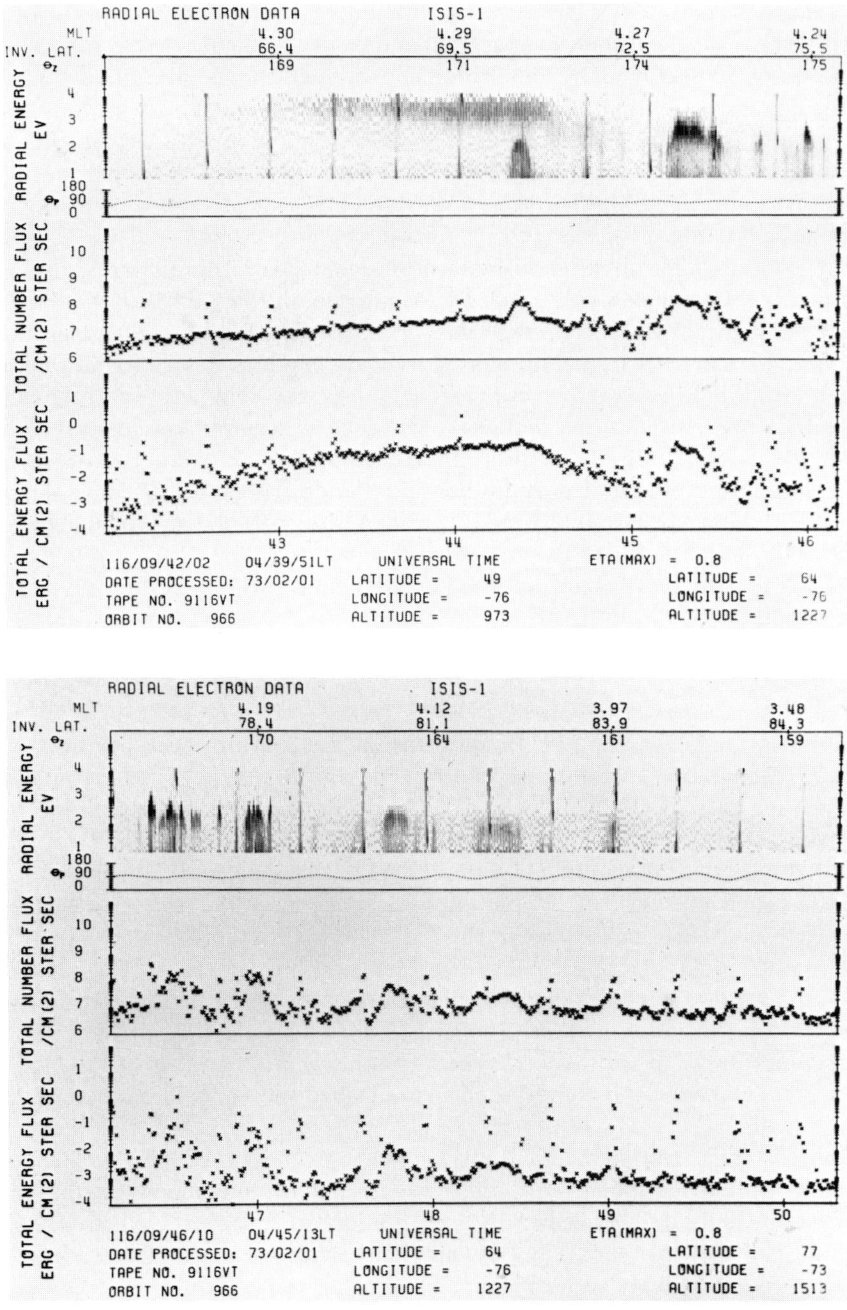

Fig. 2.29. ISIS-1 energy-time (ε-t) electron spectrogram approximately along the dawn-dusk meridian, orbit 966, on 1969, April 26. (Winningham, J. D. and Heikkila, W. J.: *J. Geophys. Res.* **79**, 949, 1974.)

2.5. Auroral Electrons and Field-Aligned Currents

2.5.1. SATELLITE OBSERVATIONS

In Section 1.3.2, it was shown that discrete arcs in the evening sector are embedded in the region of an upward field-aligned current. The distribution of the field-aligned fluxes of auroral electrons has been studied most extensively by Hoffman and Evans (1968) and Berko (1973). In their study, the alignment of the flux with respect to the geomagnetic field vector was determined by examining the ratio of the flux of 2.3 keV electrons at 0° pitch-angle to that at 60°. When this ratio was greater than or equal to 2.0, it was considered to be a field-aligned flux. Berko (1973) noted that the field-aligned flux tends to occur most frequently at 70° invariant latitude near midnight. In general, it occurs in association with high fluxes of auroral particles. However, Paschmann *et al.* (1972) noted that the pitch-angle distribution tends to be predominantly isotropic in most cases, independent of latitude and geomagnetic conditions.

2.5.2. ROCKET OBSERVATIONS

Rocket observations are useful in examining whether or not a visible arc results from a sheet of field-aligned electron fluxes, which carry upward electric currents. Indeed, Whalen and McDiarmid (1972) observed a narrowly collimated (field-aligned) beam in the vicinity of a bright loop structure. However, Arnoldy *et al.* (1974) and Arnoldy (1974) showed that a considerable amount ($\sim 10^{-6}$ A m^{-2}) of field-aligned currents in the vicinity of visible auroras was not necessarily carried by 0° pitch-angle particles. They noted also that field-aligned currents are often carried by electrons of energies from 0.5 to about 5 keV and that their spectra narrowly peak ($\sim \pm 0.5$ keV range); but those energy peaks do not necessarily coincide with the peak of isotropic monoenergetic peaks.

An extensive rocket observation, conducted by Anderson and his colleagues (Vondrak *et al.*, 1971; Park and Cloutier, 1971; Cloutier *et al.*, 1970; Cloutier *et al.*, 1973; Anderson and Cloutier, 1975; Pazich and Anderson, 1975; Spiger and Anderson, 1975; Casserly and Cloutier, 1975; and Iglesias and Anderson, 1975), showed that individual discrete auroral arcs are associated with a pair of oppositely directed field-aligned currents. They showed also that the location of the upward field-aligned current (of magnitude 2.7×10^{-5} A m^{-2}) coincides with that of the auroral arc and that there is a downward field-aligned current to the south of the auroral arc. The simultaneous particle measurements (0.5–20 keV) show, however, that the particle flux can account for a current density of 5.5×10^{-6} A m^{-2}, namely only about 17% of the current deduced from the magnetometer data. Casserly and Cloutier (1975) concluded that both the upward and downward field-aligned currents were carried primarily by electrons with energy less than 0.5 keV. It should be added here that rocket vector magnetometer measurements in the cusp region by Ledley and Farthing (1974) suggested a very intense field-aligned current of order 10^{-4} A m^{-2}.

All these observations are important in understanding the basic processes involved in accelerating auroral electrons. In Section 3.9, we shall consider a specific mechanism which can account for both the acceleration of auroral electrons and the generation of the field-aligned currents.

2.6. Auroral Particles and Atmospheric Emissions

The interaction of auroral electrons with the upper atmosphere has been studied by a number of workers, most recently by Rees (1969), Rees and Maeda (1973), Jones and Rees (1973), Rees and Jones (1973), Banks *et al.* (1974) and Kamiyama (1974). The numerical results obtained by Banks *et al.* (1974) are based on the following assumptions about the characteristics of the electron beam.

Energy spectrum:

$$F = A\ e^{-(\varepsilon-\varepsilon_0)^2/2\sigma^2}$$

Pitch-angle distribution: isotropic

The flux is normalized by the quantity called the perpendicular flux

$$F_{\text{unit}} = \pi F$$

Figure 2.30 shows the ionization rates ($cm^{-3}\ s^{-1}$) per unit flux for the spectra

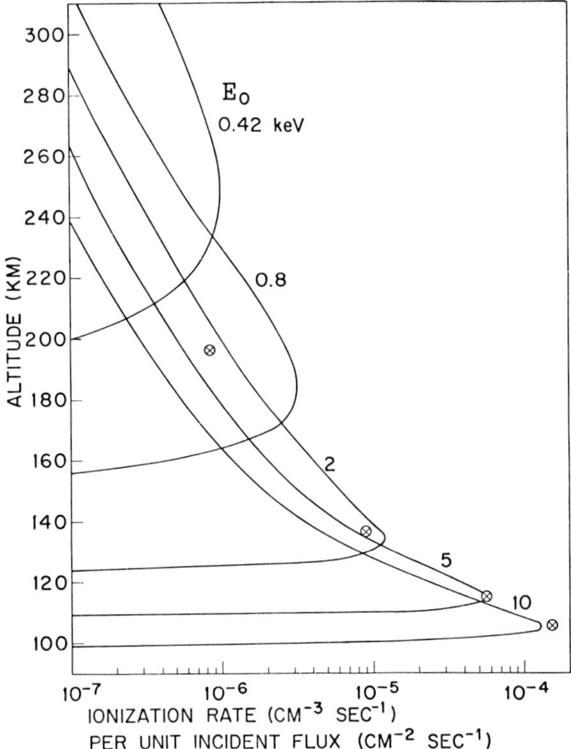

Fig. 2.30. Ionization rates per unit incident flux for Gaussian fluxes in the energy range 0.42–10 keV. (Banks, P. M., Chappell, C. R. and Nagy, A. F.: *J. Geophys. Res.* **79**, 1459, 1974.)

characterized by $\varepsilon_0 = 0.42$, 0.8, 2, 5 and 10 keV. The ionization of N_2 molecules results in the emission of the First Negative Band (1NG) through

$$N_2(X^1\,\Sigma_y^+) + \mathbf{e}^* \rightarrow N_2^+(B^2\,\Sigma_u^+) + \mathbf{e}^* + e$$

$$N_2^+(B^2\,\Sigma_u^+) \rightarrow N_2^+(X^2\,\Sigma_y^+) + h\nu$$

where **e*** denotes a primary energetic electron; for a recent rocket observation of auroral electrons and the 1NG emission, see Feldman and Doering (1975).

The emissions from atomic oxygen involve complicated chemical reactions, as well as the excitation by secondary electrons. Henriksen (1974) lists the following processes:

$$O(^3P) + e \rightarrow O(^1S) + e$$

$$O_2(X^3\,\Sigma_y) + e \rightarrow O(^1S) + O(^1D, {}^3P) + e$$

$$N_2(A^3\,\Sigma_u^+) + O(^3P) \rightarrow N_2(X^1\,\Sigma_y^+) + O(^1S)$$

$$O_2^+(X^2\,\pi_y) + e \rightarrow O(^1S) + O(^1D, {}^3P)$$

$$NO(x^2\,\pi_y) + N(^4S) \rightarrow N_2(X^1\,\Sigma_y^+) + O(^1S)$$

Figures 2.31(a) and (b) give the emission rates for both the O I 5577 Å and 6300 Å emissions for the spectra characterized by ε_0 mentioned in the above. In both estimates, the quenching by collisions of molecules and atoms is taken into account. For a recent ion composition observation by rockets see Narcisi *et al.* (1974) and Swider and Narcisi (1974). Deehr *et al.* (1973) made an extensive study of the relationship between the electron energy flux and the resulting intensity of N_2^+ 4278 Å on the basis of ESRO I/AURORAE satellite. Their result indicates that the N_2^+ 4278 Å emission rate is of order 270 R erg^{-1}.

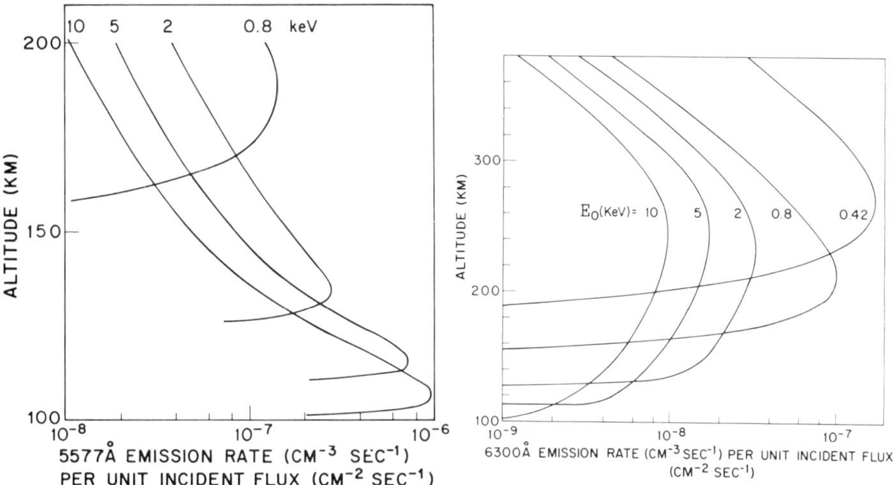

Fig. 2.31. Profiles of the 5577 Å and 6300 Å emission rates for atomic oxygen with incident Gaussian spectra, per unit incident flux. (Banks, P. M., Chappell, C. R. and Nagy, A. F.: *J. Geophys. Res.* **79**, 1459, 1974.)

Eather (1969), Eather and Mende (1971a, b, 1972), Mende and Eather (1971, 1972, 1975) and Rees and Luckey (1974) showed that spectroscopic features of auroras are quite useful in inferring spectral characteristics of auroral electrons. As we saw earlier (Section 2.4.1), the observed energy spectra of auroral electrons often have a Maxwellian form.

$$F(\varepsilon)d\varepsilon = f_0\varepsilon \exp(-\varepsilon/\alpha)d\varepsilon \text{ electrons cm}^{-2}\text{ s}^{-1}\text{ eV}^{-1}$$

Rees and Luckey (1974) showed that it is possible to estimate the value of α by measuring both the ratio of (O I 6300 Å/N_2^+ 4278 Å) and the absolute emission rate of the N_2^+ 4278 Å. With the value of α thus obtained, it is then possible to estimate the flux of electrons by knowing the emission rate of the N_2^+ 4278 Å.

Eather (1969) and Eather and Mende (1971a, b) used this method to infer energies of precipitating electrons along the auroral oval. One of the important features revealed by their study is the presence of the zone of soft electron precipitation near the poleward half of the auroral oval in the night sector, as well as in the midday part of the auroral oval. Figure 2.32 shows the intensity ratio 6300 Å/4278 Å as a function of oval coordinates (invariant latitude minus the latitude of the equatorward boundary of the oval). There is a significant increase

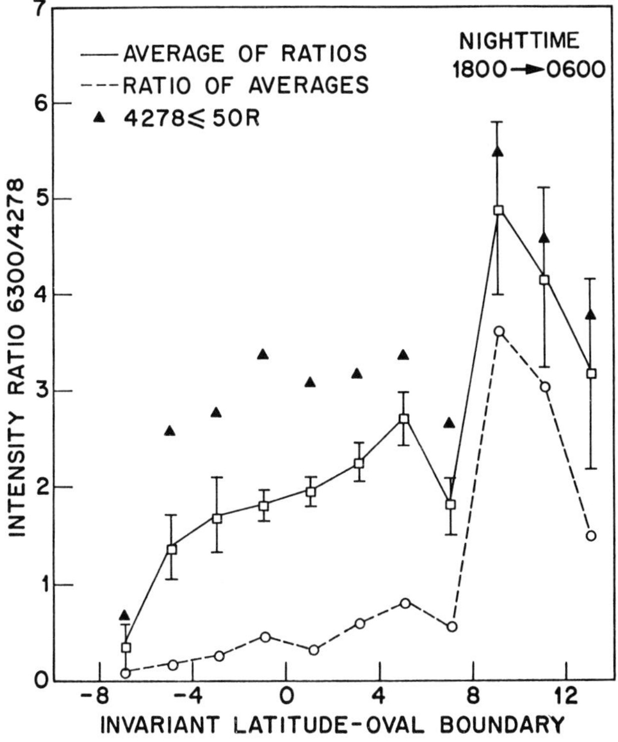

Fig. 2.32. Ratios I(6300 Å)/I(4278 Å) as a function of oval coordinates for nightside (1800–0600 MLT) oval. The solid line is the average ratio, and the dashed line is the ratio of averages. The triangles represent average of ratios for I(4278 Å) ≤ 50 *R*. (Eather, R. H. and Mende, S. B.: *J. Geophys. Res.* **77**, 660, 1972.)

of the ratio about 7° poleward of the equatorward boundary of the oval, indicating the presence of the soft precipitation region. In fact, there is a systematic increase of the average height of the aurora towards higher latitudes.

Figure 2.33 indicates the gross precipitation pattern over the entire polar region, which is inferred from the spectroscopic observations (Eather and Mende 1971a, b): in particular, the location of the nightside soft zone, the dayside soft zone and the hard zone. This result is in agreement with the satellite observations by Burch (1968) and by McDiarmid *et al.* (1975) that the average energy of auroral electrons decreases towards higher latitudes (Section 2.3).

These studies are statistical in nature. However, on the basis of an extensive airborne observation, Pike *et al.* (1976) showed that there is an oval-aligned band of enhanced red (6300 Å) emission in the night sector and that this red band extends over 2° to 10° in latitude and contains within it discrete and diffuse auroras. Further, the average pattern may be considered as the background emission pattern (often subvisual), and visual features are embedded in it. This conclusion has an important implication in inferring the origin of auroral electrons (Section 3.9).

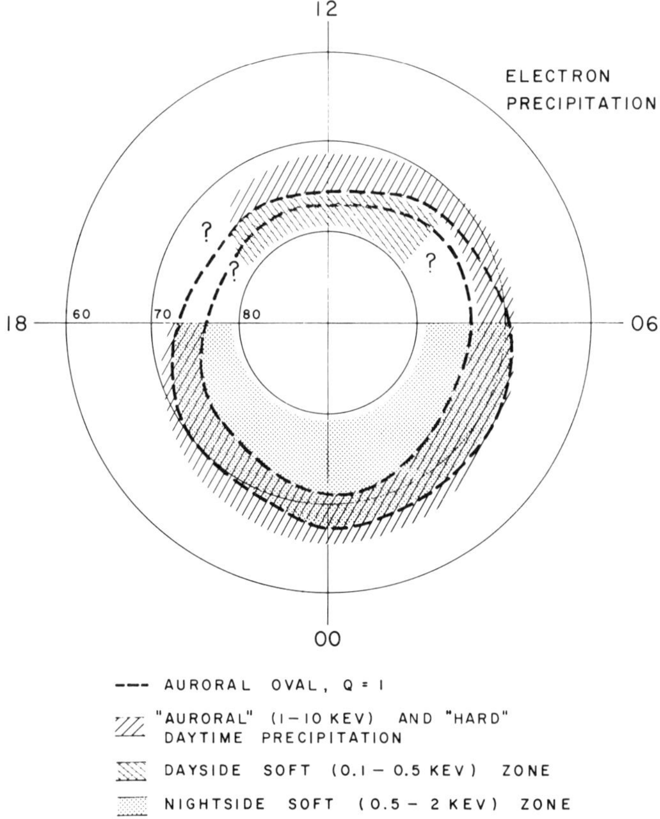

Fig. 2.33. Electron precipitation pattern over the entire polar region inferred from optical observations. (Eather, R. H. and Mende, S. B.: *J. Geophys. Res.* **77**, 660, 1972.)

2.7. Auroral Protons

The precipitation of auroral protons traditionally has been studied on the basis of hydrogen emissions (Wiens and Jones, 1969; Montbriand, 1969; Rees and Benedict, 1970; Stringer, 1971; Fukunishi and Tohmatsu, 1973; Belon *et al.*, 1974; Fukunishi, 1975; Oguti, 1973; Oguti *et al.*, 1974; for a comprehensive review, see Omholt, 1972). These optical studies showed that the hydrogen aurora appears as an oval belt, but it is displaced with respect to the oval belt of discrete auroras; it is located equatorward in the afternoon sector and overlaps the discrete auroral belt during the forenoon period. On the other hand, Fukunishi (1975) showed that in the morning sector the proton aurora is located a little poleward of the oval auroras during substorms. The Ly-α emission, observed from satellites, revealed also a similar oval belt (Chubb and Hicks, 1970; for other Ly-α observations, see Joki and Evans, 1969; Clark and Metzger, 1969; Peek, 1970; Metzger and Clark, 1971). It should be noted that a part of the hydrogen emission might arise from the excitation of terrestrial (ambient) hydrogen atoms by auroral electrons (Belon *et al.*, 1974).

Auroral hydrogen emissions observations during auroral substorms have recently been most extensively conducted at Syowa station, Antarctica. Their results will be discussed in Section 8.5.1.

Direct observations of precipitating protons by polar orbiting satellites have recently contributed considerably to our knowledge of the proton precipitation pattern over the entire polar region (Hultqvist *et al.*, 1971; Hultqvist *et al.*, 1974; Amundsen *et al.*, 1973; Amundsen *et al.*, 1975; Mizera, 1974). This subject has been most comprehensively studied by Hultqvist (1972, 1973, 1975) and Hultqvist *et al.* (1974), so that only a brief summary of their results will be presented here.

Figure 2.34 shows the precipitation pattern of 6 keV protons constructed by Riedler and Borg (1972). Figures 2.35(a) and (b) show both the electron and proton precipitation profiles, constructed by Deehr *et al.* (1973). In their review, Hultqvist (1973, 1975) and Hultqvist *et al.* (1974) emphasized that the general precipitation pattern of protons is quite similar to that of electrons. This conclusion appears to be a little different from what has been drawn from optical observations. However, Hultqvist (1972) noted that these two observations are actually compatible with each other. First of all, the *maximum* of the average keV proton precipitation occurs a little south of the maximum of the electron precipitation in the late evening. This is clearly seen in Figures 2.24(a) and 2.35. Similarly, in the morning sector the maximum of the proton precipitation is located a little poleward of the electron precipitation maximum. Secondly, large differences of the latitudinal distributions of both the proton and electron precipitation do not occur during quiet conditions, but only during disturbed conditions; for example, the results obtained by Wiens and Jones (1968a) were associated with highly disturbed conditions.

It should be noted that the so-called 'proton aurora', which was thought to be most clearly seen as a fairly uniform diffuse glow in the evening sector, is now found to be caused mainly by electrons (Lui and Anger, 1973), although at times protons can contribute significantly to the luminosity (Romick and Sharp, 1967).

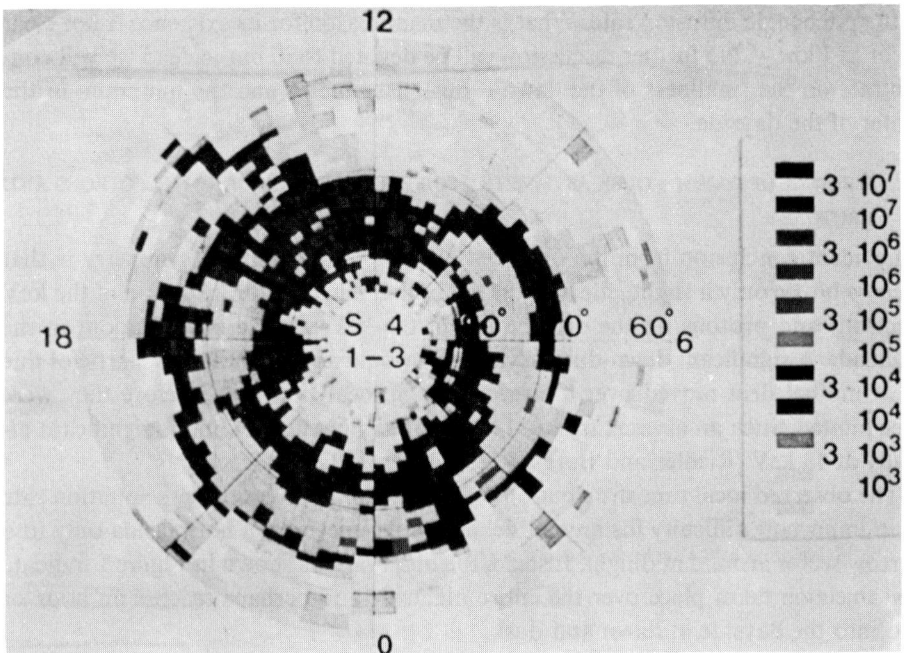

Fig. 2.34. Precipitation pattern of 6 keV protons for fairly low magnetic activity (Kp = 1 − 3), observed by the ESRO-1A satellite. (Riedler, W. and Borg, H.: *Space Research XII*, 1397, Akademie-Verlag, Berlin, 1972.)

The precipitation of protons from the ring current belt will be discussed in Section 8.5.

Rocket observations of auroral protons have so far been made much less frequently than those of auroral electrons (Miller and Shepherd, 1969; Badhwar *et al.*, 1969; Wax and Bernstein, 1970; Iglesias and Vondrak, 1974; Iglesias and Anderson, 1975). Most recently, Miller and Shepherd (1969) and Söraas *et al.* (1974) made a detailed observation of auroral protons and the Hβ emission in a post-breakup auroral glow. They showed that the observed Hβ profiles agree well with the profiles which are estimated from the observed energy spectra of auroral protons (Figure 2.36). The maximum volume emission was about 80 photons cm^{-3} s^{-1} and was located at the height of 112 ± 1 km; the computed Hβ intensity was 140 *R*, while observed intensity was 170 *R*.

2.8. Auroral Helium Ions (He^{++}, He^{+}) and Oxygen Ions (O^{+})

2.8.1. AURORAL HELIUM IONS

Measurements of the relative abundances, He^{++}/H^{+}, He^{+}/H^{+}, in auroral ions are of particular interest, because these ratios are quite different between solar wind plasma and ionospheric plasma (Axford, 1970). Whalen and McDiarmid (1972a)

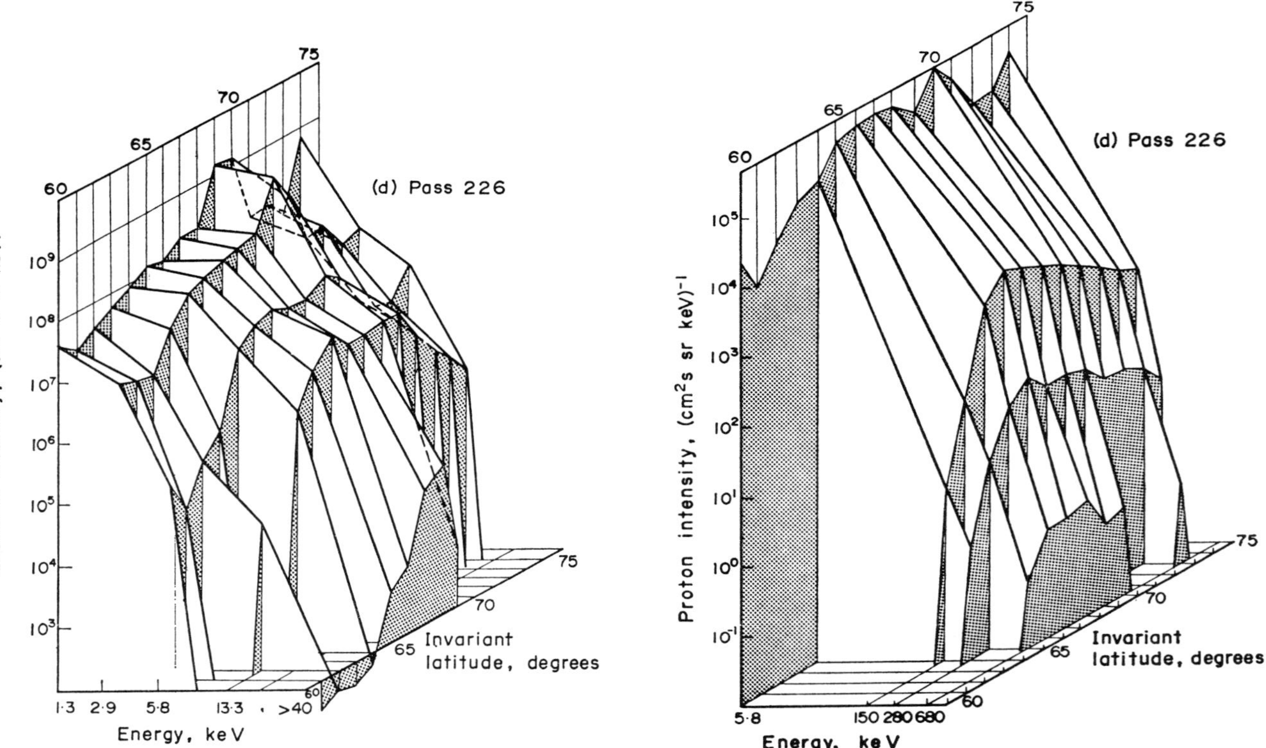

Fig. 2.35. Electron and proton energy spectra as a function of invariant latitude at 1° interval, observed by the ESRO 1/AURORAE satellite, on October 19, 1968. (Deehr *et al.*: *J. Atmosph. Terr. Phys.* **35**, 1979, 1973.)

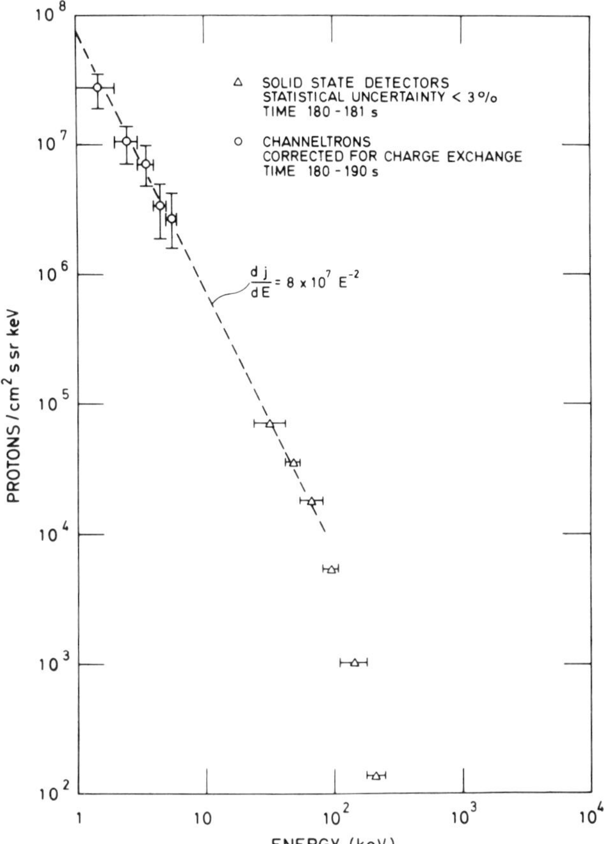

Fig. 2.36. Proton differential energy spectrum observed at an altitude of 220 km (corrected for charge exchange). (Söraas, F., Lindalen, H. R., Maseide, K., Egeland, A., Sten, T. A. and Evans, D. S.: *J. Geophys. Res.* **79**, 1851, 1974.)

found that both H^+ and He^{++} peaked at the same energy per unit charge (5 keV) and that the average ion-flux ratio He^{++}/H^+ was 3%. No detectable He^+ was found. Axford *et al.* (1972) used the metal foil sampling technique and found ^4He of order 10^6 cm^{-2} s^{-1}, but not ^3He. These observations suggest that auroral ions are of solar wind origin. On the other hand, Johnson *et al.* (1974) detected, by a satellite-borne instrument, He^+ with energies up to 1.4 keV and with the peak energy flux 0.03 erg cm^{-2} s^{-1} ster^{-1}. They suggested that He^+ was of ionospheric origin.

2.8.2. AURORAL OXYGEN IONS

It has been shown by Shelley *et al.* (1972) and Johnson *et al.* (1975) that an intense flux of order 2×10^6 cm^{-2} s^{-1} ster^{-1} of oxygen ions is often observed in the auroral region and exceeds, at times, the simultaneous proton flux during geomagnetic storms. Figure 2.37 shows parts of trajectories of the satellite (1971–089A) along which fluxes of O^+ ions were detected; the dot indicates the position of the

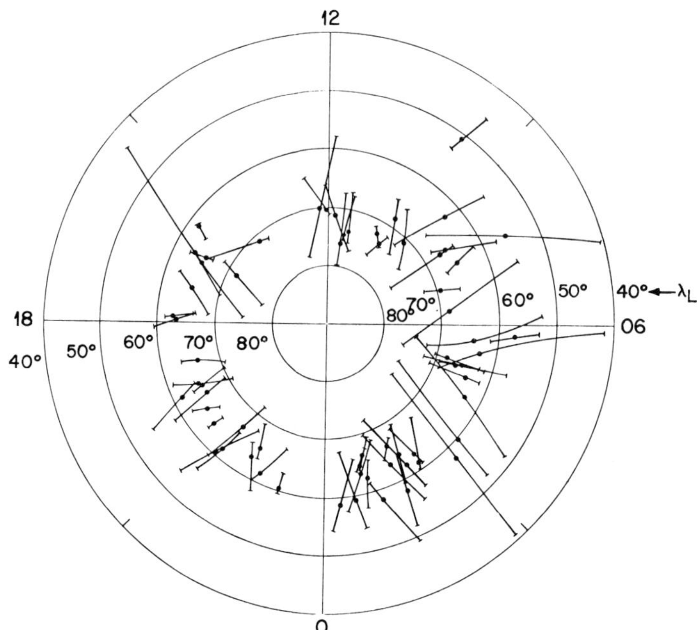

Fig. 2.37. Regions of observed oxygen ion O^+ precipitation in invariant latitude-MLT coordinates during 11 major geomagnetic storms. (Johnson, R. G., Sharp, R. D. and Shelley, E. G.: *Physics of the Hot Plasma in the Magnetosphere*. B. Hultqvist and L. Stenflo (eds.), p. 45, Plenum Press, 1975.)

maximum flux intensity on each pass segment. These observations were made during 11 geomagnetic storms of different intensities, but the O^+ ion precipitation region seems to occur along an oval-like belt. The origin of O^+ ions is at present not known. If they are proved to be of ionospheric origin, it is of interest to examine how they can be accelerated upward before they precipitate into the ionosphere.

2.9. Auroral Oval and the Polar Ionosphere

2.9.1. F REGION

The F region of the polar ionosphere is greatly disturbed by the large-scale convective ($E \times B$) motion of ionospheric plasma (Section 1.3.5) and by additional ionization caused by precipitating auroral particles. It is, however, beyond the scope of this short section to review studies of the polar ionosphere. Thus, we confine our attention to a few aspects of the ionosphere which will be referred to in later chapters (Thomas and Andrews, 1969; Buchau *et al.*, 1972; Watkins, 1976).

In the F region, there is the mid-latitude trough which separates the polar ionosphere from the rest of the ionosphere. Beyond the trough, the importance of the particle precipitation is first manifested in the region called the poleward 'wall' of the trough where the electron density rises steeply towards higher latitudes.

The trough has recently been studied by Bowman (1969), Rycroft and Thomas (1970), Rycroft and Burnell (1970), and Morse *et al.* (1971). Bates *et al.* (1973) showed, on the basis of their Chatanika radar observations, that the poleward edge of the trough can be identified with the location of a steep horizontal electron density gradient, namely a 'wall of ionization' which is approximately field-aligned, and that its position does not seem to coincide with the simultaneous location of auroras, but with the auroral location during preceding hours.

The F layer along the auroral oval shows a particular type of spread echo. Thus, this region of the ionosphere is called the F layer irregularity zone, FLIZ (Pike, 1971). Figure 2.38 shows a series of ionograms taken from an airborne ionosonde when the aircraft flew across the midday part of the oval. The first

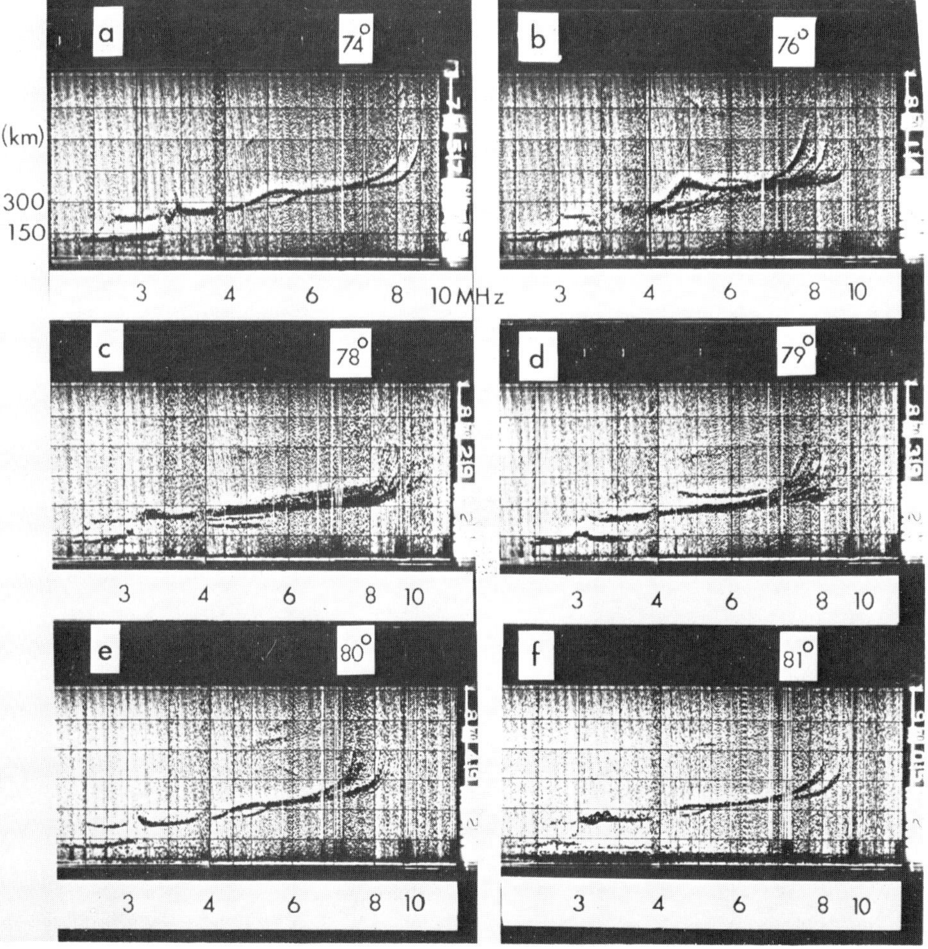

Fig. 2.38. Sequence of airborne ionograms across the auroral oval in the afternoon sector, 1957–2105 UT, 1970, May 11. The corrected geomagnetic latitude at which the ionogram was recorded is labeled. (Pike, C. P.: *J. Geophys. Res.* **76**, 7745, 1971.)

ionogram (a) shows a normal F layer echo, which was taken at invariant latitude of 74°. As the plane approached the FLIZ, it was seen first as an oblique echo, mixed in the normal echo (b). As the plane flew directly underneath the FLIZ, the oblique FLIZ echo reduced its virtual height and replaced the normal echo (c).

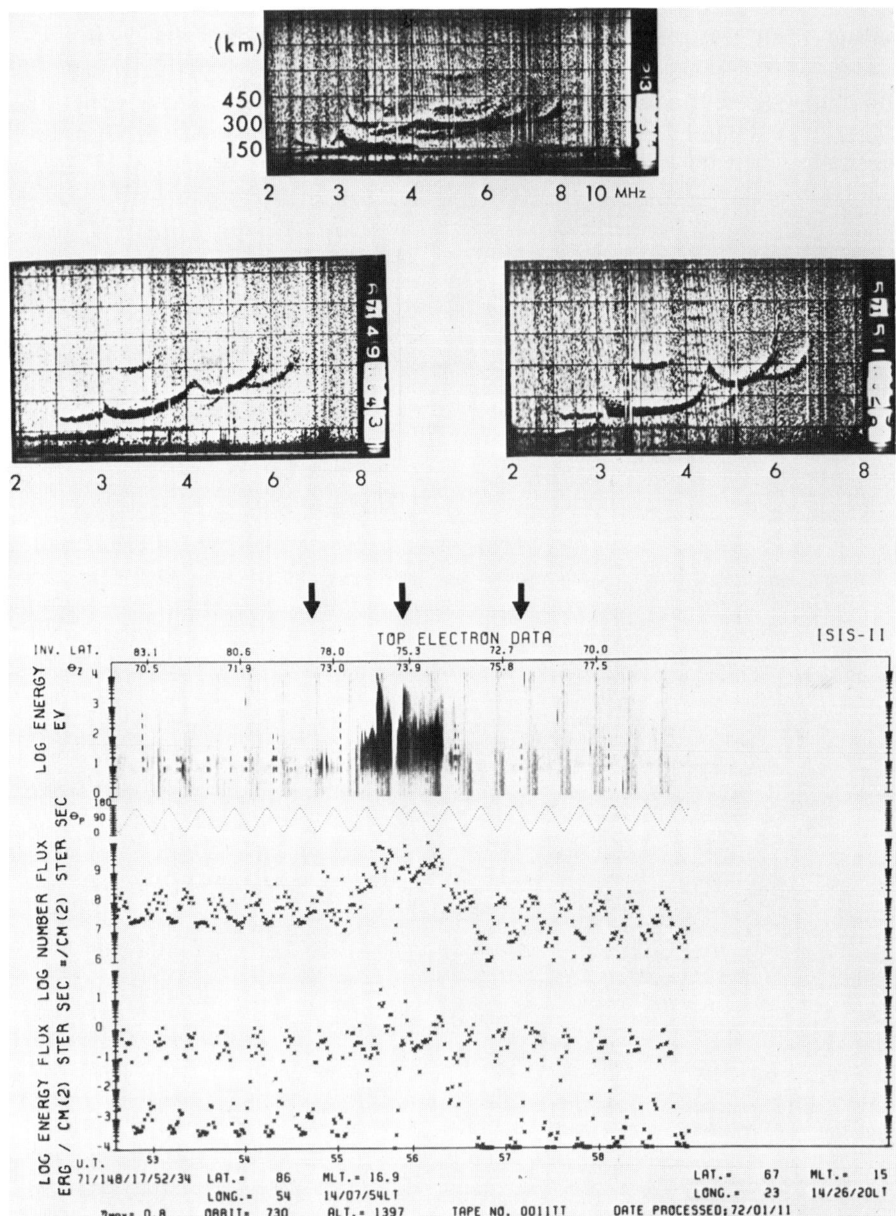

Fig. 2.39(a). Simultaneous observations of the ionosphere in the cusp region from above by the ISIS-2 satellite (SPS electron data) and from below by an airborne ionospheric sounder.

Then, as the plane flew further towards higher latitudes, moving away from the FLIZ, an oblique echo was seen again (d and e). Finally, a normal F echo was seen as the plane moved well into the polar cap (f). Pike (1971) showed that the FLIZ coincides with the F layer plasma ring proposed by Thomas and Andrews (1969); their study was based on records obtained from a sounder aboard the Allouette I satellite.

Figure 2.39(a) shows the simultaneous observation of the cusp region by an airborne sounder and by the SPS particle detector carried by the ISIS-II satellite. In the upper part there are three ionograms which were obtained at the latitudes indicated by the three arrows in the lower part of the figure, in which SPS electron data are shown. The FLIZ is clearly seen in the middle ionogram, which was obtained under the cusp precipitation region. Figure 2.39(b) shows the electron density distribution in the vicinity of the cusp region. The electron density is much higher in the cusp region than the surrounding regions.

The electron density at altitudes between 500 and 3000 km in the polar cap region is much less than that in the surrounding region. Figure 2.40 shows this tendency very clearly by showing the distribution of the points where the observed electron density was less than 100 cm^{-3} (Nelms and Chapman, 1970).

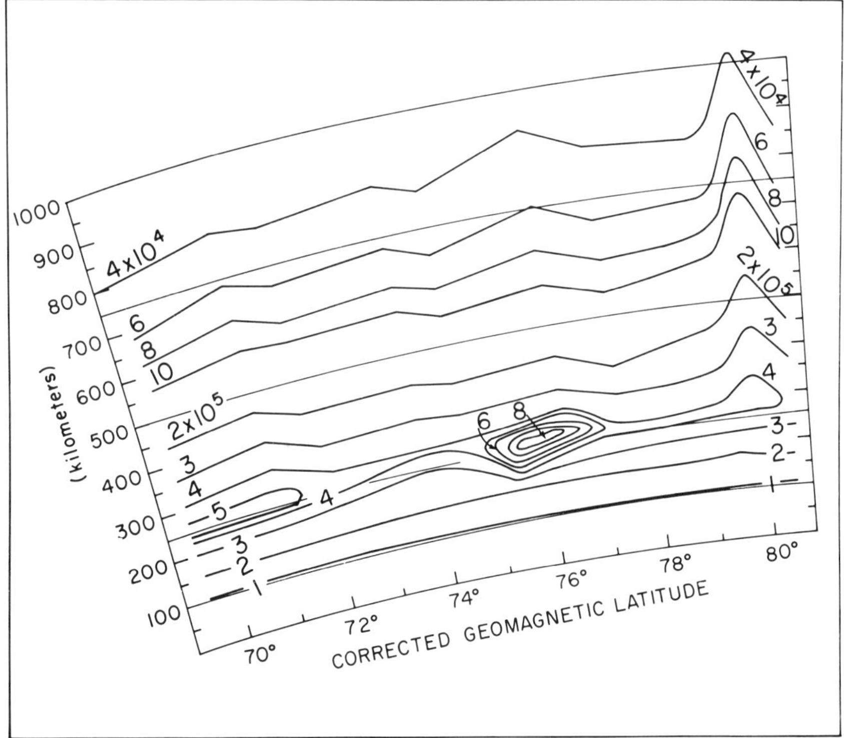

Fig. 2.39(b). Electron density profile in the vicinity of the cusp region in corrected geomagnetic latitude-altitude coordinates. (Heikkila, W. J. and Winningham, J. D.: AFCRL Report TR-74-0379, Sept. 1974.)

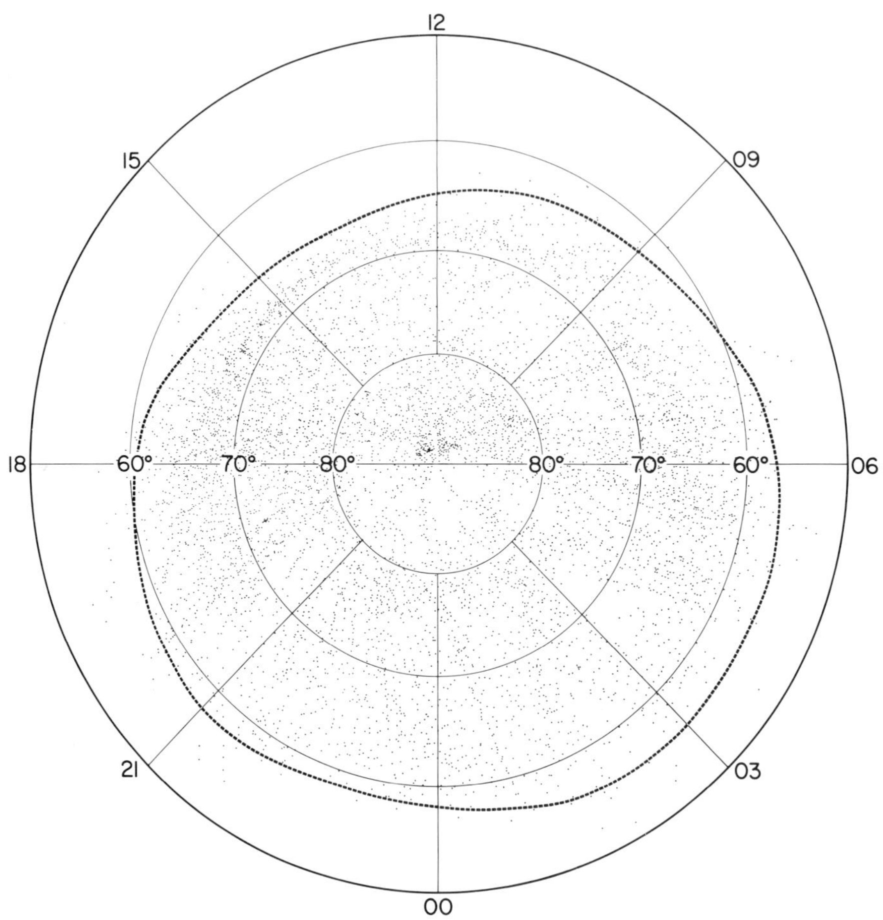

Fig. 2.40. Locations of the occurrence of low electron densities ($\leq 100/cm^3$) in the polar region, observed on Alouette 2 ionograms, in geomagnetic latitude-LT coordinates. (Nelms, G. L. and Chapman, J. H.: *The Polar Ionosphere and Magnetospheric Processes*, G. Skovli (ed.), p. 233, Gordon and Breach, 1970.)

Banks and Holzer (1968) interpreted this feature in terms of a supersonic flow of light ions (H^+, He^+) and the accompanying electrons, namely of the polar wind.

The polar F layer irregularities cause amplitude fluctuations of radio signals from satellites and radio stars. Its equatorward boundary is, however, about 10° lower than that of the auroral oval (Aarons, 1973). *In situ* measurements of electron density fluctuations in the F region have recently been made by McClure and Hanson (1973) and Dyson, McClure and Hanson (1974) on the basis of a retarding potential analyzer aboard the OGO-6. This phenomenon, scintillation, occurs in an oval-like area, but its equatorward boundary is located far equatorward of that of the auroral oval.

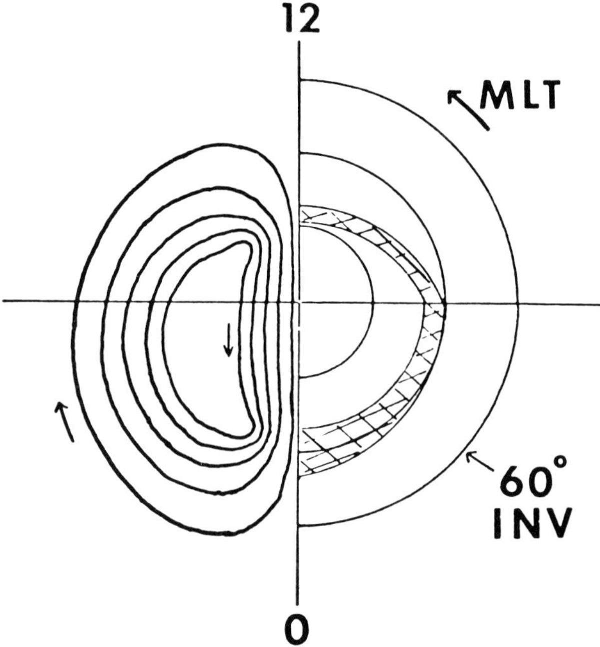

Fig. 2.41(a). Assumed plasma flow pattern and the auroral oval in computing the electron density distribution in Figure 2.42.

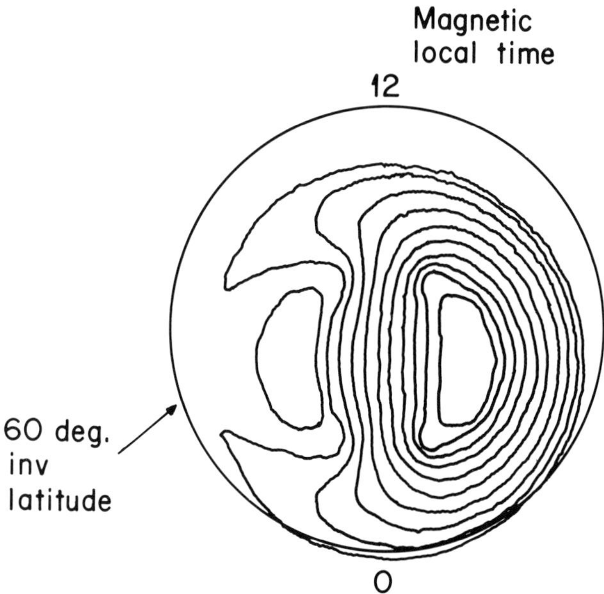

Fig. 2.41(b). Assumed plasma flow pattern (a) in invariant latitude-MLT coordinates. (Watkins, B.: Ph.D. Thesis, Univ. of Alaska, 1976.)

A large-scale distortion of the F region by the magnetospheric electric field has recently been studied by Knudsen (1974) and Watkins (1976). In particular, the latter author made a very extensive numerical study on the distribution of ionization at 300 km, by taking into the photoionization, additional ionization caused by auroral precipitation, various loss processes and the convection effect. Figures 2.41(a) and (b) show the streamlines of the convection used in his study, in the frame of the rotating earth and in the invariant latitude-MLT frame of reference, respectively. Figure 2.42 shows two examples of his results, the electron density distribution in the invariant latitude-MLT frame of reference.

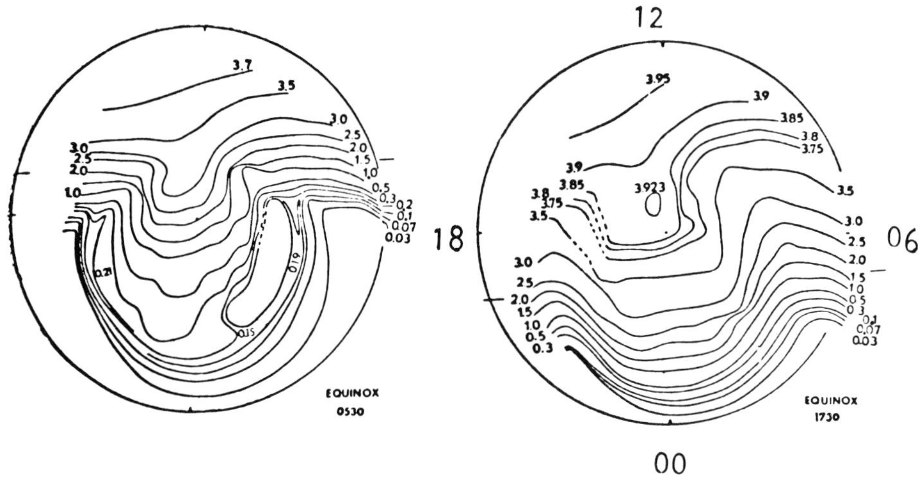

Fig. 2.42. Computed electron density contours at the F2 peak level in invariant latitude-MLT coordinates at 0530 and 1730 UT, equinox. The circle refers to the invariant latitude 60°. The two small bars either side of the circle denote the line of zero solar depression angle.

2.9.2. E REGION

On the basis of ionograms collected by an AFCRL aircraft during extremely quiet periods (Kp = 0, 1), Wagner *et al.* (1973) showed that beyond the latitude of a significant contribution of solar radiation, the E layer consists of two circumpolar belts. The first is a belt of the retarded type E (Es − r); it is characterized by a relatively thick layer (which can produce a retardation trace in the E region echo), and the layer often extends down to the D region. Poleward of this E region, there is a belt of what is called 'auroral type Es' or 'Esa'; its echo does not show the retardation. Visible auroras are often seen in the latter belt. It is quite likely that the retarded type E occurs where the quasi-circular luminous belt is located, while the auroral type Es occurs in the oval of discrete auroras (an oval-shaped luminosity). The equatorward boundary of the diffuse aurora coincides also with the poleward trough wall.

2.10. Summary

It is clear that characteristics of both auroras and auroral particles are a complicated function of local time and latitude. In this section we summarize very briefly main features studied in this chapter.

It seems reasonable to consider discrete auroral arcs (and auroral particles precipitating into them) separately from the rest of the auroral regions (and auroral particles). The latter constitute a broad background precipitation region in

TABLE 2.1
Discrete and diffuse auroras

<table>
<tr><th></th><th>Discrete Aurora (excluding the polar cap aurora)</th><th>Diffuse Aurora</th></tr>
<tr><td rowspan="5">Morphological features</td><td>A discrete aurora is a curtain-like structure and appears as a single arc or is separated from other arcs by a dark space of order a few tens of kilometers in width.</td><td>A diffuse aurora is either a broad band of luminosity at least several tens of kilometers in width or a group of discrete auroras which are closely packed along a rather narrow east-west belt. Patchy auroras develop in the diffuse aurora. The mantle aurora and the hydrogen aurora may belong to the diffuse aurora.</td></tr>
<tr><td>Predominantly an evening feature.</td><td>Consisting of an oval band and a circular band which meet together in the midnight sector.</td></tr>
<tr><td colspan="2">In the morning sector, the diffuse aurora often develops into several discrete auroras; some of the latter spread over the morning half of the polar cap and become the polar cap aurora.</td></tr>
<tr><td colspan="2">It is difficult to distinguish both types of auroras from all-sky camera photographs. The equatorward edge of the diffuse appears like a homogeneous arc.</td></tr>
<tr><td>Varying degree of conjugacy.</td><td>Appears to be conjugate.</td></tr>
<tr><td rowspan="3">Particle precipitation</td><td>Inverted V precipitation (BPS region).</td><td>A broad and relatively uniform precipitation (CPS region).</td></tr>
<tr><td>Energy spectrum is considerably variable and has a peaked component, in addition to the power law and the Maxwellian components.</td><td>Energy spectrum is relatively constant, becoming soft near the poleward and equatorward edges. The spectrum has the power law and the Maxwellian components.</td></tr>
<tr><td colspan="2">The 'trapping boundary' of the so-called 40 keV background boundary is located between the discrete and diffuse auroras.</td></tr>
<tr><td>Electric field</td><td colspan="2">The convection electric field appears to change the direction near the poleward boundary of the discrete auroral region.</td></tr>
<tr><td>Field-aligned currents</td><td>Upward current in the evening sector. Individual arcs are associated with a pair of upward and downward currents.</td><td>Weak downward current in the evening sector.</td></tr>
<tr><td>Ionosphere</td><td>Auroral type Es layer.</td><td>Retarded type Es.</td></tr>
</table>

which discrete auroras are 'embedded'. In this broad precipitation region, one of the common features is that the particles spectrum can be expressed by a combination of the power law and a Maxwellian distribution.

The broad precipitation consists of both oval and annular regions. In the oval region, the spectrum seems to be softest in the midday sector.

In the midnight sector, the spectrum is softest near the poleward boundary and becomes harder towards lower latitudes, but becomes soft again very near the equatorward boundary.

It should be noted that there are weak field-aligned currents in the broad oval precipitation region (Section 1.3.2), an inflow along the poleward half of the morning part of the oval, and an outflow along the poleward half of the afternoon part; in the equatorward half of the oval belt, this field-aligned current direction is reversed: namely, an outflow in the morning sector and an inflow in the evening sector. These field-aligned currents can also be considered as the 'background' field-aligned currents.

The region of discrete auroras may be considered to be a singular region which is embedded in the broad oval precipitation region. The precipitating auroral particles have a peaked spectrum, as well as a combination of the power law and Maxwellian components. This precipitation is closely associated with a pair of oppositely flowing field-aligned currents which are much more intense than the background field-aligned currents mentioned in the above.

The annular precipitation occurs along a broad belt whose centerline is about $L = 6$. This region partially overlaps or merges with the oval precipitation region in the midnight sector. The energy of precipitating electrons in the annular belt is, in general, considerably harder than those in the oval precipitation region.

In Table 2.1, we summarize some of the important features of discrete and diffuse auroras; note that the polar cap aurora is not included in the discrete aurora.

References

Aarons, J.: 1973, 'A Descriptive Model of F Layer High-Latitude Irregularities as Shown by Scintillation Observations', *J. Geophys. Res.* **78**, 7441.

Ackerson, K. L.: 1972, 'Observations of Charged Particle Precipitation over the Auroral Zone during Magnetic Substorm' (preprint – Dept of Phys. and Astronomy – Univ. of Iowa).

Ackerson, K. L. and Frank, L. A.: 1972, 'Correlated Satellite Measurements of Low-Energy Electron, Precipitation and Ground-Based Observations of a Visible Auroral Arc', *J. Geophys. Res.* **77**, 1128.

Akasofu, S.-I.: 1974a, 'A Study of Auroral Displays Photographed from the DMSP-2 Satellite and from the Alaska Meridian Chain of Stations', *Space Sci. Rev.* **16**, 617.

Akasofu, S.-I.: 1974b, 'Discrete, Continuous and Diffuse Auroras', *Planet. Space Sci.* **22**, 1723.

Akasofu, S.-I.: 1976, 'Recent Progress in Studies of DMSP Auroral Photographs', *Space Sci. Rev.* **19**, 169.

Akasofu, S.-I. and Yasuhara, F.: 1973, 'Red Auroras in the Morning Sector', *J. Geophys. Res.* **78**, 3027.

Akasofu, S.-I., Kimball, D. S., Buchau, J. and Gowell, R. W.: 1972, 'Alignment of Auroral Arcs', *J. Geophys. Res.* **77**, 4233.

Amundsen, R., Aarsnes, K., Lindalen, H. R. and Söraas, F.: 1973, 'High Latitude Electron and Proton Precipitation at Different Local Times and Magnetic Activity', *J. Atmosph. Terr. Phys.* **35**, 2191.

Amundsen, R., Söraas, F. and Aarsnes, K.: 1975, 'Cleft Signature in Proton Fluxes above 100 keV', *J. Geophys. Res.* **80**, 685.

Anderson, H. R. and Cloutier, P. A.: 1975, 'Simultaneous Measurements of Auroral Particles and

Electric Currents by a Rocket-Borne Instrument System: Introductory Remarks', *J. Geophys. Res.* **80**, 2146.

Andrews, M. K. and Strömman, J. R.: 1973, 'Simultaneous Observations of Low Energy Electron Fluxes and the Polar Red Emission at 6300 Å', *J. Atmosph. Terr. Phys.* **35**, 537.

Anger, C. D. and Lui, A. T. Y.: 1973, 'A Global View at the Polar Region on 18 December 1971', *Planet. Space Sci.* **21**, 873, 1973.

Anger, C. D., Sawchuk, W. and Shepherd, G. G.: 1974, 'Polar Cap Optical Aurora Seen from ISIS-2', *Magnetospheric Physics*, B. M. McCormac (ed.), p. 357, D. Reidel Publ. Co., Dordrecht-Holland.

Arnoldy, R. L.: 1970, 'Rapid Fluctuations of Energetic Auroral Particles', *J. Geophys. Res.* **75**, 228.

Arnoldy, R. L.: 1974, 'Auroral Particle Precipitation and Birkeland Currents', *Rev. Geophys. Space Phys.* **12**, 217.

Arnoldy, R. L. and Choy, L. W.: 1973, 'Auroral Electrons of Energy Less Than 1 keV Observed at Rocket Altitudes', *J. Geophys. Res.* **78**, 2187.

Arnoldy, R. L., Lewis, P. B. and Isaacson, P. O.: 1974, 'Field-Aligned Auroral Electron Fluxes', *J. Geophys. Res.* **79**, 4208.

Axford, W. I.: 1970, 'On the Origin of Radiation Belt and Auroral Primary Ions', *Particles and Fields in the Magnetosphere*, B. M. McCormac (ed.), p. 46, D. Reidel Publ. Co., Dordrecht-Holland.

Axford, W. I., Bühler, F., Chivers, H. J. A., Eberhardt, P. and Geiss, J.: 1972, 'Auroral Helium Precipitation', *J. Geophys. Res.* **77**, 6724.

Badhwar, G. D., Deney, C. L. and Kaplon, M. F.: 1969, 'Flux of Hydrogen, Helium and Electrons at Fort Churchill in 1967', *J. Geophys. Res.* **74**, 369.

Bagno, Yu. D., Osipov, K. K. and Osipov, N. K.: 1974, 'Nonstationary Ion Production by the Auroral Ionosphere', *Geomag. Aeronom.* **14**, 59.

Banks, P. M. and Holzer, T. E.: 1968, 'The Polar Wind', *J. Geophys. Res.* **73**, 6846.

Banks, P. M., Chappell, C. R. and Nagy, A. F.: 1974, 'A New Model for the Interaction of Auroral Electrons with the Atmosphere: Spectral Degradation, Backscatter, Optical Emission, and Ionization', *J. Geophys. Res.* **79**, 1459.

Bates, H. F., Belon, A. E. and Hunsucker, R. D.: 1973, 'Aurora and the Poleward Edge of the Main Ionospheric Trough', *J. Geophys. Res.* **78**, 648.

Belon, A. E., Maggs, J. E., Davis, T. N., Mather, K. B., Glass, N. W. and Hughes, G. F.: 1969, 'Conjugacy of Visual Auroras during Magnetically Quiet Periods', *J. Geophys. Res.* **74**, 1.

Belon, A. E., Romick, G. J. and Stringer, W. J.: 1974, 'Proton Precipitation during the Auroral Breakup', *Planet. Space Sci.* **22**, 735.

Berko, F. W.: 1973, 'Distributions and Characteristics of High Latitude Field-Aligned Electron Precipitation', *J. Geophys. Res.* **78**, 1615.

Berko, F. W. and Hoffman, R. A.: 1974, 'Dependence of Field-Aligned Electron Precipitation Occurrence on Season and Altitude', *J. Geophys. Res.* **79**, 3749.

Bosqued, J. M., Cardona, G. and Rème, H.: 1974, 'Auroral Electron Fluxes Parallel to the Geomagnetic Field Lines', *J. Geophys. Res.* **79**, 98.

Bowman, G. G.: 1969, 'Ionization Troughs below the F2-Layer Maximum', *Planet. Space Sci.* **17**, 777.

Boyd, J. S., Belon, A. E. and Romick, G. J.: 1971, 'Latitude and Time Variations in Precipitated Electron Energy Inferred from Measurements of Auroral Heights', *J. Geophys. Res.* **76**, 7694.

Brekke, A. and Omholt, A.: 1968, 'The Intensity Ratio I(4278)/I(5577) in Aurora', *Planet. Space Sci.* **16**, 1259.

Brekke, A. and Pettersen, H.: 1972, 'A Possible Method for Estimating Any Indirect Process in the Production of the O(^{1}S) Atoms in Aurora', *Planet. Space Sci.* **20**, 1569.

Bryant, D. A., Courtier, G. M. and Bennett, G.: 1972, 'Electron Intensities over Auroral Arcs', *Earth's Magnetospheric Processes*, B. M. McCormac, (ed.), p. 141, D. Reidel Publ. Co., Dordrecht-Holland.

Bryant, D. A., Courtier, G. M. and Bennett, G.: 1973, 'Electron Intensities over Two Auroral Arcs', *Planet. Space Sci.* **21**, 165.

Buchau, J., Whalen, J. A. and Akasofu, S.-I.: 1969, 'Airborne Observations of the Midday Aurora', *J. Atmosph. Terr. Phys.* **23**, 1021, Jan.–June.

Buchau, J., Pittenger, E. W. and Sizoo, A. H.: 1970, 'Arctic Ionosphere and Aurora Airborne Investigations', Air Force Cambridge Research Lab. AFCRL-70-0280, May 1970 Environmental Research Papers, No. 322.

Buchau, J., Gassmann, G. J., Pike, C. P., Wagner, R. A. and Whalen, J. A.: 1972, 'Precipitation Patterns in the Arctic Ionosphere Determined from Airborne Observations', *Ann. Geophys.* **28**, 443.

Burch, J. L.: 1968, 'Low-Energy Electron Fluxes at Latitudes above the Auroral Zone', *J. Geophys. Res.* **73**, 3585.

Burch, J. L.: 1972, 'Precipitation of Low-Energy Electrons at High Latitudes, Effects of Substorms, Interplanetary Magnetic Field and Dipole Hilt Angle', Goddard Space Flight Center, May.

Burrows, J. R.: 1974, 'The Plasma Sheet in the Evening Sector', *Magnetospheric Physics*, B. M. McCormac (ed.), p. 179, D. Reidel Publ. Co., Dordrecht-Holland.
Burrows, K., Stolarik, J. D. and Heppner, J. P.: 1971, 'Rocket Measurements of the Magnetic Fields Associated with Visual Aurorae', *Planet. Space Sci.* **19**, 877.
Casserly, R. T. Jr. and Cloutier, P. A.: 1975, 'Rocket-Based Magnetic Observations of Auroral Birkeland Currents in Association with a Structured Auroral Arc', *J. Geophys. Res.* **80**, 2165.
Caverly, R. S.: 1975, 'Observations from a Chain of All-Sky Cameras Coordinated with Satellite Auroral Zone Particle Precipitation', M.S. Thesis, University of Alaska, August.
Chase, L. M.: 1970, 'Energy Spectra of Auroral Zone Particles', *J. Geophys. Res.* **75**, 7128.
Choy, L. W., Arnoldy, R. L., Potter, W., Kintner, P. and Cahill, L. J. Jr.: 1971, 'Field-Aligned Particle Currents near an Auroral Arc', *J. Geophys. Res.* **76**, 8279.
Chubb, T. A. and Hicks, G. T.: 1970, 'Observations of the Aurora in the Far Ultraviolet from OGO-4', *J. Geophys. Res.* **75**, 1290.
Clark, M. A. and Metzger, P. H.: 1969, 'Auroral Lyman-Alpha Observations', *J. Geophys. Res.* **74**, 6257.
Cloutier, P. A. and Anderson, H. R.: 1975, 'Observations of Birkeland Currents', *Space Sci. Rev.* **17**, 563.
Cloutier, P. A., Anderson, H. R., Park, R. J., Vondrak, R. R., Spiger, R. J. and Sandel, B. R.: 1970, 'Detection of Geomagnetically Aligned Currents Associated with an Auroral Arc', *J. Geophys. Res.* **75**, 2595.
Cloutier, P. A., Sandel, B. R., Anderson, H. R., Pazich, P. M. and Spiger, R. J.: 1973, 'Measurement of Auroral Birkeland Currents and Energetic Particle Fluxes', *J. Geophys. Res.* **78**, 640.
Craven, J. D.: 1970, 'A Survey of Low-Energy ($E \geqslant 5$ keV) Electron Energy Fluxes over the Northern Auroral Regions with Satellite INJUN-4', *J. Geophys. Res.* **75**, 2468.
Davis, T. N.: 1962, 'The Morphology of the Auroral Displays of 1957–58, 2. Detailed Analyses of Alaska Data and Analyses of High-Latitude Data', *J. Geophys. Res.* **67**, 75.
Davis, T. N. and DeWitt, R. N.: 1963, 'Twenty-Four-Hour Observations of Aurora at the Southern Auroral Zone', *J. Geophys. Res.* **68**, 6237.
Deehr, C. S., Egeland, A., Aarsnes, K., Amundsen, R., Lindalen, H. R., Söraas, F., Dalziel, R., Smith, P. A., Thomas, G. R., Stauning, P., Borg, H., Gustafsson, G., Holmgren, L. A., Riedler, W., Raitt, J., Skovli, G., Wedde, T. and Jaeschke, R.: 1973, 'Particle and Auroral Observations from the ESRO 1/AURORAE Satellite', *J. Atmosph. Terr. Phys.* **35**, 1979.
Deehr, C. S., Winningham, J. D., Yasuhara, F. and Akasofu, S.-I.: 1976, 'Simultaneous Observations of Discrete and Diffuse Auroras by the ISIS-2 Satellite and Airborne Instruments', *J. Geophys. Res.* **81** (in press).
Denholm, J. V.: 1961, 'Some Auroral Observations inside the Southern Auroral Zone', *J. Geophys. Res.* **66**, 2105.
Derblom, H.: 1975, 'Observed Characteristics of Polar Cleft Hα and O I Emissions', *Planet. Space Sci.* **23**, 1053.
Doering, J. P., Peterson, W. K., Bostrom, C. O. and Armstrong, J. C.: 1975, 'Measurements of Low-Energy Electrons in the Day Airglow and Dayside Auroral Zone from Atmosphere Explorer C', *J. Geophys. Res.* **80**, 3934.
Duysinx, R. and Monfils, A.: 1970, 'Spatial Separation of λ 3914 and λ 5577-A Emissions in an Aurora', *J. Geophys. Res.* **75**, 2606.
Duysinx, R. and Monfils, A.: 1972, 'Auroral Spectra Recorded between 2000 and 3000 Å with a Fast Scanning Spectrometer', *Ann. Geophys.* **28**, 109.
Dyson, P. L. and Winningham, J. D.: 1974, 'Top Side Ionospheric Spread F and Particle Precipitation in the Day Side Magnetospheric Clefts', *J. Geophys. Res.* **79**, 5219.
Dyson, P. L. and Zmuda, A. J.: 1970, 'Some Ionospheric Properties at 1000 Kilometers Altitude within and near the Auroral Zone', *J. Geophys. Res.* **75**, 1893.
Dyson, P. L., McClure, J. P. and Hanson, W. B.: 1974, 'In situ Measurements of the Spectral Characteristics of F Region Ionospheric Irregularities', *J. Geophys. Res.* **79**, 1497.
Eather, R. H.: 1967, 'Auroral Proton Precipitation and Hydrogen Emissions', *Rev. Geophys.* **5**, 207.
Eather, R. H.: 1968, 'Spectral Intensity Ratios in Proton-Induced Auroras', *J. Geophys. Res.* **73**, 119.
Eather, R. H.: 1969, 'Latitudinal Distribution of Auroral and Airglow Emissions: The "Soft" Auroral Zone', *J. Geophys. Res.* **74**, 153.
Eather, R. H.: 1975, 'Advances in Magnetospheric Physics 1971–74: Aurora', one of an eight-paper U.S. National Report on magnetospheric physics to be presented through the auspices of the American Geophysical Union to the Meeting of the International Union of Geodesy and Geophysics at Grenoble, France.
Eather, R. H. and Akasofu, S.-I.: 1969, 'Characteristics of Polar Cap Auroras', *J. Geophys. Res.* **74**, 4794.

Eather, R. H. and Mende, S. B.: 1971a, 'High Latitude Particle Precipitation, and Source Regions in the Magnetosphere', Presented at Advanced Study Inst. on Magnetosphere-Ionosphere Interactions, Dalseter, Norway, April, Boston College.

Eather, R. H. and Mende, S. B.: 1971b, 'Airborne Observations of Auroral Precipitation Patterns', *J. Geophys. Res.* **76**, 1746.

Eather, R. H. and Mende, S. B.: 1972, 'Systematics in Auroral Energy Spectra', *J. Geophys. Res.* **77**, 660.

Evans, D. S.: 1968, 'The Observations of a Near Monoenergetic Flux of Auroral Electrons', *J. Geophys. Res.* **73**, 2315.

Evans, D. S.: 1971, 'Direct Observation of Temporal and Spatial Structure in Auroral Electrons', *The Radiating Atmosphere*, B. M. McCormac (ed.), p. 267, D. Reidel Publ. Co., Dordrecht-Holland.

Evans, D. S.: 1972, 'Auroral Particles and Fields', *Ann. Geophys.* **28**, 639.

Feldman, P. D. and Doering, J. P.: 1975, 'Auroral Electrons and the Optical Emissions of Nitrogen', *J. Geophys. Res.* **80**, 2808.

Feldman, P. D., Doering, J. P. and Moore, J. H.: 1971, 'Rocket Measurement of the Secondary Electron Spectrum in an Aurora', *J. Geophys. Res.* **76**, 1738.

Feldstein, Ya. I. and Starkov, G. V.: 1971, 'Auroral Oval Planetary Energetics', *J. Atmosphere Terrestrial Phys.* **33**, 197.

Frank, L. A.: 1971, 'Relationship of the Plasma Sheet, Ring Current, Trapping Boundary, and Plasmapause near the Magnetic Equator and Local Midnight', *J. Geophys. Res.* **76**, 2265.

Frank, L. A. and Ackerson, K. L.: 1971, 'Observations of Charged Particle Precipitation into the Auroral Zone', *J. Geophys. Res.* **76**, 3612.

Frank, L. A. and Ackerson, K. L.: 1972, 'Local-Time Survey of Plasma at Low Altitudes over the Auroral Zones', *J. Geophys. Res.* **77**, 4116.

Fritz, T. A.: 1970, 'Study of the High-Latitude, Outer-Zone Boundary Region for $\geq$40-keV Electrons with Satellite INJUN-3', *J. Geophys. Res.* **75**, 5387.

Fukunishi, H.: 1975, 'Dynamic Relationship between Proton and Electron Auroral Substorms', *J. Geophys. Res.* **80**, 553.

Fukunishi, H. and Tohmatsu, T.: 1973, 'Constitution of Proton Aurora and Electron Aurora Substorms Part. Meridian-Scanning Photometric System for Proton Auroras and Electron Auroras', JARE Sci., Rep. Series A No. 11, March.

Gowell, R. W. and Akasofu, S.-I.: 1969, 'Irregular Pulsations of the Morning Sky Brightness', *Planet. Space Sci.* **17**, 289.

Gustafsson, G., Feldstein, Y. I. and Shevnina, N. F.: 1969, 'The Auroral Orientation Curves for the IQSY', *Planet. Space Sci.* **17**, 1657.

Harrison, A. W. and Cairns, C. D.: 1969, 'Helium Emission (1.083μ) in Sunlit Aurora', *Planet. Space Sci.* **17**, 1213.

Heikkila, W. J.: 1972, 'The Morphology of Auroral Particle Precipitation', COSPAR, *Space Research XII*, S. A. Bowhill, L. D. Jaffe, and M. J. Roycroft (eds.), vol. 2, p. 1343, Akademie-Verlag, Berlin.

Heikkila, W. J. and Winningham, J. D.: 1971, 'Penetration of Magnetosheath Plasma to Low Altitudes Through the Dayside Magnetospheric Cusps', *J. Geophys. Res.* **76**, 883.

Heikkila, W. J., Winningham, J. D., Eather, R. H. and Akasofu, S.-I.: 1972, 'Auroral Emissions and Particle Precipitation in the Noon Sector', *J. Geophys. Res.* **77**, 4100.

Heikkila, W. J. and Winningham, J. D.: 1974, 'Auroral Data Analysis', AFCRL-TR-74-0379.

Henriksen, K.: 1974, 'Studies of the Auroral Green Line: Electron Excitation Mechanism', *The Auroral Observatory*, Tromosø, May.

Hoffman, R. A. and Berko, F. W.: 1971, 'Primary Electron Influx to Dayside Auroral Oval', *J. Geophys. Res.* **76**, 2967.

Hoffman, R. A. and Evans, D. S.: 1968, 'Field-Aligned Electron Bursts at High Latitudes Observed by OGO-4', *J. Geophys. Res.* **73**, 6201.

Hones, E. W. Jr., Asbridge, J. R., Bame, S. J. and Singer, S.: 1971, 'Energy Spectra and Angular Distributions of Particles in the Plasma Sheet and their Comparison with Rocket Measurements over the Auroral Zone', *J. Geophys. Res.* **76**, 63.

Hultqvist, B.: 1969, 'Auroras and Polar Substorms: Observations and Theory', *Rev. Geophys.* **7**, 129.

Hultqvist, B.: 1972, 'Auroral Particles', *Geofysiske Publikasjoner* **29**, 27.

Hultqvist, B.: 1973, 'Rocket and Satellite Observations of Energetic Particle Precipitation in Relation to Optical Aurora', *Symposium on Aurora and Airglow* of the IAGA Second General Scientific Assembly, Kyoto, Japan, Sept.

Hultqvist, B.: 1974, 'Rocket and Satellite Observations of Energetic Particle Precipitation in Relation to Optical Aurora', *Ann. Geophys.* **30**, 233.

Hultqvist, B.: 1975, 'The Aurora', *The Magnetosphere of the Earth and Jupiter*, V. Formisano (ed.), p. 77, D. Reidel Publ. Co., Dordrecht-Holland.

Hultqvist, B., Borg, H., Riedler, W. and Christophersen, P.: 1971, 'Observations of Magnetic-Field Aligned Anisotropy for 1 and 6 keV Positive Ions in the Upper Atmosphere', *Planet. Space Sci.* **19**, 279.

Hultqvist, B., Borg, H., Christophersen, P., Riedler, W. and Bernstein, W.: 1974, 'Energetic Protons in the keV Energy Range and Associated keV Electrons Observed at Various Local Times and Disturbance Levels in the Upper Ionosphere', NOAA Technical Report 305-SEL 29, Jan.

Iglesias, G. E. and Anderson, H. R.: 1975, 'Neutral Hydrogen Flux Measured at 100-to-200 km Altitude in an Electron Aurora', *J. Geophys. Res.* **80**, 2169.

Iglesias, G. E. and Vondrak, R. R.: 1974, 'Atmospheric Spreading of Protons in Auroral Arcs', *J. Geophys. Res.* **79**, 280.

Johnson, R. G., Sharp, R. D. and Shelley, E. G.: 1974, 'The Discovery of Energetic He^+ Ions in the Magnetosphere', *J. Geophys. Res.* **79**, 3135.

Johnson, R. G., Sharp, R. D. and Shelley, E. G.: 1975, 'Composition of the Hot Plasmas in the Magnetosphere', Review paper presented at the Nobel Symposium on the Physics of the Hot Plasma in the Magnetosphere, Kiruna, Sweden, April.

Joki, E. G. and Evans, J. E.: 1969, 'Satellite Measurements of Auroral Ultraviolet and 3914-A Radiation', *J. Geophys. Res.* **74**, 4677.

Jones, A. V.: 1971, 'Auroral Spectroscopy', *Space Sci. Rev.* **11**, 776.

Jones, R. A. and Rees, M. H.: 1973, 'Time Dependent Studies of the Aurora-1. Ion Density and Composition', *Planet. Space Sci.* **21**, 537.

Kamiyama, H.: 1974, 'Ionosphere Produced through the Precipitation of Electrons of Several Hundred Electronvolts', *Rep. Ionosph. Space Res. Japan* **28**, 129.

Kennel, C. F. and Rees, M. H.: 1972, 'Dayside Auroral-Oval Plasma Density and Conductivity Enhancements Due to Magnetosheath Electron Precipitation', *J. Geophys. Res.* **77**, 2294.

Knudsen, W. C.: 1974, 'Magnetospheric Convection and the High-Latitude F_2 Ionosphere', *J. Geophys. Res.* **79**, 1046.

Lassen, K.: 1967, 'Polar Cap Aurora', *Aurora And Airglow*, B. M. McCormac (ed.), p. 453, Reinhold Publ. Co., New York.

Lassen, K.: 1970, 'The Position of the Auroral Oval over Greenland and Spitzbergen', *Physica Norvegica* **4**, 171.

Lassen, K.: 1972, 'On the Classification of High-Latitude Auroras', *Dan. Meterol. Inst. Geophys.* Paper, R-28, Copenhagen.

Lassen, K.: 1973, 'Orientation of Auroral Arcs and Precipitation Pattern of Auroral Particles', *Dan. Meterol. Inst. Geophys.* Paper R-33, Copenhagen.

Lebeau, A. F.: 1970, 'Magnetic Field Variations in the Polar Cap', *J. Franklin Inst.* **290**, 297.

Ledley, B. G. and Farthing, W. H.: 1974, 'Field-Aligned Current Observations in the Polar Cusp Ionosphere', *J. Geophys. Res.* **79**, 3124.

Lui, A. T. Y. and Anger, C. D.: 1973, 'A Uniform Belt of Diffuse Auroral Emission Seen by the ISIS-2 Scanning Photometer', *Planet. Space Sci.* **21**, 799.

Lui, A. T. Y., Perreault, P., Akasofu, S.-I. and Anger, C. D.: 1973, 'The Diffuse Aurora', *Planet. Space Sci.* **21**, 857.

Lui, A. T. Y., Venkatesan, D., Anger, C. D., Akasofu, S.-I., Heikkila, W. J., Winningham, J. D. and Burrows, J. R.: 1976, 'Simultaneous Observations of Particle Precipitations and Auroral Emissions by the ISIS-2 Satellite in 19–24 MLT Sector', *J. Geophys. Res.* **81** (in press).

Maehlum, B. H.: 1968, 'Universal-Time Control of the Low-Energy Fluxes in the Polar Regions', *J. Geophys. Res.* **73**, 3459.

Maral, G.: 1970, 'Détermination Indirecte des Caractères Énergétiques des Précipitations Électroniques dans la Zone Aurorale Pendant la Journée', *Ann. Geophys.* **26**, 71.

Maynard, N. C. and Johnstone, A. D.: 1974, 'High-Latitude Day Side Electric Field and Particle Measurements', *J. Geophys. Res.* **79**, 3111.

McClure, J. P. and Hanson, W. B.: 1973, 'A Catalog of Ionospheric F Region Irregularity Behavior Based on OGO-6 Retarding Potential Analyzer Data', *J. Geophys. Res.* **78**, 7431.

McDiarmid, I. B., Burrows, J. R. and Wilson, M. D.: 1972, 'Solar Particles and the Dayside Limit of Closed Field Lines', *J. Geophys. Res.* **77**, 1103.

McDiarmid, I. B., Burrows, J. R. and Budzinski, E. E.: 1975, 'Average Characteristics of Magnetospheric Electrons (150 eV to 200 keV) at 1400 km', *J. Geophys. Res.* **80**, 73.

McDiarmid, I. B., Burrows, J. R. and Budzinski, E. E.: 1976, 'Particle Properties in the Dayside Cleft', *J. Geophys. Res.* **81**, 221.

McEwen, D. J. and Sivjee, G. G.: 1972, 'Rocket Measurements of Electron Influx during a Major Storm with Type A Aurora', *J. Geophys. Res.* **77**, 5523.

Mende, S. B. and Eather, R. H.: 1971, 'The Spatial Extent of Aurorae in O(5577) and N_2^+(4278) Emissions', *Planet. Space Sci.* **19**, 49.

Mende, S. B. and Eather, R. H.: 1972, 'Photometric Auroral Particle Measurements', *Earth's Magnetospheric Processes*, B. M. McCormac (ed.), p. 179, D. Reidel Publ. Co., Dordrecht-Holland.
Mende, S. B. and Eather, R. H.: 1975, 'Spectroscopic Determination of the Characteristics of Precipitating Auroral Particles', *J. Geophys. Res.* **80**, 3211.
Meng, C.-I.: 1976, 'Simultaneous Observations of Low-Energy Electron Precipitation and Optical Auroral Arcs in the Evening Sector by the DMSP Satellite', *J. Geophys. Res.* **81**, 2771.
Meng, C.-I. and Akasofu, S.-I.: 1976, 'The Relation between the Polar Cap Auroral Arc and the Auroral Oval Arc', *J. Geophys. Res.* **81**, 4004.
Metzger, P. H. and Clark, M. A.: 1971, 'Auroral Lyman Alpha Observations, 2, Event of May 1967', *J. Geophys. Res.* **76**, 1756.
Miller, J. R. and Shepherd, G. G.: 1969, 'Rocket Measurements of H Beta Production in a Hydrogen Aurora', *J. Geophys. Res.* **74**, 4987.
Miller, R. E., Fastie, W. G. and Isler, R. C.: 1968, 'Rocket Studies of Far-Ultraviolet Radiation in an Aurora', *J. Geophys. Res.* **73**, 3353.
Mishin, V. M., Saifudinova, T. I. and Zhulin, I. A.: 1970, 'A Magnetosphere Model Based on Two Zones of Precipitating Energetic Particles', *J. Geophys. Res.* **75**, 797.
Mizera, P. F.: 1974, 'Observations of Precipitating Protons with Ring Current Energies', *J. Geophys. Res.* **79**, 581.
Mizera, P. F., Croley, D. R. Jr., Morse, F. A. and Vampola, A. L.: 1975, 'Electron Fluxes and Correlations with Quiet Time Auroral Arcs', *J. Geophys. Res.* **80**, 2129.
Montbriand, L. E.: 1969, 'Morphology of Auroral Hydrogen Emissions during Auroral Substorm', Ph.D. Thesis, Univ. of Saskatchewan.
Morse, F. A., Hilton, H. H. and Mizera, P. F.: 1971, 'Polar Ionosphere: Measured Ion Density Enhancements and Soft Electron Precipitation', *J. Geophys. Res.* **76**, 6099.
Murcray, W. B.: 1969a, 'The Ratio $\rho(5577)/\rho(N_2+\text{First Negative})$ in the Aurora', *J. Geophys. Res.* **74**, 366.
Murcray, W. B.: 1969b, 'Intensity Ratio of the 6300 Å and 5577 Å O I Emissions in Quiet Aurora', *Planet. Space Sci.* **17**, 1429.
Nagata, T., Hirasawa, T., Takizawa, M. and Tohmatsu, T.: 1975, 'Antarctic Substorm Events Observed by Sounding Rockets, Ionization of the D- and E-Regions by Auroral Electrons', *Planet. Space Sci.* **23**, 1321.
Narcisi, R. S., Sherman, C., Wlodyka, L. E. and Ulwick, J. C.: 1974, 'Ion Composition in an IBC Class II Aurora, 1, Measurements', *J. Geophys. Res.* **79**, 2843.
Nelms, G. L. and Chapman, J. H.: 1970, 'The High Latitude Ionosphere: Results from the Alouette/ ISIS Topside Sounders', *The Polar Ionosphere and Magnetospheric Processes*, G. Skovli (ed.), p. 233, Gordon and Breach, Science Publ. Inc., New York.
Ogilvie, K. W.: 1968, 'Auroral Electron Energy Spectra', *J. Geophys. Res.* **73**, 2325.
Oguti, T.: 1973, 'Hydrogen Emission and Electron Aurora at the Onset of the Auroral Breakup', *J. Geophys. Res.* **78**, 7543.
Oguti, T., Fukunishi, H., Tohmatsu, T. and Nagata, T.: 1974, 'H_B Emission during Auroral Breakup', Memoirs of National Institute of Polar Research Special Issue no. 3, Nat. Inst. of Polar Res., Tokyo, March.
Omholt, A.: 1971, *The Optical Aurora*, Springer-Verlag, Berlin.
Park, R. J. and Cloutier, P. A.: 1971, 'Rocket-Based Measurement of Birkeland Currents Related to an Auroral Arc and Electrojet', *J. Geophys. Res.* **76**, 7714.
Paschmann, G., Johnson, R. G., Sharp, R. D. and Shelley, E. G.: 1972, 'Angular Distributions of Auroral Electrons in the Energy Range 0.8 to 16 keV', *J. Geophys. Res.* **77**, 6111.
Pazich, P. M. and Anderson, H. R.: 1975, 'Rocket Measurements of Auroral Electron Fluxes Associated with Field-Aligned Currents', *J. Geophys. Res.* **80**, 2152.
Peek, H. M.: 1970, 'Vacuum Ultraviolet Emission from Auroras', *J. Geophys. Res.* **75**, 6209.
Pike, C. P.: 1971, 'A Latitudinal Survey of the Daytime Polar F Layer', *J. Geophys. Res.* **76**, 7745.
Pike, C. P. and Whalen, J. A.: 1974, 'Satellite Observations of Auroral Substorms', *J. Geophys. Res.* **79**, 985.
Pike, C. P., Whalen, J. A. and Buchau, J.: 1976, 'F Layer and 6300 Å Measurements in the Midnight Sector of the Auroral Oval', *J. Geophys. Res.* **81** (in press).
Prasad, S. S. and Green, A. E. S.: 1971, 'Ultraviolet Emissions from Atomic Nitrogen in the Aurora', *J. Geophys. Res.* **76**, 2419.
Prölss, G. W.: 1974, 'Lateral Dispersion of Low-Energy Protons Precipitated in the Auroral Zones', *Planet. Space Sci.* **22**, 341.
Rearwin, S.: 1971, 'Rocket Measurements of Low-Energy Auroral Electrons', *J. Geophys. Res.* **76**, 4505.
Rearwin, S. and Hones, E. W., Jr.: 1974, 'Near-Simultaneous Measurement of Low-Energy Electrons

by Sounding Rocket and Satellite', *J. Geophys. Res.* **79**, 4322.

Reasoner, D. L. and Chappell, C. R.: 1973, 'Twin Payload Observations of Incident and Backscattered Auroral Electrons', *J. Geophys. Res.* **78**, 2176.

Reasoner, D. L., Eather, R. H. and O'Brien, B. J.: 1968, 'Detection of Alpha Particles in Auroral Phenomena', *J. Geophys. Res.* **73**, 4185.

Rees, M. H.: 1969, 'Auroral Electrons', *Space Sci. Rev.* **10**, 413.

Rees, M. H. and Benedict, P. C.: 1970, 'Auroral Proton Oval', *J. Geophys. Res.* **75**, 4763.

Rees, M. H. and Jones, R. A.: 1973, 'Time Dependent Studies of the Aurora – II. Spectroscopic Morphology', *Planet. Space Sci.* **21**, 1213.

Rees, M. H. and Luckey, D.: 1974, 'Auroral Electron Energy Derived from Ratio of Spectroscopic Emissions, 1, Model Computations', *J. Geophys. Res.* **79**, 5181.

Rees, M. H. and Maeda, K.: 1973, 'Auroral Electron Spectra', *J. Geophys. Res.* **78**, 8391.

Rees, M. H., Stewart, A. I. and Walker, J. C. G.: 1969, 'Secondary Electrons in Aurora', *Planet. Space Sci.* **17**, 1997.

Reid, G. C. and Rees, M. H.: 1961, 'The Systematic Behavior of Hydrogen Emission in the Aurora II', *Planet. Space Sci.* **5**, 99.

Riedler, W.: 1972, 'Auroral Particle Precipitation Patterns', *Earth's Magnetospheric Processes*, B. M. McCormac (ed.), p. 133, D. Reidel Publ. Co., Dordrecht-Holland.

Riedler, W. and Borg, H.: 1972, 'High-Latitude Precipitation of Low-Energy Particles as Observed by ESRO IA', *Space Research XII*, 1397, Akademie-Verlag, Berlin.

Romick, G. J. and Brown, N. B.: 1971, 'Midday Auroral Observations in the Oval, Cusp Region, and Polar Cap', *J. Geophys. Res.* **76**, 8420.

Romick, G. J. and Sharp, R. D.: 1967, 'Simultaneous Measurements of an Incident Hydrogen Flux and the Resulting Hydrogen Balmer Alpha Emission in an Auroral Hydrogen Arc', *J. Geophys. Res.* **72**, 4791.

Romick, G. J., Belon, A. E. and Stringer, W. J.: 1974, 'Photometric Measurements of H-Beta in the Aurora', *Planet. Space Sci.* **22**, 725.

Rossberg, L.: 1971, 'Newly Observed Premidnight Asymmetry in the 40 keV Electron Flux Profile', *J. Geophys. Res.* **76**, 6980.

Royrvik, O.: 1976, 'Pulsating Aurora: Local and Global Morphology', Ph.D. Thesis, University of Alaska.

Russell, C. T., Chappell, C. R., Montgomery, M. D., Neugebauer, M. and Scarf, F. L.: 1971, 'O5 Observations of the Polar Cusp on November 1, 1968', *J. Geophys. Res.* **76**, 6743.

Rycroft, M. J. and Burnell, S. J.: 1970, 'Statistical Analysis of Movements of the Ionospheric Trough and the Plasmapause', *J. Geophys. Res.* **75**, 5600.

Rycroft, M. J. and Thomas, J. O.: 1970, 'The Magnetospheric Plasmapause and the Electron Density Trough at the Alouette I Orbit', *Planet. Space Sci.* **18**, 65.

Sandford, B. P.: 1964, 'Aurora and Airglow Intensity Variations with Time and Magnetic Activity at Southern High Latitudes', *J. Atmosph. Terr. Phys.*, **26**, 749.

Sandford, B. P.: 1967, 'High Latitude Night-Sky Emissions', *Aurora and Airglow*, B. M. McCormac (ed.), p. 443, Reinhold Publ. Co., New York.

Schunk, R. W., Raitt, W. J. and Banks, P. M.: 1975, 'Effect of Electric Fields on the Daytime High-Latitude E and F Regions', *J. Geophys. Res.* **80**, 3121.

Sharp, R. D. and Johnson, R. G.: 1968, 'Some Average Properties of Auroral Electron Precipitation as Determined from Satellite Observations', *J. Geophys. Res.* **73**, 969.

Sharp, R. D., Johnson, R. G., Shelley, E. G. and Harris, K. K.: 1974, 'Energetic O^+ Ions in the Magnetosphere', *J. Geophys. Res.* **79**, 1844.

Sharp, W. E. and Hays, P. B.: 1974, 'Low-Energy Auroral Electrons', *J. Geophys. Res.* **79**, 4319.

Shelley, E. G., Johnson, R. G. and Sharp, R. D.: 1972, 'Satellite Observations of Energetic Heavy Ions during a Geomagnetic Storm', *J. Geophys. Res.* **77**, 6104.

Shepherd, G. G. and Thirkettle, F. W.: 1973, 'Magnetospheric Dayside Cusp; A Topside View of Its 6300-Angström Atomic Oxygen Emission', *Science* **180**, 737.

Shepherd, G. G., Anger, C. D., Brace, L. H., Burrows, J. R., Heikkila, W. J., Hoffman, J., Maier, E. J. and Whitteker, J. H.: 1973, 'An Observation of Polar Aurora and Airglow from the ISIS-II Spacecraft', *Planet. Space Sci.* **21**, 819.

Snyder, A. L., Buchau, J. and Akasofu, S.-I.: 1972, 'Formation of Auroral Patches in the Midday Sector during a Substorm', *Planet. Space Sci.* **20**, 1116.

Snyder, A. L. Jr. and Akasofu, S.-I.: 1974, 'Major Auroral Substorm Features in the Dark Sector by a USAF DMSP Satellite', *Planet. Space Sci.* **22**, 1511.

Snyder, A. L. Jr. and Akasofu, S.-I.: 1976, 'Auroral Oval Photographs from the DMSP-8531 and 10533 Satellites', *J. Geophys. Res.* **81**, 1799.

Snyder, A. L., Akasofu, S.-I. and Davis, T. N.: 1974, 'Auroral Substorms Observed from above the North Polar Region by a Satellite', *J. Geophys. Res.* **79**, 1393.

Snyder, A. L., Akasofu, S.-I. and Kimball, D. S.: 1975, 'The Continuity of the Auroral Oval in the Afternoon Sector', *Planet. Space Sci.* **23**, 225.

Söraas, F.: 1972, 'ESRO IA/B Observations at High Latitudes of Trapped and Precipitating Protons with Energies above 100 keV', *Earth's Magnetospheric Processes*, B. M. McCormac (ed.), p. 121, D. Reidel Publ. Co., Dordrecht-Holland.

Söraas, F., Lindalen, H. R., Maseide, K., Egeland, A., Sten, T. A. and Evans, D. S.: 1974, 'Proton Precipitation and the H Emission in a Post-Breakup Auroral Glow', *J. Geophys. Res.* **79**, 1851.

Spiger, R. J. and Anderson, H. R.: 1975, 'Electron Currents Associated with an Auroral Band', *J. Geophys. Res.* **80**, 2161.

Stenbaek-Nielsen, H. C., Wescott, E. M., Davis, T. N. and Peterson, R. W.: 1973, 'Differences in Auroral Intensity at Conjugate Points', *J. Geophys. Res.* **78**, 659.

Stoffregen, W.: 1969, 'Transient Emissions on the Wavelength of Helium I, 5876 Å Recorded during Auroral Break-Up', *Planet. Space Sci.* **17**, 1927.

Stringer, W. J.: 1971, 'The Relationship of Auroral Hydrogen Emissions to Auroral Morphology', Ph.D. Thesis, University of Alaska, May.

Swider, W. and Narcisi, R. S.: 1974, 'Ion Composition in an IBC Class 11 Aurora, 2, Model', *J. Geophys. Res.* **79**, 2849.

Thomas, J. O. and Andrews, M. K.: 1969, 'The Trans-Polar Exospheric Plasma, 3: A Unified Picture', *Planet. Space Sci.* **17**, 433.

Tulinov, V. F., Zhuchenko, Yu. M., Lipovetskiy, V. A., Tulinov, G. F. and Feygin, V. M.: 1973, 'Soft Electron Fluxes in the Upper Atmosphere of the Polar Region', *Geomag. Aeronom.* **13**, 437.

Vasyliunas, V. M.: 1974, 'Summary of Magnetospheric Cleft Symposium', *EOS*, Feb.

Venkatarangan, P., Burrows, J. R. and McDiarmid, I. B.: 1975, 'On the Angular Distributions of Electrons in 'Inverted V' Substructures', *J. Geophys. Res.* **80**, 66.

Vij, K. K., Venkatesan, D. and Anger, C. D.: 1975, 'Investigation of Electron Precipitation during an Auroral Substorm by Rocket-Borne Detectors', *J. Geophys. Res.* **80**, 3205.

Viswanathan, K. S. and Rajappa, N.: 1969, 'A Note on Auroral Electrons', *Planet. Space Sci.* **17**, 1051.

Vondrak, R. R., Anderson, H. R. and Spiger, R. J.: 1971, 'Rocket-Based Measurement of Particle Fluxes and Currents in an Auroral Arc', *J. Geophys. Res.* **76**, 7701.

Wagner, R. A. and Pike, C. P.: 1971, 'A Discussion of Arctic Ionograms', Paper pres. at AGARD Technical Meeting in Lindau/Harz, Germany in Sept.

Wagner, R. A., Snyder, A. L. and Akasofu, S.-I.: 1973, 'The Structure of the Polar Ionosphere during Exceptionally Quiet Periods', *Planet. Space Sci.* **21**, 1911.

Walker, J. C. G. and Rees, M. H.: 1968, 'Ionospheric Electron Densities and Temperatures in Aurora', *Planet. Space Sci.* **16**, 459.

Watkins, B.: 1976, 'A Computer Model of the Polar F-Region Ionosphere', Ph.D. Thesis, University of Alaska.

Wax, R. L. and Bernstein, W.: 1970, 'Rocket-Borne Measurements of Hβ Emissions and Energetic Hydrogen Fluxes during an Auroral Breakup', *J. Geophys. Res.* **75**, 783.

Westerlund, L. H.: 1969, 'The Auroral Electron Energy Spectrum Extended to 45 eV', *J. Geophys. Res.* **74**, 351.

Whalen, B. A. and McDiarmid, I. B.: 1968, 'Direct Measurement of Auroral Alpha Particles', *J. Geophys. Res.* **73**, 2307.

Whalen, B. A. and McDiarmid, I. B.: 1972a, 'Further Low-Energy Auroral-Ion Composition Measurements', *J. Geophys. Res.* **77**, 1306.

Whalen, B. A. and McDiarmid, J. B.: 1972b, 'Observations of Magnetic-Field-Aligned Auroral Electron Precipitation', *J. Geophys. Res.* **77**, 191.

Whalen, B. A., Miller, J. R. and McDiarmid, I. B.: 1971a, 'Energetic Particle Measurements in a Pulsating Aurora', *J. Geophys. Res.* **76**, 978.

Whalen, B. A., Miller, J. R. and McDiarmid, I. B.: 1971b, 'Sounding Rocket Observations of Particle Precipitation in a Polar-Cap Electron Aurora', *J. Geophys. Res.* **76**, 6847.

Whalen, B. A., Green, D. W. and McDiarmid, I. B.: 1974, 'Observations of Ionospheric Ion Flow and Related Convective Electric Fields in and near an Auroral Arc', *J. Geophys. Res.* **79**, 2835.

Whalen, J. A., Buchau, J. and Wagner, R. A.: 1971, 'Airborne Ionospheric and Optical Measurements of Noontime Aurora', *J. Atmosph. Terr. Phys.* **33**, 661.

Whalen, J. A. and Pike, C. P.: 1973, 'F-Layer and 6300-A Measurements in the Day Sector of the Auroral Oval', *J. Geophys. Res.* **78**, 3848.

Wiens, R. H. and Jones, Vallance A.: 1969, 'Studies of Auroral Hydrogen Emissions in West-Central Canada, 3. Proton and Electron and Electron Auroral Ovals', *Can. J. Phys.* **47**, 1493.

Winningham, J. D.: 1972, 'Characteristics of Magnetosheath Plasma Observed at Low Altitudes in the Dayside Magnetospheric Cusps', *Earth's Magnetospheric Processes*, B. M. McCormac (ed.), p. 68, D. Reidel Publ. Co., Dordrecht-Holland.

Winningham, J. D. and Heikkila, W. J.: 1974, 'Polar Cap Auroral Electron Fluxes Observed with ISIS-1', *J. Geophys. Res.* **79**, 949.

Winningham, J. D., Akasofu, S.-I., Yasuhara, F. and Heikkila, W. J.: 1973, 'Simultaneous Observations of Auroras from the South Pole Station and of Precipitating Electrons by ISIS-1', *J. Geophys. Res.* **78**, 6579.

Zaytseva, S. A., Starkov, G. V. and Shevnina, N. F.: 1973, 'Orientation of Auroral Arcs', *Geomag. Aeronom.* **13**, 579.

Zwick, H. H. and Shepherd, G. G.: 1973, 'Upper Atmospheric Temperatures from Doppler Line Widths – V. Auroral Electron Energy Spectra and Fluxes Deduced from the 5577 and 6300 Å Atomic Oxygen Emissions', *Planet. Space Sci.* **21**, 605.

CHAPTER 3

DISTRIBUTION OF PLASMAS IN THE MAGNETOSPHERE

The purpose of this chapter is to review briefly present knowledge on the distribution of plasmas in the magnetosphere and to attempt to find the link between magnetospheric plasmas and auroral particles. The distribution of auroral particles was discussed in the previous chapter. In this chapter, we shall examine, first, plasma characteristics in five domains: the plasma mantle, the polar cusp, the plasma sheet, the Van Allen belt (including the ring current belt) and the plasmasphere. In the second part we examine how plasma particles in different domains contribute to different auroral regions. The last part of this chapter is devoted to a detailed discussion of acceleration processes of auroral particles.

3.1. Five Plasma Domains

An intensive survey of the magnetosphere during the last decade has revealed that it consists of several plasma domains which are characterized by plasmas of different energy characteristics. Figure 3.1 shows the present view of the distribution of plasmas in the magnetosphere.

The streaming plasma of magnetosheath origin just inside the magnetopause was first convincingly shown by Akasofu *et al.* (1973) who named the domain occupied by this particular plasma 'the boundary layer'. Their finding was soon confirmed and extended by Rosenbauer *et al.* (1975) who named the domain 'the plasma mantle'. In this book, we adopt the latter name for the regions which are occupied by the streaming magnetosheath-like plasma. The discovery of this plasma domain has an important implication in understanding the origin of plasma in the plasma sheet. It should also be noted that the plasma mantle coincides with the layer of energetic electrons (>40 keV) discovered by Meng and Anderson (1970, 1975).

The presence of the plasma sheet, a sheet-like distribution of plasma, centered around the midplane of the magnetotail and called the 'neutral sheet', was discovered by Bame *et al.* (1967), and has been studied by a number of workers (Bame, 1968; Gringauz, 1969; Frank, 1971b, c; Nishida and Lyon, 1972).

There is little doubt that plasma particles from the inner or central plasma sheet (CPS) contribute significantly in exciting the diffuse auroral luminosity, after they are precipitated into the ionosphere by various plasma processes. Plasma particles in the upper (in the northern hemisphere) and the lower (in the southern hemisphere) boundary layers of the plasma sheet (BPS) appear to be responsible for the background luminosity in the poleward half of the broad oval

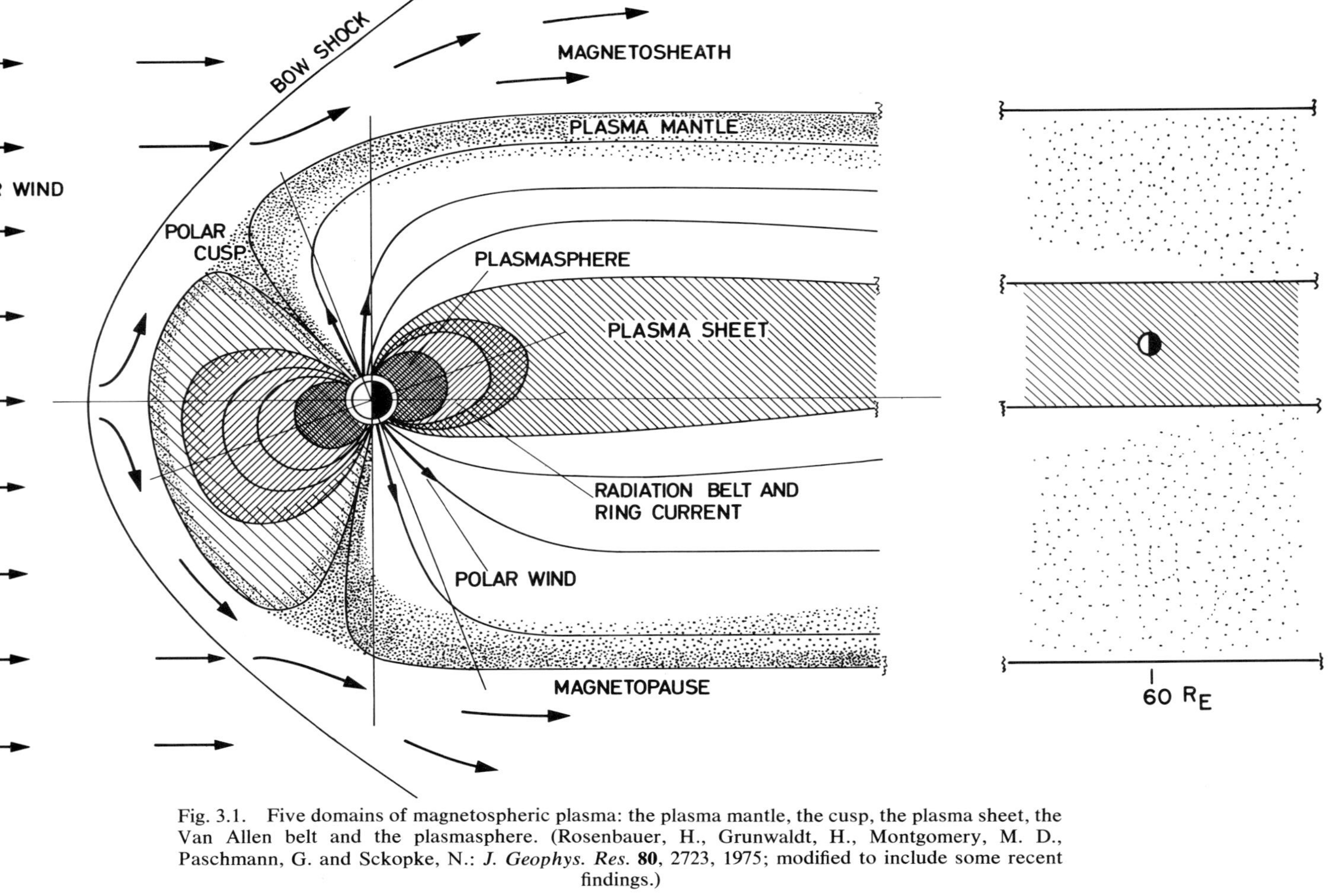

Fig. 3.1. Five domains of magnetospheric plasma: the plasma mantle, the cusp, the plasma sheet, the Van Allen belt and the plasmasphere. (Rosenbauer, H., Grunwaldt, H., Montgomery, M. D., Paschmann, G. and Sckopke, N.: *J. Geophys. Res.* **80**, 2723, 1975; modified to include some recent findings.)

band (Section 2.10). In these boundary layers, a particular mechanism develops and accelerates electrons toward the Earth. The accelerated electrons cause discrete auroral arcs. The thinness of discrete arcs (~ 500 m) suggests that the acceleration mechanism must occur in narrowly confined regions which are embedded in the boundary layers.

The presence of the ring current belt had long been inferred from studies of geomagnetic storms. The ring current protons were first identified by Frank (1967). DeForest and McIlwain (1971) showed most convincingly that they are injected from the plasma sheet. This belt coincides also with the Van Allen outer belt, where energetic electrons are injected from the plasma sheet and are then subsequently trapped. It has been suggested that both the protons and electrons from the Van Allen belt contribute significantly to auroral luminosity in the diffuse annular belt (Section 2.1).

The presence of plasma in the exosphere was inferred by Storey (1953) on the basis of the propagation characteristics of whistlers. Carpenter (1963) showed that this plasma region is rather sharply bounded. This particular plasma domain is called the plasmasphere, and the plasma distribution has since then been directly observed by satellites (Taylor *et al.*, 1968; Chappell *et al.*, 1970). It has been suggested that the mid-latitude red arc results from an interaction of plasmas in the plasmasphere and in the ring current belt.

3.2. Plasma Mantle

Magnetosheath-like plasmas have recently been found at several regions in the magnetosphere, and the regions occupied by them have different names: the boundary layer, the plasma mantle, the lobe plasma, etc.; see Figure 3.2. As

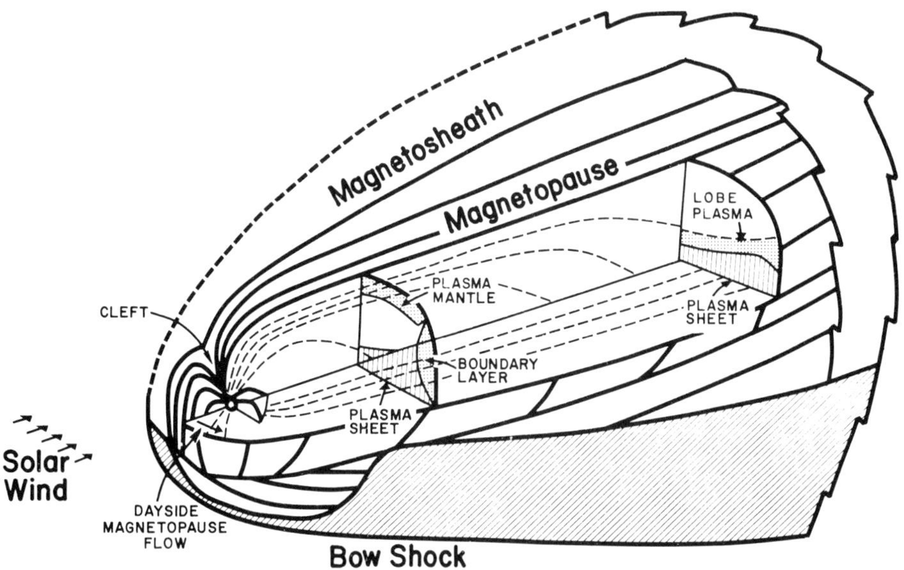

Fig. 3.2. Three regions of the plasma mantle, together with the cleft (cusp) and the plasma sheet. (Courtesy of Freeman, J. W., Hill, H. K. and Hardy, D. A.)

mentioned earlier, these plasma regions are altogether called the plasma mantle. Figure 3.2 indicates their general locations. Their common characteristics are that they have magnetosheath-like energy spectra and flow in the anti-solar direction. In this section, we examine plasma characteristics in different regions of the plasma mantle.

3.2.1. FRONTAL REGION OF THE MAGNETOSPHERE

The plasma mantle in the frontal portion of the magnetosphere was extensively studied by Rosenbauer *et al.* (1975) and Haerendel and Paschmann (1975). In Figure 3.3 the observed locations of the plasma mantle along satellite (HEOS) trajectories are projected onto the X-Z plane. The plasma mantle forms a thick (0.5–4 R_E) layer just inside the magnetopause. They found also that the plasma flows in the anti-solar direction with a speed of about 100–200 km s^{-1}, and this speed is always less than the concurrent flow speed in the nearby magnetosheath. Figure 3.4 shows an example of contours for the proton velocity distribution

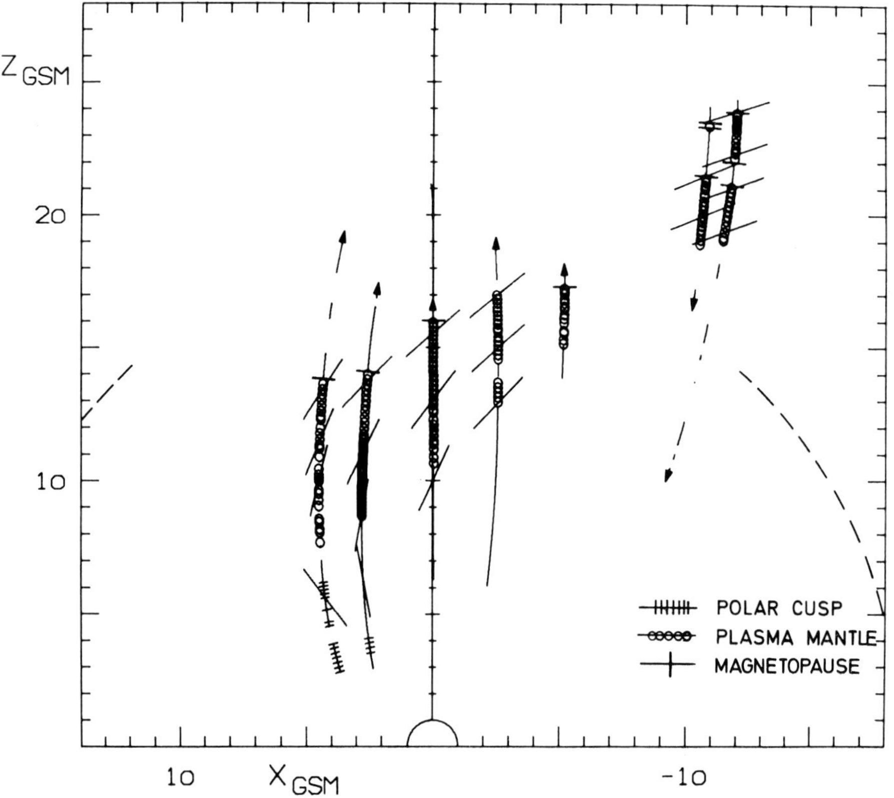

Fig. 3.3. Geometry of the plasma mantle in the noon-midnight meridian. Parts of HEOS-2 satellite orbits along which the plasma mantle was traversed are shown. (Rosenbauer, H., Grunwaldt, H., Montgomery, M. D., Paschmann, G. and Sckopke, N.: *J. Geophys. Res.* **80**, 2723, 1975.)

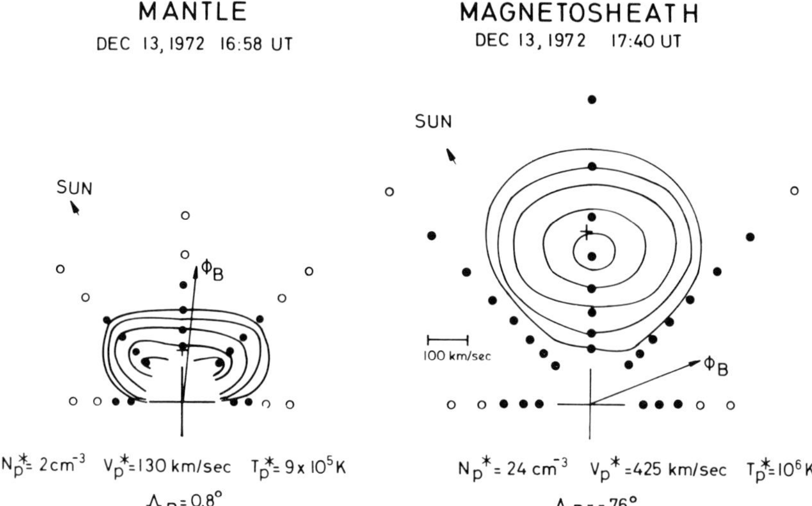

Fig. 3.4. Contours of the proton velocity distribution function, in the spacecraft equatorial plane, in the mantle and magnetosheath. The direction of the magnetic field projected onto the spacecraft equatorial plane is shown by the arrow marked Φ_B; its inclination with respect to the plane is given by Λ_B. (Rosenbauer, H., Grunwaldt, H., Montgomery, M. D., Paschmann, G. and Sckopke, N.: *J. Geophys. Res.* **80**, 2723, 1975.)

function (lines of constant density in velocity space), together with the direction of the local magnetic field and of the Sun, for both the mantle plasma and the magnetosheath plasma. The plasma parameters for both locations are also given.

3.2.2. MAGNETOTAIL ($X > -30\ R_E$)

Figure 3.5 shows parts of the IMP-6 trajectories (the X-Y plane projection) along which the maximum counting rates of protons occurred in the energy range from a little less than 140 eV to a little more than 1.68 keV for all Kp values. The numbers 1, 2, 3 and 4 refer to different channels for the above energy range, and each number represents one of the energy ranges which indicated the maximum counting rate for an hour interval along a satellite trajectory. The highest energy for the entire energy range corresponds to about the mean energy of solar wind protons. Thus, this energy range is most suitable in detecting the plasma mantle. Indeed, it can be seen that protons of this particular energy range are distributed approximately along the magnetopause, delineating the plasma mantle along the dayside magnetopause, and also along the dawn-dusk magnetopause. The presence of the mantle is more apparent during quiet periods than during disturbed periods.

Figure 3.6 shows an example of traverse of the magnetosheath and the plasma mantle by the Vela satellite at a geocentric distance of 18 R_E. Three boxes in the left side of the figure show, from the top, the proton density, temperature and bulk speed. Four boxes on the right-hand side show four samples of proton data, taken in the magnetosheath (a), the transition region from the magnetosheath to the

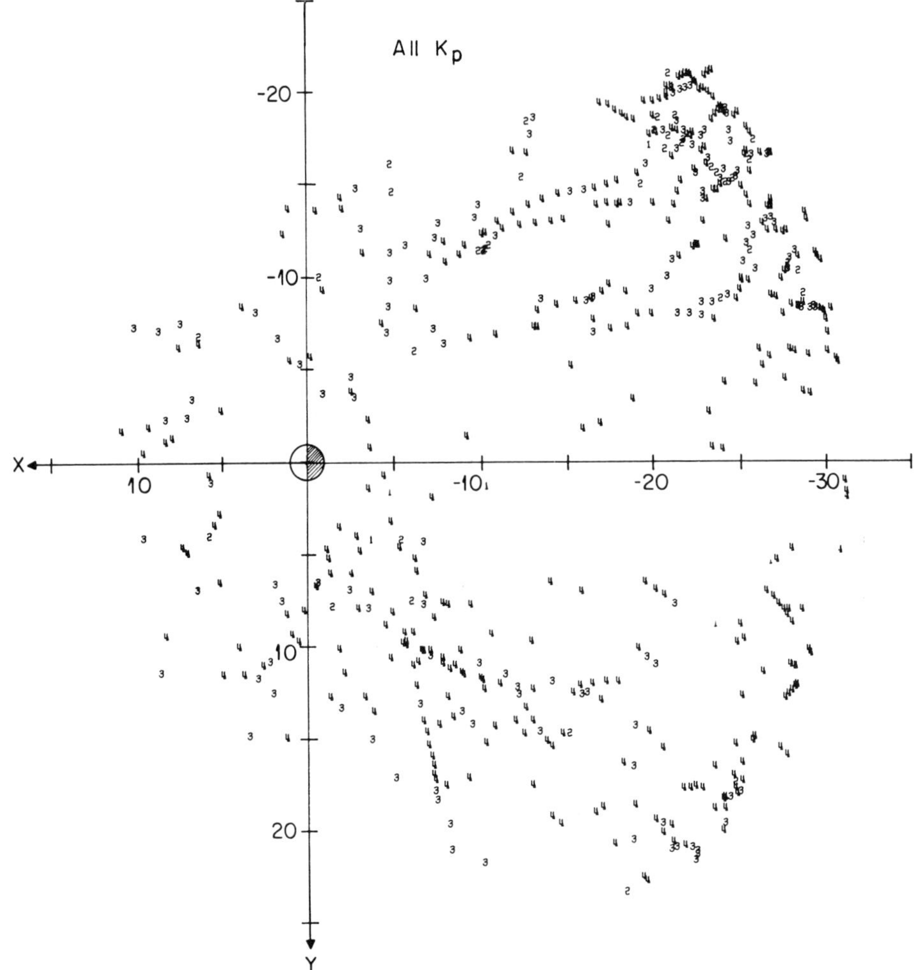

Fig. 3.5. Parts of the IMP-6 trajectories (the X-Y plane projection) along which the maximum counting rate of protons occurred in the energy range from a little less than 140 eV to a little more than 1.68 keV for all Kp values. (Courtesy of Hones, E. W. Jr., Akasofu, S.-I., Lui, A. T. Y. and Bame, S. J.)

plasma mantle (b), and in the plasma mantle (c, d), respectively. In the magnetosheath, one can see a large spin modulation of the counting rate, indicating a rapid flow of plasma (~ 300 km s^{-1}) in the anti-sunward direction. As the satellite crossed the magnetopause, the rate of spin modulation decreased sharply, but the modulation remained long after the crossing, indicating the presence of mantle plasma.

The Vela 5 and 6 satellite observations suggest that the mantle plasma tends to occupy four corners in the so-called 'Vela sphere' (since the trajectories of Vela satellites lie on a surface of sphere of radius 18 R_E, namely, the four regions of $|dZ_{SM}| \geq 5\ R_E$ and angular distances from midnight $\geq 50°$, namely, $\Phi_{SM} < 130°$ and $\Phi_{SM} > 230°$). It should be noted that Vela satellite trajectories do not cover the

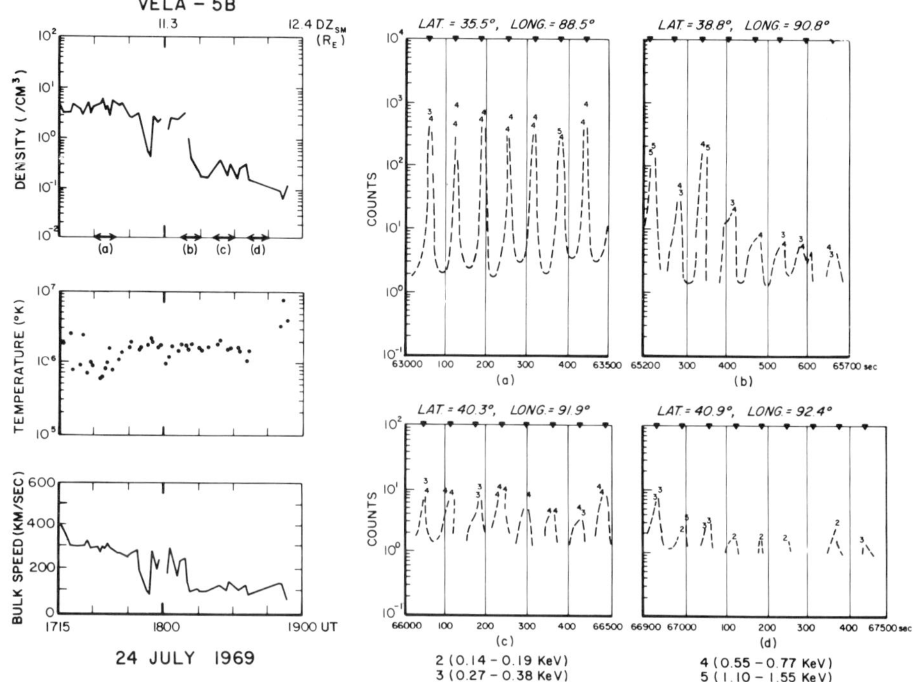

Fig. 3.6. Plasma mantle traversed by the Vela-5 B satellite from the magnetosheath on the duskside of the magnetotail on 1969, July 24. Four short periods, (a), (b), (c) and (d), are chosen to show how the spin modulation of counting rate changed during the traverse. The energy range for each channel indicated at the peak of the spin modulation is given at the bottom. (Akasofu, S.-I., Hones, E. W. Jr., Bame, S., Asbridge, J. R. and Lui, A. T. Y.: *J. Geophys. Res.* **78**, 7257, 1973.)

entire sphere of radius 18 R_E, so it is quite likely that there is mantle plasma in the upper and lower high latitude lobe regions; see Figure 3.2.

3.2.3. MAGNETOTAIL AT THE LUNAR DISTANCE ($X \sim -60\ R_E$)

Hardy *et al.* (1975) showed that at the lunar distance the mantle plasma is found over the entire range in Φ_{SM}, rather than only in the four corners at the Vela satellite distance, namely over the entire high latitude lobe just above (in the northern hemisphere) and below (in the southern hemisphere) the plasma sheet; see Figure 3.2. However, it should be noted again that latitudinal coverage by a detector on the lunar surface is very limited. It may well be that at the lunar distance or perhaps a little beyond, the mantle plasma flows in the entire high latitude lobe region (see Figure 3.1). In Section 3.2.4 we shall see that at $X \sim 1000\ R_E$ plasma particles are streaming in the anti-solar direction in the entire magnetotail. Spectral characteristics of plasmas in the plasma mantle and plasma sheet at the lunar distance ($\sim 60\ R_E$) are shown in Figure 3.7.

Hardy *et al.* (1975) found also that there is an interesting correlation between the B_y component of the interplanetary magnetic field (IMF) and quadrants in the

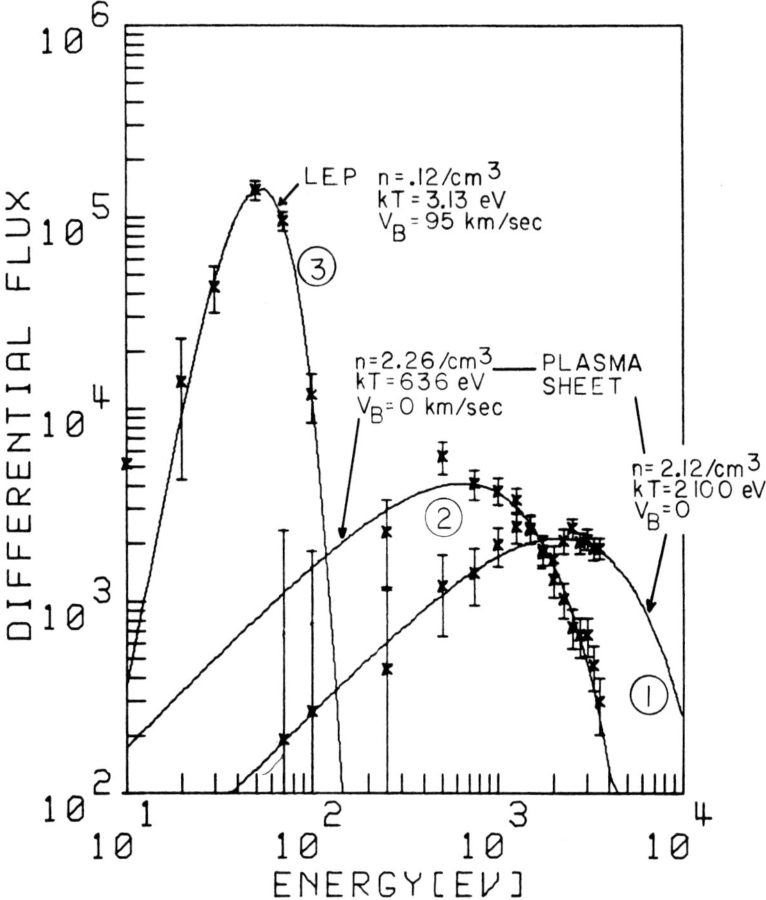

Fig. 3.7. Comparative differential flux spectra for plasma in the plasma mantle (LEP) and the plasma sheet at the lunar distance. (Hardy, D. A., Hills, H. K. and Freeman, J. W.: *Geophys. Res. Lett.* **2**, 169, 1975.)

tail cross-section (viewing toward the Sun), in which the mantle plasma is found. The mantle plasma is found very frequently in the quadrants II and IV when the B_y component is negative and in the quadrants I and III when the B_y component is positive. They suggested that this is because the IMF lines are connected to the northern afternoon and the southern morning sectors when the B_y component is negative and to the northern morning and southern afternoon sectors when the B_y component is positive. They suggested also that magnetosheath plasma enters from the dayside magnetopause through the merging region and then is convected downward in the northern hemisphere and upward in the southern hemisphere, namely toward the plasma sheet in both hemispheres, as it flows downstream in the anti-solar direction. The required electric field (dawn-to-dusk) is of order 93 to 223 kV which is a little greater than the generally accepted value of 40 to 80 kV (on the basis of the measurement across the polar cap; see Section 1.3.4).

3.2.4. DISTANT MAGNETOTAIL ($X \sim -500 \sim -1000\ R_E$)

The structure of the magnetotail at 500–1000 R_E has been a matter of dispute for many years. Does the magnetosphere maintain a cohesive structure at such a great distance? Or, is it filamentary or turbulent? The focal point of discussion is how to interpret plasma and magnetic field data obtained by the Pioneer 7 space probe at 900–1000 R_E downstream of solar wind, since the tail-like condition was observed intermittently for about 6 days (Ness *et al.*, 1967; Wolfe *et al.*, 1967; Fairfield, 1968; Intriligator *et al.*, 1969; Walker *et al.*, 1975; Villante, 1975; Villante and Lazarus, 1975).

According to the last authors, the magnetotail can most easily be identified by a low density region, with a large variability, with the magnetic field reversal from radial to anti-radial by a vector rotation in the north-south plane. The field vector has the northward component at the time of minimum field magnitude in 8 of 11 examples. Figure 3.8 shows an example of the midplane crossing. Note that all the

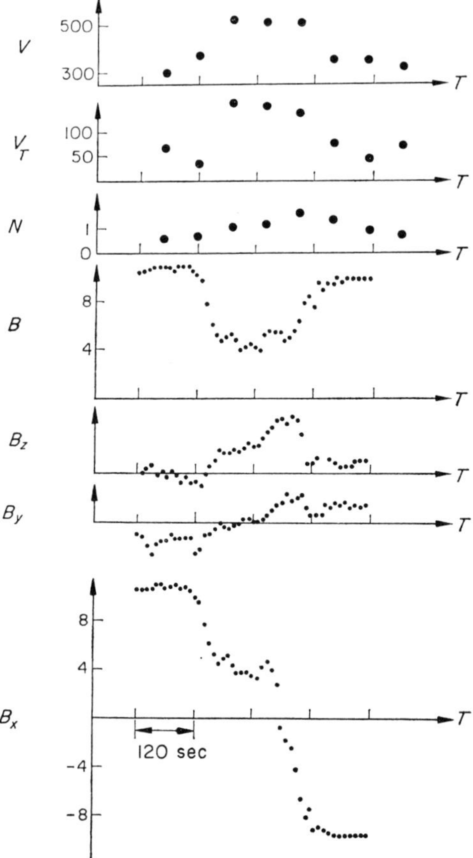

Fig. 3.8. Plasma parameters and the magnetic field across the midplane of the magnetotail at $X \simeq -1000\ R_E$. (Villante, O: *Planet. Space Sci.* **23**, 723, 1975.)

parameters vary rather systematically during the crossing and that the plasma particles are streaming in the anti-solar direction *throughout* the magnetotail.

Walker *et al.* (1975) noted that variations in the flow direction of the solar wind are sufficient to explain the intermittent observations of tail-like plasma and that the width of the tail is of order 60–90 R_E. Within the tail, the velocity of plasma flow is usually within 100 km s^{-1} of the magnetosheath velocity. Thus, they concluded: "many of the observations are understandable through a relatively simple description of the plasma, and we feel that a characterization of the tail as being 'filamentary' or 'turbulent' implies a more fragmented structure than is shown by our observations."

3.3. Polar Cusp

Heikkila and Winningham (1971) and Frank (1971b, c) found that magnetosheath-like plasma is present in a rather limited region in the dayside magnetosphere. The energy spectra of both protons and electrons are remarkably similar to those in the magnetosheath. Figure 3.9 shows an example of such a comparison. It can be seen that the spectra of electrons in the magnetosheath, plasma mantle and cusp are essentially identical. Frank (1971b, c) inferred that there is a funnel-like region centered around the Chapman-Ferraro neutral point which is connected to the dayside ionosphere, so that magnetosheath plasma can enter the magnetosphere through it. Heikkila (1972) proposed that magnetosheath plasma enters the magnetosphere in the form of a wedge. He called this particular region 'cleft'.

McDiarmid *et al.* (1972) showed, however, that the cusp plasma is most often observed in the closed, rather than open, field line region. They found that energetic electrons in the cusp region have a pancake-like pitch-angle distribution peaking at 90°. Further, they showed that solar electrons do not penetrate into the cusp region. These findings suggest that magnetosheath plasma can enter the magnetosphere across the magnetopause.

A variety of plasma instabilities appears to take place in the cusp region. Russell *et al.* (1971), Scarf *et al.* (1972) and Fredricks *et al.* (1973) observed intense ULF magnetic field fluctuations and VLF electric field noise which they interpreted in terms of drift instabilities and associated wave-particles and wave-wave interactions. D'Angelo *et al.* (1974) observed also intense magnetic field fluctuations which they interpreted in terms of the Kelvin-Helmholtz instability.

3.4. Plasma Sheet

The origin of plasma in the plasma sheet has long been discussed only in terms of the average energy characteristics (*cf.* Hill, 1974). There has been little study of day-to-day changes of its energy characteristics. Thus in the first subsection the plasma sheet will be examined during a prolonged quiet condition. This observation shows how quickly plasma sheet particles are replenished. Another important feature of the plasma is that the average energy of protons and electrons depends considerably on locations in the magnetosphere. We shall deal with this subject in

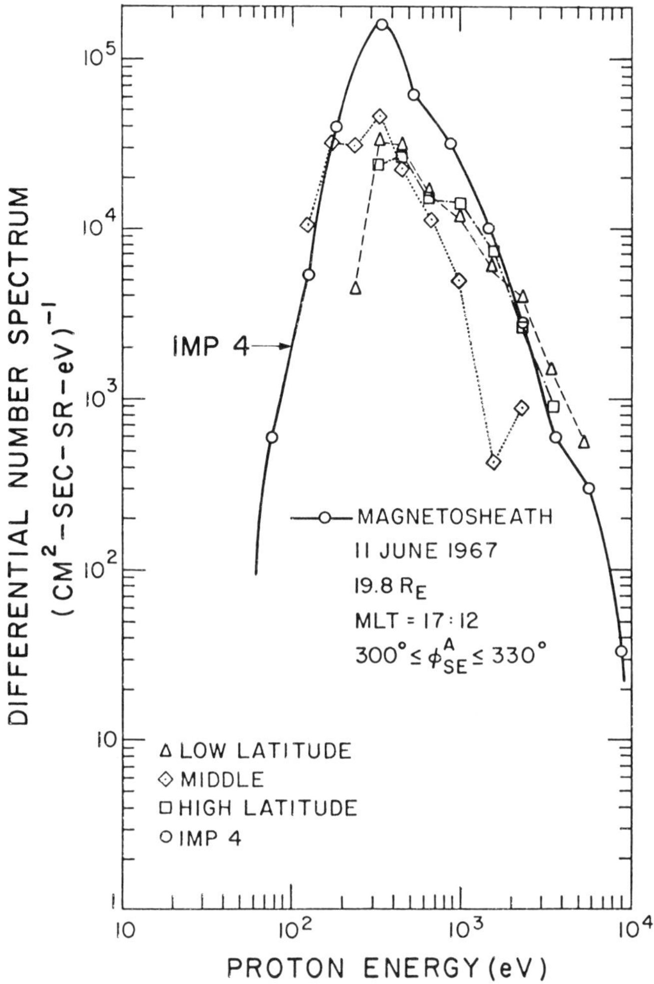

Fig. 3.9. Absolute comparison of proton spectra obtained in the magnetosheath (IMP-4) and in the cusp region (ISIS-1). (Heikkila, W. J. and Winningham, J. D.: *J. Geophys. Res.* **76**, 883, 1971.)

the second subsection. The third subsection deals with equilibrium and stability conditions of the plasma sheet.

3.4.1. PLASMA SHEET DURING VERY QUIET PERIODS

A very soft magnetosheath-like proton spectrum in the plasma sheet has been observed by Vela satellites from time to time (*cf.* Hones *et al.*, 1971). However, it was only recently that the occurrence of such a soft spectrum was systematically examined. Figure 3.10 shows parts of the trajectories of the Vela satellites along which soft spectra were observed. One of the softest spectra was measured between 1112 UT and 1220 UT, on 1965, October 17 at $dZ_{SM} \sim 0.2\, R_E$, and was

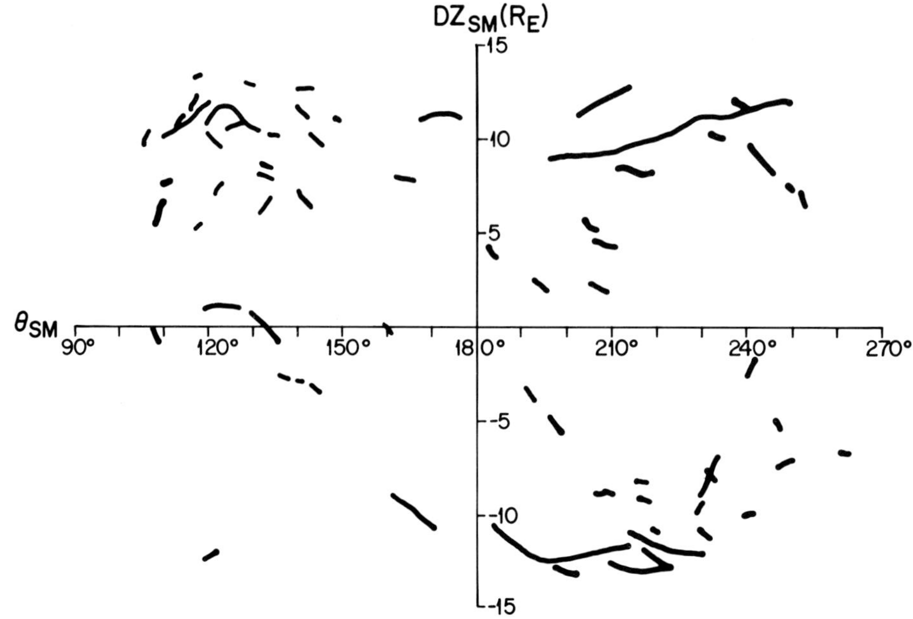

Fig. 3.10. Map showing parts of Vela orbits through the magnetotail, along which the detectors observed nonstreaming, magnetosheath-like protons. (Akasofu, S.-I., Hones, E. W. Jr., Bame, S. J., Asbridge, J. R. and Lui, A. T. Y.: *J. Geophys. Res.* **78**, 7257, 1973.)

reported by Hones *et al.* (1971); this event occurred after a very prolonged quiet period, as indicated by the following list of Kp values:

October 14	$2_0, 3-, 1_0, 2+, 1+, 1_0, 0_0, 1_0$
October 15	$1-, 1-, 0+, 1-, 0_0, 0_0, 0+, 0_0$
October 16	$0_0, 0_0, 1-, 0+, 0+, 0+, 1-, 1-$
October 17	$0_0, 0_0, 0_0,$ *1 −*, *1 +*

The italicized values of Kp are those for the period during which the proton measurements were made. Similar protons were observed also on the following days:

(a) 1969	October 31	$1-, 2_0, 2_0, 1_0, 0+, 2-$
	November 1	$1_0, 1-, 0+,$ *0 +*
(b) 1970	May 9	$2-, 1_0, 1-, 0+, 1-, 1_0, 1-, 1-$
	May 10	$0_0, 1-, 0+, 1-, 0+, 0+,$ *0 +*, $0+$
(c) 1970	June 22	$1+, 0+, 0+,$ *1_0*, *1_0*
(d) 1970	January 25	$1_0, 1-, 0+, 0_0, 0+, 0+,$ *1 −*, *1_0*

A similar softening of the proton spectra was observed frequently by the IMP-6 satellite; at times the proton energy became as low as that of magnetosheath protons even in the inner part of the plasma sheet. Hardy *et al.* (1975) noted also that a 'cold plasma' appears at the lunar distance during quiet periods.

These observations suggest that after a prolonged (40–70 h) quiet period, the energy of protons in the plasma sheet becomes less than 1 keV even in the central part ($|Z_{SM}| < 1\ R_E$) of the sheet. This fact implies that even during a quiet period the plasma sheet particles are replenished by magnetosheath plasma. Further, since the soft spectra are observed more frequently at greater Z_{SM} values (namely in the upper and the lower boundary layers of the plasma sheet), the replenishment process is likely to begin first in the upper and lower boundary regions and perhaps also first at great distances and then gradually at smaller distances.

On the other hand, a single substorm seems to be able to energize such cool plasma particles to a very high energy (~ 20 keV). The energized plasma is seen during the recovery phase of substorms (Section 6.6). During such a period, the entire plasma in the plasma sheet is energized, and the average energy of protons often becomes greater than 25 keV even in the upper or lower boundary of the plasma sheet.

Although the energization of plasma sheet particles during substorms will be discussed in detail in Section 6.6, it is instructive to see a very sudden hardening of the energy spectrum of plasma sheet protons even during a very weak substorm. On 1970, February 20, the Vela 5A satellite was moving upward through the magnetotail near local midnight between 0600 UT and 1200 UT (Figure 3.11). It encountered a relatively low, varying flux of only moderately energetic protons from 0600 UT to $\sim$ 1015 UT. From $\sim$ 0940 UT the particle flux decreased until the

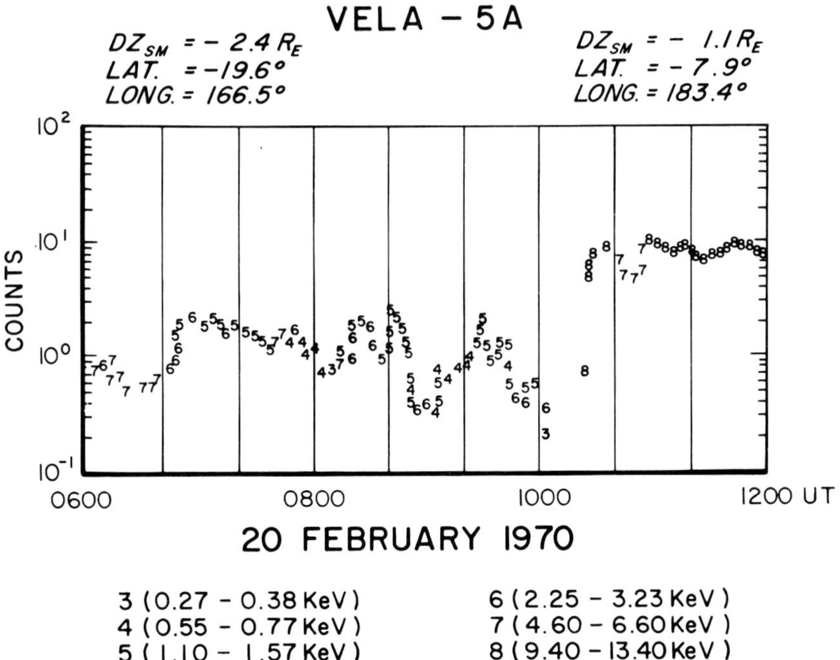

Fig. 3.11. Sudden energization of protons in the plasma sheet during a very weak substorm. (Akasofu, S.-I., Hones, E. W. Jr., Bame, S., Asbridge, J. R. and Lui, A. T. Y.: *J. Geophys. Res.* **78**, 7257, 1973.)

time when a short data gap started at 1015 UT; at that time the proton flux had dropped below detectable intensities. But the electron spectrum was measurable and was very similar to the tail lobe electron spectra. Then, the proton flux increased suddenly at ~ 1024 UT, and a relatively hard electron spectrum (peak flux at $\varepsilon_e \sim 1$ keV) was also encountered. The dropout of plasma starting at ~ 0940 UT and the sudden reappearance of a much hotter plasma closely resemble the typical behavior of the plasma sheet during substorms (Sections 6.2 and 6.6). Indeed, the Fort Yukon all-sky camera showed that auroras became slightly active near the northern horizon at about 0950 UT; thus, a weak substorm or substorm-like activity occurred along a very contracted oval at dp. lat. 72° in the midnight sector. Since such a weak substorm can considerably energize plasma sheet particles, it is difficult to see the replenishing process during a moderately active period.

Except during a prolonged quiet period and during substorms, however, plasma sheet particles are hottest near the neutral sheet, and there is an indication, at least statistically, of a *gradual but substantial* cooling toward the upper and lower boundaries.

It is interesting to note in this connection that the average energy of electrons is substantially softer in the upper and lower boundary regions of the plasma sheet than in the region surrounding the midplane of the plasma sheet (Hones, 1968). Figure 3.12 shows the average energy of electrons at different parts of the Vela sphere. This result has an important implication in identifying plasma domains which are responsible for supplying auroral particles (Section 3.8.1).

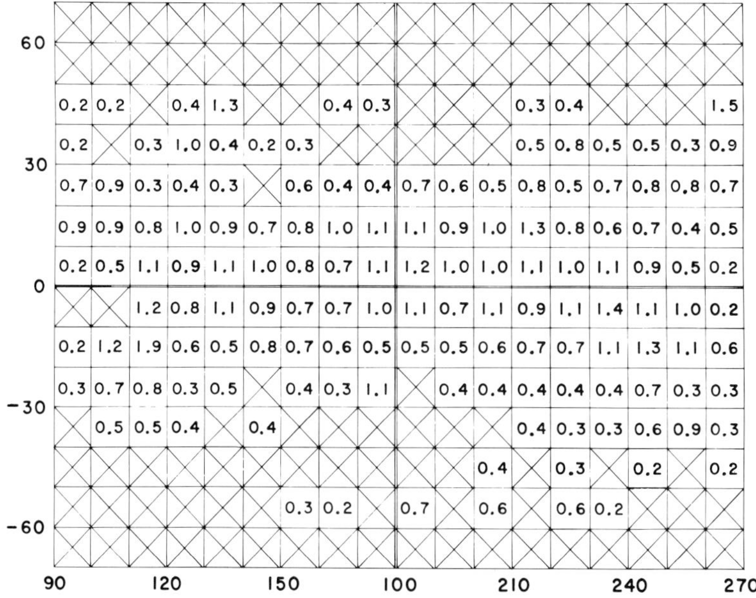

Fig. 3.12. Average energy of electrons at different regions of the plasma sheet at $X \simeq -18\ R_E$ (Hones, E. W. Jr.: *Physics of the Magnetosphere*, R. L. Corovillano, J. F. McClay and H. R. Radoski (eds.), p. 392, D. Reidel Publ. Co., Dordrecht-Holland, 1968.)

3.4.2. AVERAGE ENERGY CHARACTERISTICS

(a) $-10\, R_E > X > -30\, R_E$

The average energy of protons in the plasma sheet is known to be about 6 keV. Figure 3.13 shows the distribution of the energy density of plasma electrons (100 eV–18 keV), in units of eV cm^{-3} $ster^{-1}$, mapped in solar magnetospheric coordinates on a rectangular projection of the back of the Vela sphere, i.e. the spherical surface of $r \sim 18\, R_E$ on which the Vela satellites move. Plasma sheet protons (100 eV–18 keV) typically have energies ~6 times as great as simultaneously observed plasma electrons. Thus, Figure 3.13 indicates also the energy density distribution of protons if the numbers shown are multiplied by ~6 (Hones, 1972).

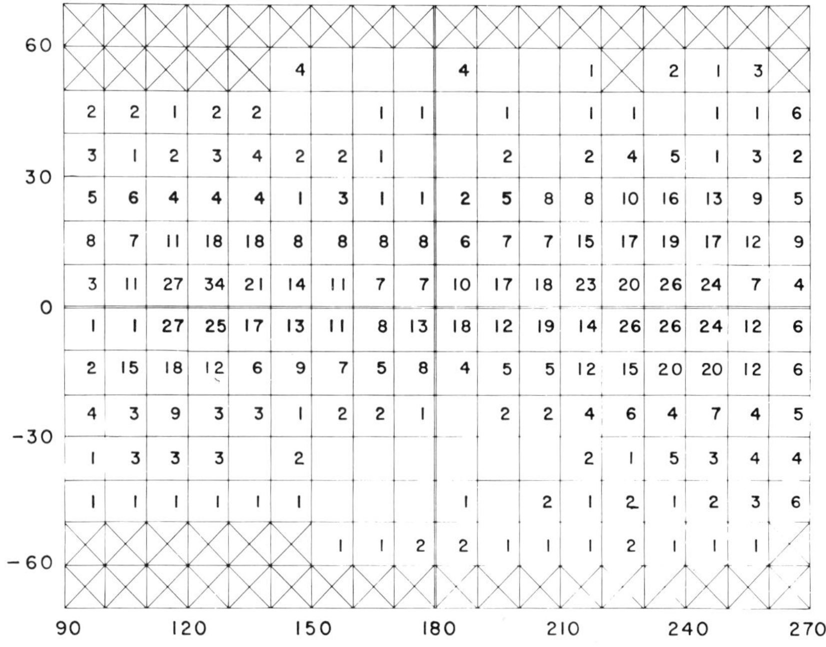

Fig. 3.13. Distribution of the energy density of plasma electrons (100 eV–18 keV), in units of eV cm^{-3} $ster^{-1}$, mapped in solar magnetospheric coordinates on a rectangular projection of the back of the Vela sphere. (Hones, E. W. Jr.: *Solar-Terrestrial Relations Conference*, August 28–September 1, 1972, University of Calgary, Calgary, Canada.)

(b) *Plasma Sheet at the Lunar Distance* ($X \sim -60\, R_E$)

Plasma characteristics in the plasma sheet at the lunar distance have been studied by a number of workers (Nishida and Lyon, 1972; Burke and Reasoner, 1973; Rich *et al.*, 1973). Nishida and Lyon (1972) showed that the cross-sectional shape (in the Y-Z of plane) of the plasma sheet at the lunar distance is similar to that at 18 R_E. Rich *et al.* (1973) showed that the thickness of the plasma sheet is of order 5 R_E, and the typical plasma parameters are $n = 0.10 + 0.05\ cm^{-3}$, $\kappa T_e = 200 \pm 50$ eV, $\kappa T_i = 2.5 \pm 0.75$ keV.

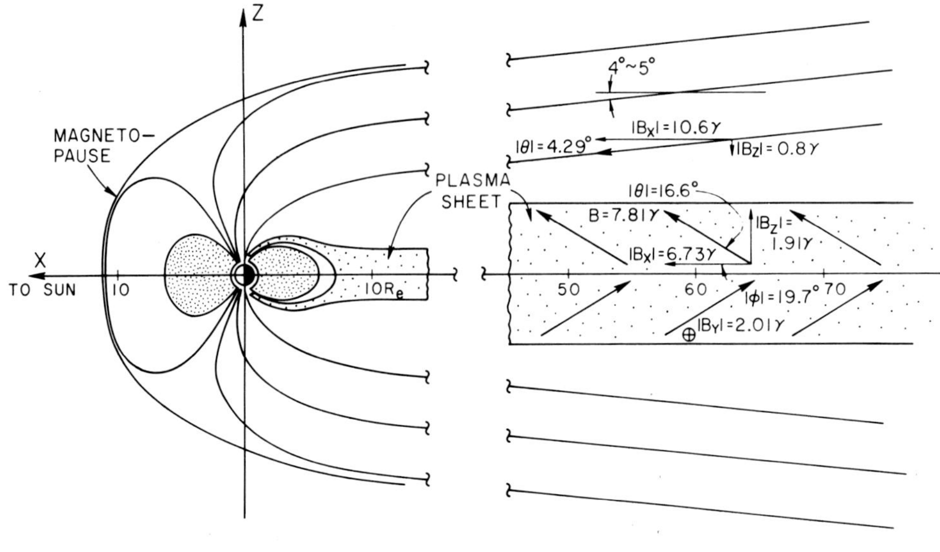

CROSS-SECTION X-Z PLANE

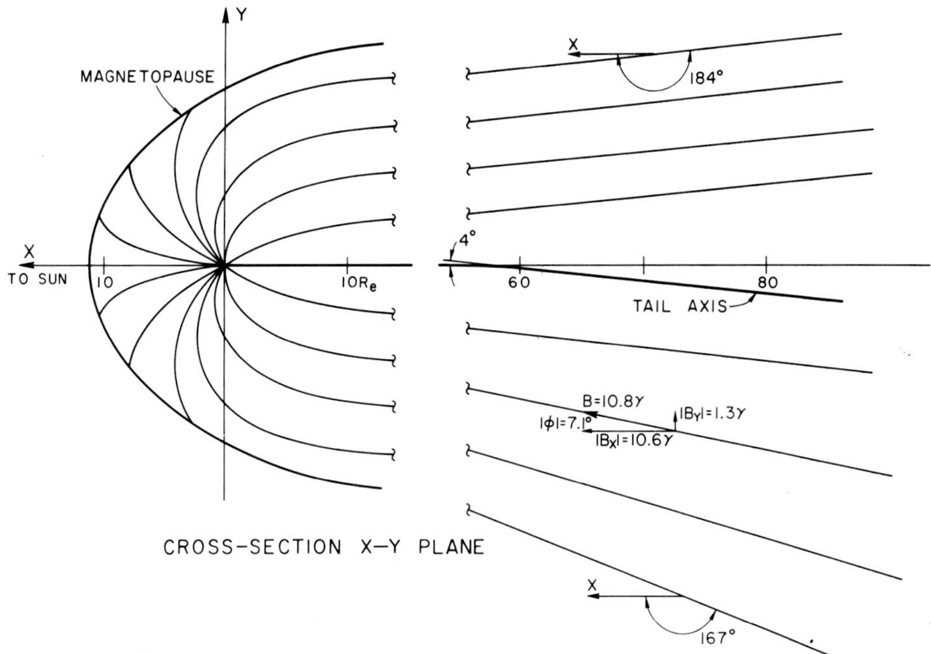

Fig. 3.14. Configuration of the plasma sheet and of the magnetic field at the lunar distance: (a) the *X*-*Z* plane and (b) the *X*-*Y* plane. (Meng, C.-I. and Anderson, K. A.: *J. Geophys. Res.* **79**, 5143, 1974.)

Hardy *et al.* (1975) noted that the differential energy spectra of plasma sheet particles at the lunar distance show a peak below 3 keV, although the average energy at the Vela distance (18 R_E) is known to be 6 keV (Bame, 1968). It may well be that this difference indicates a decrease of the average energy at greater geocentric distances. However, it must be noted that these measurements were made during different periods.

The distribution of energetic electrons and the magnetic field configuration in the magnetotail at the lunar distance have most recently been studied by Meng (1971), Meng and Mihalov (1972a, b) and Meng and Anderson (1974). They showed that the occurrence frequency of the energetic electrons is highest near the solar magnetospheric equator and falls off gradually away from the equatorial plane. The occurrence frequency is much higher on the dawnside than on the duskside. This tendency was noted earlier at the Vela distance by Montgomery (1968).

The magnetic field configuration obtained by Meng and Anderson (1974) is shown in Figures 3.14(a) and (b). In the high latitude lobes, the magnetic field lines are systematically and orderly distributed. The field lines show a clear indication of longitudinal divergence ($\sim 7° \times 2 = 14°$) and of latitudinal divergence ($4° \sim 5° \times 2 = 8° \sim 10°$) away from the aberrated tail axis.

Inside the plasma sheet, the field lines are considerably more deviated from the aberrated tail axis. The longitudinal and latitudinal deviations are about 20° and 17°, respectively. The average B_z value is about 1.9 γ. They compared the field configuration for low Kp (Kp $\leq 1+$) and high Kp (Kp $\geq 2-$) and found that the general pattern does not change significantly, except that in the plasma sheet the latitudinal deviation $|\theta|$ decreases from 17.2° for the low Kp values to 15.8° for the high Kp values. This feature will be discussed further in Section 6.4.

3.5. Origin and Dynamics of the Plasma Sheet

3.5.1. MERIDIONAL CONVECTION OF MAGNETOSPHERIC PLASMA

There are now several important facts available in considering the source of plasma sheet particles.

(1) At a geocentric distance of 18 R_E, the energy of protons becomes soft first in the upper and the lower boundary layers as an extended quiet period begins. If a very quiet magnetic condition (Kp = 0) lasts for 40 h or more, the average energy becomes about 1 keV or less even in the inner plasma sheet. One of the most plausible interpretations of this phenomenon is that the plasma in the plasma sheet is continuously replenished by fresh magnetosheath particles.

(2) At a geocentric distance of 18 R_E, the average energy of plasma particles is significantly less near the upper and lower boundary layers than in the vicinity of the midplane.

(3) At the lunar distance, the plasma mantle is located just above the plasma sheet over the entire longitude range, and perhaps over the entire high latitude lobe.

(4) At a very great distance ($\sim 1000\, R_E$) downstream, there is often a high

speed plasma flow (within 100 km s^{-1} of the magnetosheath velocity) throughout the magnetotail.

(5) Hardy *et al.* (1975) suggested that the mantle plasma is convected toward the plasma sheet by the $(E \times B)$ drift.

In the original theory of magnetospheric convection by Axford and Hines (1961), it was assumed that the same magnetospheric plasma repeats a convective motion in the equatorial plane. A new pattern of convection which emerges from the above study is as follows: Solar wind plasma particles enter the magnetosphere through the merging region and blow through the high latitude lobe region along the entire length of the magnetotail. However, some of them become plasma sheet particles after being convected toward the plasma sheet and 'trapped' in it. Then, they participate in the earthward motion (Section 1.3.5) and depart from the magnetosphere across the dayside magnetopause.

Therefore, the convective motion is a meridional one, rather than an equatorial one, and further, most of the particles leave the magnetosphere after a single circulation. Thus, if a quiet period lasts for more than a day, a significant part of plasma in the plasma sheet is replenished by fresh magnetosheath plasma which is convected into the plasma sheet from the plasma mantle. Note that the dawn-to-dusk electric field appears to be always present, even when the interplanetary magnetic field has a large northward component (Section 4.4.2). On the other hand, even a single weak substorm can energize protons in the plasma sheet to 25 keV or even more (Section 3.4.1). Therefore, the opportunity of observing magnetosheath-like protons in the inner plasma sheet does not occur very frequently.

3.5.2. ELECTRIC CURRENTS, MAGNETIC FIELD STRUCTURE AND THE BALANCE OF STRESSES IN THE PLASMA SHEET

(a) *Two-Dimensional Maxwell-Vlasov and Macroscopic Approaches*

It was suggested in the previous section that plasma sheet particles are continuously replenished by mantle plasma particles even during quiet periods. Therefore, strictly speaking, the magnetotail must always be in dynamic equilibrium. However, during quiet periods, the velocity of the cross-tail electric field is rather weak (~10 kV), so that the resulting $(E \times B)$ drift speed is of order 20 km s^{-1}. Such a low speed would not contribute significantly to equilibrium conditions in the magnetotail. Thus, it is generally believed that the magnetotail is nearly in static equilibrium during a quiet period, namely,

$$\boldsymbol{j} \times \boldsymbol{B} = \nabla p$$

This requires, however, that the current-carrying layer should be rather broad, since there does not seem to be a sufficient pressure gradient in the magnetotail to counterbalance a large Lorentz force, if the current is confined in a very thin layer (Rich *et al.*, 1972). Indeed, Bowling and Wolf (1974) showed that there is a layer of nearly uniform current density, approximately 2.3 to 2.6 R_E thick, in the central region of the plasma sheet. However, this problem needs further study.

The above equation has been studied for a two-dimensional situation by a number of workers, including Soop and Schindler (1973), Toichi (1972), Kan (1973) and Schindler (1975). The magnetic potential A_y associated with $\boldsymbol{B}(B_x, 0, B_z)$ is related to the current density j_y by

$$\frac{\partial^2 A_y}{\partial x^2} + \frac{\partial^2 A_y}{\partial z^2} = -4\pi(j_{y+} + j_{y-})$$

where $\pm$ signs are for ions and electrons, respectively. This equation may be rewritten as

$$\frac{\partial^2 A_y}{\partial x^2} + \frac{\partial^2 A_y}{\partial z^2} = 4\pi n_0|e|\left(1 + \frac{T_+}{T_-}\right) V_- \exp\left(-\frac{eV_- A_y}{\kappa T_-}\right)$$

for the distribution of ions and electrons of the form

$$f = n_0\left(\frac{m}{2\kappa T}\right)^{3/2} \exp(-\eta^2) \exp\left(-\frac{\varepsilon}{\kappa T}\right) \exp\left(2\eta \frac{p}{\sqrt{2m\kappa T}}\right)$$

where η denotes a dimensionless parameter, and

$$p = mv + eA_y$$

Toichi (1972) solved the above equation under the following boundary conditions after rewriting the first equation in a dimensionless form:

$$\frac{\partial^2 A_y^*}{\partial x^{*2}} + \frac{\partial^2 A_y^*}{\partial z^{*2}} = e^{-2A_y^*}$$

$$A_y^* = \log\cosh(z^*) \qquad \text{at } x^* = \infty$$

$$A_y^* = \log\{\cosh(z^*) - \alpha\} \qquad \text{at } x^* = 0$$

$$B_x^* = -\frac{\partial A_y^*}{\partial z^*} = \begin{cases} -1 & \text{at } z^* = \infty \\ +1 & \text{at } z^* = -\infty \end{cases}$$

where

$$x^* = x/\lambda, \qquad z^* = z/\lambda, \qquad A_y^* = A_y/\lambda B_0, \qquad B_x^* = B_x/B_0,$$

$$\lambda = (T_-/V_-)\sqrt{\kappa/2\pi n_0 e^2(T_+ + T_-)}, \quad B_0 = \sqrt{8\pi n_0 \kappa(T_+ + T_-)},$$

and α is an arbitrary parameter. Figure 3.15 shows the configuration of the magnetic field in the plasma sheet for $\alpha = 0.6$ and $\alpha = 0.4$, respectively; for the other parameters, see Table 3.1. Toichi suggested that model 1 corresponds to a quiet condition and model 2 to the pre-substorm condition. He showed also that the B_z component is smaller in model 2 than in model 1. Figures 3.16(a) and (b) show the distribution of the number density n as a function of X and Z, respectively, for the two models.

(b) *Kinetic Approaches*

Particle orbits in the vicinity of the midplane of the magnetotail have been studied by Speiser (1965, 1967), Alfvén (1968), Cowley (1971, 1973), Sonnerup (1971), Dungey (1972, 1975), Eastwood (1972, 1974, 1975) and Pudovkin and Tsyganenko (1973). Assuming a simple neutral sheet geometry (where the B_z component is null), Alfvén (1968) showed that both protons and electrons are convected toward

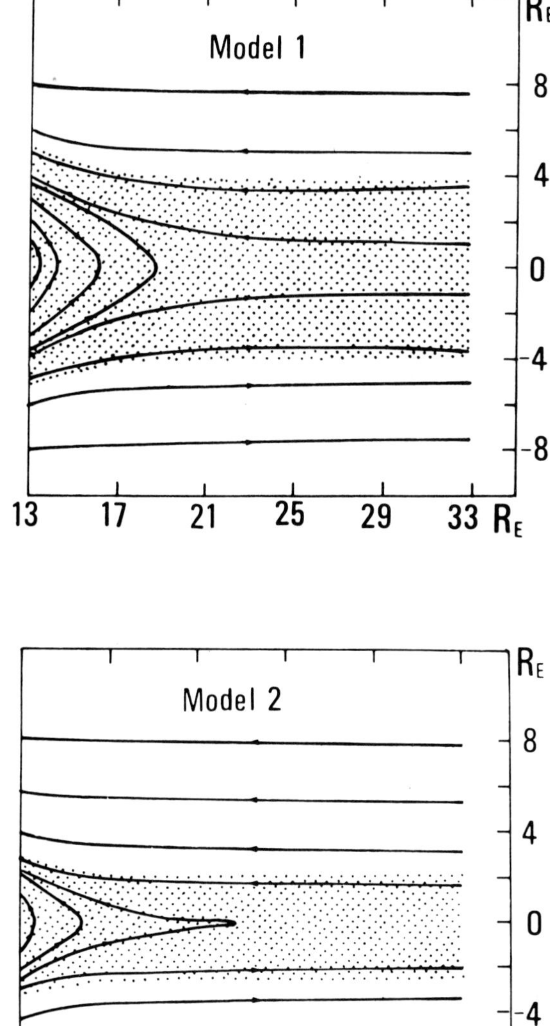

Fig. 3.15. Computed configurations of the plasma sheet for Models 1 and 2. (Toichi, T.: *Cosmic Electrodyn.* **3**, 81, 1972.)

the neutral sheet, from both the northern and southern hemispheres, and that the electrons drift toward the dawnside of the magnetopause and the protons toward the duskside magnetopause; their drift motions are a meandering type, across the neutral sheet. Figure 3.17(a) shows the trajectory of hot protons and electrons which enter the neutral sheet from the northern hemisphere. Alfvén (1968) and

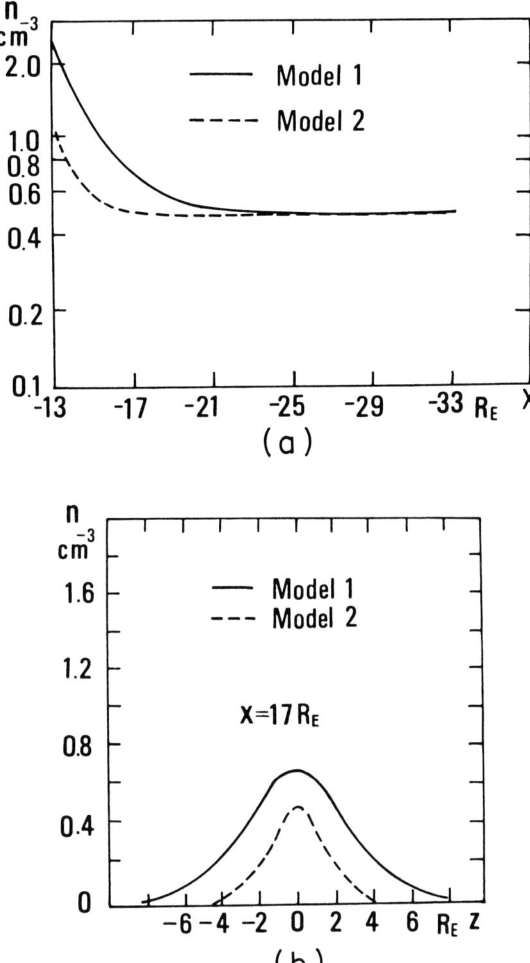

Fig. 3.16. Distribution of the number density n as a function of X and Z. (Toichi, T.: *Cosmic Electrodyn.* **3**, 81, 1972).

TABLE 3.1
Magnetotail parameters

	Model 1 (Quiet state)	Model 2 (Pre-substorm state)
λ (km)	$4\,R_E$	$2\,R_E$
n_0(cm^{-3})	0.5	0.5
$T_+ = 2T_-$(keV)	1.0	1.0
α	0.6	0.4
$B_0(\gamma)$	17.0	17.0
V(km s^{-1})	3.5	6.8
$\Phi(\gamma R_E^2)$	1800	480
N_T (particles)	7.5×10^{29}	3.3×10^{29}
J_T(A)	4.2×10^6	3.7×10^6

Alfvén and Fälthammar (1971) found a simple relationship between the potential drop $\Phi = Ed$ and the other related quantities in this situation:

$$\Phi = \frac{B_0^2}{4\pi n_e d}$$

where d denotes the dawn-dusk dimension of the magnetotail.

Cowley (1971, 1973) extended considerably Alfvén's study, particularly by noting that such drift motions give rise to charge separation and space charges, distorting Alfvén's simple potential pattern; see Figure 3.17(b). The convective motion should thus take place along the disturbed potential contours. For example, magnetosheath protons enter from the dawnside magnetopause and move toward the duskside magnetopause. Similarly, electrons are convected toward the duskside magnetopause, but accelerated along the neutral sheet toward the dawnside magnetopause. Such an electron beam may carry much of the sheet current in the dawn-to-dusk direction. This layer of the electron beam is much narrower than the layer in which the protons meander. This problem was recently discussed in terms of the stationary Vlasov theory by Bornatici and Schindler (1974).

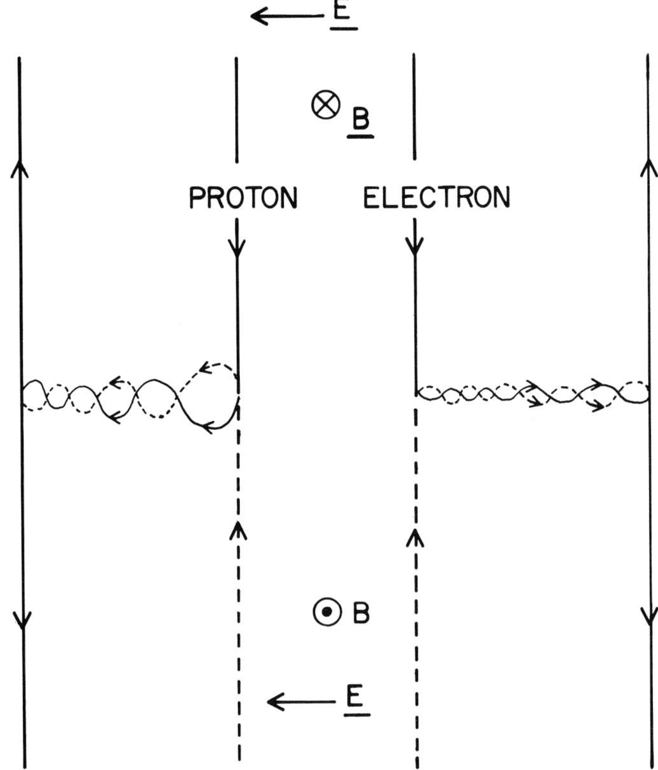

Fig. 3.17(a). Schematic diagram showing the trajectories of protons and electrons in the vicinity of the neutral sheet in Alfvén's model. (Cowley, S. W. H.: *Cosmic Electrodyn.* **3**, 448, 1973; slightly modified.)

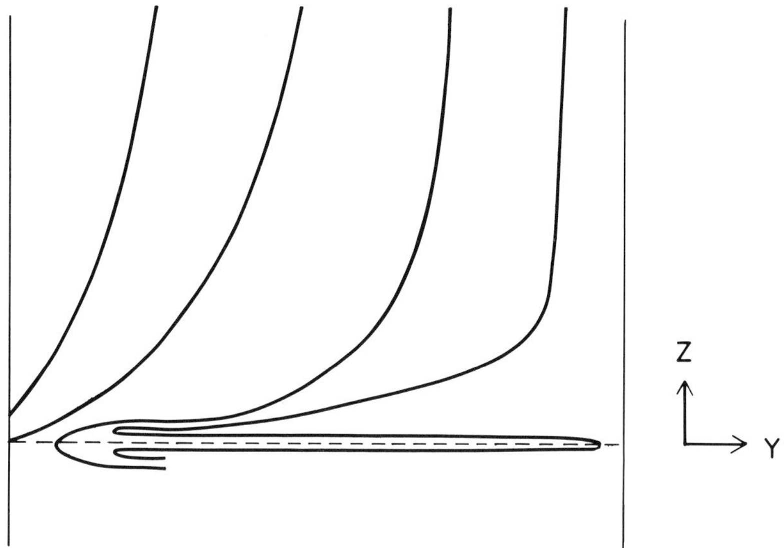

Fig. 3.17(b). Electric equipotentials in Cowley's neutral sheet model. (From Dungey, J. W.: *Space Sci. Rev.* **17**, 173, 1975.)

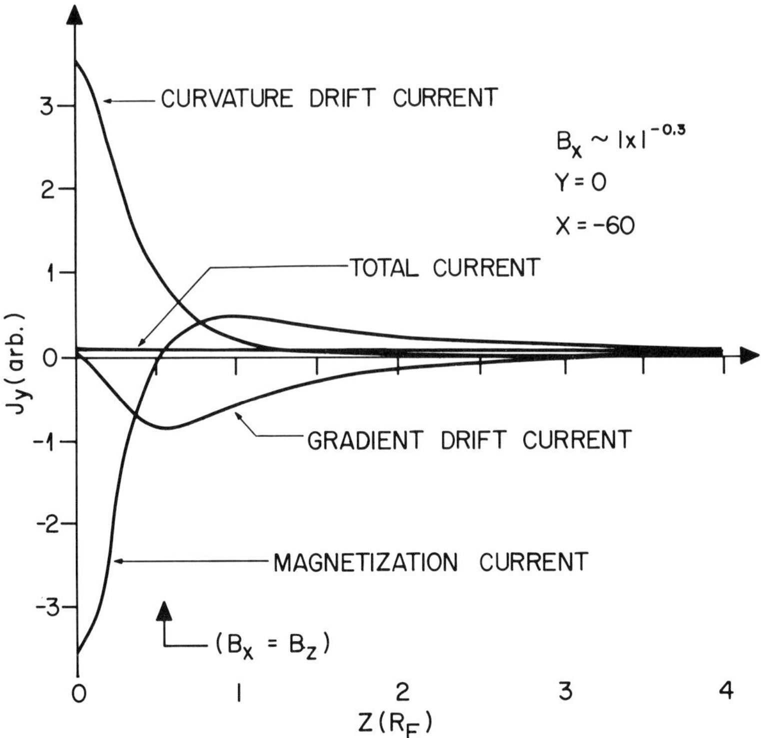

Fig. 3.18(a). Distributions of the three components of the cross-tail current and of the total current as a function of the distance from the midplane at $X = -60\ R_E$ and $Y = 0$.

Bird and Beard (1972a, b) and Huang (1973) examined the electric currents which arise from the gradient drift, curvature drift and gyration of plasma particles in the plasma sheet, by providing a first approximation magnetic field distribution and plasma distribution for a bisected cylindrical geometry. The magnetic field due to the electric current is computed. Then, on the basis of the newly computed magnetic field, the current distribution is computed again, and so on. They showed that both the magnetic field and plasma distribution converge rapidly to a self-consistent result. Figure 3.18(a) shows the current distribution for the different sources and also the resultant current intensity.

Note that the resultant current is rather small and varies very slowly as a function of Z (the distance from the midplane), although the individual component currents are intense and vary rapidly. Figure 3.18(b) shows the magnetic field components in the magnetotail from all the source currents and the resultant field.

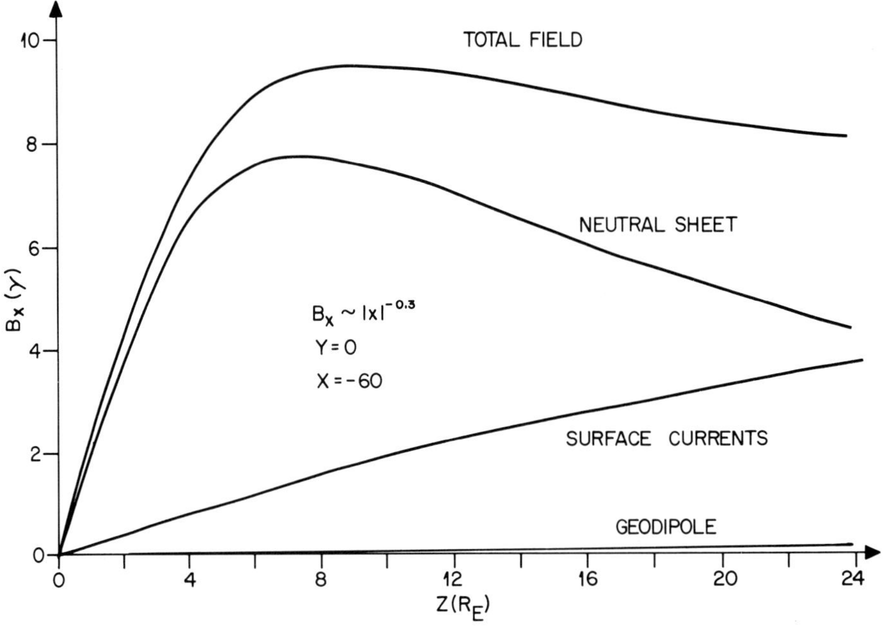

Fig. 3.18(b). Distribution of the magnetic fields produced by the cross-tail current, the surface (magnetopause) current, as well as the Earth's dipole field as a function of the distance Z from the midplane at $X = -60\ R_E$. (Bird, M. K. and Beard, D. B.: *Planet. Space Sci.* **20**, 2057, 1972.)

3.6. Van Allen Belts

3.6.1. RING CURRENT BELT: THE QUIET TIME BELT

Figure 3.19 shows a three-dimensional 'map' of the energy density of protons (1–872 keV) on quiet days, constructed on the basis of the S^3 satellite data (Smith *et al.*, 1975). One of the important features revealed by this presentation is that the

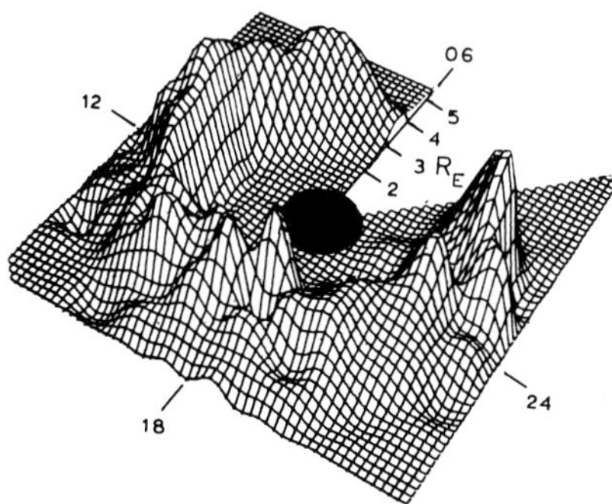

Fig. 3.19. Three-dimensional 'map' of the energy density of protons on quiet days. (Smith, P. H., Hoffman, R. A. and Bewtra, N. K.: *EOS* **56**, 618, 1975.)

maximum energy density for protons of energies from 1 to 872 keV lies very close to $L = 3.5$ at all local times.

The presence of such a proton belt, together with the plasma sheet, causes a significant distortion of the geomagnetic field (Lee and Cahill, 1975). Smith *et al.* (1975) noted that the proton belt in Figure 3.25 would produce about a 15 γ depression of the horizontal component on the Earth's surface and a maximum

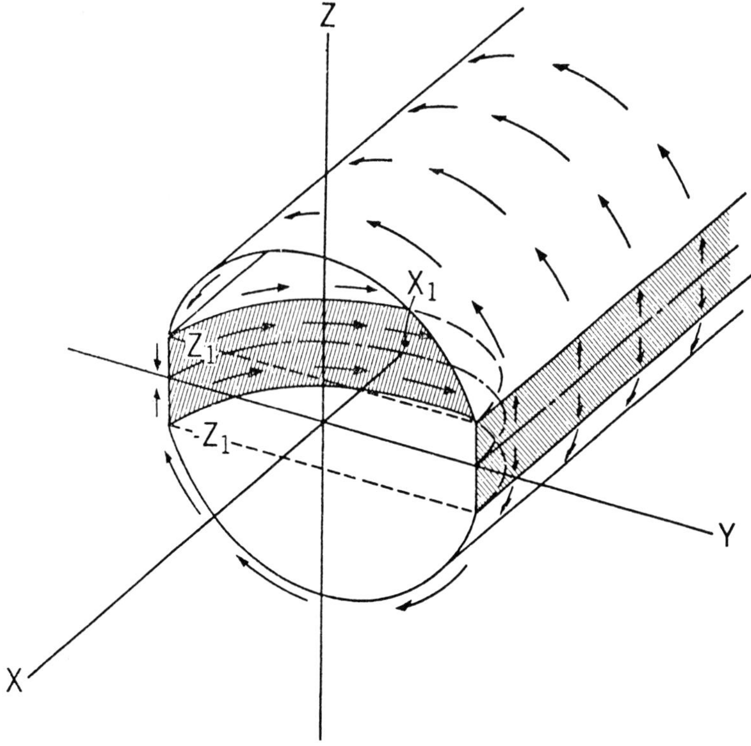

Fig. 3.20. Schematic view of the ring and cross-tail current system near its inner edge. (Sugiura, M. and Poros, D. J.: *Planet. Space Sci.* **21**, 1763, 1973.)

depression at $L = 3.5$ of about 37 γ (or 5% of the main field). Sugiura (1972, 1973a, b) and Sugiura and Poros (1973) showed that the magnetic field distortion in the night sector can be represented well by a circular disk current of a finite thickness, illustrated in Figure 3.20. They suggested that the westward cross-tail current is continuous from the magnetotail to a geocentric distance of about 3 R_E. Thus, in their view, the ring current is a part of the cross-tail current which flows along circular arcs concentric with the Earth. The magnetic field structure beyond a geocentric distance of 5 R_E has most recently been studied by Hedgecock and Thomas (1975). They showed that the tail-like structure becomes evident even at a geocentric distance of about 7 R_E (Figure 3.21). Plasma characteristics near the inner edge of the plasma sheet were studied by Schield and Frank (1970).

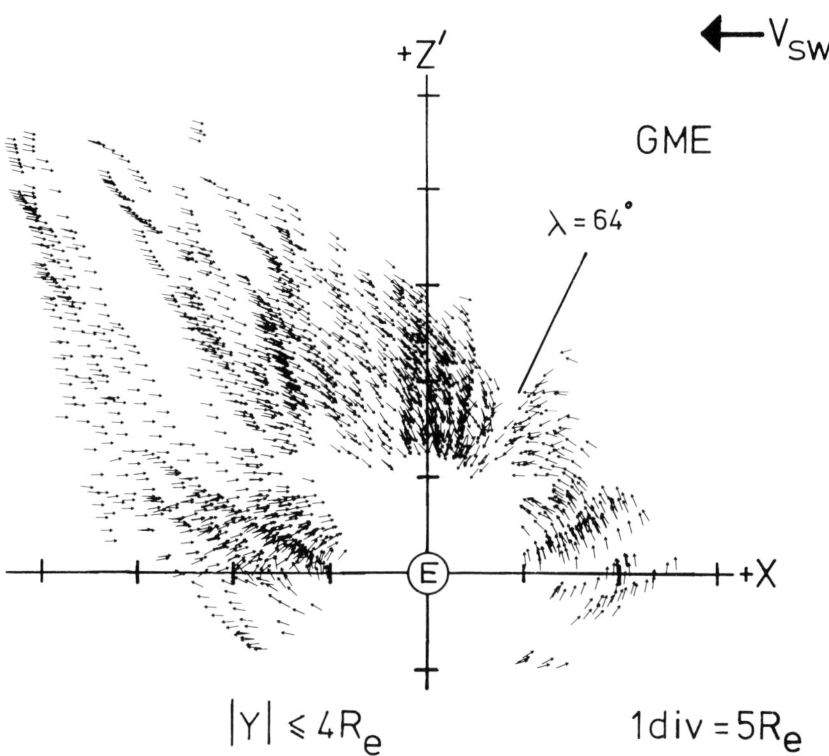

Fig. 3.21. Orientation of the magnetic field vectors in the X-Z plane at $|X| > 5\ R_E$ and $Y \leq 4\ R_E$. (Hedgecock, P. C. and Thomas, B. T.: *Geophys. J. Roy. Astron. Soc.* **41**, 391, 1975.)

3.6.2. ELECTRON BELT

The classical outer Van Allen belt is where electrons of energies of order 5–500 keV are relatively stably trapped. It is an important source of auroral electrons which produces a part of the annular auroral belt. The nature of the electron belt has been discussed in a number of review papers and books (*cf.* Hess, 1968; Schulz, 1975; Schulz and Lanzerotti, 1975). It should be added, however, that it is substorm processes which energize plasma sheet electrons to about 5 keV and inject them into the trapping region. Thus, the magnetospheric substorm is a part of source processes for the Van Allen belt. This subject will be dealt with in detail in Section 8.2.1.

3.7. Plasmasphere

The composition and the number density distribution of individual constituents in the plasmasphere have been extensively studied by ion mass spectrometers. Figure 3.22 shows a typical number density distribution for H^+, He^+ and O^+ as a function of L; note that the plasmapause was located at $L = 4.9$ in this particular example. The location of the plasmapause varies, however, as a function of local time. Particularly, it extends outward in the 1500–2200 LT sector, and thus this region is called the 'bulge region'. Figure 3.23 shows the average location of the plasmapause. Many workers have examined how the plasmapause might be related to the mid-latitude trough of the F region. Taylor and Walsh (1972) demonstrated that the projection of the plasmapause onto the ionosphere, along the geomagnetic field lines, coincides with the region of a pronounced light ion trough (LIT) in both H^+ and He^+. Figure 3.24 shows the projected locations of the light ion trough (dots) and the location of the plasmapause determined by ground-based VLF techniques.

It is generally understood that the plasma in the plasmasphere is supplied from the ionosphere. The upward flux of ions along the geomagnetic field lines is estimated to be of order 3×10^8 ions cm^{-2} s^{-1} in the daytime (Park, 1970). On the

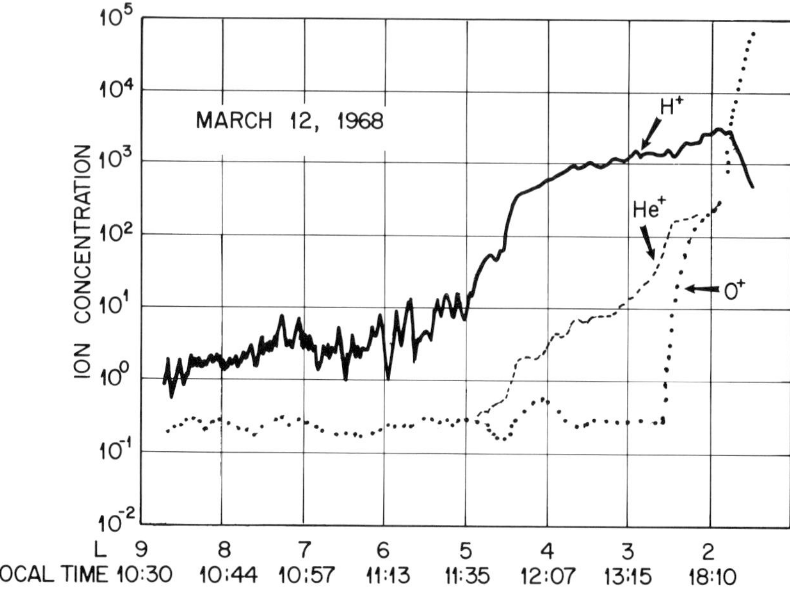

Fig. 3.22. Distribution of H^+, He^+ and O^+ densities as a function of L (and LT), observed by the OGO-5 satellite. (Chappell, C. R., Harris, K. K., and Sharp, G. W.: *J. Geophys. Res.* **75**, 50, 1970.)

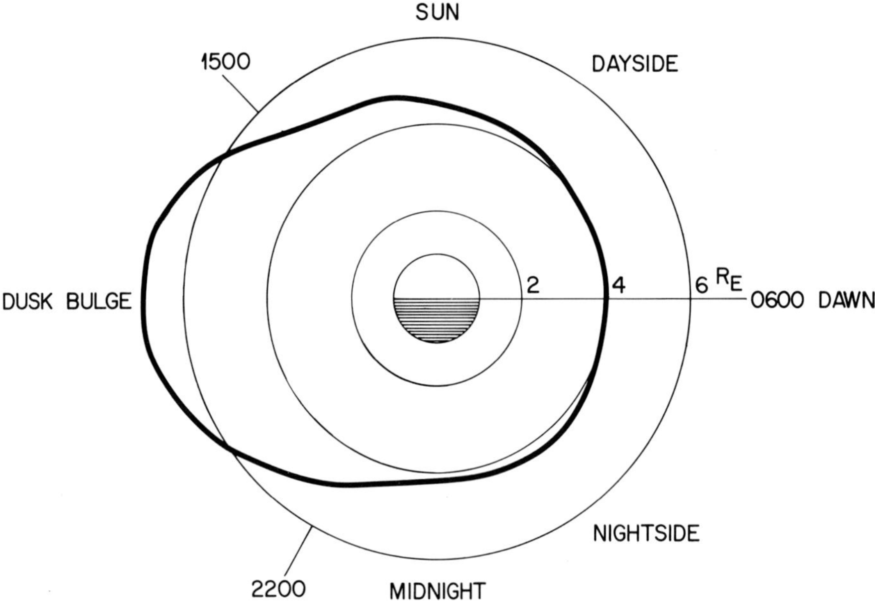

Fig. 3.23. Geometry of the plasmapause in the equatorial plane.

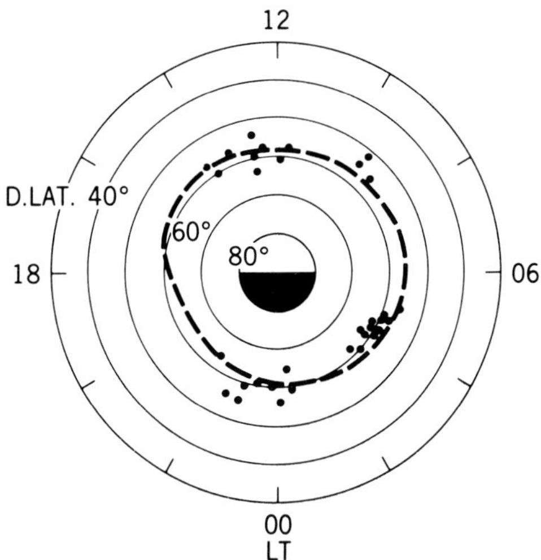

Fig. 3.24. Comparison of plasmapause locations determined independently from the light ion trough (dots) and the ground-based VLF method (dashed line). (Taylor, H. A. Jr. and Walsh, W. J.: *J. Geophys. Res.* **77**, 6716, 1972.)

other hand, the plasmasphere loses its plasma when the geomagnetic tubes of force reach the magnetopause.

The plasmasphere exhibits a variety of phenomena during disturbed periods (Grebowsky, 1970, 1971; Chappell, 1972, 1974; Chappell *et al.*, 1970, 1971; Taylor *et al.*, 1971; Chen and Grebowsky, 1974; Grebowsky *et al.*, 1974). Among them are: (1)

an earthward shift of the plasmapause, (2) the formation of a plasma tail from the plasmasphere, and (3) an interaction with the ring current belt, resulting in the mid-latitude red arc. These subjects will be discussed in Section 8.6.

3.8. Magnetospheric Plasmas and Auroral Particles

In Section 2.10 it was proposed that auroral particles precipitating into discrete auroral arcs should be considered separately from the rest of the auroral particles which produce the broad oval belt and the annular belt. Discrete auroral arcs occupy only very limited regions and thus can be considered as a singular region within the broad oval region. This implies that a particular type of acceleration process develops *within* the broad oval belt, namely within the background precipitation region.

Therefore, we examine first the contribution of plasma particles from the five plasma domains to the broad oval and annular precipitation belts. The following two subsections are concerned with this subject. We shall discuss possible acceleration mechanisms of arc-producing auroral particles in Section 3.9.

3.8.1. OVAL BELT

On the basis of a simple geometrical consideration, one can infer that the oval belt is produced by the precipitation of particles mainly from three plasma domains: the plasma mantle, the polar cusp and the plasma sheet. In the following we shall list some of the supporting observations for this inference.

(1) The dayside part of the oval belt is essentially a red band, suggesting that magnetosheath-like plasma particles precipitate there from either or both the plasma mantle and the polar cusp (Section 3.3).

(2) It was shown in Section 2.6 that the average energy of electrons precipitating into the nightside part of the broad oval belt decreases toward the poleward boundary of the belt. Hones *et al.* (1971) showed that the electron spectrum softens toward the upper and the lower boundaries of the plasma sheet, namely from about 1.2 keV near the midplane to about 400 eV near the boundaries (Section 3.4.1 and Figure 3.12).

(3) Lassen (1974) showed that the upper boundary and the inner edge of the plasma sheet coincide, respectively, with the poleward and equatorward boundaries of the auroral oval which are determined by a statistical study of auroral arc alignment (Figure 3.25).

(4) The equatorward half of the broad oval belt is occupied by the diffuse aurora and is connected to the CPS precipitation region. Choy *et al.* (1971) found an excellent agreement between an electron spectrum obtained by a rocket in the diffuse auroral region and the simultaneous spectrum obtained by a geosynchronous satellite (Figure 3.26); the agreement is likely to be more than a mere coincidence.

(5) Near the equatorward boundary of the diffuse aurora, the average energy of electrons softens (Winningham *et al.*, 1975). Near the inner edge of the plasma sheet, the average energy of electrons also softens (Schield and Frank, 1970).

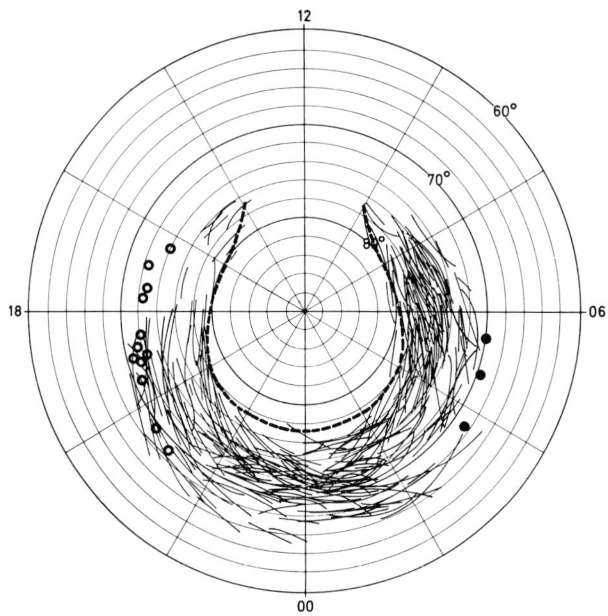

Fig. 3.25. Location of auroral arcs observed from Greenland in December 1965 and January 1966 at quiet times (Kp = 0.1), together with the projected inner edge of the plasma sheet (open circles and dots) and projected outer boundary of the plasma sheet at $X = -18\ R_E$ (Lassen, K.: *J. Geophys. Res.* **79**, 3857, 1974.)

Thus, it is quite likely that the projection of the inner boundary of the plasma sheet, along the geomagnetic field lines, coincides with the equatorward boundary of the diffuse aurora.

3.8.2. ANNULAR BELT

There is little doubt that energetic protons and electrons in the Van Allen belt contribute significantly to the annular belt. After being injected into the belt, the protons drift westward and the electrons drift eastward. It is during these drifting motions that a variety of wave-particle interactions play the major role in precipitating both the protons and electrons into the polar upper atmosphere. In Section 8.5 we shall examine their motions in the Van Allen belt and the subsequent precipitation processes.

3.9. Acceleration Processes of Arc-Producing Auroral Electrons

3.9.1. INTRODUCTION

In the previous section it was shown that the upper and lower boundaries of the plasma sheet contain magnetosheath-like plasma and that these regions may be connected to the poleward half of the broad oval belt by the geomagnetic field

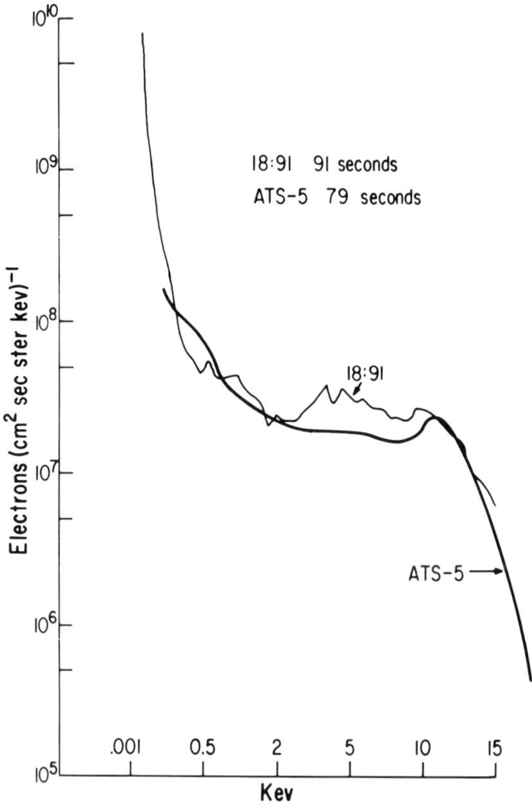

Fig. 3.26. Comparison of auroral electron spectra obtained nearly simultaneously by a rocket and by the synchronous ATS-5 satellite. (Choy, L. W., Arnoldy, R. L., Potter, W., Kintner, P. and Cahill, L. J. Jr.: *J. Geophys. Res.* **76**, 8279, 1971.)

lines. It is *within* this region that auroral arcs develop. The most important difference in characteristics between auroral particles which precipitate into the 'background' region (namely, the broad oval belt) and those which precipitate into auroral arcs is that the former have an exponential or a power law spectrum and a Maxwellian spectrum, while the latter have an additional 'monoenergetic' or a sharply peaked component. There are thus at least two precipitation processes, one for discrete arcs and the other for the background region. The latter process will be discussed in Section 8.5.2.

Evans (1974, 1975) showed most convincingly that a simple way to produce such a 'monoenergetic' component from hot Maxwellian plasma is to have a potential drop along geomagnetic field lines. An interesting by-product of the potential drop is that it constitutes a 'barrier' for upgoing electrons, particularly for secondary electrons. They will be 'reflected' downward at the barrier and reappear as precipitating electrons. By several numerical models Evans demonstrated how closely one can reproduce some of the spectra observed by rockets and satellites. Figure 3.27(a) shows an example of his spectrum, together with one

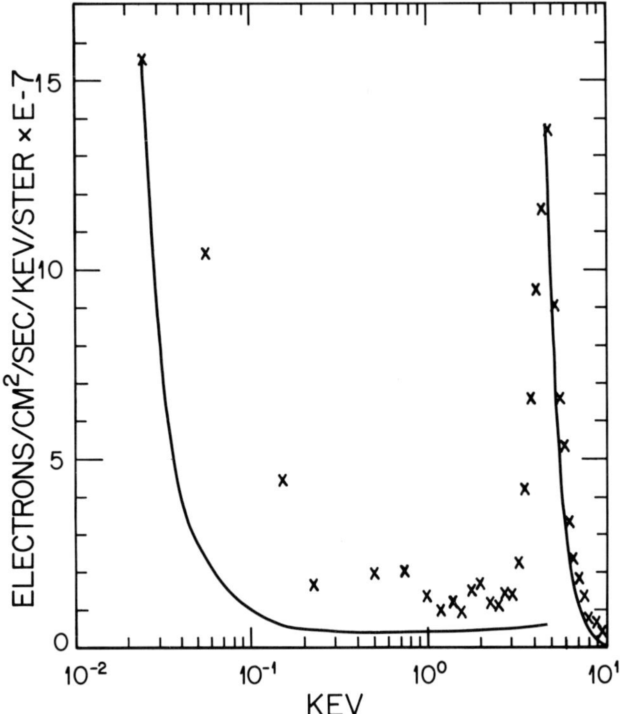

Fig. 3.27(a). Comparison of an observed auroral electron energy spectrum (0° pitch-angle) and the computed spectrum by assuming that energetic electrons are accelerated through a field-aligned potential drop. The lower energy electron fluxes are calculated from the secondary and backscattered electrons produced from the atmosphere by the primary electron beam. (Evans, D. S.: *Physics of the Hot Plasma in the Magnetosphere*, Hultqvist, B. and Stenflo, L. (eds.), p. 319, Plenum Press, 1975.)

of the spectra observed by Arnoldy and Choy (1973). Figure 3.29(b) shows another example of his spectrum which reproduces well the spectrum observed by Frank and Ackerson (1971).

One of the central questions in auroral physics is then how a potential drop of a few kilovolts can develop along the geomagnetic field lines which are embedded in the upper and lower boundary layers of the plasma sheet. What conditions make a particular group of field lines unique and singular in developing such a potential drop? There are several important clues in searching for the solution to this problem.

(1) Auroral arcs tend to appear predominantly in the *poleward half* of the broad oval belt in the evening and midnight sectors (Sections 2.1 and 2.2).

(2) In the poleward half of the oval belt in the evening sector, there is a weak upward current (Section 1.3.2).

(3) An auroral arc is associated with a pair of intense currents, an upward current within the arc and a downward current a little on the equatorward side (Section 2.5.2).

(4) A part of the upward current appears to be carried by a downward flux of electrons of energies less than 500 eV (Section 2.5.2).

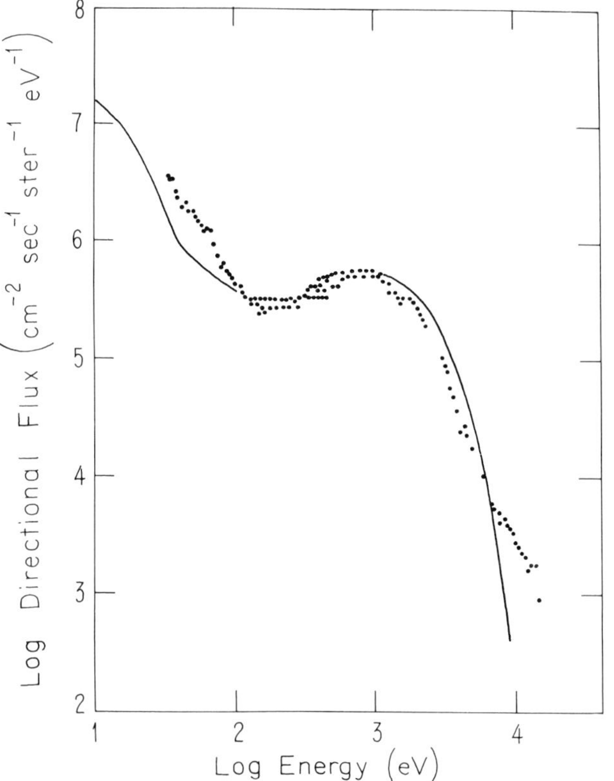

Fig. 3.27(b). Comparison of an observed auroral electron energy spectrum (dots) and the computed spectrum by assuming a 400 V potential drop along a magnetic field line. (Evans, D. S.: *J. Geophys. Res.* **79**, 2853, 1974.)

(5) Gurnett (1972) suggested that there is a particular type of potential structure associated with the inverted V structure. It is shown in Figure 3.30. A similar potential configuration was suggested earlier by Carlqvist and Boström (1970). From a simple geometrical consideration illustrated in the figure, the potential difference Φ_s between the satellite altitude and the auroral altitude is of order 6 kV. Thus, electrons in the inverted V structure can acquire energy of order 6 kV below an altitude of about 1000 km.

(6) There are oppositely directed plasma flows (parallel and antiparallel to an auroral arc). The potential distribution in Figure 3.28 can cause the observed $(E \times B)$ drift motion (Hallinan and Davis, 1970; Hallinan, 1976).

(7) There is some evidence of the presence of the parallel (pointing downward) electric field of ~10–24 mV m^{-1} even in rocket altitudes (Mozer and Fahleson, 1970; Kelley *et al.*, 1971; Bering *et al.*, 1973; Kelley *et al.*, 1975).

(8) There are a number of attempts to infer the presence of potential drop along geomagnetic field lines and its location on the basis of differences of time variations of both precipitating electrons and protons (Bryant *et al.*, 1969; Rème and Bosqued, 1971; Johnstone, 1971; Johnstone and Davis, 1974; Maynard *et al.*,

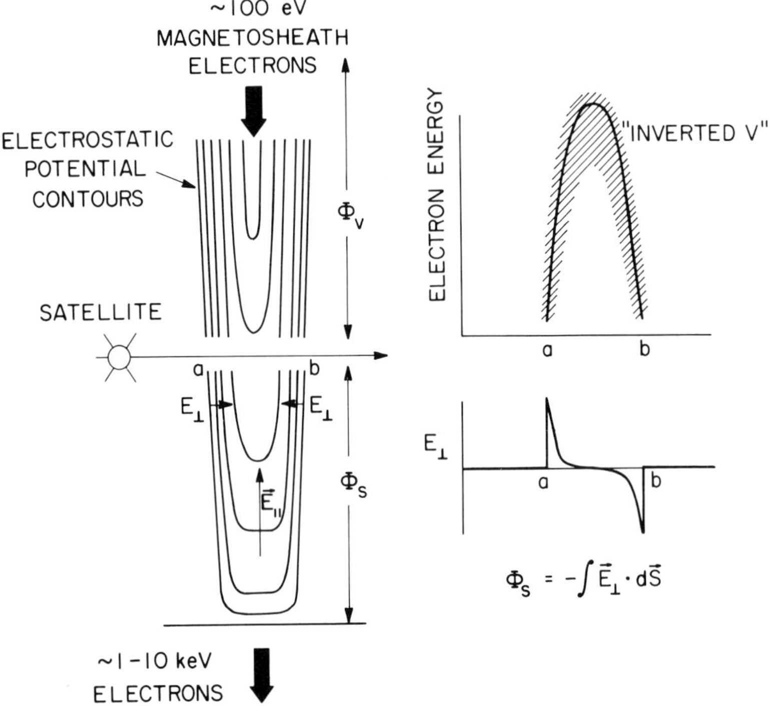

Fig. 3.28. Schematic electrostatic potential contours in the vicinity of an 'inverted V' precipitation region, deduced from both electric field and particle observations which are indicated on the right-hand side. (Gurnett, D. A.: *Critical Problems of Magnetospheric Physics*, Dyer, E. R. (ed.), p. 123, National Academy of Sciences, U.S.A., 1972.)

1973; Hall and Bryant, 1974; Bryant *et al.*, 1975). Most of these observations suggest field-aligned acceleration mechanisms below a few earth radii in altitude.

3.9.2. AURORAL ARCS AND THEIR TOPOLOGICAL RELATION WITH THE MAGNETOSPHERIC STRUCTURE

One of the important clues in an attempt to understand the acceleration processes of arc-producing electrons is the location of an auroral arc with respect to the boundary of the open and closed field lines. Auroral arcs are often multiple, so that we shall consider first an auroral arc which delineates the poleward boundary of the oval. An extensive study of DMSP photographs shows that the poleward boundary of the evening part of the oval is delineated most often by a single bright arc (Figures 2.1 and 2.2). Even in the morning sector, the poleward boundary of the diffuse aurora is much brighter than the rest and develops a distinct arc structure (namely, an omega band) during active periods.

The question then is where the poleward boundary arc is located with respect to the boundary of the open and closed field lines. However, in order to answer this question, one must be able to determine the boundary of the open and closed field lines.

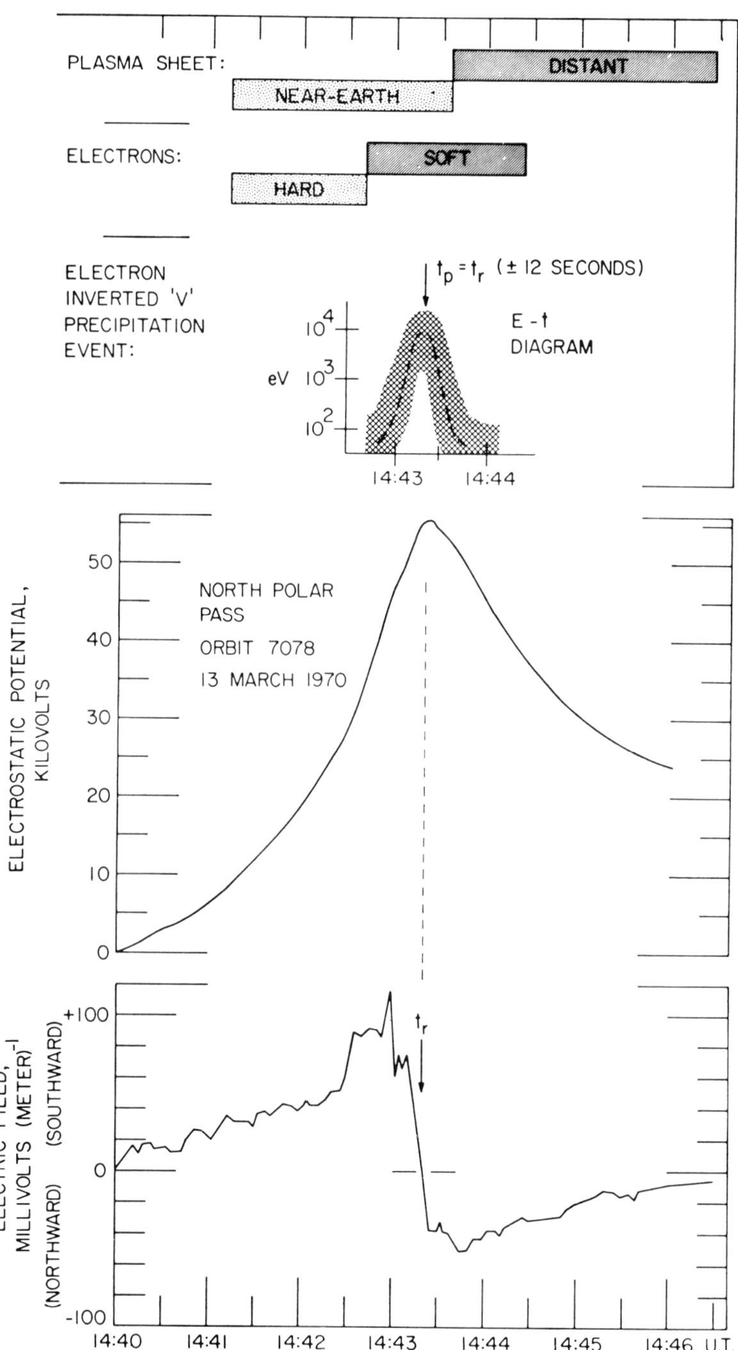

Fig. 3.29. Plasma and electric field distributions in the vicinity of a typical 'inverted V' precipitation region. (Frank, L. A. and Gurnett, D. A.: *J. Geophys. Res.* **76**, 6829, 1971.)

Perhaps one of the important clues to this question is the fact that the peak of the inverted V structure appears to coincide with the point of the reversal of the polar electric field (Gurnett and Frank, 1973). Figure 3.29 shows an example of such an observation in the evening sector. Since an auroral arc is expected to occur at or very near the peak of the inverted V structure and since the electric field reversal is expected to occur at or very near the boundary of the open and closed field lines, one can infer that the polewardmost arc appears very near this boundary. Swift and Gurnett (1973) showed also that the reversal location coincided with the location of a discrete auroral arc.

3.9.3. ARC ENERGIZATION SYSTEM

As stressed earlier, a discrete auroral arc is a unique feature in the polar upper atmosphere and occurs within a wide oval belt of particle precipitation. A discrete auroral arc is thus a localized feature; it must be produced by a localized energization system. This localized energization system will be referred to as the arc energization system which must include the energy source, the acceleration region and the electromotive force for the field-aligned current connected to the arc. The arc energization system is embedded in the large-scale magnetosphere-ionosphere coupling system. Its V-shaped electric equipotential configuration was discussed in Section 3.9.1.

One of the interesting features of the discrete auroral arc is that it is associated with field-aligned currents. Satellite and rocket observations have revealed the existence of localized field-aligned currents connected to a discrete auroral arc (Sections 2.5.2 and 3.9.1), embedded in large-scale field-aligned currents connected to the auroral oval (Section 1.3.2). The large-scale field-aligned current is always present; the localized current is observed only in the presence of a discrete arc. The former may be termed the oval field-aligned current; the latter, the arc field-aligned current.

It should be noted that the arc field-aligned current system cannot be a minor perturbation of the oval field-aligned current system. This is because the current in the arc system is about ten times more intense than in the oval. This, together with the fact that the arc energization system is a very localized feature, indicates that there must be a very concentrated energy supply mechanism which can maintain the V-shaped electric potential, drive the arc field-aligned currents and accelerate auroral electrons.

Kan (1975) and Kan and Akasofu (1976) suggested that an earthward flow of plasma in a very thin layer near the outer surface of the plasma sheet can not only generate the V-shaped potential, but also drive the arc field-aligned current. The energy source of these processes is the kinetic energy of ions in the suggested thin layer of the plasma flow which may be generated in the vicinity of the X-line, namely the merging and reconnection regions (Section 1.4.2). When the earthward flow of plasma is shocked electrostatically, a parallel electric field is produced, giving rise to a field-aligned potential jump in the kilovolt range (Kan, 1975). Figure 3.30(a) illustrates the proposed arc energization system. Since the plasma flow can be treated as a thin sheet flow, the equipotential surface of the electrostatic shock

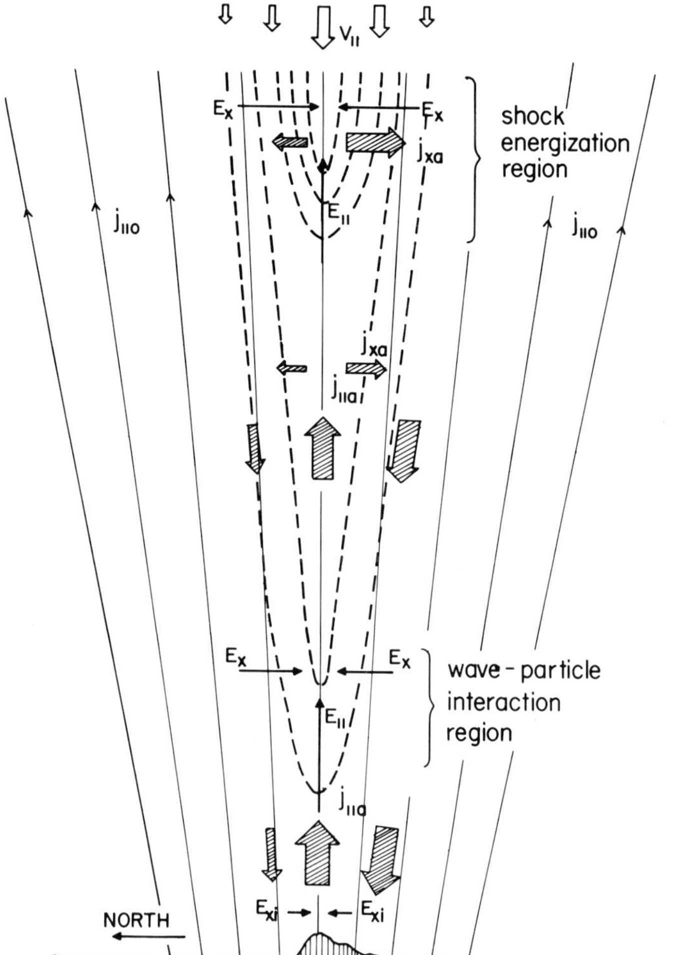

Fig. 3.30(a). Schematic configuration of the arc energization system. The parallel electric field $E_{\|}$ appears both in the shock energization region and in the wave particle interaction region. The arc field-aligned current $j_{\|a}$ is driven by the mechanism indicated in Figure 3.30(b) both in and below the shock energization region. The distribution of arc luminosity is indicated at the bottom.

must be two-dimensional ($\partial/\partial y = 0$; the y-axis is taken to be parallel to the arc, the x-axis is perpendicular pointing to the north and the z-axis is parallel to $\boldsymbol{B}$) and can be expected to be V-shaped. As soon as the V-shaped potential configuration develops, the plasma flow becomes sheared.

In addition to shear flow, an electric current across the magnetic field results from the polarization drift due to spatial variations of the electric field and/or magnetic field along the guiding center trajectory of the gyrating particle, as the plasma flows through the V-shaped potential region.

$$\left(j_{\perp} \simeq \frac{\boldsymbol{B}}{B^2} \times (\rho(\boldsymbol{V} \cdot \nabla)\boldsymbol{V})\right)_x \simeq \left(\frac{\rho}{B^2}\right) V_z B_z \frac{\partial V_y}{\partial z} = \left(\frac{\rho}{B^2}\right) V_z \frac{\partial E_x}{\partial z}$$

Since the polarization drift is proportional to the mass-to-charge ratio of a particle, the polarization current is carried predominantly by the ions in the plasma. Figure 3.30(b) illustrates schematically this situation. Since $\nabla \times \boldsymbol{E} = 0$ and thus,

$$\frac{\partial E_x}{\partial z} = \frac{\partial E_z}{\partial x}$$

$j_{\perp_x}$ is proportional to $\partial E_z / \partial x$. The electric current thus produced in the arc energization region can drive $j_{\perp_x}$ which closes in the ionosphere by the Pedersen current.

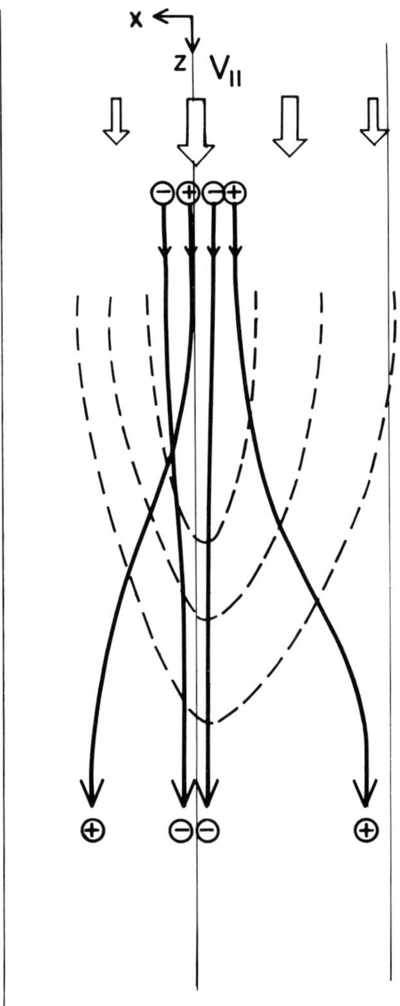

Fig. 3.30(b). Guiding center trajectories of ions and electrons (solid curves with arrow head) in the energization region. The parallel plasma flow component at the top of the energization region is indicated by $V_\parallel$ and represented by unshaded boldface arrows.

On the other hand, the electrons in the plasma flow stream down (along the field line which passed through the center of the V-shaped configuration) and are accelerated, constituting the upward-directed current in the arc field-aligned current system.

The accelerated electron beam interacts with the ambient plasma in the topside ionosphere at several thousand kilometers (< a few Earth radii). This is the wave-particle interaction region shown in Figure 3.20(a). Anomalous resistivity (Section 3.9.4(a)) and/or the double layer (Section 3.9.4(b)) may be generated in the wave-particle interaction region. Thus, an upward parallel electric field may also be produced in the wave-particle interaction region which is located below the arc energization region. Note that the shock energization region also serves as the dynamo which generates the electrical energy dissipated in the double layer and/or in the anomalous resistivity in the wave-particle interaction region. The average electron beam energy is degraded after the beam passes through the wave-particle interaction region. It is not clear at present whether or not the energy degradation process in the wave-particle interaction region is sufficient to account for the observation that a part of the upward arc field-aligned current is carried by particles with energy less than 500 eV.

It is important to note that the proposed arc energization system is consistent with the observation on the auroral kilometric radiation reported by Gurnett (1974). Observations show that the kilometric radiation originates from below 3 R_E in the auroral region. Gurnett (1974) suggested that it is probably generated by plasma instabilities due to intense field-aligned electron beams with energies of tens of kilovolts interacting with the ambient plasma. Thus, the existence of an acceleration region above the wave-particle interaction region is required for this mechanism to operate. The proposed energization system places this acceleration region in the plasma sheet. In this connection, it may be noted that intense field-aligned electron beams have been observed by the ATS-6 satellite at ~ 6.6 R_E and ~ 11° geometric latitude (McIlwain, 1975) and that the intense arc field-aligned current may have been detected by the OGO-5 satellite at 5 to 7 R_E (Sugiura, 1975; Section 1.3.2).

It may be noted that in a steady state, the curl $\boldsymbol{E} = 0$ condition requires that the equipotential surface be closed. Outside the energization region, the equipotential surface must be parallel to the magnetic surface. Hence, energization regions must exist in pair in a steady state and are magnetically conjugate in the magnetosphere. Thus, the equipotential surface must extend from an energization region in one hemisphere to an energization region in the opposite hemisphere as suggested by A. T. Y. Lui (private communication, 1975). Figure 3.30(c) illustrates the equipotential surfaces (dashed curves) enclosing a pair of conjugate shock energization regions in the plasma sheet (one in each hemisphere). The formation of auroral arcs has also been discussed by Atkinson (1970), Ogawa and Sato (1971), Sato and Holzer (1973) and Holzer and Sato (1973). Atkinson (1970) showed that magnetic flux tubes are connected to the auroral oval and if high conductive strips are produced in the ionosphere by some mechanism, they undergo a highly nonlinear oscillation that produces standing waves (auroral arcs). He considered then the field-aligned currents generated by the waves and their

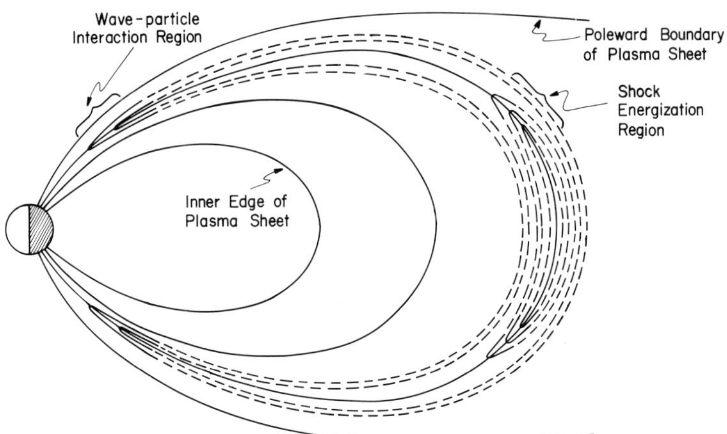

Fig. 3.30(c). Closed equipotential contours (dashed curves) in the plasma sheet. (Courtesy of Kan, J. R. and Akasofu, S.-I.)

closure mechanism (the polarization current) in the ionosphere. Sato and Holzer (1973) and Holzer and Sato (1973) considered growth of a weak ionospheric electron density perturbation (that is, narrow in latitude and broad in local time) which is embedded in the magnetosphere-ionosphere system. They showed that such a perturbation can grow exponentially in time, driving a current through the magnetosphere-ionosphere system. They also suggested that the associated current driven instabilities could cause both pitch-angle scattering of trapped particles and anomalous resistivity.

3.9.4. POSSIBLE PROCESSES IN THE WAVE-PARTICLE INTERACTION REGION

(a) *Current Driven Instabilities*

There are three current driven instabilities which are known to cause anomalous resistivities. They are the Buneman two-stream instability, the ion-cyclotron instability and the ion-acoustic instability. Figure 3.31 shows the critical drift velocity as a function of the electron-to-ion temperature ratio T_e/T_i for the three instabilities. Swift (1965) was the first to point out that current driven instabilities may play an important role in causing the potential drop. Kindel and Kennel (1971) examined the current density for the three instabilities for the upper ionospheric conditions and showed that the ion-cyclotron instability requires the smallest electron drift speed among the three. On the basis of their results, Kan (1974) estimated the magnitude of the electric fields which arise from the three instabilities as a function of the electron density $n\,(\mathrm{cm}^{-3})$ when the drift speed exceeds the critical value. They are given by

$$E_{\|\mathrm{B}} \simeq 2.6\times 10^4 j_\| / \sqrt{n} A^{1/3} \qquad \text{for the Buneman instability,}$$

$$E_{\|\mathrm{ic}} \simeq 6.5\times 10^2 j_\| (\Omega_e/\omega_{pe}) \sqrt{nA} \qquad \text{for the ion-cyclotron instability,}$$

$$E_{\|\mathrm{ia}} \simeq 6\times 10^3 j_\| / \sqrt{n} \qquad \text{for the ion-acoustic instability,}$$

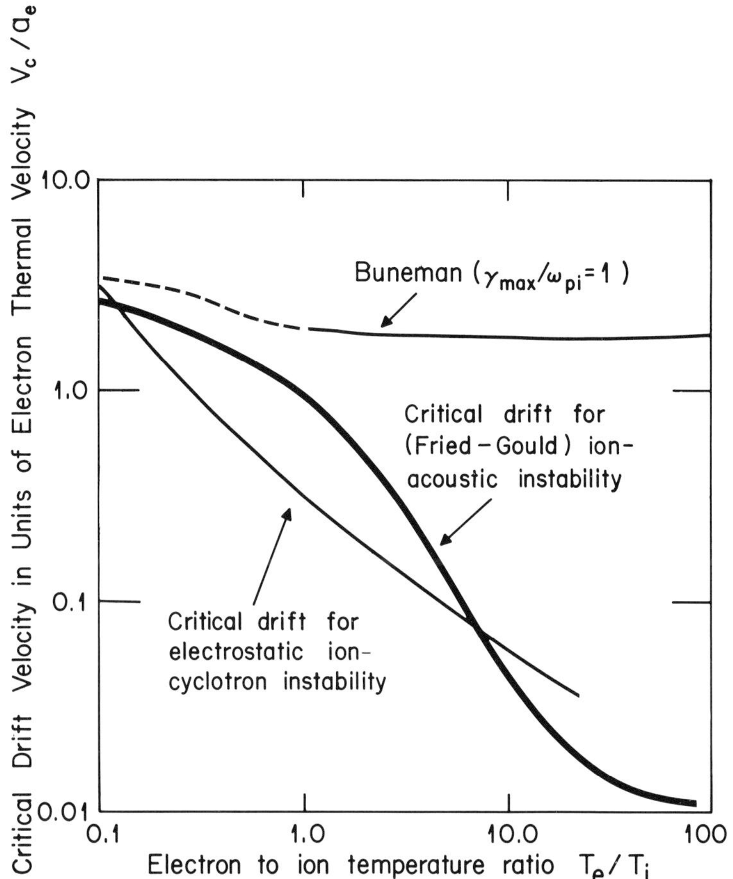

Fig. 3.31. Critical drift speeds V_c (normalized to the electron thermal speed a_e) for the ion-cyclotron waves, ion-acoustic waves and the Buneman instability, as a function of electron to ion temperature ratio, T_e/T_i, in hydrogen plasma. (Kindel, J. M. and Kennel, C. F.: *J. Geophys. Res.* **76**, 3055, 1971).

where E is given in volt m^{-1}, $j_{\|}$ in A m^{-2}; A is the atomic mass number. Figure 3.32 shows the magnitude of the resulting electric field for a particular distribution of electron density in the upper part of the ionosphere. The curves 1, 2 and 3 correspond to $j_{\|}$ (200 km) $= 3 \times 10^9$, 10^{10} and 10^{11} electrons $cm^{-2}\,s^{-1}$, respectively. Each curve indicates the most important instabilities under the given conditions, as well as the important positive ions (H^+ or O^+) for ion-cyclotron instability; see also Chmyrev (1971).

It can be seen that for $j_{\|}$ (200 km) $= 10^{10}$ electrons $cm^{-2}\,s^{-1}$, the total potential drop between 10^3 km and 2×10^4 km is about 100 kV; in this particular case, the ion-acoustic instability plays the most important role. The current driven instabilities have recently been studied by Volosevich and Liperovskiy (1972) and Papadopoulos and Coffey (1974).

Scarf *et al.* (1973, 1975) and Fredricks *et al.* (1973) showed that field-aligned currents, observed by the transverse component of the magnetic field, are often

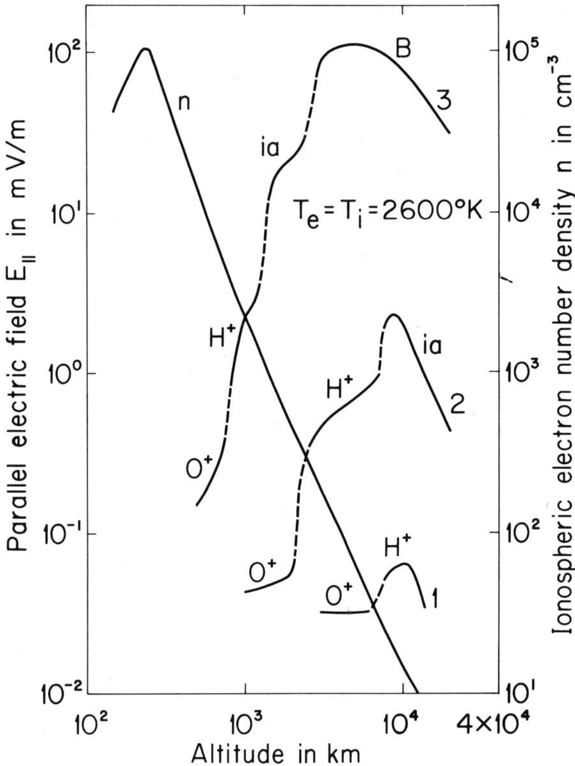

Fig. 3.32. Parallel electric field profiles due to anomalous resistivities. Curves 1, 2 and 3 correspond to $j_\parallel$ (at 200 km altitude) = 3×10^9, 10^{10} and 10^{11} electrons cm^{-2} s^{-1}, respectively. Symbols O^+, H^+, ia and B along each curve denote the dominant instability in different altitude range; O^+ stands for O^+-cyclotron instability, H^+ for H^+-cyclotron instability, ia for the ion-acoustic instability and B for the Buneman two-stream instability. (Kan, J. R.: *EOS* **55**, 1007.)

associated with intense bursts of electrostatic emissions of frequencies of order of 600 Hz–10 kHz with amplitudes over 100 mV m^{-1} and suggested that these waves are capable of becoming resistive to the field-aligned currents. Figure 3.33 is an example of their observations. Perkins (1968) suggested that a monoenergetic beam of electrons causes unstable electrostatic waves in the upper ionosphere and that the resulting stochastic process accelerates a few electrons to ~40–100 keV.

(b) *Double Layer*

In a low pressure discharge tube, one can often observe a particular type of plasma sheath, called the double layer, in which a very high electric field can develop. The simplest double layer can be observed in a discharge tube in which the cross-section of the tube changes discontinuously. Figure 3.34 shows the distribution of potential, electric field and space charge in a typical plane double layer. Alfvén (1958) was the first to point out its possible significance in magnetospheric physics.

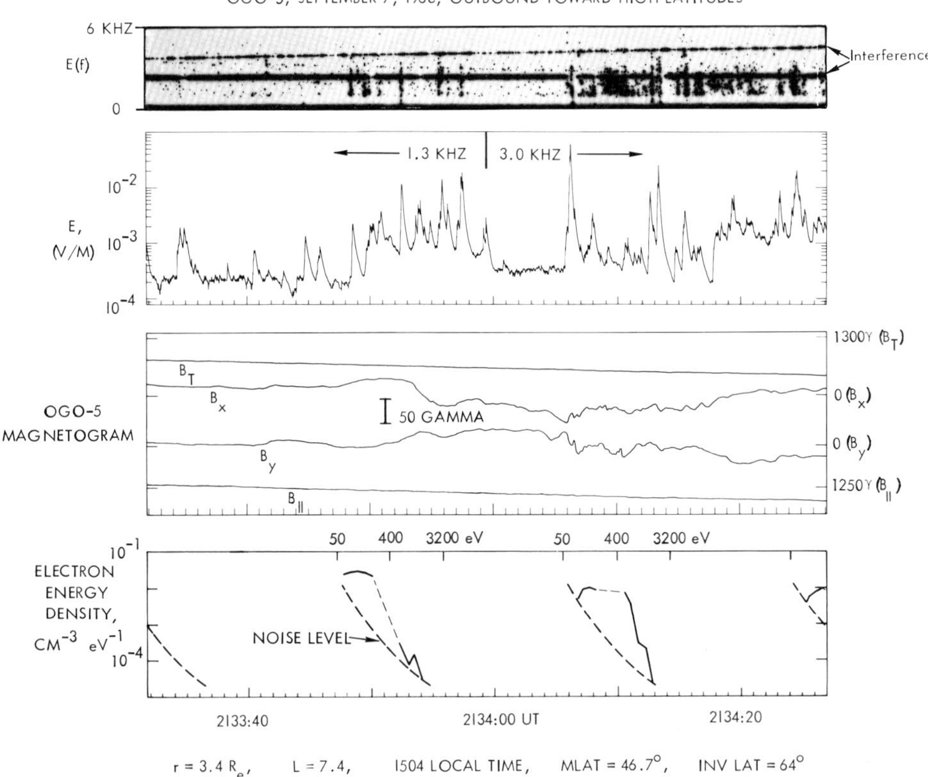

Fig. 3.33. Intense magnetospheric electric field noise burst associated with field-aligned currents (deduced from magnetic field data in the middle row). (Scarf, F. L., Fredricks, R. W., Green, I. M., and Russell, C. T.: *J. Geophys. Res.* **77**, 2274, 1972.)

Block (1972a, b, 1975) listed the following characteristics for a double layer:

(1) The density is lower than in the surrounding plasma.

(2) There is a substantial deviation from quasi-neutrality.

(3) The total net space charge integrated over the double layer is very small or negligible.

(4) The thickness of the double layer is less than a mean free path of the plasma particles. Usually, the thickness is of order 50 Debye lengths.

(5) The electric field within the double layer is very much stronger than any field that can possibly exist in the undisturbed plasma.

(6) The potential difference across the layer is of order kT/e or larger.

A stational double layer can be thought of as a vacuum region that divides the plasma into two. Electrons from one of the plasma regions are accelerated by the double layer, but positive ions are reflected there. On the other hand, electrons from the other plasma regions are reflected, but positive ions are accelerated by it. Block suggested that the most likely altitude for stable double layers is 1000–2000 km on closed field lines; for more recent works on the double layer, see Swift (1975) and Block and Fälthammar (1975).

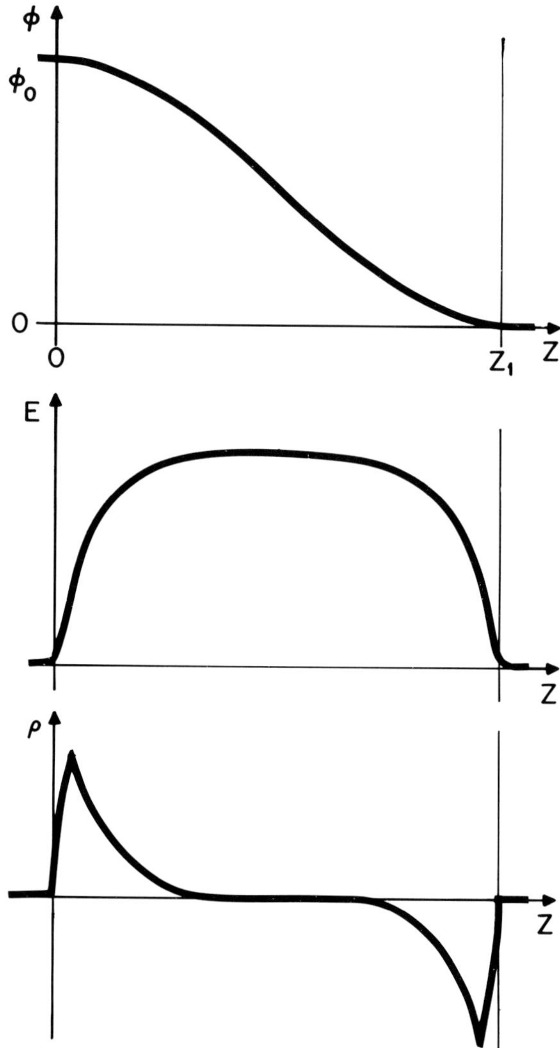

Fig. 3.34. Electric potential, the electric field and space charge distributions in a double layer. (Block, L. P.: *Physics of the Hot Plasma in the Magnetosphere*, Hultqvist, B. and Stenflo, L. (eds.), p. 229, Plenum Press, 1975.)

References

Akasofu, S.-I., Hones, E. W. Jr., Bame, S. J., Asbridge, J. R. and Lui, A. T. Y.: 1973, 'Magnetotail and Boundary Layer Plasmas at a Geocentric Distance of $\sim 18\,R_E$: Vela 5 and 6 Observations', *J. Geophys. Res.* **78**, 7257.

Alekseyev, I. I. and Shabanskiy, V. P.: 1971, 'Model of the Magnetospheric Field', *Geomag. Aeronom.* **11**, 480.

Alfvén, H.: 1958, 'On the Theory of Magnetic Storms and Aurorae', *Tellus* **10**, 104.

Alfvén, H.: 1968, 'Some Properties of Magnetospheric Neutral Surfaces', *J. Geophys. Res.* **73**, 4379.

Alfvén, H. and Fälthammar, C.-G.: 1971, 'A New Approach to the Theory of Magnetosphere', *Cosmic Electrodyn.* **2**, 78.

Amundsen, R., Søraas, F. and Aarsnes, K.: 1975, 'Cleft Signature in Proton Fluxes above 100 keV', *J. Geophys. Res.* **80**, 685.

Anderson, K. A., Chase, L. M., Lin, R. P., McCoy, J. E. and McGuire, R. E.: 1972, 'Solar-Wind and Interplanetary Electron Measurements on the Apollo 15 Subsatellite', *J. Geophys. Res.* **77**, 4611.

Armstrong, T. P. and Krimigis, S. M.: 1968, 'Observations of Protons in the Magnetosphere and Magnetotail with Explorer 33', *J. Geophys. Res.* **73**, 143.

Armstrong, J. C., Akasofu, S.-I. and Rostoker, G.: 1975, 'A Comparison of Satellite Observations of Birkeland Currents with Ground Observations of Visible Aurora and Ionospheric Currents', *J. Geophys. Res.* **80**, 575.

Arnoldy, R. L. and Choy, L. W.: 1973, 'Auroral Electrons of Energy Less Than 1 keV Observed at Rocket Altitude', *J. Geophys. Res.* **78**, 2187.

Arnoldy, R. L., Hendrickson, R. A. and Winckler, J. R.: 1975, 'Echo 2: Observations at Fort Churchill of a 4-keV Peak in Low-Level Electron Precipitation', *J. Geophys. Res.* **80**, 2316.

Atkinson, G.: 1970, 'Auroral Arcs: Result of the Interaction of a Dynamic Magnetosphere with the Ionosphere', *J. Geophys. Res.* **75**, 4746.

Atkinson, G.: 1973, 'A Proposed Experiment to Determine if Field-Aligned Currents Close by Polarization Currents', *J. Geophys. Res.* **78**, 7549.

Axford, W. I. and Hines, C. O.: 1961, 'A Unifying Theory of High-Latitude Geophysical Phenomena and Geomagnetic Storms', *Can. J. Phys.* **39**, 1433.

Babic, M. and Torven, S.: 1974, 'Current Limiting Space Charge Sheaths in a Low Pressure Arc Plasma', *TRITA-EPP*-7402, Royal Inst. Technology, Stockholm.

Bame, S. J.: 1968, 'Plasma Sheet and Adjacent Regions', *Earth's Particles and Fields*, B. M. McCormac (ed.), p. 873, Reinhold, New York.

Bame, S. J., Asbridge, J. R., Felthauser, H. E., Hones, E. W. Jr. and Strong, I. B.: 1967, 'Characteristics of the Plasma Sheet in the Earth's Magnetotail ', *J. Geophys. Res.* **72**, 113.

Barsukov, V. M., Van'yan, L. L. and Pudovkin, M. I.: 1972, 'Possible Acceleration of Auroral Electrons by the Electric Field of Pulsations', *Geomag. Aeronom.* **12**, 504.

Beach, R., Cresswell, G. R., Davis, T. N., Hallinan, T. J. and Sweet, L. R.: 1968, 'Flickering, a 10-cps Fluctuation within Bright Auroras', *Planet. Space Sci.* **16**, 1525.

Bering, E. A., Kelley, M. C. and Mozer, F. S.: 1973, 'Split Langmuir Probe Measurements of Current Density and Electric Fields in an Aurora', *J. Geophys. Res.* **78**, 2201.

Berko, F. W.: 1973, 'Distributions and Characteristics of High-Latitude Field-Aligned Electron Precipitation', *J. Geophys. Res.* **78**, 1615.

Berko, F. W. and Hoffman, R. A.: 1974, 'Dependence of Field-Aligned Electron Precipitation Occurrence on Season and Altitude', *J. Geophys. Res.* **79**, 3749.

Berko, F. W., Hoffman, R. A., Burton, R. K. and Holzer, R. E.: 1975, 'Simultaneous Particle and Field Observations of Field-Aligned Currents', *J. Geophys. Res.* **80**, 37.

Bernstein, W. and Wax, R. L.: 1970, 'Impulsive Precipitation Events during an Auroral Breakup', *J. Geophys. Res.* **75**, 3915.

Bird, M. K.: 1975, 'Solar Wind Access to the Plasma Sheet along the Flanks of the Magnetotail', *Planet. Space Sci.* **23**, 27.

Bird, M. K. and Beard, D. B.: 1972a, 'The Self-Consistent Geomagnetic Tail under Static Conditions', *Planet. Space Sci.* **20**, 2057.

Bird, M. K. and Beard, D. B.: 1972b, 'Self-Consistent Description of the Magnetotail Current System', *J. Geophys. Res.* **77**, 4864.

Block, L. P.: 1972a, 'Acceleration of Auroral Particles by Electric Double Layers', *Earth's Magnetospheric Processes*, B. M. McCormac (ed.), p. 258, D. Reidel Publ. Co., Dordrecht-Holland.

Block, L. P.: 1972b, 'Potential Double Layers in the Ionosphere', *Cosmic Electrodynamics* **3**, 346.

Block, L. P.: 1975, 'Double Layers', *Physics of the Hot Plasma in the Magnetosphere*, B. Hultqvist and L. Stenflo (eds.), p. 229, Plenum Press, New York.

Block, L. P. and Fälthammar, C.-G.: 1968, 'Effects of Field-Aligned Currents on the Structure of the Ionosphere', *J. Geophys. Res.* **73**, 4807.

Block, L. P. and Fälthammar, C.-G.: 1968, 'Mechanisms That May Support Magnetic-Field-Aligned Electric Fields in the Magnetosphere', Dept. of Plasma Phys. Rep. *TRITA-EPP*-75-22, Royal Inst. Technology, Stockholm.

Bornatici, M. and Schindler, K.: 1974, 'A Self-Consistent Approach to Adiabatic Neutral Sheets', *J. Geophys. Res.* **79**, 529.

Bosqued, J. M. and Rème, H.: 1974, 'Evidence of Energy Dependent Mechanisms for the Precipitation of Auroral Electrons', *Planet. Space Sci.* **22**, 1279.

Bowling, S. B.: 1974, 'The Influence of the Direction of the Geomagnetic Dipole on the Position of the Neutral Sheet', *J. Geophys. Res.* **79**, 5155.

Bowling, S. B. and Wolf, R. A.: 1974, 'The Motion and Magnetic Structure of the Plasma Sheet near 30 R_E', *Planet. Space Sci.* **22**, 673.

Boyd, J. S., Davis, T. N., Brown, N. B., Hallinan, T. J. and Wallis, D. D.: 1972, 'Observations of Fast Auroral Waves', *Planet. Space Sci.* **20**, 437.

Brace, L. H. and Theis, R. F.: 1974, 'The Behavior of the Plasmapause at Mid-Latitudes: ISIS-1 Langmuir Probe Measurements', *J. Geophys. Res.* **79**, 1871.

Brekke, A.: 1972, 'The Frequency of Pulsating Aurora and Its Relationship to Other Characteristic Parameters', *Planet. Space Sci.* **20**, 285.

Brinton, H. C., Pickett, R. A. and Taylor, H. A. Jr..: 1968, 'Thermal Ion Structure of the Plasmasphere', *Planet. Space Sci.* **16**, 899.

Bryant, D. A., Smith, M. J. and Courtier, G. M.: 1975, 'Distant Modulation of Electric Intensity during an Expansion Phase of the Auroral Substorm', *Planet. Space Sci.* **23**, 867.

Burke, W. J. and Reasoner, D. L.: 1973, 'Observations of Plasma Flow in the Neutral Sheet at Lunar Distance during Two Magnetic Bays', *J. Geophys. Res.* **78**, 6790.

Burrows, J. R.: 1961, 'A Review of the Magnetospheric Characteristics of Solar Flare Particles', National Res. Council of Canada.

Burrows, J. R.: 1974, 'The Plasma Sheet in the Evening Sector', *Magnetospheric Physics*, B. M. McCormac (ed.), p. 179, D. Reidel Publ. Co., Dordrecht-Holland.

Burrows, J. R. and McDiarmid, I. B.: 1972, 'Trapped Particle Boundary Regions', *Critical Problems of Magnetospheric Physics*, Proceedings of the COSPAR, AGA, URSI Symposium, Madrid, Spain, May 11–13, 1972. E. R. Dyer (ed.), p. 83, National Academy of Sciences, Washington, D.C.

Carlqvist, P. and Boström, R.: 1970, 'Space-Charge Regions above the Aurora', *J. Geophys. Res.* **75**, 7140.

Carpenter, D. L.: 1963, 'Whistler Evidence of a 'Knee' in the Magnetospheric Ionization Density Profile', *J. Geophys. Res.* **68**, 1675.

Chappell, C. R.: 1972a, 'Recent Satellite Measurements of the Morphology and Dynamics of the Plasmasphere', *Rev. Geophys. Space Phys.* **10**, 951.

Chappell, C. R.: 1972b, 'Thermal Ions in the Magnetosphere', *Earth's Magnetospheric Processes*, B. M. McCormac (ed.), p. 280, D. Reidel Publ. Co., Dordrecht-Holland.

Chappell, C. R.: 1974, 'Detached Plasma Regions in the Magnetosphere', *J. Geophys. Res.* **79**, 1861.

Chappell, C. R., Harris, K. K. and Sharp, G. W.: 1970, 'The Morphology of the Bulge Region of the Plasmasphere', *J. Geophys. Res.* **75**, 3848.

Chappell, C. R., Harris, K. K. and Sharp, G. W.: 1971, 'The Dayside of the Plasmasphere', *J. Geophys. Res.* **76**, 7632.

Chase, L. M.: 1968, 'Spectral Measurements of Auroral-Zone Particles', *J. Geophys. Res.* **73**, 3469.

Chase, L. M.: 1969, 'Evidence that the Plasma Sheet Is the Source of Auroral Electrons', *J. Geophys. Res.* **74**, 348.

Chase, L. M., McGuire, R. E., McCoy, J. E., Lin, R. P., Anderson, K. A. and Hones, E. W. Jr.: 1974, 'Plasma and Energetic Particles in the Magnetotail at 60 R_E', *J. Geophys. Res.* **79**, 4779.

Chen, A. J. and Grebowsky, J. M.: 1974, 'Plasma Tail Interpretations of Pronounced Detached Plasma Regions Measured by OGO-5', *J. Geophys. Res.* **79**, 3851.

Chmyrev, V. M.: 1971, 'Excitation of Ion-Acoustic Waves in the Magnetosphere', *Geomag. Aeronom.* **11**, 545.

Choe, J. Y. and Beard, D. B.: 1974, 'The Near Earth Magnetic Field of the Magnetotail Current', *Planet. Space Sci.* **22**, 609.

Choy, L. W., Arnoldy, R. L., Potter, W., Kintner, P. and Cahill, L. J. Jr.: 1971, 'Field-Aligned Particle Currents near an Auroral Arc', *J. Geophys. Res.* **76**, 8279.

Coppi, B., Laval, G. and Pellat, R.: 1966, 'Dynamics of the Geomagnetic Tail', *Phys. Rev. Lett.* **16**, 1207.

Courtier, G. M. and Bryant, D. A.: 1974, 'Electron Precipitation in a Non-Uniform Glow Aurora', *Planet. Space Sci.* **22**, 1067.

Courtier, G. M., Bennett, G. and Bryant, D. A.: 1971, 'Pitch-Angle Diffusion of Electrons in a Glow Aurora', *J. Atmosph. Terr. Phys.* **33**, 847.

Cowley, S. W. H.: 1971, 'The Adiabatic Flow Model of a Neutral Sheet', *Cosmic Electrodyn.* **2**, 90.

Cowley, S. W. H.: 1973, 'A Self-Consistent Model of a Simple Magnetic Neutral Sheet System Surrounded by a Cold, Collisionless Plasma', *Cosmic Electrodynamics* **3**, 448.

Cresswell, G. R.: 1968, 'Fast Auroral Waves', *Planet. Space Sci.* **16**, 1453.

D'Angelo, N.: 1969, 'Role of the Universal Instability in Auroral Phenomena', *J. Geophys. Res.* **74**, 909.

D'Angelo, N.: 1973, 'Ultralow Frequency Fluctuations at the Polar Cusp Boundaries', *J. Geophys. Res.* **78**, 1208.
D'Angelo, N., Bahnsen, A. and Rosenbauer, H.: 1974, 'Wave and Particle Measurements at the Polar Cusp', *J. Geophys. Res.* **79**, 3129.
Davies, C. M.: 1969, 'The Structure of the Magnetopause', *Planet Space Sci.* **17**, 333.
DeForest, S. E. and McIlwain, C. E.: 1971, 'Plasma Clouds in the Magnetosphere', *J. Geophys. Res.* **76**, 3587.
Dungey, J. W.: 1972, 'Theory of Neutral Sheets', *Earth's Magnetospheric Processes*, B. M. McCormac (ed.), p. 210, D. Reidel Publ. Co., Dordrecht-Holland.
Dungey, J. W.: 1975, 'Neutral Sheets', *Space Sci. Rev.* **17**, 173.
Eastwood, J. W.: 1972, 'Consistency of Fields and Particle Motion in the "Speiser" Model of the Current Sheet', *Planet. Space Sci.* **20**, 1555.
Eastwood, J. W.: 1974, 'The Warm Current Sheet Model and Its Implications on the Temporal Behaviour of the Geomagnetic Tail', *Planet. Space Sci.* **22**, 1641.
Eastwood, J. W.: 1975, 'Some Properties of the Current Sheet in the Geomagnetic Tail', *Planet. Space Sci.* **23**, 1.
Evans, D. S.: 1974, 'Precipitating Electron Fluxes Formed by a Magnetic Field Aligned Potential Difference', *J. Geophys. Res.* **79**, 2853.
Evans, D. S.: 1975, 'Evidence for the Low Altitude Acceleration of Auroral Particles', *Physics of the Hot Plasma in the Magnetosphere*, B. Hultqvist and L. Stenflo (eds.), p. 319, Plenum Press, New York.
Fairfield, D. H.: 1968, 'Simultaneous Measurements on Three Satellites and the Observation of the Geomagnetic Tail at 1000 R_E', *J. Geophys. Res.* **73**, 6179.
Frank, L. A.: 1967, 'On the Extraterrestrial Ring Current during Geomagnetic Storms', *J. Geophys. Res.* **72**, 2753.
Frank, L. A.: 1970, 'Further Comments Concerning Low Energy Charged Particle Distributions within the Earth's Magnetosphere and its Environs.', *Particles and Fields in the Magnetosphere*, B. M. McCormac (ed.), p. 319, D. Reidel Publ. Co., Dordrecht-Holland.
Frank, L. A.: 1971a, 'Relationship of the Plasma Sheet, Ring Current, Trapping Boundary, and Plasmapause near the Magnetic Equator and Local Midnight', *J. Geophys. Res.* **76**, 2265.
Frank, L. A.: 1971b, 'Comments on a Proposed Magnetospheric Model', *J. Geophys. Res.* **76**, 2512.
Frank, L. A.: 1971c, 'Plasma in the Earth's Polar Magnetosphere', *J. Geophys. Res.* **76**, 5202.
Frank, L. A. and Ackerson, K. L.: 1971, 'Observations of Charged Particle Precipitation into the Auroral Zone', *J. Geophys. Res.* **76**, 3612.
Frank, L. A. and Gurnett, D. A.: 1971, 'Distributions of Plasmas and Electric Fields over the Auroral Zones and Polar Caps', *J. Geophys. Res.* **76**, 6829.
Fredricks, R. W., Scarf, F. L. and Russell, C. T.: 1973, 'Field-Aligned Currents, Plasma Waves, and Anomalous Resistivity in the Disturbed Polar Cusp', *J. Geophys. Res.* **78**, 2133.
Grebowsky, J. M.: 1970, 'Model Study of Plasmapause Motion', *J. Geophys. Res.* **75**, 4329.
Grebowsky, J. M.: 1971, 'Time-Dependent Plasmapause Motion', *J. Geophys. Res.* **76**, 6193.
Grebowsky, J. M., Tulunay, Y. K. and Chen, A. J.: 1974, 'Temporal Variations in the Dawn and Dusk Midlatitude Trough and Plasmapause Position', *Planet. Space Sci.* **22**, 1089.
Gringauz, K. I.: 1969, 'Low Energy Plasma in the Earth's Magnetosphere', *Rev. Geophys.* **7**, 339.
Gudkova, V. A., Zelenyy, L. M. and Liperovskiy, V. A.: 1973, 'Dynamics of Longitudinal Currents in the Magnetosphere', *Geomag. Aeronom.* **13**, 272.
Gurnett, D. A.: 1972, 'Electric Field and Plasma Observations in the Magnetosphere', *Critical Problems of Magnetospheric Physics*, Proc. joint COSPAR/IAGU/URSI Symposium, Madrid, Spain, 11–13 May, 1972, E. R. Dyer (ed.), p. 123, National Academy of Sciences, Washington, D.C.
Gurnett, D. A.: 1974, 'The Earth as a Radio Source: Terrestrial Kilometric Radiation', *J. Geophys. Res.* **79**, 4227.
Gurnett, D. A.: 1975, 'The Earth as a Radio Source: The Nonthermal Continuum', *J. Geophys. Res.* **80**, 2751.
Gurnett, D. A. and Frank, L. A.: 1972a, 'VLF Hiss and Related Plasma Observations in the Polar Magnetosphere', *J. Geophys. Res.* **77**, 172.
Gurnett, D. A. and Frank, L. A.: 1972b, 'ELF Noise Bands Associated with Auroral Electron Precipitation', *J. Geophys. Res.* **77**, 3411.
Gurnett, D. A. and Frank, L. A.: 1973, 'Observed Relationships between Electric Fields and Auroral Particle Precipitation', *J. Geophys. Res.* **78**, 145.
Gurnett, D. A. and Frank, L. A.: 1974, 'Thermal and Suprathermal Plasma Densities in the Outer Magnetosphere', *J. Geophys. Res.* **79**, 2355.
Gurnett, D. A. and Mosier, S. R.: 1969, 'VLF Electric and Magnetic Fields Observed in the Auroral

Zone with the Javelin 8.46 Sounding Rocket', *J. Geophys. Res.* **74**, 3979.
Haerendel, G. and Paschmann, G.: 1975, 'Entry of Solar Wind Plasma into the Magnetosphere', *Physics of the Hot Plasma in the Magnetosphere*, B. Hultqvist and L. Stenflo (eds.), p. 23, Plenum Press, New York.
Hall, D. S. and Bryant, D. A.: 1974, 'Collimation of Auroral Particles by Time-Varying Acceleration', *Nature* **251**, 402.
Hallinan, T. J.: 1976, 'Spiral-Like Auroral Forms: Observations and a Proposed Theory', Ph.D. Thesis, University of Alaska.
Hallinan, T. J. and Davis, T. N.: 1970, 'Small-Scale Auroral Arc Distortions', *Planet. Space Sci.* **18**, 1735.
Hardy, D. A., Hills, H. K. and Freeman, J. W.: 1975, 'A New Plasma Regime in the Distant Geomagnetic Tail', *Geophys. Res. Lett.* **2**, 169.
Hasegawa, A.: 1971, 'Dynamics of Auroral Formation', *The Radiating Atmosphere*, B. M. McCormac (ed.), p. 384, D. Reidel Publ. Co., Dordrecht-Holland.
Haskell, G. P. and Southwood, D. J.: 1972, 'Comments on Paper by J. R. Sharper and W. J. Heikkila', "Fermi Acceleration of Auroral Particles",' *J. Geophys. Res.* **77**, 6926.
Hedgecock, P. C. and Thomas, B. T.: 1975, 'HEOS Observations of the Configuration of the Magnetosphere', *Geophys. J. R. Astron. Soc.* **41**, 391.
Heikkila, W. J.: 1972, 'The Morphology of Auroral Particle Precipitation', COSPAR, *Space Research XII*, S. A. Bowhill, L. D. Jaffe and M. J. Rycroft (eds.), vol. 2, 1343, Akademie-Verlag, Berlin.
Heikkila, W. J. and Winningham, J. C.: 1971, 'Penetration of Magnetosheath Plasma to Low Altitudes through the Dayside Magnetospheric Cusps', *J. Geophys. Res.* **76**, 883.
Hess, W. N.: 1968, *The Radiation Belt and Magnetosphere*, Blaisdell Publ. Co., Walthalm, Mass.
Hewson-Browne, R. C. and Kendall, P. C.: 1971, 'A Constant-Pressure, Tail-Like, Extrusion of Plasma into the Confined Magnetic Field of a Line Dipole', *Planet. Space Sci.* **19**, 869.
Hill, T. W.: 1974, 'Origin of the Plasma Sheet', *Rev. Geophys. Space Phys.* **12**, 379.
Hoffman, R. A. and Laaspere, T.: 1971, 'A Comparison of VLF Auroral Hiss with Precipitating Low-Energy Electrons Using Simultaneous Data from Two OGO-4 Experiments', *Goddard Space Flight Center, X-646-71-319*, Aug.
Holzer, T. E. and Sato, T.: 1973, 'Quiet Auroral Arcs and Electrodynamic Coupling between the Ionosphere and the Magnetosphere', 2, *J. Geophys. Res.* **78**, 7330.
Hones, E. W. Jr.: 1968, 'Review and Interpretation of Particle Measurements Made by the Vela Satellites in the Magnetotail', *Physics of the Magnetosphere*, R. Carovillano, J. F. McClay and H. R. Radoski (eds.), p. 392, D. Reidel Publ. Co., Dordrecht-Holland.
Hones, E. W. Jr.: 1972, 'Substorm Phenomena in the Distant Magnetosphere', a paper presented at the Solar-Terrestrial Relations Conference, August 28–September 1, 1972, University of Calgary, Calgary, Canada.
Hones, E. W. Jr., Asbridge, J. R., Bame, S. J. and Singer, S.: 1971, 'Energy Spectra and Angular Distributions of Particles in the Plasma Sheet and Their Comparison with Rocket Measurements over the Auroral Zone', *J. Geophys. Res.* **76**, 63.
Hruska, A.: 1968, 'Interaction of Plasma Waves and Particles in the Nonuniform Magnetosphere and Acceleration of Auroral Particles', *Planet. Space Sci.* **16**, 1297.
Huang, Y. H.: 1973, 'The Magnetic Field in the Earth's Tail', *Planet. Space Sci.* **21**, 528.
Hultqvist, B.: 1971, 'On the Production of a Magnetic-Field-Aligned Electric Field by the Interaction between the Hot Magnetospheric Plasma and the Cold Ionosphere', *Planet. Space Sci.* **19**, 749.
Intriligator, D. S., Wolfe, J. H., McKibbin, D. D. and Collard, H. R.: 1969, 'Preliminary Comparison of Solar Wind Plasma Observations in the Geomagnetospheric Wake at 1000 and 500 Earth Radii', *Planet. Space Sci.* **17**, 321.
Johnstone, A. D.: 1971, 'Correlation between Electron and Proton Fluxes in Postbreakup Aurora', *J. Geophys. Res.* **76**, 5259.
Johnstone, A. D. and Davis, T. N.: 1974, 'Low-Altitude Acceleration of Auroral Electrons during Breakup Observed by a Mother-Daughter Rocket', *J. Geophys. Res.* **79**, 1416.
Jorgensen, T. S.: 1968, 'Interpretation of Auroral Hiss Measured on OGO-2 and at Byrd Station in Terms of Incoherent Cerenkov Radiation', *J. Geophys. Res.* **73**, 1055.
Kan, J. R.: 1973, 'On the Structure of the Magnetotail Current Sheet', *J. Geophys. Res.* **78**, 3773.
Kan, J. R.: 1974, 'Energization Processes for Auroral Primary Electrons', *EOS* **55**, 1007.
Kan, J. R.: 1975, 'Energization of Auroral Electrons by Electrostatic Shock Waves', *J. Geophys. Res.* **80**, 2089.
Kan, J. R. and Akasofu, S.-I.: 1976, 'Energy Source and Mechanism for Accelerating the Electrons and Driving the Field-Aligned Currents of the Discrete Auroral Arc', *J. Geophys. Res.* **81** (in press).

Karlson, E. T.: 1970, 'On the Equilibrium of the Magnetopause', *J. Geophys. Res.* **75**, 2438.
Kelley, M. C. and Mozer, F. S.: 1972a, 'A Satellite Survey of Vector Electric Fields in the Ionosphere at Frequencies of 10 to 500 Hertz, 1, Isotropic, High-Latitude Electrostatic Emissions', *J. Geophys. Res.* **77**, 4158.
Kelley, M. C. and Mozer, F. S.: 1972b, 'A Satellite Survey of Vector Electric Fields in the Ionosphere at Frequencies of 10 to 500 Hertz, 2, The Electric Component of ELF Hiss', *J. Geophys. Res.* **77**, 4174.
Kelley, M. C. and Mozer, F. S.: 1972c, 'A Satellite Survey of Vector Electric Fields in the Ionosphere at Frequencies of 10 to 500 Hertz, 3, Low-Frequency Equatorial Emissions and Their Relationship to Ionospheric Turbulence', *J. Geophys. Res.* **77**, 4183.
Kelley, M. C. and Mozer, F. S.: 1972d, 'A Technique for Making Dispersion Relation Measurements of Electrostatic Waves', *J. Geophys. Res.* **77**, 6900.
Kelley, M. C., Mozer, F. S. and Fahleson, U. V.: 1970, 'Measurements of the Electric Field Component of Waves in the Auroral Ionosphere', *Planet. Space Sci.* **18**, 847.
Kelley, M. C., Mozer, F. S. and Fahleson, U. V.: 1971, 'Electric Fields in the Nighttime and Daytime Auroral Zone', *J. Geophys. Res.* **76**, 6054.
Kelley, M. C., Pedersen, A., Fahleson, U. V., Jones, D. and Köhn, D.: 1974, 'Active Experiments Stimulating Waves and Particle Precipitation with Small Ionospheric Barium Releases', *J. Geophys. Res.* **79**, 2859.
Kelley, M. C., Haerendel, G., Kappler, H., Mozer, F. S. and Fahleson, U. V.: 1975, 'Electric Field Measurements in a Major Magnetospheric Substorm', *J. Geophys. Res.* **80**, 3181.
Kennel, C. F., Fredricks, R. W. and Scarf, F. L.: 1970, 'High Frequency Electrostatic Waves in the Magnetosphere', *Particles and Fields in the Magnetosphere*, B. M. McCormac (ed.), p. 257, D. Reidel Publ. Co., Dordrecht-Holland.
Kindel, J. M. and Kennel, C. F.: 1971, 'Topside Current Instabilities', *J. Geophys. Res.* **76**, 3055.
Knight, S.: 1973, 'Parallel Electric Fields', *Planet. Space Sci.* **21**, 741.
Kurth, W. S., Baumback, M. M. and Gurnett, D. A.: 1975, 'Direction-Finding Measurements of Auroral Kilometric Radiation', *J. Geophys. Res.* **80**, 2764.
Kuznetsov, S. N., Sosnovets, E. N., Tserskaya, L. V., Tel'tsov, M. V. and Khorosheva, O. V.: 1972, 'Electron and Proton Acceleration in the Outer Regions of the Magnetosphere during Polar Substorms', *Geomag. Aeronom.* **12**, 426.
Lassen, K.: 1974, 'Relation of the Plasma Sheet to the Nighttime Auroral Oval', *J. Geophys. Res.* **79**, 3857.
Laval, G. and Pellat, R.: 1970, 'Particle Acceleration by Electrostatic Waves Propagating in an Inhomogeneous Plasma', *J. Geophys. Res.* **75**, 3255.
Lee, Y. C. and Cahill, L. J. Jr.: 1975, 'Quiet Time Inflation of the Inner Magnetosphere in the Afternoon and Evening Quadrants', *J. Geophys. Res.* **80**, 1003.
Lemaire, J. and Scherer, M.: 1973, 'Plasma Sheet Particle Precipitation: a Kinetic Model', *Planet. Space Sci.* **21**, 281.
Lemaire, J. and Scherer, M.: 1974, 'Ionosphere-Plasmasheet Field-Aligned Currents and Parallel Electric Fields', *Planet. Space Sci.* **22**, 1485.
Lim, T. L. and Laaspere, T.: 1972, 'An Evaluation of the Intensity of Cerenkov Radiation from Auroral Electrons with Energies down to 100 eV', *J. Geophys. Res.* **77**, 4145.
Maeda, K.: 1975, 'A Calculation of Auroral Hiss with Improved Models for Geoplasma and Magnetic Field', *Planet. Space Sci.* **23**, 843.
Maehlum, B. N. and Moestue, H.: 1973, 'High Temporal and Spatial Resolution Observations of Low Energy Electrons by a Mother-Daughter Rocket in the Vicinity of Two Quiescent Auroral Arcs', *Planet. Space Sci.* **21**, 1957.
Maggs, J. E. and Davis, T. N.: 1968, 'Measurements of the Thicknesses of Auroral Structures', *Planet. Space Sci.* **16**, 205.
Markham, T. P.: 1970, 'Artificial Conjugate Auroral and Afterglow Spectra', *Planet. Space Sci.* **18**, 731.
Maynard, N. C. and Johnstone, A. D.: 1974, 'High-Latitude Day Side Electric Field and Particle Measurements', *J. Geophys. Res.* **79**, 3111.
Maynard, N. C., Bahnsen, A., Christophersen, P., Egeland, A. and Lundin, R.: 1973, 'An Example of Anticorrelation of Auroral Particles and Electric Fields', *J. Geophys. Res.* **78**, 3976.
McDiarmid, I. B. and Budzinski, E. E.: 1968, 'Search for Low-Altitude Acceleration Mechanisms during an Auroral Substorm', *Can. J. Phys.* **46**, 911.
McDiarmid, I. B., Burrows, J. R. and Wilson, M. D.: 1972, 'Solar Particles and the Dayside Limit of Closed Field Lines', *J. Geophys. Res.* **77**, 1103.
McIlwain, C. E.: 1975, 'Auroral Electron Beams near the Magnetic Equator', *Physics of the Hot Plasma in the Magnetopause*, B. Hultqvist and L. Stenflo (eds.), p. 91, Plenum Press, New York.

Meng, C.-I.: 1971, 'Energetic Electrons in the Magnetotail at 60 R_E', *J. Geophys. Res.* **76**, 862.
Meng, C. I. and Anderson, K. A.: 1970, 'A Layer of Energetic Electrons (>40 keV) near the Magnetopause', *J. Geophys. Res.* **75**, 1827.
Meng, C.-I. and Anderson, K. A.: 1974, 'Magnetic Field Configuration in the Magnetotail near 60 R_E', *J. Geophys. Res.* **79**, 5143.
Meng, C.-I. and Anderson, K. A.: 1975, 'Characteristics of the Magnetopause Energetic Electron Layer', *J. Geophys. Res.* **80**, 4237.
Meng, C.-I. and Mihalov, J. D.: 1972a, 'Average Plasma-Sheet Configuration near 60 Earth Radii', *J. Geophys. Res.* **77**, 1739.
Meng, C.-I. and Mihalov, J. D.: 1972b, 'On the Diamagnetic Effect of the Plasma Sheet near 60 R_E', *J. Geophys. Res.* **77**, 4661.
Montgomery, M. D.: 1968, 'Observations of Electrons in the Earth's Magnetotail by Vela Launch 2 Satellites', *J. Geophys. Res.* **73**, 871.
Mozer, F. S.: 1968, 'Rapid Variations of Auroral Particle Fluxes', *J. Geophys. Res.* **73**, 999.
Mozer, F. S. and Fahleson, U. V.: 1970, 'Parallel and Perpendicular Electric Fields in an Aurora', *Planet. Space Sci.* **18**, 1563.
Murayama, T. and Simpson, J. A.: 1968, 'Electrons within the Neutral Sheet of the Magnetospheric Tail', *J. Geophys. Res.* **73**, 891.
Ness, N. F., Scearce, C. S. and Cantarano, S.: 1967, 'Probable Observations of the Geomagnetic Tail at 10^3 Earth Radii by Pioneer 7', *J. Geophys. Res.* **72**, 3769.
Nishida, A. and Lyon, E. F.: 1972, 'Plasma Sheet at Lunar Distance: Structure and Solar-Wind Dependence', *J. Geophys. Res.* **77**, 4086.
Obertz, P.: 1973, 'Two-Dimensional Problem of the Shape of the Magnetosphere', *Geomag. Aeronom.* **13**, 758.
O'Brien, B. J.: 1970, 'Considerations That the Source of Auroral Energetic Particles Is Not a Parallel Electrostatic Field', *Planet. Space Sci.* **18**, 1821.
O'Brien, B. J. and Reasoner, D. L.: 1971, 'Measurements of Highly Collimated Short-Duration Bursts of Auroral Electrons and Comparison with Existing Auroral Models', *J. Geophys. Res.* **76**, 8258.
Ogawa, T. and Sato, T.: 1971, 'New Mechanisms of Auroral Arcs', *Planet. Space Sci.* **19**, 1393.
Oguti, T.: 1969, 'Conjugate Point Problems', *Space Sci. Rev.* **9**, 745.
Oguti, T.: 1971, 'Electric Current in and near Aurora', *Rep. Ionsph. Space Res. Japan*, **25**, 109.
Oguti, T.: 1974a, 'Rotational Deformations and Related Drift Motions of Auroral Arcs', *J. Geophys. Res.* **79**, 3861.
Oguti, T.: 1974b, 'Identification of Hiss-Emitting Auroral Activity', *Rep. Ionosph. Space Res. Japan*, **28**, 124.
Ossakow, S. L.: 1968, 'Anomalous Resistivity along Lines of Force in the Magnetosphere', *J. Geophys. Res.* **73**, 6366.
Papadopoulos, K. and Coffey, T.: 1974, 'Anomalous Resistivity in the Auroral Plasma', *J. Geophys. Res.* **79**, 1558.
Park, C. G.: 1970, 'Whistler Observations of the Interchange of Ionization between the Ionosphere and the Protonsphere', *J. Geophys. Res.* **75**, 4249.
Paschmann, G.: 1972, 'Angular Distributions of Precipitating Electrons', *Earth's Magnetospheric Processes*, B. M. McCormac (ed.), p. 168, D. Reidel Publ. Co., Dordrecht-Holland.
Paschmann, G., Grünwaldt, H., Montgomery, M. D., Rosenbauer, H. and Sckopke, N.: 1973, 'Plasma Observations in the High-Latitude Magnetosphere', Max-Planck-Institut für Physik und Astrophysik, Seventh ESLAB Symposium, 22–25 May.
Pazich, P. M.: 1972, 'Rocket-Based Measurement of Auroral Particle Fluxes Associated with Field-Aligned Currents', Thesis, Rice University, Houston, Texas, Nov.
Perkins, F. W.: 1968, 'Plasma-Wave Instabilities in the Ionosphere over the Aurora', *J. Geophys. Res.* **73**, 6631.
Pudovkin, M. I. and Tsyganenko, N. A.: 1973, 'Particle Motion and Currents in the Neutral Sheet of the Magnetospheric Tail', *Planet. Space Sci.* **21**, 2027.
Rearwin, S. and Hones, E. W. Jr.: 1974, 'Near-Simultaneous Measurement of Low-Energy Electrons by Sounding Rocket and Satellite', *J. Geophys. Res.* **79**, 4322.
Rème, H. and Bosqued, J. M.: 1971, 'Evidence near the Auroral Ionosphere of a Parallel Electric Field Deduced from Energy and Angular Distributions of Low-Energy Particles', *J. Geophys. Res.* **76**, 7683.
Rich, F. J., Vasyliunas, V. M. and Wolf, R. A.: 1972, 'On the Balance of Stresses in the Plasma Sheet', *J. Geophys. Res.* **77**, 4670.
Rich, F. J., Reasoner, D. L. and Burke, W. J.: 1973, 'Plasma Sheet at Lunar Distance: Characteristics and Interactions with the Lunar Surface', *J. Geophys. Res.* **78**, 8097.

Roederer, J. G.: 1970, *Dynamics of Geomagnetically Trapped Radiation*, Springer-Verlag, Berlin.
Roldugin, V. K., Belen'kaya, B. N. and Mal'tseva, N. F.: 1971, 'Electric Fields in the Ionosphere during Pulsating Auroras', *Geomag. Aeronom.* **11**, 691.
Rosenbauer, H., Grünwaldt, H., Montgomery, M. D., Paschmann, G. and Sckopke, N.: 1975, 'Heos 2 Plasma Observations in the Distant Polar Magnetosphere: The Plasma Mantle', *J. Geophys. Res.* **80**, 2723.
Rosenberg, T. J.: 1968, 'Correlated Bursts of VLF Hiss, Auroral Light and X-Rays', *Planet. Space Sci.* **16**, 1419.
Russell, C. T.: 1972a, 'Noise in the Geomagnetic Tail', *Planet. Space Sci.* **20**, 1541.
Russell, C. T.: 1972b, 'Magnetic and Electric Waves in Space', *Earth's Magnetospheric Processes*', B. M. McCormac (ed.), p. 39, D. Reidel Publ. Co., Dordrecht-Holland.
Russell, C. T. and Holzer, R. E.: 1970, 'AC Magnetic Fields', *Particles and Fields in the Magnetosphere*, B. M. McCormac (ed.), p. 195, D. Reidel Publ. Co., Dordrecht-Holland.
Russell, C. T., Chappell, C. R., Montgomery, M. D., Neugebauer, M. and Scarf, F. L.: 1971, 'OGO-5 Observations of the Polar Cusp on November 1, 1968, *J. Geophys. Res.* **76**, 6743.
Sato, T. and Holzer, T. E.: 1973, 'Quiet Auroral Arcs and Electrodynamic Coupling between the Ionosphere and the Magnetosphere, 1', *J. Geophys. Res.* **78**, 7314.
Sato, T. and Ogawa, T.: 1972, 'Discussion of Paper by G. Atkinson, "Auroral Arcs: Result of the Interactions of a Dynamic Magnetosphere with the Ionosphere",' *J. Geophys. Res.* **77**, 1994.
Scarf, F. L.: 1975, 'Characteristics of Instabilities in the Magnetosphere Deduced from Wave Observations', *Physics of the Hot Plasma in the Magnetosphere*, B. Hultqvist and L. Stenflo (eds), p. 271, Plenum Press, New York.
Scarf, F. L.: 'Characteristics of Instabilities in the Magnetosphere Deduced from Wave Observations', Space Sci. Dept., TRW Systems Group, One Space Park, Redondo Beach, California.
Scarf, F. L. and Fredricks, R. W.: 1972, 'Electrostatic Waves in the Magnetosphere', *Earth's Magnetospheric Processes*, B. M. McCormac (ed.), p. 329, D. Reidel Publ. Co., Dordrecht-Holland.
Scarf, F. L., Fredricks, R. W. and Crook, G. M.: 1968, 'Detection of Electromagnetic and Electrostatic Waves on OV3-3', *J. Geophys. Res.* **73**, 1723.
Scarf, F. L., Kennel, C. F., Fredricks, R. W., Green, I. M. and Crook, G. M.: 1970, 'AC Fields and Wave Particle Interactions', *Particles and Fields in the Magnetosphere*, B. M. McCormac (ed.), p. 275, D. Reidel Publ. Co., Dordrecht-Holland.
Scarf, F. L., Fredricks, R. W., Green, I. M. and Russell, C. T.: 1972, 'Plasma Waves in the Dayside Polar Cusp', *J. Geophys. Res.* **77**, 2274.
Scarf, F. L., Fredricks, R. W., Russell, C. T., Kivelson, M., Neugebauer, M. and Chappell, C. R.: 1973, 'Observation of a Current-Driven Plasma Instability at the Outer Zone-Plasma Sheet Boundary', *J. Geophys. Res.* **78**, 2150.
Scarf, F. L., Frank, L. A., Ackerson, K. L. and Lepping, R. P.: 1974, 'Plasma Wave Turbulence at Distant Crossings of the Plasma Sheet Boundaries and the Neutral Sheet', *Geophys. Res. Lett.* **1**, 189.
Scarf, F. L., Fredricks, R. W., Russell, C. T., Neugebauer, M., Kivelson, M. and Chappell, C. R.: 1975, 'Current-Driven Plasma Instabilities at High Latitudes', *J. Geophys. Res.* **80**, 2030.
Schield, M. A. and Frank, L. A.: 1970, 'Electron Observations between the Inner Edge of the Plasma Sheet and the Plasmasphere', *J. Geophys. Res.* **75**, 5401.
Schindler, K.: 1972, 'A Self-Consistent Theory of the Tail of the Magnetosphere', *Earth's Magnetospheric Processes*, B. M. McCormac (ed.), p. 200, D. Reidel Publ. Co., Dordrecht-Holland.
Schindler, K.: 1974, 'A Theory of the Substorm Mechanisms', *J. Geophys. Res.* **79**, 2803.
Schindler, K.: 1975, 'Plasma and Fields in the Magnetotail', *Space Sci. Rev.* **17**, 589.
Schindler, K. and Ness, N. F.: 1972, 'Internal Structure of the Geomagnetic Neutral Sheet', *J. Geophys. Res.* **77**, 91.
Schindler, K. and Soop, M.: 1968, 'Stability of Plasma Sheaths', *Phys. of Fluids* **11**, 1192.
Schubert, G., Sonett, C. P., Smith, B. F., Schwartz, K. and Colburn, D. S.: 1975, 'Using the Moon to Probe the Geomagnetic Tail Lobe Plasma', *Geophys. Res. Lett.* **2**, 277.
Schultz, M.: 1975, 'Geomagnetically Trapped Radiation', *Space Sci. Rev.* **17**, 481.
Schultz, M. and Lanzerotti, L. J.: 1975, *Particle Diffusion in the Radiation Belts*, Springer-Verlag, Berlin.
Sharber, J. R. and Heikkila, W. J.: 1972, 'Fermi Acceleration of Auroral Particles', *J. Geophys. Res.* **77**, 3397.
Sharp, R. D., Carr, D. L., Johnson, R. G. and Shelley, E. G.: 1971, 'Coordinated Auroral-Electron Observations from a Synchronous and a Polar Satellite', *J. Geophys. Res.* **76**, 7669.
Siscoe, G. L.: 1972a, 'Consequences of an Isotropic Static Plasma Sheet in Models of the Geomagnetic Tail', *Planet Sci.* **20**, 937.

Siscoe, G. L.: 1972b, 'On the Plasma Sheet Contribution to the Force Balance Requirements in the Geomagnetic Tail', *J. Geophys. Res.* **77**, 6230.

Smith, P. H., Hoffman, R. A. and Bewtra, N. K.: 1975, 'Ring Current Distribution as Observed by S^3 at Quiet Times and during Magnetic Storms', *EOS* **56**, 618.

Sonnerup, B. U. Ö.: 1971, 'Adiabatic Particle Orbits in Magnetic Null Sheet', *J. Geophys. Res.* **76**, 8211.

Sonnerup, B. U. Ö. and Cahill, L. J. Jr.: 1968, 'Explorer 12 Observations of the Magnetopause Current Layer', *J. Geophys. Res.* **73**, 1757.

Soop, M. and Schindler, K.: 1973, 'A Self-Consistent Two-Dimensional Approach to Magnetospheric Structures', *Astrophys. Space Sci.* **20**, 287.

Sozou, C. and Windle, D. W.: 1969, 'Non-Linear Symmetric Inflation of a Magnetic Dipole', *Planet. Space Sci.* **17**, 999.

Speiser, T. W.: 1965, 'Particle Trajectories in Model Current Sheets, 1, Analytical Solutions', *J. Geophys. Res.* **70**, 4219.

Speiser, T. W.: 1967, 'Particle Trajectories in Model Current Sheets, 2, Applications to Auroras Using a Geomagnetic Tail Model', *J. Geophys. Res.* **72**, 3919.

Speiser, T. W.: 1968, 'On the Uncoupling of Parallel and Perpendicular Particle Motion in a Neutral Sheet', *J. Geophys. Res.* **73**, 1112.

Spitz, A. L. and Watson, M. D.: 1972, 'Auroral Photometric Observations at Geomagnetically Conjugate Points', *Planet. Space Sci.* **20**, 1337.

Spreiter, J. R. and Alksne, A. Y.: 1969, 'Effect of Neutral Sheet Currents on the Shape and Magnetic Field of the Magnetosphere Tail', *Planet. Space Sci.* **17**, 233.

Stenbaek-Nielsen, H. C.: 1974, 'Indications of a Longitudinal Component in Auroral Phenomena', *J. Geophys. Res.* **79**, 2521.

Stenbaek-Nielsen, H. C., Davis, T. N. and Glass, N. W.: 1972, 'Relative Motion of Auroral Conjugate Points during Substorms', *J. Geophys. Res.* **77**, 1844.

Storey, L. R. O.: 1953, 'An Investigation of Whistling Atmospherics', *Phil. Trans. Roy. Soc.* **A246**, 113.

Sugiura, M.: 1972, 'Equatorial Current Sheet in the Magnetosphere', *J. Geophys. Res.* **77**, 6093.

Sugiura, M.: 1973a, 'Quiet Time Magnetospheric Field Depression at 2.3–3.6 R_E', *J. Geophys. Res.* **78**, 3182.

Sugiura, M.: 1973b, 'Magnetospheric Field Morphology at Magnetically Quiet Times', *Radio Sci.* **8**, 921.

Sugiura, M.: 1975, 'Identifications of the Polar Cap Boundary and the Auroral Belt in the High-Altitude Magnetosphere: A Model for Field-Aligned Currents', *J. Geophys. Res.* **80**, 2057.

Sugiura, M. and Poros, D. J.: 1973, 'A Magnetospheric Field Model Incorporating the OGO-3 and 5 Magnetic Field Observations', *Planet. Space Sci.* **21**, 1763.

Swift, D. W.: 1965, 'A Mechanism for Energizing Electrons in the Magnetosphere', *J. Geophys. Res.* **70**, 3061.

Swift, D. W.: 1970, 'Particle Acceleration by Electrostatic Waves', *J. Geophys. Res.* **75**, 6324.

Swift, D. W.: 1975, 'On the Formation of Auroral Arcs and Acceleration of Auroral Electrons', *J. Geophys. Res.* **80**, 2096.

Swift, D. W. and Gurnett, D. A.: 1973, 'Direct Comparison between Satellite Electric Field Measurements and Visual Aurora', *J. Geophys. Res.* **78**, 7306.

Swift, D. W. and Kan, J. R.: 1975, 'A Theory of Auroral Hiss and Implications on the Origin of Auroral Electrons', *J. Geophys. Res.* **80**, 985.

Taylor, H. A. Jr. and Walsh, W. J.: 1972, 'The Light-Ion Trough, the Main Trough, and the Plasmapause', *J. Geophys. Res.* **77**, 6716.

Taylor, H. A. Jr., Brinton, H. C., Pharo, M. W. II and Rahman, N. K.: 1968, 'Thermal Ions in the Exposphere: Evidence of Solar and Geomagnetic Control', *J. Geophys. Res.* **73**, 5521.

Taylor, H. A. Jr., Grebowsky, J. and Walsh, W. J.: 1971, 'Structured Variations of the Plasmapause: Evidence of a Corotating Plasma Tail', *J. Geophys. Res.* **76**, 6806.

Taylor, H. E. and Perkins, F. W.: 1971a, 'Correction', *J. Geophys. Res.* **76**, 6210.

Taylor, H. E. and Perkins, F. W.: 1971b, 'Auroral Phenomena Driven by the Magnetospheric Plasma', *J. Geophys. Res.* **76**, 272.

Toichi, T.: 1972, 'Two-Dimensional Equilibrium Solution of the Plasma Sheet and Its Application to the Problem of the Tail Magnetosphere', *Cosmic Electrodyn.* **3**, 81.

Vesecky, J. F. and Frankel, M. S.: 1975, 'Observations of a Low-Frequency Cutoff in Magnetospheric Radio Noise Received on Imp 6', *J. Geophys. Res.* **80**, 2771.

Villante, U.: 1975, 'Some Remarks on the Structure of the Distant Neutral Sheet', *Planet. Space Sci.* **23**, 723.

Villante, U. and Lazarus, A. J.: 1975, 'Double Streams of Protons in the Distant Geomagnetic Tail', *J. Geophys. Res.* **80**, 1245.

Volosevich, A. V. and Liperovskiy, V. A.: 1972, 'Threshold of Appearance of Anomalous Resistance for Field-Aligned Currents in the Magnetosphere', *Geomag. Aeronom.* **12**, 670.

Walker, R. J., Villante, U. and Lazarus, A. J.: 1975, 'Pioneer 7 Observations of Plasma Flow and Field Reversal Regions in the Distant Geomagnetic Tail', *J. Geophys. Res.* **80**, 1238.

Walker, R. J. and Farley, T. A.: 1972, 'Spatial Distribution of Energetic Plasma Sheet Electrons', *J. Geophys. Res.* **77**, 4650.

Wescott, E. M., Rieger E. P., Stenbaek-Nielsen, H. C., Davis, T. N., Peek, H. M. and Bottoms, P. J.: 1975, 'The $L = 6.7$ Quiet Time Barium Shaped Charge Injection Experiment "Chachalaca",' *J. Geophys. Res.* **80**, 2738.

Whalen, B. A. and McDiarmid, I. B.: 1973, 'Pitch Angle Diffusion of Low-Energy Auroral Electrons', *J. Geophys. Res.* **78**, 1608.

Whalen, B. A., Verschell, H. J. and McDiarmid, I. B.: 1975, 'Correlations of Ionospheric Electric Fields and Energetic Particle Precipitation', *J. Geophys. Res.* **80**, 2137.

Wilhelm, K. and Munch, J. W.: 1975, 'Geomagnetic Micropulsations Caused by Field-Aligned Current Sheets and the Polar Electrojet', Max-Planck-Institut für Aeronomie, April.

Willis, D. M.: 1970, 'The Electrostatic Field at the Magnetopause', *Planet. Space Sci.* **18**, 749.

Willis, D. M.: 1971, 'Structure of the Magnetopause', *Rev. Geophys. Space Phys.* **9**, 953.

Winckler, J. R., Arnoldy, R. L. and Hendrickson, R. A.: 1975, 'Echo 2: A Study of Electron Beams Injected into the High-Latitude Ionosphere from a Large Sounding Rocket', *J. Geophys. Res.* **80**, 2083.

Winningham, J. D., Yashura, F., Akasofu, S.-I. and Heikkila, W. J.: 1975, 'The Latitudinal Morphology of 10-eV to 10-keV Electron Fluxes during Magnetically Quiet and Disturbed Times in the 2100–0300 MLT Sector', *J. Geophys. Res.* **80**, 3148.

Wolfe, J. H., Silva, R. W., McKibbin, D. D. and Mason, R. H.: 1967, 'Preliminary Observations of a Geomagnetospheric Wake at 1000 Earth Radii', *J. Geophys. Res.* **72**, 4577.

CHAPTER 4

RESPONSES OF THE MAGNETOSPHERE TO INTERPLANETARY DISTURBANCES

4.1. Interplanetary Disturbances

Large-scale interplanetary disturbances, their origins and their effects on the magnetosphere, have been discussed in a number of reviews (*cf.* S.T.P., Chapter 7). In studying magnetospheric substorms, it is important, first of all, to identify characteristics of the response of the magnetosphere to various interplanetary disturbances, and to distinguish them from substorm-associated changes. This is because these interplanetary disturbances are almost always present even during a fairly quiet period. Thus, when magnetospheric quantities are monitored by satellite-borne instruments or by ground-based instruments, the response of the magnetosphere to these disturbances is recorded as 'noise' in addition to the substorm 'signal'. For example, a weak shock wave compresses the magnetosphere, and its effect is observed as an increase of the magnetic field of order of 10 γ throughout the magnetosphere. This amount is relatively small on the ground where the field intensity is large ($\sim$50000 γ). However, it is relatively a very large change in the magnetotail, since the magnetotail field intensity in the high latitude lobe is only of order 10–30 γ. Thus, one must identify the compression effects in the records, by monitoring the solar wind pressure, and distinguish them from substorm-associated changes. In the following, we discuss three elementary interplanetary disturbances and their magnetospheric responses.

(i) Interplanetary discontinuities which are associated with changes in pressure $(p+(B^2/8\pi))$:

(a) Weak shock waves

(b) Tangential and rotational discontinuities

(ii) Interplanetary disturbances which are associated with changes in the EW component of the interplanetary magnetic field (IMF)

(iii) Interplanetary disturbances which are associated with changes in the NS component (the B_z component) of the IMF.

4.2. Interplanetary Pressure Disturbances and Magnetospheric Responses

On the basis of the Chapman-Ferraro theory of the formation of the magnetosphere, it is not difficult to infer that an enhanced solar wind pressure after the passage of an interplanetary shock wave compresses the magnetosphere, causing an increase of the magnetic field in the magnetospheric cavity. The three-dimensional theory predicts that the amount of the increase ΔH at the equator on the ground is of order

$$\Delta H = 26.1 \times 10^4(\sqrt{p_2} - \sqrt{p_1})\gamma$$

where p_1 (dyne cm^{-2}) and p_2 denote the solar wind pressure before and after the passage of the shock wave, respectively (*cf.* S.T.P., p. 587). However, actual observations indicate that the numerical factor in the above equation is of order 13.1×10^4 (Siscoe *et al.*, 1968; Mikerina and Ivanov, 1971).

Gosling *et al.* (1967) and Burlaga and Ogilvie (1967) showed that some interplanetary tangential discontinuities, as well as interplanetary shock waves, are associated with solar wind pressure changes. Ivanov and Mikerina (1971) showed that a rotational discontinuity observed on 1967, October 28 was also associated with a sudden pressure change. Some of such interplanetary discontinuities are responsible for the sudden compression of the magnetosphere and are recorded as sudden impulses (si). When a sudden impulse is followed by storminess, it is called the storm sudden commencement (ssc).

Changes of the magnetic field in the magnetotail at the time of ssc's and si's have been studied by Patel (1968, 1972), Sugiura *et al.* (1968) and Ondoh (1970). The most noticeable change is a sudden increase of the field magnitude. Ondoh (1970) showed that the magnitude of the sudden increase is greater in the dayside magnetosphere than that on the Earth's surface, but is slightly smaller in the nightside magnetosphere. Figure 4.1 shows his results.

Sugiura *et al.* (1968) examined the ssc of the 1966, July 8 storm in detail by monitoring the propagation of the interplanetary shock wave by two satellites, IMP-3 and Explorer 33, and concluded that the observed magnetic field increase in the tail cannot simply be due to an increased lateral pressure along the magnetopause. Both Sugiura *et al.* (1968) and Ondoh (1970) suggested that the

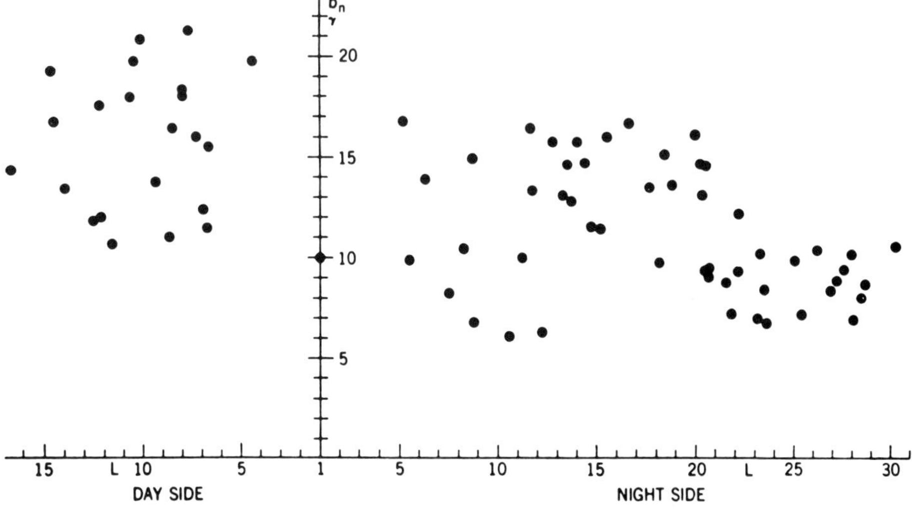

Fig. 4.1. Normalized magnitude of the increase of the magnetic field as a function of L in both the dayside and nightside magnetosphere. (Ondoh, T.: *J. Radio Research Lab.* **17**, 199, 1970.)

transfer of the magnetic flux from the dayside magnetosphere to the magnetotail may account for the observed increase. Patel (1972) showed also that the average propagation speed of the compression effect in the magnetosphere is of order 760 km s^{-1} with a range of 550–1230 km s^{-1}.

Ondoh (1970) showed that the rise time is shorter in the dayside magnetosphere than that on the ground (~200 s), but is longer in the nightside magnetosphere. He showed also that the dayside increase can be explained in terms of the simple compression, namely of the shortening of the magnetopause distance in the dayside. An inspection of Patel's study (1972) indicates also that changes of the magnetotail field are not limited to the magnitude and that both the θ and ϕ components vary appreciably during the passage of interplanetary shock waves. Patel and Cahill (1974) showed also that a definite polarization pattern of the disturbance vector is seen in high latitudes in the Van Allen belt region at the time of ssc's; see also Kaufmann and Walker (1974). Patel (1972) showed also that negative sudden impulses are associated with a decrease of the magnitude of the magnetotail field.

The geocentric distance D of the magnetopause is given by

$$D = (f^2 B_0^2 / 2\pi k \rho V_s^2)^{1/6}$$

where

$B_0 = 0.32$ G,
$\rho =$ the mass density of the solar wind,
$V_s =$ the solar wind speed,

and the factor k is a measure of how efficiently solar wind particles transfer their momentum to the magnetosphere ($k = 1$ for inelastic collisions and $k = 2$ for elastic collisions) and the factor ($f = 1 \sim 1.54$) relates the geomagnetic field just inside the magnetopause to the undisturbed dipole field at that point. Fairfield (1971) found, on the basis of a large number of magnetopause crossings, that the observed value of the magnetopause distance and the estimated distance (by using the observed values of ρ and V_s^2) agree when f^2/k is taken to be 1.4.

Lui *et al.* (1976) examined the effect of the solar wind pressure (nV_s^2) on the radius R_T of the magnetotail. Figure 4.2 shows their results. It can be seen that R_T is a very sensitive function of nV_s^2.

Other reported magnetospheric responses to a sudden interplanetary pressure change are:

(i) Burch (1973b) reported the occurrence of a widespread precipitation of protons (0.7–7 keV) in predawn hours during the initial phase of geomagnetic storms.

(ii) Vorob'yev (1974) reported a brief increase of auroral activity at the time of ssc's; see also Roldugin (1974).

(iii) Leinbach *et al.* (1970), Brown *et al.* (1972) and Brown and Driatsky (1973) reported a complicated change of the distribution of cosmic noise absorption at the time of ssc's.

(iv) Brown (1974) reported a marked decrease of cosmic noise absorption at the time of a negative sudden impulse which happened to occur during a substorm.

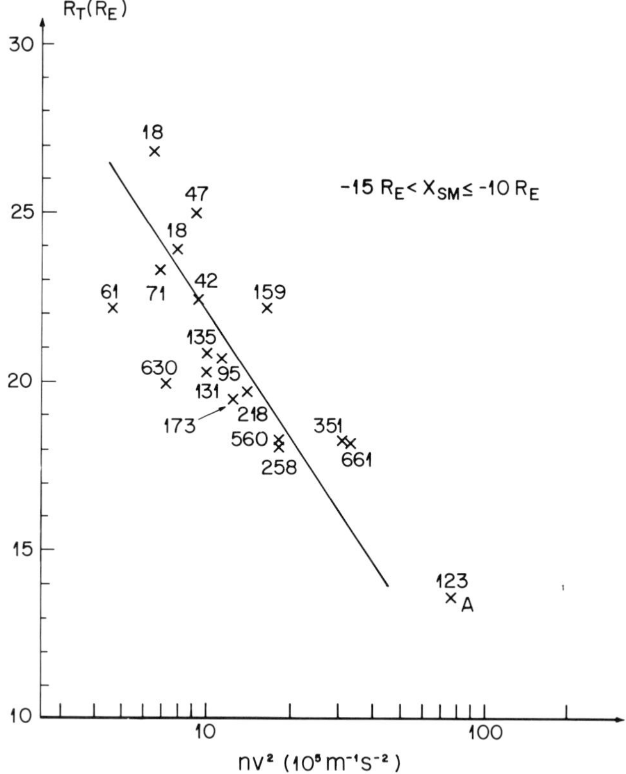

Fig. 4.2. Radius of the magnetotail at $-15\ R_E < X < -10\ R_E$ as a function of solar wind pressure nV_s^2 (10^5 m^{-1} s^{-2}). (Courtesy of Lui, A. T. Y. and Akasofu, S.-I.)

(v) Harang (1969) reported a marked increase of VLF emissions at the time of the ssc of the great magnetic storm of 1967, May 25–26.

(vi) Lin *et al.* (1973) reported a complicated plasma flow at the synchronous distance at the time of the sudden commencement of the 1967, February 15 storm (2348 UT). In particular, they observed a westward flow of about 50 km s^{-1} (corresponding to a radial electric field of 10 mV m^{-1}).

One of the complex features in studying magnetotail changes during the compression is that an intense compression triggers a magnetospheric substorm without delay (Schieldge and Siscoe, 1970; Kawasaki *et al.*, 1971). Figure 4.3 shows an example of the substorms which are triggered by interplanetary shock waves. The arrival of the shock wave in the vicinity of the magnetosphere (to the location of the satellite) is signaled by a sudden increase of the field magnitude B_{IMF}. The corresponding ground magnetic records from several auroral zone stations show the occurrence of large positive or negative bays immediately after the arrival of a positive pulse (ssc). Kawasaki *et al.* (1971) showed that IGY ssc's of magnitude greater than 20 γ had a very high probability (90%) of triggering substorms. Figure 4.4 shows a few examples of the simultaneous magnetic records from Honolulu and College. It can be clearly seen that intense negative

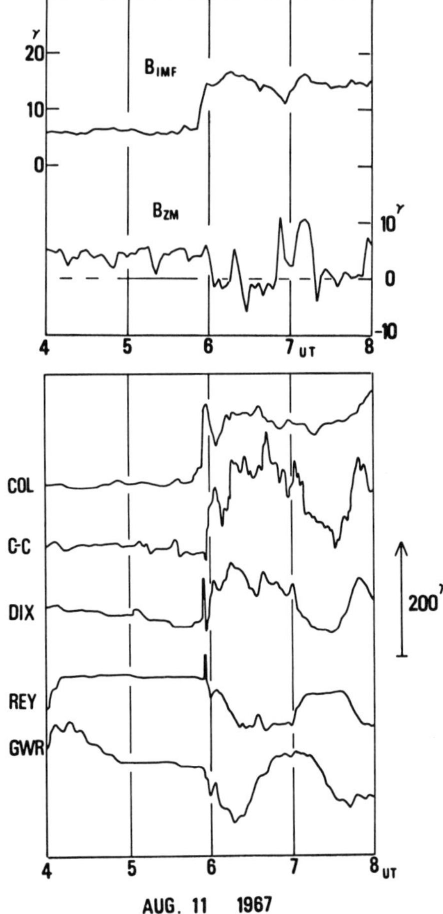

Fig. 4.3. Magnetospheric substorm triggered by an IMF shock wave. The IMF parameters ($|B|$ and B_z) are shown in the upper panel. The lower panel shows the *H* component magnetic records from several auroral zone stations. An intense substorm occurred immediately after the ssc caused by the shock wave. (Courtesy of Iijima, T. and Kokubun, S.)

bays occurred at College almost precisely at the time of the ssc's observed at Honolulu. Such a high probability of triggering substorms makes it difficult to study the response of the magnetosphere to sudden pressure changes alone.

Kawasaki *et al.* (1971) showed also that on the average, 49% of IGY ssc's were associated with substorms, but only 4% during the IQSY. They attributed this difference to the magnitude of ssc's. They showed also that the triggering probability did not depend on how the interplanetary magnetic field varied across shock waves or tangential discontinuities. On the other hand, Burch (1972a) and Iijima (1973) noted that the triggering probability is high when the interplanetary magnetic field has a southward component at the time of ssc's. As we shall discuss in Section 4.4.7, it is possible to monitor properly substorm activity at magnetic stations along the auroral zone only when the IMF has an appreciable southward

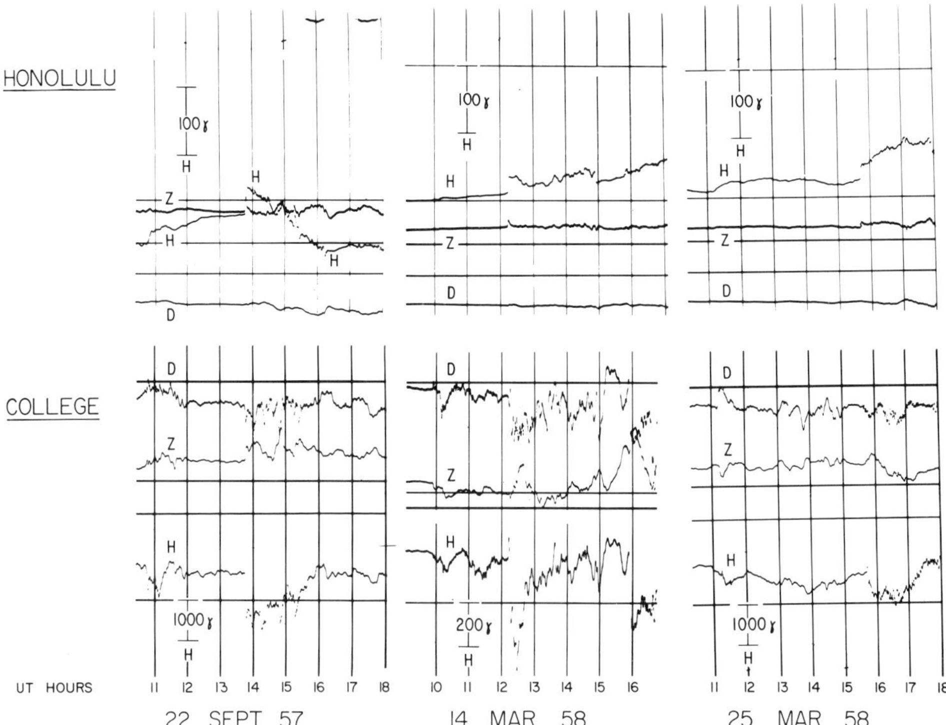

Fig. 4.4. Example of substorms (the College magnetic records) triggered by interplanetary shock waves (see ssc's in the Honolulu magnetic records). (Kawasaki, K., Akasofu, S.-I., Yasuhara, F. and Meng, C.-I.: *J. Geophys. Res.* **76**, 6781, 1971.)

component. Thus, the problem may at least partially be related to the observing probability rather than the true triggering probability. A possible mechanism for triggering substorms by an interplanetary shock wave was theoretically investigated by Lyatskiy and Mal'tsev (1971).

4.3 Changes of the IMF EW Component and Magnetospheric Responses

4.3.1. MERGING OF THE GEOMAGNETIC FIELD LINES WITH THE IMF EW COMPONENT

It has so far been assumed that the IMF has only the NS component. In this section, we examine how the topology of the magnetosphere changes if the EW component is added. This new situation corresponds to the case in which a magnetic field is added along the z-axis in a two-dimensional merging configuration (Section 1.4.2). Thus, the system becomes no longer two-dimensional (Cowley, 1973). For example, if there is no z component, all the field lines cross perpendicularly the z-z' line. However, this is no longer the case. Figure 4.5 illustrates this situation. A field line moves first down toward the diffusion region (1-1′ to 4-4′). As it enters the diffusion region, it splits into two and also becomes

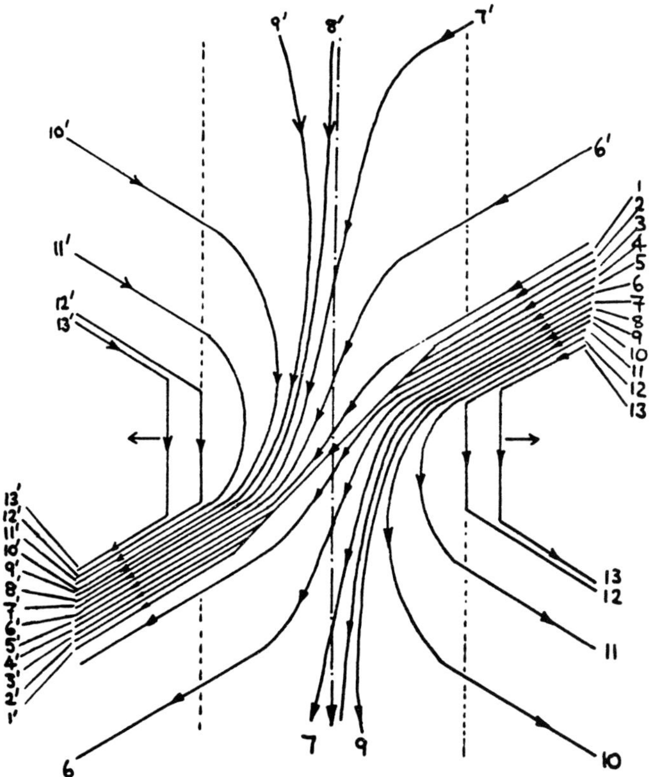

Fig. 4.5. Reconnection of nonparallel field lines. The X-line is shown by the dash-dot line. (Cowley, S. W. H.: *Radio Sci.* **8**, 903, 1973.)

extended along the neutral line (z-z'). Thus, once a field line enters into the diffusion region, its identity is lost, although its ends (where the condition $\boldsymbol{E} + \boldsymbol{V} \times \boldsymbol{B} = 0$ still holds) can be identified. As the field lines move deeper into the diffusion region, the splitting increases (6-6′, 7-7′). When it reaches the neutral line, one end of the two split lines (8-8′) extends to infinity (because the only magnetic field component along the z-z' line is the newly added component). As they move out of the diffusion region, their extension along the z-z' line starts to diminish (9-9′, 10-10′). Then they reach the boundary of the diffusion region and flow into the convection region (10-10′ to 13-13′).

Before applying the above feature of the interplanetary EW component on the magnetosphere, it is necessary to learn about two types of neutral point. In the simplest situation, the two types of neutral point can be formed by adding a northward component to the Earth dipole field. Then, the Earth's field is confined to a spherical cavity, and two neutral points (one in each hemisphere) are formed at points along the dipole axis. Following Cowley (1973), the northern neutral point is called type A neutral point and the southern point is called type B point. The difference between the two points can easily be recognized by observing the field line configuration near the two points. When the two points are observed along the dipole axis as shown in Figure 4.6, the field lines converge toward the northern point (type A), while they diverge away from the southern point (type B).

Now returning to our problem, consider that the IMF has only a westward component. The field line configuration projected onto the y-z plane is shown in Figure 4.7(a). There is a type A neutral point in the northern hemisphere, and a type B neutral point in the southern hemisphere (Figure 4.7b). The two neutral points are connected by two field lines; they are called the X-lines.

Cowley showed schematically by three series of diagrams how this particular merging can take place in magnetospheric situations. In Figure 4.8(a), a westward-directed field line is convected toward the magnetosphere (i). This field line undergoes the following three changes: the splitting (ii), the extension along the line connecting the two neutral points (iii), and finally reaching to the neutral points (iv). Figure 4.8(b) shows the corresponding outward convection of a closed field line which merges with an IMF line to form an open field line, after leaving the diffusion region. Figure 4.8(c) shows how the merging proceeds. Figure 4.9 shows the configuration of the merged field lines on the magnetopause, viewed from the Sun, for both the eastward and westward components.

Now, we examine how the merging of the EW component of the IMF with the geomagnetic field alters the polar electric field and the convection pattern which

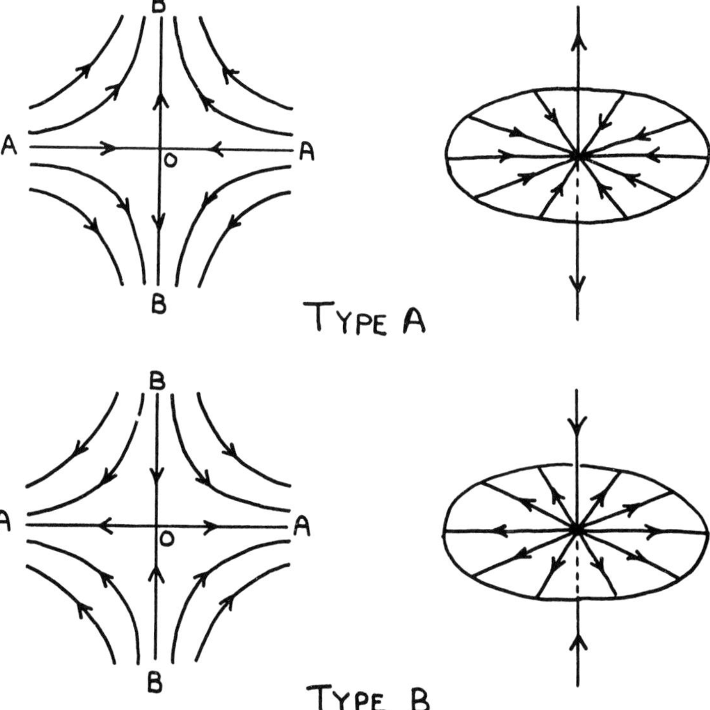

Fig. 4.6. Field-line configurations near the two types of neutral points. (Cowley, S. W. H.: *Radio Sci.* **8**, 903, 1973.)

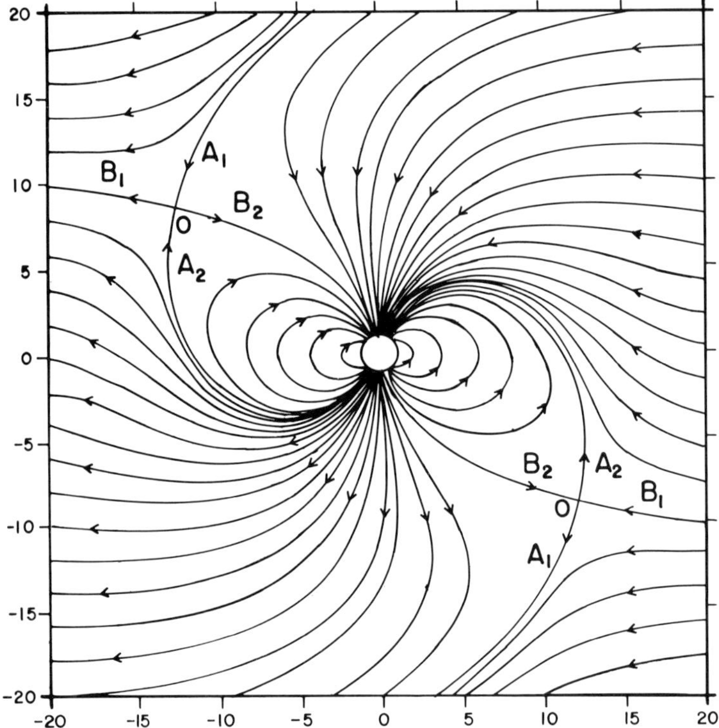

Fig. 4.7(a). Field lines in the Y-Z plane when the geomagnetic field lines and a uniform westward IMF merge. (Cowley, S. W. H.: *Radio Sci.* **8**, 903, 1973.)

were studied only for the NS component in Sections 1.3.4 and 1.3.5. The solar wind-magnetosphere dynamo in the new situation can be understood qualitatively by adding a westward component to b_{norm}, the normal component to the magnetopause (Section 1.4.1). Then, the Lorentz force gives rise to additional deflection to solar wind protons and electrons: the protons are deflected northward and electrons southward in both hemispheres. Negative space charge tends to accumulate just inside the magnetopause in the northern hemisphere, and positive space charge in the southern hemisphere (Stern, 1973).

For the solar wind-magnetosphere dynamo with the NS component of the IMF, the charge distribution in the polar ionosphere can be expressed by $\rho = \rho_0 \sin \psi$ where $\psi = 15° \times \text{LMT}$, reckoned from midnight (Section 1.3.4); this situation can be understood by assuming that the morning half of the oval is positively charged and the evening half is negatively charged. Now, when more negative space charges are added to the northern auroral oval in the day sector, the potential contour lines will no longer be symmetric with respect to the noon-midnight meridian. The negative space charge region expands toward the morning sector. On the other hand, in the southern hemisphere, the positive space charge region expands toward the evening sector. It is not difficult to infer that the opposite effect will occur when the interplanetary magnetic field has an

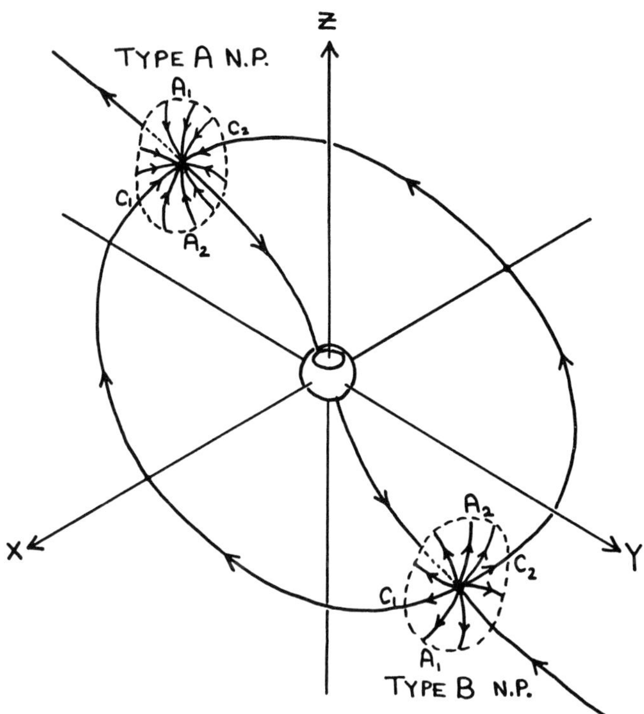

Fig. 4.7(b). Three-dimensional arrangement of the neutral points (Types A and B), the field lines in their vicinity, and the field-line ring which connects the two neutral points for the case shown in Figure 4.7(a). (Cowley, S. W. H.: *Radio Sci.* **8**, 903, 1973.)

eastward component. Leont'yev and Lyatskiy (1974) and Volland (1975) computed the expected electric field and current distributions for both the toward and away sector days.

4.3.2. OBSERVATIONS

Far from the Sun in interplanetary space, the configuration of the interplanetary magnetic field lines resemble an Archimedean spiral. Thus, magnetic field vector $\boldsymbol{B}$ has a radial component B_r, either inward (toward the Sun) or outward (away from the Sun), and an azimuthal component B_ϕ. Interplanetary space is often divided into alternating spiral sectors, in each of which the radial component is directed either outward or inward (*cf.* S.T.P., pp. 26–31). The azimuthal component B_ϕ (namely, the EW component) is directed westward in the toward sector and eastward in the away sector.

Heppner (1972) showed that the electric field distribution becomes asymmetric with respect to the noon-midnight meridian when the EW component has a non-zero value, and that the asymmetry is clearly controlled by it. Figure 4.10 shows schematically the field distributions for both the away (eastward component) and the toward (westward component) sectors. Heppner (1972) summarized

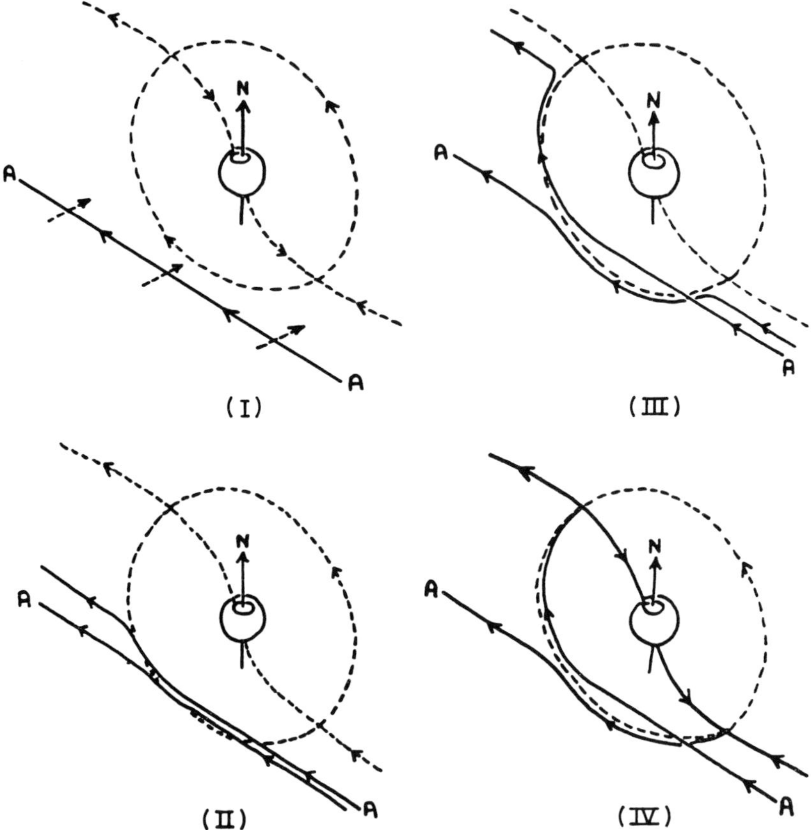

Fig. 4.8(a). Deformation of an IMF field line as it is convected into the dayside diffusion region.

his results schematically by inferring the convection pattern in the northern polar region for both sectors. They are shown in Figure 4.11. Such an asymmetry of the polar electric field and convection pattern is expected to cause a significant difference in the distribution of the electric current in the polar ionosphere and thus in the daily magnetic variation at polar cap stations between away and toward days.

Differences of the daily magnetic variations at Thule, Resolute Bay and Godhavn between away days and toward days were first noted by Svalgaard (1968, 1972, 1975). This was soon confirmed by Mansurov (1969), Jørgensen *et al.* (1972), Friis-Christensen *et al.* (1972), Sumaruk and Feldstein (1973) and Campbell and Matsushita (1973). Further, Friis-Christensen *et al.* (1972) confirmed that it is the EW component (the B_y or B_ϕ component) which is responsible for the change. In order to examine the differences of the daily variation between the two types of days, the daily (X and Y) records are superposed for several away (A) and toward (T) days; the results are shown in Figure 4.12.

Svalgaard (1973) proposed that such changes of the daily magnetic variations could be explained by an additional circular current superposed on the S_q^p current, a

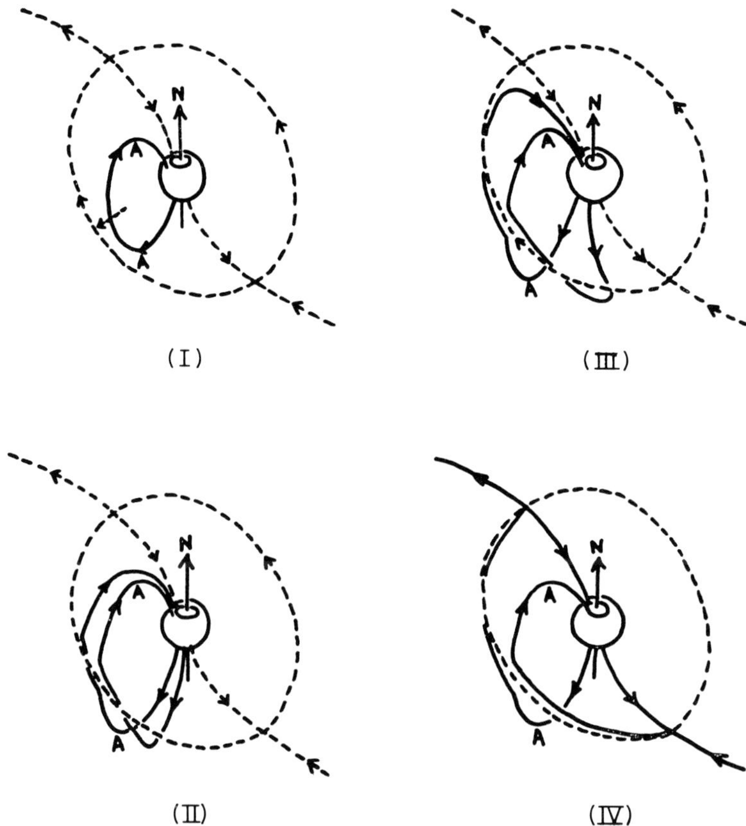

Fig. 4.8(b). Deformation of a geomagnetic field line as it is convected into the dayside diffusion region.

counter-clockwise one (viewed from above the north polar region) during away days and clockwise one during toward days. On the basis of the OGO-2, -4 and -6 satellites, Langel (1974) concluded that the current is located in the ionosphere.

Friis-Christensen and Wilhjelm (1975) have recently made the most extensive analysis of polar cap magnetic variations by grouping daily variations in terms of $B_y > 0\ \gamma$, $B_y \simeq 0\ \gamma$, $B_y < 0\ \gamma$, as well as in terms of $B_z > 0\ \gamma$, $B_z \simeq 0\ \gamma$ and $B_z < 0\ \gamma$. Their results are shown in Figure 4.13. It is at present not certain whether the major changes in the figure can all be explained in terms of the modified convection pattern proposed by Heppner and of the additional current proposed by Svalgaard.

The IMF B_y component is not necessarily uniform in each interplanetary sector. There are some days when the B_y component varies considerably, changing the sign quite rapidly (periods of less than 30 min). Kawasaki *et al.* (1973) showed that the polar cap current responds quite clearly to such rapid variations of the B_y component. Figure 4.14 shows the IMF B_y component and the corresponding Alert Y component. The two curves are slightly shifted in order to allow for the propagation time of the observed interplanetary signal at the location

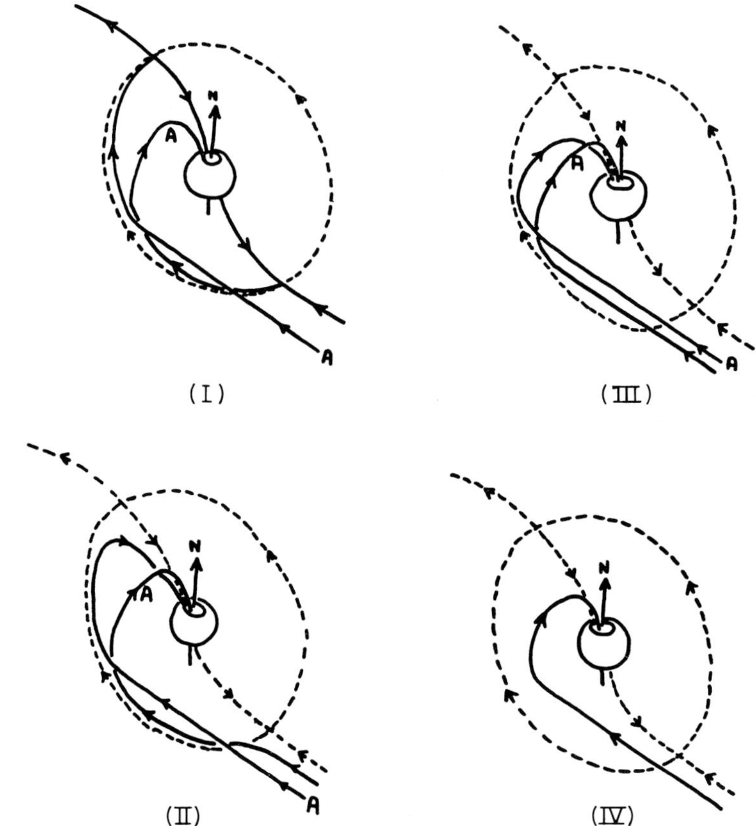

Fig. 4.8(c). Merging process of the IMF field line in Figure 4.8(a) and the geomagnetic field line in Figure 4.8(b). (Cowley, S. W. H.: *Radio Sci.* **8**, 903, 1973.)

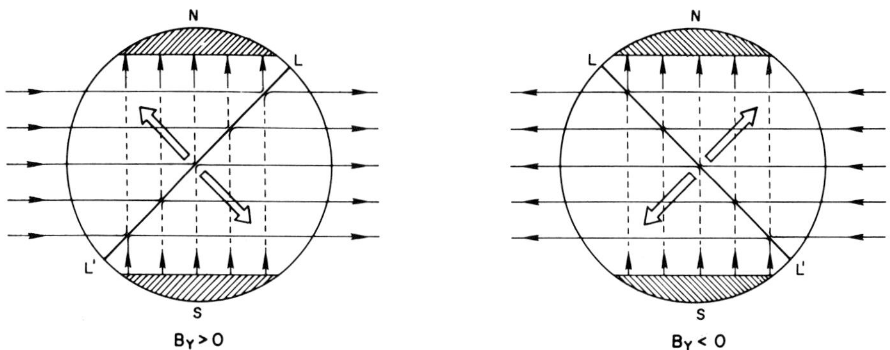

Fig. 4.9. Geometry of the merged field lines on the magnetopause, viewed from the Sun, for the cases of $B_y > 0$ and $B_y < 0$. (Mozer, F. S., Gonzalez, W. D., Bogott, F., Kelley, M. C. and Schutz, S.: *J. Geophys. Res.* **79**, 56, 1974.)

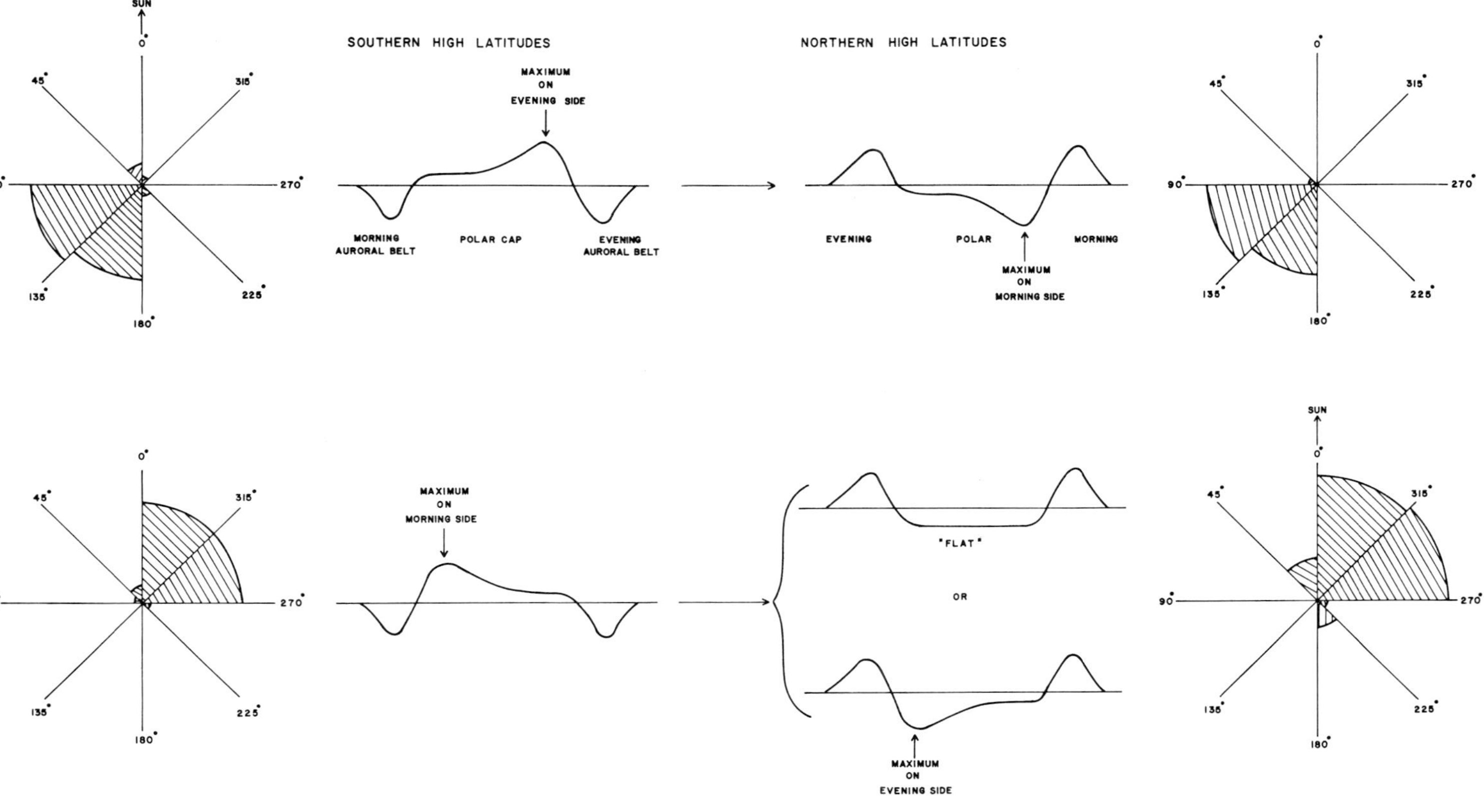

Fig. 4.10. Anti-correlated distribution of the dawn-dusk component of the electric field along the dawn-dusk meridian in both the northern and southern polar caps for different IMF ϕ angles. (Heppner, J. P.: *J. Geophys. Res.* **77**, 4877, 1972.)

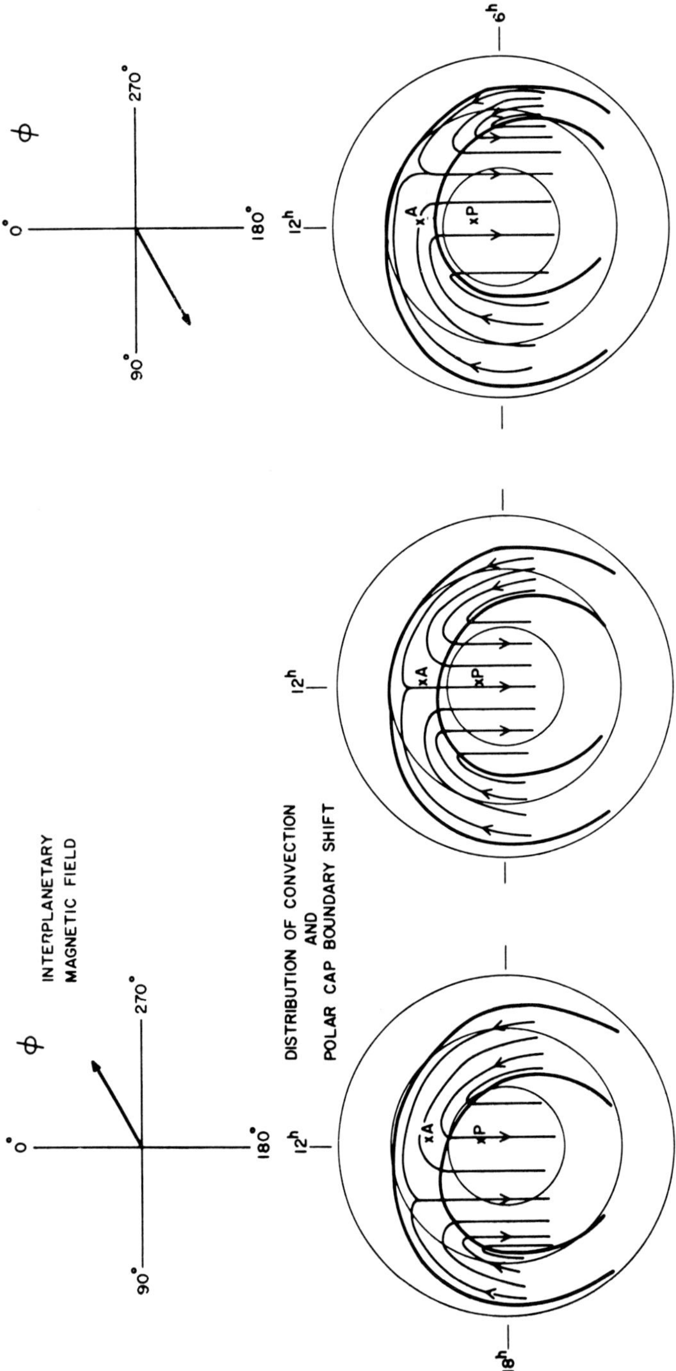

Fig. 4.11. Schematic drawings of the convection pattern in the northern ionosphere for $270° < \phi < 0°$, $\phi = 0°$ and $90° < \phi < 180°$. (Heppner, J. P.: *J. Geophys. Res.* **77**, 4877, 1972.)

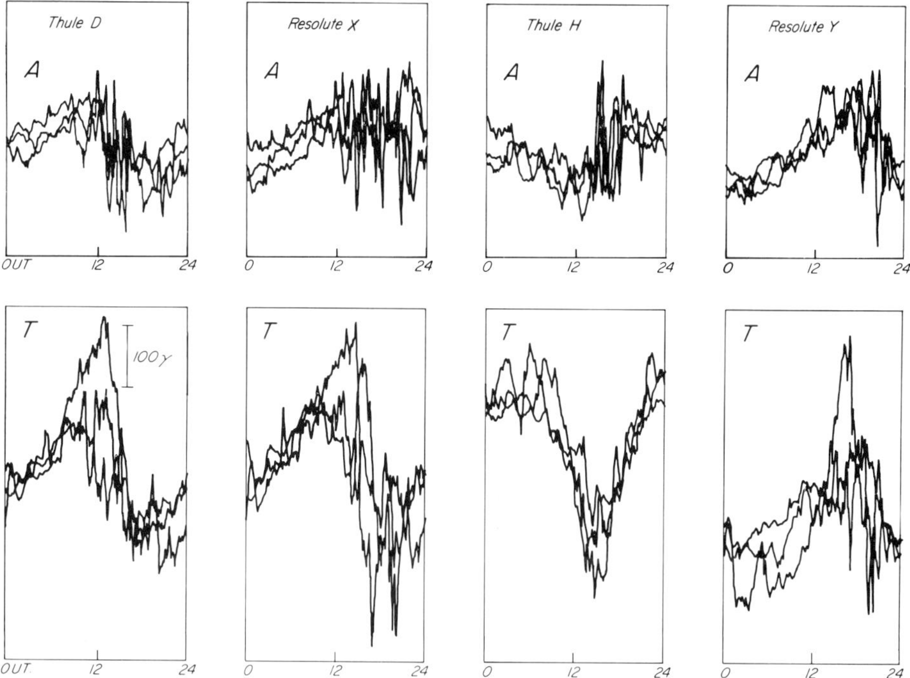

Fig. 4.12. Superposed Thule (the D and H components) and Resolute (the X and Y components) daily magnetic records for several 'toward (T)' and 'away (A)' days, respectively.

of the IMP-3 satellite to the magnetopause. The two curves correlate remarkably well, even for fluctuations of time-scales of order 20 min.

Starkov *et al.* (1973) did not find any significant difference in the occurrence of auroras for the different sectors. Burch (1973) reported that the AL index peaks for toward sectors in the spring and for away sectors in the fall. Bhargava and Rangarajan (1975) found that there is an interesting relationship between the low latitude magnetic field and the polarity of the interplanetary magnetic field. They found that a change from an away sector to a toward sector is associated with a large depression in the H component field.

Kawasaki *et al.* (1973) pointed out that since the magnitude of the IMF is relatively steady, changes of the EW component are often associated with corresponding changes of the NS component (see Section 9.1.2). In a simple situation, a vector of the constant magnitude rotates clockwise or counter-clockwise in the plane which is perpendicular to the interplanetary 'garden-hose' field. Thus when the EW component decreases, the NS component increases. As we shall study in the next section, changes of the NS component cause a particular type of geomagnetic variation in the polar cap region. Therefore, one must be extremely cautious in distinguishing magnetospheric responses to changes of the EW component of the IMF from responses to changes of the NS component. Berthelier *et al.* (1974) statistically eliminated effects of changes of the NS component in their study of the EW modulation effects.

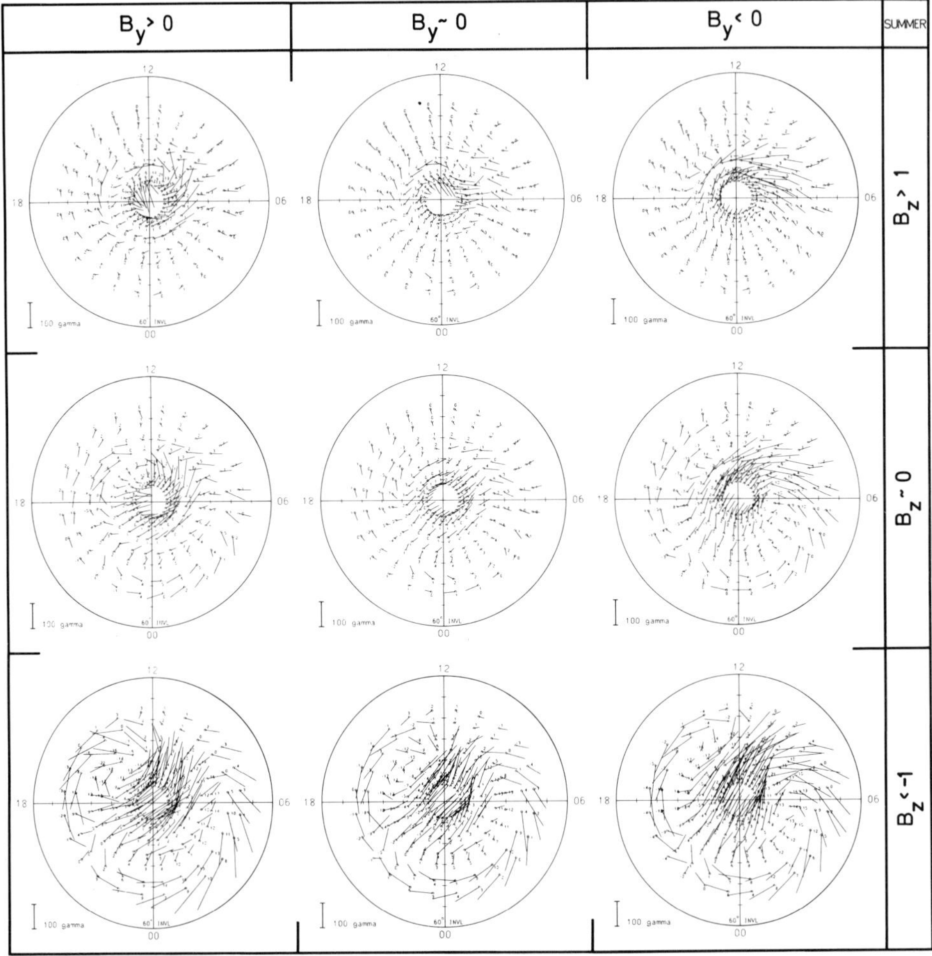

Fig. 4.13. Daily magnetic variation vectors in high latitudes for $B_y > 0$, $B_y \simeq 0$, $B_y < 0$ and $B_z < -1\ \gamma$, $B_z \sim 0\ \gamma$, $B_z > 1\ \gamma$ for summer conditions. (Friis-Christensen, E. and Wilhjelm, J.: *J. Geophys. Res.* **80**, 1248, 1975.)

4.4. Changes of the IMF NS Component and Magnetospheric Responses

4.4.1. INTRODUCTION

Merging between the geomagnetic field and the IMF is greatly enhanced when the latter has a southward component ($B_z < 0$). Figure 4.15 shows, by a block-diagram, some of the proposed phenomena associated with the enhanced merging. This subject has been extensively studied in the past, but most of the studies were rather qualitative and require re-examination. The complexity of the problem can easily be realized by the following example. From Equation (12) in Section 1.4.3, it is not immediately obvious whether or not an enhanced

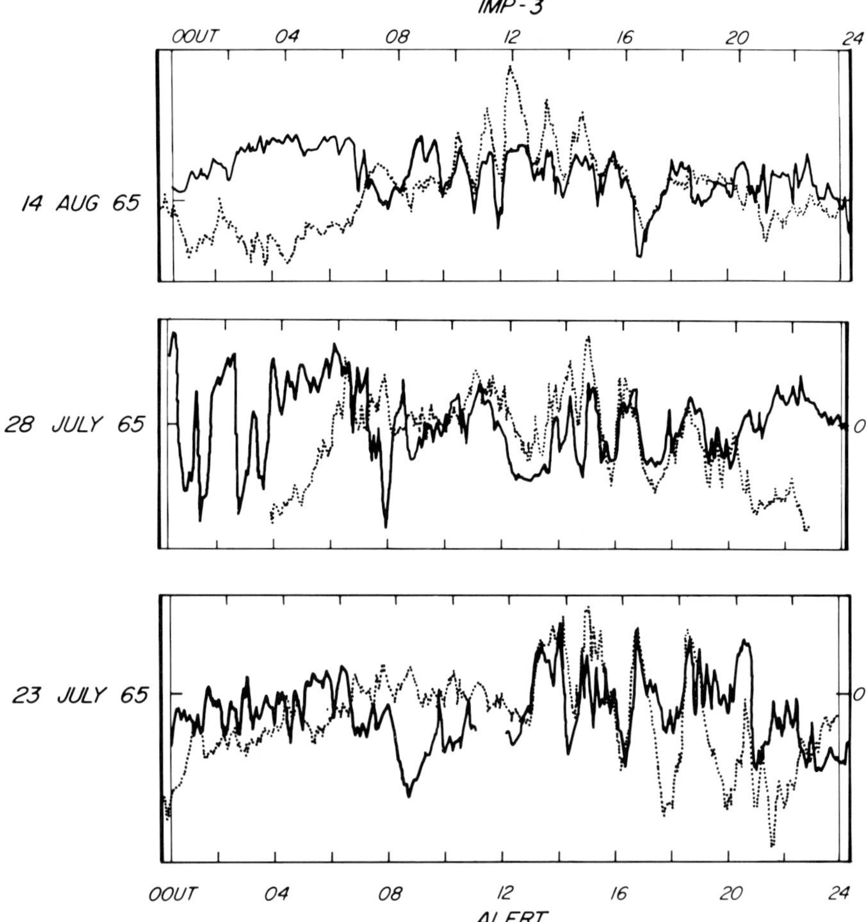

Fig. 4.14. Short-period fluctuations of the IMF B_y component (solid curves) and the corresponding Y component magnetic records from Alert. (Kawasaki, K., Akasofu, S.-I., Yasuhara, F. and Meng, C.-I.: *Planet. Space Sci.* **21**, 1743, 1973.)

merging rate Φ_D would result in an increase of the magnetotail field B_T, because it is not known how R_T varies as a function of Φ_D, even if we assume that L remains unchanged. In Equation (14), it is not immediately obvious how d_1 and d_2 vary individually as Φ_D increases. Thus, as an example, we may write

$$B_T = f_1(p_{s\perp}, \Phi_D, R_T, \Sigma, \ldots)$$
$$\cdots\cdots\cdots\cdots$$

where $p_{s\perp}$ and Σ denote solar wind pressure and substorm effects, respectively. Thus, when one wishes to examine effects of substorms (Σ) on B_T, one must keep in mind that it is rather rare to find periods when B_z and thus Φ_D remain nearly constant in time and thus that one must be cautious in distinguishing magnetospheric responses associated with B_z (or Φ_D) changes from the substorm effects Σ.

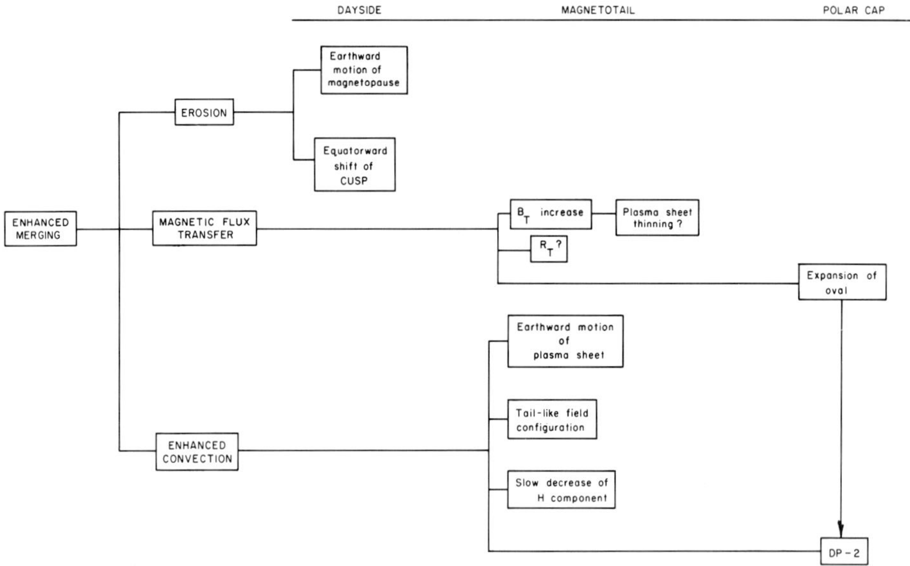

Fig. 4.15. Expected and/or observed responses of the magnetosphere to an enhanced merging of the IMF with geomagnetic field lines on the dayside magnetopause.

Conversely, in studying effects of the IMF B_z component on various magnetospheric parameters, it is important to monitor carefully substorm activity, so that one can exclude substorm periods. Therefore, when one intends to establish the relationship

$$B_T = f_2(B_z(\mathrm{IMF}))$$

one must be certain that

$$p_{s\perp} = \text{const.}$$

$$\Sigma = 0$$

Unfortunately such an elementary precaution has not been taken in many studies in this particular subject, resulting in a considerable confusion. Thus, in this section, we must review critically past studies of the relationships between the B_z component and various magnetospheric parameters by examining whether or not the claimed relationships can be justified. We shall see that many of the claimed relationships between magnetospheric parameters and Φ_D are indeed inaccurate simply because they were contaminated by substorm effects ($\Sigma \neq 0$). Further, one of the most important subjects in magnetospheric physics is to uncover the relationship between Φ_D and Σ. As we shall see in Section 5.2, Σ is found to be a very complicated function of Φ_D, although many workers have considered that each substorm is related to each enhancement of Φ_D.

McPherron (1970, 1972, 1973) proposed that the magnetospheric substorm has an additional phase, called a 'growth phase', which is initiated by the southward turning of the IMF vector. McPherron *et al.* (1973c) and McPherron *et al.* (1973b) proposed further that the southward turning of the IMF vector causes a particular chain of processes which lead to the onset of the expansive phase of substorms and that the proposed growth phase features are manifestations of such a chain of processes. McPherron *et al.* (1973c) state: "A southward turning of the interplanetary magnetic field is accompanied by erosion of the dayside magnetosphere, flux transport to the geomagnetic tail, and thinning and inward motion of the plasma sheet, . . . the expansion phase of substorms can originate near the inner edge of the plasma sheet as a consequence of rapid plasma sheet thinning".

Their conclusion implies that the direct cause of *each* substorm can be tracked back to *each* southward turning of the IMF vector. We shall see in Section 5.2 that a series of magnetospheric substorms is generated by a single southward turning and that many substorms occur during the periods when the IMF is directed northward or even after the northward turning. In fact, Caan *et al.* (1975) have recently concluded that substorms tend to occur at about the time of the northward turning (Section 4.4.4(c)).

Therefore, a significant percentage of substorms is not immediately preceded by the IMF southward turning. This is one of the most important reasons why the responses of the magnetosphere to the north-south component of the IMF should be identified and distinguished from substorm phenomena.

We shall see also in this chapter that many of the proposed growth phase features occur clearly during the expansive phase; many workers misidentified the onset time of the expansive phase. Thus, after reviewing most of the papers which deal with the proposed growth phase, it has become apparent that the proposed growth phase features are either responses of the magnetosphere to the southward turning of the IMF vector (actually, more accurately to $\partial B_z/\partial t < 0$, with B_z either positive or negative), expansive phase features or even recovery phase features.

Equations (12), (13), (14) and (15) in Section 1.4.3 provide the relationships between the potential drop in the dayside magnetosphere Φ_D and various magnetospheric quantities in a steady state. However, it is important to examine how the magnetosphere responds to an enhanced merging Φ_D as a function of time. For this purpose, let us define here the production rate Φ_N of reconnected (or closed) field lines along the nightside X-line, in addition to the production rate Φ_D of merged (or open) field lines along the dayside X-line.

Note that the potential drop Φ_D across the magnetotail is equal to the production rate of open field lines, which is equal to the total amount of IMF flux brought to the magnetopause per unit time, namely $V_s B_i w$, where V_s, B_i and w denote the solar wind speed, the IMF intensity and the length of the X-line. In this book, the quantities Φ_D and Φ_N are often referred to as the merging and reconnection rate, respectively. The conventional definition of the merging rate M (dimensionless) is given in Section 1.4.2.

Consider then the quantity S defined by

$$S = \int_0^T (\Phi_D - \Phi_N)\, dt$$

where the time $t = 0$ is reckoned from the time when the IMF begins to decrease after having a large northward component ($B_z \gtrsim +5\ \gamma$) for an extended period, say 6–12 h.

Figure 4.16 illustrates schematically time variations of Φ_D, Φ_N, $(\Phi_D - \Phi_N)$ and $S = \int_0^\tau (\Phi_D - \Phi_N)\, dt$ when the north-south component of the IMF B_z varies from a large positive value to a negative value for about 2 hours and then back to a large positive value again. Note that an extended period of a large positive B_z value is the initial condition.

In constructing Figure 4.16, it is assumed that both Φ_D and Φ_N have a finite value even when the B_z component has a large positive value. This is because the magnetotail (a product of the solar wind-magnetosphere dynamo) appears to be a permanent feature, and thus there must always be a finite amount of open flux, even when B_z has a large positive value for an extended period. Meng and Anderson (1974) examined extensively magnetotail data at the lunar distance (60 R_E) for both low Kp values (Kp$\leq$1+) and high Kp values (Kp$\geq$2−) and showed that there is no significant difference in the structure of the magnetotail for both conditions.

Under such a condition, one would expect to have a steady state, and both Φ_D and Φ_N would have the same minimum value. Indeed, the auroral oval has the minimum size, and its noon and midnight locations are 80°–82° and 72°–74° in invariant latitudes, respectively. The fact that such a condition can occur suggests strongly that there is a finite amount of magnetic energy in the magnetotail which is not available as substorm energy. Thus, the magnetic energy in the magnetotail

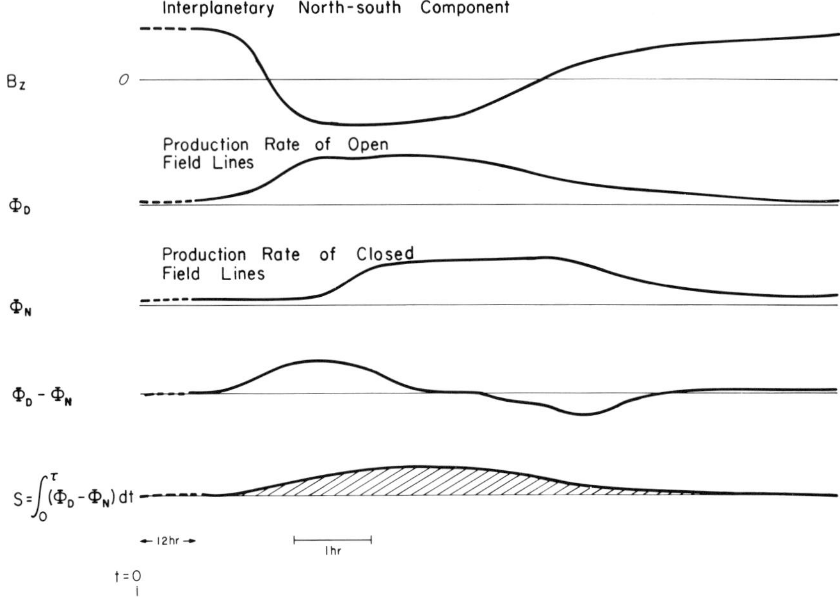

Fig. 4.16. Hypothetical change of the B_z component of the IMF as a function of time and the expected changes of the quantities Φ_D, Φ_N, $(\Phi_D - \Phi_N)$ and $S = \int (\Phi_D - \Phi_N)\, dt$ (Akasofu, S.-I.: *Planet. Space Sci.* **23**, 1349, 1975.)

during an extended period of a large positive B_z value cannot be dissipated as substorm energy. This may also be inferred from the fact that the auroral oval does not shrink to a point even during a very quiet period or most intense substorms. Thus, when the oval contracts to its minimum size, the magnetosphere is in a sort of ground state.

Now, let us suppose that B_z begins to decrease after such a quiet condition. As B_z begins to decrease, Φ_D begins to increase immediately. Let us suppose also that B_z reaches a minimum value in less than one hour after the southward turning and maintains a steady state value for about two hours.

This immediate response of Φ_D to the B_z change can be inferred from the fact that the equatorward motion of the cusp or of the midday aurora follows closely B_z changes; this cusp motion will be discussed in detail in Section 4.4.3(b). On the other hand, it is not at present accurately known how rapidly the midnight portion of the oval responds to the initial B_z change. This depends on how quickly the information on a change of Φ_D can reach the nightside X-line and the production rate Φ_N begins to respond to it. The delay time τ_m can be estimated by

$$\tau_m = \frac{B_P(A_1 - A_0)}{\Phi_D} \lesssim 40 \text{ min}$$

where

$$\Phi_D = V_s B_i w = (400 \text{ km s}^{-1}) \times (5\ \gamma) \times (15\ R_E)$$

and B_P denotes the magnetic field intensity in the polar ionosphere; A_0 and A_1 are the area of the minimum oval and an expanded oval, respectively (A_0 and A_1 are assumed to be bounded by the latitude circles of 72° and 65°, respectively).

Because of this delay, the quantity $(\Phi_D - \Phi_N)$ should increase initially. Note that this quantity is directly proportional to the rate of change of the polar cap area, namely the area bounded by the auroral oval. Thus, when $(\Phi_D - \Phi_N)$ is positive, the oval expands. However, Φ_N begins to increase about 40 min after $t = 0$, (that is, when the IMF signal reaches the nightside X-line) and eventually a new steady state $\Phi_D = \Phi_N$ will be reached. Then, the quantity $(\Phi_D - \Phi_N)$ becomes null, and the auroral oval ceases to expand.

Suppose then the B_z begins to increase after maintaining a large negative value for about 2 h. The quantity Φ_D responds immediately to the B_z change and begins to decrease. However, again, the corresponding Φ_N variation will be delayed for about 40 min, until the new information on B_z can reach the nightside X-line. During this period, the quantity $(\Phi_D - \Phi_N)$ becomes negative. If a large B_z value is maintained for a prolonged period, both Φ_D and Φ_N reach the same minimum value.

Meanwhile, the quantity $S = \int_0^T (\Phi_D - \Phi_N)\, dt$ increases until $(\Phi_D - \Phi_N)$ becomes null, and then begins to decrease. Note that the area of the polar cap is proportional to the quantity S. Thus, the auroral oval expands until S reaches the maximum value and then begins to contract poleward. Eventually, S will become null after a prolonged period of a large B_z value (that is, when both Φ_D and Φ_N reach the same minimum value again).

It should be noted that the quantity S is equal to the amount of open magnetic flux at a time reckoned from $t = 0$ (after a prolonged period of a large positive B_z value).

$$S = B_P(A_1 - A_0)$$

Thus, the amount of S can be monitored, if one can observe continuously the area of the auroral oval. The change in shape of the auroral oval will be discussed further in detail in Section 4.4.6(b). The significance of the quantity S on substorms will be discussed in Sections 5.1 and 5.2.

4.4.2. MERGING WITH THE IMF OF AN ARBITRARY ANGLE

It is rather rare to find times when the IMF field vector is directed purely in either the north-south or east-west direction. Thus, it is necessary to extend the analysis in Section 1.4.2 to include the merging when the direction of the IMF vector is arbitrary. Let θ be the angle between the transverse component of the IMF vector ($|\boldsymbol{B}_i| = \sqrt{B_y^2 + B_z^2}$) and the geomagnetic field vector ($\boldsymbol{B}_g$). The X-line is perpendicular to the line which connects the two vectors $\boldsymbol{B}_i$ and $\boldsymbol{B}_g$. As noted in Section 1.4.2, the merging rate for an anti-parallel situation is given by

$$M_i < kV_A$$

where $k = (1 + \sqrt{2})$ and V_A is the Alfvén wave speed. Sonnerup (1974) showed that when the two fields are not anti-parallel, V_A does not change, but B_i in $V_A = B_i/\sqrt{4\pi\rho}$ should be replaced by the component perpendicular to the X-line. Thus,

$$M_i < (1 + \sqrt{2}) V_A \sin \alpha$$

where α denotes the angle between the X-line and the $\boldsymbol{B}_i$ vector (see Figure 4.17(a)). Let $B_\parallel$ be a common component of $\boldsymbol{B}_i$ and $\boldsymbol{B}_g$ along the X-line.

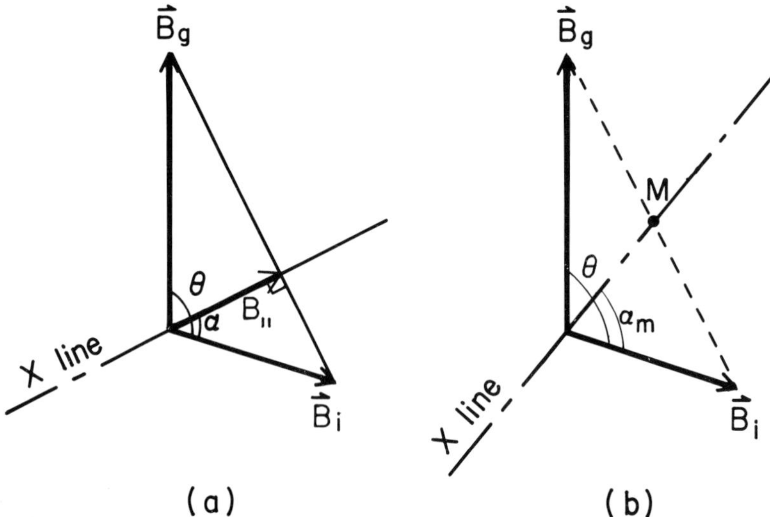

Fig. 4.17. Relationship between the geomagnetic field vector B_g, the IMF vector B_i and the X-line: (a) Sonnerup's model (*J. Geophys. Res.* **79**, 1546, 1974), (b) Kan and Su's model. (Courtesy of Kan, J. R.)

Now, since

$$B_{\parallel} = B_g \cos(\theta - \alpha) = B_i \cos \alpha$$

$$\tan \alpha = \frac{B_i/B_g - \cos \theta}{\sin \theta}$$

and thus

$$\text{if } B_g = B_i, \qquad \alpha = \theta/2$$

Noting that the electric field along the X-line is given by

$$E_{\parallel} = V_s B_i \sin \alpha$$

we have

$$E_{\parallel} \leq (1 + \sqrt{2}) V_A B_i \sin^2 \alpha$$

where V_s denotes the solar wind velocity.

Thus, we have

$$E_{\parallel} \leq (1 + \sqrt{2}) V_A B_i \frac{(B_i/B_g - \cos \theta)^2}{1 + (B_i/B_g)^2 - 2(B_i/B_g) \cos \theta}$$

If $B_i = B_g$,

$$E \leq (1 + \sqrt{2}) V_A B_i \sin^2(\theta/2)$$

On the other hand, Gonzalez and Mozer (1974) obtained

$$E = \frac{\Phi_D}{\pi R} = \frac{2 V_s B_i F(B_M/B_g, \theta)}{\sin \theta/2}$$

where $B_M = qB_i$

$q =$ the amplification factor

$R =$ the geocentric distance to the dawn or dusk magnetopause ($= 15\, R_E$)

$B_g = 70\ \gamma$

$$F(B_M/B_g, \theta) = 0 \quad \text{for } \cos\theta \geq B_i/14\ \gamma \text{ and } B_i < 14\ \gamma$$

$$= \frac{B_i - 14 \cos \theta}{(B_i^2 - 28 B_i \cos \theta + 196)^{1/2}} \quad \text{for } B_i < 14\ \gamma \text{ and } \cos \theta < B_i/14\ \gamma$$

$$= \sin(\theta/2) \qquad \text{for } B_i \geq 14\ \gamma$$

Figure 4.18 shows the calculated relationship between Φ_D and B_z by taking B_y as the parameter. Mozer *et al.* (1974) and Gonzalez and Mozer (1974) tested their model calculation of the potential by comparing the estimated electric field E_p (by using observed variations of the IMF and assuming the size of the polar cap to be 3000 km) and the observed variations of the electric field in the polar cap. Figure 4.19 shows hourly averages of the estimated electric field and the mean of the dawn-to-dusk component of the electric fields observed at Thule and Resolute Bay. There is a general similarity in their time variations, although the estimated electric field is about three times greater than the observed one. They noted that if the IMF B_i is amplified near the apex of the magnetosphere by a factor of 5 ($q = 5$ in their estimate, namely $B_M = 5B_i$), the average ratio of the observed field to the estimated field is 0.35.

Kan and Su (1976) pointed out that Sonnerup's results lead to a much weaker

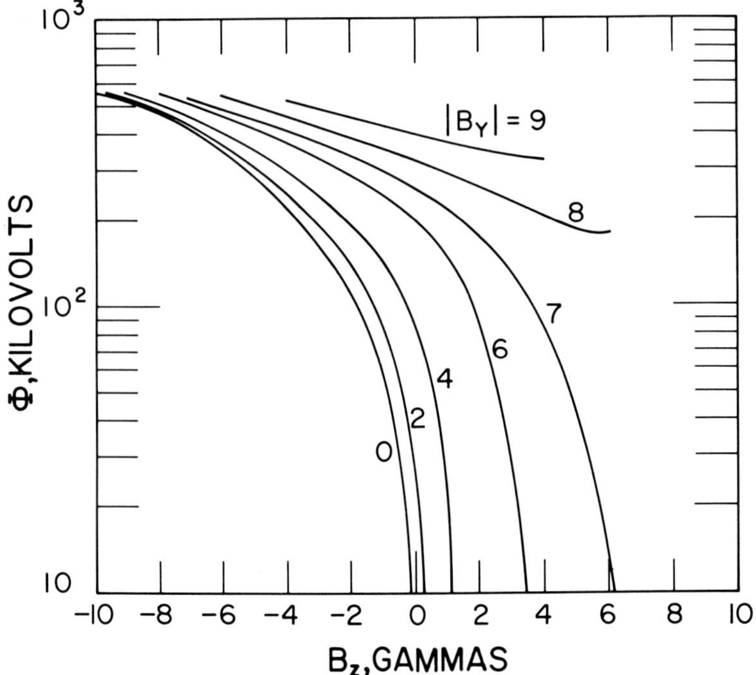

Fig. 4.18. Electric potential along the *X*-line as a function of the B_z component for different values of the $|B_y|$ component. (Gonzalez, W. D. and Mozer, F. S.: *J. Geophys. Res.* **79**, 4186, 1974.)

electric field than the observed value for a small value of θ. They suggested that the *X*-line makes an angle α_m with OB_i (see Figure 4.17(b)) and satisfies the following relation,

$$\begin{aligned} \sin \alpha_m &= \frac{\sin \theta}{[1 + (B_i/B_g)^2 + 2(B_i/B_g) \cos \theta]^{1/2}} \qquad && \theta \leq \theta_c \\ &= 1 && \theta > \theta_c \end{aligned}$$

where

$$\theta_c = (\pi/2) + \sin^{-1} (B_i/B_g), \text{ which is greater than or equal to } \pi/2.$$

Since $\alpha_m > 0$ for $\theta > 0$, merging can proceed at all angles between $\boldsymbol{B}_i$ and $\boldsymbol{B}_g$ except when they are exactly parallel.

The maximum reconnection electric field can then be written as

$$E_{\|\max} = \begin{cases} k_0 V_{A0} B_i \dfrac{\sin^2 \theta}{1 + (B_i/B_g)^2 + 2(B_i/B_g) \cos \theta} & \theta \leq \theta_c \\ k_0 V_{A0} B_i & \theta > \theta_c \end{cases}$$

Figure 4.20(a) shows $E_{\|\max}$ as a function of θ for various values of (B_i/B_g) plotted in solid curves, together with Sonnerup's results (dashed curves).

There is so far little statistical study on the relationship between the observed dawn-dusk electric field and the interplanetary magnetic field. Figure 4.20(b) is

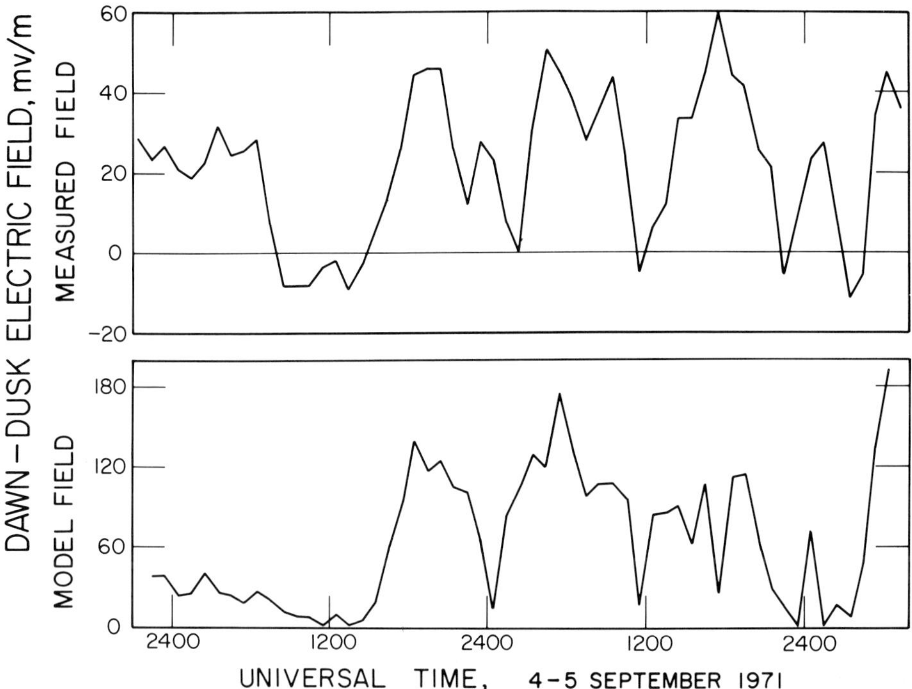

Fig. 4.19. Observed electric field in the polar cap and the computed electric field on the basis of the observed IMF. (Gonzalez, W. D. and Mozer, F. S.: *J. Geophys. Res.* **79**, 4186, 1974.)

constructed on the basis of the published electric field records by Heppner and the simultaneous interplanetary magnetic field record.

It shows a comparison between the averaged polar cap electric field measured by OGO-6 (Heppner, 1972) and the model electric fields as functions of the angle θ. The open circles represent the measured electric field averaged over the polar cap; the solid circles represent the calculated electric field on the basis of the equation obtained by Kan and Su; and the crosses represent the calculated field by Sonnerup's equation.

4.4.3. EROSION OF THE DAYSIDE MAGNETOSPHERE

(a) *Magnetopause Motion*

The earthward motion of the magnetopause associated with an enhanced merging has been discussed in terms of 'erosion' of the dayside magnetosphere. The motion becomes obvious, however, only after examining changes of the balance between the solar wind pressure and the Lorentz force associated with the electric current which flows at the magnetopause at the time when an enhanced merging takes place. In considering this problem, it is instructive to note that for the closed magnetospheric model (Chapman-Ferraro magnetosphere), the solar wind pressure is balanced by the Lorentz force associated with the Chapman-Ferraro

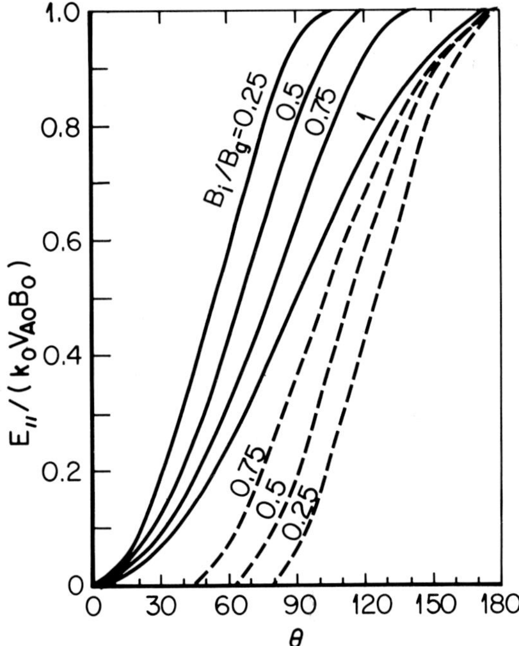

Fig. 4.20(a). Electric field associated with merging between the geomagnetic field (B_g) and the IMF (B_i) for different values of the ratio B_i/B_g as a function of the angle θ between $\boldsymbol{B}_i$ and $\boldsymbol{B}_g$. The solid curves are based on Kan and Su's estimates and the dashed curves on Sonnerup's estimates.

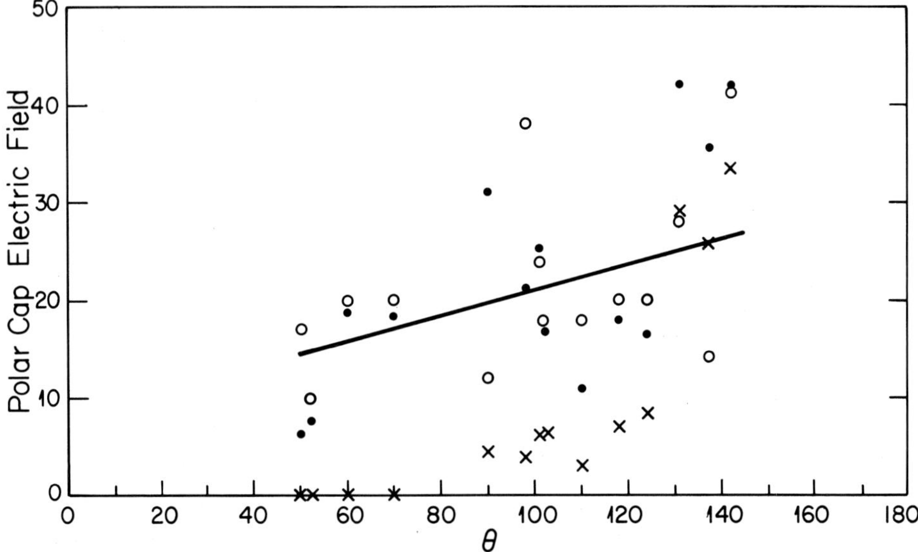

Fig. 4.20(b). Polar cap electric field (mV m^{-1}) measured by the OGO-6 (○), merging electric fields predicted by Kan and Su's model (●) and by Sonnerup's model (×). (Courtesy of Kan, J. R.)

current (which flows around the neutral point). When the solar wind pressure is enhanced, the magnetopause is pushed toward the Earth until the Chapman-Ferraro current increases to the point at which the corresponding Lorentz force can counterbalance the enhanced solar wind pressure. Suppose that the IMF and the geomagnetic field are anti-parallel across the magnetopause, without merging. This is an extended case of the Chapman-Ferraro model in which the solar wind pressure term p is replaced by $(p + B^2/8\pi)$. If merging is allowed in this situation, the quantity $\nabla \times \boldsymbol{B} (= 4\pi \boldsymbol{j})$ will be reduced. Then, the Lorentz force $\boldsymbol{j} \times \boldsymbol{B}$ is reduced and thus the magnetopause is pushed toward the Earth until a new balance is attained, namely until $\boldsymbol{j} \times \boldsymbol{B}$ becomes large enough for the same solar wind pressure. Because of the complexity involved, a simple superposition of the dipole field and the IMF (Forbes and Speiser, 1971; Hill and Rassbach, 1975) cannot accurately predict the location of the magnetopause.

Kan and Akasofu (1974) showed that B_i and the geocentric distance D of the magnetopause is related (Section 1.4.3) by

$$|B_i| D_0^3 = 2[(D_0/D)^3 - 1]$$

where D_0 denotes the geocentric distance of the magnetopause when $B_i = 0$.

Suppose that the interplanetary (or the magnetosheath) magnetic field just outside the magnetopause has initially a magnitude of 40 γ and is directed northward ($D_0 = 10\, R_E$) and that it turns southward. If the solar wind pressure does not change during the southward turning, the magnetopause moves earthward by a distance of $2\, R_E$, from $10\, R_E$ to $8\, R_E$ in geocentric distance, if the IMF B_z is $-57\, \gamma$.

Fairfield (1971) examined effects of the IMF B_z component on the magnetopause distance and found that the subsolar distances were $10.5\, R_E$ and $(11.6 \pm 2.0) R_E$ for the southward-directed and northward-directed fields. Thus, the IMF B_z does not seem to affect seriously the magnetopause location. Maezawa (1974) showed statistically that the magnetopause is located about 0–2 R_E closer to the Earth for southward-oriented IMF fields than for northward-oriented fields (Figure 4.21). In his study, the effect of changing solar wind pressure is eliminated by normalizing the observed magnetopause distance by the simultaneous solar wind pressure data.

Aubry *et al.* (1970) and Aubry *et al.* (1971) examined an inward motion of the magnetopause, observed by the OGO-5 satellite, which occurred at about 1840 UT on 1968, March 27. The general situation at that time is illustrated in Figure 4.22. They associated the observed inward motion to the southward turning of the IMF vector which was observed, by the Explorer 35 satellite, at the lunar distance near the Sun-Earth line; this southward-turning signal was supposed to reach the vicinity of the magnetopause at about 1724–1729 UT. Indeed, this was confirmed by the Explorer 33 satellite, which was located in the magnetosheath at that time. After the 'normal' crossing of the magnetopause at about 1700 UT, OGO-5 moved toward the Earth. Between 1730 and 1840 UT (between B and C in Figure 4.22), the magnetopause was 'oscillating' and began to move inward at 1840 UT.

The authors examined solar wind parameters and noted that the observed inward motion could not be associated with an increase of the solar wind pressure

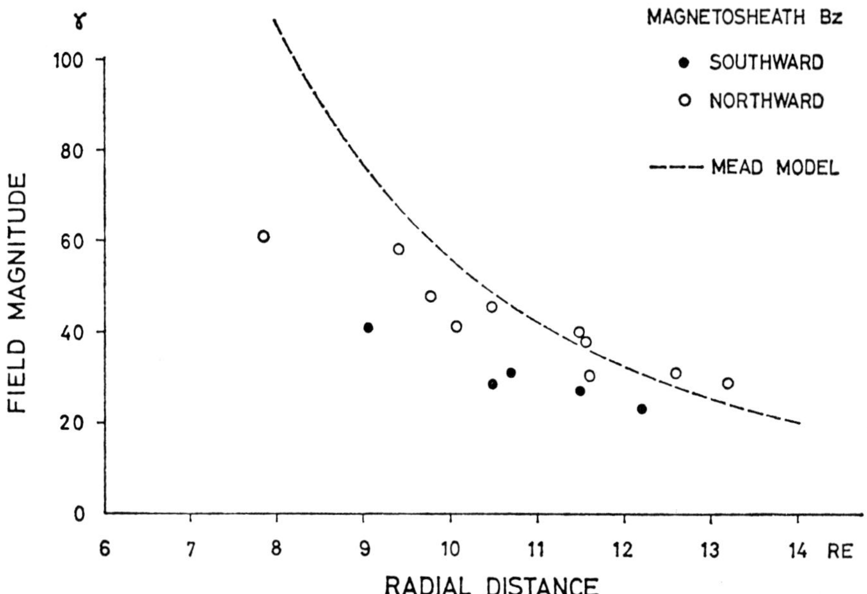

Fig. 4.21. Locations of the magnetopause for the northward B_z component (open circles) and the southward B_z component (dots). (Maezawa, K.: *Planet. Space Sci.* **22**, 1443, 1974.)

and proposed that it must be due to the erosion of the magnetopause which is caused by an enhanced merging associated with the arrival of the southward-directed IMF.

Aubry *et al.* (1971, 1972) examined the simultaneous magnetic record from Sodankyla and claimed that a substorm began about one hour after the inward motion. Thus, they concluded that this inward motion occurred prior to the substorm and that it was a growth phase feature. However, the AE index was as large as 500 γ at 1815 UT, as an intense substorm was in progress at that time. Sodankyla is a subauroral zone station, located in the evening sector at that time, and is not a suitable station for monitoring substorm activity. Thus it is not possible to claim that the observed inward motion is a growth phase feature.

Holzer and Reid (1975) and Reid and Holzer (1975) examined the time dependence of the erosion by using an electrical circuit analogy, in which the ionosphere is represented by a resistance and the magnetosphere by a set of inductances and a capacitance. The dawn-dusk electric field across the polar cap, induced by the solar wind-magnetosphere dynamo in the ionosphere, is represented by the voltage applied by a generator. Changes in the rate of dayside merging lead to changes in the potential drop across the magnetospheric capacitor ($\Delta q = C\Delta\Phi_D$) and, after a characteristic time delay, to changes in the voltage V applied by the generator (Figure 4.23). The governing equations are

$$\mathrm{d}q/\mathrm{d}t = i_2 - i_1$$
$$L\,\mathrm{d}i_1/\mathrm{d}t = -V_1 + q/C - i_1R_1$$
$$L\,\mathrm{d}i_2/\mathrm{d}t = V_2 - q/C - i_2R_2$$

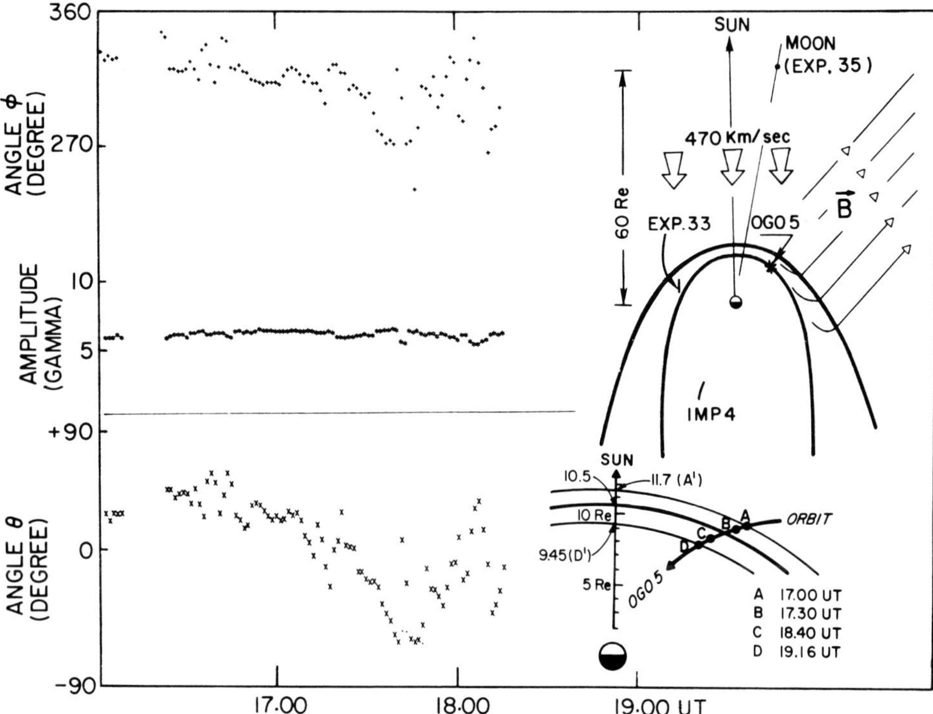

Fig. 4.22. Earthward drift of the dayside magnetopause, which occurred after the southward turning of the IMF vector on 1968, March 27. The IMF data (Explorer 35) are shown on the left-hand side. The successive locations of the magnetopause are shown in the lower half of the right-hand side. (Aubry, M. P., Russell, C. T. and Kivelson, M. G.: *J. Geophys. Res.* **75**, 7018, 1970.)

where

$$L = 90 \text{ H}$$
$$C = 150 \text{ F}$$
$$R = 0.17\ \Omega$$
$$LC = 1.35 \times 10^4 \text{ s}^2$$
$$L/R = 530 \text{ s}$$

and the subscripts 1 and 2 refer to the northern and southern hemispheric quantities, respectively. In the magnetosphere-ionosphere system represented here, there are two time constants, τ_1 and τ_2: the time τ_1 needed for magnetosheath plasma to travel along the freshly merged field lines to the ionosphere and the time τ_2 required for the foot of a merged field line to travel one quarter of the total noon-midnight dimension of the polar cap. The potential drop Φ_D is assumed to have a form $\Phi_{RM} \sin(\theta/2)$; the associated merging electric field is denoted by E_R.

Reid and Holzer (1975) examined the transient response of the magnetospheric electric field E_M and the applied ionospheric electric field E_A to a step function-like change in the reconnection electric field E_R for two sets of the parameters: (i)

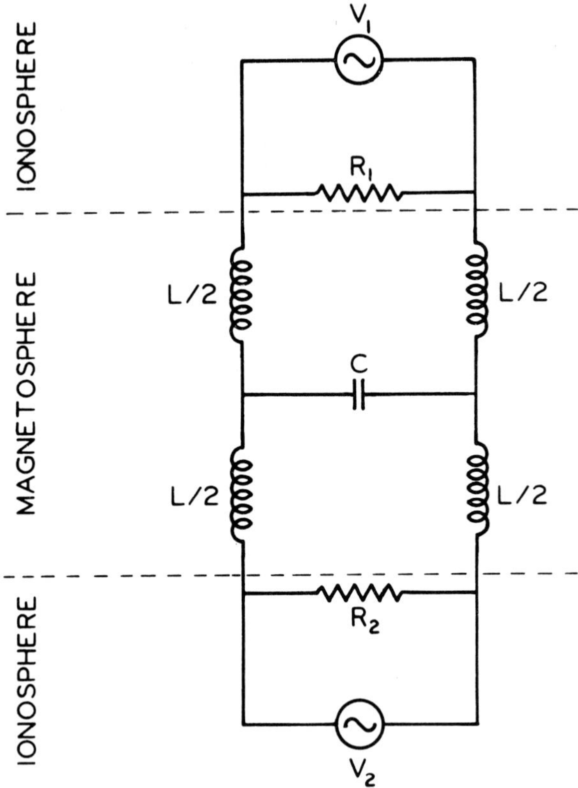

Fig. 4.23. Equivalent electric circuit used in examining transient motions of the magnetopause associated with the southward turning of the IMF vector. The magnetosphere is represented by inductors and a capacitor, and the ionosphere is represented by resistors and voltage (or current) generators. (Holzer, T. E. and Reid, G. C.: *J. Geophys. Res.* **80**, 2041, 1975.)

$\tau_1 = 7.5$ min, $\tau_2 = 30$ min, $\Phi_{RM} = 100$ kV, and (ii) $\tau_1 = 5$ min, $\tau_2 = 15$ min, $\Phi_{RM} = 200$ kV. Their results are shown in Figures 4.24(a) and (b). Here, the change ΔD in the magnetopause position is assumed to be given by

$$\Delta D \approx \tau_2 \Delta E_R / B_{MP}$$

where B_{MP} ($\simeq 50\ \gamma$) denotes the magnetic field strength just inside the magnetosphere.

(b) *Cusp Motion*

An enhanced merging and the subsequent 'transfer' of the merged field lines from the dayside magnetosphere to the magnetotail are expected to shift the cusp location equatorward until a new steady state can be reached. This phenomenon was studied by Burch (1972b, 1973c, 1974) who examined the location of the cusp precipitation, observed by a polar orbiting satellite, OGO-4. He showed that after

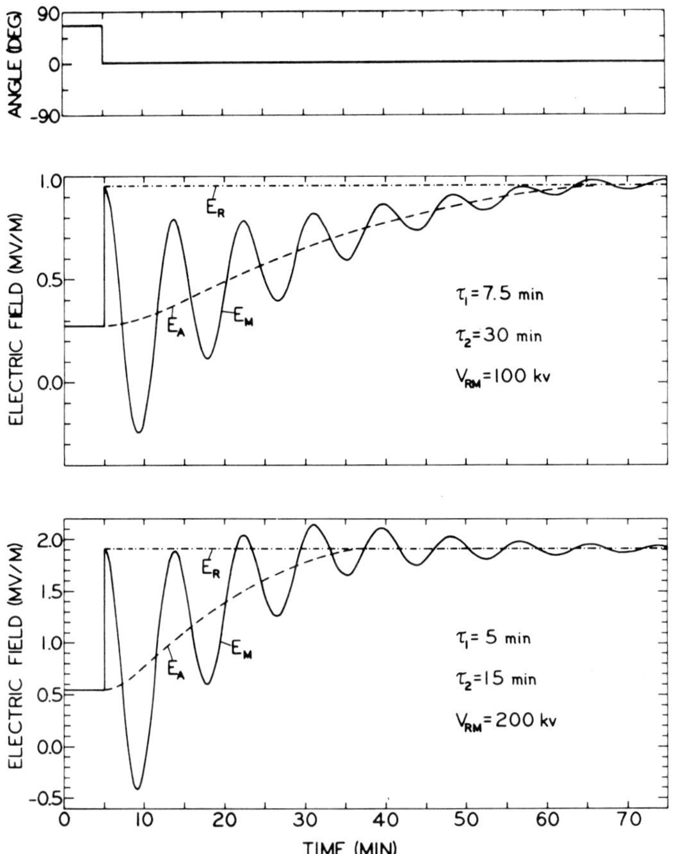

Fig. 4.24(a). Transient response of the magnetospheric electric field E_M and the applied ionospheric electric field E_A to a change in the merging electric field E_R associated with a change in the IMF angle.

a sharp southward turning of the IMF vector, the cusp location is found at progressively lower latitudes (Figure 4.25). Figure 4.26 shows also both the poleward and equatorward locations of the cusp as a function of the 45-min average of the interplanetary B_z component preceding the satellite observation. Burch (1974) noted also that the width of the cusp (the distance between the poleward and equatorward boundaries) increases as the B_z component increases (larger positive values). The equatorward shift of the cusp at a great geocentric distance during a southward IMF condition was reported by Kivelson *et al.* (1973).

Recently, Kamide *et al.* (1976) examined the scatter of the points in Figure 4.25 and found that it can be reduced considerably by grouping the points into two: quiet times and substorm times. They showed that the IMF B_z component is an important factor in controlling the latitude of the cusp, but that the cusp location is further shifted equatorward during substorms. This will be discussed further in Section 9.2.3.

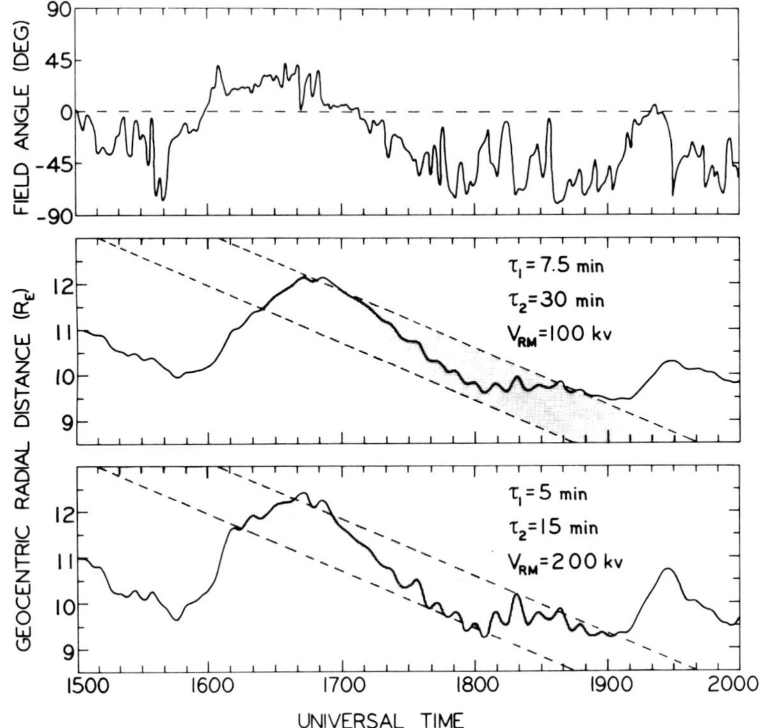

Fig. 4.24(b). Application of Reid and Holzer's model to the observed earthward shift of the magnetopause, which is illustrated in Figure 4.22. From the top: the observed change of the IMF θ angle, the estimated magnetopause location (solid lines) for two sets of the parameters (middle and bottom), and the relative location of the OGO-5 satellite (the shade indicates uncertainty in the location). (Reid, G. C. and Holzer, T. E.: *J. Geophys. Res.* **80**, 2050, 1975.)

4.4.4. MAGNETIC FLUX TRANSFER TO THE MAGNETOTAIL

(a) *High Latitude Lobe Field B_T*

In Section 1.4.3, we noted that the potential drop Φ_D along the X-line is related to the magnetic field intensity B_T in the high latitude lobe of the magnetotail by

$$\Phi_D = B_T\left(\frac{\pi}{2}\right)R_T^2\left(\frac{V_s}{L}\right) \quad \text{or} \quad B_T = \left(\frac{2}{\pi}\right)\left(\frac{L}{V_s}\right)\frac{1}{R_T^2}\Phi_D$$

However, as mentioned in Section 4.4.1, it is not immediately obvious from the above equation how B_T varies as a function of Φ_D, unless R_T is known as a function of Φ_D. Further, as we examined in Section 4.2, B_T is quite sensitive to changes of the solar wind pressure ($p_{s\perp}$), so that great care is needed in examining B_T changes associated with the IMF B_z. Further, undoubtedly B_T varies considerably during substorms, so that substorm activity should be monitored carefully in this particular study.

Meng and Colburn (1974) made an extensive study of the relationship between

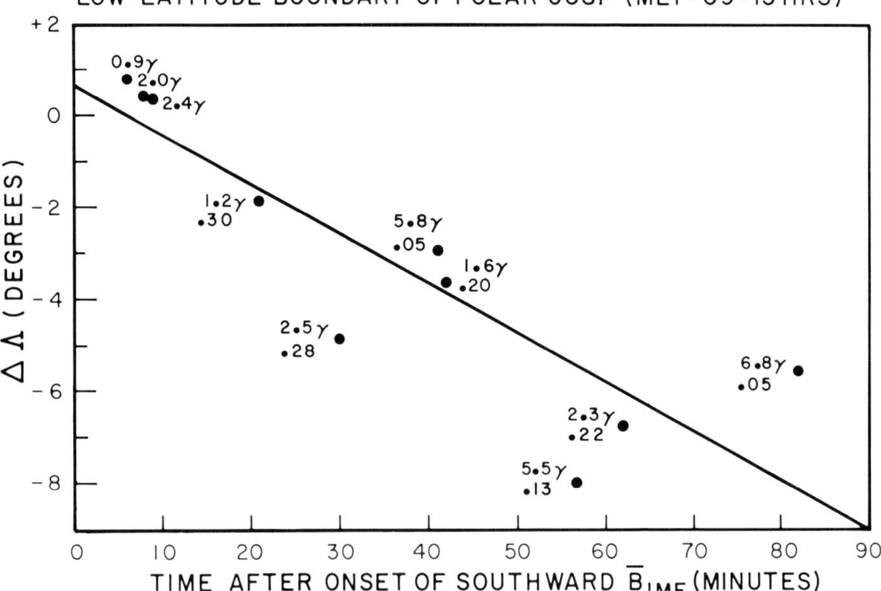

Fig. 4.25. Differences between the latitudes of the observed lower boundary of polar cusp and the expected boundaries for 10 cases that followed sharp onsets of southward IMF. The average southward component of IMF during the period between the onset of southward field and the observation is given for each point, together with the ratio of southward flux eroded into the tail to the flux brought to the front of the magnetosphere. (Burch, J. L.: *J. Geophys. Res.* **77**, 6696, 1972.)

the high latitude lobe field and the IMF B_z component. Most of their examples are not free from 'substorm contamination'. They showed, however, a few examples which are not associated with a significant change in the AE index. Figure 4.27 shows one of their examples. There are three increases in $F(\approx B_T)$, corresponding to three southward turnings ($B_z = -2 \sim -4\ \gamma$). The amount of the increase in F was of order 10% in all three cases, indicating an increase of 20% in the magnetic energy density in the magnetotail.

This aspect was also studied in detail by Iijima and Nagata (1972). Figure 4.28 shows one of their examples: from the top, the PCM index (the magnitude of the horizontal magnetic disturbance field in the polar cap region), the four ground magnetic indices (the AU, and AL, Dst and Asy indices), the magnetic pressure ($B_T^2/8\pi$), and the maximum value of the shearing stress component in the Maxwell stress tensor. There were three prolonged periods ($\gtrsim 2$ h) of the southward component during the period of consideration. Each period was associated with a significant increase of both magnetic pressure and shear stress in the magnetotail; the largest increase occurred during the second southward turning. Magnetic pressure doubled during that period, thus resulting in an increase of 40% in B_T. Iijima and Nagata (1972) identified the three corresponding growth phase periods, as indicated in the AL index. It is, however, difficult to identify the onset time of the expansive phase from ground magnetic records alone (see Section 4.4.7). Indeed, during the second event, all-sky photographs from Fort Yukon

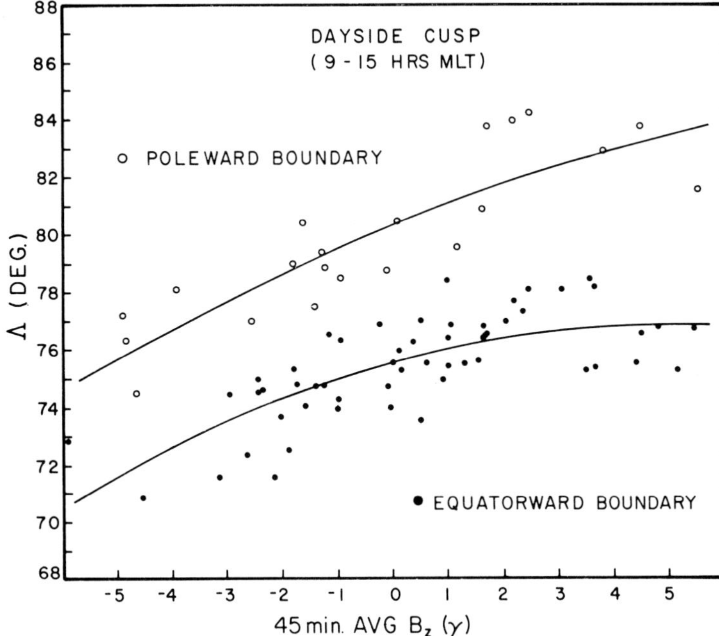

Fig. 4.26. Locations of the poleward (○) and equatorward (●) boundaries of dayside cusp electron precipitation for 45 minute B_z averages. (Burch, J. L.: *Radio Sci.* **8**, 955, 1973.)

showed a typical expansive phase feature at about 1150 UT on that day, although Iijima and Nagata inferred that the growth phase lasted from about 1015 UT to 1230 UT.

Iijima (1972) examined also changes of magnetic pressure for 64 isolated substorms and demonstrated that magnetic pressure increases until a sharp increase in the AL index occurs. He concluded that the period of this increase of magnetic pressure and of the associated slow growth of the AL index should be identified with a growth phase. His results will be discussed further in Section 4.5.2.

Aubry and McPherron (1971) also examined changes of B_T associated with the southward turning of the IMF vector. Figure 4.29 shows, from the top: the magnetotail field parameters, the total magnitude $F = |B|$ (which is nearly equal to B_T in the high latitude lobe), the Z_{SM} $(=B_z)$ component in the magnetotail, and the latitude θ and longitude ϕ (in solar magnetospheric coordinates).

They identified the onset time of two substorms during this period, at 1230 and 1610 UT, respectively. According to their determination of the onset times, the two increases of F occurred between 1100 and 1230 UT and between 1415 and 1610 UT, respectively, namely prior to their substorm onset. In the bottom of the figure, the corresponding AE index is shown, to point out some of the problems associated with their conclusions. At about 1110 UT, the AE index began to increase, reaching the maximum intensity of order 750 γ at as early as 1215 UT. There is no doubt that the expansive phase began well before 1230 UT, which was

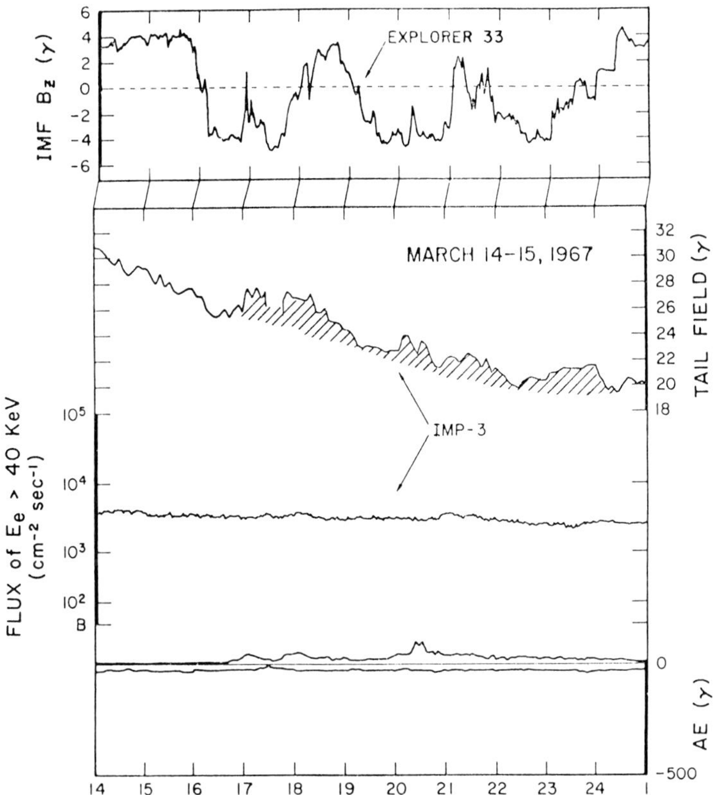

Fig. 4.27. Increases of the magnetotail lobe field after southward turnings of the IMF vector. (Meng, C.-I. and Colburn, D. S.: *J. Geophys. Res.* **79**, 1831, 1974.)

the time identified as the onset time of the expansive phase by Aubry and McPherron (1971). In this particular case, the southward turning of the IMF vector, the increase of F and the substorm began at about the same time, making it difficult to identify the cause of the increase of F. However, it is clearly incorrect to ascribe the observed increase of F to a growth phase feature.

The AE index began to decrease at about 1250 UT; there was then a fairly quiet period between 1400 and 1540 UT. At 1545 UT, a new increase of the AE index began. However, Aubry and McPherron (1971) identified the onset time of the expansive phase to be as late as 1610 UT. First of all, F began to increase at about 1415 UT, well before the sudden southward turning of the IMF vector (~ 1515 UT). The increase was enhanced at about the time when the new AE index increase began. Thus there is again no justification to conclude that this particular increase was a growth phase feature. It appears that they assumed *a priori* that the expansive phase of a substorm must be associated with a decrease in B_T. However, as noted in Section 4.4.1, B_T is a function of several parameters, such as Φ_D, $p_{s\perp}$, Σ, R_T, etc. In fact, a detailed study of B_T changes shows that the decrease begins sometime during substorms, but not necessarily at the onset time of the

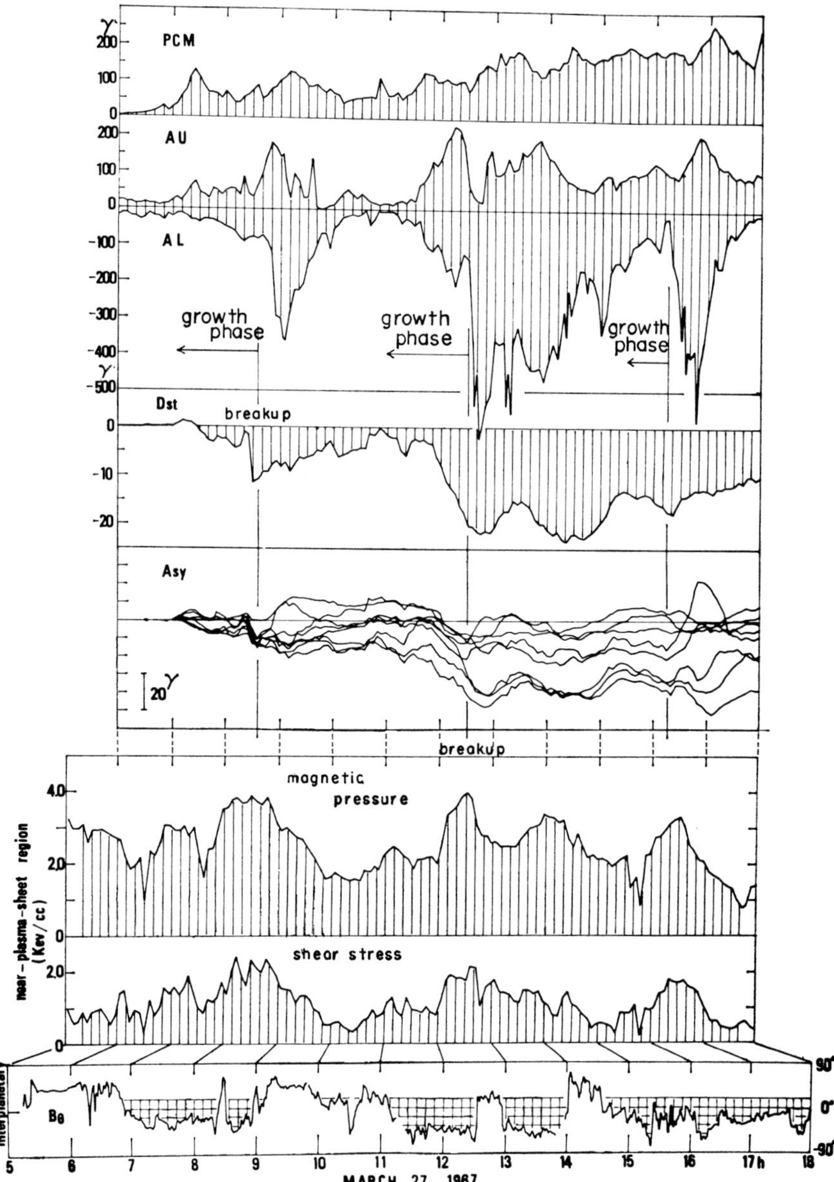

Fig. 4.28. Relationship between geomagnetic activity and IMF changes. Geomagnetic activity is expressed by the polar cap magnetic disturbance index (PCM), the AU and AL indices, and the Dst and Asy indices. The middle panel shows magnetic pressure ($B_T^2/8\pi$) and shear stress in the high latitude lobe. The IMF θ angle is shown in the bottom panel. (Iijima, T. and Nagata, T.: *Planet. Space Sci.* **20**, 1095, 1972.)

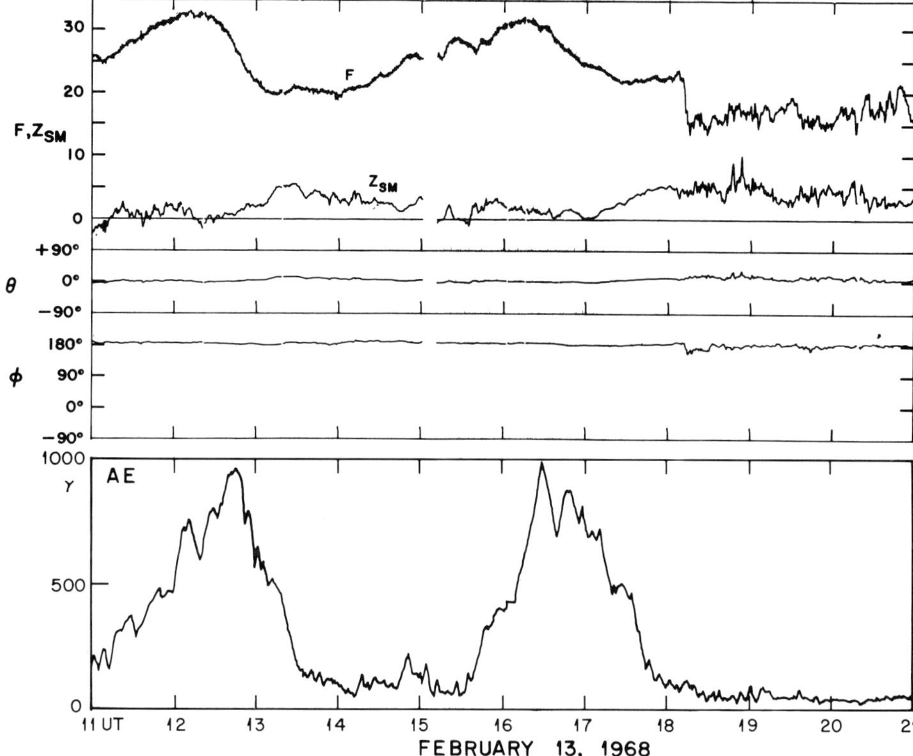

Fig. 4.29. Changes of the magnetic field ($F = B$, $Z(=B_z)$, θ and ϕ) in the magnetotail during substorms (expressed in terms of the AE index). (Fairfield, D. H. and Ness, N. F.: *J. Geophys. Res.* **75**: 7032, 1970.)

expansive phase (Meng and Colburn, 1974). One possibility for the complexity is that B_T begins to decrease when the reconnection rate Φ_N in the nightside becomes greater than the merging rate Φ_D in the dayside. If this is indeed the case, B_T may not decrease if Φ_D is greater than Φ_N throughout the substorm.

McPherron (1973) and Caan *et al.* (1975) examined further this feature for 20 'isolated' substorms. The IMF component (SM), the total lobe magnetic energy density ($F^2/8\pi$), the mid-latitude H component change and the AE index are examined by assuming that the onset time of a substorm can be determined from the time of a sharp onset of mid-latitude positive bay. Figure 4.30 shows their results. It can be clearly seen that the magnetic energy density in the high latitude lobe increases substantially prior to the sharp onset of mid-latitude positive bays. This increase was interpreted as a result of the enhanced merging, caused by the southward component of the IMF (the top diagram), prior to the onset of the substorms. However, the average AE index in the bottom of the diagram indicates that it reached more than 400 γ prior to the onset time of the sharp positive bays, in spite of the claim that the middle latitude signature can determine the onset time of the expansion phase. It was more than 300 γ even two hours before the onset. For these obvious reasons, it is difficult to accept their claim that the

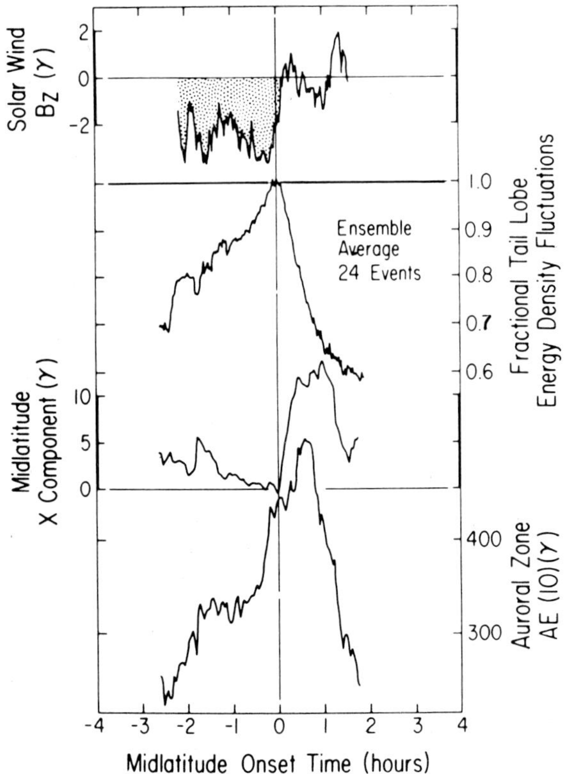

Fig. 4.30. Superimposed epoch averages of the IMF B_z, tail lobe magnetic energy ($B_T^2/8\pi$), the mid-latitude positive bay and the AE index. (Caan, M. N., McPherron, R. L. and Russell, C. T.: *J. Geophys. Res.* **80**, 191, 1975.)

observed increase in the tail lobe magnetic energy is purely an effect of the enhanced merging associated with the southward component of the IMF. As noted earlier, it is also puzzling to see that the expansive phase appears to begin at about the time of the northward turning of the IMF vector.

As noted earlier, B_T is very sensitive to changes of the solar wind pressure (Section 4.2). Arnoldy (1971) reported a large increase of B_T at 0809 UT on 1967, April 1, after the southward turning of the IMF vector. However, the event coincided also with the ssc of the 1967, April 1 storm. Thus it is quite likely that the observed increase of B_T was mainly due to the compression effect, rather than to an enhanced merging.

A large increase of magnetic energy density in the high latitude lobe, which began at about 0330 UT on 1968, June 22, was examined by Caan *et al.* (1973). Figure 4.31 shows, from the top, the solar wind density, velocity, the interplanetary B_z component, the tail lobe field, and the AE index. The authors identified their growth phase to be the period between 0330 and 0535 UT. However, it is quite obvious that the large increase of solar wind pressure (which began at 0330 UT) contributed significantly to the increase of ($B_T^2/8\pi$). Indeed, they

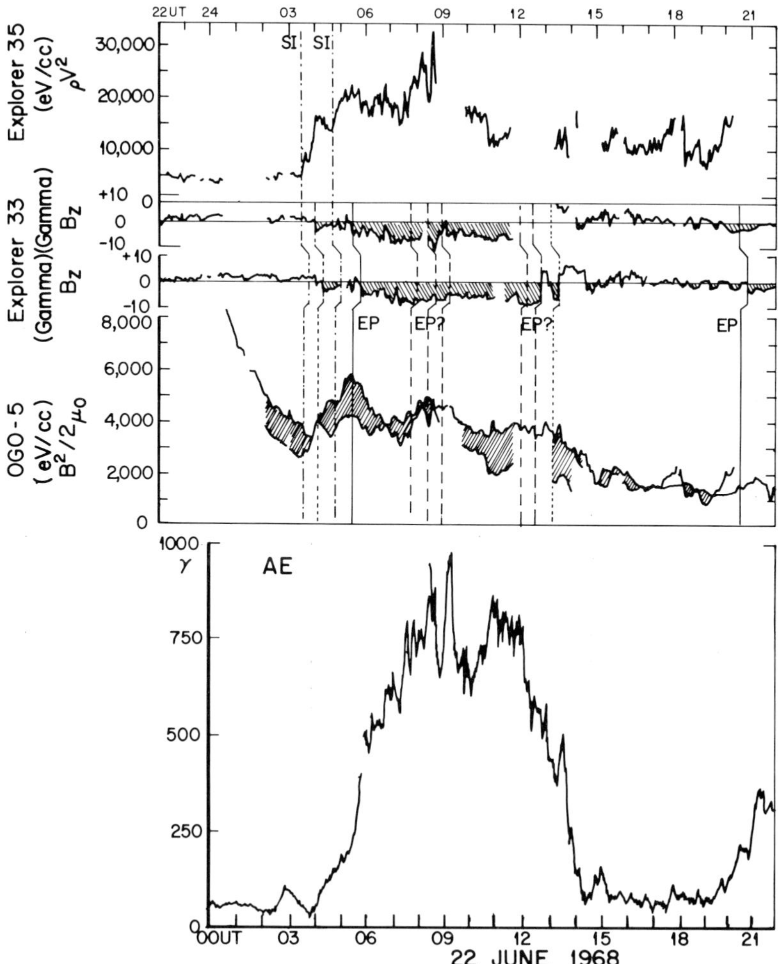

Fig. 4.31. Solar wind dynamic pressure (Explorer 35), the IMF B_z component (Explorer 33), tail lobe magnetic energy density ($B_T^2/8\pi$) and the AE index. The lobe magnetic energy density is calculated by taking into account solar wind pressure variations. When the measured energy exceeds the calculated energy, the difference is indicated by the shade. (Caan, M. N., McPherron, R. L. and Russell, C. T.: *J. Geophys. Res.* **78**, 8087, 1973.)

attempted to subtract the effect of the increased solar wind pressure by using the formulas devised by Olson and Pfitzer (1974) and concluded that the southward B_z component increased $B_T^2/8\pi$ by a factor of about 2 (100%). However, their formula must be 'calibrated' before such a use. As mentioned in Section 4.2, there is a disagreement by a factor of 2 even between the estimated value of ΔH and the observed value on the ground. We found earlier that the southward component of $2 \sim 4\ \gamma$ can cause an increase of $B_T^2/8\pi$ of order 20%, not as much as $\sim 100\%$.

(b) '*Radius*' *of the Magnetotail* R_T

If the general shape of the magnetopause is controlled by the solar wind pressure alone, the earthward shift of the apex of the magnetopause (located at a geocentric distance of D_0) will be associated with a reduction of the 'radius' of the magnetotail. This is because the radius R_T of the magnetotail is approximately given by $R_T \simeq 1.5\ D_0$.

The situation will be much more complicated for an enhanced merging caused by the southward turning of the IMF vector. An increase of Φ_D will cause an enhancement of the cross-tail current, which in turn increases B_T. Atkinson and Unti (1969) suggested that an enhanced tail current causes a large expansion of the magnetotail, as well as an earthward shift of the apex of the magnetopause, and that R_T is quite sensitive to the intensity of the cross-tail current.

Lui *et al.* (1976) examined R_T at $0\ R_E > X > -30\ R_E$ for various interplanetary conditions $B_z > 1\ \gamma$, $-1\ \gamma < B_z < +1\ \gamma$, and $B_z < -1\ \gamma$ and found that there is no clear indication that the IMF B_z component affects the radius R_T. Figure 4.32 shows their results.

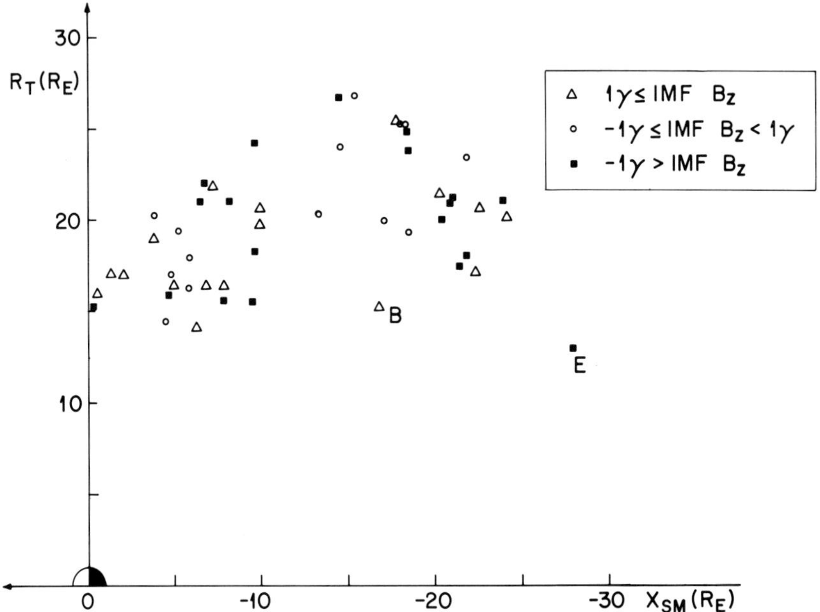

Fig. 4.32. Locations of the magnetopause in the X-Z plane for different values of the IMF B_z component. (Courtesy of Lui, A. T. Y. and Akasofu, S.-I.)

(c) *Plasma Sheet Thinning*

Aubry and McPherron (1971) suggested that an enhanced merging and thus an enhanced convective motion of plasma would reduce the thickness of the plasma sheet. However, their study of magnetotail responses to the southward turning of the IMF was criticized in the previous subsection on the basis of an inadequate determination of the substorm onset time. Russell (1974) argued also that the thickness of the plasma sheet should be reduced when merging along the dayside

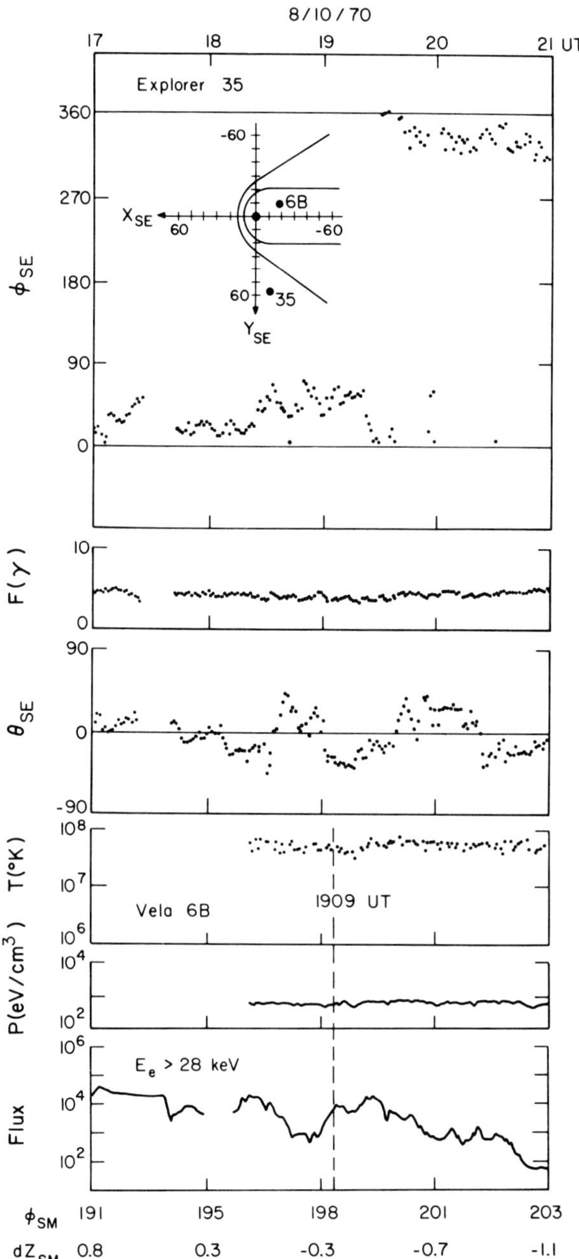

Fig. 4.33. Relationship between the southward turning of the IMF vector (top three panels) and plasma sheet variations (bottom three panels). (Lui, A. T. Y., Hones, E. W. Jr., Venkatesan, D., Akasofu, S.-I. and Bame, S. J.: *J. Geophys. Res.* **80**, 929, 1975.)

X-line proceeds faster than reconnection along the X-line in the magnetotail (implying that the transferred magnetic flux 'squeezes' the plasma sheet) and further that a growth phase lasts until the plasma sheet has become thin enough to allow the formation of a new neutral line in the magnetotail. Thus, the problem is again related to the proposed growth phase. Such inferences were, however, questioned by Meng and Colburn (1974) who examined both magnetic field and particle data in the magnetotail in conjunction with the corresponding variations of the B_T component.

Lui *et al.* (1975) examined recently plasma sheet behavior after nineteen sharp southward turnings of the IMF vector. They found that of these, twelve showed a large decrease of plasma pressure (the thinning) and that all of these twelve events were associated with substorms. On the other hand, seven cases showed only a slight plasma sheet thinning (if any) and there was no indication of substorms at the standard auroral zone stations. Figure 4.33 shows one of their examples; the IMF turned southward at 1901 UT and 2027 UT, remaining southward for 35 min and 55 min, respectively. Yet there are no significant decreases of the plasma pressure during this whole interval at Vela 6B, located near local midnight and less than 1 R_E from the neutral sheet. No polar magnetic substorms are observed on the ground. Very small decreases of plasma pressure can be noted at 1915 UT and 2048 UT. They last for 13 min and 20 min, respectively. The time separation between the decreases is 1 h 33 min, comparable to the time separation between the southward turnings (1 h 26 min). These pressure decreases could be the response of the plasma sheet (very near the neutral sheet) to these two sudden southward turnings of the IMF. The recovery of the first small plasma pressure decrease is accompanied by a moderate enhancement of energetic electron flux.

Lui *et al.* (1975) concluded that the southward IMF component alone does not reduce significantly plasma pressure in the plasma sheet. When a large decrease of plasma pressure is observed, it is associated with a substorm. Indeed, Akasofu *et al.* (1971b) found that a sudden complete thinning of the plasma sheet occurs within a few minutes of the onset of the auroral substorm (indicated by a sudden brightening of an auroral arc in the midnight sector). We shall discuss in detail the thinning as a substorm feature in Section 6.3.

4.4.5. ENHANCED CONVECTION IN THE PLASMA SHEET

(a) *Earthward Advance of the Plasma Sheet*

Jaggi and Wolf (1973) predicted that the potential Φ_D is related to the minimum penetration depth L_G of the plasma sheet by

$$L_G = 7.93\left(\frac{31\ \gamma\eta\mu}{\Phi_D\langle\sigma_p\rangle}\right)^{1/3}$$

Thus an enhanced Φ_D would cause a deeper penetration of the plasma sheet toward the Earth (Section 1.4.3). Yasuhara *et al.* (1976) examined 32 IMF southward turnings in an attempt to find whether the plasma injection observed at the synchronous distance is closely related in time to the southward turning of the IMF vector.

Figure 4.34 shows an example of their results. After the northward turning of the IMF vector at 0237 UT, the B_z component was as large as $+4\ \gamma$ and became negative at about 0450 UT (Figure 4.34(a)). The AE index was quite low during the period when B_z was positive. At about 0500 UT the AE index began to increase. Akasofu *et al.* (1973b) suggested that such a variation arises from the expansion of the auroral oval and of the associated enhancement of the S_q^p current, rather than from the growth of an entirely new current system DP-2, as suggested by Nishida (1968); see Section 4.4.6(c). Indeed, the AL index was almost null during this AE increase, an important feature of the expanding S_q^p current system. The corresponding ATS-5 plasma data shows no indication of plasma injection during this slow growth of the AE index (Figure 4.34(b)).

At 0635 UT, the AE index increased suddenly. An intense plasma injection was observed at the ATS-5 location at about the *same time.* Note that electrons and protons over the entire detector energy range appeared almost simultaneously at the satellite location. The satellite local time was 2320 LT. The simultaneity of

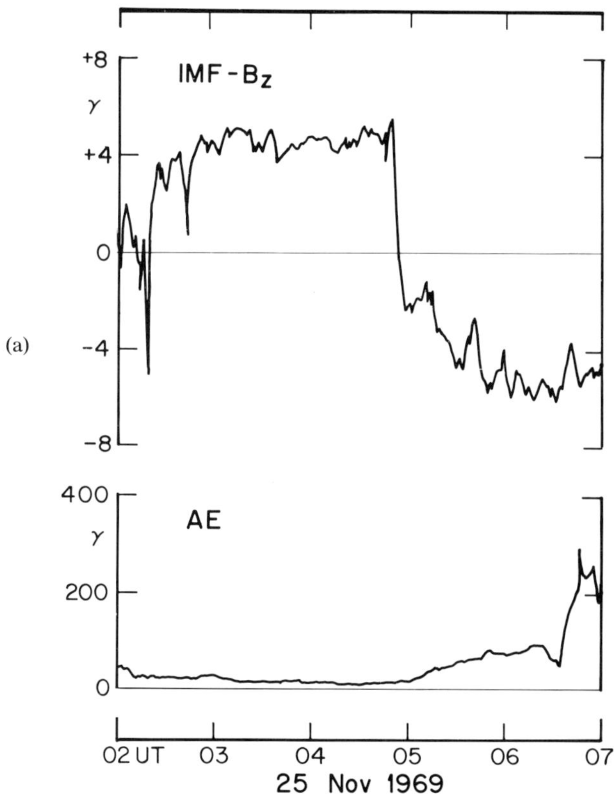

Fig. 4.34(a, b). Plasma injection which occurred simultaneously with a sudden enhancement of the AE index, but 90 min after the southward turning of the IMF vector. (Courtesy of Yasuhara, F., Akasofu, S.-I. and McIlwain, C. E.)

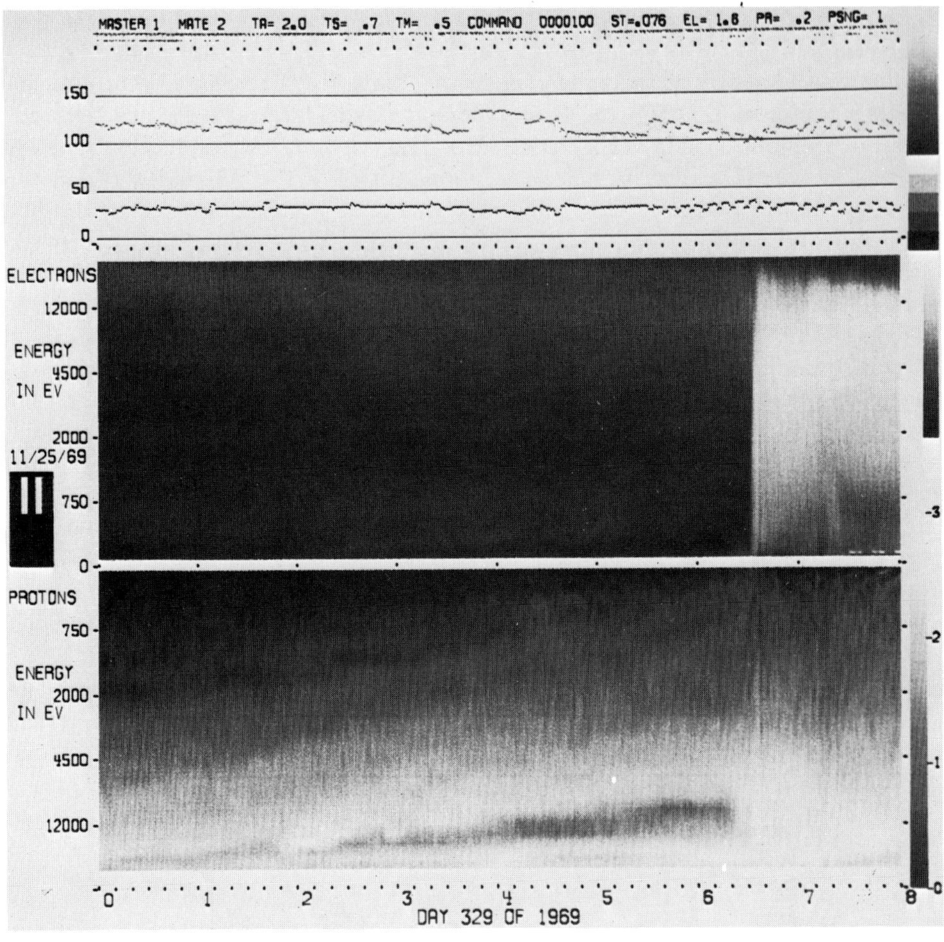

Fig. 4.34(b).

the sudden growth of the AE index and the plasma injection suggests strongly that the plasma injection was directly related to some substorm processes. There are at least two possibilities in interpreting the observed delay. The first one is that the time constant of the penetration of the enhanced electric field deep into the inner magnetosphere ($X \simeq -6\ R_E$) is of order of one to two hours or that it takes about one or two hours for the inner edge of the plasma sheet to reach the geocentric distance from its quiet-time location, say, $\sim 10\ R_E$; see Shelley *et al.* (1971). The second possibility is that there is no *direct* relationship between the southward turning of the IMF vector and the plasma injection. In fact, some plasma injections take place even after northward turnings (Section 8.2.1).

The simultaneity of the plasma injection and substorm onset for all the 32 events seems to rule out the first possibility, since there is no reason why the arrival time of the front of the advancing plasma sheet to the geosynchronous distance should coincide almost exactly with the substorm onset.

The plasma injection at the substorm onset suggests that a sudden change of the distribution of the electric field occurs at the onset time of substorms. More specifically, some substorm processes make it possible for an intense electric field to penetrate into the inner magnetosphere, allowing the plasma sheet to advance closer toward the Earth. We shall discuss this problem in Section 8.2.2.

(b) *Tendency Toward the Tail-Like Field*

Effects of the IMF B_z component on the near-Earth plasma sheet have been provided by an extensive study of two substorms, observed by the OGO-5 satellite, which occurred on 1968, August 15 (Aubry *et al.*, 1972; McPherron *et al.*, 1973c); Figure 4.35. Here, we examine the second substorm on that day. A southward-turning of the IMF vector took place at about 0630 UT. The authors determined the onset of the expansive phase to be 0714 UT and noted that the magnetic field vector changed from a dipole-like one to a tail-like one between 0650 and 0730 UT. Figure 4.36 shows the projection of the magnetic field vectors between 0650 and 0730 UT onto the X-Z_{GSM} plane. It can be seen that between 0650 and 0713 UT the projected vectors increased in magnitude, while the angle with respect to the equatorial plane was reduced. The authors concluded that this can be explained in terms of an increase of the magnetotail current and/or of the earthward motion of the plasma sheet, and that vector changes occurred prior to the onset of the expansive phase. However, a large increase of the AE index began at about 0630 UT and recorded a 500 γ level at the time (0714 UT) which they identified as the onset time of the expansive phase. Thus it is incorrect to ascribe this particular field observation as a growth phase feature.

Nishida and Nagayama (1975) found that after the southward turning of the B_z component, the tendency toward the tail-like configuration is accentuated first at a geocentric distance of 6.6 R_E with a delay of about 10 min and later at a distance of 80 R_E with a delay of about 30 min.

(c) *Slow Decrease of the H Component at the Synchronous Distance and in Low Latitudes*

McPherron (1970) proposed that prior to the onset of the expansive phase, the H component of the magnetic field gradually decreases at the synchronous distance. He ascribed this phenomenon to an earthward motion of the plasma sheet and/or an increase of the magnetotail current during his proposed growth phase. Figure 4.35 serves to illustrate this particular phenomenon. The H component magnetic records from the ATS-1 satellite show two large decreases at about 0340 and 0645 UT, each prior to the onset time of each substorm, determined by McPherron *et al.* (1973a). The first decrease began about an hour before their onset time, and the second one about 45 min before their onset time. However, it can be seen that the AE index began to increase at about the time, or even earlier, than the onset time of the two decreases, reaching a 500 γ level. Thus, again, it is incorrect to ascribe these two decreases to a growth phase feature.

As we shall discuss in great detail in Section 8.2.1, ring current protons are injected into the trapping regions at the onset time of a substorm. These protons form a partial ring current and produce the diamagnetic effect.

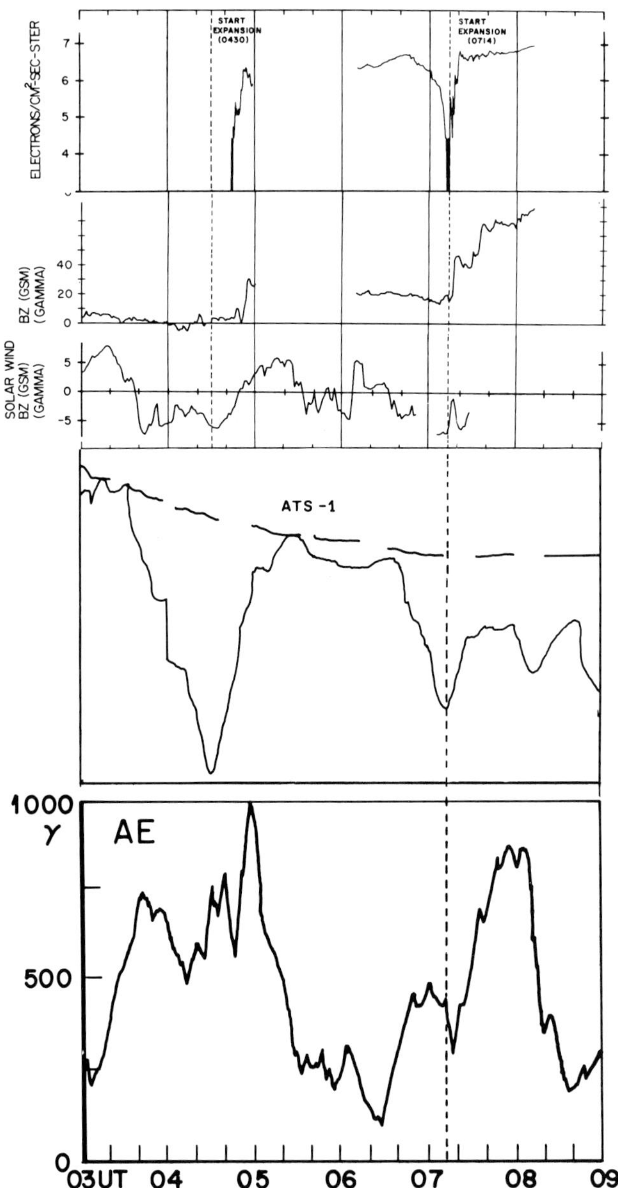

Fig. 4.35. Simultaneous observations of substorms by the OGO-5 satellite (energetic electron fluxes and the B_z component) in the magnetotail and by the ATS-1 satellite (the H component magnetic field) at the synchronous distance, together with the IMF B_z component and the AE index. (McPherron, R. L., Aubry, M. P., Russell, C. T. and Coleman, P. J. Jr.: *J. Geophys. Res.* **78**, 3068, 1973.)

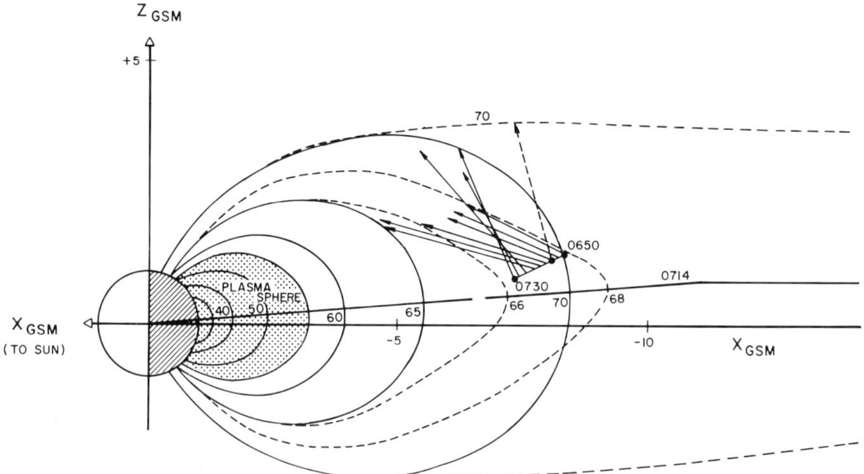

Fig. 4.36. Magnetic field vectors, projected onto the noon-midnight plane, between 0650 and 0730 UT, 1968, August 15. (McPherron, R. L., Aubry, M. P., Russell, C. T. and Coleman, P. J. Jr.: *J. Geophys. Res.* **78**, 3068, 1973.)

Since an intense substorm began at about the onset time of the two decreases, it is quite likely that a significant part of these two decreases was caused by diamagnetism of ring current protons which were injected into the trapping region during the two substorms. Thus they are rather poor examples to illustrate effects of the southward component of the IMF on the magnetic field at the synchronous distance. Further, the injected protons drift westward, and thus their effects can be seen in the evening sector but not in the morning sector. Indeed, McPherron (1970) showed that the H component decrease during the claimed growth phase occurs only in the evening sector. However, this is precisely an expansive phase feature.

The drifting protons can also cause a profound effect on the distribution of energetic electrons in the trapping region. This is because diamagnetism produced by the protons alters considerably their drift paths. Since energetic electrons tend to drift along contours of $B = \text{constant}$, the diamagnetic effect of the protons shifts the drift paths toward the Earth, so that the trapping boundary also shifts toward the Earth. Therefore, if a satellite is located in the outer skirt of the trapping region (where the energetic electron flux sharply drops outward), a drastic decrease of the flux of energetic electrons can take place (Lezniak *et al.* 1968; Lezniak and Winckler, 1968; Pfitzer and Winckler, 1969; Pfitzer *et al.*, 1969; and Winckler, 1970). This phenomenon will be discussed in detail in Section 8.4.

A large reduction of fluxes of energetic electrons has indeed been observed in association with the H component decrease at the synchronous distance. Figure 4.37 shows an example of such a decrease on 1966, December 25. A large decrease of the flux and of the H component began at about 0600 UT. This feature has also been described by some as a growth phase feature. However, this particular event began soon after a large increase of the AE index ($\sim 400\ \gamma$).

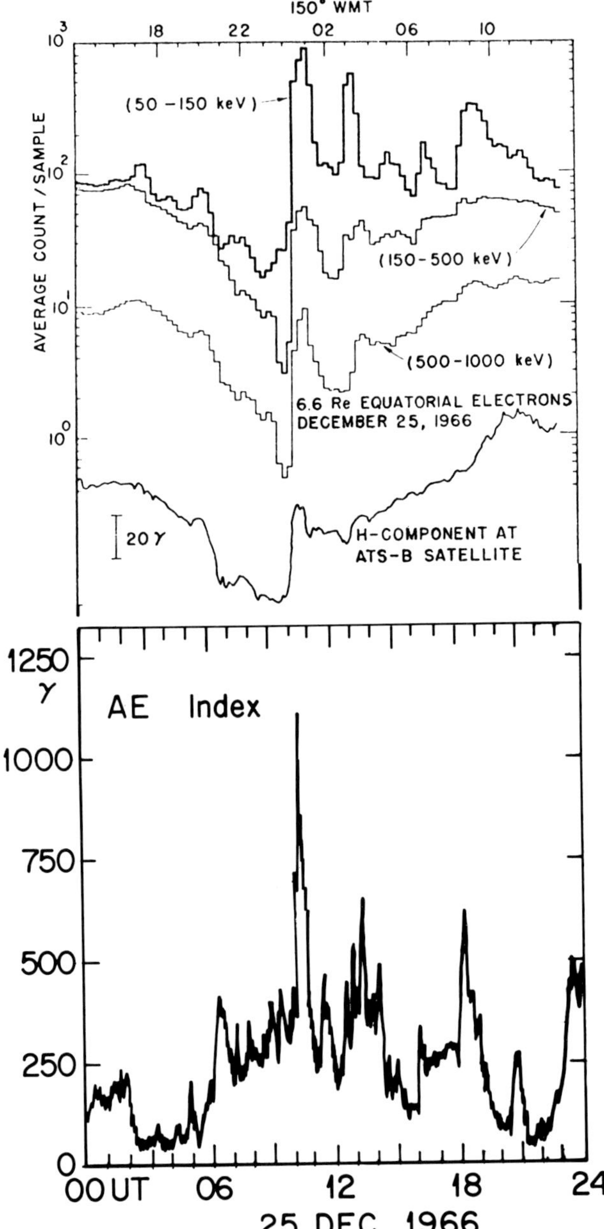

Fig. 4.37. Energetic electron fluxes and the H component magnetic field changes observed by the ATS-1 satellite at the synchronous distance. (Courtesy of Winckler, J. R.) The AE index is shown at the bottom.

Iijima and Nagata (1972) and Kokubun and Iijima (1975) showed also that the H component magnetic field in low latitudes decreases prior to the onset of the expansive phase or during their proposed growth phase. Figure 4.28 shows an example of their results. It is most clearly seen during the second southward turning of the IMF. However, the Dst index shows a decrease of as much as 20 γ even prior to the onset time of their expansive phase. Since the Dst index is derived from the longitudinal average of the H component, some stations should have shown even larger decreases. This is indeed the case as indicated by the Asy index; note also that both positive and negative changes were observed, and they are important features of the magnetospheric substorm (Section 7.2.3).

Parks and Pellat (1972) and Parks *et al.* (1974) showed that the quantity defined by

$$\xi = \tan^{-1} B_y/B_z = \tan^{-1} E_z/E_y$$

correlates well with the flux of energetic electrons at the synchronous distance. Here, B_y and B_z denote the IMF east-west and north-south components, respectively, and the electric field components E_y and E_z are obtained from the equation $\boldsymbol{E} + \boldsymbol{V} \times \boldsymbol{B} = 0$.

4.4.6. POLAR CAP PHENOMENA

(a) *Dawn-to-Dusk Electric Field in the Polar Region*

In Section 4.4.2, we examined the general relationship between the IMF and the dawn-to-dusk electric field. In this subsection, we examine first an example of the electric field observations by balloon-borne probes (Mozer, 1971).

Figure 4.38 shows a simultaneous observation of westward electric fields at Churchill, Yellowknife and Uranium City on 1969, August 8. A large westward electric field grew at both Yellowknife and Uranium City at about 1700 UT on that day, well before the onset of a negative bay (0740 UT) at three Canadian stations, including Great Whale River. The figure shows also the corresponding IMF; it turned southward rather abruptly at 0630 UT. It also shows the corresponding magnetic records from several stations in the polar cap, Godhavn, Narssarssuaq, Mould Bay, College and Alert, where a distinct disturbance began at about 0630 UT. It may well be that these disturbance features can be described as the DP-2 variation which will be discussed in subsection (c). As we shall see in Section 7.7.3, Mozer (1971) also examined 15-min averages of the electric field data obtained during 19 balloon observations; the onset time $T = 0$ of the substorms was determined by the onset time of negative bays. He suggested that the increase of the westward electric field prior to $T = 0$ is a growth phase feature. However, as was mentioned earlier in Section 4.4.1, it is more appropriate to attribute the increase to the southward turning of the IMF vector.

Figure 4.39(a) shows the distribution of approximate ($E \times B$) drift vectors, observed by the INJUN-5 satellite, along three successive passages over the northern and southern polar region on 1970, February 24 (Gurnett and Akasofu, 1974). An intense dawn-to-dusk electric field was observed at about the time the satellite entered the southern polar cap (1515 UT), after leaving the northern polar

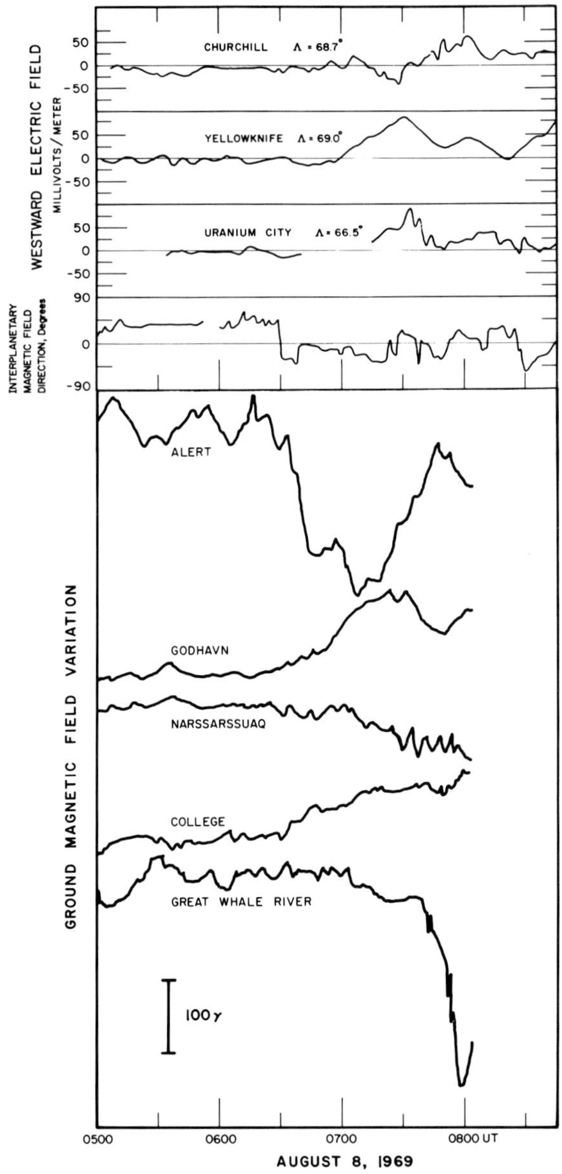

Fig. 4.38. Simultaneous measurements of the westward electric field at Churchill, Yellowknife and Uranium City, together with the IMF θ component and the H component magnetic variations in high latitudes – Alert, Godhaven, Narssarssuaq, College and Great Whale River. (Mozer, R. S.: *J. Geophys. Res.* **76**, 7595, 1971.)

cap at about 1435 UT. The corresponding IMF turned southward between 1322 and 1419 UT. If the growth of the electric field is assumed to be associated with the southward turning (which reached the apex of the magnetopause at about 1402 UT), the electric field did not grow for about 28 min after the turning, since there was no significant electric field during the first orbit (~1430 UT); Figure 4.39(b). A large increase of the AE index began at about 1530 UT.

(a)

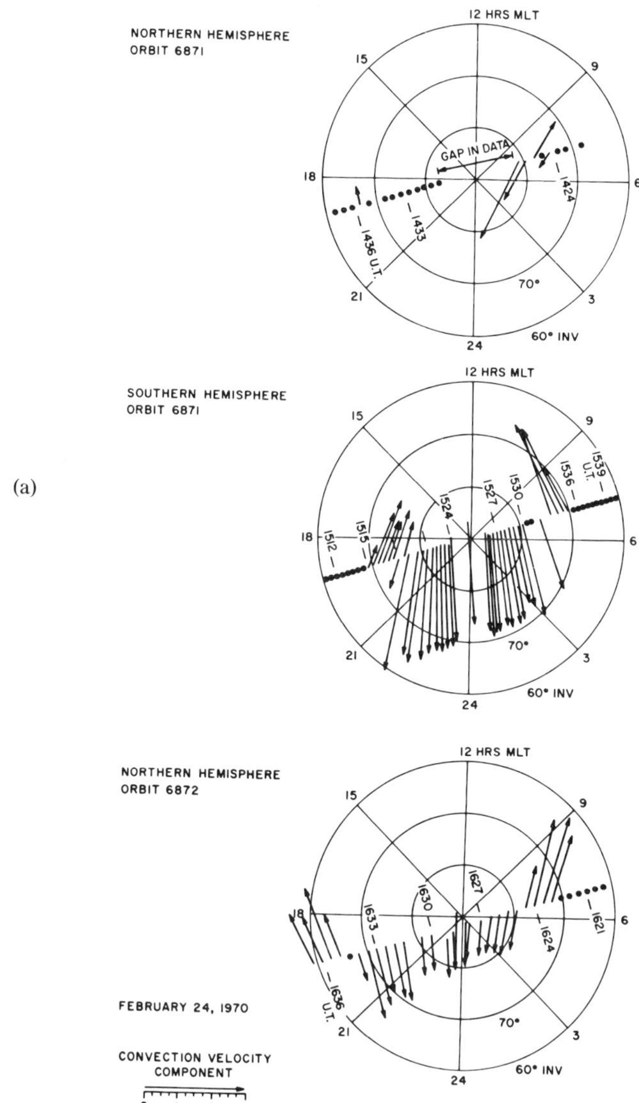

Fig. 4.39(a, b). Sudden enhancement of the polar cap electric field and the associated IMF and geomagnetic field changes. (Gurnett, D. A. and Akasofu, S.-I.: *J. Geophys. Res.* **79**, 3197, 1974.)

(b) *Expansion of the Auroral Oval*

It was found in Section 1.4.3 that the potential Φ_D is related to the noon-midnight meridian dimension d_1 and the dawn-dusk dimension d_2 of the auroral oval by

$$\Phi_D = \frac{d_1 d_2 V_s B_P}{L}$$

where V_s denotes the solar wind speed, L the length of the magnetotail and B_P the magnetic field intensity in the polar ionosphere.

(b)

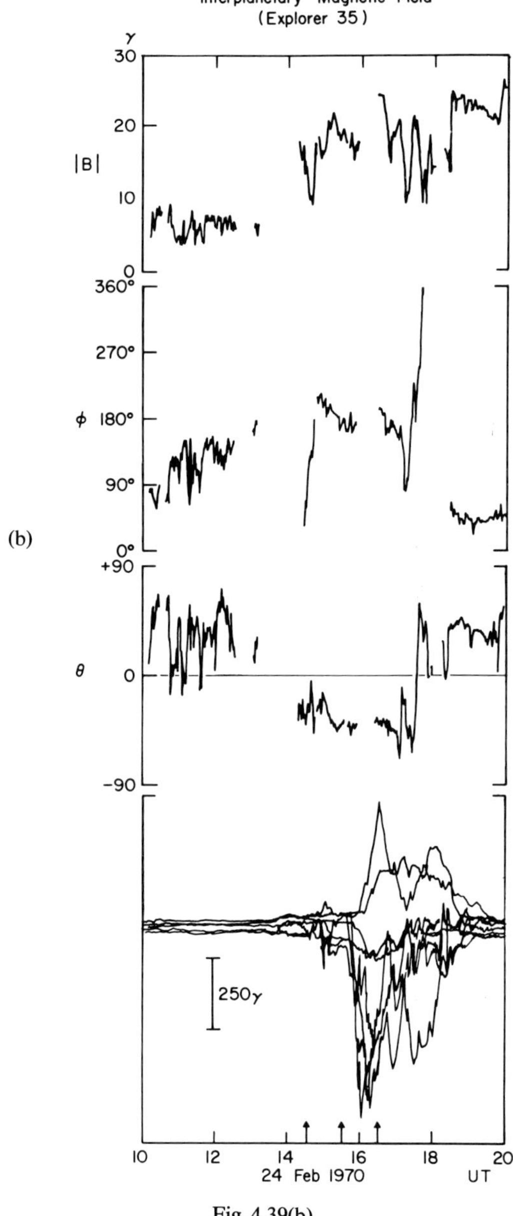

Fig. 4.39(b).

Holzworth and Meng (1975) showed that the auroral oval can often be approximated by a circle ($d_1 \simeq d_2$) and examined the relationship between the radius $d(d_1 = d_2)$ and the B_z component. Their result is shown in Figure 4.40. It is, however, not well known at present how fast the auroral oval *as a whole* responds to changes of the B_z component. Pike *et al.* (1974) showed, on the basis of a

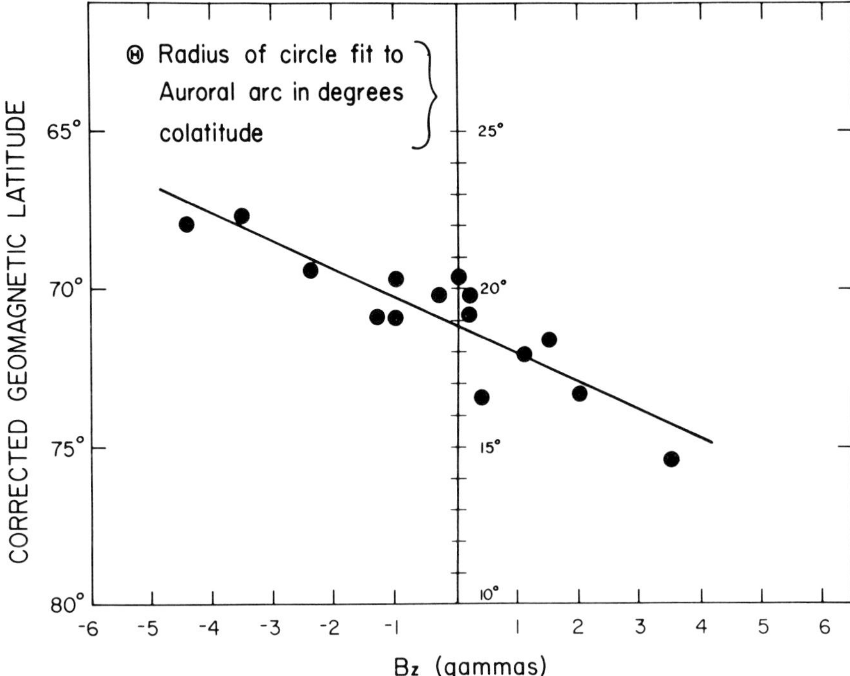

Fig. 4.40. Midnight locations of the auroral oval as a function of the IMF B_z component. (Holzworth, R. H. and Meng, C.-I.: *Geophys. Res. Lett.* **2**, 277, 1975.)

simultaneous observation of the oval by an airborne camera and the Alaska meridian chain of cameras, that the equatorward shift of the oval, after the southward turning of the IMF vector, was observed nearly simultaneously in the 1500 and 1930 MLT sectors. It is also quite likely that the equatorward motion of auroras and the simultaneous increase of the westward electric field, observed by Kelley *et al.* (1971), are associated with the southward turning of the IMF. The expansion and contraction of the auroral oval associated with changes of the IMF will further be discussed in Section 4.4.7.

(c) *DP-2 Variation*

It has been proposed by Nishida (1968a, b) that there is a new type of magnetic variation, called the DP-2 variation, which is distinct from the polar magnetic substorm variations. He showed that the growth and decay of the DP-2 field closely follows changes of the B_z component of the IMF in a very particular way. Figure 4.41 shows an example of the relationship between DP-2 variation and B_z. Note that negative B_z values are plotted above the base line ($B_z = 0$) and positive B_z values below it. In order to illustrate the correlation, some of the points at which dB_z/dt varies from a positive value to a negative value are connected by straight lines, regardless of the signs of B_z values at those points. Then, the areas enclosed by the B_z (t = time) curve and those straight lines are shaded. With such a diagram in hand, Nishida could illustrate that the DP-2 variation can easily be

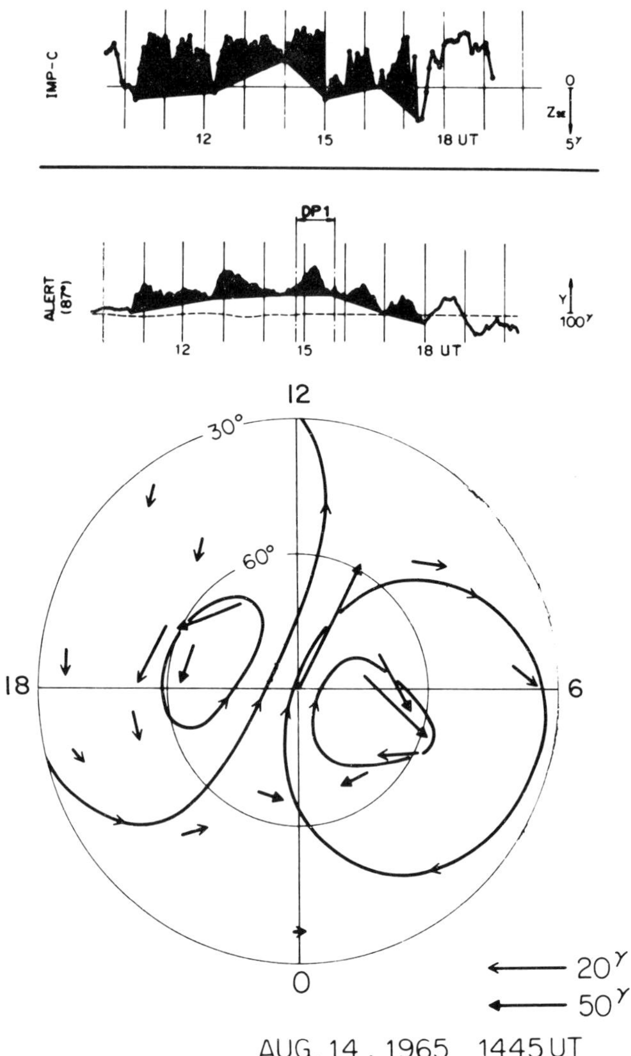

Fig. 4.41. Modulation of the S_q^p variations by the IMF B_z component, namely, the DP-2 variations, and its equivalent current system. (Nishida, A.: *Planet. Space Sci.* **19**, 205, 1971.)

identified in magnetic records from polar cap stations and a magnetic equatorial station, Huancayo. In Figure 4.41, the *Y* component record from Alert is shown, and the identified DP-2 variations are also shaded. He showed that the equivalent current system of the DP-2 variation consists mainly of two vortices or the so-called 'twin vortex mode'. Further, Nishida (1968a, b) suggested that the DP-2 current spreads down to the equator and that there occurs an eastward zonal current. However, on the basis of an incoherent scatter radar observation at Jicamarca, (Peru), Matsushita and Balsley (1972) questioned the existence of such an extended current; see also Nishida (1973b) and Matsushita and Balsley (1973).

Akasofu *et al.* (1973b) and Leont'yev and Lyatskiy (1974) showed that a significant part, if not all, of the DP-2 variation can reasonably be explained by the combined effect of the equatorward expansion of the permanent current system S_q^p and its enhancement. The expansion of the oval is now found to be controlled by the north-south component of the IMF (see (c)). They concluded thus that the DP-2 variation arises from a *modulation* of the permanently existing S_q^p current system by the IMF, rather than by an intermittent growth of a new type of current system.

In order to demonstrate this possibility, Akasofu *et al.* (1973b) assumed that the oval shifts from the dp. lat. 80° latitude circle to the dp. lat. 78° circle and that the S_q^p current is doubled from the quiet day value. The distribution of the computed vectors for such a combination is illustrated in Figure 4.42. Considering the uncertainty in choosing a proper baseline, the resemblance with the observed vector distribution is enough to account for a significant part, if not all, of the DP-2 variations.

Kokubun (1971) was the first to recognize that a twin-vortex current system grows in the polar region after the southward turning of the interplanetary

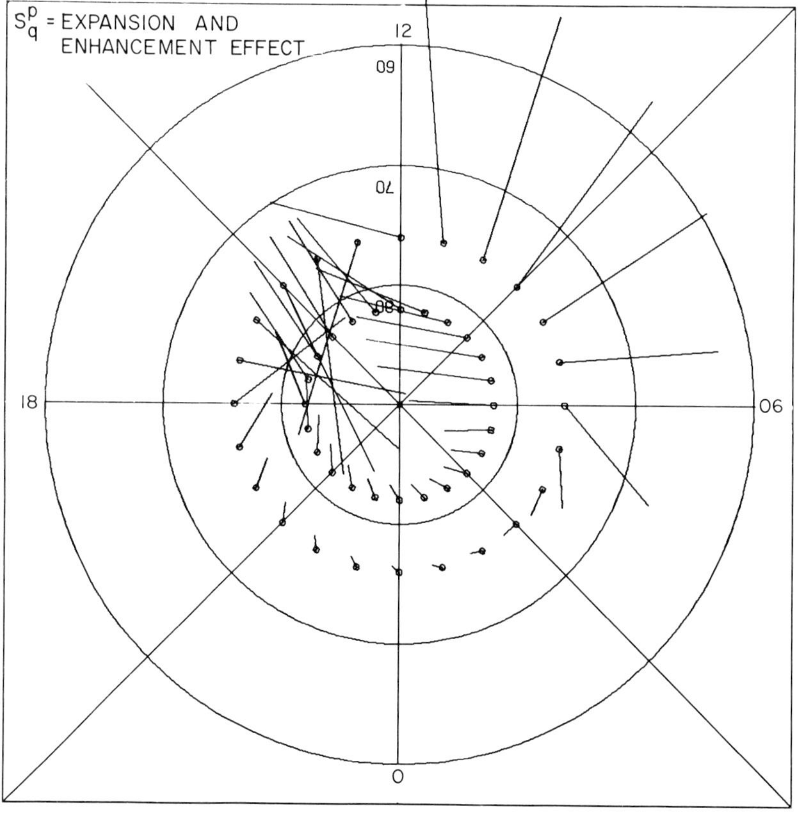

Fig. 4.42. Computed modulation of the S_q^p variations by assuming an expansion of the oval and enhancement of the S_q^p current system (including the field-aligned currents). (Akasofu, S.-I., Yasuhara, F. and Kawasaki, K.: *Planet. Space Sci.* **21**, 2232, 1973.)

magnetic field. When the growth of this current system is observed at the auroral zone stations in the midnight sector, it appears as a slowly growing negative bay. He ascribed this phenomenon to an enhanced convection and suggested that it is a manifestation of the energy storing process in the magnetotail.

This particular feature has then been extensively studied by Nishida (1971), Nishida and Kokubun (1971), Iijima (1972, 1974), Iijima and Nagata (1972), Iijima and Kokubun (1973), Pudovkin *et al.* (1970), Pudovkin and Troshichev (1972), Troshichev *et al.* (1974b) and others. Figure 4.43 shows an example of such a current system constructed by Iijima (1972). A striking resemblance between this twin-vortex current system and the DP-2 current system should be noted. Indeed, it is quite likely that they are essentially the same phenomenon. Troshichev *et al.* (1974a), Troshichev *et al.* (1974b) and Tverskaya and Khorosheva (1974) found that two basic current systems, DP_{11} and DP_{12}, appear alternately on a disturbed day.

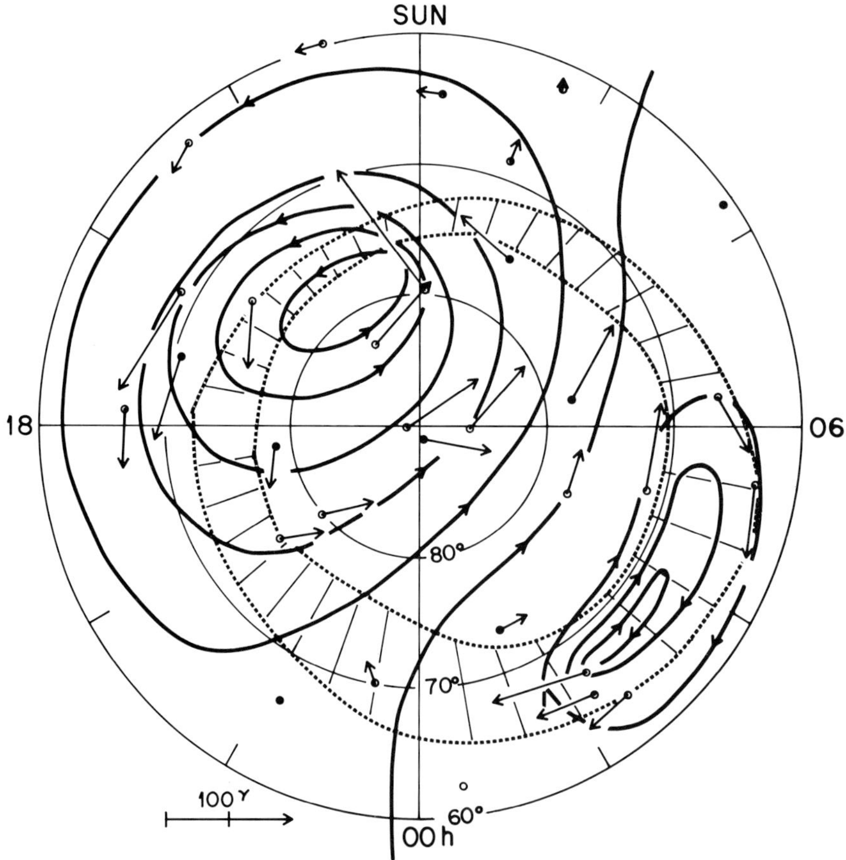

Fig. 4.43. Equivalent current system in the polar region for the proposed growth phase during a substorm on 1965, September 12–13. Current intensity between two adjacent lines is 3.1×10^4 A. (Iijima, T.: *Rep. Ionosph. Space Res. Japan* **26**, 267, 1972.)

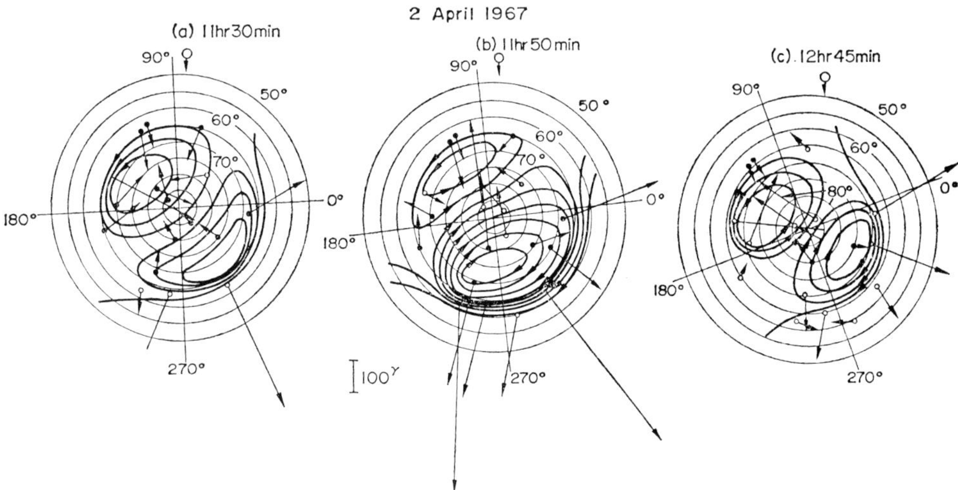

Fig. 4.44. Equivalent current system during a 'growth phase', the expansive phase and the recovery phase during a substorm on 1967, April 2. (Troshichev, O. A., Kuznetsov, B. M. and Pudovkin, M. I.: *Planet. Space Sci.* **22**, 1403, 1974.)

Their DP_{11} is identical to the substorm current system. They noted that the DP_{12} current system appears during the proposed growth phase and also during the recovery phase. An example of their analysis is shown in Figure 4.44. It can be seen that the DP_{12} current system is essentially identical to the DP-2 current system. It should be noted, however, that such a distinction is somehow arbitrary, since both are most likely to co-exist (Kawasaki and Akasofu, 1972; see also Nishida, 1973).

Burch (1973a) reported that the ratio of DP-2 type activity at Vostok to that at Thule was greater by a factor of about 1.75 for toward sectors than for away sectors for periods near the 1967 spring and fall equinoxes.

4.4.7. CORRELATION BETWEEN B_s AND THE AE INDEX

(a) *Arnoldy's Study of the Relationship between B_s and the AE Index*

Arnoldy (1971) gave an empirical equation between the AE index and solar wind parameters:

$$\begin{aligned}\mathrm{AE} = &-0.26(\Sigma B_s\tau)_0 - 0.91(\Sigma B_s\tau)_1\\ &-0.33(\Sigma B_s\tau)_2 + 0.12p\end{aligned}$$

where B_s and p denote the magnitude of southward-directed component of IMF and the dynamic pressure of the solar wind, respectively. The quantity $(\Sigma B_s\tau)$ is a summation over one hour of the product of ten-minute average values of B_s and time increment τ; the subscripts 0, 1 and 2 for the summation refer to the lead time in hours of the summation with respect to the AE index. The second term, the summation of B_s over the hour preceding the value of AE, is the most significant term. In Figure 4.45 the actual AE index (hourly) is compared with the AE index given by the above empirical equation. The agreement between the two curves is good, suggesting that the quantity $(\Sigma B_s\tau)_1$ is of fundamental importance in substorm processes.

Similar, but finer time-resolution studies (~5.5 min) have recently been made by Foster *et al.* (1971), Rostoker *et al.* (1972) and Meng *et al.* (1973). The results obtained by Foster *et al.* (1971) are shown in Figure 4.46. They noted that the AE index shows a slow growth prior to a sharp growth and that the latter occurs at about the time when the IMF B_z component reaches the minimum value. Meng *et al.* (1973) noted: (i) The increase of the AE index is observed, on the average, at about 15 min or so after the onset of the southward IMF event. (ii) The fine time-scale cross-correlation computation between the $-B_z(=B_s)$ component of the IMF and the AE index showed a significant correlation; the location of the peak value of the coefficient varied from −10 min to −90 min, the most probable location being about −40 min (Ivanov and Mikerina (1973) found the time lag ranges from 1 h 10 min to 4 h 40 min). (iii) This time lag between southward turning of the IMF and the AE index should not be interpreted as being the duration of a growth phase (see also Hirshberg and Holzer, 1975). (iv) The termination of the southward IMF (northward turning of the field) is frequently associated with the start of the recovery of the AE activity.

Arnoldy (1971) and Foster *et al.* (1971) showed also that other solar wind parameters, such as the number density, temperature, velocity, and the magnitude of the IMF, do not have a significant correlation with the AE index (Figure 4.47).

The above studies indicate clearly that the southward component of the IMF has a definite correlation with the AE index. In order to understand basic physical processes involved in this correlation, it is necessary, however, to understand what the AE index actually represents.

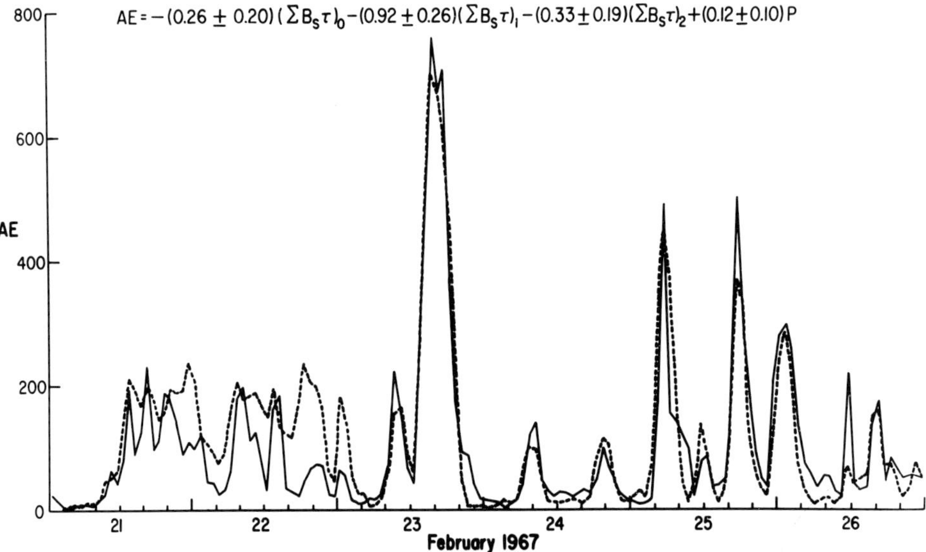

Fig. 4.45. Comparison of the empirical AE index deduced from the IMF B_s component and solar wind parameters and the actual AE index. (Arnoldy, R. L.: *J. Geophys. Res.* **76**, 5189, 1971.)

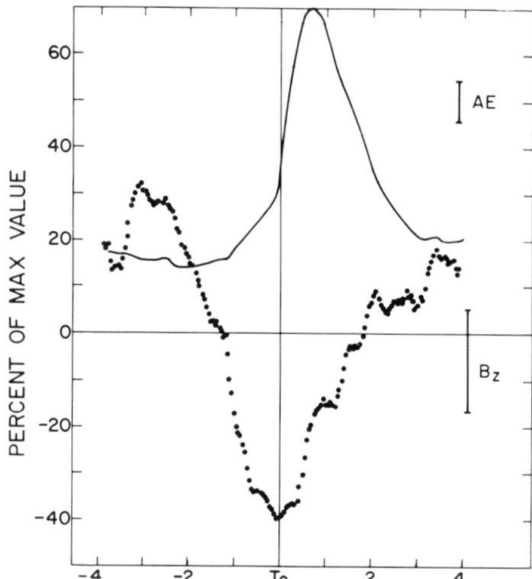

Fig. 4.46. Average percentage of the maximum value of the AE index and the IMF B_z component during 54 isolated substorms. T_o was defined as the time of the sharp onset of the negative bay observed in the H component of the auroral zone station in the midnight sector. (Foster, J. C., Fairfield, D. H., Ogilvie, K. W. and Rosenberg, T. J.: *J. Geophys. Res.* **76**, 6971, 1971.)

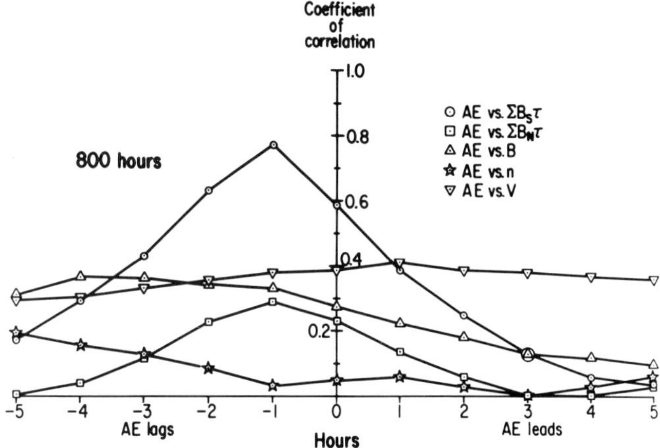

Fig. 4.47. Linear correlation coefficient of various interplanetary parameters with the AE index for lead or lag time up to 5 h. (Arnoldy, R. L.: *J. Geophys. Res.* **76**, 5189, 1971.)

(b) *The AE Index*

The AE index is devised to monitor the intensity of the electrojet on the basis of horizontal component magnetic records from a set of auroral zone magnetic stations which are fairly uniformly distributed in longitude (Davis and Sugiura, 1966; Allen and Krochl, 1975). By superposing these records, the upper envelope of the superposed records provides the AU index and the lower envelope provides the AL index. The AE index is defined as the distance between the upper and lower envelopes:

$$AE = AU + AL$$

Thus the AE index is based on the assumption that at least one of the auroral zone magnetic stations is capable of monitoring the westward electrojet, which is assumed to be most intense along the auroral *zone* in the midnight or the early morning sector, and that at least one of them is capable of monitoring the eastward electrojet which is assumed to be the most intense along the auroral *zone* in the afternoon sector. Unfortunately, however, the latitude of the electrojets can be quite variable, and the westward electrojet flows along the auroral *oval* which changes its size (Section 4.4.6(b)). In particular, a serious problem arises when the auroral oval contracts poleward of the auroral zone in the midnight sector, since all the AE stations are then located well outside the auroral oval during such a period.

As we saw in Sections 1.4.3 and 4.4.6, the B_z component of the IMF controls the size (namely, d_1 and d_2) of the auroral oval. When the B_z component has a large northward component, the auroral oval contracts poleward, well beyond the latitude of the auroral zone. In such occasions, the auroral zone stations are no longer capable of monitoring properly geomagnetic activity which occurs along such a contract oval (Feldstein and Starkov, 1968; Akasofu *et al.*, 1971a; Akasofu *et al.*, 1973a). Figures 4.48(a) and (b) show two examples of auroral substorms observed over Inuvik (inv. lat. ~70°) during a prolonged period of a large positive IMF B_z component. The corresponding magnetic records from College show little sign of substorm activity at those times, despite the fact that College was then located in the midnight sector. In Section 5.2 we shall see also several DMSP photographs which show substorm activity along a contracted oval when the IMF B_z component has a large positive component. Figure 4.49 shows a detailed analysis of a substorm which occurred well poleward of College. The substorm was associated with an intense poleward expansion of the oval and also of the growth of the electrojet (~250 γ). *In such an occasion, however, it has been thought that geomagnetic and auroral activities are absent over the entire polar region. Actually, the auroral zone stations become simply incapable of monitoring geomagnetic and auroral activities when the B_z component has a large northward component.*

On the other hand, when the B_z component has an appreciable southward component, the oval expands down to the latitude of the auroral zone in the midnight sector. Thus, *it is only in such occasions when the AE index becomes meaningful and when substorm activity can properly be monitored in terms of the AE index.* A close correlation between the AE index and the IMF B_z, such as that shown in Figure 4.50, is due mostly, if not entirely, to this fact.

Therefore, *if we rely only on the AE index or auroral zone magnetic records as an indicator of substorm activity, substorms appear to occur mostly when the B_z component has the southward component or particularly after the southward turning. Further, for the latter case, the magnetospheric response to the southward turning may appear to precede each substorm onset. These are the major reasons for overemphasizing the roles of the IMF B_z component on substorm processes.*

It has also been argued by some workers that weak substorms during the period of the northward IMF should not be regarded as substorms or should not be considered in an initial study. However, we shall see in Section 5.2 that there is no major qualitative difference between substorms with $B_z > 0$ and those with

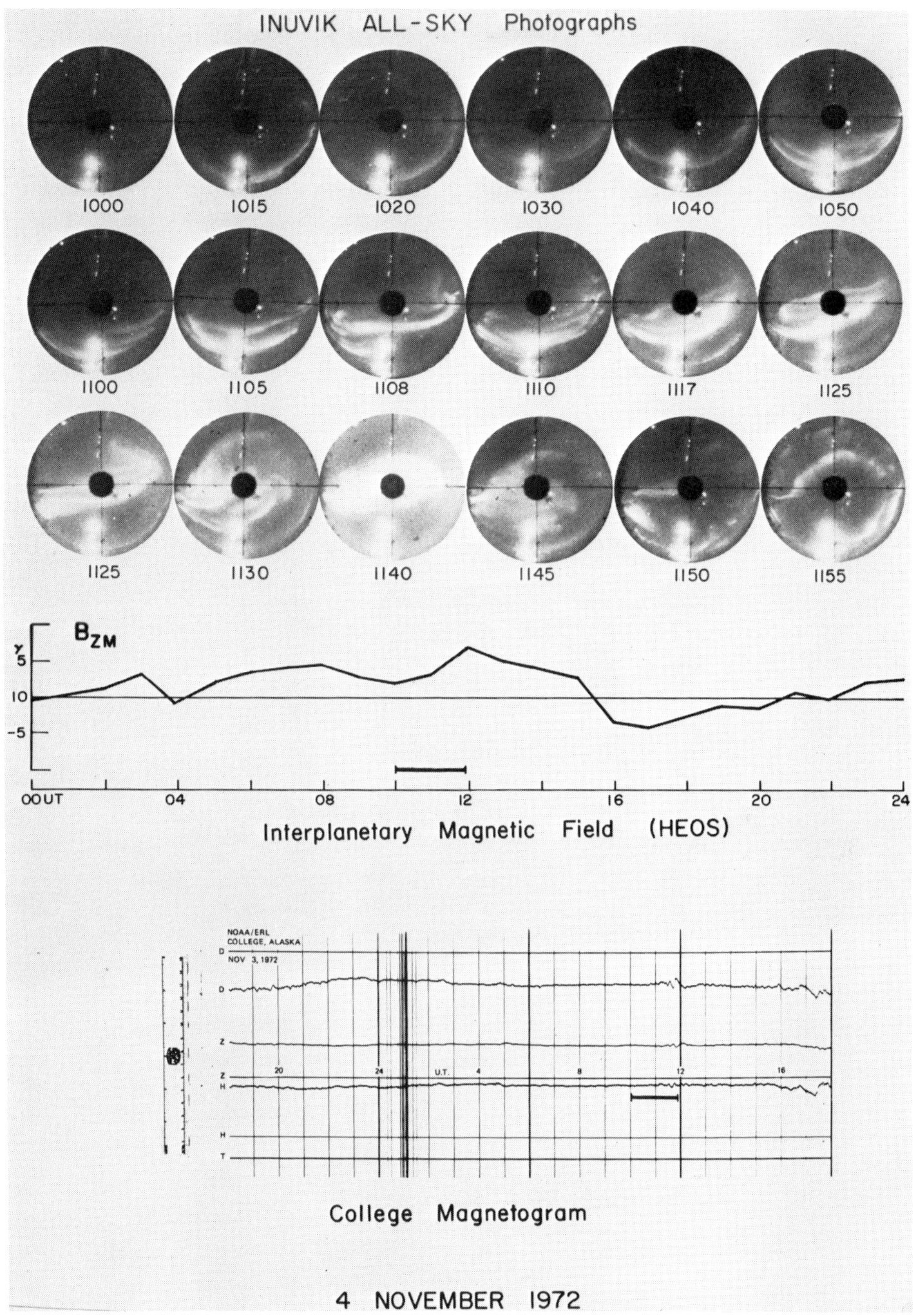

(a)

Fig. 4.48(a, b). Auroral substorms observed along contracted auroral oval. They tend to occur during periods when the IMF B_z component has a large northward component. Note further that the College magnetic records show no obvious indication of substorms between 10 and 12 UT. (Akasofu, S.-I.: *Planet. Space Sci.* **23**, 1349, 1975.)

INUVIK ALL-SKY Photographs
0800
0815
0830
0832
0840
0842
0843
0845
0850
0853
1035
1036
1038
1040
1045
1050
B_{ZM}
γ
5
0
-5
00 UT
04
08
12
16
20
24
Interplanetary Magnetic Field (HEOS)
NOAA/ERL
COLLEGE, ALASKA
NOV 4, 1972
D
Z
H
T
20
24
U.T.
4
8
12
16
College Magnetogram
5 NOVEMBER 1972

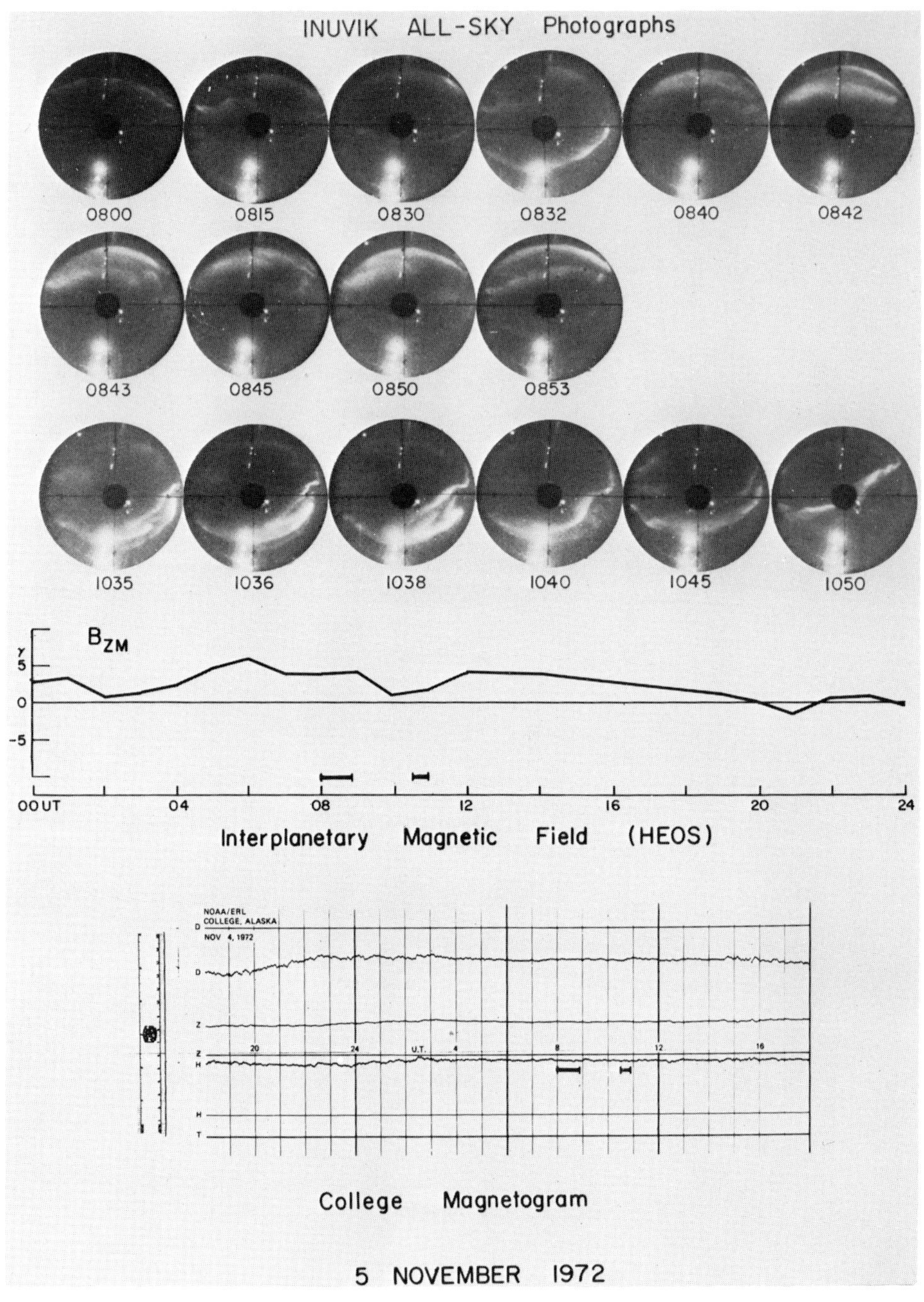

(b)

Fig. 4.48(b)

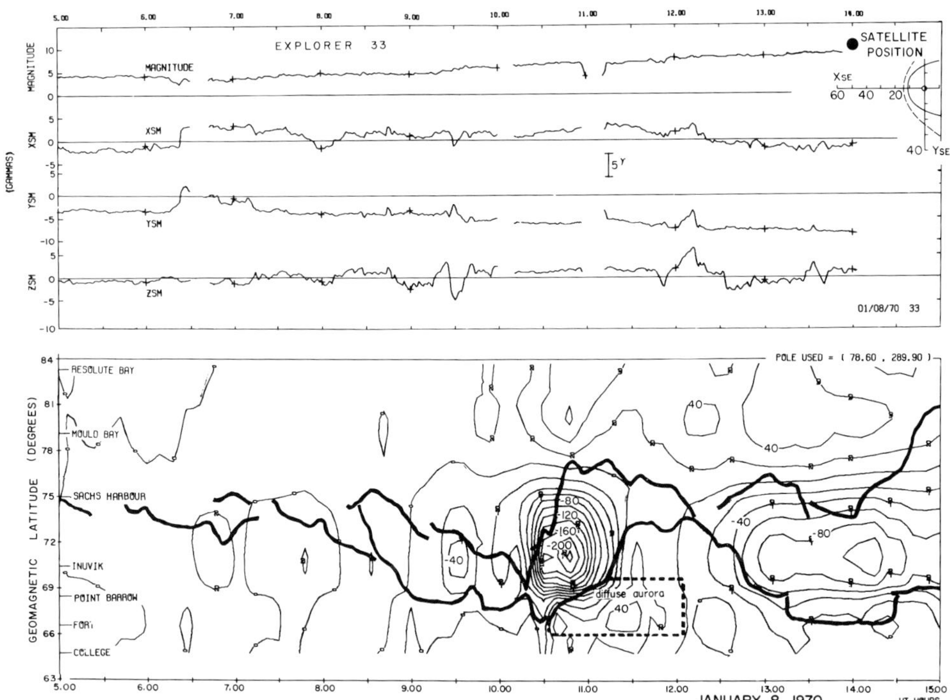

Fig. 4.49. Detailed features of a magnetospheric substorm which occurred along a contracted oval on 1970, January 8. In the upper panel, the IMF parameters are shown. In the lower panel, the high latitude and low latitude boundaries of the auroral oval are shown by thick curves. Note a sudden poleward expansion of the oval at 1020 UT. The thin lines show the iso-intensity contours of the *H* component, deduced from magnetic records obtained from the Alaska meridian chain of observatories. (Kamide, Y. and Akasofu, S.-I.: *J. Geophys. Res.* **79**, 3755, 1974.)

$B_z < 0$. This feature was also seen in Figures 4.48 and 4.49. Many substorm quantities vary continuously for a wide range of B_z, from a large positive value to a large negative value. Thus there is no reason to group substorms into two, one with $B_z > 0$ and the other with $B_z < 0$. Secondly, by confining attention to substorms with $B_z < 0$ alone, one is bound to ignore a large number of substorms. McPherron *et al.* (1973c) recently noted that their growth phase model refers only to the first substorm after the southward turning of the interplanetary magnetic field. This is obviously because only the first substorm after the southward turning has their proposed growth phase. It is also interesting to note that a rocket observation of weak substorms, conducted by Johnstone *et al.* (1974), showed also that there is no basic difference between weak and intense substorms.

Another major reason for identifying incorrectly growth phase features is simply due to the lack of an adequate examination of magnetic records. Records from subauroral zone stations should not be used in monitoring substorm activity, since they are incapable of accurately monitoring substorm activity. It is also not possible to determine accurately (± 10 min) the onset time of substorms on the

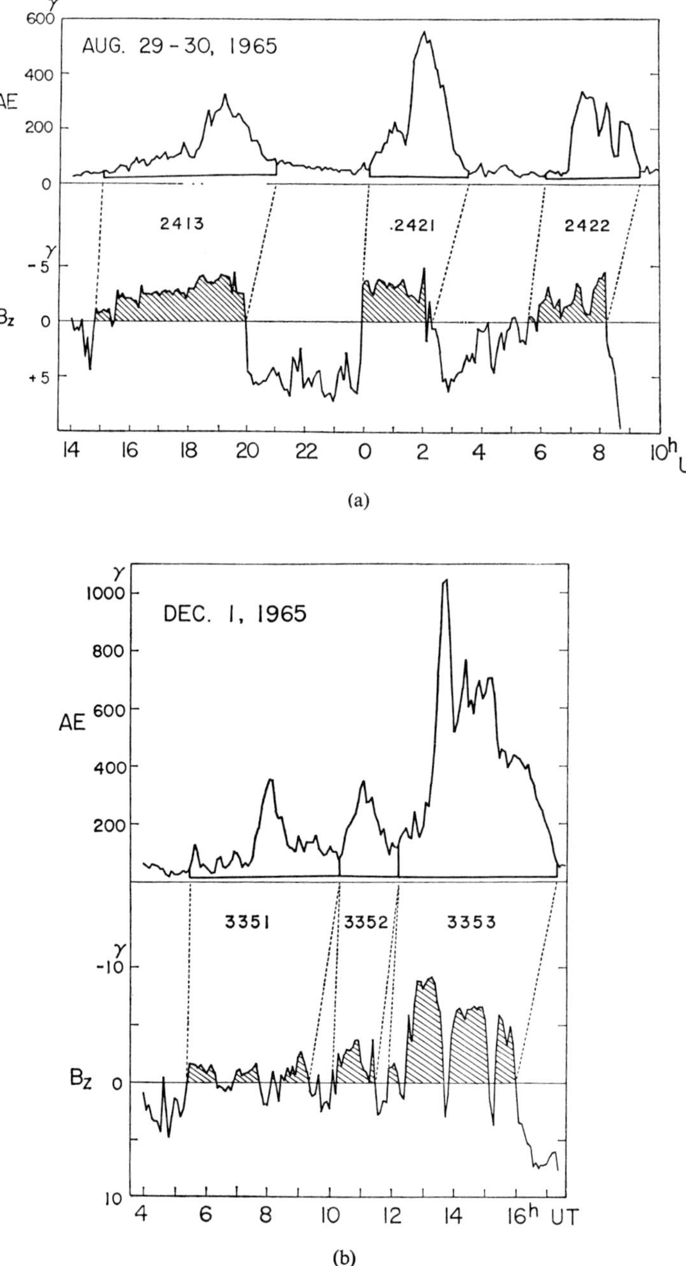

(a)

(b)

Fig. 4.50. Correlation between the AE index and the IMF B_z component. (Murayama, T. and Hakamada, K.: *Planet. Space Sci.* **23**, 75, 1975.)

basis of auroral zone magnetic records; this is particularly the case in the evening sector because of the complexity associated with westward traveling surges which have a finite propagation time. Unfortunately, there are several papers which claim to find growth phase features which occur only 10–15 min prior to the onset of the expansion phase, which was determined from auroral zone magnetic records.

Still another reason for the confusion is our inability to define the baseline in satellite magnetic records. When substorms occur frequently, it is difficult to determine whether a particular magnetospheric quantity observed is recovering toward its baseline value during the recovery phase or whether a new change is beginning to take place after the end of the previous substorm. In Sections 6.3 and 6.6 we shall see magnetic field observation in the magnetotail during successive substorms. During the recovery phase (*not after*) of each substorm, the B_z component slowly decreases, so that the field configuration becomes 'tail-like' after increasing rapidly at an early epoch. On the other hand, Aubry *et al.* (1972) suggested that such a tendency occurs during a growth phase (Section 4.4.5(b)).

Further, the AE index is not appropriate in identifying different *phases* of the substorm. The onset time of the expansive and recovery phases of the magnetospheric substorm can, so far, be determined only on the basis of auroral data. The expansive phase of the magnetospheric substorm is defined to begin at the time of a sudden brightening of an auroral arc and to end at the time when the poleward expanding bulge reaches the highest latitude. Magnetic records from the auroral zone cannot tell when the bulge reaches the highest latitude. The intensity of a negative bay observed at an auroral zone station depends on its distance from the station, as well as the total current intensity, density distribution and width of the westward electrojet. Although it is a rather common practice to identify the onset time of substorms by the so-called 'sharp onset' of a negative bay, this practice is applicable only when magnetic records are available from the station which is located very near the auroral arc which brightens at the onset time of the expansive phase (Akasofu and Snyder, 1972; Tsurutani and Bogott, 1972). Similarly, there is no basis for determining the onset of the recovery phase by the maximum epoch of a negative bay or of the AE index. The recovery of a negative bay can be caused by the movement of the electrojet away from the station. The AE index or magnetic records from the auroral zone can tell only different *epochs* (early or late) of substorms.

All these cautions are rather elementary. However, they have often been ignored. Since a number of magnetospheric phenomena have been discussed in the past on the basis of inadequate use of ground magnetic records, it is quite natural to have some confusion in trying to piece together satellite observations during different substorms. For example, in many cases, claimed growth phase features were actually expansive phase features and sometimes even recovery phase features of previous substorms.

It has also been claimed that one of the most important indications of the proposed growth phase is a gradual development of a negative bay at auroral zone stations in the midnight sector or a gradual development in the AE index, before its sudden enhancement or the so-called 'sharp onset'. Indeed, a large number of papers have been published on a growth phase during the past several years by making use of this particular feature. However, at least in some cases, the gradual development in the AE index can actually be substorm effects along a contracted oval (Kamide *et al.*, 1975). It was shown also in Section 4.4.6(b) that the modulation of the S_q^p current by an enhanced merging rate can give rise to a particular type of magnetic field variations, first identified by Nishida (1968a, b) as the DP-2 variations. Since these magnetic variations will be reflected in the AE index, AE index changes should be treated carefully.

The onset time of positive bays observed in middle latitudes has also been used extensively in determining $T = 0$ (McPherron *et al.*, 1973). However, as shown by Akasofu (1968, p. 175 and Figure 126), this onset time can differ considerably at stations in the evening sector. Thus, one must be cautious in using this auxiliary feature.

4.4.8. SUMMARY

There is little doubt that the north-south component B_z of the IMF plays a vital role in magnetospheric processes by controlling the merging rate between the IMF and the geomagnetic field. In an attempt to identify various magnetospheric responses to the IMF B_z component, it is very important to remember that most of the magnetospheric quantities are a function of many parameters. Therefore, a painful analysis is needed in examining how each parameter contributes to a particular magnetospheric quantity. Without such an analysis, a phenomenologi-

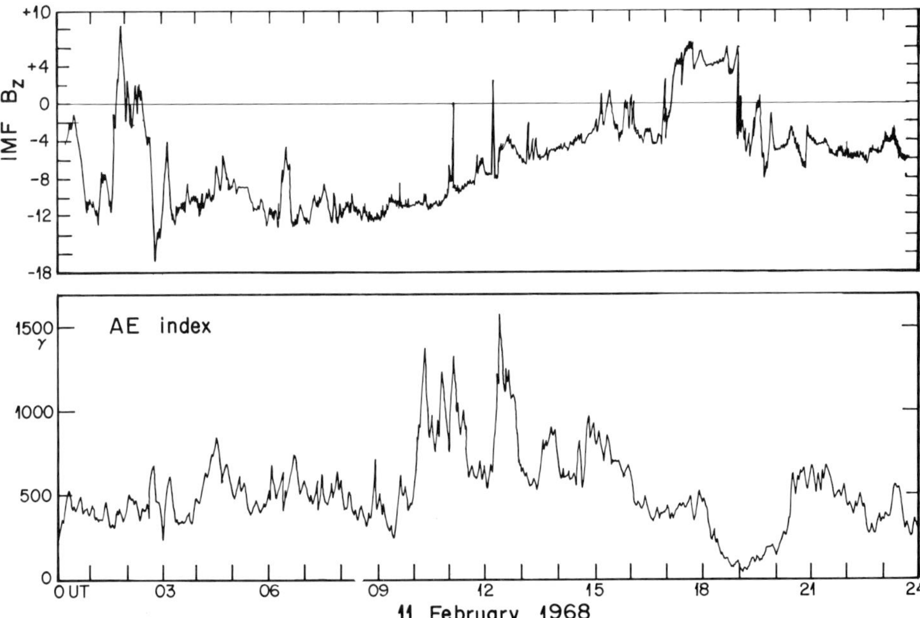

Fig. 4.51. Occurrence of intense magnetospheric substorms, between 09 and 13 UT on 1968, February 11, without any obvious decrease of the IMF B_z component except that they occurred during a prolonged period of a large negative B_z component.

cal theory is of little worth. As we saw in this chapter, we have just begun to separate the response of the magnetosphere to the B_z component from substorm effects. There is still much uncertainty in this particular study. Note in particular that there is no indication that the direct cause of *each* substorm can be tracked back to *each* southward turning of the IMF vector. So long as the magnetosphere has a sufficient amount of energy to dissipate, it generates magnetospheric substorms (see also Section 5.2). For example, in Figure 4.51, it is quite clear that a series of intense ($\sim 1500\ \gamma$) substorms between 09 and 16 UT occurred without an increase of the B_s ($=-B_z$) one or two hours prior to their onsets.

In studying magnetospheric substorms, one has to determine how various magnetospheric quantities observed at different locations vary as a function of substorm time T, and then examine how they are related to each other. Ideally, such a study could be achieved if there were a large number of satellites, observing simultaneously all the desired quantities at a large number of locations. Since this is impossible, one has to compile satellite data observed at different locations and at different times, and analysed by different workers. Such a task will be possible only if there is the identical time frame of reference, namely the identical substorm time. For this purpose, it is essential to determine accurately the substorm onset time $T = 0$ which is defined as the time when an auroral arc in the midnight sector suddenly brightens. Unfortunately, auroral dates are not always available, so that several auxiliary methods have been devised, mostly without justification and 'calibration'. This fact has been the source of confusion in substorm studies during the last several years.

References

Akasofu, S.-I.: 1968, *Polar and Magnetospheric Substorms*, D. Reidel Publ. Co., Dordrecht-Holland.

Akasofu, S.-I. and Snyder, A. L.: 1972, 'Comments on the Growth Phase of Magnetospheric Substorms', *J. Geophys. Res.* **77**, 6275.

Akasofu, S.-I., Wilson, C. R., Snyder, L. and Perreault, P. D.: 1971a, 'Results from a Meridian Chain of Observatories in the Alaskan Sector (1)', *Planet. Space Sci.* **19**, 477.

Akasofu, S.-I., Hones, E. W. Jr., Montgomery, M. D., Bame, S. J. and Singer, S.: 1971b, 'Association of Magnetotail Phenomena with Visible Auroral Features', *J. Geophys. Res.* **76**, 5985.

Akasofu, S.-I., Perreault, P. D., Yasuhara, F. and Meng, C.-I.: 1973a, 'Auroral Substorms and the Interplanetary Magnetic Field', *J. Geophys. Res.* **78**, 7490.

Akasofu, S.-I., Yasuhara, F. and Kawasaki, K.: 1973b, 'A Note on the DP-2 Variation', *Planet. Space Sci.* **21**, 2232.

Aleksandrova, I. A. and Zaytseva, S. A.: 1972, 'Polar-Cap Auroras and the Intensity of the DCF field', *Geomag. Aeronom.* **12**, 310.

Allen, J. H. and Krochl, H. W.: 1975, 'Spatial and Temporal Distributions of Magnetic Effects of Auroral Electrojets as Derived from AE Indices', *J. Geophys. Res.* **80**, 3667.

Arnoldy, R. L.: 1971, 'Signature in the Interplanetary Medium for Substorms', *J. Geophys. Res.* **76**, 5189.

Atkinson, G. and Unti, T.: 1969, 'Two-Dimensional Chapman-Ferraro Problem with Neutral Sheet, 3, Implied Magnetospheric Flows and Their Time Dependence', *J. Geophys. Res.* **74**, 6275.

Aubry, M. P. and McPherron, R. L.: 1971, 'Magnetotail Changes in Relation to the Solar Wind Magnetic Field and Magnetospheric Substorms', *J. Geophys. Res.* **76**, 4381.

Aubry, M. P., Russell, C. T. and Kivelson, M. G.: 1970, 'On Inward Motion of the Magnetopause before a Substorm', *J. Geophys. Res.* **75**, 7018.

Aubry, M. P., Kivelson, M. G. and Russell, C. T.: 1971, 'Motion and Structure of the Magnetopause', *J. Geophys. Res.* **76**, 1673.

Aubry, M. P., Kivelson, M. G., McPherron, R. L., Russell, C. T. and Colburn, D. S.: 1972, 'Outer Magnetosphere near Midnight at Quiet and Disturbed Times', *J. Geophys. Res.* **77**, 5487.

Berthelier, A., Berthelier, J. J. and Guerin, C.: 1974, 'The Effect of the East-West Component of the Interplanetary Magnetic Field on Magnetospheric Convection as Deduced from Magnetic Perturbations at High Latitudes', *J. Geophys. Res.* **79**, 3187.

Bhargava, B. N. and Rangarajan, G. K.: 1975, 'Interplanetary Magnetic Field Polarity and Low-Latitude Geomagnetic Field', *Planet. Space Sci.* **23**, 929.

Brown, R. R.: 1974, 'On Decreases in Ionospheric Absorption Associated with Negative Sudden Impulses in the Geomagnetic Field', *J. Geophys. Res.* **79**, 1113.

Brown, R. R. and Driatsky, V. M.: 1973, 'Further Studies of Ionospheric and Geomagnetic Effects of Sudden Impulses', *Planet. Space Sci.* **21**, 1931.

Brown, R. R., Leinbach, H., Akasofu, S.-I., Driatsky, V. M. and Schmidt, R. J.: 1972, 'Quadruple Conjugate Pair Observations of the Sudden Commencement Absorption Event on June 17, 1965', *J. Geophys. Res.* **77**, 5602.

Burch, J. L.: 1972a, 'Preconditions for the Triggering of Polar Magnetic Substorms, by Storm Sudden Commencements', *J. Geophys. Res.* **77**, 5629.

Burch, J. L.: 1972b, 'Precipitation of Low-Energy Electrons at High Latitudes: Effects of Interplanetary Magnetic Field and Dipole Tilt Angle', *J. Geophys. Res.* **77**, 6696.

Burch, J. L.: 1973a, 'Effects of Interplanetary Magnetic Sector Structure on Auroral Zone and Polar Cap Magnetic Activity', *J. Geophys. Res.* **78**, 1047.

Burch, J. L.: 1973b, 'High-Latitude Proton Precipitation and Light Ion Density Profiles during the Magnetic Storm Initial Phase', *J. Geophys. Res.* **78**, 6569.

Burch, J. L.: 1973c, 'Rate of Erosion of Dayside Magnetic Flux Based on a Quantitative Study of the Dependence of Polar Cusp Latitude on the Interplanetary Magnetic Field', *Radio Sci.* **8**, 955.

Burch, J. L.: 1974, 'Observations of Interactions between Interplanetary and Geomagnetic Fields', *Rev. Geophys. Space Physics* **12**, 363.

Burlaga, L. F. and Ogilvie, K. W.: 1969, 'Causes of Sudden Commencements and Sudden Impulses', *J. Geophys. Res.* **74**, 2815.

Caan, M. N., McPherron, R. L. and Russell, C. T.: 1973, 'Solar Wind and Substorm-Related Changes in the Lobes of the Geomagnetic Tail', *J. Geophys. Res.* **78**, 8087.

Caan, M. N., McPherron, R. L. and Russell, C. T.: 1975, 'Substorm and Interplanetary Magnetic Field Effects on the Geomagnetic Tail Lobes', *J. Geophys. Res.* **80**, 191.

Campbell, W. H. and Matsushita, S.: 1973, 'Correspondence of Solar Field Sector Direction and Polar Cap Geomagnetic Field Changes for 1965', *J. Geophys. Res.* **78**, 2079.

Coroniti, F. V. and Kennel, C. F.: 1972, 'Changes in Magnetospheric Configuration during the Substorm Growth Phase', *J. Geophys. Res.* **77**, 3361.

Cowley, S. W. H.: 1973, 'A Qualitative Study of the Reconnection between Earth's Magnetic Field and an Interplanetary Field of Arbitrary Orientation', *Radio Sci.* **8**, 903.

Davis, T. N. and Sugiura, M.: 1966, 'Auroral Electrojet Activity Index AE and Its Universal Time Variations', *J. Geophys. Res.* **71**, 785.

Fairfield, D. H.: 1971, 'Average and Unusual Locations of the Earth's Magnetopause and Bow Shock', *J. Geophys. Res.* **76**, 6700.

Feldstein, Y. I. and Starkov, G. V.: 1968, 'Auroral Oval in the IGY and IQSY Period and a Ring Current in the Magnetosphere', *Planet. Space Sci.* **16**, 129.

Forbes, T. G. and Speiser, T. W.: 1971, 'Mathematical Models of the Open Magnetosphere: Application to Dayside Auroras', *J. Geophys. Res.* **76**, 7542.

Foster, J. C., Fairfield, D. H., Ogilvie, K. W. and Rosenberg, T. J.: 1971, 'Relationship of Interplanetary Parameters and Occurrence of Magnetospheric Substorms', *J. Geophys. Res.* **76**, 6971.

Friis-Christensen, E. and Wilhjelm, J.: 1975, 'Polar Cap Currents for Different Directions of the Interplanetary Magnetic Field in the *Y-Z* plane', *J. Geophys. Res.* **80**, 1248.

Friis-Christensen, E., Lassen, K., Wilhjelm, J., Wilcox, J. M., Gonzalez, W. and Colburn, D. S.: 1972, 'Critical Component of the Interplanetary Magnetic Field Responsible for Large Geomagnetic Effects in the Polar Cap', *J. Geophys. Res.* **77**, 3371.

Gonzalez, W. D. and Mozer, F. S.: 1974, 'A Quantitative Model for the Potential Resulting from Reconnection with an Arbitrary Interplanetary Magnetic Field', *J. Geophys. Res.* **79**, 4186.

Gosling, J. T., Asbridge, J. R., Bame, S. J., Hundhausen, A. J. and Strong, I. B.: 1967, 'Discontinuities in the Solar Wind Associated with Sudden Geomagnetic Impulses and Storm Commencements', *J. Geophys. Res.* **72**, 3357.

Gurnett, D. A. and Akasofu, S.-I.: 1974, 'Electric and Magnetic Field Observations during a Substorm on February 24, 1970', *J. Geophys. Res.* **79**, 3197.

Hansen, A. M., D'Angelo, N. and Bahsen, A.: 1974, 'On the Magnetotail Boundary Layer and the Azimuthal Component of the Interplanetary Magnetic Field', Danish Space Research Institute, Jan. 1974.

Harang, L.: 1969, 'Bursts of VLF-Emission Simultaneous with Sharp Dip of Absorption and SC in Geomagnetic H', *Planet. Space Sci.* **17**, 1565.

Heppner, J. P.: 1972, 'Polar Cap Electric Field Distributions Related to the Interplanetary Magnetic Field Direction', *J. Geophys. Res.* **77**, 4877.

Hill, T. W. and Rassbach, M. E.: 1975, 'Interplanetary Magnetic Field Direction and the Configuration of the Dayside Magnetosphere', *J. Geophys. Res.* **80**, 1.

Hirshberg, J. and Holzer, T. E.: 1975, 'Relationship between the Interplanetary Magnetic Field and "Isolated Substorms",' *J. Geophys. Res.* **80**, 3553.

Holzer, T. E. and Reid, G. C.: 1975, 'The Response of the Dayside Magnetosphere-Ionosphere System to Time-Varying Field Line Reconnection at the Magnetopause, 1, Theoretical Model', *J. Geophys. Res.* **80**, 2041.

Holzworth, R. H. and Meng, C.-I.: 1975, 'Mathematical Representation of the Auroral Oval', *Geophys. Res. Lett.* **2**, 377.

Iijima, T.: 1972, 'Development of Weak Electrojets during the Growth Phase of Magnetic Substorms', *Rep. Ionosph. Space Res. Japan* **26**, 267.

Iijima, T.: 1973, 'Interplanetary and Ground Magnetic Conditions Preceding SSC-Triggered Substorms', *Rep. Ionosph. Space Res. Japan* **27**, 205.

Iijima, T.: 1974, 'Development of Polar Magnetic Substorms', *Rep. Ionosph. Space Res. Japan* **28**, 69.

Iijima, T. and Kokubun, S.: 1973, 'Geomagnetic S_q^p Variation on an Extremely Quiet Day', *Rep. Ionosph. Space Res. Japan* **27**, 195.

Iijima, T. and Nagata, T.: 1972, 'Signatures for Substorm Development of the Growth Phase and Expansion Phase', *Planet. Space Sci.* **20**, 1095.

Ivanov, K. G. and Mikerina, N. V.: 1971, 'Rotational Discontinuity in the Solar Wind and Magnetic Disturbance on Earth', *Geomag. Aeronom.* **11**, 868.

Ivanov, K. G. and Mikerina, N. V.: 1973, 'Southern Component of the Interplanetary Magnetic Field and Magnetospheric Substorms', *Geomag. Aeronom.* **13**, 412.

Jaggi, R. K. and Wolf, R. A.: 1973, 'Self-Consistent Calculation of the Motion of a Sheet of Ions in the Magnetosphere', *J. Geophys. Res.* **78**, 2852.

Johnstone, A. D., Boyd, J. S. and Davis, T. N.: 1974, 'Study of a Small Magnetospheric Substorm', *J. Geophys. Res.* **79**, 1403.

Jørgensen, T. S., Friis-Christensen, E. and Wilhjelm, J.: 1972, 'Interplanetary Magnetic-Field Direction and High-Latitude Ionospheric Currents', *J. Geophys. Res.* **77**, 1976.

Kamide, Y.: 1974, 'Association of DP and DR Fields with the Interplanetary Magnetic Field Variation', *J. Geophys. Res.* **79**, 49.

Kamide, Y. and Akasofu, S.-I.: 1974, 'Latitudinal Cross Section of the Auroral Electrojet and Its Relation to the Interplanetary Magnetic Field Polarity', *J. Geophys. Res.* **79**, 3755.

Kamide, Y., Akasofu, S.-I., DeForest, E. and Kisabeth, J. L.: 1975, 'Weak and Intense Substorms', *Planet. Space Sci.* **23**, 579.

Kamide, Y., Burch, J. L., Winningham, J. D. and Akasofu, S.-I.: 1976, 'Dependence of the Latitude of the Cleft on the Interplanetary Magnetic Field and Substorm Activity', *J. Geophys. Res.* **81**, 698.

Kan, J. R. and Akasofu, S.-I.: 1974, 'A Model of the Open Magnetosphere', *J. Geophys. Res.* **79**, 1379.

Kan, J. R. and Su, S. Y.: 1976, 'An Upper Limit on Magnetopause Reconnection', *J. Geophys. Res.* (submitted).

Kaufmann, R. L. and Walker, D. N.: 1974, 'Hydromagnetic Waves Excited during an SSC', *J. Geophys. Res.* **79**, 5187.

Kawasaki, K. and Akasofu, S.-I.: 1972, 'Geomagnetic Disturbances in the Polar Cap: S_p^q and DP-2', *Planet. Space Sci.* **20**, 1163.

Kawasaki, K., Akasofu, S.-I., Yasuhara, F. and Meng, C.-I.: 1971, 'Storm Sudden Commencements and Polar Magnetic Substorms', *J. Geophys. Res.* **76**, 6781.

Kawasaki, K., Yasuhara, F. and Akasofu, S.-I.: 1973, 'Short-Period Interplanetary and Polar Magnetic Field Variations', *Planet. Space Sci.* **21**, 1743.

Kelley, M. C., Starr, J. A. and Mozer, F. S.: 1971, 'Relationship between Magnetospheric Electric Fields and the Motion of Auroral Forms', *J. Geophys. Res.* **76**, 5269.

Kivelson, M. G., Russell, C. T., Neugebauer, M., Scarf, F. L. and Fredricks, R. W.: 1973, 'Dependence of the Polar Cusp on the North-South Component of the Interplanetary Magnetic Field', *J. Geophys. Res.* **78**, 3761.

Kokubun, S.: 1971, 'Polar Substorm and Interplanetary Magnetic Field', *Planet. Space Sci.* **19**, 697.

Kokubun, S. and Iijima, T.: 1975, 'Time-Sequence of Polar Magnetic Substorms', *Planet. Space Sci.* **23**, 1483.

Kovner, M. S.: 1972, 'Dependence of the Position of the Magnetopause on the Orientation of the Interplanetary Magnetic Field', *Geomag. Aeronom.* **12**, 823.

Langel, R. A.: 1974, 'Variation with Interplanetary Sector of the Total Magnetic Field Measured at the OGO-2, 4, and 6 Satellites', *Planet. Space Sci.* **22**, 1413.

Langel, R. A.: 1975, 'Relation of Variations in Total Magnetic Field at High Latitude with the Parameters of the Interplanetary Magnetic Field and with DP2 Fluctuations', *J. Geophys. Res.* **80**, 1261.

Leinbach, H., Schmidt, R. J. and Brown, R. R.: 1970, 'Conjugate Observations of an Electron Precipitation Event Associated with the Sudden Commencement of a Magnetic Storm', *J. Geophys. Res.* **75**, 7099.

Leont'yev, S. V. and Lyatskiy, W. B.: 1974, 'Electric Field and Currents Connected with Y-Component of Interplanetary Magnetic Field', *Planet. Space Sci.* **22**, 811.

Leont'yev, S. V., Lyatskiy, V. B. and Mal'tsev, Yu. P.: 1974, 'DP2 Current System and Its Place in the Development of a Substorm', *Geomag. Aeronom.* **14**, 91.

Lezniak, T. W. and Winckler, J. R.: 1968, 'Structure of the Magnetopause at 6.6 R_E in Terms of 50- to 150-keV Electrons', *J. Geophys. Res.* **73**, 5733.

Lezniak, T. W., Arnoldy, R. L., Parks, G. K. and Winckler, J. R.: 1968, 'Measurement and Intensity of Energetic Electrons at the Equator at 6.6 R_e', *Radio Sci.* **3**, 710.

Lin, C.-A., Young, D. T. and Wolf, R. A.: 1973, 'Magnetospheric Plasma Motion during a Sudden Commencement', *Planet. Space Sci.* **21**, 1713.

Lui, A. T. Y., Hones, E. W. Jr., Venkatesan, D., Akasofu, S.-I. and Bame, S. J.: 1975, 'Response of the Plasma Sheet at $\sim 18\, R_E$ to Sudden Southward Turnings of the Interplanetary Magnetic Field', *J. Geophys. Res.* **80**, 929.

Lui, A. T. Y. and Akasofu, S.-I.: 1976, 'Cross-Sectional Area of the Near-Earth Nightside Magnetosphere', *J. Geophys. Res.* (submitted).

Lyatskiy, V. B. and Mal'tsev, Yu. P.: 1971, 'Shock Wave in the Tail of the Magnetosphere as the Reason for the Active Phase of a Magnetospheric Substorm', *Geomag. Aeronom.* **11**, 872.

Maezawa, K.: 1974, 'Dependence of the Magnetopause Position on the Southward Interplanetary Magnetic Field', *Planet. Space Sci.* **22**, 1443.

Mansurov, S. M.: 1969, 'New Evidence of a Relationship between Magnetic Fields in Space and on Earth', *Geomag. Aeronom.* **9**, 622.

Matsushita, S. and Balsley, B. B.: 1972, 'A Question of DP-2', *Planet. Space Sci.* **20**, 1259.

Matsushita, S. and Balsley, B. B.: 1973 'Comments on Nishida's Reply', *Planet. Space Sci.* **21**, 1260.

Matsushita, S., Tarpley, J. D. and Campbell, W. H.: 1973, 'IMF Sector Structure Effects on the Quiet Geomagnetic Field', *Radio Science* **8**, 963.

McPherron, R. L.: 1970, 'Growth Phase of Magnetospheric Substorms', *J. Geophys. Res.* **75**, 5592.

McPherron, R. L.: 1972, 'Substorm Related Changes in the Geomagnetic Tail: The Growth Phase', *Planet. Space Sci.* **20**, 1521.

McPherron, R. L.: 1973, 'Satellite Studies of Magnetospheric Substorms on August 15, 1968, 1. State of the Magnetosphere', *J. Geophys. Res.* **78**, 3044.

McPherron, R. L.: 1974, 'Critical Problems in Establishing the Morphology of Substorms in Space', *Magnetospheric Physics*, B. M. McCormac (ed.), p. 335, D. Reidel, Publ. Co., Dordrecht-Holland.

McPherron, R. L., Parks, G. K., Colburn, D. S. and Montgomery, M. D.: 1973a, 'Satellite Studies of Magnetospheric Substorms on August 15, 1968, 2. Solar Wind and Outer Magnetosphere', *J. Geophys. Res.* **78**, 3054.

McPherron, R. L., Russell, C. T., Kivelson, M. G. and Coleman, P. J. Jr.: 1973b, 'Substorms in Space: The Correlation between Ground and Satellite Observations of the Magnetic Field', *Radio Sci.* **8**, 1058.

McPherron, R. L., Russell, C. T. and Aubry, M. P.: 1973c, 'Satellite Studies of Magnetospheric Substorms on August 15, 1968, 9. Phenomenological Model for Substorms', *J. Geophys. Res.* **78**, 3131.

Meng, C.-I. and Anderson, K. A.: 1974, 'Magnetic Field Configuration in the Magnetotail near 60 R_E', *J. Geophys. Res.* **79**, 5143.

Meng, C.-I. and Colburn, D. S.: 1974, 'Magnetotail Variations Associated with the Southward Interplanetary Magnetic Field', *J. Geophys. Res.* **79**, 1831.

Meng. C.-I., Tsurutani, B., Kawasaki, K. and Akasofu, S.-I.: 1973, 'Cross-Correlation Analysis of the AE Index and the Interplanetary Magnetic Field B_z Component', *J. Geophys. Res.* **78**, 617.

Mikerina, N. V. and Ivanov, K. G.: 1971, 'Geomagnetic Effects of Discontinuities and Models of Magnetosphere Deformations', *Geomag. Aeronom.* **11**, 776.

Mishin, V. M., Bazarzhapov, A. D., Nemtsova, E. I., Popov, G. V. and Shelomentsev, V. V.: 1973a, 'The Effect of Interplanetary Magnetic Field on Magnetospheric Convection and Electric Fields in the Ionosphere', Siberian Inst. of Terrestrial Magnetism, Ionosphere and Radio Wave Propagation.

Mishin, V. M., Bazarzhapov, A. D., Saifudinova, T. I., Urbanovich, V. D. and Shelomentsev, V. V.: 1973b, 'Development of Magnetic Substorm. 1.', *Sibizmir Report 7-73.*

Moldovanu, A.: 1974, 'Geomagnetic Effects of Interplanetary Sector Structure', *Planet. Space Sci.* **22**, 193.

Moldovanu, A. and Bradu, E.-B.: 1973, 'Comparison of the Sectorial Structure of the Interplanetary Magnetic Field and the Occurrence of SC and SI Events', *Gerlands Beitr. Geophysik, Leipzig* **82**, 171.

Mozer, F. S.: 1971, 'Origin and Effects of Electric Fields during Isolated Magnetospheric Substorms', *J. Geophys. Res.* **76**, 7595.

Mozer, F. S., Gonzalez, W. D., Bogott, F., Kelley, M. C. and Schutz, S.: 1974, 'High-Latitude Electric Fields and the Three-Dimensional Interaction between the Interplanetary and Terrestrial Magnetic Fields', *J. Geophys. Res.* **79**, 56.

Murayama, T. and Hakamada, K.: 1975, 'Effects of Solar Wind Parameters on the Development of Magnetospheric Substorms', *Planet. Space Sci.* **23**, 75.

Nishida, A.: 1968a, 'Geomagnetic D_p2 Fluctuations and Associated Magnetospheric Phenomena', *J. Geophys. Res.* **73**, 1795.

Nishida, A.: 1968b, 'Coherence of Geomagnetic DP2 Fluctuations with Interplanetary Magnetic Variations', *J. Geophys. Res.* **73**, 5549.

Nishida, A.: 1971, 'DP2 and Polar Substorm', *Planet. Space Sci.* **19**, 205.

Nishida, A.: 1973a, 'Discussion on Geomagnetic Disturbances in the Polar Cap: S_q^p and DP-2 by K. Kawasaki and S.-I. Akasofu', *Planet. Space Sci.* **21**, 691.

Nishida, A.: 1973b, 'Reply to "A Question of DP-2" by S. Matsushita and B. B. Balsley', *Planet. Space Sci.* **21**, 1255.

Nishida, A.: 1975, 'Interplanetary Field Effect on the Magnetosphere', *Space Sci. Rev.* **17**, 353, 1975.

Nishida, A. and Kokubun, S.: 1971, 'New Polar Magnetic Disturbances: S_q^p, SP, DPC, and DP2', *Rev. Geophys. Space Phys.* **9**, 417.

Nishida, A. and Nagayama, N.: 1975, 'Magnetic Field Observations in the Low Latitude Magnetotail during Substorms', *Planet. Space Sci.* **23**, 1119.

Olson, W. P. and Pfitzer, K. A.: 1974, 'A Quantitative Model of the Magnetospheric Magnetic Field', *J. Geophys. Res.* **79**, 3739.

Ondoh, T.: 1970, 'Magnetospheric Sudden Impulses', *J. Radio Res. Lab.* **17**, 199.

Parks, G. K. and Pellat, R.: 1972, 'Correlation of Interplanetary Space B_z Field Fluctuations and Trapped Particle Redistribution', *J. Geophys. Res.* **77**, 266.

Parks, G. K., Gardner, E. H., Lin, C. S., Newell, R. E. and Wehrenberg, P. J.: 1974, 'Interplanetary Control of Magnetospheric Dynamics', *Correlated Interplanetary and Magnetospheric Observations*, D. E. Page (ed.), p. 49, D. Reidel Publ. Co., Dordrecht-Holland.

Patel, V. L.: 1968, 'Sudden Impulses in the Geomagnetotail', *J. Geophys. Res.* **73**, 3407.

Patel, V. L.: 1972, 'Sudden Impulses in the Geomagnetotail and the Vicinity', *Planet. Space Sci.* **20**, 1127.

Patel, V. L. and Cahill, L. J. Jr.: 1974, 'Magnetic Field Variations of the SI in the Magnetosphere and Correlated Effects in Interplanetary Space', *Planet. Space Sci.* **22**, 1117.

Perona, G. E.: 1972a, 'A Theory on the Latitude and Local Time Distribution of Precipitating Electrons during a Sudden Commencement', *Earth's Magnetospheric Process*, B. M. McCormac (ed.), p. 351, D. Reidel Publ. Co., Dordrecht-Holland.

Perona, G. E.: 1972b, 'Theory on the Precipitation of Magnetospheric Electrons at the Time of a Sudden Commencement', *J. Geophys. Res.* **77**, 101.

Pfitzer, K. A. and Winckler, J. R.: 1969, 'Intensity Correlations and Substorm Electron Drift Effects in the Outer Radiation Belt Measured with the OGO 3 and ATS-1 Satellites', *J. Geophys. Res.* **74**, 5005.

Pfitzer, K. A., Lezniak, T. W. and Winckler, J. R.: 1969, 'Experimental Verification of Drift-Shell Splitting in the Distorted Magnetosphere', *J. Geophys. Res.* **74**, 4687.

Pike, C. P., Meng, C.-I., Akasofu, S.-I. and Whalen, J. A.: 1974, 'Observed Correlations between Interplanetary Magnetic Field Variations and the Dynamics of the Auroral Oval and the High-Latitude Ionosphere', *J. Geophys. Res.* **79**, 5129.

Pudovkin, M. I. and Troshichev, O. A.: 1972, 'On the Types of Current Patterns of Weak Geomagnetic Disturbances at the Polar Caps', *Planet. Space Sci.* **20**, 1773.

Pudkovkin, M. I., Isaev, S. I. and Zaitzeva, S. A.: 1970, *Ann. Geophys.* **26**, 761.

Reid, G. C. and Holzer, T. E.: 1975, 'The Response of the Day Side Magnetosphere-Ionosphere System to Time-Varying Field Line Reconnection at the Magnetopause, 2, Erosion Event of March 27, 1968', *J. Geophys. Res.* **80**, 2050.

Roldugin, V. K.: 1974, 'Optical Effects of SC and SI at Loparskaya and Mirny', *Geomag. Aeronom.* **14**, 75.

Rossberg, L.: 1974, 'Remarks on the Growth Phase of Substorms', *Magnetospheric Physics*, B. M. McCormac (ed.), p. 367, D. Reidel Publ. Co., Dordrecht-Holland.

Rossberg, L., Riedler, W., Skovli, G., Stauning, P. and Thomas, G. R.: 1974, 'Interplanetary Magnetic

Field Direction and Auroral Zone Disturbances', *Correlated Interplanetary and Magnetospheric Observations*, D. E. Page (ed.), p. 85, D. Reidel Publ. Co. Dordrecht-Holland.
Rostoker, G., Lam, H.-L. and Hume, W. D.: 1972, 'Response Time of the Magnetosphere to the Interplanetary Electric Field', *Can. J. Phys.* **50**, 544.
Russell, C. T.: 1974, 'The Solar Wind and Magnetospheric Dynamics', *Correlated Interplanetary and Magnetospheric Observations*, D. E. Page (ed.), p. 3, D. Reidel Publ. Co., Dordrecht-Holland.
Russell, C. T. and McPherron, R. L.: 1973, 'The Magnetotail and Substorms', *Space Sci. Rev.* **15**, 205.
Russell, C. T., McPherron, R. L. and Coleman, P. J. Jr.: 1971, 'Magnetic Field Variations in the Near Geomagnetic Tail Associated with Weak Substorm Activity', *J. Geophys. Res.* **76**, 1823.
Schieldge, J. P. and Siscoe, G. I.: 1970, 'A Correlation of Occurrence of Simultaneous Sudden Magnetospheric Compressions and Geomagnetic Bay Onsets with Selected Geophysical Indices', *J. Atmosph. Terr. Phys.* **32**, 1819.
Shelley, E. G., Johnson, R. G. and Sharp, R. D.: 1971, 'Plasma Sheet Convection Velocities, Inferred from Electron Flux Measurements at Synchronous Altitude', *Radio Sci.* **6**, 305.
Siscoe, G. L., Formisano, V. and Lazarus, A. J.: 1968, 'Relation between Geomagnetic Sudden Impulses and Solar Wind Pressure Changes, An Experimental Investigation', *J. Geophys. Res.* **73**, 4869.
Snyder, A. L. and Akasofu, S.-I.: 1972, 'Observations of the Auroral Oval by the Alaskan Meridian Chain of Stations', *J. Geophys. Res.* **77**, 3419.
Sonnerup, B. U. Ö.: 1974, 'Magnetopause Reconnection Rate', *J. Geophys. Res.* **79**, 1546.
Starkov, G. V., Fel'dshteyn, Y. I. and Shevnina, N. F.: 1973, 'Relationship between the Sector Structure of the Interplanetary Magnetic Field and Auroral Activity', *Geomag. Aeronom.* **13**, 807.
Stern, D. P.: 1973, 'A Study of the Electric Field in an Open Magnetospheric Model', *J. Geophys. Res.* **78**, 7292.
Sugiura, M., Skillman, T. L., Ledley, B. G. and Heppner, J. P.: 1968, 'Propagation of the Sudden Commencement of July 8, 1966, to the Magnetotail', *J. Geophys. Res.* **73**, 6699.
Sumaruk, P. V. and Fel'dshteyn, Ya. I.: 1973, 'Seasonal Changes of Variations of the Z Component in the Polar Region in Connection with the Sign of the YSE Component of the Interplanetary Magnetic Field', *Geomag. Aeronom.* **13**, 469.
Svalgaard, L.: 1968, 'Sector Structure of the Interplanetary Magnetic Field and Daily Variation of the Geomagnetic Field at High Latitudes', Pap. 6. Dansk. Meteorol. Inst. Geophys. Charlottenlund, Denmark.
Svalgaard, L.: 1972, 'Interplanetary Magnetic-Sector Structure, 1926–1971', *J. Geophys. Res.* **77**, 4027.
Svalgaard, L.: 1973, 'Polar Cap Magnetic Variations and Their Relationship with the Interplanetary Magnetic Sector Structure', *J. Geophys. Res.* **78**, 2064.
Svalgaard, L.: 1974, 'The Relation between the Azimuthal Component of the Interplanetary Magnetic Field and the Geomagnetic Field in the Polar Caps', *Correlated Interplanetary and Magnetospheric Observations*, D. E. Page (ed.), p. 61, D. Reidel Publ. Co., Dordrecht-Holland.
Svalgaard, L.: 1975, 'On the Use of Godhavn H Component as an Indicator of the Interplanetary Sector Polarity', *J. Geophys. Res.* **80**, 2717.
Tashkinova, L. G. and Tverskoy, B. A.: 1972, 'Oscillation of DP1 and DP2 Current Systems', *Geomag. Aeronom.* **12**, 509.
Troshichev, O. A., Kuznetsov, B. M. and Pudovkin, M. I.: 1974a, 'The Current Systems of the Magnetic Substorm Growth and Explosive Phases', *Planet. Space Sci.* **22**, 1403.
Troshichev, O. A., Pudovkin, M. L., Pegov, L. A. and Grafe, A.: 1974b, 'Some Remarks on the DP-2 Current System', *Gerlands Beitr. Geophys.* **83**, 275.
Tsurutani, B. and Bogott, F.: 1972, 'Onset of Magnetospheric Substorms', *J. Geophys. Res.* **77**, 4677.
Tsurutani, B. T. and Meng, C.-I.: 1972, 'Interplanetary Magnetic-Field Variations and Substorm Activity', *J. Geophys. Res.* **77**, 2964.
Tverskaya, L. V. and Khorosheva, O. V.: 1974, 'Some Characteristics of the Development of Gigantic DP-2 Variations during Magnetic Storms', *Geomag. Aeronom.* **14**, 86.
Unti, T. and Atkinson, G.: 1968, 'Two-Dimensional Chapman-Ferraro Problem with Neutral Sheet, 1, The Boundary', *J. Geophys. Res.* **73**, 7319.
Volland, H.: 1975, 'Differential Rotation of the Magnetospheric Plasma as Cause of the Svalgaard-Mansurov Effect', *J. Geophys. Res.* **80**, 2311.
Vorob'yev, V. G.: 1974, 'SC-Associated Effects in Auroras', *Geomag. Aeronom.* **14**, 72.
Winckler, J. R.: 1970, 'The Origin and Distribution of Energetic Electrons in the Van Allen Radiation Belts', *Particles and Fields in the Magnetosphere*, B. M. McCormac (ed.), p. 332, D. Reidel Publ. Co., Dordrecht-Holland.
Yasuhara, F., Akasofu, S.-I. and McIlwain, C. E.: 1976, 'Plasma Injection at the Geosynchronous Distance', *J. Geophys. Res.* (submitted).

MAGNETOSPHERIC SUBSTORMS: INTRODUCTION

It is believed that the magnetospheric substorm results when the energy stored in the magnetotail is explosively released toward the inner magnetosphere and finally deposited as heat energy in the upper atmosphere (Dungey, 1961; Axford, 1969). We have learned in the last four chapters that the solar wind-magnetosphere interaction constitutes a magnetohydrodynamic (MHD) dynamo and that its efficiency is regulated by the north-south component (B_z) of the interplanetary magnetic field (IMF). When the IMF has a large ($\sim 5\ \gamma$) northward component for a prolonged period (6 h or more), the efficiency of the dynamo becomes minimum. The corresponding state may be considered to be the 'ground state' of the magnetosphere. However, even in this situation, solar wind particles are continuously captured by the magnetosphere as they flow in the anti-solar direction in the high latitude lobe of the magnetotail and then are convected toward the dayside magnetopause through the plasma sheet. Thus, when such a quiet condition lasts for more than 48 h, a significant part of the plasma sheet particles can be replaced by magnetosheath particles (Sections 3.4 and 3.5).

When the magnitude of the IMF north-south component decreases from about $5\ \gamma$, the dynamo efficiency increases. Then, the quantity $S = \int_0^T (\Phi_D - \Phi_N)\, dt$ defined in Section 4.4.1 becomes finite and positive, and excess energy is being accumulated in the magnetotail. In this situation, the magnetosphere is no longer in the ground state and tends to release sporadically its accumulated energy in the form of kinetic energy of auroral particles and the joule heat energy in the ionosphere, as well as in the form of kinetic energy of particles in the plasma sheet. The magnetospheric substorm appears to be this energy release process.

It is known that the Earth's lower atmosphere tends to maintain heat exchange by a sporadic vortex formation process called *cyclogenesis*, which is much more efficient in exchanging warm and cold air masses than the simple meridional Hadley cell-like convection. There is a certain analogy between cyclogenesis and the process associated with the substorm. Cyclogenesis energy is accumulated through the differential heating between the equatorial and polar regions by solar (black body) radiation. The ultimate source of energy of the magnetospheric substorm is the kinetic energy of solar wind plasma. Substorm energy is accumulated when the rate of merging Φ_D of the interplanetary and geomagnetic field lines at the front of the magnetosphere becomes greater than the rate of reconnection Φ_N of the merged field lines at the anti-sunward end of the magnetotail (that is, when there is a differential merging rate $((\Phi_D - \Phi_N) > 0)$ and thus the quantity S becomes positive). In this situation, the magnetosphere does not simply increase (enhance) the reconnection rate Φ_N (the plasma convection).

Instead, the magnetosphere releases sporadically the accumulated energy so long as $S > 0$. The manifestation of this process is the magnetospheric substorm. There have been disagreements, among workers, as to how the magnetosphere achieves such a sudden increase of Φ_N. It has been suggested by a number of workers that a new magnetic neutral line is formed in the near-Earth plasma sheet, where Φ_N becomes considerably greater than that along the X-line at the anti-solar end of the plasma sheet. However, we shall see in Chapter 6 that there is no obvious indication of the new neutral line formation in the near-Earth plasma sheet. It is suggested thus that the magnetosphere achieves the energy release in a two-step process, first by deflating the plasma sheet during the *expansive* phase and then enhancing Φ_N along the X-line at the anti-solar end of the plasma sheet during the *recovery* phase.

In the first section (Section 5.1) of this chapter, we classify magnetospheric disturbances into two; reversible and irreversible disturbances. The magnetospheric substorm belongs to the latter. In Sections 5.2 and 5.3, we examine two important quantities of the magnetospheric substorm, substorm energy and intensity, as well as the substorm function Σ, namely how substorms are related to the merging rate Φ_D along the dayside X-line. In particular, it will be shown in Section 5.2 that when the magnetosphere is 'pulsed' once by the IMF southward turning, a series of magnetospheric substorms, with decreasing intensity, are generated. Section 5.4 presents an introductory review of theories of the magnetospheric substorm.

In Section 5.5, the direction of our study in the later half of this book is outlined. In the last section of this chapter, we shall examine how a magnetospheric *storm* is related to magnetospheric *substorms*.

5.1. A New Classification of Magnetospheric Disturbances

It is convenient to classify magnetospheric disturbances into the following:

(1) Reversible or quasi-reversible disturbances; no heat production,

(2) Irreversible disturbances; a sudden heat production.

5.1.1. Reversible or Quasi-Reversible Disturbances

A reversible disturbance is defined to be the one for which the magnetosphere returns to a quiet-time configuration after the responsible interplanetary disturbance is removed. There is no significant increase of heat production from the quiet-time value in the polar cap circuit and thus no sudden large reduction in the amount of S (or of the amount of excess energy accumulated in the magnetotail). We have so far found three interplanetary disturbances which cause reversible or quasi-reversible disturbances:

(i) *A Weak Interplanetary Shock Wave*

A weak interplanetary shock wave simply compresses the magnetosphere (Section 4.2). After the solar wind returns to a pre-shock condition, the magnetosphere expands, returning to a quiet-time configuration. There is no significant increase of heat production by this process.

(ii) *The East-West Component of the IMF*

In Section 4.3, it was found that the east-west component of the IMF causes a specific response of the magnetosphere. Svalgaard (1973) showed that a counterclockwise current occurs in the polar cap region (observing from above the north geomagnetic pole) when the IMF has a westward component (that is, in the away sector), and a clockwise current occurs from an eastward component (that is, in the toward sector). In Section 4.3 it was shown that part of these circular currents arise from the asymmetry of the electric field distribution (namely, a slight modulation of the space charge distribution along the oval), which is caused by the IMF EW component. If this is indeed the case, this particular magnetospheric response is not strictly a reversible process (since there will be an appreciable increase of joule heat production), but is a quasi-reversible process.

(iii) *The North-South Component of the IMF*

In Section 4.4, it was shown that when the time derivative of the north-south component (B_z) of the IMF becomes negative (namely, $\partial B_z/\partial t < 0$), there occurs a specific response in the magnetosphere.

The expansion of the auroral oval and thus of the polar S_q^p current circuit causes a particular type of magnetic variation. It was pointed out in Section 4.4.6 that the DP-2 variations, identified first by Nishida (1968), can be understood in terms of a combination of an expansion of the circuit and a slight increase of the total current intensity. That is to say, the DP-2 variations arise from a slight modulation of the S_q^p current, instead of the development of an entirely new current system, by changes of the B_z component of the IMF. Thus, the response of the magnetosphere to changes of the B_z component of the IMF is not strictly a reversible process. However, since the increased heating is rather small compared with what will be described as an irreversible process, we may classify this modulation as a quasi-reversible process. Another reason for identifying these responses as a reversible process is that when $\partial B_z/\partial t$ becomes positive, S can decrease without sudden heat production in the polar ionosphere.

5.1.2. IRREVERSIBLE DISTURBANCES

Here we identify the magnetospheric substorm as an irreversible phenomenon. Indeed, there occurs a large increase of joule heating in the polar cap ionosphere during relatively short periods (~3 hours). The quantity S and thus the magnetic energy accumulated in the magnetotail are considerably reduced during substorms.

5.2. Substorm Energy ε_Σ and Substorm Function $\Sigma = \Gamma(\Phi_D)$

5.2.1. GROUND STATE OF THE MAGNETOSPHERE

In the introduction of the present chapter, it was noted that the ultimate source of energy for substorms is the kinetic energy of solar wind particles which is accumulated in the magnetotail by the solar wind-magnetosphere dynamo process and that the substorm is associated with the conversion of this stored energy into heat energy in the polar ionosphere. The efficiency of the dynamo is regulated by the B_z component of the IMF which controls the rates of merging Φ_D and Φ_N. In Section 4.4.1, it was suggested that during a prolonged period of a large positive B_z value ($\sim +5\ \gamma$), both Φ_D and Φ_N reach the same minimum value, and the quantity $S = \int (\Phi_D - \Phi_N)\,dz$ becomes null. During such a period, the oval contracts to its minimum size, the midday and midnight portions of the oval being located at about $80° \sim 82°$ and $72° \sim 74°$, respectively. The total area of the minimum size oval A_0 is about $7.4 \times 10^6\ \text{km}^2$. Further, *there is no indication of substorm activity along the minimum size oval.*

Therefore, it may be considered that the magnetosphere reaches its ground state during such a period. On the other hand, it is known that the magnetotail is always present even during a very quiet period; correspondingly, the auroral oval does not shrink to a point. Therefore, even when $S = 0$ (during a prolonged period of a large positive B_z value), there is a considerable amount of open flux and stored energy in the magnetotail. However, this magnetotail energy is not available for substorms. In the next subsection, we shall see a number of DMSP photographs which indicate these important points.

5.2.2. SUBSTORM FUNCTION $\Sigma = \Gamma(\Phi_D)$

We now know how the merging rates Φ_D and Φ_N are related to the magnetic energy $\varepsilon_{\Sigma 0}$ available for substorms:

$$\varepsilon_{\Sigma 0} = \frac{L}{\pi^4 R_T^2}\left(\int (\Phi_D - \Phi_N)\,d\tau\right)^2$$

$$= \frac{L}{\pi^4 R_T^2} S^2$$

The above equations can be obtained from Equations (5) and (7) in Section 1.4.3 by noting that the magnetic flux in the northern (or the southern) half of the magnetotail ($B_T \pi R_T^2$) should be equal to $S = B_p(A_1 - A_0)$; see also Section 4.4.1.

The next question is how substorms (Σ) are related to the merging rate Φ_D or the energy $\varepsilon_{\Sigma 0}$. As mentioned in Section 4.4, a number of workers have attempted to prove that the substorm function Σ is a delta function, namely

$$\Sigma = \delta(t - \tau_1)$$

The delta function is supposed to be triggered by a step function-like change of the IMF B_z component, namely

$$B_z = U(t) \qquad \begin{aligned} U &= +5\ \gamma \quad t \leq 0 \\ &= -5\ \gamma \quad t \geq 0 \end{aligned}$$

Here τ_1 denotes a time constant which indicates a time delay of the substorm onset after the southward turning of the IMF vector at $t = 0$. This implies that each substorm is directly related to an increase of Φ_D, namely a southward turning of the IMF vector. In Section 4.4, we discussed in great detail how and why a number of workers reached such a conclusion. It was shown that their inferences were based on auroral and magnetic observations along the auroral zone stations which are incapable of monitoring substorm activity when the IMF vector turns northward and thus the auroral oval contracts well poleward of auroral zone stations. In the following, we examine the relationship between auroral activity and the IMF B_z component on the basis of auroral photographs taken from DMSP satellites and the simultaneous IMF data (Akasofu, 1975).

Figure 5.1 shows several DMSP photographs which were taken in the later half of 1972, October 28. The IMF data on that day show that the B_z component turned northward at about 1010 UT and remained positive until about 1930 UT. Some of the DMSP photographs taken during this period (orbits 275, 276, 277 and 278) show clearly intense substorm activity. All of them show typical substorm features and show no qualitative difference from substorms observed during orbits 279 and 280 when the B_z component was negative. It can easily be inferred that the quantity $S = \int (\Phi_D - \Phi_D)\,dt$ became minimum at about 14–16 UT but remained finite, and that a part of the remaining energy was dissipated by the

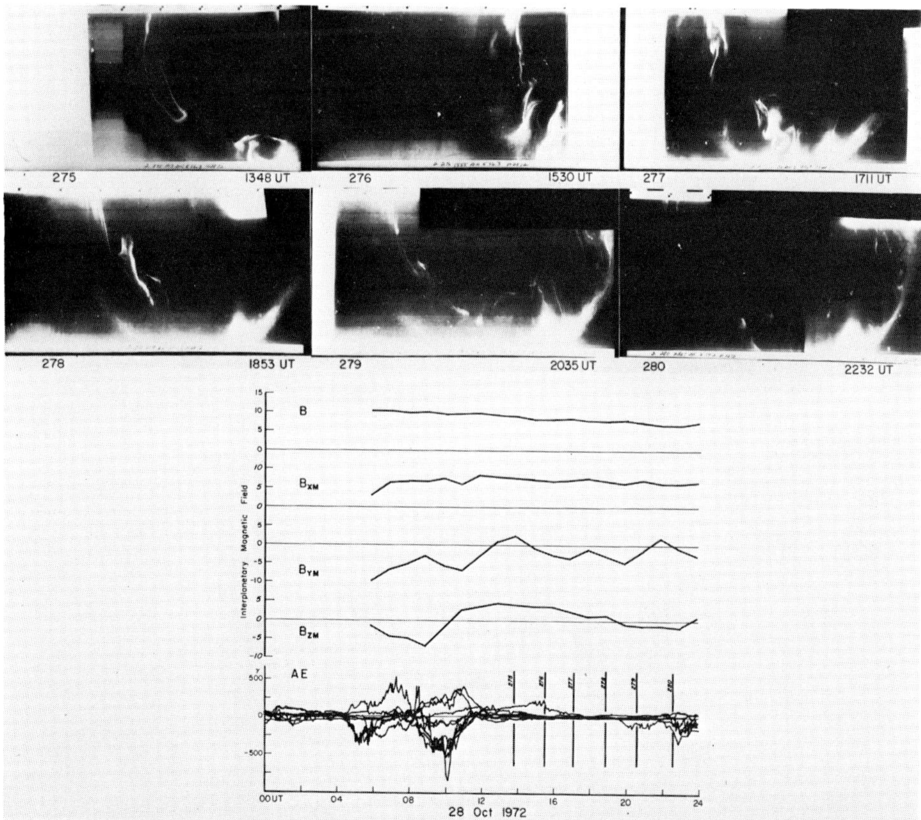

Fig. 5.1. Occurrence of auroral substorms during a prolonged period of a large positive B_z component. From the top: DMSP-2 photographs, the IMF parameters and the AE index. (Akasofu, S.-I.: *Planetary Space Sci.* **23**, 1349, 1975.)

substorm at about 15 UT (orbit 276). Then, excess energy began to accumulate again after 16 UT, since the B_z component began to decrease at that time.

Figure 5.2 shows a series of DMSP photographs which were taken on 1972, 24–25 December. They are some of the most illuminating photographs for studying the role of the IMF on substorm processes, although a similar situation can be found at other periods. At about 08 UT on December 24, the B_z component became positive, and a prolonged period of the northward field began. It can be seen that auroras were quite active at least until about 1410 UT (orbits 640, 641, 642 and 643), in spite of large positive B_z values during that period. One of the most interesting features of auroral activity during this period is that the size of the auroral oval was gradually decreasing after each substorm, indicating that the quantity S and the excess energy $\varepsilon_{\Sigma 0}$ were diminishing. The oval at orbit 643 was considerably smaller than that at orbit 640. The oval became even smaller and quite dim at orbit 645. During the next two orbits (646 and 647), the oval

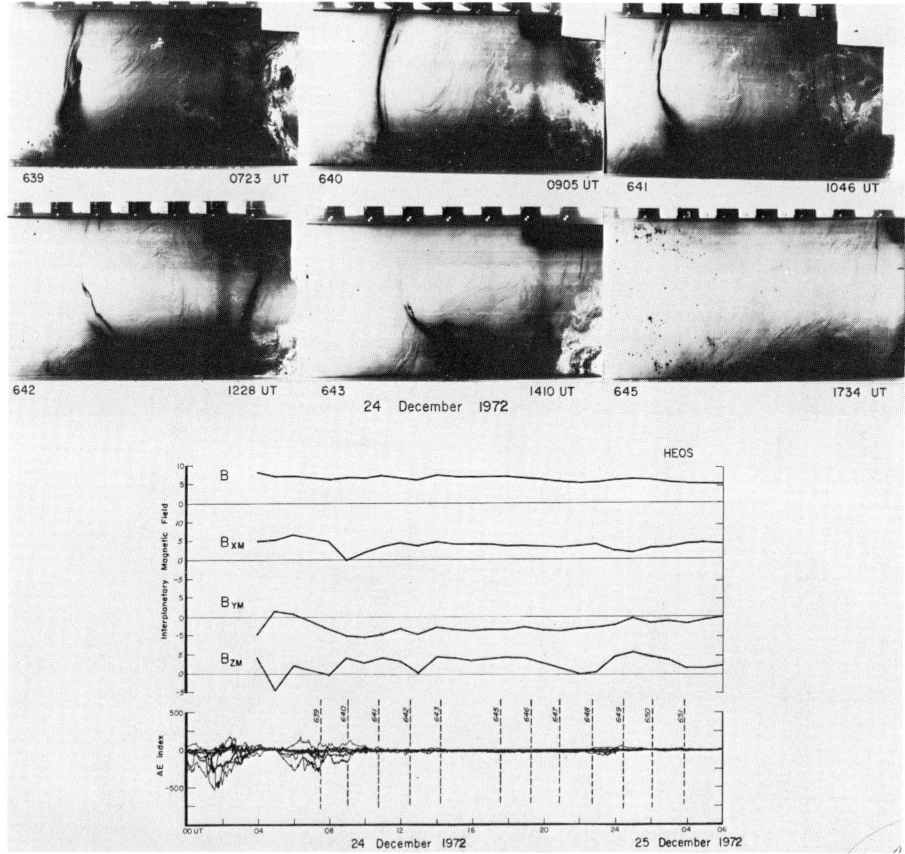

Fig. 5.2(a). Occurrence of auroral substorms during a prolonged period of a large positive B_z component. From the top: DMSP-2 photographs, the IMF parameters and the AE index. (Akasofu, S.-I.: *Planetary Space Sci.* **23**, 1349, 1975.)

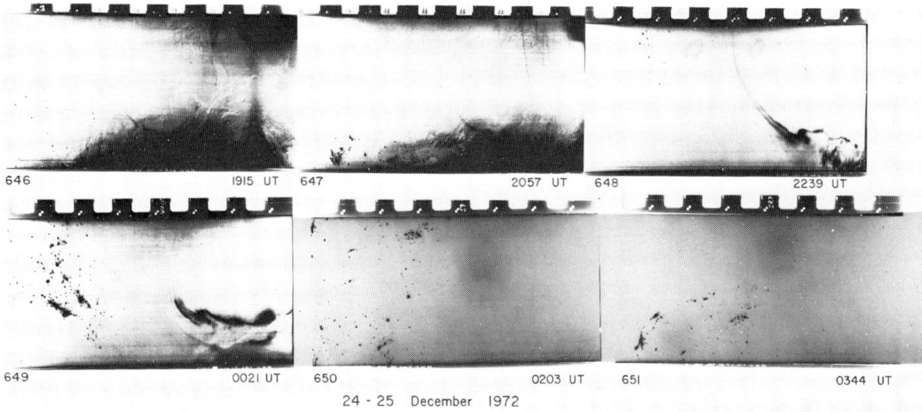

Fig. 5.2(b). Occurrence of auroral substorms during a prolonged period of a large positive B_z component. From the top: DMSP-2 photographs, the IMF parameters and the AE index. (Akasofu, S.-I.: *Planetary Space Sci.* **23**, 1349, 1975.)

contracted even further poleward. It is likely that the quantity S became almost null at about 20–21 UT. The fact that the oval was very small during orbit 647 suggests that a small decrease of the B_z component after a prolonged period of a large positive value contributes only a small amount to the quantity S. However, during orbit 648, a typical substorm feature was observed, indicating that a finite amount of the excess energy was being accumulated during the period after 21 UT.

The B_z component was positive throughout the day of 1972, December 26, except for a short period between 1945 and 2130 UT. The auroral oval was very dim during orbits 650 and 651. Although it is not shown here, the B_z component had small positive values between 04 and 13 UT on that day. A dim contracted oval was seen around 12–14 UT, together with weak substorm activity; its midnight latitude was about 70°. It is likely that S had a small positive value during that period.

A prolonged period of large positive B_z values occurred also on December 27. Figure 5.3(a) shows DMSP photographs, the interplanetary magnetic data and the AE index between 18 UT, December 26, and 24 UT, December 27. A dim, small oval was observed during orbits 674 and 675. Then weak substorm activity was observed during orbits 676 and 677, although the AE index had no typical substorm features. The DMSP photograph at orbit 679 shows a bright arc in the afternoon part of the oval; the B_z component was negative for about one hour before that time. After orbit 679, the oval became too dim to study in detail in DMSP photographs during the rest of the day. However, auroras were seen over Sachs Harbour (one of the Alaska meridian chain stations, at inv. lat. ~74°) at least until 12 UT. Figures 5.3(b) and (c) show two other examples of a substorm along a contracted oval.

The above results may be summarized schematically in Figure 5.4 which shows, from the top: a hypothetical change of the IMF B_z component, the

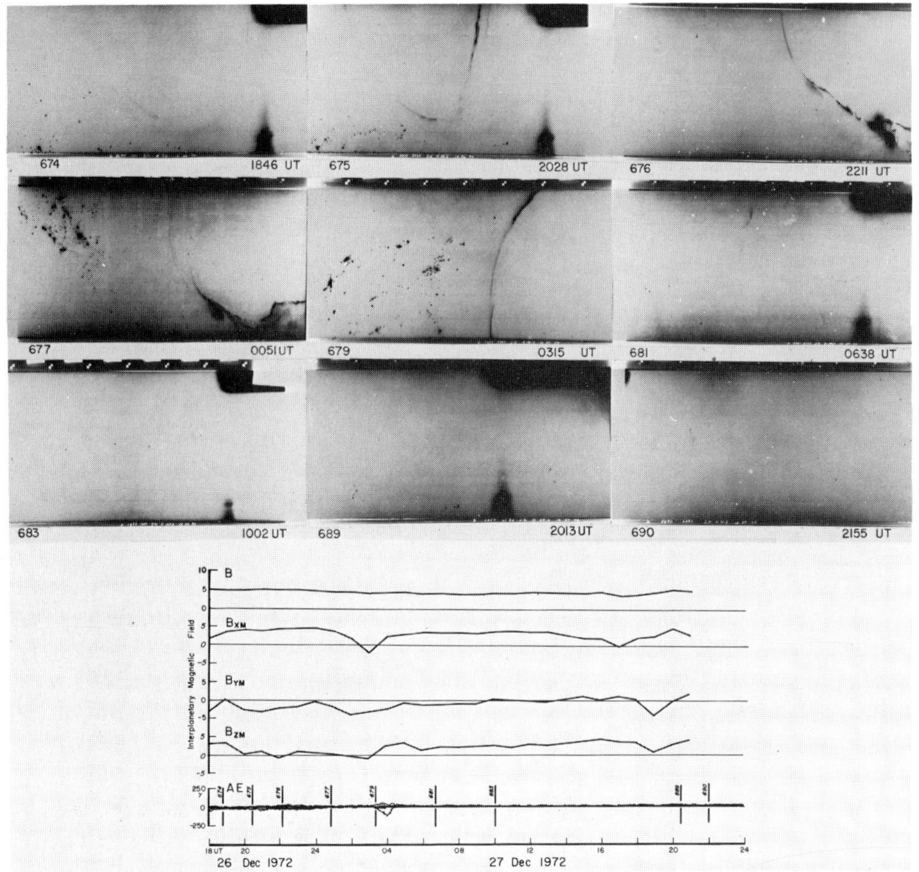

Fig. 5.3(a). Auroral oval during a prolonged period of a large positive B_z component. (Akasofu, S.-I.: *Planetary Space Sci.* **23**, 1349, 1975.)

corresponding changes of the auroral oval, auroral activity and the excess energy in the magnetotail. After a prolonged period of a large positive B_z component, the auroral oval contracts to its minimum size. After a sudden decrease in the B_z component and the subsequent southward turning, the auroral oval rapidly expands equatorward. The magnetic energy in the magnetotail increases simultaneously. One or two intense substorms occur along the expanded oval.

Then, the B_z component begins to increase slowly and turns northward, after remaining southward for a few hours. During this period, several substorms occur successively, although they cannot be observed at auroral zone stations. After each substorm, the excess energy decreases. This occurs as long as the excess energy is available in the magnetotail. Eventually the excess energy is completely dissipated, and the auroral oval contracts to its minimum size.

Such an ideal situation does not occur very often. In particular, a prolonged period (~ 12 h) of a large northward component ($\sim +5\ \gamma$) occurs infrequently,

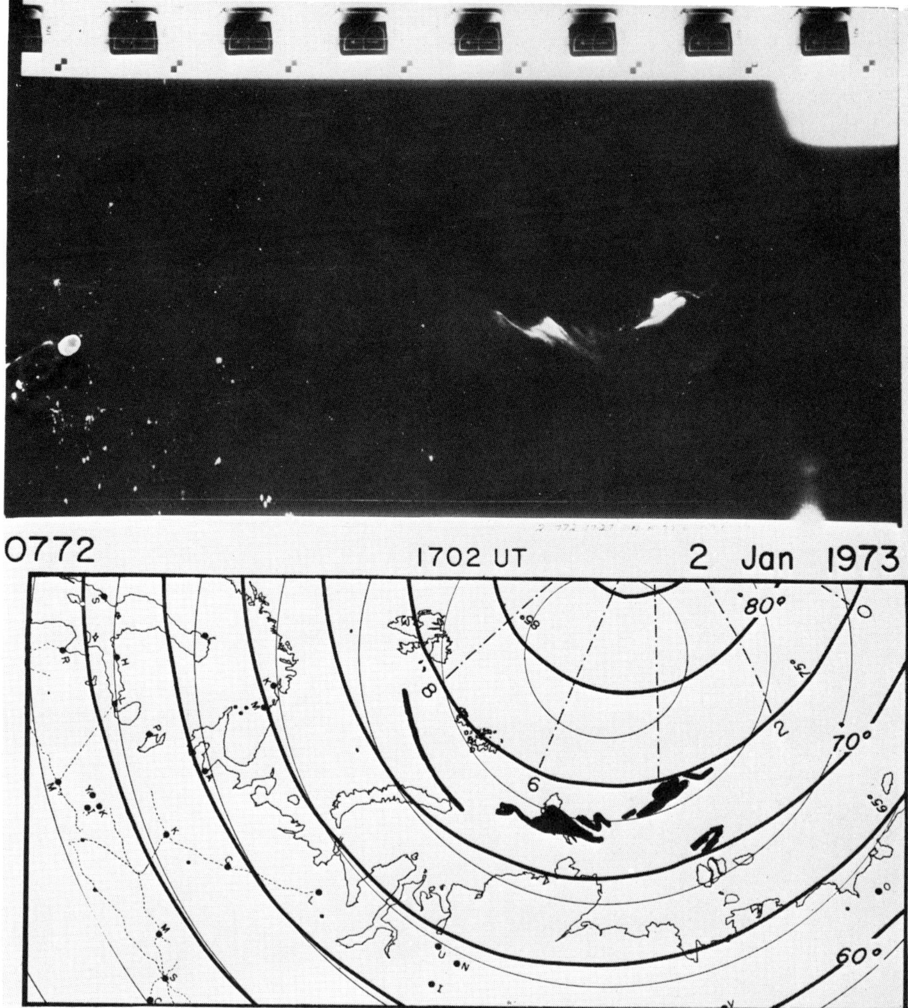

Fig. 5.3(b). Auroral substorms along contracted ovals. (Akasofu, S.-I.: *Space Sci. Rev.* **22**, 1511, 1974.)

perhaps at most a few times a month. Thus, the magnetosphere almost always has the excess energy.

Now, on the basis of the results summarized in Figure 5.4, we attempt to find the substorm function Σ. Here the magnetosphere is considered to be a 'black box' which is 'pulsed' by the interplanetary signal.

As noted in Section 4.4, the magnetosphere can accumulate the excess energy only when Φ_D is greater than Φ_N. Since Φ_N follows Φ_D with a certain time delay, say $\tau_0 \simeq 45$ min $-$ 1 h, the energy accumulation lasts only for the period τ_0. Suppose, for simplicity, that the IMF B_z component varies like a step function; it changes from $+5\ \gamma$ to $-5\ \gamma$ at $t = 0$. Thus, the magnetosphere has a 'shutter'

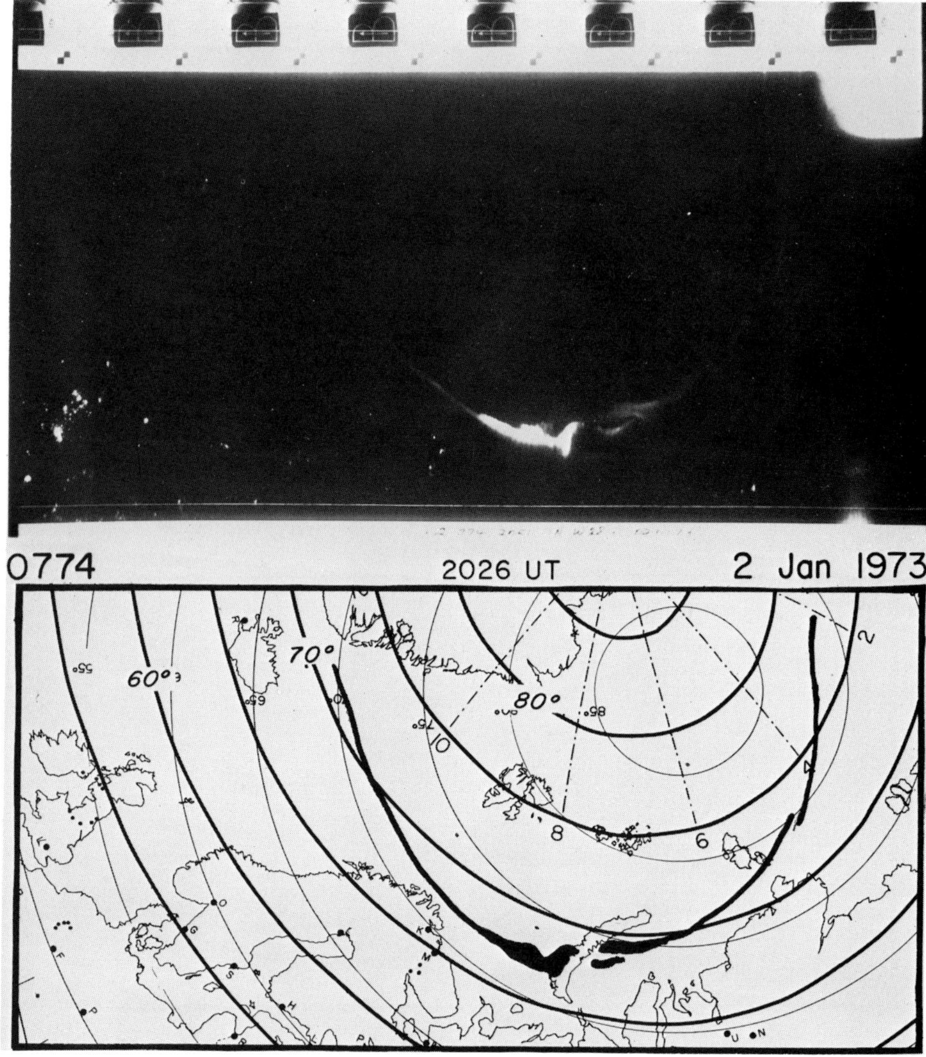

Fig. 5.3(c). Auroral substorms along contracted ovals. (Akasofu, S.-I.: *Space Sci. Rev.* **22**, 1511, 1974.)

which is triggered to open when the time derivative of the IMF B_z component $\partial B_z/\partial t$ becomes negative, but closes after $t \simeq \tau_0$. After the magnetosphere is 'pulsed' by this input, substorms appear as a series of pulses of diminishing intensity (say, $\tau_4 \simeq 6$ h). For simplicity, we assume that the first pulse appears τ_1 minutes after the southward turning of the IMF vector (or more precisely after $\partial B_z/\partial t$ becomes negative) and that each pulse lasts for τ_2 with an interval of τ_3. In summary, we can conclude that the substorm function $\Sigma(\Phi_D)$ is a series of pulses of diminishing intensity (Akasofu, 1976). Figure 5.5 illustrates schematically this situation in a block diagram form.

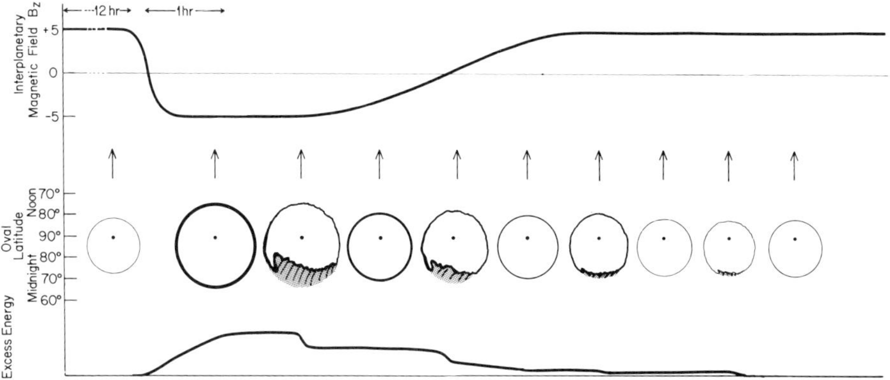

Fig. 5.4. Hypothetical change of the IMF B_z component and associated changes of the auroral oval, the occurrence of auroral substorms and changes in the amount of excess magnetic energy in the magnetotail. (Akasofu, S.-I.: *Space Sci. Rev.* **19**, 169, 1976.)

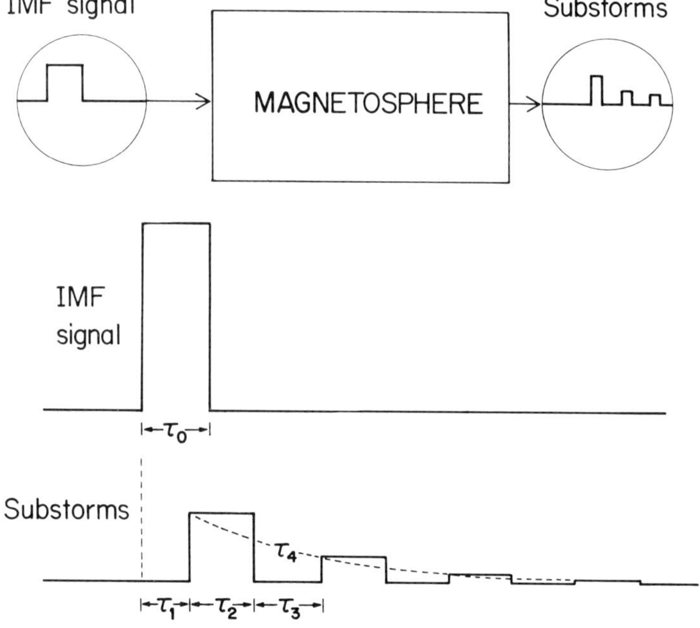

Fig. 5.5. Series of magnetospheric substorms resulting as a response of the magnetosphere to an impulsive IMF B_z 'signal'. The magnetosphere may be considered to be a 'black box'.

5.3. Substorm Intensity

The relationship between $\varepsilon_{\Sigma 0}$ and S, which is given by

$$\varepsilon_{\Sigma 0} = \frac{L}{\pi^4 R_{\mathrm{T}}^2} S^2$$

provides an important clue in finding the relationship between the energy ε_Σ

associated with individual substorms and the energy $\varepsilon_{\Sigma 0}$ stored in the magnetotail. Substorm energy ε_{Σ} may be written as

$$\varepsilon_{\Sigma} \approx \varepsilon_{k} + \varepsilon_{j} + \varepsilon_{r}$$

where ε_k, ε_j and ε_r denote the energy associated with the total kinetic energy of auroral particles, the joule heat energy associated with the auroral electrojet, and the total kinetic energy of ring current particles, respectively.

It is commonly assumed that in electron precipitation, about one ionization in 50 leads to the emission of a 3914 Å photon. Let us suppose that the average intensity of the 3914 Å emission in the auroral bulge is 1 kR. This requires an emission of 10^9 photon cm^{-2} s^{-1}. The ion production rate must then be of order 5×10^{10} ions cm^{-2} s^{-1}. Since the energy lost by an electron for each ionization collision is about 35 eV, the total energy flux required for the above emission is given by (35 eV ion^{-1})(5×10^{10} ions cm^{-2} s^{-1}) = 1.75×10^{12} eV cm^{-2} s^{-1} = 2.8 erg cm^{-2} s^{-1}; note that if this energy is carried by 5 keV electrons, the required flux should be of order 2.8×10^8 electrons cm^{-2} s^{-1}. Assuming that the auroral bulge has an area of $A = 3.4 \times 10^{16}$ cm^2 (the north-south dimension $\simeq 15°$ and the east-west dimension $\simeq 120°$), the total energy carried by the precipitation electrons is thus of order 10^{17} erg s^{-1}. Then, assuming the duration of a substorm to be 2 h (7.2×10^3 s), the total kinetic energy injected into one polar upper atmosphere is of order 7.2×10^{20} erg, the total energy ε_k being 1.4×10^{21} erg for both hemispheres.

The energy dissipated by joule heating in the electrojet region is given by $EjA = 1.8 \times 10^{10}\ W \simeq 10^{17}$ erg s^{-1}, where $E \simeq 10$ mV m^{-1}, $j \simeq 300$ A km^{-1} and the area A. The total joule heat produced in one polar upper atmosphere is then 7.2×10^{20} erg, the total energy ε_j being 1.4×10^{21} erg for both hemispheres.

The ring current energy can be estimated from the magnetic field ΔB produced by the belt (cf. S.T.P., p. 637) in low latitudes. For $\Delta B \simeq 50\ \gamma$, the total kinetic energy ε_r of the ring current belt is of order 2×10^{22} erg. Assuming that the above energy is injected during a period of 2 h = 7.2×10^3 s, the injection rate is of order 2.8×10^{18} erg s^{-1}.

Summary

	Auroral precipitation	Joule heat dissipation	Ring current injection	Total
Rate (erg/sec)	10^{17} [a]	10^{17} [a]	2.8×10^{18}	3.0×10^{18}
Total (erg)	$\varepsilon_k = 1.4 \times 10^{21}$	$\varepsilon_j = 1.4 \times 10^{21}$	$\varepsilon_r = 2 \times 10^{22}$	$\varepsilon_{\Sigma} = 2.3 \times 10^{22}$

[a]One hemisphere.

The purpose of the following two subsections is to examine whether there is a simple relationship between $\varepsilon_{\Sigma 0}$ and ε_k and also between $\varepsilon_{\Sigma 0}$ and ε_j (Akasofu and Kamide, 1976).

5.3.1. KINETIC ENERGY OF AURORAL PARTICLES

The DMSP (noon-midnight meridian) satellites can photograph approximately one half of the auroral oval over the antarctic region during the period between 15 and 18 UT once a day. For the purpose of this study, 30 DMSP photographs are chosen which show clearly an oval with substorm features. We assume then that (i) the equatorward boundary of the auroral oval gives the boundary of the oval area A_q just prior to substorm onset, (ii) the brightest (the saturation level) area A_k in the photograph gives a measure of the total kinetic energy of auroral particles, and (iii) the auroral behavior is symmetric with respect to the noon-midnight meridian. It may be noted that the intensity of solar flares is measured in terms of the area of the brightest Hα emission. In this sense, A_k may be considered to be a measure of the intensity of substorms. Several examples of DMSP photographs used in this study are shown in Figure 5.6. The scaling became inaccurate for large ovals, since the photographs did not cover the entire afternoon-evening part of the oval. Thus some extrapolation became inevitable.

With the above definitions, it may be stated that (i) the stored energy $\varepsilon_{\Sigma 0}$ just prior to the onset of a substorm is given by

$$\varepsilon_{\Sigma 0} = \frac{LB_{\mathrm{P}}^2}{\pi^4 R_{\mathrm{T}}^2}(A_q - A_0)^2$$

(ii) ε_k of auroral particles is proportional to A_k (since the kinetic energy flux deposited in the polar upper atmosphere should be proportional to A_k).

Thus, if there is any relationship between the stored energy $\varepsilon_{\Sigma 0}$ and the kinetic energy ε_k, there should also be a relationship between A_q and A_k. This is indeed the case, and the relationship is shown in Figure 5.7 together with the curve expressed by

$$A_k = 0.05(A_q - A_0)^2$$

where $A_0 = 3.7 \times 10^6\ \mathrm{km}^2$ (since we are dealing with only one half of the oval). It can be clearly seen, in spite of a number of admittedly crude assumptions and approximations, that A_k is proportional to $(A_q - A_0)^2$ and thus that ε_k is proportional to $\varepsilon_{\Sigma 0}$. Following the definition of the intensity of solar flares, A_k may be chosen as a measure of the intensity of substorms. With this definition, the intensity of substorms is proportional to the excess energy $\varepsilon_{\Sigma 0}$ stored in the magnetotail.

Since the number of available photographs was limited, it was not possible to select photographs which show only the maximum epoch of substorms. Obviously, part of the reason for the considerable scatter of points in Figure 5.7 arises from this fact. However, examining a large number of photographs which could not be used for scaling (because for example, the midday part of the oval was missing), an extremely large expanding bulge does not occur very frequently. One of the main reasons for the scatter appears to be the difference in the intensity of substorms for a given value of A_q. If this is indeed the case, there can be an order of magnitude difference in substorm intensity for a given A_q. This point will be substantiated in the following subsection.

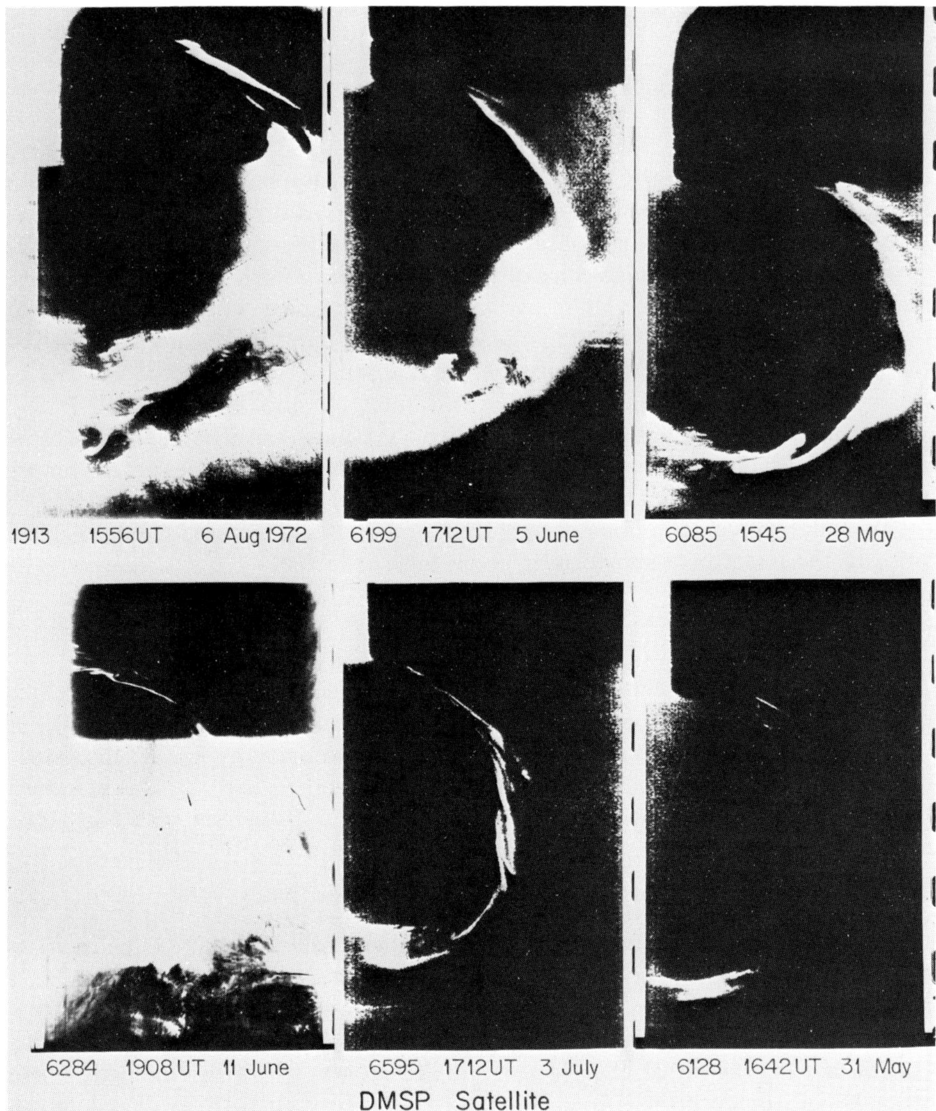

Fig. 5.6. DMSP photographs showing the relationship between the intensity of auroral substorms (estimated from the size of the area covered by brightest auroras) and the size of the auroral oval.

5.3.2. JOULE HEAT ENERGY OF THE AURORAL ELECTROJET

It is also of great interest to examine relationships between the power P_j dissipated in the auroral ionosphere as joule heat and the amount of the excess energy $\varepsilon_{\Sigma 0}$. It can be estimated in terms of the total current I of the westward electrojet and may be expressed as

$$P_j = I^2/\Sigma_c$$

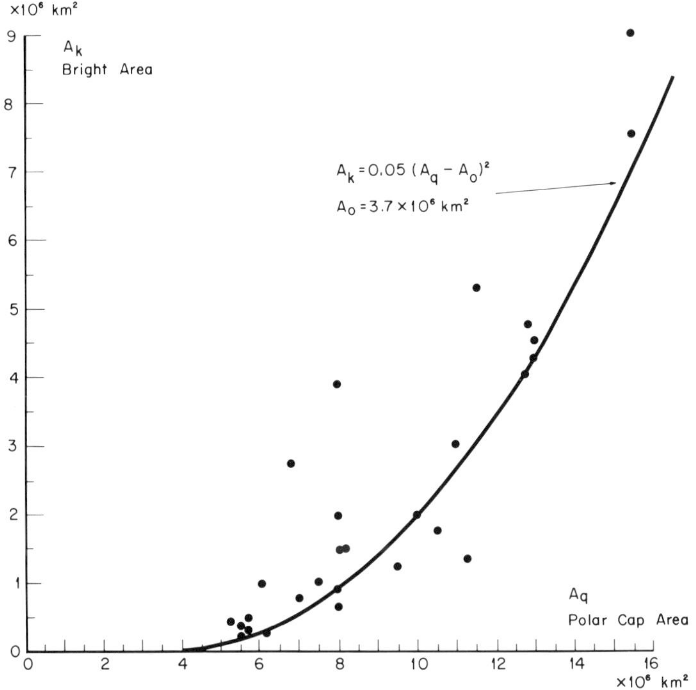

Fig. 5.7. Relationship between the intensity of auroral substorms (estimated from the size of the area covered by brightest auroras) and the area of the auroral oval. (Akasofu, S.-I. and Kamide, Y.: *Planetary Space Sci.* **24**, 223, 1976.)

where Σ_c denotes the height-integrated Cowling conductivity. The energy ε_j of the westward electrojet dissipated in time τ is then given by

$$\varepsilon_j = \int^{\tau} P_j \, dt = I^2\tau/\Sigma_c$$

Let us define here α as the ratio between the energy stored in the magnetosphere and that dissipated as joule heat ε_j in the polar ionosphere:

$$\varepsilon_j = \alpha\varepsilon_{\Sigma 0}$$

Here, let us say that the center of the electrojet is located at colatitude θ_c, and the oval is assumed to be a circle (cf. Holzworth and Meng, 1975). A_c and A_0 denote the area of the circles of radius d_c and d_0 or colatitude θ_c and θ_0, respectively. Then, rewriting the expression given by Gonzalez and Mozer (1974), we have

$$\varepsilon_j = \frac{I^2}{\Sigma_c}\tau = \alpha\frac{V_s\tau B_p^2}{\pi^4 R_T^2}(A_c - A_0)^2$$

and

$$I^2 = \alpha\Sigma_c\frac{V_s B_p^2}{\pi^4 R_T^2}(A_c - A_0)^2$$

or

$$I = \sqrt{\alpha \Sigma_c V_s} \frac{B_p}{\pi R_T} (d_c^2 - d_0^2)$$

$$= \sqrt{\alpha \Sigma_c V_s} \frac{B_p}{\pi R_T} \beta^2 (\theta_c^2 - \theta_0^2)$$

where θ_c and θ_0 are the colatitudes in the midnight sector which correspond to d_c and d_0, respectively, and

$$\beta = \theta_c / d_c = \theta_0 / d_0$$

Here, we use typical values of the parameters as follows: $\beta = 0.81$, $B_p = 0.6$ G, $V_s = 400$ km s^{-1}, $R_T = 15\ R_E$ and $\Sigma_c = 50$ mho.

The two values (I, θ_c) have already been obtained for a large number of substorms at their maximum epoch by Kamide and Akasofu (1974) on the basis of magnetic records from the Alaska meridian chain of stations. Note that the current density J(A m^{-1}) is given (Kamide and Brekke, 1975) by

$$J(\text{A m}^{-1}) = 2 \times 10^3\ \Delta H(\gamma)$$

In Figure 5.8, we show the relation between I and θ_c. The figure shows also two

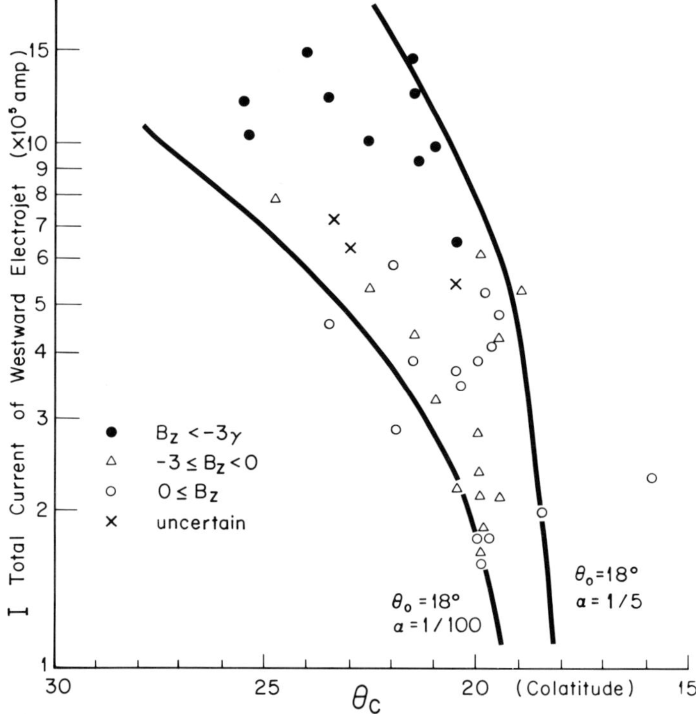

Fig. 5.8. Relationship between the total current intensity of the westward auroral electrojet and the colatitude of the auroral oval in the midnight sector. (Akasofu, S.-I. and Kamide, Y.: *Planetary Space Sci.* **24**, 223, 1976.)

curves which are calculated by using the last equation with the following parameters:

$$\alpha = 1/5, \quad \theta_0 = 18°, \quad \text{and}$$
$$\alpha = 1/100, \quad \theta_0 = 18°$$

It can be seen that the observed points are well bounded by the two curves. Thus, it may be concluded that the quantity S^2 is proportional to the energy ε_j, and that ε_j lies between 1/5 and 1/100 of $\varepsilon_{\Sigma 0}$ (and A_q); this is in agreement with the results obtained in the previous section by using the DMSP photographs.

On the basis of the present study, the following results were obtained:

(1) The magnetosphere is in the 'ground state' during a prolonged period (~6–12 h) of a large (~5 γ) northward IMF. The auroral oval has its minimum size (the area of about $A_0 = 7.4 \times 10^6$ km^2) and there is no observable substorm feature. It is interesting that the relationships between $\varepsilon_{\Sigma 0}$ and ε_k and between $\varepsilon_{\Sigma 0}$ and ε_j lead to approximately the same value of A_0. Since there is no report that the magnetotail disappears during such a period, it may be concluded that the magnetic energy stored in the magnetotail in the ground state is not available for substorms.

(2) The main part of the magnetosphere may be considered to be a 'black box' which produces a series of pulses after it is pulsed once. Thus the substorm function Σ is represented by a series of pulses.

(3) The area A_k covered by brightest auroras and the area A_q bounded by the oval has a simple relation which is given by

$$A_k = 0.05(A_q - A_0)^2$$

Following the definition of the intensity of solar flares, A_k may be chosen as a measure of the intensity of substorms. Since $(A_1 - A_0)^2$ is proportional to the excess energy $\varepsilon_{\Sigma 0}$ stored in the magnetotail, it may be concluded that the intensity of substorms is proportional to $\varepsilon_{\Sigma 0}$.

(4) The joule heat energy ε_j lies between 1/5 and 1/100 of the excess energy $\varepsilon_{\Sigma 0}$, indicating that there is a difference in substorm intensity for a given value of $\varepsilon_{\Sigma 0}$.

5.4. Time-Dependent Merging

5.4.1. STABILITY OF THE PLASMA SHEET

Schindler (1972) and Schindler and Soop (1968) showed that the stability criteria for a two-dimensional plasma sheet may be written in the form

$$\frac{\mathrm{d}a}{\mathrm{d}x} > 0 \text{ stable}$$

$$\left.\begin{array}{l}\dfrac{\mathrm{d}a}{\mathrm{d}x} < 0 \\ \varepsilon = L_z/L_x \to 0\end{array}\right\} \text{unstable}$$

where a denotes the distance between the neutral sheet and the magnetopause

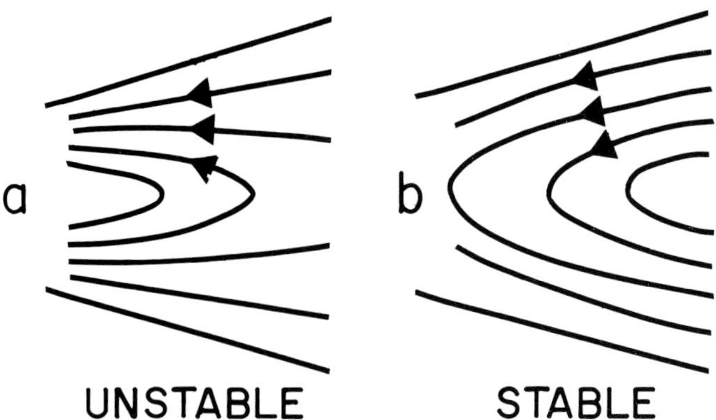

Fig. 5.9. Unstable and stable magnetic field configurations in a sheet of plasma. (Schindler, K: *J. Geophys. Res.* **79**, 2803, 1974.)

and is a function of distance from the Earth. The equation $z = a(x)$ determines the geometry of the magnetopause; it is also assumed that the variation of the plasma and field quantities along the tail is small: namely, the characteristic length L_x for those quantities along the x-axis is large compared with the characteristic length L_z along the z-axis. Figure 5.9 shows schematically both the stable and unstable configurations. Since the magnetotail has a flaring configuration and B_z is positive, Schindler concluded that the magnetotail is unstable. Schindler (1974) noted that for the observed magnetic field configuration in the magnetotail, the ion-tearing instability is most likely to occur (Section 5.4.3).

5.4.2. NUMERICAL SIMULATION OF RECONNECTION

Sonnerup (1970) examined the magnetic field configuration in the vicinity of the magnetic X-line for different merging rates, proposing the leading wave (OL) and the trailing wave (OT) in separating the inflow and outflow regions (Section 1.4.2). For a small merging rate ($M_1 = v_1/V_A$), both OL and OT make small angles with respect to the center line of the outflow region. However, for the maximum merging rate $M_1 = (1 + \sqrt{2})$, these angles become maximum.

Fukao and Tsuda (1973) examined how the merging develops as a function of time by using a particular type of initial magnetic field (the magnetic potential A) and the flow (the velocity potential ψ configurations). Both A and ψ are related to $\boldsymbol{B}$ and $\boldsymbol{U}$ by

$$B_x = \frac{\partial A}{\partial y}, \quad B_y = -\frac{\partial A}{\partial x}$$

$$U_x = \frac{\partial \psi}{\partial y}, \quad U_y = -\frac{\partial \psi}{\partial x}$$

The set of MHD equations (MKS) for an incompressible fluid,

$$\nabla \cdot \boldsymbol{U} = 0$$

$$\nabla \cdot \boldsymbol{B} = 0$$

$$\nabla \times \boldsymbol{E} = -\frac{\partial \boldsymbol{B}}{\partial t}$$

$$\frac{1}{\mu} \nabla \times \boldsymbol{B} = \boldsymbol{J}$$

$$\boldsymbol{E} + \boldsymbol{U} \times \boldsymbol{B} = \boldsymbol{J}/\sigma$$

$$\rho\left\{\frac{\partial \boldsymbol{U}}{\partial t} + (\boldsymbol{U} \cdot \nabla)\boldsymbol{U}\right\} = \boldsymbol{J} \times \boldsymbol{B} - \nabla \mathrm{p} - \rho\nu\nabla \times \nabla \times \boldsymbol{U}$$

are combined to give the set of nonlinear simultaneous equations

$$\frac{\partial A}{\partial t} = \left[-\frac{\partial \psi}{\partial y}\frac{\partial}{\partial x} - \frac{\partial \psi}{\partial x}\frac{\partial}{\partial y}\right] A + R_m^{-1}\nabla^2 A$$

$$\nabla^2 \frac{\partial \psi}{\partial t} = -\left[\frac{\partial \psi}{\partial y}\frac{\partial}{\partial x} - \frac{\partial \psi}{\partial x}\frac{\partial}{\partial y}\right]\nabla^2\psi$$

$$+\left[\frac{\partial A}{\partial y}\frac{\partial}{\partial x}\frac{\partial A}{\partial x}\frac{\partial}{\partial y}\right]\nabla^2 A + R^{-1}\nabla^4\psi$$

where distances are normalized by L, the size of the whole domain, U by V_A (the Alfvén wave speed) and t by L/V_A. The magnetic Reynolds number R_m and the Reynolds number R are taken to be 5×10^3, and the maximum fluid speed along the x-axis is given by $U = 1.0$. Their numerical results are shown in Figure 5.10. The formation of the X-type neutral line is clearly seen at $T = 6 \times 10^{-2}$. In our magnetospheric situation, L/V_A may be of order a few seconds. Thus, their results suggest that the X-type neutral line can be formed in a very short period under the given initial conditions.

5.4.3. PLASMA PROCESSES

As mentioned in Section 5.1 the magnetospheric substorm may be considered to be the process by which the magnetosphere tends to reduce efficiently the accumulated energy in the magnetotail. Thus, there is no doubt that the overall condition of the magnetosphere determines when and where the enhanced reconnection should take place. Yeh and Axford (1970) expressed this situation by saying that the reconnection rate is determined by the boundary conditions around the reconnection region. However, in order to understand various substorm phenomena, it is necessary to 'translate' the above hydromagnetic description of the magnetospheric substorm in terms of specific plasma processes which are directly related to the causes of the magnetospheric substorm.

The critical problem here is what plasma processes allow an explosive reconnection, if the reconnection process is indeed the process which converts the magnetic energy into the substorm energy. We noted earlier that Schindler (1974) proposed that ion-tearing mode instability plays.a crucial role.

Actually, this has been a long-standing problem, not only in magnetospheric physics, but also in solar and astrophysics in general. In the diffusion region of

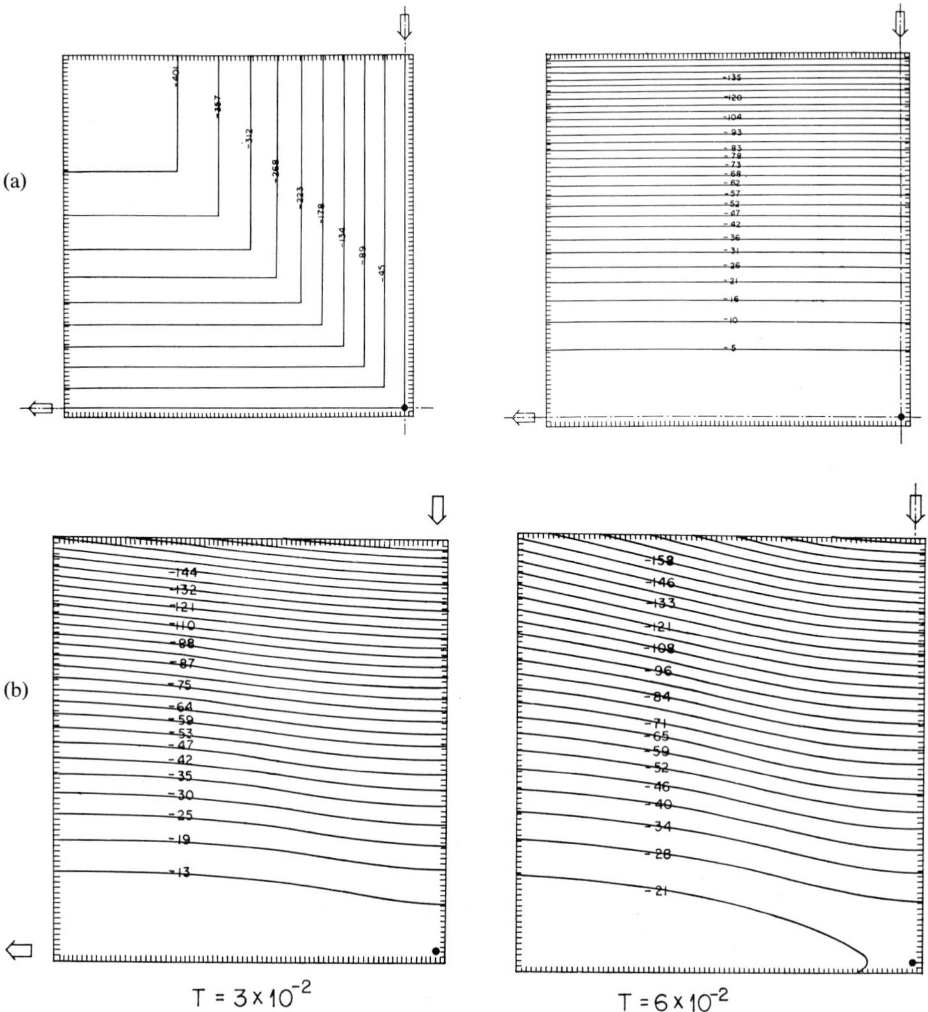

Fig. 5.10(a, b). Development of an X-type neutral line. (a) The initial flow and the magnetic field distribution at $T = 0$ are shown in the left- and right-hand side, respectively. (b) Magnetic field configurations at $T = 3 \times 10^{-2}$ and 6×10^{-2}. (Fukao, S. and Tsuda, T.: *Planet. Space Sci.* **21**, 1151, 1973.)

reconnection (Section 1.4.2), the magnetic field $\boldsymbol{B}$ is given by

$$\frac{\partial \boldsymbol{B}}{\partial t} = \frac{1}{\sigma} \nabla^2 \boldsymbol{B}$$

Thus, the exponential decay time t is given by

$$t \sim \mathscr{L}^2 \sigma$$

where $\mathscr{L}$ denotes a characteristic length of the system and σ the electrical conductivity. Thus, in making the time t short, either $\mathscr{L}$ or σ should be reduced.

This problem was noted earlier in Section 1.4.2(a), where it was pointed out that the characteristic time for Sweet-Parker's model of merging is as long as 10^{14} s. It was pointed out also that this long characteristic time means a very small value of the merging rate $M = 6.5 \times 10^{-8}$. It means also a very small energy production rate. For example, assuming the dimension of the diffusion region to be as large as $\mathscr{L}(=1000\ R_E) \times w(=$ dawn-dusk width of the magnetotail $= 30R_E)$, the energy production rate is only of order

$$(B^2/8\pi)\mathscr{L}wv = 6.3 \times 10^{13}\ \text{erg s}^{-1}$$

where $B = 10^{-4}$ G. It is for this reason that Petschek (1964) and Sonnerup (1970) proposed that MHD waves play an important role in enhancing the merging rate. Indeed, their merging rates ($M = 0.03$ for Petschek's model and $M \simeq 1 + \sqrt{2}$ for Sonnerup's model; see Section 1.4.2) are sufficient for a rapid reconnection. However, we shall see in Section 9.2.2(b) that the magnetic field configuration and plasma flow pattern predicted by Petschek do not seem to agree with the observed ones. Sonnerup's mechanism cannot produce a sufficient amount of energy, because his model is applicable for a high β condition (Section 1.4.2(c)).

As a mechanism for reducing the scale length $\mathscr{L}$, various tearing mode instabilities have been proposed (Furth *et al.*, 1963; Coppi *et al.*, 1966), while various wave-particle interaction mechanisms have been considered in reducing σ (Scarf *et al.*, 1974; Syrovatskii, 1972; Bowers, 1973). Speiser (1970) proposed also that a new effective conductivity should be defined in the diffusion region. Piddington (1967) suggested that the ionosphere may contribute considerably in enhancing the reconnection by discharging the space charges which develop in the (ideal) current pinch (the neutral sheet). In the following, we shall review some of the above studies; for a recent theoretical review of the subject, see Schindler (1975).

Schindler (1974) suggested that the instability of the plasma sheet, examined in Section 5.4.1, is associated with the ion-tearing mode instability. In most of the magnetotail conditions, motions of electrons in the plasma sheet can be described in terms of the guiding center motion (namely, gyroscopic). However, if the gyroperiod of protons $\tau(= 2\pi m/eB_z)$ associated with the B_z component becomes equal to the characteristic time of the instability, the instability will grow. This criterion can be written as

$$\frac{\zeta \Omega_i}{2\pi\gamma} \lesssim 1$$

where Ω_i denotes the ion gyro-frequency with respect to the B_z component, γ the growth rate of the ion-tearing mode instability and $\zeta > 1$ a constant. Figure 5.11 shows the stability diagram for this particular instability. It can be seen that the thinner the plasma sheet is, the smaller L_z (the thickness of the plasma sheet) is and/or the smaller the normal component B_z is, the more quickly the instability tends to grow. Schindler suggested that for $B_z \simeq 1\ \gamma$, the instability sets in when the thickness of the plasma sheet becomes a fraction of an Earth radius.

In order for this ion-tearing mode instability to be important in substorm processes, it has to grow in a nonlinear manner, namely to macroscopic amplitudes. That is to say, the tearing mode instability may produce a number of

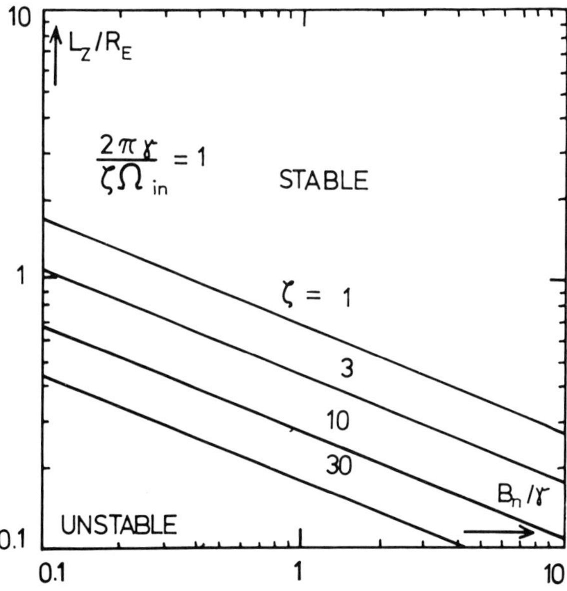

Fig. 5.11. Stability diagram for the ion-tearing mode instability: γ denotes the growth rate of the ion-tearing model, Ω_{in} the ion gyro-frequency associated with the normal magnetic field component B_n, and ζ is an unknown numerical factor which is expected to lie in the range as specified. (Schindler, K: *J. Geophys. Res.* **79**, 2803, 1974.)

small 'magnetic loops' along the midplane, but each loop has to grow and combine to result in a large-scale reconfiguration of the magnetic field in the magnetotail. Figure 5.12 shows schematically how the tearing mode instability might eventually grow in its nonlinear growth stage (Schindler, 1974). At present, however, there is no theoretical treatment of the nonlinear development of the tearing mode instability. Schindler and Ness (1972) and Bowling (1975) showed that during periods of enhanced geomagnetic activity ($Kp \geq 3$), magnetometer records (rapid fluctuations in both the magnitude and the direction of the field) from the vicinity of the midplane at the lunar distance are statistically consistent with what one expects from a small-scale closed loop-like structure. Therefore, it is possible that such loop-like structures result from the ion-tearing mode instability. However, it is at present not certain whether they eventually form much larger-scale loops in the way suggested in Figure 5.12. Speiser (1973) showed that most of the magnetic field observations are consistent with a wavy current sheet in which small-scale loops are embedded (Figure 5.13). These studies have an important implication in interpreting changes of the B_z component during substorms (Section 6.2).

There have also been a few suggestions that an anomalous resistivity might develop in the cross-tail current region, resulting in reduction of the effective conductivity and eventually in disruption of the current. Syrovatskii (1972) and Bowers (1973) considered the growth of ion-acoustic waves in the cross-tail current, but the basic requirement for the growth of this particular wave is $T_e > T_+$, which is not satisfied in the distant plasma sheet. Recently, Scarf *et al.* (1974) detected intense ELF electromagnetic noise (electron whistler mode waves,

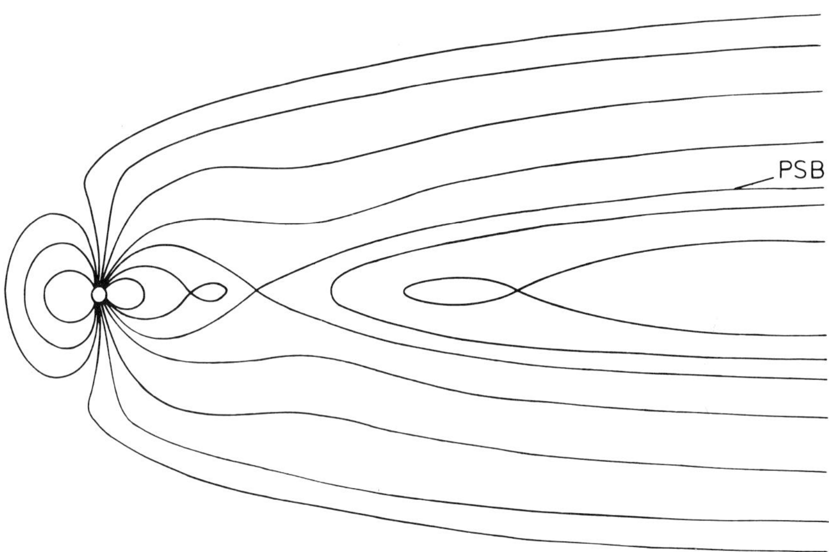

Fig. 5.12. Schematic diagram showing the possible growth of the ion-tearing mode instability, resulting in a large-scale change of the magnetic field configuration. (Schindler, K.: *J. Geophys. Res.* **79**, 2803, 1974.)

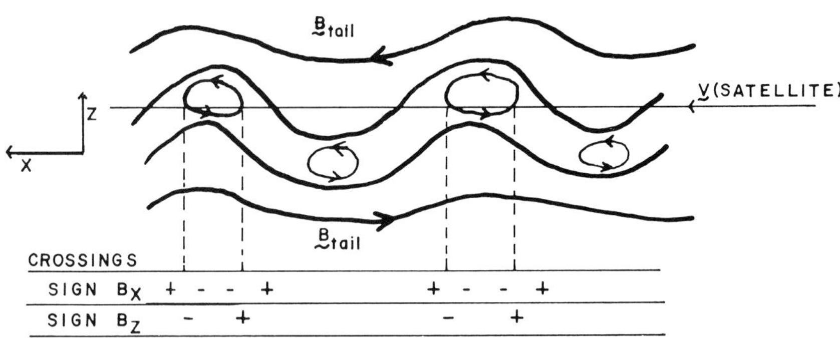

Fig. 5.13. Schematic model of a wavy current sheet, to explain typical variations of the magnetic field near the midplane of the magnetotail. (Speiser, T. W.: *Radio Sci.* **8**, 973, 1973.)

the lowest mode of $(n+1/2)\Omega_e \simeq 140$ Hz) in the distant plasma sheet ($X \sim -35\ R_E$) during magnetospheric substorms. Figure 5.14 shows their results. The concurrent magnetic and plasma flow data suggested to the authors that the satellite was close to the magnetic neutral line.

It is important to recall in this respect that the two magnetic field lines which cross at the X-line become perpendicular to each other when the rate of merging (or reconnection) becomes maximum. For such a magnetic field configuration, however, $\nabla \times \boldsymbol{B} = 0$ and thus there is no electric current along the X-line (Section 1.4.2). Therefore, the complete blocking of the magnetotail current along a narrow strip may indeed be equivalent to the highest rate of reconnection. However, as we shall see in Sections 6.2 and 7.2, there is no indication of disruption of the

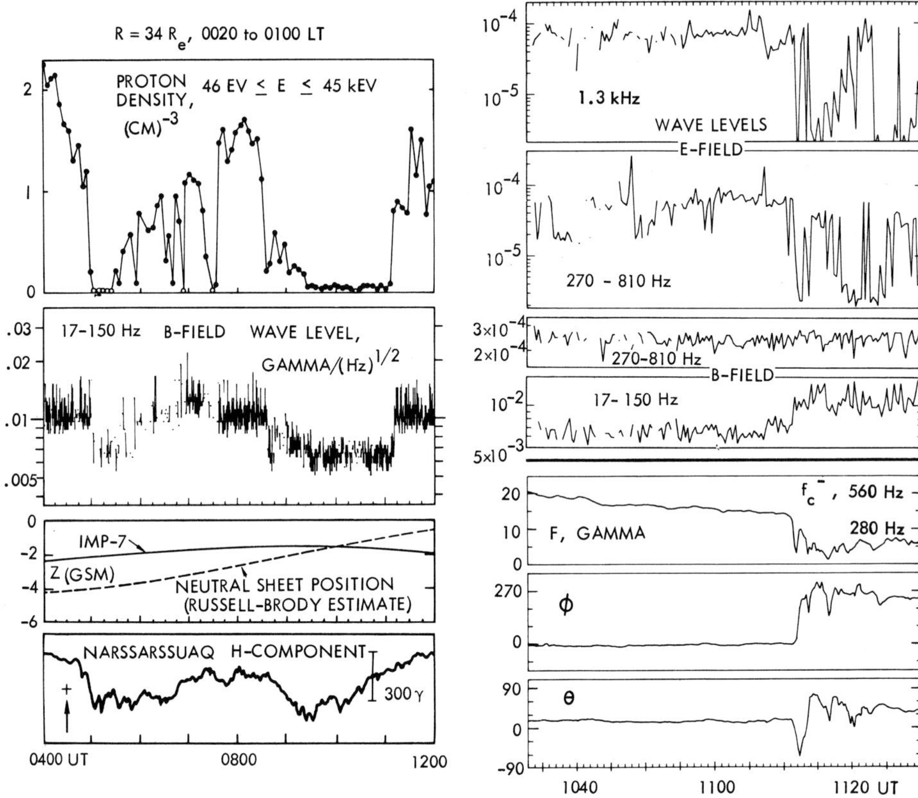

Fig. 5.14. Intense ELF electromagnetic noise observed in the plasma sheet during two substorms on 1972, October 28. The electric field noise observations are shown in the top three panels on the right-hand side, the magnetic field noise observations in the next two panels. The other panels show the relevant data during the event. (Scarf, F. L., Frank, L. A., Ackerson, K. L. and Lepping, R. P.: *Geophys. Res. Lett.* **1**, 189, 1974.)

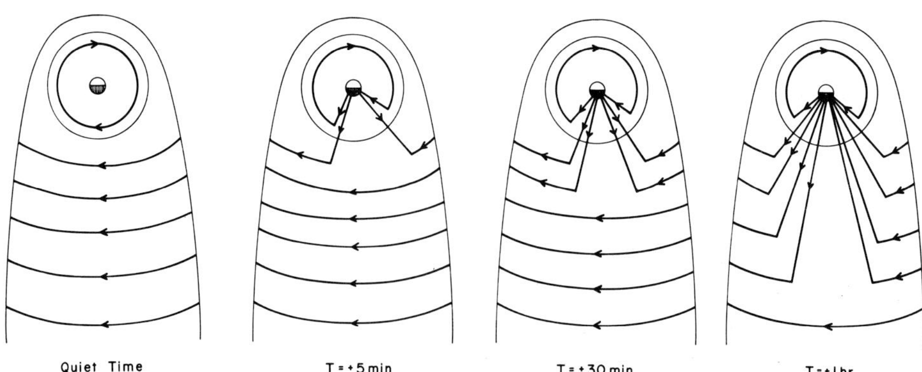

Fig. 5.15. Schematic diagram showing how the cross-tail current might be disrupted during the course of a magnetospheric substorm. (Akasofu, S.-I.: *Solar-Terrestrial Physics*, E. R. Dyer, (ed.), p. 131, D. Reidel Publ. Co., 1972.)

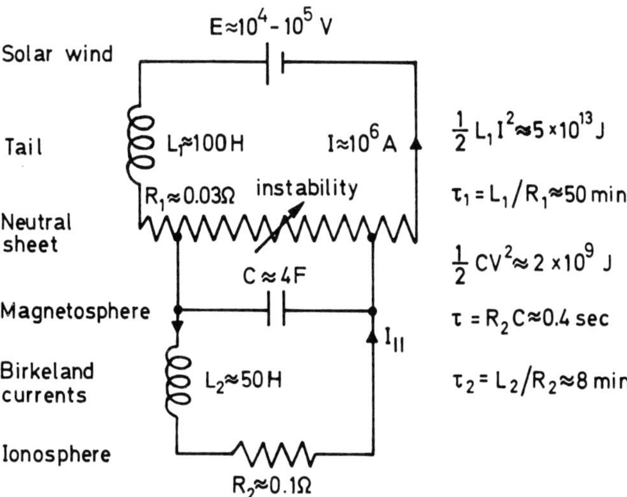

Fig. 5.16(a). Equivalent circuit for the substorm current system. At the substorm onset, an instability occurs in the cross-tail current, increasing considerably the electric resistance and thus causing the diversion of the current to the polar ionosphere.

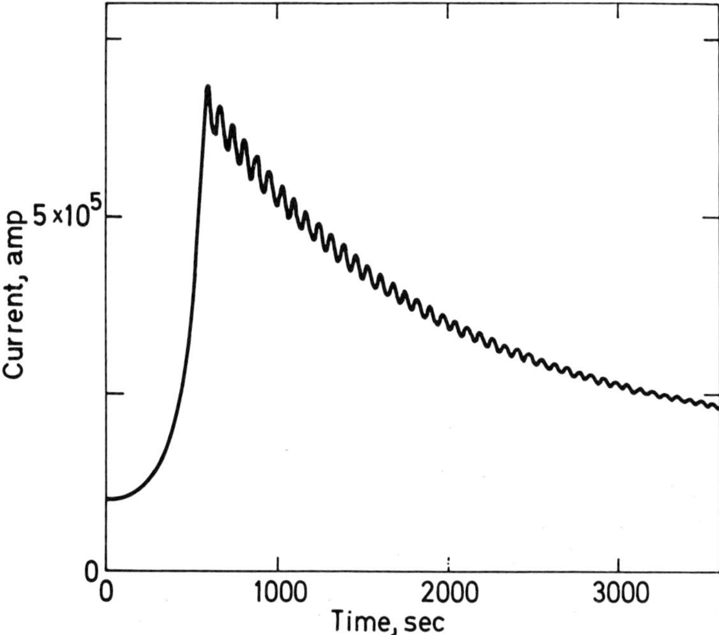

Fig. 5.16(b). Time variations of the current to the ionosphere in the circuit in Figure 5.16(a), after a sudden increase of the resistance in the cross-tail current. (Boström, R: *Critical Problems of Magnetospheric Physics*, E. R. Dyer (ed.), p. 139, National Academy of Sciences, Washington, D.C., 1972.)

cross-tail current along a narrow ($<10\,R_E$) strip in the distant tail, so that the disruption may take place only at distances $X = -6 \sim -10\,R_E$. In any event, the blocked part of the cross-tail current must find a new circuit, and the obvious one would be the polar cap circuit. Indeed, Atkinson (1967), Akasofu (1972) and McPherron *et al.* (1973) have considered how the cross-tail current might be diverted to the polar ionosphere. Figure 5.16 shows an example of such a phenomenological model.

Speiser (1970) suggested that the effective conductivity in the diffusion region should be redefined when reconnection takes place in collisionless plasma. In such a situation the dissipated magnetic energy from the reconnection process can be carried away by accelerated particles, rather than going into particle (ohmic) heating. On the basis of his extensive study of trajectories of particles in the midplane of the magnetotail, he showed that the effective conductivity σ_i can be written as

$$\sigma_i = \frac{ne^2}{m}\tau = \left(\frac{Ln^2e^3}{2mE}\right)^{1/2} \simeq 7\times 10^{-2}\ \text{mho m}^{-1}$$

where L is the length of the acceleration region, and τ the lifetime of the particle in the system, instead of the collision interval. The value of σ_i thus defined may be compared with the collisional conductivity $\simeq 10^7$ mho m^{-1} in the plasma sheet.

5.4.4. SUBSTORM TIME CONSTANTS

The magnetospheric substorm is a transient phenomenon with a lifetime of order 2–3 h. Its growth and decay may be most clearly manifested in the westward electrojet. The electrojet grows rapidly during the expansive phase, which lasts typically 30 min, and then decays slowly during the next 1–2 h, the recovery phase. It is reasonable to infer that such transient behavior of the magnetosphere arises from a particular set of large-scale magnetospheric quantities. Further, since we are here dealing with the electric current in the magnetosphere, it is reasonable to consider an equivalent circuit for the magnetospheric substorm; note that we have already used the concept of an equivalent circuit in Section 4.4.3(b). Here, we describe first the equivalent circuit for the magnetospheric substorm proposed by Fälthammar and Boström (Boström, 1972, 1974).

They assumed that the westward electrojet results from the disruption of a part of the magnetotail current which flows in a single-loop coil of radius 10 R_E and length (along the $-X$ axis) 10 R_E. The resistance of the loop is small, only of order 0.03 Ω, and its inductance L is of order 100 H (Figure 5.16(a)). The total current I in the loop is of order $(1/2)\times(30\ \text{mA m}^{-1})\times 10\,R_E \simeq 10^6$ A. The current is supplied by the solar wind-magnetosphere dynamo (Section 1.4.1); its electromotive force is of order 10^4–10^5 V. Thus, the magnetic energy in this inductive circuit is $(1/2)L_1I^2 = 5\times 10^{13} J = 5\times 10^{20}$ erg; note that the above value is smaller than for the energy released during a single substorm. Akasofu (1969) suggested that such a current disruption might cause a large voltage ($\sim$50 kV) to be developed in the circuit; see also Alfvén and Carlqvist (1967).

In this situation, it is assumed that the disruption of the loop current is caused by an anomalous resistivity which grows in the cross-tail current. This portion of the circuit is indicated by a variable resistor in the figure. When the resistance is suddenly increased, the loop current finds its own way to the ionosphere. The resistance of the ionosphere is of order 0.1 Ω and the inductance associated with the added portion of the circuit is of order $L_2 \simeq 50\ H$.

The magnetospheric circuit has also capacitance, since displacement current can flow in the plasma sheet. Charging the capacitor is equivalent to setting plasma in motion. Fälthammar and Boström estimated the capacitance from the kinetic energy of order 2×10^9 J = 2×10^{16} erg (plasma flow of speed 10 km s^{-1} in the plasma sheet of a volume of order 100 R_E^3 and number density 1 cm^{-3}). A capacitor charged to $V = 30$ kV has the same energy $(1/2)\ CV^2$ if its capacitance is 4 F.

Such a current circuit has three time constants (Figure 5.16(b)),

$$\tau_1 = L_1/R_1 \simeq 50 \text{ min}$$

$$\tau = R_2 C \simeq 0.4 \text{ s}$$

$$\tau_2 = L_2/R_2 \simeq 8 \text{ min}$$

Fälthammar and Boström suggested that τ_1 corresponds to the decay time of a substorm. Further, this circuit provides small oscillations of time constant $\tau = R_2C \simeq 0.4$ s, and they suggested that those oscillations may correspond to the Pi2 micropulsations (Section 7.5.2).

As noted in Section 5.2.2, however, the magnetosphere produces a series of substorms with diminishing intensity, after it is 'pulsed' by the IMF B_z component. Thus, the equivalent circuit will be much more complicated than the one described above.

5.4.5. SHORT REVIEW OF THEORIES OF THE MAGNETOSPHERIC SUBSTORM

During the last decade, major theoretical efforts on the magnetospheric substorm have been concentrated on the search of suitable instability mechanisms either at the outer surface of the ring current belt (Swift, 1967; Liu, 1970) or at the inner edge of the plasma sheet (Atkinson, 1971; Coroniti and Kennel, 1972; Kropotkin, 1972).

Swift (1967) examined the growth of the interchange instability at the outer surface of the ring current belt. The ionosphere plays an important role in inhibiting the growth of the instability by depolarizing the space charges associated with it. He suggested that when the inhibiting effect of the ionosphere is suddenly reduced by the growth of the acoustic wave instability (resulting from field-aligned currents) and thus the ring current belt is de-coupled from the ionosphere, the interchange instability grows rapidly. In supporting his theory, Swift noted that during an early epoch of the expansive phase the brightened arc remains stationary, but ripples and waves appear along it and that their growth time agrees with the predicted one. Liu (1970) examined also the possibility of growth of the interchange and the drift flute mode instabilities at the surface of the ring current belt and suggested that the drift flute mode instability can grow when the density gradient exceeds a critical limit; the ionosphere is assumed to be perfectly conducting.

The first indication of substorm, a sudden brightening of a discrete arc, occurs near the poleward boundary of the diffuse aurora (or near the equatorward boundary of the belt of discrete auroras). Note that the equatorward boundary of the diffuse aurora corresponds to the projection (along the geomagnetic field lines) of the inner edge of the plasma sheet (Section 3.8.1). Therefore, it appears that the direct cause of substorms should be found well within the plasma sheet.

Nevertheless, Swift and Liu developed mathematically essential 'building blocks' for substorm theories, in particular the growth of an instability and the associated field-aligned currents, the subsequent growth of a current driven instability and the resulting decoupling between the magnetosphere and the ionosphere which leads to an explosive onset of the substorm.

It has been suggested by Kennel (1969) that the pitch-angle distribution of plasma near the inner edge of the advancing plasma sheet may become strongly anisotropic and thus an intense precipitation may occur as a result of the growth of instabilities associated with the pitch-angle anisotropy. Atkinson (1971) suggested that as plasma is removed from the inner edge of the plasma sheet by such instabilities, its diamagnetic effects disappear, allowing inward collapse of tail-like magnetic field lines to the dipole-like form. Atkinson suggested that the inward collapse occurs much faster than the large-scale convection, resulting in a build up of dipole-like field lines on the nightside. He suggested further that the cross-tail current must be diverted to the ionosphere. However, we shall see in Section 8.2 that hot plasma is not removed from the region suggested by Atkinson. Instead, plasma pressure increases, in spite of the fact that there occurs an appreciable increase of the B_z component.

Coroniti and Kennel (1972) consider that the electrical conductivity of the ionosphere has a peak near the equatorward boundary of the auroral oval and thus such a non-uniformity of the conductivity tends to result in an equatorward polarization electric field. During a quiet condition, the space charges associated with the electric field can freely escape into the magnetosphere, generating field-aligned currents. When the convection electric field is enhanced, the field-aligned currents will also be increased to the point at which electrostatic ion-cyclotron waves begin to develop. The resulting anomalous resistivity (Section 3.9.4) will disrupt the perfect electrical communication between the ionosphere and the magnetosphere. Thus, the free closure of the Hall current in the magnetosphere is prevented, resulting in a large build-up of the space charges near the equatorward boundary of the oval. The resulting polarization electric field drives an intense westward current, the growth of the westward electrojet. At the same time, the polarization electric field causes a rapid motion of plasma toward the magnetopause in the dawn sector, which generates a rarefaction wave (the slow hydromagnetic rarefaction wave), propagating in the anti-solar direction. Coroniti and Kennel (1972) suggested that when the rarefaction wave arrives at the X-line at the anti-solar end of the magnetotail, it triggers enhanced reconnection or that a new neutral line, perhaps stimulated by the rarefaction wave, forms much closer to the Earth and then propagates down the tail.

As we shall discuss in Section 9.2, the first possibility in their idea has some attractive features; in particular, it does not require the neutral line formation during an early epoch of the substorm, since we find no clear indication of the large-scale change of the magnetic field structure which is predicted by a number of workers, namely the formation of a new magnetic neutral line in the near-Earth plasma sheet. Further, it appears that their mechanism may work near the boundary between the diffuse and discrete auroras, rather than near the inner edge of the plasma sheet.

Kropotkin (1972) called attention to the presence of a sharp density gradient near the inner edge of the plasma sheet and showed that it results from three mechanisms: (i) the convective motion of plasma, (ii) the diffusion of plasma across the magnetic field lines in the high gradient region and (iii) particle losses in that region due to instabilities which cause the particle precipitation. After obtaining a steady state boundary condition, Kropotkin (1972) examined the possibility of the growth of the flute type instability at the boundary as the cause of the onset of the expansion phase.

As mentioned already, the region of the auroral arc, which brightens first at $T = 0$, is connected by geomagnetic field lines to a region within the plasma sheet (perhaps, $X \simeq -5 \sim -15\ R_E$), rather than to the inner boundary of the plasma sheet or the outer surface of the ring current belt. The theories proposed by Coppi *et al.* (1966) and Schindler (1974) consider a tearing mode instability within the plasma sheet. Schindler's theory was discussed in Section 5.4.3.

5.5. Magnetospheric Substorms

In an attempt to analyze a vast amount of both satellite and ground-based observations during magnetospheric substorms and to synthesize them systematically, it is logical to begin our study by reviewing observations which give supporting evidence for the neutral line formation and the enhanced reconnection. This is because a large number of satellite observations have been interpreted in terms of a new neutral line formation.

This approach does not necessarily imply that the reconnection theory has already been established. On the contrary, we shall see in the next chapter that some of the claimed evidence that the reconnection process takes place in the near-Earth plasma sheet is doubtful.

5.5.1. NEUTRAL LINE FORMATION – ENHANCED RECONNECTION OR PLASMA SHEET DEFLATION – ENHANCED RECONNECTION

The formation of a new magnetic neutral line in the near-Earth plasma sheet should give rise to: (i) a specific change of the magnetic field configuration (in particular, the $\boldsymbol{B}$ vector reversal beyond the distances of the newly formed neutral line), (ii) a specific change of the plasma flow pattern (toward and away from the Sun at locations inside and outside the neutral line, respectively), and (iii) the associated change of the configuration of the plasma sheet, namely the phenomenon called 'thinning'. We shall review carefully the observations which claim to find the above features in the first part of Chapter 6.

At the end of Chapter 6, we shall finish the review by concluding that the magnetosphere enhances the reconnection rate Φ_N in a two-step process, first by deflating the plasma sheet during the *expansive* phase and then enhancing Φ_N along the nightside X-line during the *recovery* phase, instead of forming a new neutral line in the near-Earth plasma sheet at $T = 0$.

5.5.2. ENHANCEMENT OF THE AURORAL OVAL CIRCUIT CURRENT

During substorms, it appears that the cross-tail current in the magnetotail is disrupted and that the disrupted current will find its way to the polar ionosphere, enhancing the current in the auroral oval circuit. The enhanced current in the oval circuit should cause specific magnetic perturbations in the magnetosphere (Figure 5.17). One of the purposes of Chapter 7 is to analyze magnetometer records from satellites at various locations, as well as from ground stations, and then to examine whether these observed magnetic field perturbations can be explained in terms of the magnetic field of an enhanced current along the auroral oval circuit, namely in terms of an enhanced S_q^p current.

An enhanced oval current forms concentrated electric currents along the

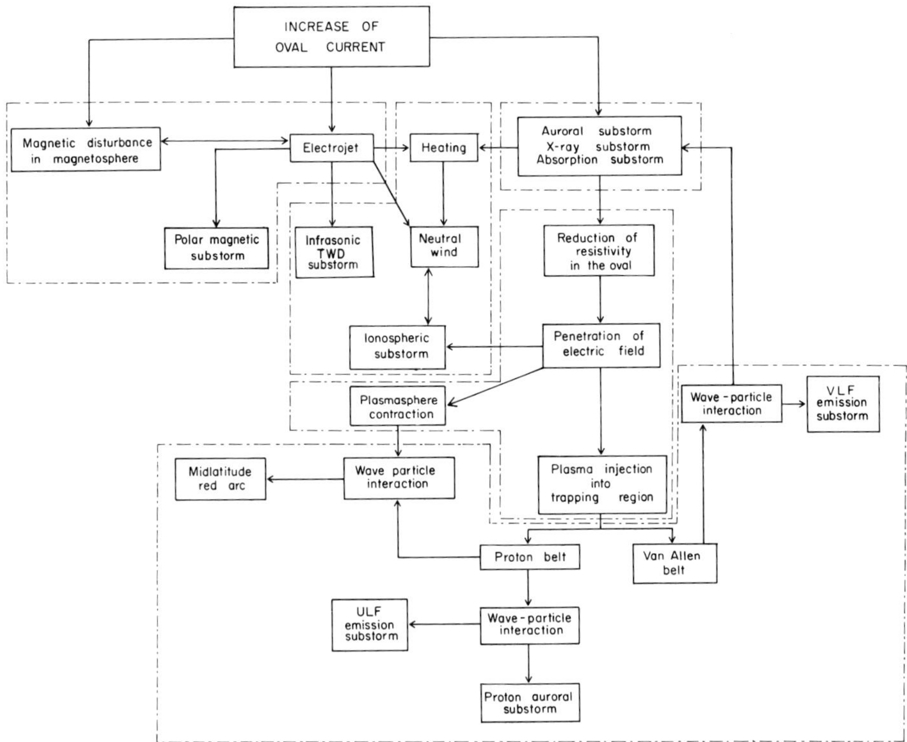

Fig. 5.17. Cause-effect relationship among some of the major features of the magnetospheric substorm.

auroral oval, the auroral electrojets. In the later half of Chapter 7, we shall examine in detail the auroral electrojet, and its relationship to the field-aligned currents and ionospheric electric fields.

Precipitating auroral particles seriously disturb the polar ionosphere by causing ionization. Most of the kinetic energy of auroral particles is eventually converted into thermal energy of ionospheric plasma. The power of the electrojet is dissipated as joule heat in the ionosphere. Thus the excess or stored energy in the magnetotail is eventually converted into heat energy in the polar ionosphere by these two processes. The heating of the polar upper atmosphere is expected to cause various phenomena, such as neutral wind, infrasonic waves and traveling wave disturbances (Figure 5.17).

5.5.3. PENETRATION OF THE CONVECTION ELECTRIC FIELD INTO THE INNER MAGNETOSPHERE AND THE RESULTING PLASMA INJECTION

One of the important processes associated with the magnetospheric substorm is a sudden penetration of the convection electric field into the inner magnetosphere, resulting in the injection of plasma from the plasma sheet into the trapping region. Protons thus injected become ring current protons, and the electrons are fed into the outer Van Allen belt. This process may result from an enhanced conductivity (or a reduced resistivity) in the auroral oval, which allows the primary (the polar dawn-dusk) electric field to penetrate deeper into mid-latitudes by short-circuiting the polarization space charges which tend to accumulate near the front of the plasma sheet. Injected plasma particles drift under a complicated combination of magnetic and electric fields. This chain of processes will be one of the major subjects in Chapter 8; see Figure 5.17.

In this view, however, the ionosphere does not play an active role in the injection processes. On the other hand, as suggested by Coroniti and Kennel (1972), it may play a crucial role in triggering processes of substorms. This subject will be discussed further in Section 9.2.2.

Ring current protons and Van Allen belt electrons are subject to plasma wave-particle interactions by which their pitch-angles can be altered significantly. As a result, some of them are dumped into the upper atmosphere. Particular ULF and VLF emissions have been identified as the responsible plasma waves for scattering protons and electrons. The wave-particle interaction processes in the magnetosphere have been extensively studied, both theoretically and observationally. Since this particular area in magnetospheric physics is now relatively well studied, details are left to a number of excellent review articles (cf. Fredricks, 1975; Gendrin, 1975).

The penetrated electric field is expected to cause also a large-scale deformation of the plasmasphere. Further, the interaction between cold plasma in the plasmasphere and hot plasma in the ring current develops a particular type of plasma instability. It has been suggested that the mid-latitude red arc results from such a process. The penetrating convection electric field can cause an intense ($E \times B$) drift motion of ionospheric plasma. These subjects will also be discussed in Chapter 8.

5.6. Geomagnetic and Magnetospheric Storms

Akasofu and Chapman (1963a) found that a geomagnetic storm period can be identified as the period when intense substorms occur frequently (cf. S.T.P., pp. 599–615). Since the excess energy available for substorms is closely related to the IMF B_z component, it is reasonable to expect that the B_z component is one of the important parameters which control the development of geomagnetic storms.

During the early part of the last decade, however, IMF data were not available. After an extensive analysis of the development of geomagnetic storms, Akasofu and Chapman (1963a) concluded: "The variety of development of the storms seems to suggest some intrinsic differences between the solar streams far beyond what we would expect from a mere difference between their pressures. The nature of their intrinsic difference is, at present, not known." It is now clear that one of the 'unknown' parameters which were sought by them is the IMF B_z component.

The development of geomagnetic storms and their relation to the IMF and other data have been studied by Hirshberg and Colburn (1969), Hirshberg *et al.* (1970), Grafe (1972), Burton *et al.* (1975), and most recently by Perreault (1974).

Here, we examine two geomagnetic storms which were studied by Perreault (1974). Figure 5.18 shows, from the top, the magnitude of the IMF magnitude $|B|$, the B_z component, the combined H component magnetic record from the standard auroral zone stations (the AE index) and the combined H component magnetic record from the standard low latitude stations, for the 1967, January 13–14 storm. In spite of its great intensity, the development of this particular storm was quite simple, except that the sudden commencement (ssc) was not a step function-like change. The main phase began at about 20 UT, when all the low latitude H component records began to show a rapid decrease. The storm reached the maximum epoch at about 02 UT on January 14, and then began to decay. The ring current became almost symmetric at about 10 UT.

The corresponding high latitude records show an intense substorm activity for about two hours after the ssc. Then the auroral zone became quiet for about 6 h, until about 20 UT, when an intense substorm activity began. The substorm activity appeared to cease rather abruptly at about 10 UT on January 14.

The magnitude of the IMF $|B|$ increased suddenly at the time of ssc and gradually decreased throughout the entire storm period, after reaching the maximum value at about 15 UT, January 13. By comparing the time variation of $|B|$ and of the storm intensity, it may be concluded that the magnitude $|B|$ did not play a major role as a parameter which controls the development of the main phase.

The B_z component had some rapid changes during the first few hours of the storm, attaining unusually large positive values for several hours. Then it began to decrease and became negative at about 20 UT. It remained negative until about 09 UT, when it increased very suddenly and became positive.

There is little doubt that the B_z component played an important role in the development of the main phase. The onset of the main phase, manifested by the decrease of the low latitude H component and also by the growth of the substorm activity, coincided roughly with the southward turning of the IMF $\boldsymbol{B}$ vector. On

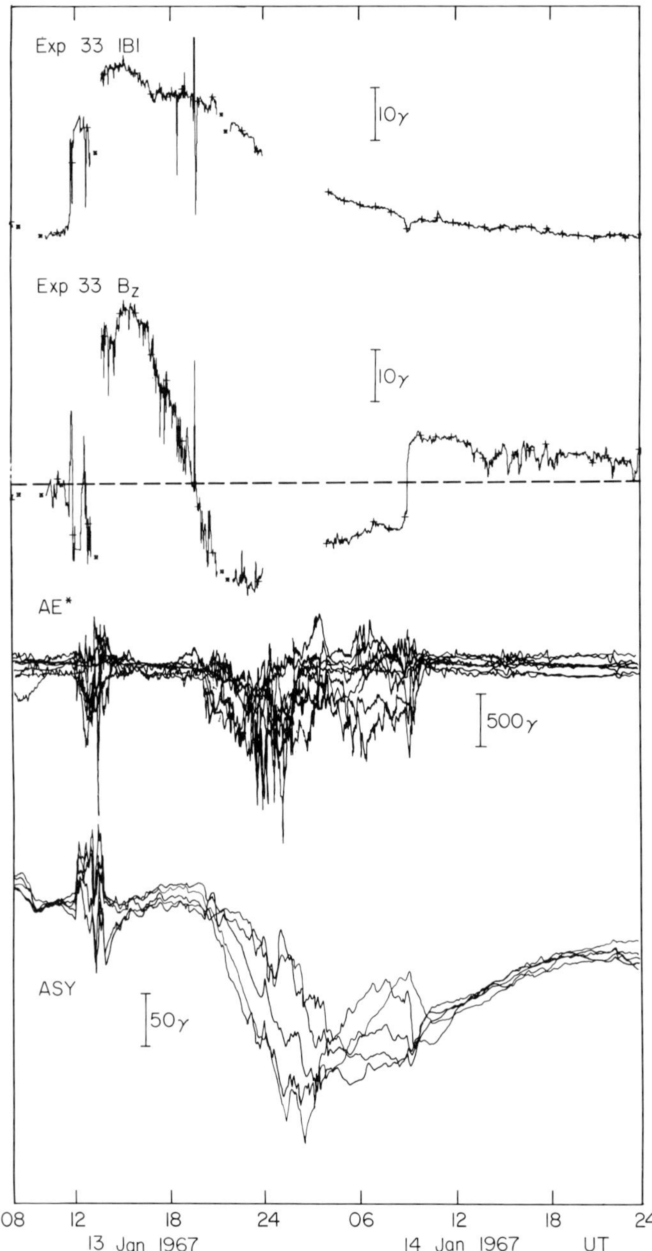

Fig. 5.18. IMF magnetic field changes ($|B|$, B_z) and the AE (AU, AL) and the Dst-Asy index during the geomagnetic storm of 1967, January 13–14. (Perreault, P. D.: Ph.D. Thesis, University of Alaska, August 1974.)

the other hand, the northward turning of the B_z component coincided roughly with the time when the ring current became symmetric and the substorm activity (expressed in terms of the AE index) ceased rather abruptly.

The substorm activity was too intense to distinguish individual substorms in the AE index during the storm. However, this does not necessarily mean that the corresponding auroral activity was continuous, as Akasofu and Chapman (1963b)

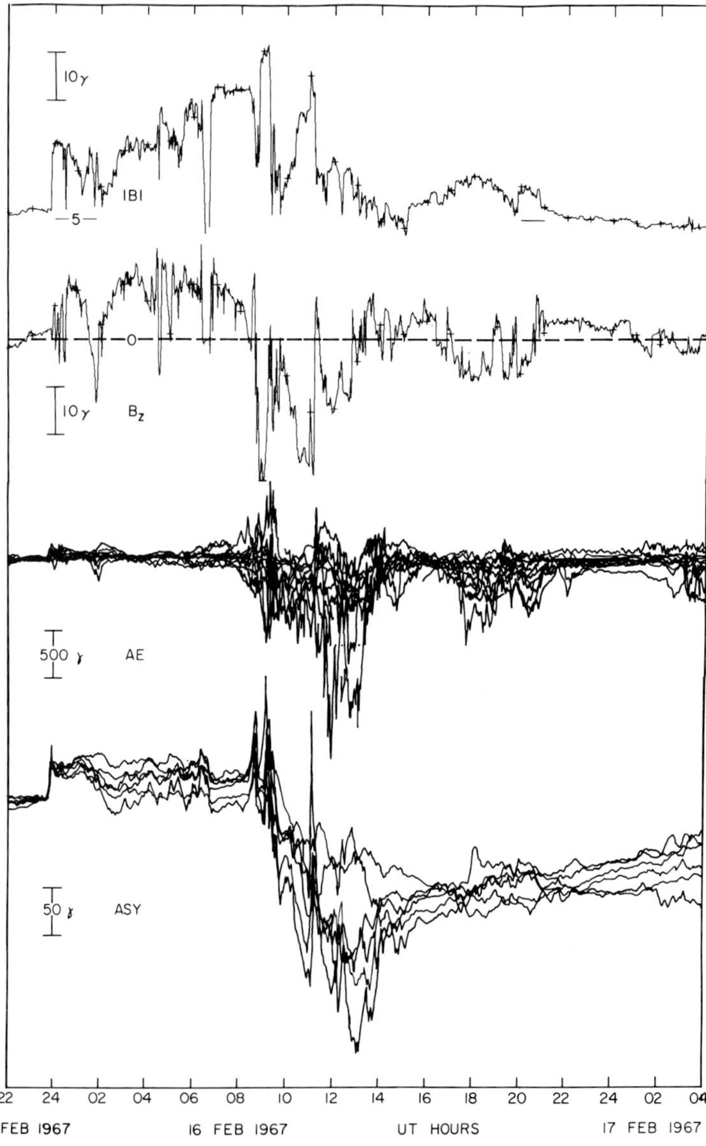

Fig. 5.19. IMF magnetic field changes ($|B|$, B_z) and the AE (AU, AL) and the Dst-Asy index during the geomagnetic storm of 1967, February 15–17. (Perreault, P. D.: Ph.D. Thesis, University of Alaska, August 1974.)

showed in their study of the great storm of 1958, February 11, one of the most intense magnetic storms during the last few decades.

Figure 5.19 shows both the interplanetary and ground magnetic field data for the magnetic storm of 1967, February 14–17. The development of this storm was more typical than that of 1967, January 13–14, although the IMF changes were more complex in this storm than in the previous one.

A step function-like increase of the *H* component, indicating the storm onset and a sudden compression of the magnetosphere, was recorded at all low latitude stations. The ssc was followed by a quasi-steady state, the initial phase, which lasted until about 09 UT. This relatively quiet period was suddenly disrupted by two large impulsive changes. The main phase began to develop soon after the end of the second impulse. There is little doubt that the magnetosphere was suddenly compressed at those times. Although it is not shown here, there occurred a sudden increase of helium (He^{++}), from a few percent to as large as about 25% at that time, doubling the solar wind pressure (Hirshberg *et al.*, 1970). The ring current became fairly symmetric at about 16 UT on February 16, although this tendency was interrupted between 18 and 22 UT. Again, there is no doubt that the B_z component played the major role in controlling the growth of the main phase.

Burton *et al.* (1975) obtained an empirical equation for the relationship between the *Y* component of the interplanetary electric field and the Dst index:

$$\frac{d}{dt}\mathrm{Dst}_0 = F(E) - a\,\mathrm{Dst}_0$$

where

$$\mathrm{Dst}_0 = \mathrm{Dst} - b(P)^{1/2} + c$$

$a = 3.6 \times 10^{-5}\ \mathrm{s}^{-1}$ (measure of ring current decay)

$b = 0.20\ \gamma\ (\mathrm{eV\ cm}^{-3})^{-1/2}$ (measure of the response to dynamic pressure changes in the solar wind)

$c = 20\ \gamma$ (measure of the quiet day currents)

$$F(E) = 0 \quad \text{for } E_y < 0.50\ \mathrm{mV\ m}^{-1}$$
$$= d(E_y - 0.5) \quad \text{for } E_y > +0.50\ \mathrm{mV\ m}^{-1}$$
$$d = -1.5 \times 10^{-3}\ \gamma(\mathrm{mV\ m}^{-1})^{-1}\ \mathrm{s}^{-1}$$
$$P = nV_s^2\ 10^{-2}\ \mathrm{eV\ cm}^{-3}$$

References

Akasofu, S.-I.: 1969, 'Magnetospheric Substorm as a Discharge Process', *Nature* **221**, 1020.

Akasofu, S.-I.: 1972, 'Magnetospheric Substorms: A Model', *Solar-Terrestrial Physics*, E. R. Dyer (ed.), p. 131, D. Reidel Publ. Co., Dordrecht-Holland.

Akasofu, S.-I.: 1975, 'The Roles of the North-South Component of the Interplanetary Magnetic Field on Large-Scale Auroral Dynamics Observed by the DMSP Satellite', *Planet. Space Sci.* **23**, 1349.

Akasofu, S.-I.: 1976, 'Recent Progress in Studies of DMSP Auroral Photographs', *Space Sci. Rev.* **19**, 169.

Akasofu, S.-I. and Chapman, S.: 1963a, 'Magnetic Storms: The Simultaneous Development of the Main Phase (DR) and of Polar Magnetic Substorms (DP)', *J. Geophys. Res.* **68**, 3155.

Akasofu, S.-I. and Chapman, S.: 1963b, 'The Lower Limit of Latitude (U.S. Sector) of Northern Quiet Auroral Arcs, and Its Relation to Dst (H)', *J. Atmospheric Terrestr. Phys.* **25**, 9.

Akasofu, S.-I., and Kamide, Y.: 1976, 'Substorm Energy', *Planet. Space Sci.* **24**, 223.

Alfvén, H. and Carlqvist, P.: 1967, 'Currents in the Solar Atmosphere and a Theory of Solar Flares', *Solar Phys.* **1**, 220.

Atkinson, G. J.: 1967, 'An Approximate Flow Equation for Geomagnetic Flux Tubes and Its Application to Polar Substorms', *J. Geophys. Res.* **72**, 5373.

Atkinson, G. J.: 1970, 'Auroral Arcs: Result of the Interactions of a Dynamic Magnetosphere with the Ionosphere', *J. Geophys. Res.* **75**, 4746.

Atkinson, G. J.: 1971, 'Magnetospheric Flows and Substorms', Review paper presented at the Advanced Study Inst. on Magnetosphere-Ionosphere Interactions, Dalseter, Norway, April 14–23.

Axford, W. I.: 1969, 'Magnetospheric Convection', *Rev. Geophys.* **7**, 421.

Boström, R.: 1972, 'Magnetosphere-Ionosphere Coupling', *Critical Problems of Magnetospheric Physics*, E. R. Dyer (ed.), p. 139, National Academy of Sciences, Washington, D.C.

Boström, R.: 1974, 'Ionosphere-Magnetosphere Coupling', *Magnetospheric Physics*, B. M. McCormac (ed.), p. 45, D. Reidel Publ. Co., Dordrecht-Holland.

Bowers, E. C.: 1973, 'A Short Wavelength Instability in the Neutral Sheet of the Earth's Geomagnetic Tail', *Astrophys. Space Sci.* **24**, 349.

Bowling, S. B.: 1975, 'Transient Occurrence of Magnetic Loops in the Magnetotail', *J. Geophys. Res.* **80**, 4741.

Burton, R. K., McPherron, R. L. and Russell, C. T.: 1975, 'An Empirical Relationship between Interplanetary Conditions and Dst', *J. Geophys. Res.* **80**, 4204.

Coppi, B., Laval, G. and Pellat, R.: 1966, 'Dynamics of the Geomagnetic Tail', *Phys. Rev. Lett.* **16**, 1207.

Coroniti, F. V. and Kennel, C. F.: 1972, 'Magnetospheric Substorms', *Cosmic Plasma Physics*, K. Schindler (ed.), p. 15, Plenum Press, New York.

Dungey, J. W.: 1961, 'Interplanetary Magnetic Field and the Auroral Zones', *Phys. Rev. Lett.* **6**, 47.

Fredricks, R. W.: 1975, 'Wave-Particle Interactions and Their Relevance to Substorms', *Space Sci. Rev.* **17**, 449.

Fukao, S. and Tsuda, T.: 1973, 'Re-Connection of Magnetic Lines of Force: Evolution in Incompressible MHD Fluids', *Planet. Space Sci.* **21**, 1151.

Furth, H. P., Killeen, J. and Rosenbluth, M. N.: 1963, 'Finite-Resistivity Instabilities of a Sheet Pinch', *Phys. Fluids* **6**, 459.

Gendrin, R.: 1975, 'Waves and Wave-Particle Interactions in the Magnetosphere: A Review', *Space Sci. Rev.* **18**, 145.

Gonzalez, W. D. and Mozer, F. S.: 1974, 'A Quantitative Model for the Potential Resulting from Reconnection with an Arbitrary Interplanetary Magnetic Field', *J. Geophys. Res.* **79**, 4186.

Grafe, A.: 1972, 'About the Connection between Equatorial Ring Current and Polar Electrojet', *Planet. Space Sci.* **20**, 183.

Hirshberg, J. and Colburn, D. S.: 1969, 'Interplanetary Field and Geomagnetic Variations – A Unified View', *Planet. Space Sci.* **17**, 1183.

Hirshberg, J., Alksne, A., Colburn, D. S., Bame, S. J. and Hundhausen, A. J.: 1970, 'Observation of a Solar Flare Induced Interplanetary Shock and Helium Enriched Driven Gas', *J. Geophys. Res.* **75**, 1.

Holzworth, R. H. and Meng, C.-I.: 1975, 'Mathematical Representation of the Auroral Oval', *Geophys. Res. Lett.* **2**, 377.

Kamide, Y. and Brekke, A.: 1975, 'Auroral Electrojet Current Density Deduced from the Chatanika Radar and from the Alaska Meridian Chain of Magnetic Observatories', *J. Geophys. Res.* **80**, 587.

Kennel, C. F.: 1969, 'Consequences of a Magnetospheric Plasma', *Rev. Geophys.* **7**, 379.

Kropotkin, A. P.: 1972, 'On the Physical Mechanism of the Magnetospheric Substorm Development', *Planet. Space Sci.* **20**, 1245.

Liu, C. S.: 1970, 'Low-Frequency Drift Instabilities of the Ring Current Belt', *J. Geophys. Res.* **75**, 3789.

McPherron, R. L., Russell, C. T. and Aubry, M. P.: 1973, 'Satellite Studies of Magnetospheric Substorms on August 15, 1968, 9. Phenomenological Model for Substorms', *J. Geophys. Res.* **78**, 3131.

Nishida, A.: 1968, 'Geomagnetic D_p^2 Fluctuations and Associated Magnetospheric Phenomena', *J. Geophys. Res.* **73**, 1795.

Perreault, P. D.: 1974, 'On the Relationship between Interplanetary Magnetic Fields and Magnetospheric Storms and Substorms', Ph.D. Thesis, University of Alaska, August.

Petschek, H. E.: 1964, 'Magnetic Field Annihilation', *AAS-NASA Symposium on the Physics of Solar Flares*, NASA Spec. Pub. SP-50, 425.

Piddington, J. H.: 1967, 'Magnetic Field Annihilation in Current Pinches', *Planet. Space Sci.* **15**, 733.

Sato, T.: 1974, 'Response Theory of Solar Wind-Magnetosphere-Ionosphere Interactions', GRL-74-01, August.

Scarf, F. L., Frank, L. A., Ackerson, K. L. and Lepping, R. P.: 1974, 'Plasma Wave Turbulence at Distance Crossings of the Plasma Sheet Boundaries and the Neutral Sheet', *Geophys. Res. Lett.* **1**, 189.

Schindler, K.: 1972, 'A Self-Consistent Theory of the Tail of the Magnetosphere', *Earth's Magnetospheric Processes*, B. M. McCormac (ed.), p. 200, D. Reidel Publ. Co., Dordrecht-Holland.

Schindler, K.: 1974, 'A Theory of the Substorm Mechanism', *J. Geophys. Res.* **79**, 2803.

Schindler, K.: 1975, 'Plasma and Fields in the Magnetospheric Tail', *Space Sci. Rev.* **17**, 589.

Schindler, K. and Ness, N. F.: 1972, 'Internal Structure of the Geomagnetic Neutral Sheet', *J. Geophys. Res.* **77**, 91.

Schindler, K. and Soop, M.: 1968, 'Stability of Plasma Sheaths', *Phys. of Fluids*, **11**, 1192.

Sonnerup, B. U. Ö.: 1970, 'Magnetic-Field Re-Connection in a Highly Conducting Incompressible Fluid', *J. Plasma Phys.* **4**, 161.

Sonnerup, B. U. Ö.: 1974, 'The Reconnecting Magnetosphere', *Magnetospheric Physics*, B. M. McCormac (ed.), p. 22, D. Reidel Publ. Co., Dordrecht-Holland.

Speiser, T. W.: 1970, 'Conductivity without Collisions or Noise', *Planetary Space Sci.* **18**, 613.

Speiser, T. W.: 1973, 'Magnetospheric Current Sheets', *Radio Sci.* **8**, 973.

Svalgaard, L.: 1973, 'Polar Cap Magnetic Variations and Their Relationship with the Interplanetary Magnetic Sector Structure', *J. Geophys. Res.* **78**, 2064.

Swift, D. W.: 1967, 'The Possible Relationship between the Auroral Breakup and the Interchange Instability of the Ring Current', *Planetary Space Sci.* **15**, 1225.

Syrovatskii, S. I.: 1972, 'Origin of the Geomagnetic Tail and Neutral Sheet', *Critical Problems of Magnetospheric Physics*, E. R. Dyer (ed.), p. 35, National Academy of Sciences, Washington, D.C.

Yeh, T. and Axford, W. I.: 1970, 'On the Re-Connection of Magnetic Field Lines in Conducting Fluids', *J. Plasma Phys.* **4**, 207.

CHAPTER 6

MAGNETOTAIL PHENOMENA DURING MAGNETOSPHERIC SUBSTORMS

6.1. Introduction

The purpose of this chapter is to examine first of all whether satellite observations in the magnetotail can be interpreted in terms of the formation of a new magnetic neutral line and the subsequent enhancement of reconnection along the scheme outlined in Section 4.4.1. For this purpose it may be worthwhile to describe briefly the most widely conjectured model of magnetotail phenomena during the magnetospheric substorm. The model is well illustrated in Figure 6.1, but similar models have been proposed by a number of other workers (*cf.* Hones, 1973, 1975; McPherron *et al.*, 1973b). These workers suggest: A southward turning of the interplanetary magnetic field is accompanied by erosion of the dayside magnetosphere, flux transport to the geomagnetic tail, and a gradual thinning and inward motion of the plasma sheet. As a result, at about $T = -30$ min, "The plasma sheet has become very thin over its whole length beyond $X \simeq -10$ to $-15\ R_E$. At $T = 0$, a neutral line forms at $X \simeq -10$ to $15\ R_E$. Reconnection of oppositely directed field lines starts there and produces rapid flow of plasma both earthward and anti-earthward from its location" (Hones, 1973). Thus, "The expansion phase of substorms can originate near the inner edge of the plasma sheet as a consequence of rapid plasma sheet thinning. At this time a portion of the inner edge of the tail current is 'short circuited' through the ionosphere. This process is consistent with the formation of a neutral point in the near-tail region" (McPherron *et al.*, 1973). The neutral line remains at the same location up to 1 h or more and then shifts far outward in the tail to cause the plasma sheet expansion in a late epoch of substorm. In a review paper, Schindler (1975) summarized these conjectures: "There are a number of observational facts that seem to be consistent with the concept of spontaneous merging. A number of authors interpreting experimental results have independently concluded that the break-up of large substorms is in fact associated with the formation of a neutral line in the near-Earth part of the tail"; see also Akasofu (1974b).

However, we shall conclude at the end of this chapter that there is no definite indication of the formation of a new magnetic neutral line in the near-Earth plasma sheet which leads to the large-scale change of the magnetic field configuration suggested in Figure 6.1. Therefore, we are confronted with the problem of finding a new mechanism for the plasma sheet thinning and for the auroral bulge formation during the expansive phase, without invoking an enhanced reconnection.

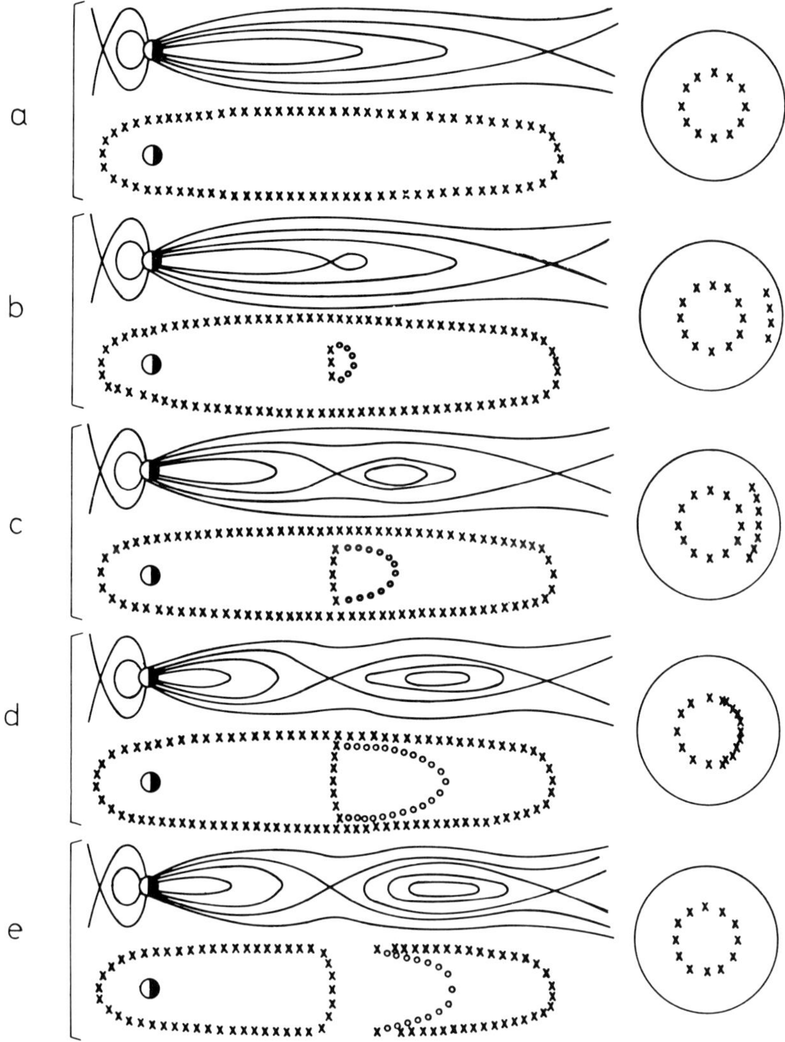

Fig. 6.1. Most widely conjectured model of magnetotail variations during the magnetospheric substorm. (Courtesy of Vasyliunas, V. M.)

6.2. *B* Vector Dipping

The most certain indication of the formation of a new magnetic neutral line in the magnetotail may perhaps be the occurrence of the ***B*** *vector reversal* ($|\boldsymbol{B}| = B_z < 0$). Magnetic field variations in the magnetotail during substorms have been studied extensively by Camidge and Rostoker (1970), Nishida (1973), Nishida and Nagayama (1973), Nishida and Hones (1974), Nishida and Nagayama (1975) and others. In Section 5.4.3, we noted Schindler's suggestion that a number of small magnetic loops (formed by ion-tearing mode instability) may eventually combine to form the large-scale change of the magnetic field structure (Figure 6.1).

The first systematic study of the B_z component changes in the plasma sheet was made by Nishida and Nagayama (1975). They reported that the occurrence of the B_z component reversal is most apparent at geocentric distances $-25\,R_E > X > -35\,R_E$, where the B_z component becomes of order -5γ, and the reversal lasts for 1 h or so. At $X < -35\,R_E$, the magnitude of the negative B_z component becomes smaller than that at $-25\,R_E > X > -35\,R_E$. Figure 6.2(a) shows a collection of the three component changes at distances $-25\,R_E > X > -35\,R_E$. The negative B_z values are shaded for emphasis.

In an attempt to prove that the occurrence of the B_z component reversal is not due to the dipping of the neutral sheet or to the formation of turbulent loops of magnetic field lines in the plasma sheet, Nishida and Nagayama plotted the B_z component as a function of Z and concluded that the occurrence of the B_z component reversal occurs more frequently than the positive B_z component at $Z < 5\,R_E$. They also called attention to the fact that the disappearance of the negative B_z component is rather abrupt, taking place in less than 20 min. Nishida and Nagayama (1975) also suggested that the length of the neutral line is often limited in longitude and that it does not extend to the magnetopause. Burke and Reasoner (1973) also reported the B_z component reversal during substorms at the lunar distance.

Since the above observations are crucial in determining whether a new magnetic neutral line is formed deep in the plasma sheet, it is worthwhile to re-examine carefully the data studied by Nishida and Nagayama (1973) with higher time resolution data. Figures 6.2(b), (c), (d) and (e) present the results of this new effort. In Figure 6.2(b), it can be seen that the negative B_z component was associated with a large increase of the magnitude of the field (B_T) and a sharp drop of energetic electron fluxes, as well as a large increase of the AE index. These features are common to almost all the examples presented by Nishida and Nagayama (1973).

There is little doubt that these observed features occur when the boundary of the thinning plasma sheet is passing the location of the satellite (Section 6.3). The magnetic field vector tends to dip toward the equatorial plane near the boundary of the thinning plasma sheet. In most cases, this dip angle is very small (~10°); in this respect, Figure 6.2(b) shows an exceptionally large dip angle (~45°) for a few minutes. Such a slight *dipping* of the magnetic field vector should not be confused with the ***B** vector reversal* and is hardly a conclusive proof of the neutral line formation. The examples provided by Burke and Reasoner (1973) show also that the dip angle is of order 20–30°. Thus, the problem in the early studies was that the authors examined only the B_z component and did not distinguish between the $\boldsymbol{B}$ vector reversal ($|\boldsymbol{B}| = B_z < 0$) and the B_z component reversal (namely, the dipping, $B_z < 0$, $|\boldsymbol{B}| \gg |B_z|$).

In the following example, we shall see in detail how the $\boldsymbol{B}$ vector changes during a similar event. Figure 6.3(a) shows the simultaneous measurements of the energetic electrons ($E > 18$ keV) and the magnetic field (in the solar magnetospheric coordinates) from the IMP-6 satellite and the AE index during a thinning event on 1971, September 27. The occurrence of a substorm was roughly indicated by an enhancement of the AE index starting at ~2025 UT. The energetic electron

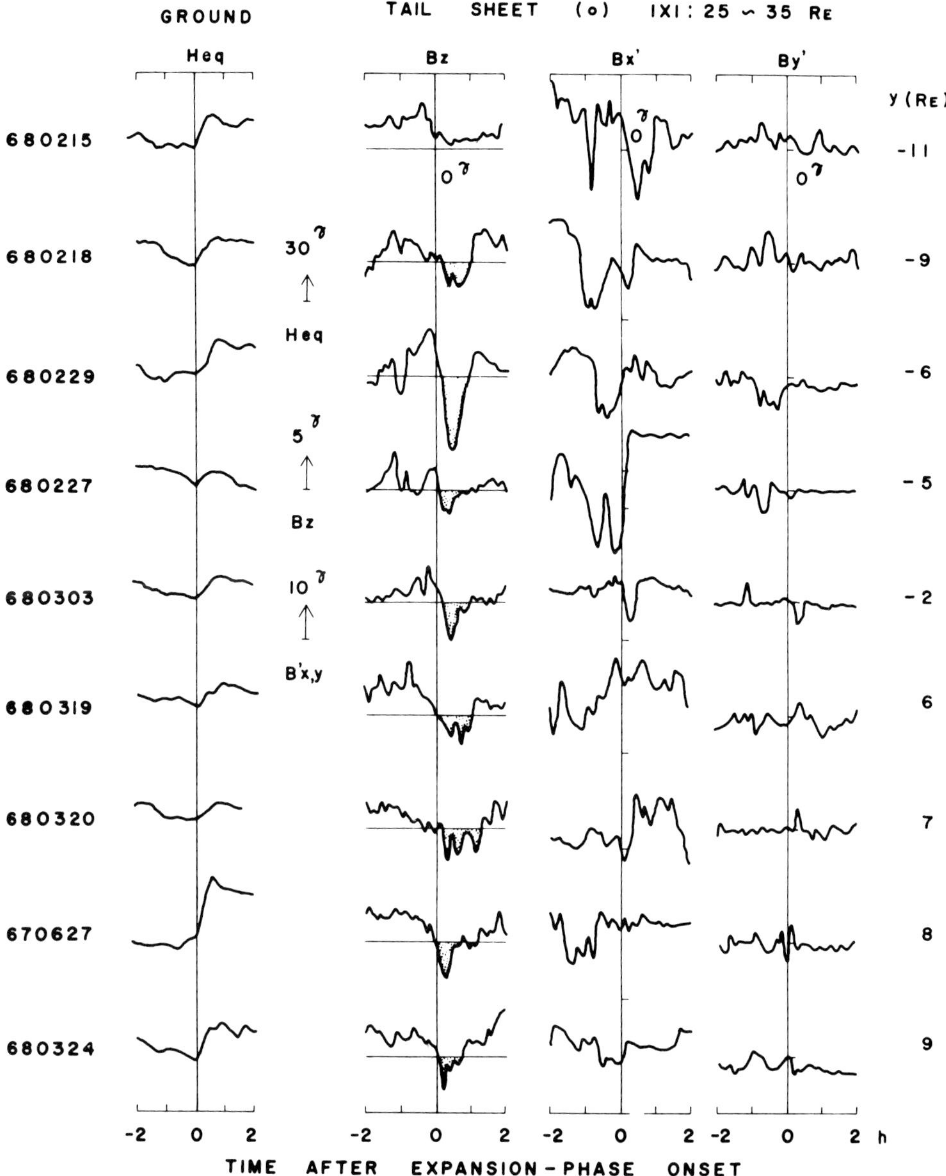

Fig. 6.2(a). Magnetic field (B_z, B_x, B_y) variations in the magnetotail during magnetospheric substorms. The onset time of the substorms is assumed to be the onset time of positive bays in low latitudes (right column). (Nishida, A. and Nagayama, N.: *J. Geophys. Res.* **78**, 3782, 1973.)

flux started to decrease at ~2004 UT, and a substantial drop occurred at 2021 UT. The B_z component was negative, starting at 2005 UT, and reached its minimum of $-2\,\gamma$ at ~2055 UT. The angle $\alpha = \tan^{-1}(B_z/|B_x|)$ at this time was about $-5°$, indicating that the magnetic field vector at this time dipped only very slightly southward. The angle α is used since it shows precisely the variation of the

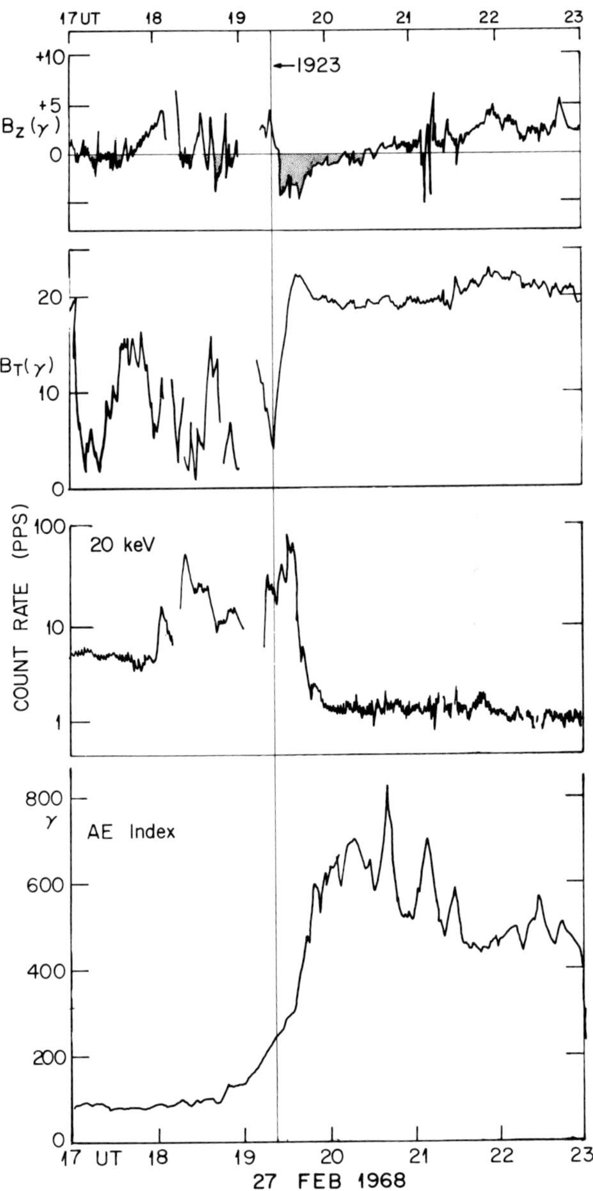

Fig. 6.2(b, c, d, e). High time resolution data of Figure 6.2(a), together with the simultaneous energetic electron data and the AE index. (Courtesy of Lui, A. T. Y., Meng, C.-I. and Akasofu, S.-I.)

magnetic field orientation on the plane parallel to the noon-midnight meridian plane. The B_z component started to increase after 2055 UT while the energetic electron flux was still decreasing. The slow decrease of energetic electron flux is an indication of partial thinning, i.e., the plasma sheet boundary approached the satellite but did not completely overtake it. At 2209 UT, the energetic electron flux

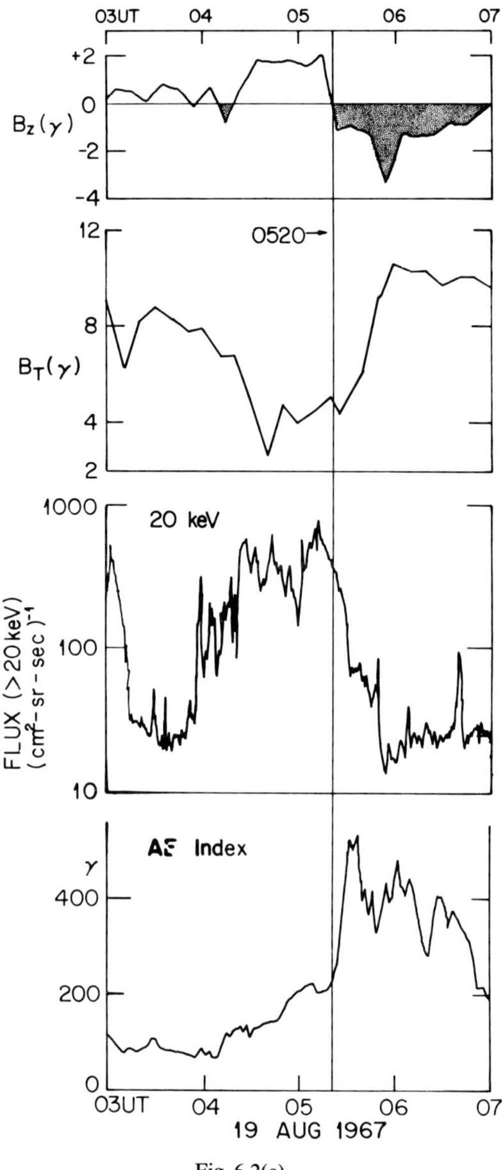

Fig. 6.2(c).

started to recover abruptly. At this time, the B_z component had already increased to about 2 γ. The complete recovery of the energetic electron flux occurred at about 2235 UT, which was immediately followed by further increases in the B_z component. The angle α of the magnetic field vector was as large as 68° for a brief period at the full recovery of the energetic electron flux (2235 UT).

The variation of the magnetic field during this interval is further elucidated by the vector plots in Figure 6.3(b). The projection of the observed magnetic field

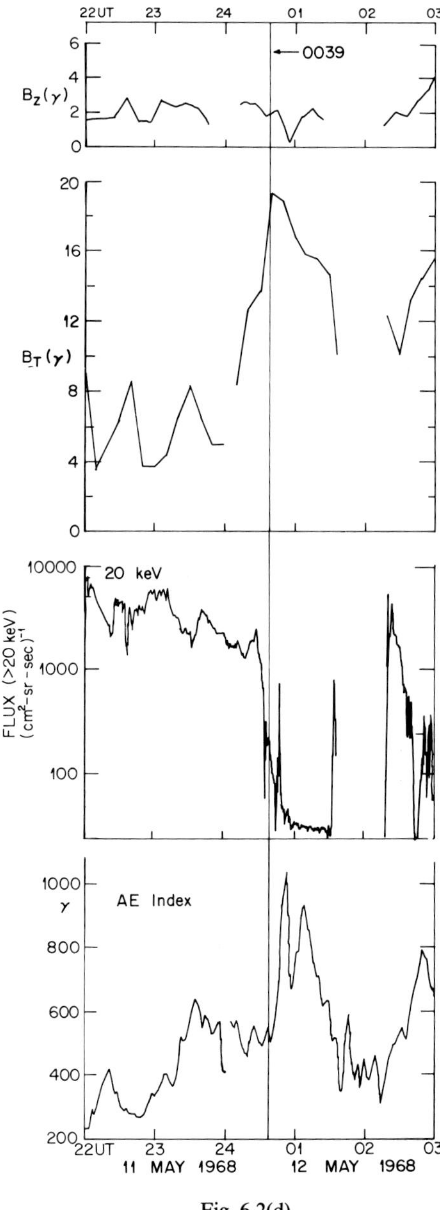

Fig. 6.2(d).

vectors on the X-Z plane is drawn with the time axis oriented horizontally and vertically in the upper and lower presentations, respectively. Each vector is constructed from magnetic field measurements averaged over 15.36 s. For clarity purposes, only one vector from every eighth data point is selected for plotting, and this corresponds to a spacing of about 2 min between successive vectors. It is clear from the upper presentation in Figure 6.3(b) that a slight southward dipping

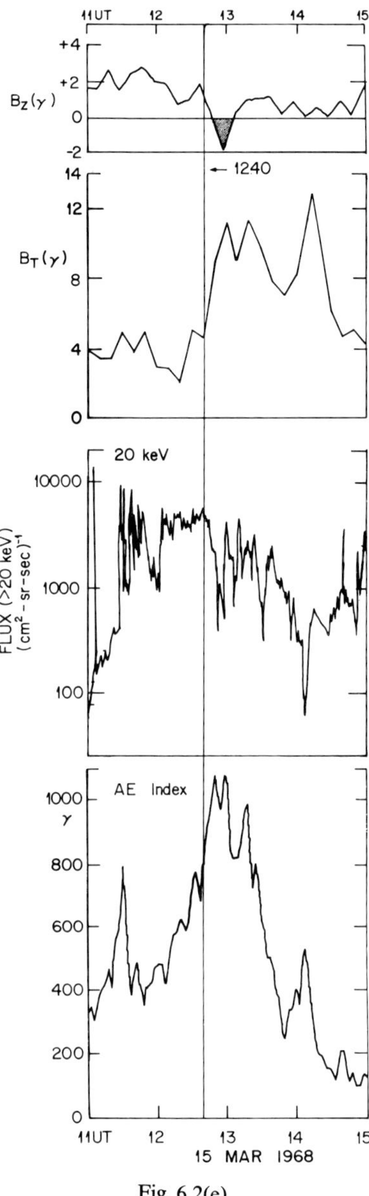

Fig. 6.2(e).

of the magnetic field vector was observed from 2020 to 2050 UT during the thinning of the plasma sheet. The start of the southward dipping was associated with an increase in the magnitude of the magnetic field, as shown in the lower plot. The plasma sheet thinning can be recognized from the lower vector plot by noting the increased magnitude of the magnetic field vectors. Northward rotation of the magnetic field vector was seen from 2210 to 2300 UT, namely during the

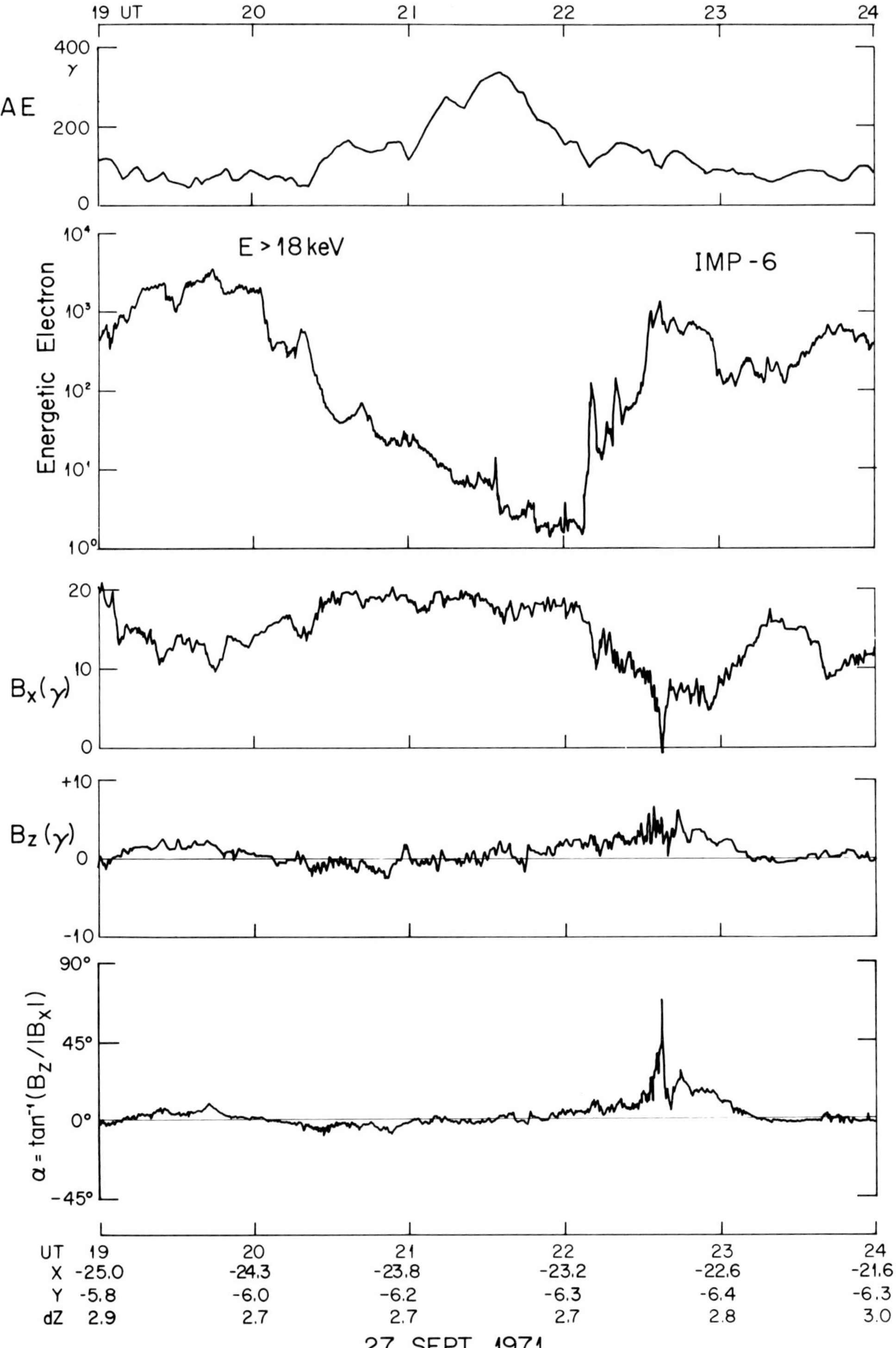

Fig. 6.3(a). Thinning and the subsequent expansion of the plasma sheet were observed by the IMP-6 satellite during a substorm on 1971, September 27. The hourly location of the satellite is given in solar magnetospheric coordinates at the bottom of the figure. The dZ coordinate is the estimated distance from the neutral sheet. During the early part of the plasma sheet thinning interval, negative B_z was detected and was associated with a slight southward dipping of the magnetic field vector as indicated by the variation of the angle α of the magnetic field.

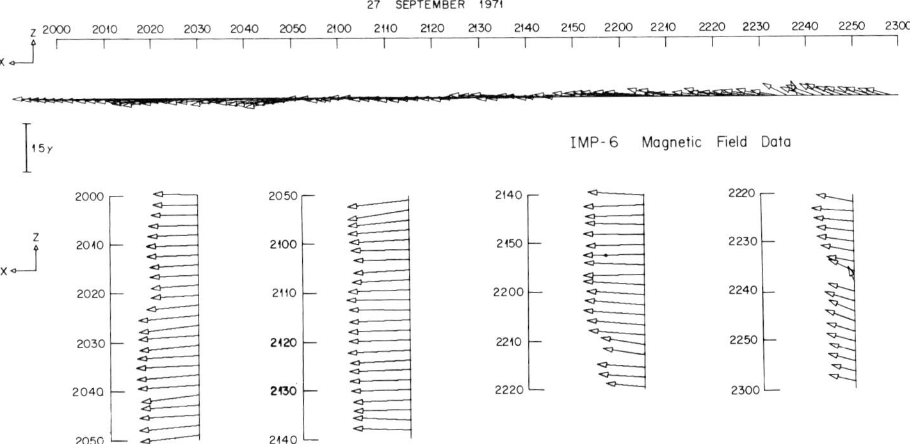

Fig. 6.3(b). Magnetic field vectors projected on the X-Z plane (solar magnetospheric coordinates) are plotted for the substorm period on 1971, September 27. The X component is plotted horizontally, and the Z component is plotted vertically. The time axis is oriented horizontally in the upper presentation and vertically in the lower presentation. The scale for the magnetic field magnitude is given at the left of the figure. The spacing between successive vectors is about 2 min. Slight southward dipping is observed during plasma sheet thinning, and large northward rotation is observed during plasma sheet expansion.

expansion of the plasma sheet. Note that the magnitude of the magnetic field vectors was considerably decreased by diamagnetism of the plasma sheet.

It may be argued that the dipping results from the formation of a new magnetic neutral line inside the location of the satellite, since it has been suggested that the plasma sheet would be pinched most seriously at the location where the neutral line is being formed (see Figure 6.1). However, it can be shown that such a dipping is common even at $X > -10\ R_E$.

In the following example, the satellite was inbound in the near-Earth region. This was indicated also by a rather steady increase in the magnitude of B_x given in Figure 6.3(c). Thinning of the plasma sheet at the IMP-6 location was detected by a gradual decrease of energetic electrons starting at 0800 UT in association with an increase in the AE index. Simultaneous with the decrease in energetic electrons was the decrease in B_z. The magnetic field was observed to dip southward from 0802 to 0900 UT. The largest negative B_z value and α was $-9\ \gamma$ and $-11°$, respectively. There was a brief recovery of energetic electrons from 0842 to 0852 UT. Active auroras were observed over Canada during this growth phase-like period. A very transient, but large, enhancement of AE was observed at ~0900 UT and was accompanied by a rapid decrease of energetic electrons. The final recovery of the energetic electrons occurred at 0905 UT when the AE index started to decrease and the B_z component was positive and increasing rapidly.

The vector plots showing the magnetic field variations during this thinning event are given in Figure 6.3(d). The change in the scale for the magnetic field magnitude in the vector plots is necessary to accommodate the large values of the field components in the near-Earth region. It is important to note that southward dipping of the magnetic field observed during this thinning event is very similar to

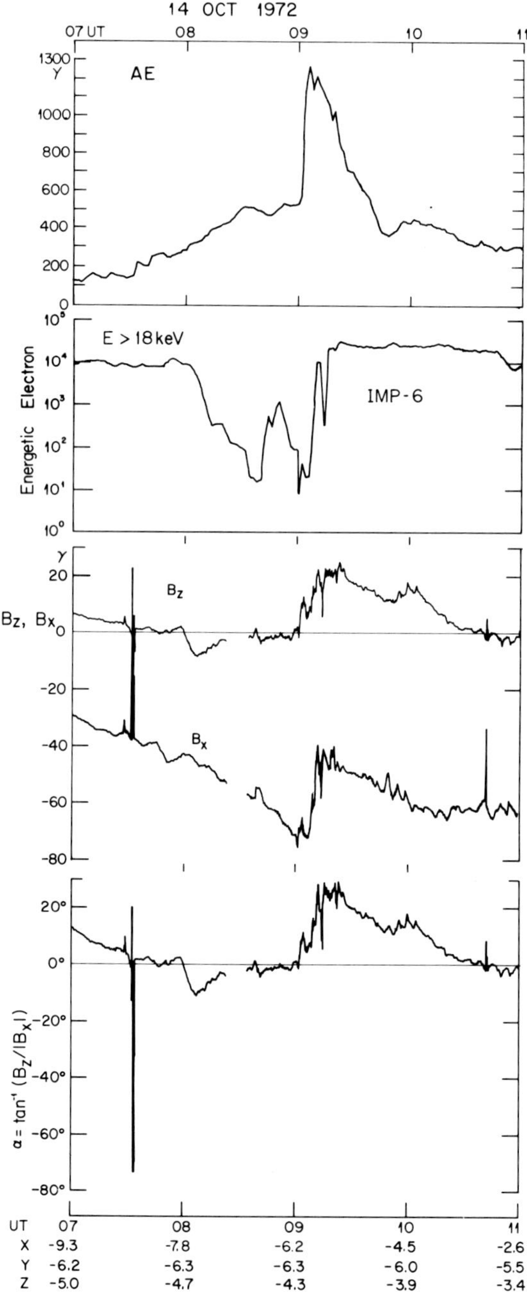

Fig. 6.3(c). Thinning of the plasma sheet detected in the near-Earth region during a substorm on 1972, October 14. The large fluctuation of all magnetic field parameters at 0732 UT was probably noise. Note that a slight southward dipping of the magnetic field was observed during plasma sheet thinning.

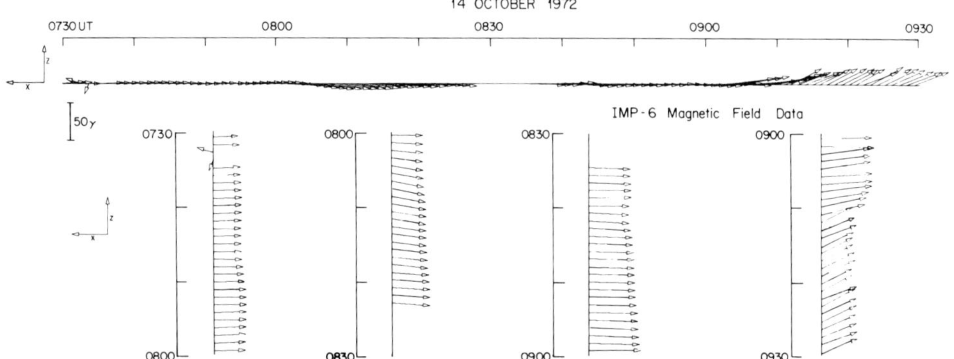

Fig. 6.3(d). Plots showing the slight southward dipping and large northward rotation during thinning and expansion of the plasma sheet, respectively, on 1972, October 14. Note that the observed changes in the magnetic field direction are essentially the same as those observed in the distant magnetotail at $X = -25$ to -30 R_E.

the southward dipping observed during the three thinning events previously shown. However, the location of the satellite during this thinning event was at $X \simeq -6.2$ to -7.8 R_E, too close to the Earth to interpret the observed southward dipping to be due to the formation of a magnetic neutral line earthward of the satellite.

We have so far shown that the negative B_z component during plasma sheet thinning results from a slight southward dipping of the magnetic field near the boundary of the thinning plasma sheet, rather than from the reversal of the total $\boldsymbol{B}$ vector. However, magnetic field changes in the vicinity of the midplane (the so-called neutral sheet) are most crucial in determining whether the southward dipping near the plasma sheet boundary implies the reversal of the B_z ($\simeq B$) component in the midplane.

Here, we examine three events in which the IMP-6 satellite was at $X \simeq -25$ to -30 R_E and was in the vicinity of the midplane during the substorm expansive phase. In this distance range, the suggested formation of a near-Earth neutral line between $X \simeq -10$ and -20 R_E would result in the $\boldsymbol{B}$ vector pointing predominantly southward.

Therefore, validity of such a conjecture can be examined by a study of magnetic field variations when the satellite is located near the midplane by noting particularly the sign of B_z component at the midplane crossing. This approach avoids complications introduced by tilting and flapping motions of the midplane.

Figure 6.4(a) shows two intervals, 0310–0335 UT and 0455–0545 UT, in which the energetic electron flux was decreasing, indicating thinning of the plasma sheet. The two events were clearly accompanied by simultaneous enhancements in the AE index. The changing sign of the B_x component and the small values of B_T observed by the satellite during this period showed that the satellite was near the midplane. The satellite crossed the midplane four times during the first thinning event and at least ten times during the second event.

The B_z component was negative when the energetic electron flux first started

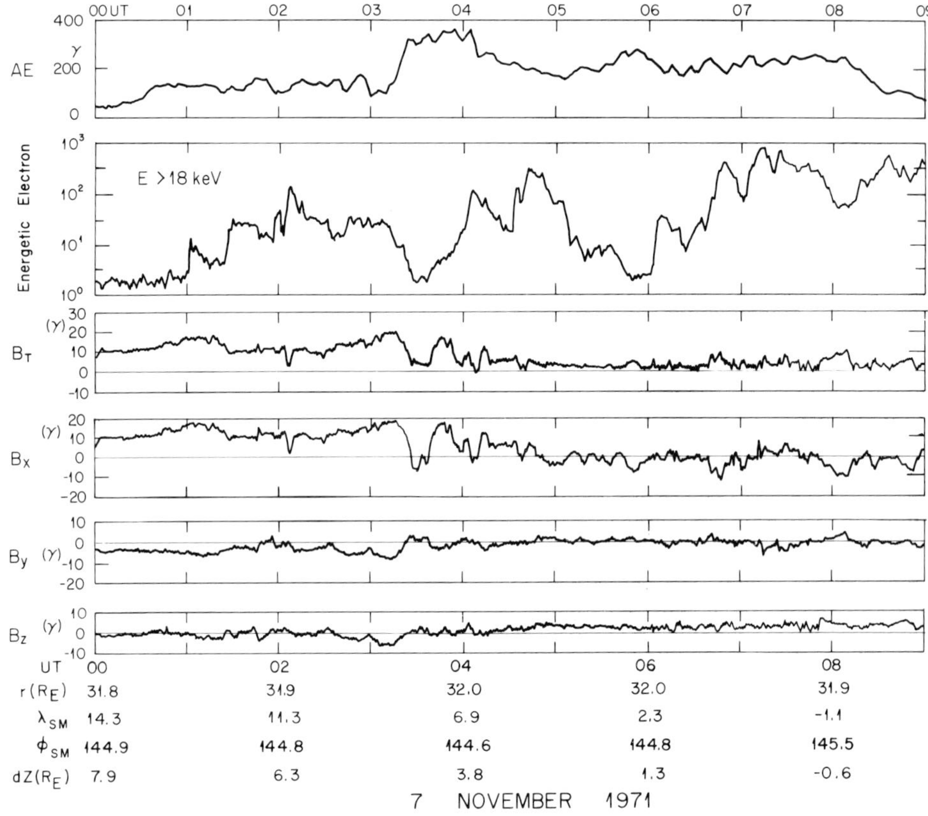

Fig. 6.4(a). Two intervals of plasma sheet thinning, 0310–0335 UT and 0455–0545 UT, during substorms were observed by the IMP-6 satellite on 1971, November 7. The satellite crossed the midplane (the so-called neutral sheet) four times during the first interval and ten times during the second interval of plasma sheet thinning. The B_z component during the midplane crossings was predominantly positive. (Courtesy of Lui, A.T.Y., Meng, C.-I. and Akasofu, S.-I.)

to decrease. However, the total magnetic field magnitude was large ($\sim 20\ \gamma$) at that time, indicating that the plasma sheet became very thin. The satellite was approaching the midplane during the period when the energetic electron flux continued to decrease. The satellite crossed the midplane at about 0320 UT. During this crossing, as well as at the time of the following three crossings, the B_z component was predominantly positive, although a very brief period of negative B_z occurred around 0337 UT after the midplane crossings, when the magnitude of the total magnetic field was relatively large. The actual orientation of the magnetic field vector on the X-Z plane can be seen in Figure 6.4(b). The crossings of the midplane are readily seen in the lower vector plot and the B_z component was positive during these crossings.

During the second thinning, the B_z component was positive ($\sim +5\ \gamma$) and very steady. Therefore, there was no indication that the total magnetic field vector was

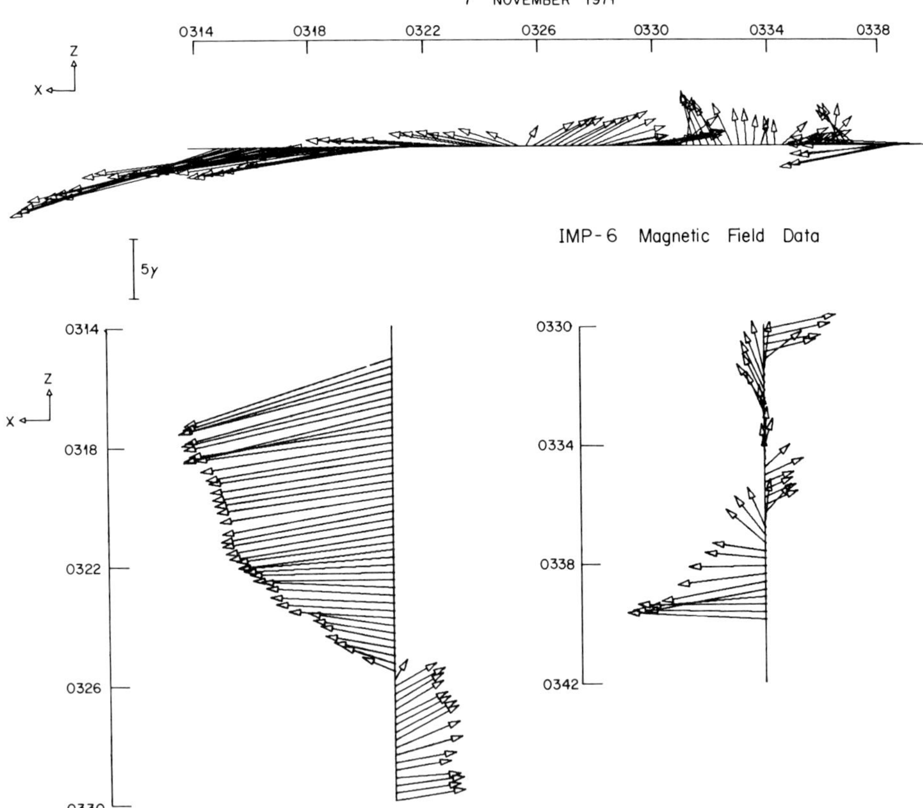

Fig. 6.4(b). Plot showing the magnetic field vectors projected on the X-Z plane during midplane crossings in the first interval of plasma sheet thinning on 1971, November 7.

pointing southward during the two events, as one would expect from the formation of a neutral line at $X > -25\ R_E$.

In the next example (1971, November 7), the AE index started to increase at about 0950 UT and was accompanied by a decrease in the energetic electron flux at the IMP-6 location (see Figure 6.4(c)). The simultaneous magnetic field measurements indicate that the satellite was near the midplane between 0950 and 1025 UT. Crossing of the midplane occurred at least eleven times within this brief plasma sheet thinning period.

The magnetic field vectors on the X-Z plane during this period of interest are shown in Figure 6.4(d). The crossing of the midplane can be readily seen from the lower vector plot and the sign of B_z during each crossing can be determined. The vector plots indicate that only four out of the eleven crossings show southward B_z. The other seven crossings show northward B_z. Therefore, the B_z component was predominantly positive during this period, *opposite* to what one would expect from the formation of a neutral line in the near-Earth plasma sheet.

In the next example (1971, October 30), the energetic electron flux started to

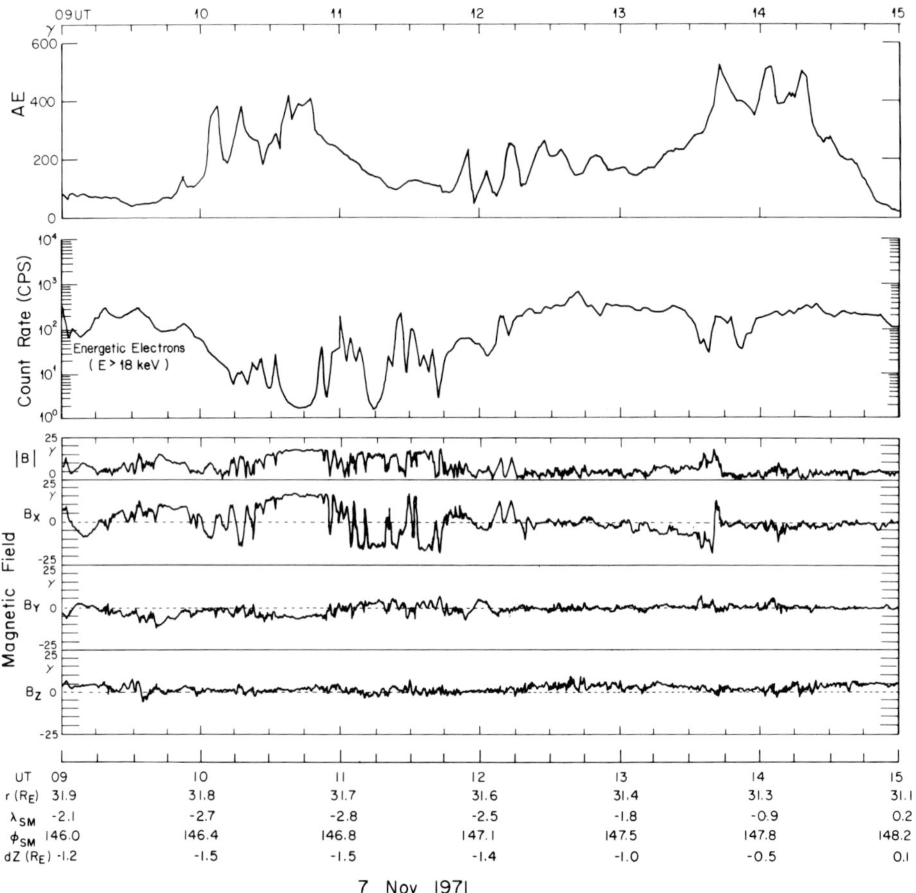

Fig. 6.4(c). The IMP-6 satellite was near the midplane between 0950 and 1025 UT during thinning of the plasma sheet on 1971, November 7.

decrease at about 1440 UT and an enhanced AE index was simultaneously observed (see Figure 6.5(a)). The observed magnitude of the magnetic field was low at that time, testifying that the satellite was in the plasma sheet. Thinning of the plasma sheet lasted until about 1507 UT. During this interval, the B_z component was negative for only about 3 min during an *early epoch of thinning*, from 1444 to 1447 UT (see Figure 6.5(b)), and was positive for the rest of the time (namely, as long as 24 min). Again, it is difficult to explain the observation in terms of the neutral line formation in the near-Earth plasma sheet. Note that this is an intense substorm since the AE index at the maximum epoch was about 900 γ.

In addition to the above study of individual events, a statistical study of midplane crossings at $X < -20\ R_E$ during substorm expansive phase over the same period of data from the IMP-6 satellite was conducted. A total of 90 midplane crossing events have been identified in which the sign of B_z at the crossing can be determined. Results show that 81% of crossings indicate northward B_z, as

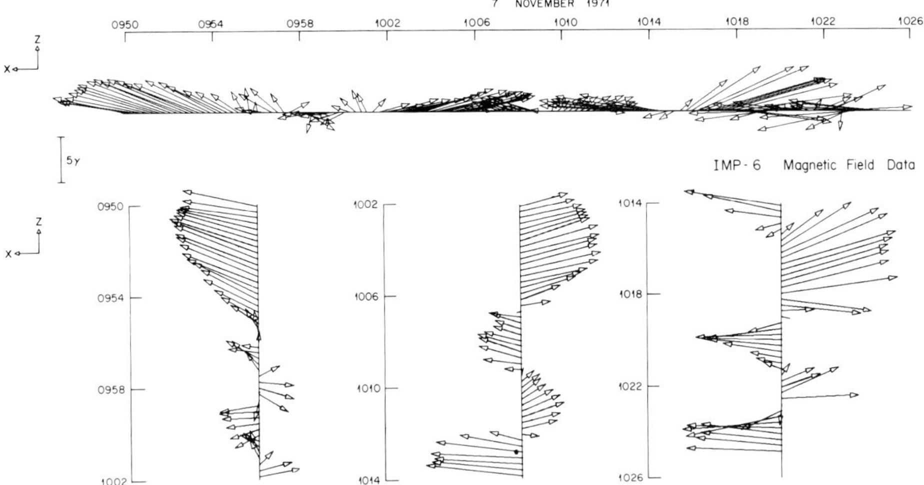

Fig. 6.4(d). Projection of the magnetic field vectors on X-Z plane during thinning of the plasma sheet between 0950 and 1025 UT on 1971, November 7.

compared with only 19% showing southward B_z. Further, *only one out of twenty* groups of crossings shows predominance of southward B_z; only three show equal frequencies of southward and northward B_z, and sixteen show a clear majority of northward B_z. This ratio between northward and southward B_z at midplane crossings is far larger than expected from neutral line formation.

These studies, together with the conclusions reached by Meng and Anderson (1974), Schindler and Ness (1972, 1974), Bowling (1975) and Speiser (1973), suggest strongly that the large-scale change of the magnetic field structure illustrated in Figure 6.1 does not seem to take place within the lunar distance for most substorms. Most of the B_z component changes appear to arise from the dipping of the vector or from small-scale loops. It may be noted also that Yasuhara *et al.* (1975) examined the B_z component change in the magnetotail by assuming a complete disruption of the cross-tail current in a narrow width of order 7–8 R_E along the length of the tail. Note that such a complete disruption would be sufficient to generate an electrojet of intensity of 10^6 A along both the northern and southern ovals. However, they found that the above complete disruption would cause a large positive B_z component ($\sim +25\ \gamma$) just inside the disruption region and also a large negative B_z component ($\sim -25\ \gamma$) just outside. Since such a large change has not been reported at distances $X \simeq -10 \sim -60\ R_E$, it is likely that the disruption occurs only at distances $X \simeq -6 \sim -10\ R_E$ or beyond the moon (see Section 7.2).

6.3. Plasma Sheet Thinning

One of the most dramatic changes in the plasma sheet during substorms is plasma sheet thinning. This feature has been studied most extensively by Hones and his colleagues at a geocentric distance of about 18 R_E, on the basis of data taken by the Vela satellites whose orbits have a fixed geocentric distance of about 17–18 R_E

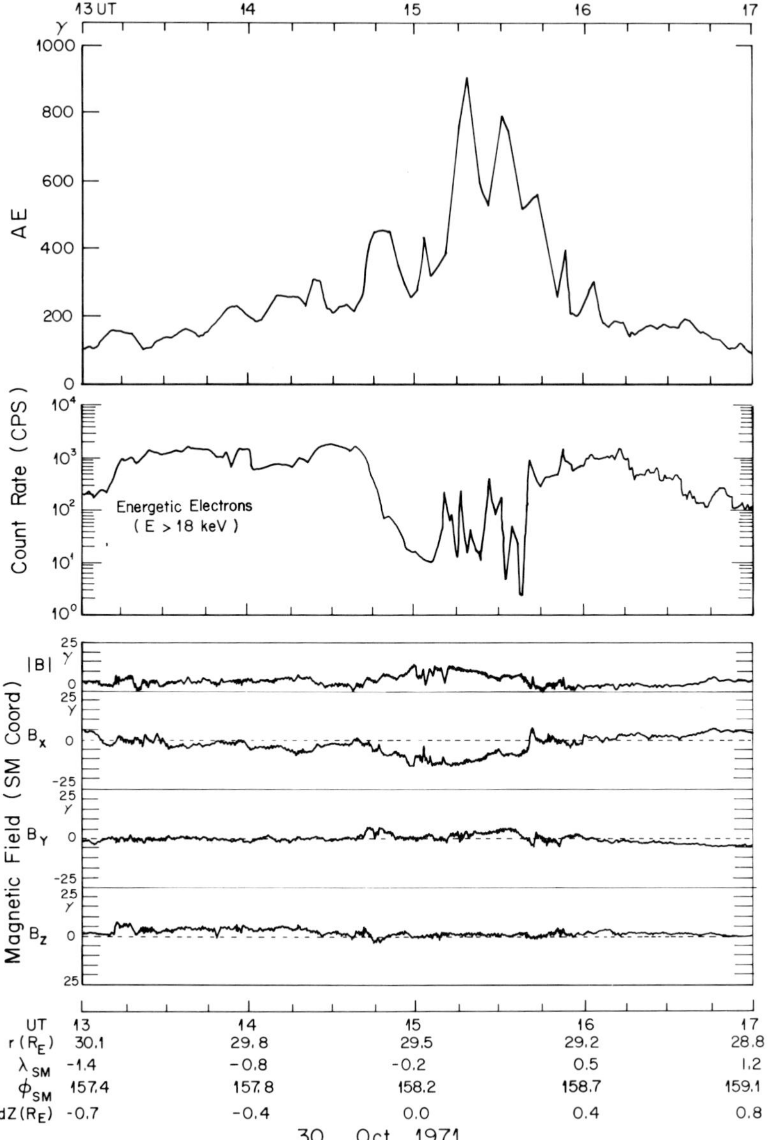

Fig. 6.5(a). Midplane crossings during thinning of the plasma sheet on 1971, October 30. (Courtesy of Lui, A. T. Y., Meng, C.-I. and Akasofu, S.-I.)

(Hones *et al.*, 1967; Hones *et al.*, 1971c; Akasofu *et al.*, 1971b; Meng *et al.*, 1971), and also at other locations (Lazarus *et al.*, 1968; Fairfield and Ness, 1970; Aubry and McPherron, 1971; Nishida and Lyon, 1972; Nishida and Hones, 1974; Lui *et al.*, 1975a).

In understanding the cause of plasma sheet thinning, it is important to know how the plasma sheet deforms during thinning. Fortunately, the profile of thinning has now become fairly clear, because of extensive satellite observations during

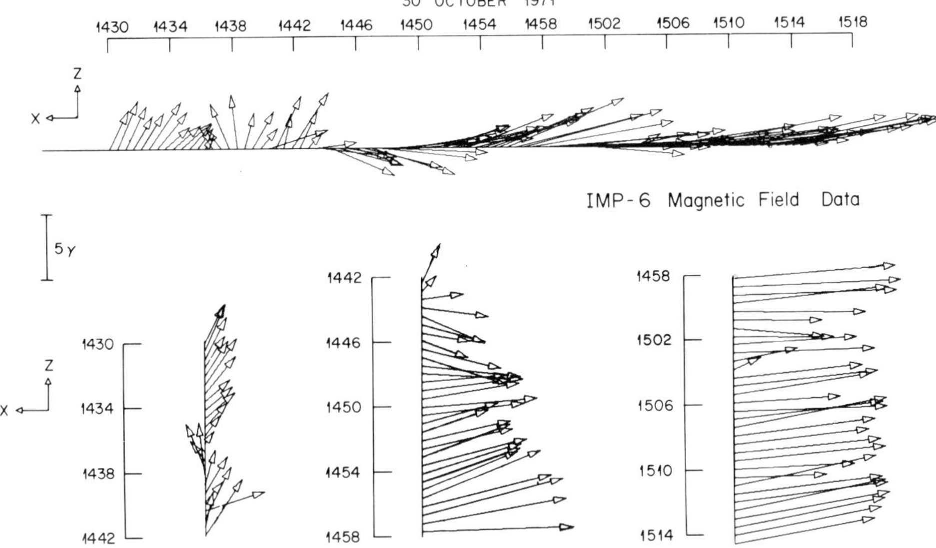

Fig. 6.5(b). Vector plots showing that B_z was positive during the midplane crossings.

the last decade. In this subsection, we review these studies briefly. Another important problem of the thinning is its onset time with respect to the onset time of the expansive phase. McPherron *et al.* (1973b) suggested that thinning begins well before (30 min to 1 h) the onset of the expansive phase, and thus it can be regarded as a *direct cause* of the hypothetical neutral line formation.

6.3.1. PROFILE OF THINNING

(a) *Z-Axis Dependence*

The distribution of plasma as a function of Z during thinning has been studied by Hones *et al.* (1971c) on the basis of records from two Vela satellites located at different distances from the midplane of the magnetotail. Figure 6.6(a) shows an example of their observations; both the Vela 3A and 4A satellites were in the morning sector during this period, crossing the midplane at different times, at about ~ 12 UT for Vela 4A and ~ 21 UT for Vela 3A (Figure 6.6(b)). During the first substorm, 12–15 UT, thinning was observed at the Vela 3A location but not at the Vela 4A location. On the other hand, during the second and third substorms, 20–24 UT and 02–05 UT, thinning was observed at the Vela 4A location but not at the Vela 3A location. Therefore, thinning is more apparent at a greater Z and may not occur near the midplane. During the last substorm, 07–10 UT, both satellites were near the southern boundary, observing the same thinning.

This observation also shows that thinning is not caused by compression of the plasma sheet, since there is no indication of an increase of plasma density near the midplane while thinning is in progress near the boundary.

Among at least four possibilities of the plasma sheet profile during thinning

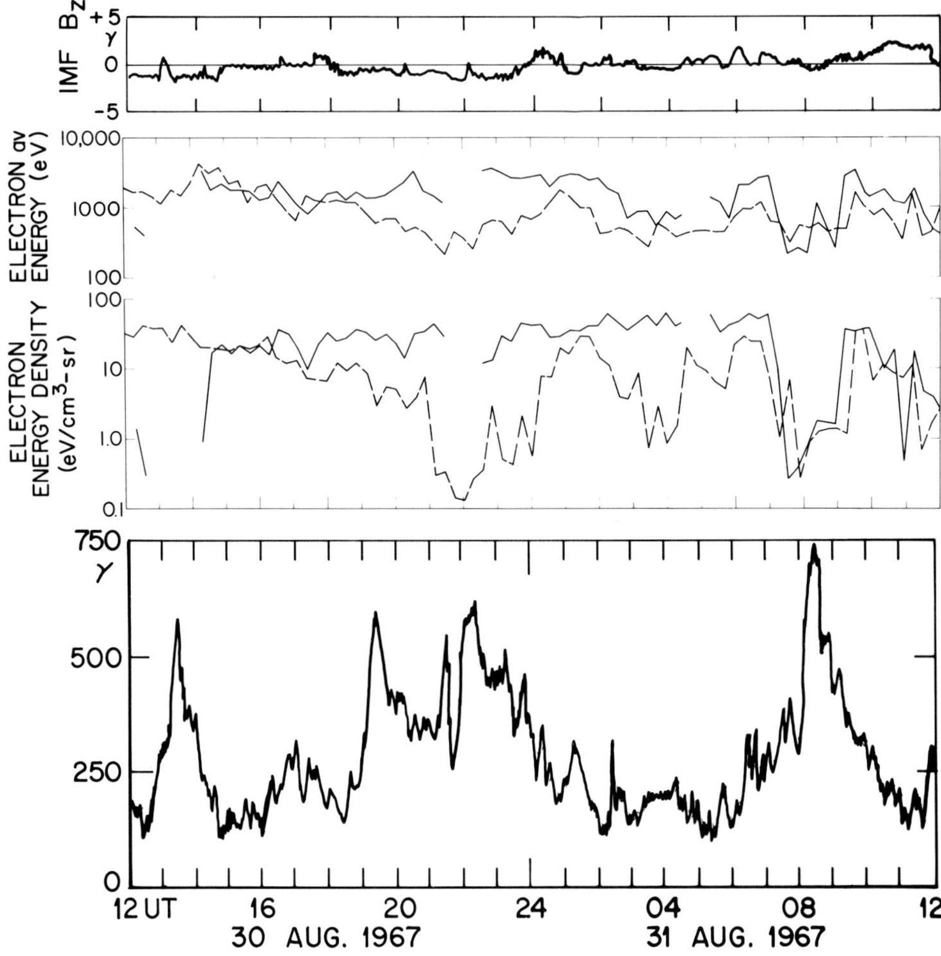

Fig. 6.6(a). Plasma sheet thinning observed by a pair of Vela satellites at $X \simeq -18\ R_E$. The IMF B_z component, the AE index and the trajectory in solar magnetospheric latitude-longitude coordinates of the two satellites, Vela 3A and 4A, are also shown in (b). (Hones, E. W. Jr., Asbridge, J. R. and Bame, S. J.: *J. Geophys. Res.* **76**, 4402, 1971.)

which are illustrated in Figure 6.7, Hones *et al.* (1971c) concluded that either the profile (b) or (d) or both are most likely. Unfortunately, the Vela satellites do not carry a magnetometer, so that it is difficult to examine thinning in the vicinity of the midplane of the plasma sheet. Meng and Akasofu (1971) examined thinnings when a satellite was crossing the midplane. Three such examples are shown in Figure 6.8. The times of the neutral sheet crossings of the satellite can be noted by observing the time when the angle ϕ changes by 180°. The time T_1 denotes the onset time of substorms determined from auroral zone magnetic records, and the time T_2 gives the time of reappearance of the plasma sheet observed as a sharp increase of the electron flux. It can be seen that during the period between T_1 and T_2, the magnetic field increased significantly while the electron flux reached the

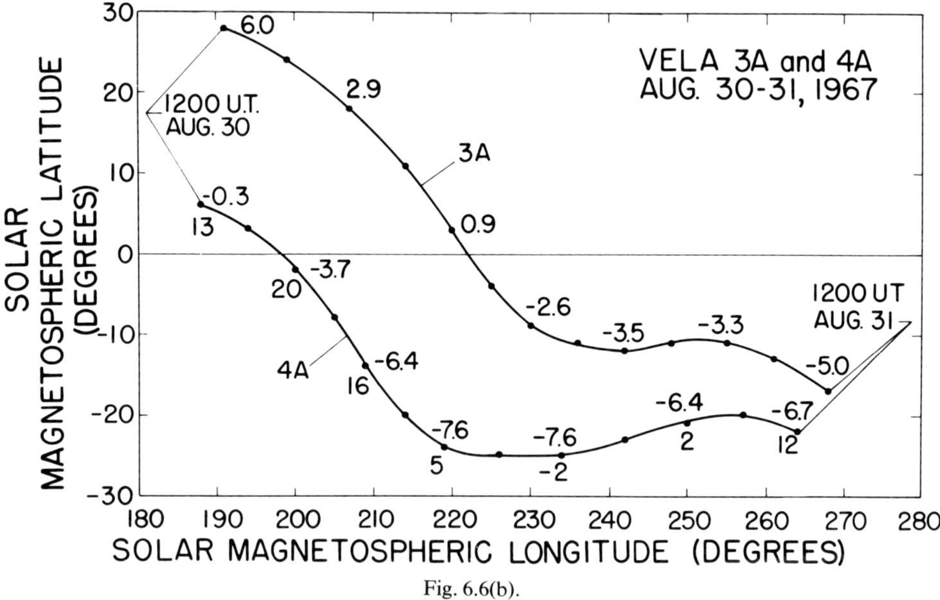

Fig. 6.6(b).

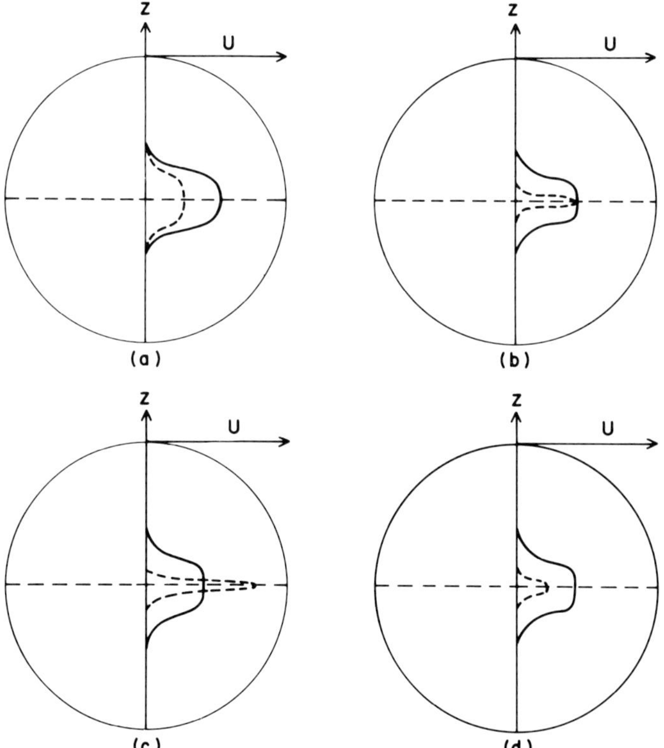

Fig. 6.7. Four possibilities of plasma distribution during thinning. Either (b) or (d) appears to be the actual change. (Hones, E. W. Jr., Asbridge, J. R. and Bame, S. J.: *J. Geophys. Res.* **76**, 4402, 1971.)

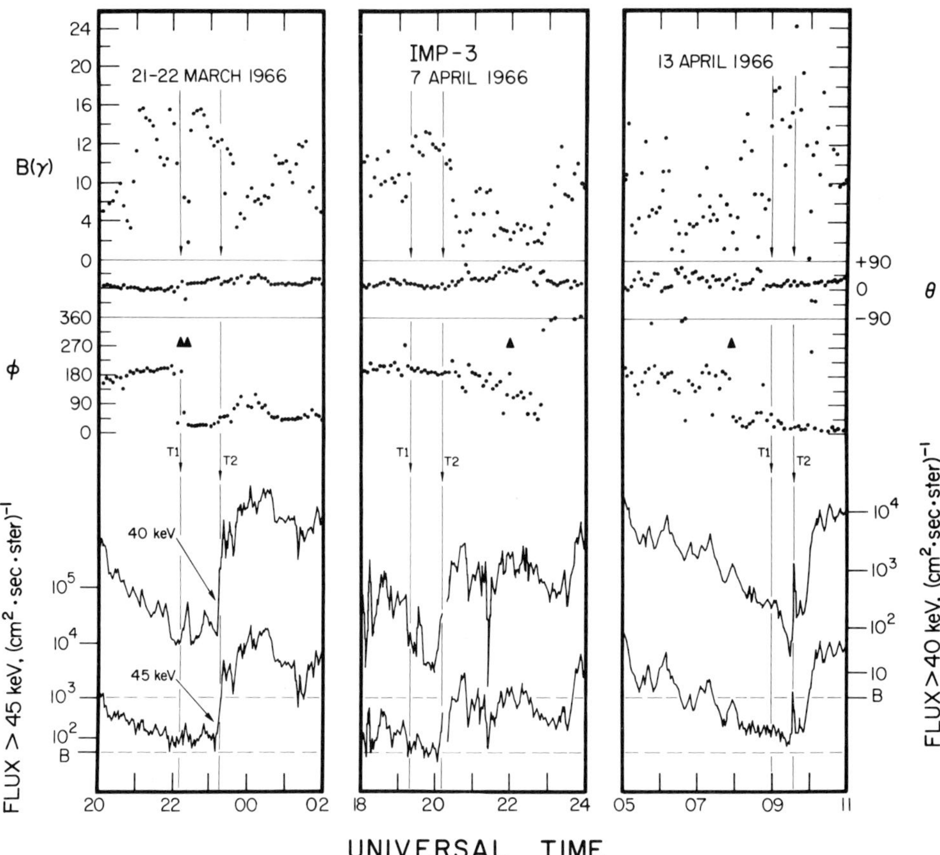

Fig. 6.8. Magnetic field and energetic electron flux variations at the midplane of the magnetotail during magnetospheric substorms. (Meng, C.-I. and Akasofu, S.-I.: *J. Geophys. Res.* **76**, 4679, 1971.)

background value. Thus, thinning seems to occur even on the midplane, although it may not be a complete thinning.

(b) *Profile in the Y-Z Plane*

The thinning profile in the Y-Z plane was studied by Lui *et al.* (1975). Figure 6.9 indicates locations on the Vela sphere (at a geocentric distance of 17–18 R_E), where the satellites observed a complete thinning (or the so-called 'dropout'), namely, the plasma density fell below the detector threshold. It can clearly be seen that the plasma sheet becomes thinnest in the midnight sector and that a complete thinning does not occur at small Z values near the magnetopause.

Thinning does not seem to take place uniformly over the entire Y-Z cross-section of the plasma sheet (Akasofu *et al.*, 1971b). When thinning is observed in the evening sector, together with the simultaneous ground all-sky records in the same meridian, it tends to occur a little before a westward traveling surge traverses the field of view of the camera. Figure 6.10 shows this situation schematically.

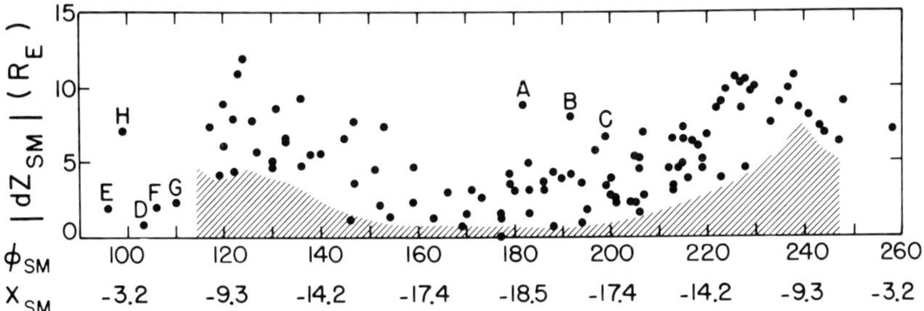

Fig. 6.9. Locations where complete plasma thinnings (or so-called 'plasma dropouts') are observed by the Vela satellites, in solar magnetospheric longitude-d$Z(R_E)$ coordinates. The region where plasma dropouts were not observed is shaded. (Lui, A.T. Y., Hones, E. W. Jr., Venkatesan, D., Akasofu, S.-I. and Bame, S. J.: *J. Geophys. Res.* **80**, 4649, 1975.)

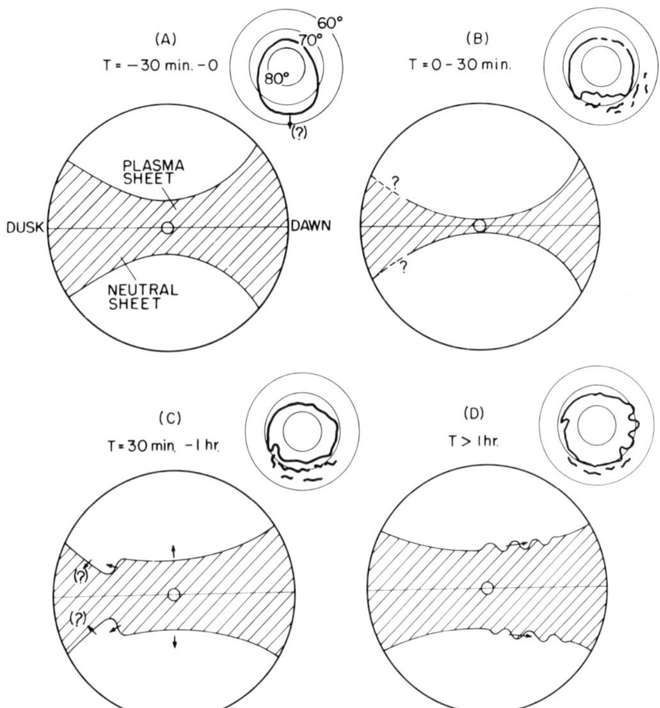

Fig. 6.10. Schematic diagram showing variations of the cross-section of the plasma sheet during the magnetospheric substorm at different phases of the magnetosphere substorm (indicated by the auroral substorm pattern). (Akasofu, S.-I., Hones, E. W. Jr., Montgomery, M. D., Bame, S. J. and Singer, S.: *J. Geophys. Res.* **76**, 5985, 1971.)

(c) *Profile in the X-Y Plane*

The thinning profile in the X-Y plane has been studied by Lui and Akasofu (1976). Figure 6.11 shows portions of the trajectories of the IMP-6 satellite along which thinnings (defined in terms of a decrease of proton counting rate by an order of magnitude or greater) were observed. It can be seen that thinning occurs over the entire plasma sheet in the dark sector.

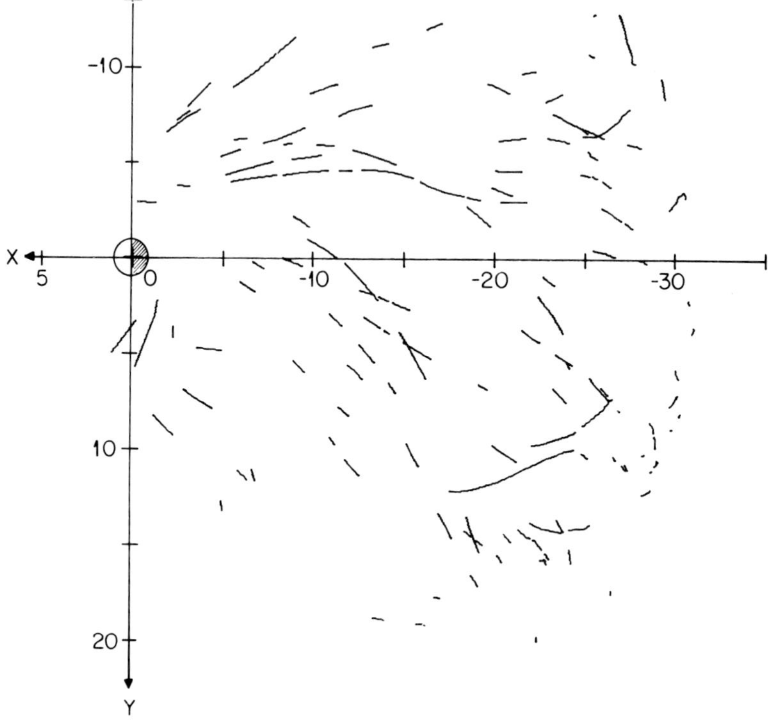

Fig. 6.11. Parts of the trajectories of the IMP-6 satellite in the X-Y plane, along which plasma sheet thinnings were observed. (Courtesy of Lui, A. T. Y. and Akasofu, S.-I.)

(d) *Profile in the X-Z Plane*

The thinning profile in the X-Z plane was also studied by Lui *et al.* (1975). Figure 6.12 shows their results. There are at least two features to be noted in the figure. The first is that thinning may occur even in the 'horn' of the plasma sheet, although it is difficult to distinguish thinning from an inward motion of the plasma sheet in

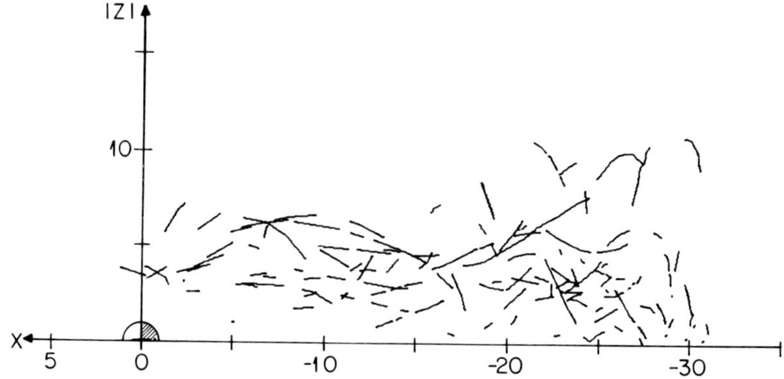

Fig. 6.12. Parts of the trajectories of the IMP-6 satellite in the X-Z plane, along which plasma sheet thinnings were observed. (Courtesy of Lui, A. T. Y. and Akasofu, S.-I.)

this particular region. The second feature is that plasma dropouts do not occur in the vicinity of the X axis, particularly at $X > -25\ R_E$.

6.3.2. TIMING OF THINNING

As mentioned earlier, the timing of thinning with respect to the onset time of the expansive phase is crucial in understanding the cause. It will be shown here that at a geocentric distance of 17–18 R_E in the midnight sector, thinning occurs almost simultaneously with the substorm onset.

Figure 6.13 shows an example of the thinning of the plasma sheet observed at about 19.1 R_E (by Vela 4B) on 1969, September 18. The thinning began at 0514

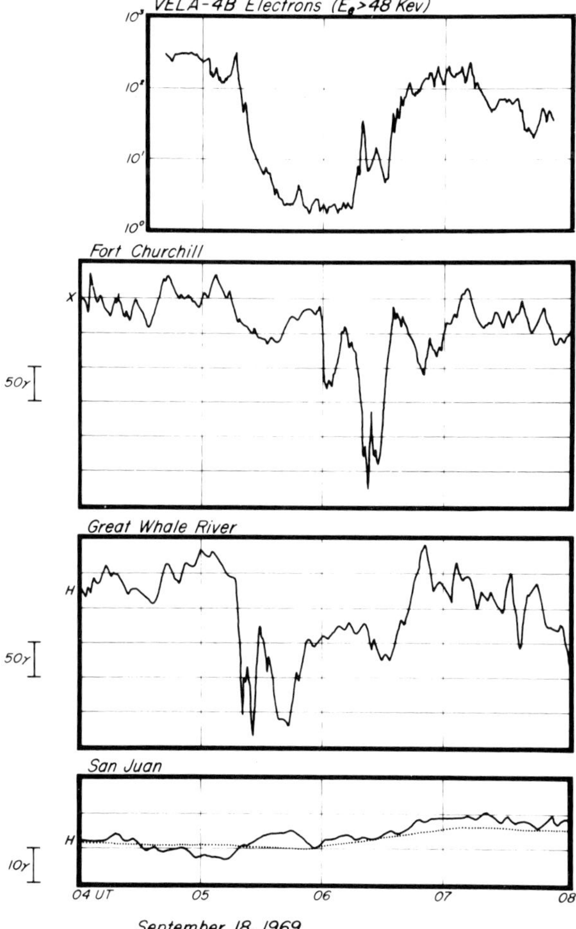

Fig. 6.13. Sudden decrease of energetic electron fluxes (indicating the onset of plasma sheet thinning) at about 0517 UT on 1969, September 18, observed by the Vela 4B satellite at $X \simeq -18\ R_E$, and the simultaneous ground magnetic records in the midnight sector (Churchill, Great Whale River and San Juan). (Hones, E. W. Jr., Akasofu, S.-I., Bame, S. J. and Singer, S.: *J. Geophys. Res.* **76**, 8241, 1971.)

(30 s) UT; it was observed as a sharp decrease of energetic electron flux; Vela 4B was located about 1.3 R_E below the estimated location of the midplane of the magnetotail. The onset time of this thinning coincided very closely with a sudden brightening of the auroral arc near the southern horizon of Churchill, which occurred between 0513 and 0514 UT. Figure 6.14 shows selected all-sky photographs from Churchill for this period. The onset of the corresponding negative bay began at 0518 UT at Great Whale River and 0514 UT at Churchill; the bay was much more sharply defined at Great Whale River than at Churchill. The auroral activity subsided temporarily about 30 min after the onset, and the new activity began between 0555 and 0557 UT. The negative bay was well-defined at Churchill. The thinning continued without interruption during this period.

Figure 6.15 shows another example of the thinning which began between 2237 and 2245 UT on 1967, May 1, observed as a sharp decrease of the energy density in the plasma sheet. The corresponding all-sky photograph from Syowa, Antarctica (conjugate to Leirvogur) showed a sudden brightening of an arc at 2239 (30 s) UT; Figure 6.16. It should particularly be noted that the thinning at $X = -18\ R_E$ was in progress during the poleward expansive motion of auroras, from the equatorward horizon to the poleward horizon at Syowa.

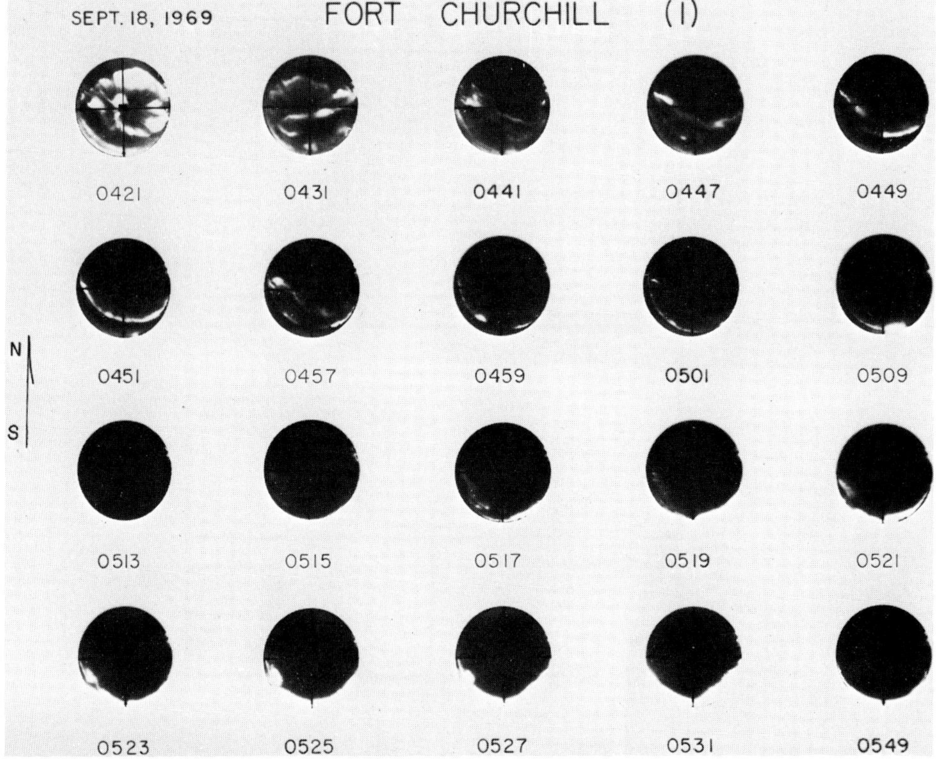

Fig. 6.14(1, 2). All-sky photographs taken from Churchill, Canada, at the time of the plasma sheet thinning shown in Figure 6.13. (Hones, E. W. Jr., Akasofu, S.-I., Bame, S. J. and Singer, S.: *J. Geophys. Res.* **76**, 8241, 1971.)

Fig. 6.14(2).

Figure 6.17 shows another example of the simultaneous occurrence of plasma sheet thinning (a Vela satellite) and of the expansive phase. We shall confine our attention to the substorm which began at about 0959 UT. The College all-sky camera photographs showed that an arc in the southern sky of College suddenly brightened at 1000 UT. The poleward expansive motion can clearly be seen in the subsequent photographs. As the simultaneous interplanetary magnetic data indicate, this particular substorm began more than two hours after the southward turning, without preceding thinning of the plasma sheet.

A very deep plasma thinning began at about 1004 UT, about 4 min after the sudden brightening. This delay appears to be caused by an initial increase of plasma fluxes at the onset, which will be discussed in Section 6.8.1.

The plasma thinning was manifested by a sudden decrease of plasma pressure by a factor of 60 in a few minutes and also by a decrease of plasma density by

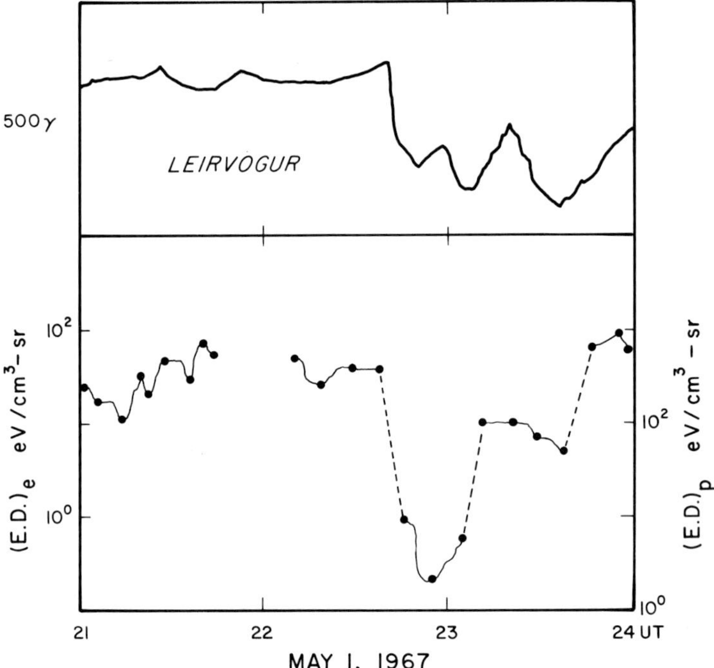

Fig. 6.15. Sudden decrease of the energy density of electrons at about 2241 UT, 1967, May 1, observed by the Vela 3B satellite at $X = -18\,R_E$, indicating the onset of plasma sheet thinning. The simultaneous ground magnetic record (the H component) from Leirvogur, then a midnight auroral zone station, is also shown. (Hones, E. W. Jr., Akasofu, S.-I., Bame, S. J. and Singer, S.: *J. Geophys. Res.* **76**, 8241, 1971.)

about one order of magnitude. It should be noted that there was *no* gradual change of any of the plasma parameters before 1000 UT. This is a clear-cut example which indicates three vital features of plasma sheet thinning:

(i) Plasma sheet thinning does not follow, in any obvious way, southward turning of the interplanetary magnetic field, since it began *very suddenly* at about 2 h after the southward turning of the IMF vector. In Section 4.4.4(c), we noted already that the southward turning alone does not cause thinning of the plasma sheet.

(ii) The onset time of thinning coincides, within a few minutes of accuracy, with the substorm onset $T = 0$.

(iii) Thinning continues until about the maximum epoch (~1030 UT) of the substorm, in spite of the fact that auroras were very active in the polar region.

There are numerous other examples which indicate that plasma sheet thinning at $X \approx -18\,R_E$ begins at the onset of negative bays, regardless of whether the bays start suddenly or gradually (Akasofu *et al.*, 1971b).

Buck *et al.* (1973) reported that the speed of the thinning plasma sheet boundary was about 4 km s^{-1} during a substorm on 1968, August 15.

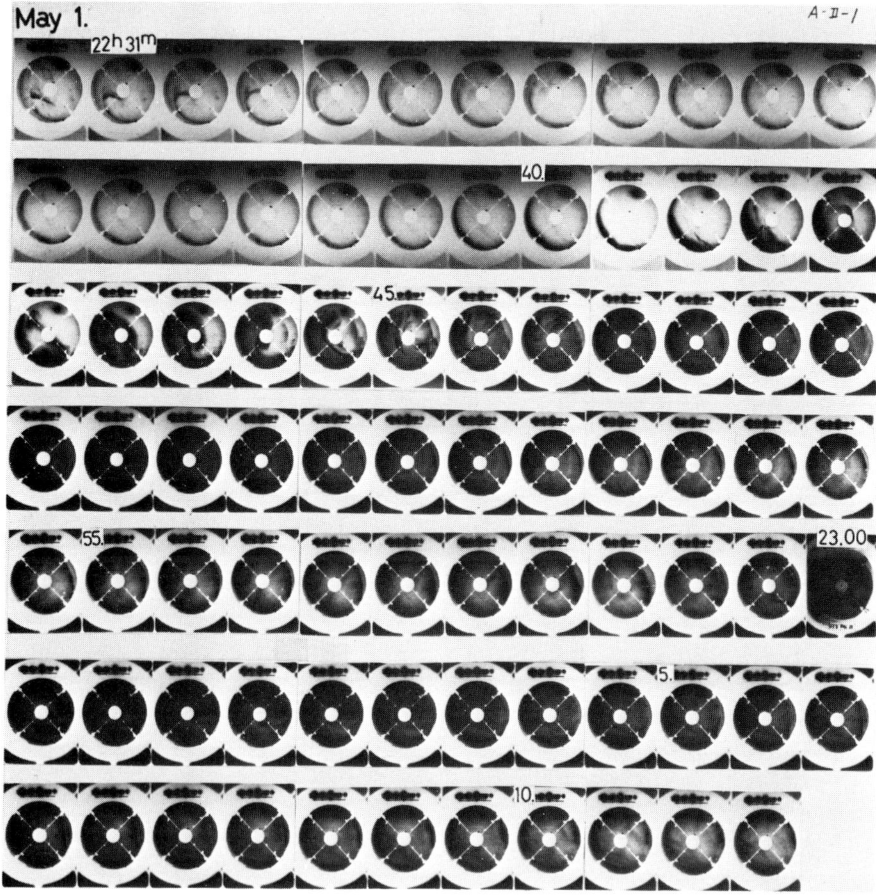

Fig. 6.16. All-sky photographs taken from Syowa station, Antarctica, during the plasma sheet thinning shown in Figure 6.15. (Hones, E. W. Jr., Akasofu, S.-I., Bame, S. J. and Singer, S.: *J. Geophys. Res.* **76**, 8241, 1971.)

6.3.3. THINNING AND THE B_z COMPONENT 'REVERSAL'

Nishida and Hones (1974) examined thinnings observed by Vela satellites at a geocentric distance of 17–18 R_E and the simultaneous magnetic field variations observed by Explorer 34 at geocentric distances greater than 25 R_E. They showed that both the thinnings observed at the Vela distance and the B_z component reversal occur at the Explorer 34 satellite nearly simultaneously. Figure 6.18(a) shows an example of such simultaneous observations. The B_z component became suddenly negative at Explorer 34 at 0444 UT, and the thinning began at 0429 ~ 0438 UT at Vela. Although it is not shown there, B became very large (~ 20 γ) at that time, indicating that the negative B_z component resulted from the dipping rather than the reversal. The onset of low latitude bays was at about 0437 UT, while auroral zone negative bays in the evening sector began at about

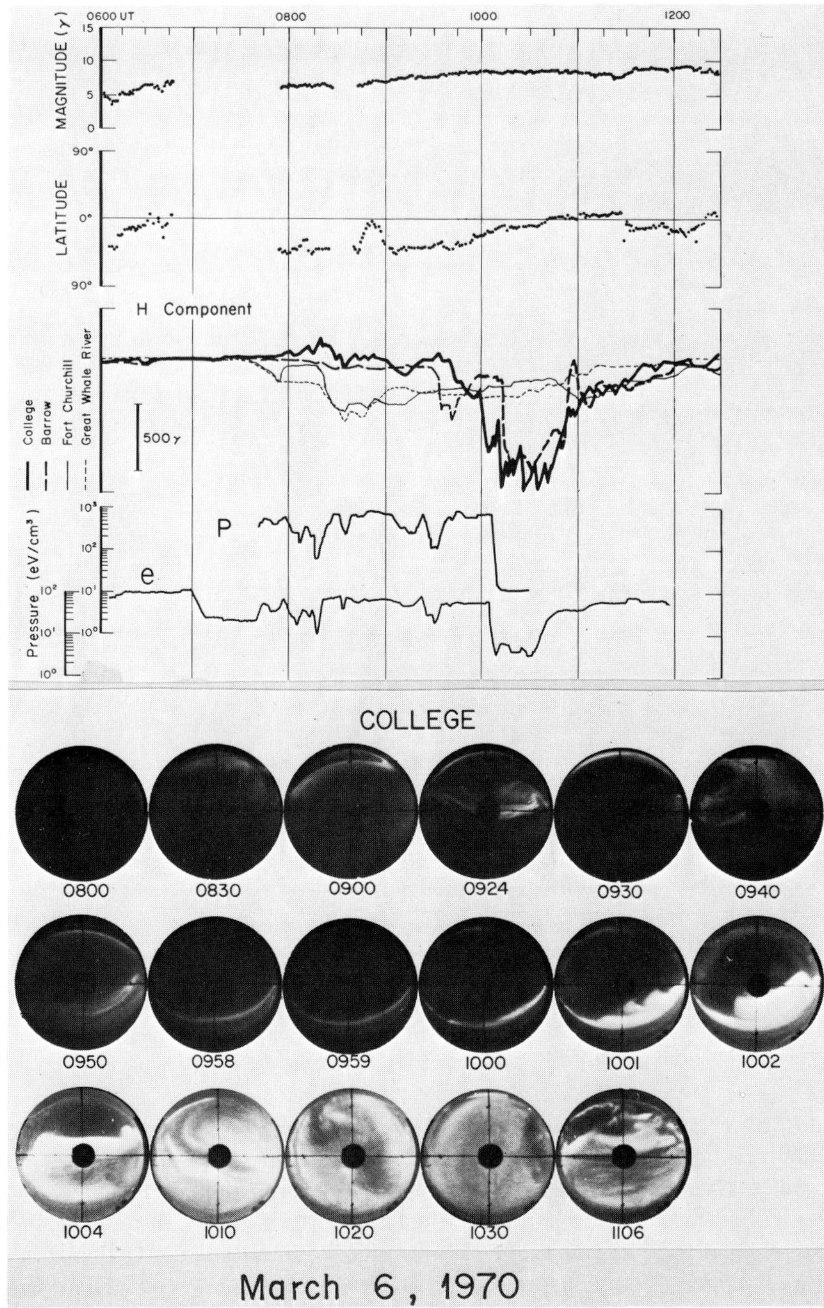

Fig. 6.17. Simultaneity of the onset of plasma sheet thinning and of magnetospheric substorms. From the top: the IMF ($|B|$ and θ), the H component magnetic records from four auroral zone stations, proton and electron pressure data in the plasma sheet (Vela 5A at $X \simeq -18\, R_E$) and the simultaneous College all-sky camera data. (Courtesy of Hones, E. W. Jr., Akasofu, S.-I. and Perreault, P.)

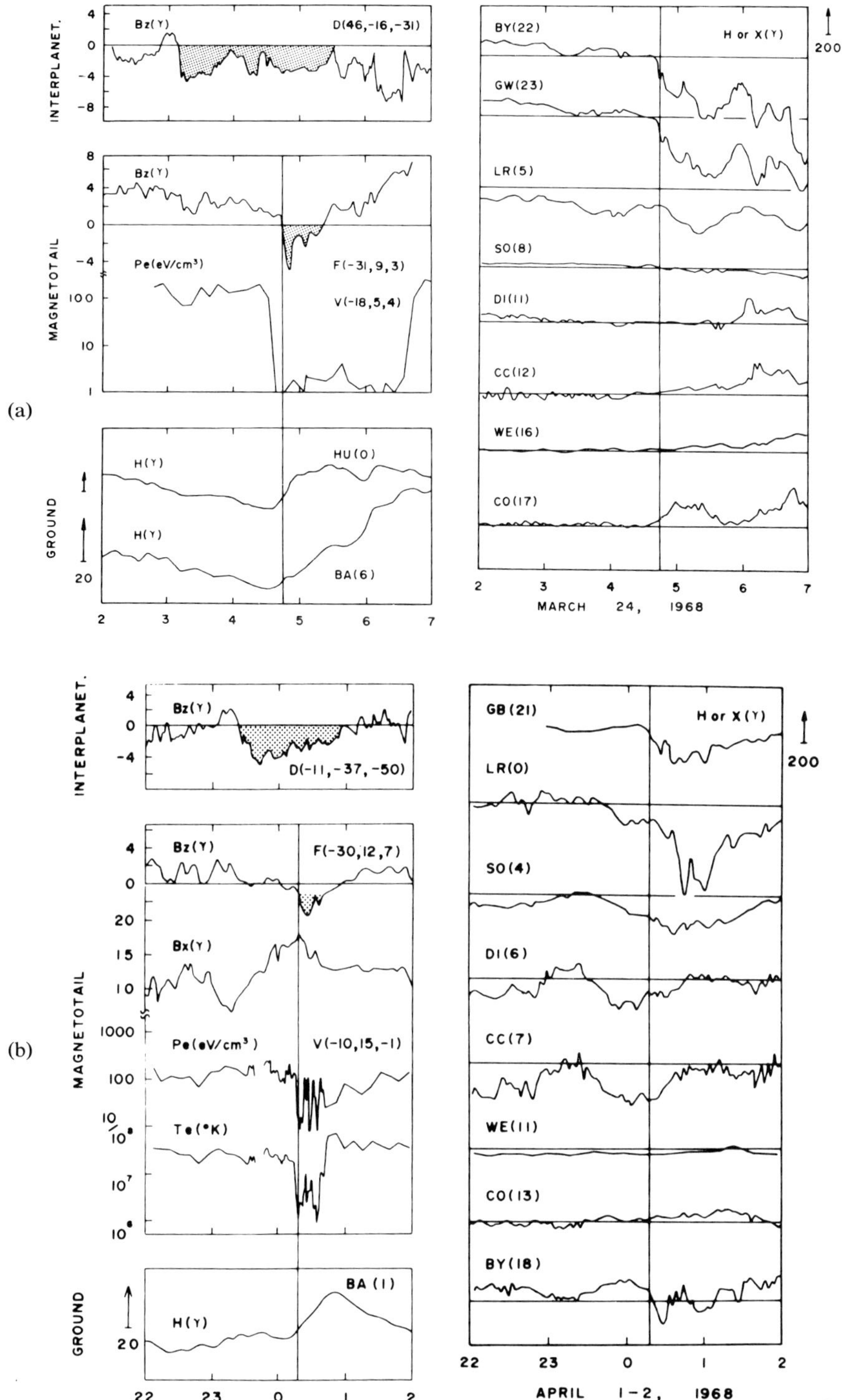

Fig. 6.18(a, b). Simultaneous observation of the magnetic field (Explorer 34), and of plasma energy density. The IMF B_z component and low and high latitude magnetic records are also shown. (Nishida, A. and Hones, E. W. Jr.: *J. Geophys. Res.* **79**, 535, 1974.)

0440 UT. There is little doubt that the thinning began at the onset of the expansive phase, so that there was a slight delay of the onset of the $\boldsymbol{B}$ vector dipping, at least in this particular example. For such a delayed case, they suggested that the thinning can start before the neutral line formation. However, it seems that the distance between the two satellites (13 R_E) was too great to make such a conclusion. The authors also suggested that the expansion of the plasma sheet (Section 6.3.3) and the onset of the positive B_z component are often simultaneous. However, this claimed relation is rather tenuous, as is seen in Figure 6.18(a).

Figure 6.18(b) shows another example of the correlation between a negative B_z component and thinning of the plasma sheet. It can be seen that the B_x component was very large at the time when the B_z component became negative, indicating again that the negative change was caused by the dipping. Note that in this example both phenomena occurred simultaneously.

6.3.4. SUMMARY

The above extensive study of plasma sheet observations reveals the following general features on the substorm-time variations of the magnetic field within and in the 'horn' of the plasma sheet:

(i) During plasma sheet thinning, the magnetic field dips slightly southward. The dipping becomes temporarily large near the crossing time of the plasma sheet boundary.

(ii) Southward dipping of the magnetic field occurs also in the near-Earth region at $X > -10\ R_E$ during plasma sheet thinning.

(iii) Large southward dippings of the magnetic field occur predominantly at locations with large Z values, rather than at those with low Z values.

(iv) The statistical study on the B_z component at midplane (or the so-called neutral sheet) from $X \simeq -20$ to $-32\ R_E$ during plasma sheet thinning at the substorm expansive phase shows that northward B_z is observed in 81% of crossings and southward B_z in only 19% of crossings.

(v) Northward rotations of magnetic field giving rise to positive B_z are frequently observed during a later stage of plasma sheet thinning.

(vi) Northward rotations of the magnetic field at plasma sheet expansion are considerably larger than southward dipping of the magnetic field during plasma sheet thinning.

One can construct the magnetic field configuration change in the magnetotail during the plasma sheet thinning on the basis of the above set of observations. This is shown schematically in Figure 6.19, with points A to G illustrating magnetic field changes at various locations of the magnetosphere and the magnetotail during plasma sheet thinning. The magnetic field observations of Fairfield (1973) and Cummings and Coleman (1968) are incorporated to construct the field variations at points A and B, respectively.

At the onset of thinning (Stage 1), the inclination of the magnetic field at the inner portion of the plasma horn (location A) decreases while the field dips slightly southward at the outer portion of the plasma horn (location C). The field near the equatorial plane becomes dipole-like (location B), in association with the

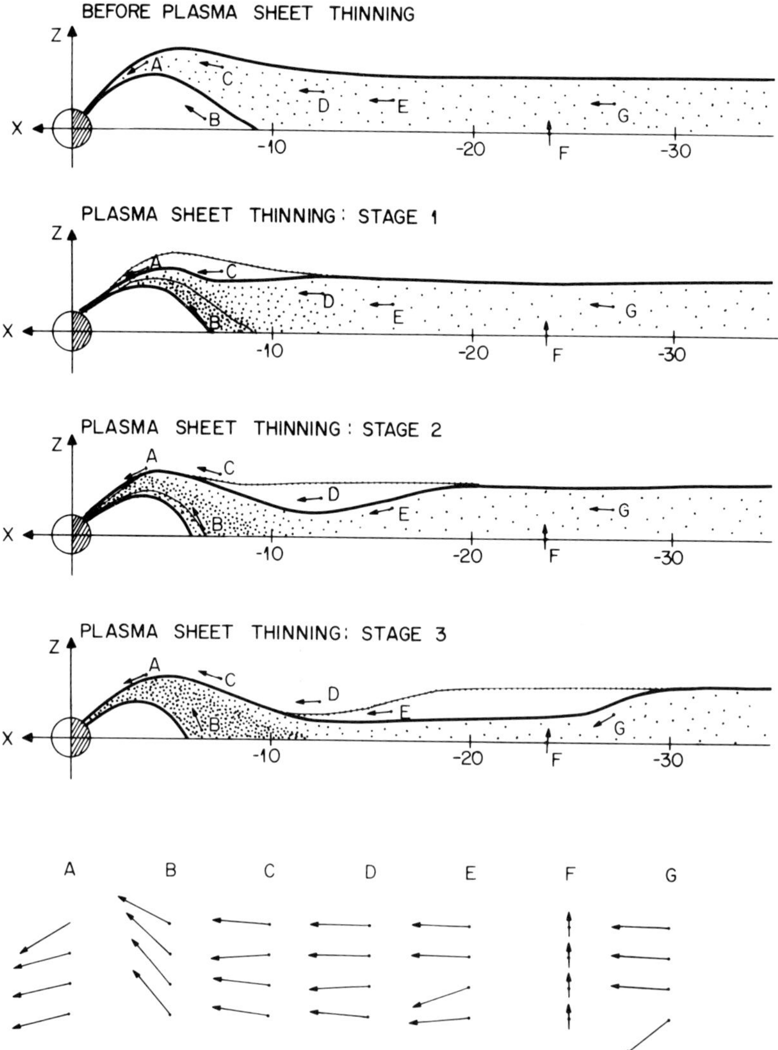

Fig. 6.19. A schematic diagram illustrating the configurational changes in the magnetosphere and the magnetotail during plasma sheet thinning at the substorm expansion phase. The thin lines mark the plasma sheet boundary in the previous stage. The dotted region is the plasma sheet and its horn. The variations of the magnetic field direction at each of the four stages are further illustrated at the bottom of the figure. (Courtesy of Lui, A. T. Y., Meng, C.-I. and Akasofu, S.-I.)

earthward injection of the plasma sheet particles (Vasyliunas, 1968; DeForest and McIlwain, 1971). As the plasma sheet thinning proceeds tailward (Stage 2), the field at C starts to rotate slightly northward while northward rotations of the field at A and B continue. Southward dipping of the field occurs at locations D and E at this time; the largest dipping is experienced near the time of the crossing of the boundary of the plasma sheet. The third stage depicts the plasma sheet thinning reaching $X = -30\ R_E$. Positive B_z values are observed at the midplane (location F), in spite of a large southward dipping occurring further down the tail (location G). It is suggested that the above sequence of processes proceeds rapidly (with a tailward propagation speed of a few hundred km s^{-1}), so that it is difficult to find the progress on the basis of the AE index which cannot determine accurately ($\pm$ 10 min) the onset time of the expansion phase.

Further, most important of all, it should be noted that thinning proceeds during the entire period of the expansive phase. There is no indication of the production of hot plasma at $X = -15 - -30\ R_E$, in spite of the fact that the aurora is most active in the polar region during this period.

If a new magnetic neutral line is expected to form in the near-Earth plasma sheet, one would expect an intense anti-sunward flow of hot plasma from the expected neutral line region, as well as the $\boldsymbol{B}$ vector reversal. On the contrary to these expectations, the plasma sheet simply thins (namely, deflates), without the $\boldsymbol{B}$ vector reversal.

This set of observations constitutes a strong objection to the hypothesis that a magnetic neutral line is formed in the near-Earth plasma sheet at the substorm expansive phase, causing the large-scale changes of the magnetic field structure envisaged by Nishida and Nagayama (1973) and Nishida and Hones (1974).

6.4. Magnetotail Field B_T and Radius R_T

6.4.1. DECREASE OF THE MAGNETOTAIL LOBE FIELD B_T

There have been several reports which indicate that during substorms the tail lobe field decreases, suggesting that a part of the accumulated energy in the magnetotail is converted into substorm energy. Iijima (1972) examined this feature for 64 isolated substorms. Figure 6.20 shows his result. He examined magnetic field

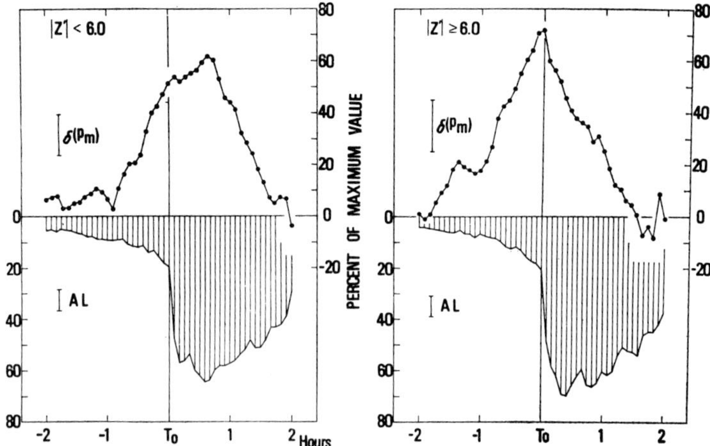

Fig. 6.20. Relationship between changes of magnetic pressure ($B_T^2/8\pi$) and the AL index for 64 isolated substorms; T_0 is defined to be a sudden increase of the AL index. (Iijima, T.: *Rep. Ionosphere Space Res. Japan* **26**, 149, 1972.)

variations for low Z values (near the midplane including the plasma sheet region) and for high Z values (at great distances from the midplane). The results are shown in terms of percentage $\Delta P_{\mathrm{m}} = \Delta (B_{\mathrm{T}}^2/8\pi)$ from the quiet day level. It can be seen that magnetic pressure begins to decrease about 45 min after the onset of the expansive phase at a small distance from the midplane (by assuming that it is given by a sharp increase of the AL index), but soon after at a great distance from the midplane. By assuming also that the onset time of the expansive phase coincides with the onset time of a positive bay in low latitudes, Caan *et al.* (1975) found that B_{T} begins to decrease at the expansion onset. Further, the initial increase in ΔP_{m} was described as a growth phase feature by Iijima (1972) and Caan *et al.* (1975). As pointed out in Section 4.4.4(a), the onset time of B_{T} decrease does not coincide with the onset time of the expansive phase.

It is not difficult to understand why B_{T} may not decrease even if the magnetic energy in the magnetotail is assumed to be dissipated during the expansive phase. Note that the entire magnetotail is in the state of pressure balance. If the magnetic pressure ($B_{\mathrm{T}}^2/8\pi$) in the high latitude lobe decreases, the solar wind pressure will compress the magnetotail until $B_{\mathrm{T}}^2/8\pi$ can again balance the solar wind pressure. As a result, B_{T} tends to remain unchanged so long as the solar wind pressure does not change. Therefore, it is not possible to examine changes of magnetic energy in the magnetotail in terms of B_{T} (see also Section 9.2.2).

Indeed, it will be shown shortly that R_{T} decreases during an early epoch of substorms (Section 6.4.2). Therefore, there is no particular reason why the magnetotail field should begin to decrease at the onset time of the expansive phase. It is unfortunate that without realizing this complexity, a number of papers have been published on the assumption that the onset of the expansive phase can be identified as the time when B_{T} begins to decrease.

Further, in this particular study it is important to examine the simultaneous plasma data. This is because the plasma sheet expands during a later epoch of substorms and may engulf a satellite (Section 6.6). In this case, the diamagnetic effect of the expanding plasma reduces considerably the local magnetic field intensity. Without plasma data, it is difficult to distinguish the suggested decrease of B_{T} from that caused by diamagnetism of the plasma sheet. It is likely that the decrease of B_{T}, studied by Iijima and Caan *et al.*, was mostly the diamagnetic effect.

In summary, the tail lobe field B_{T} is a function of a number of parameters (Sections 1.4.3 and 4.4.4), and thus one cannot assume that B_{T} begins to decrease at the onset time of the expansive phase and that the amount of the dissipated magnetic energy is estimated simply from the observed B_{T} decrease.

6.4.2. DECREASE OF THE RADIUS OF THE MAGNETOTAIL R_{T}

Recently Maezawa (1975) showed that the magnetotail radius R_{T} in the region $-20\ R_{\mathrm{E}} > X > -70\ R_{\mathrm{E}}$ increases during 1–2 h *before* the onset time of substorms and recovers within about one hour after the substorm onset. On the other hand, Lui *et al.* (1976) showed that the magnetotail radius in the region $0\ R_{\mathrm{E}} > X > -30\ R_{\mathrm{E}}$ decreases rapidly shortly after the substorm onset and recovers about one

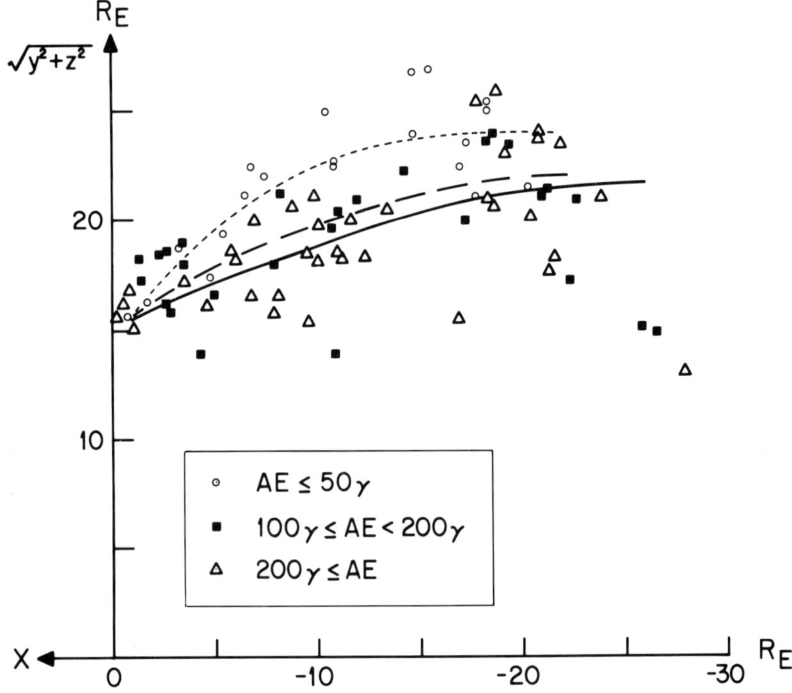

Fig. 6.21. Locations of the magnetopause, in X-$\sqrt{Y^2+Z^2}$ coordinates, for three different ranges of the AE index. (Courtesy of Lui, A. T. Y. and Akasofu, S.-I.)

hour after the onset. Figure 6.21 shows the locations of the magnetopause for three different ranges of the AE index. It can be seen that the cross-section of the tail becomes smaller during substorms than during quiet times. This reduction in R_T is likely due to the fact that the magnetic flux in the high latitude lobe begins to decrease during an early epoch of a substorm, but the external solar wind pressure remains constant.

6.5. Auroral Bulge

Many features described in this section have been considered to be indications of an enhanced reconnection, namely reconnection of *open* field lines and the subsequent contraction of the newly formed closed field lines. We have seen in Section 6.2, however, that there is no conclusive indication of the formation of a new magnetic neutral line in the near-Earth plasma sheet and that thinning may be interpreted as deflation of the plasma sheet. Actually, most of the features in this section can also be *qualitatively* interpreted as indications of the contraction of stretched (but *closed*) field lines in the plasma sheet during thinning, provided that the diversion of the cross-tail current to the polar upper atmosphere occurs during the expansive phase (Sections 6.5.2 and 7.2).

6.5.1. AURORAL BULGE

The formation of the auroral bulge in the midnight sector is one of the most interesting features of the magnetospheric substorm. Figure 6.22(a) shows a typical bulge observed by the DMSP-2 satellite. Figure 6.22(b) shows details of auroral features near the front of the expanding bulge. In many cases the expansion appears as a rapid poleward motion (of speed 300–500 m s^{-1}) of an active arc, but in other cases it appears as a continuous formation of new arcs at the front.

One of the important features of the auroral bulge is that *within* the bulge, active arc segments move very rapidly equatorward (Davis, 1962; Snyder and

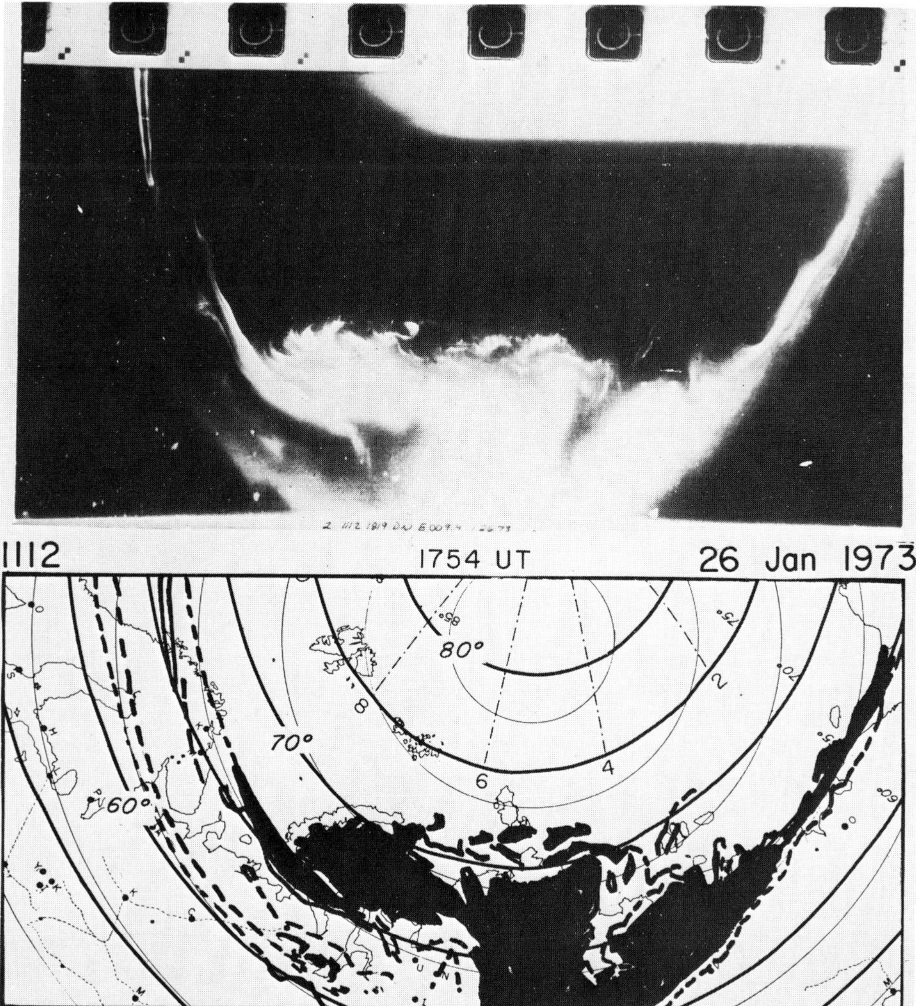

Fig. 6.22(a). Typical poleward expanding bulge, photographed by the DMSP-2 satellite. (Akasofu, S.-I.: *Space Sci. Rev.* **16**, 617, 1974.)

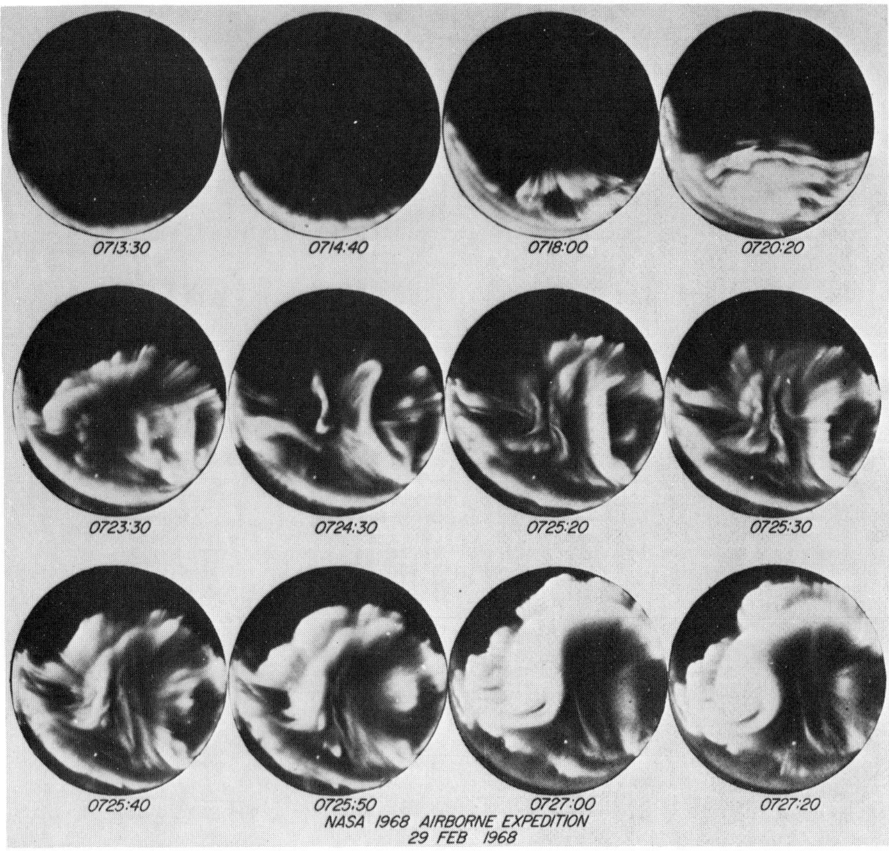

Fig. 6.22(b). Poleward expanding auroral bulge, photographed by an airborne camera. (Akasofu, S.-I.: *Planet. Space Sci.* **20**, 821, 1972.)

Akasofu, 1972; Vorobjev *et al.*, 1975). Figures 6.23(a) and (b) show some examples of this observation. Vorobjev *et al.* (1975) reported also that arcs within the auroral oval shift equatorward.

Characteristics of auroral particles in the auroral bulge have also been studied by a number of workers (Burrows and McDiarmid, 1972; Frank and Ackerson, 1972; Hruska, 1973; Hruska *et al.*, 1972; Hoffman and Burch, 1973; Winningham *et al.*, 1975; Rossberg, 1971, 1974; Rossberg *et al.*, 1974; Deehr *et al.*, 1973). Figure 6.24 shows an example of the bulge observation by the ISIS-1 satellite. It can be seen that it is mainly the BPS region (Section 2.4.1(a)) which expands and contracts during substorms and that the CPS region is relatively stable. It should be noted that the onset of the expansive phase is signaled by a sudden brightening of an auroral arc which is located near the equatorward boundary of the belt of discrete auroras and thus near the poleward edge of the diffuse aurora. Thus, the substorm begins near the boundary of the CPS and BPS regions, either at the ionospheric end or in the equatorial plane. It is of utmost importance to identify this boundary region in the equatorial plane. In this respect, it should be noted that

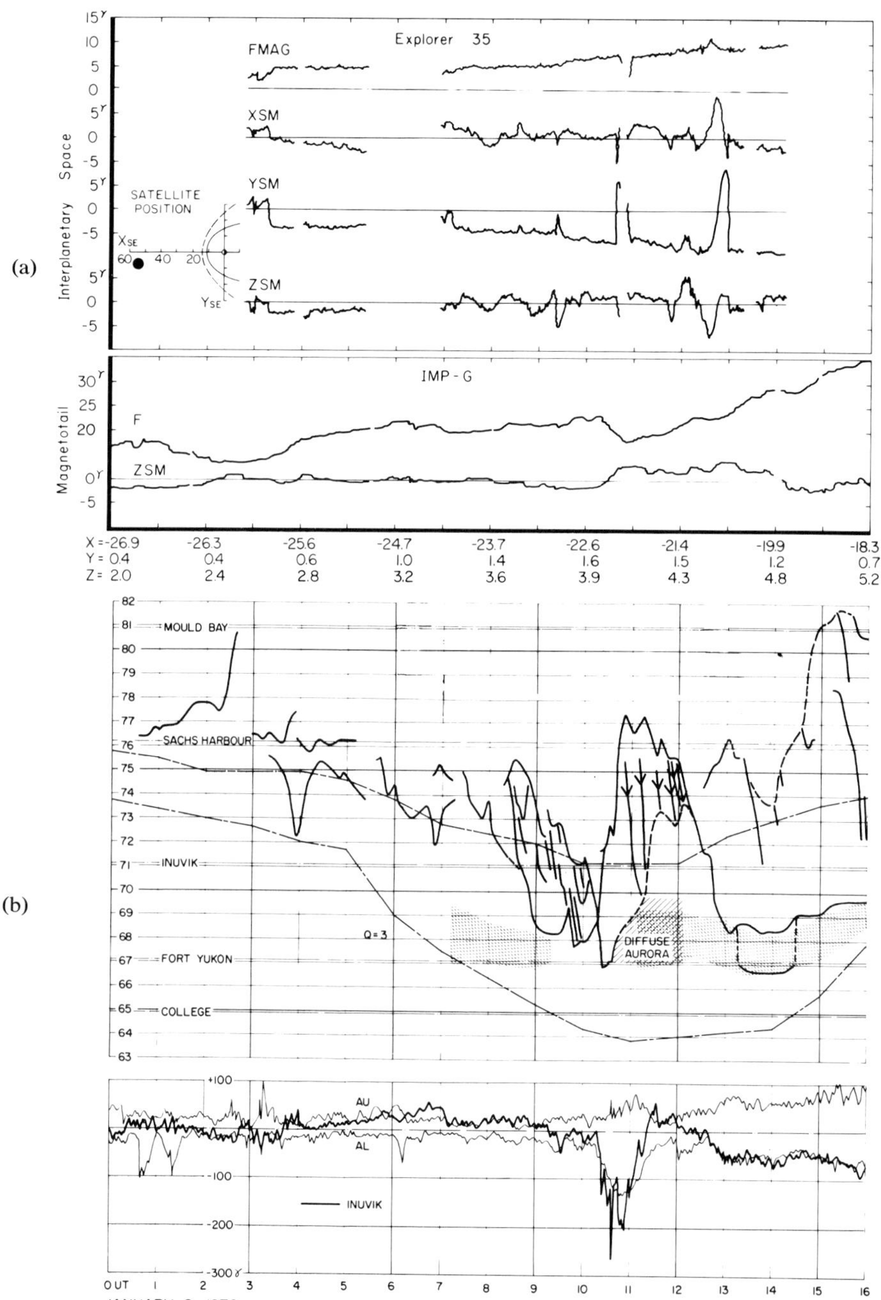

Fig. 6.23(a, b). Auroral oval observed from the Alaska meridian chain of all-sky cameras and the concurrent IMF ($|B|, X, Y, Z$), the magnetic field changes in the magnetotail (courtesy of Fairfield, D. H.) and the AU and AL indices. The equatorward motion of auroras during the expansive phase is indicated by arrows. (Snyder, A. L. and Akasofu, S.-I.: *J. Geophys. Res.* **77**, 3419, 1972.)

Explorer 33
FMAG
XSM
YSM
ZSM
SATELLITE POSITION
Interplanetary Space
IMP-G
Magnetotail
F
ZSM
X =-23.3 -22.1 -20.8 -19.3 -17.6 15.7 -13.5 -10.9
Y =- 1.4 -1.2 -0.8 -0.1 0.5 0.8 0.7 0.4
Z = 3.5 3.9 4.5 4.9 5.2 5.3 5.5 5.5
SACHS HARBOUR
INUVIK
FORT YUKON
COLLEGE
DIFFUSE AURORA
DIFFUSE
Q = 3
AU
AL
INUVIK
JANUARY 5, 1970

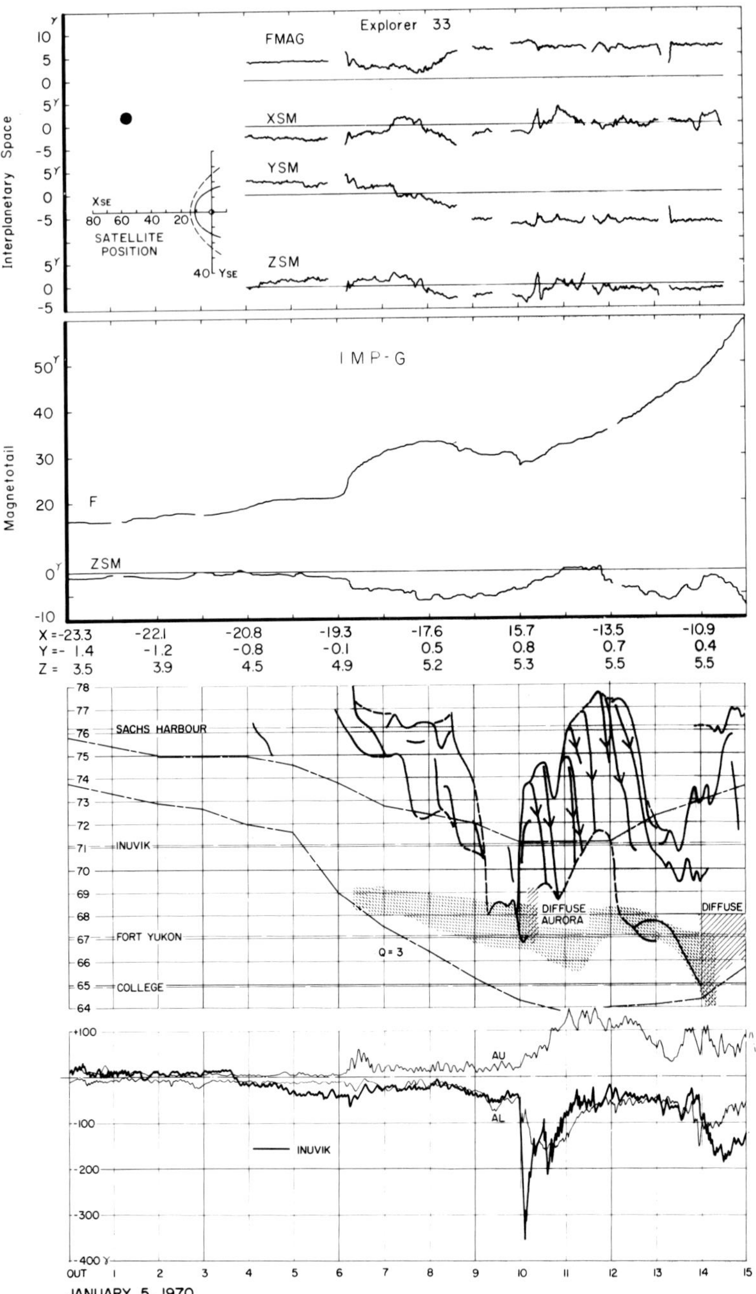

Fig. 6.23(b).

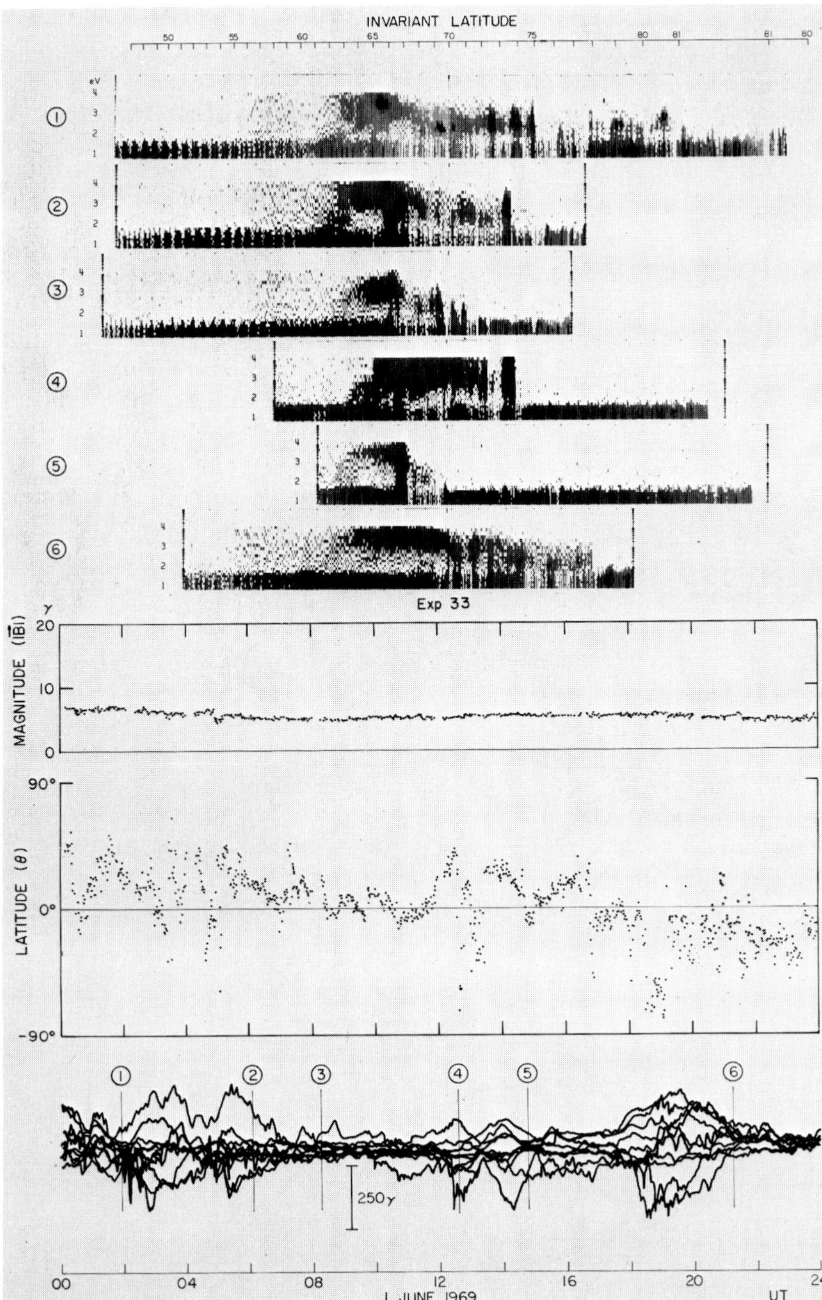

Fig. 6.24. Electron spectrograms taken approximately along the midnight meridian by the ISIS-1 satellite on 1969, June 1. The IMF data ($|B|$, θ) and the AE index are shown together with the times of the satellite passage. (Winningham, J. D., Yasuhara, F., Akasofu, S.-I. and Heikkila, W. J.: *J. Geophys. Res.* **80**, 3148, 1975.)

the boundary of the diffuse and discrete auroras is commonly observed well equatorward of College (dp. lat. 65°). If one assumes that a new magnetic neutral line is responsible for the sudden brightening or a sudden formation of an arc near the boundary, one must also assume that the neutral line can be formed at a geocentric distance of as close as 4–6 R_E.

6.5.2. AURORAL PARTICLES IN THE BULGE

Axford (1969) suggested that the poleward expanding bulge in the midnight sector results from the fact that the newly reconnected field lines contract rapidly toward the Earth and therefore plasma particles are accelerated by either the betatron process or the Fermi process or both; as a result, a hot plasma region of a dipolar shape grows rapidly, and the poleward expanding bulge may be considered to be the projection of this expanding region onto the ionosphere. Indeed, Kivelson *et al.* (1973) found a large increase of energetic electrons of pitch-angles near 90° in the equatorial plane at the synchronous distance but no appreciable increase of

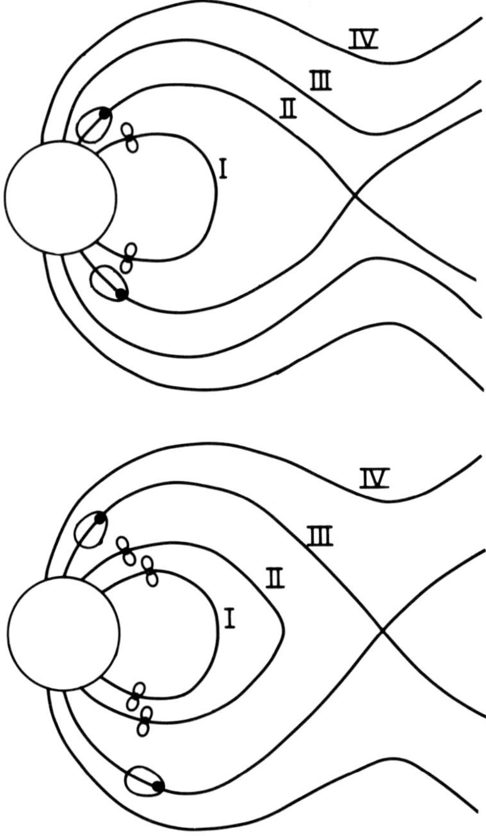

Fig. 6.25(a). Schematic diagram showing changes of the pitch-angle distribution, from the pancake-type to the cigar-type, as the magnetic field configuration changes during substorms (assuming the reconnection model).

electrons of small pitch-angles, indicating that the betatron mechanism is the predominant feature (Section 8.4).

The most common method to test whether a particular geomagnetic field line is open or closed is to examine the presence of the trapped particles. They have the pitch-angle distribution which peaks at 90°, namely the so-called 'pancake' distribution. Note that it is difficult to find any other process whereby untrapped particles would have this particular pitch-angle distribution. On the other hand, the 'cigar' pitch-angle distribution is believed to be caused by a potential drop along the field lines (Section 3.9).

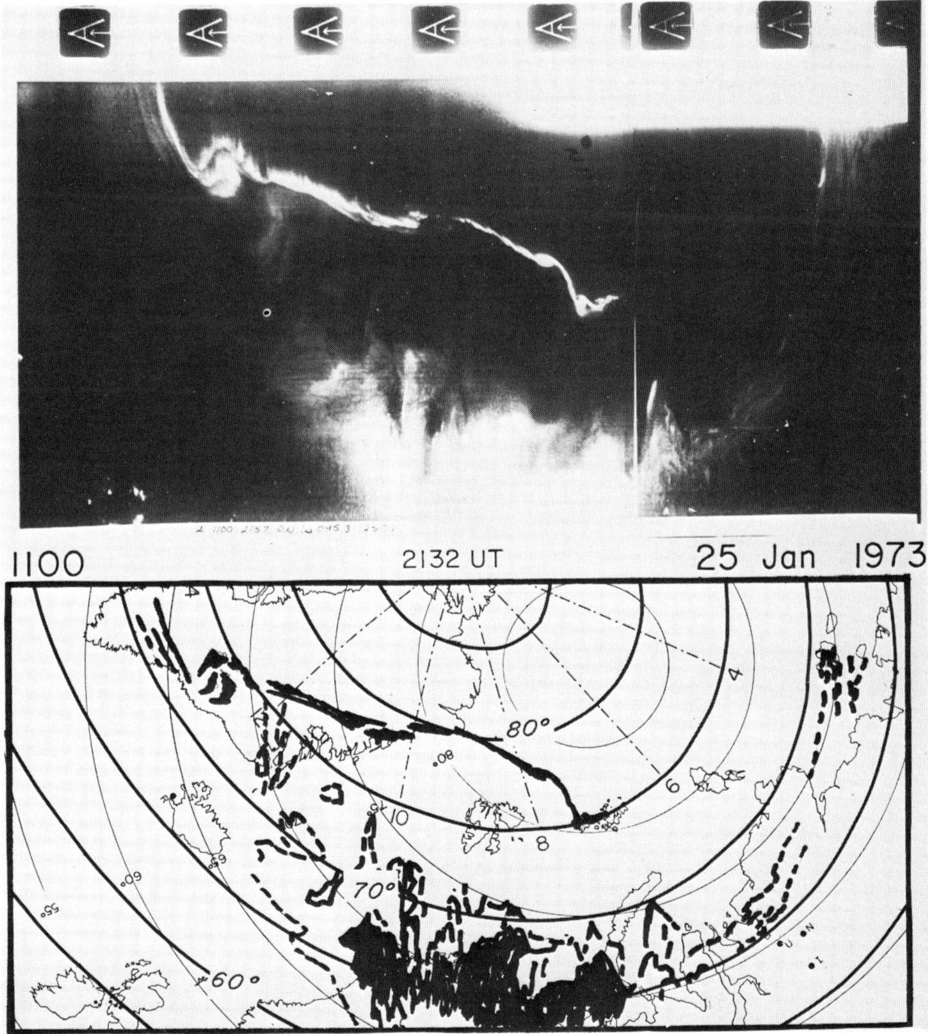

Fig. 6.25(b). Maximum epoch of the expansion phase. Note the development of dark area between the poleward advancing discrete arc and the diffuse aurora, which is not seen during an early epoch (see Figure 6.22(a)). (Akasofu, S.-I.: *Space Sci. Rev.* **16**, 617, 1974.)

Field lines which originate well within the polar cap are believed to be open (Section 1.1). Thus, if one finds the pancake pitch-angle distribution behind the advancing front of the poleward expanding bulge, one can infer that the geomagnetic field lines in the bulge are closed. Figure 6.25(a) shows this situation schematically; for a different interpretation, see Section 7.2.5(c).

It is common to observe a dark area between the poleward advancing arc and the poleward boundary of the diffuse aurora during the expansive phase; Figure 6.25(b) shows an example of DMSP photographs which show this particular feature. The reason for the presence of the dark area between the advancing arc and the diffuse aurora is not because there are no particles, but because the particles have a pancake distribution above the atmosphere, and so there is little precipitating flux. In order to substantiate this, Figure 6.26 shows an example of SPS and EPD data on 1969, July 12, taken from the ISIS-1 (polar orbiting) satellite. These data were taken approximately along the midnight meridian and thus show the midnight (latitudinal) cross-section of the precipitation region; for details of the format of the SPS detector, see Section 2.4.1.

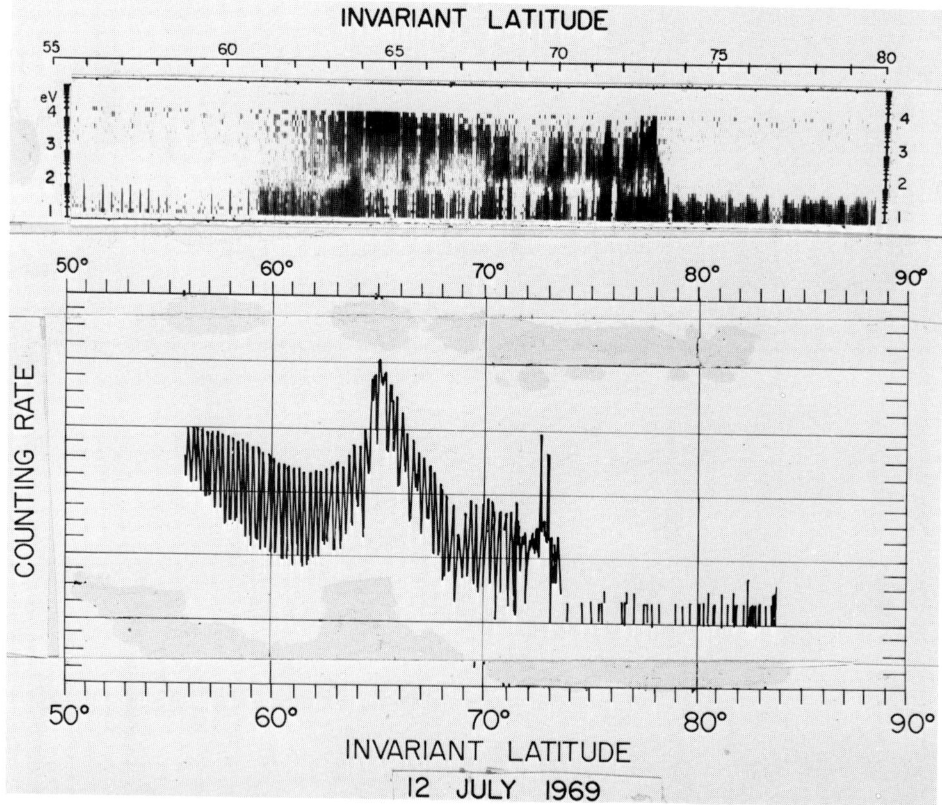

Fig. 6.26. Simultaneous SPS (10 eV–10 keV) and EPD (>22 keV) electron data, taken from the ISIS-1 satellite, approximately along the midnight meridian. (The SPS data, courtesy of W. J. Heikkila and the EPD data of J. R. Burrows.)

The SPS data show clearly a high flux at the high latitude boundary of the precipitation region, indicating the presence of the poleward advancing arc. However, within the bulge energetic electrons (EPD data) show a double modulation of flux, indicating that the pitch-angle distribution is peaked at 90°. This feature was discussed by Hruska *et al.* (1972).

6.5.3. INCREASE OF THE MAGNETIC FIELD COMPONENT B_z

Fairfield and Ness (1970) showed that a significant increase of the B_z component magnitude occurs in the magnetotail. Figure 6.27 shows an example of this increase. It can be seen that a distinct increase of the B_z component was observed during each of two successive substorms. Note that the magnitude of the B_z component was about 5 γ; the satellite was at $X = -29\ R_E$ and $Y = 11\ R_E$ at that time. It should be noted, however, that the B_z component shows a definite sign of increase in a late epoch of both substorms (at about the time when the AE index reached the highest values). This indicates that the increase occurred after the field magnitude (F) decreased rather sharply, namely when the satellite was

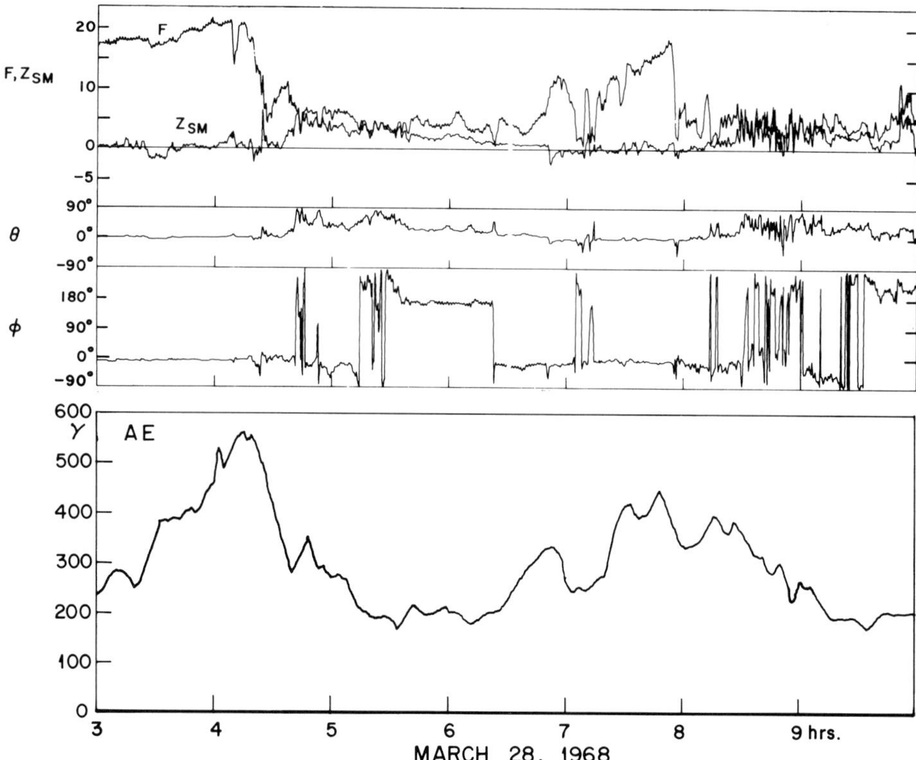

Fig. 6.27. Magnetic field variations ($|B| = F$, Z, θ and ϕ) in the magnetotail during two magnetospheric substorms on 1968, March 18 (the AE index). (Fairfield, D. H. and Ness, N. F.: *J. Geophys. Res.* **77**, 611, 1972.)

engulfed by the expanding plasma sheet during a late epoch of the substorm (Section 6.6); the decrease of the field magnitude is caused by diamagnetism of the expanding plasma sheet. Thus, the B_z component increase is mostly confined *within* the expanding plasma sheet (see also Section 6.6).

Figure 6.28 is another good example in which the B_z component increases during each of four successive substorms. Again, the increase was observed in the expanding plasma sheet. However, note also that the B_z component begins to decrease well before the end of each substorm, while the B_x component increases.

Figures 6.27 and 6.28 serve to illustrate one of the important questions related to the fundamental mechanism of the magnetospheric substorm. It has been widely believed that the magnetospheric substorm is the process by which the magnetosphere relaxes by itself from the stressed condition exerted by the solar wind, and that as a result the magnetic field configuration in the magnetotail tends to return from a tail-like configuration to the *original* Earth's dipole configuration; the stretched field lines contract rapidly toward the Earth after they are

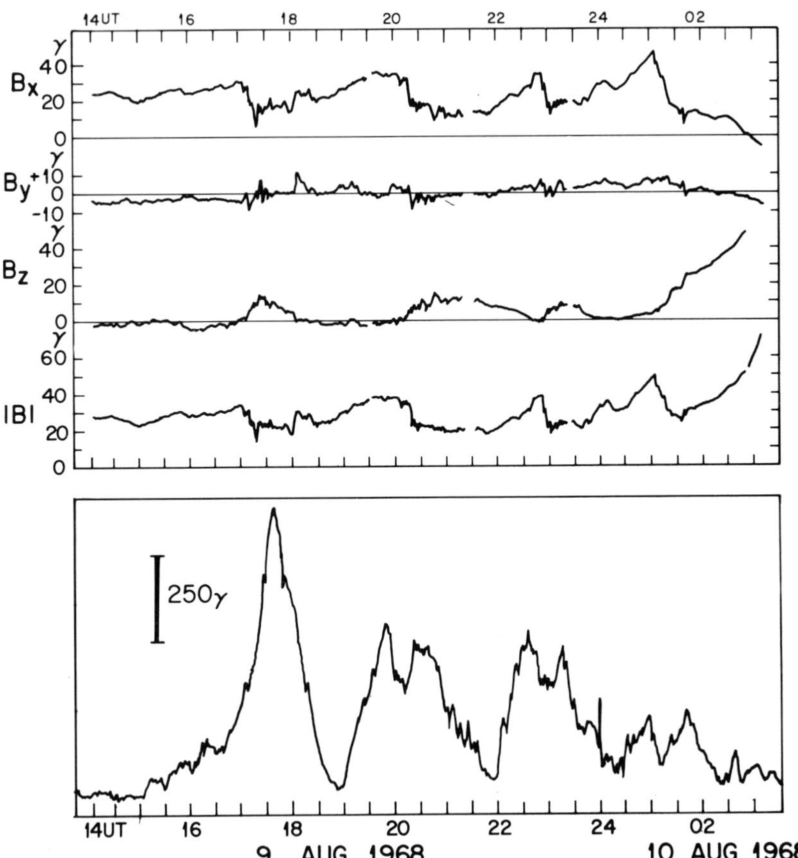

Fig. 6.28. Magnetic field variations (B_x, B_y, B_z, $|B|$) in the magnetotail during several successive magnetospheric substorms. (Russell, C. T., McPherron, R. L. and Coleman, P. J. Jr.: *J. Geophys. Res.* **76**, 1823, 1971.)

reconnected, resulting in more or less dipolar field lines. Indeed, it has been believed that the observed increase of the B_z component discussed above is an important indication of this tendency.

There are two problems associated with this interpretation. First it should be recalled that the magnitude of the increased B_z component is about 5 γ at a geocentric distance of 30 R_E or greater where the Earth's dipole field should be 1 γ or less. Secondly, it can be seen that the B_z component begins to decrease well before the end of each substorm, after reaching a peak value, while the B_x component begins to increase. Thus, the field configuration is becoming tail-like again during the recovery phase.

These two features suggest that the magnetospheric substorm is not simply a manifestation of the relaxation process of the magnetosphere, from a stressed ('unstable') configuration to the original Earth's dipole ('stable') configuration. It is more reasonable to interpret this chain of phenomena in terms of the diversion of the cross-tail current, plasma sheet deflation and the subsequent inflation (Section 9.2). Figure 6.5 presents another example which shows very clearly the tendency for the magnetotail structure to become tail-like during the recovery phase. This result raises another problem in interpreting substorm processes. It has been proposed by McPherron *et al.* (1973b) that the magnetic field configuration becomes more tail-like during their proposed growth phase. This and the earlier examples show that the tendency toward the tail-like structure is not unique to the proposed growth phase. Then how can one distinguish the proposed growth phase from the recovery phase of a previous substorm?

6.5.4. INCREASE OF COSMIC RAY PROTON CUT-OFF

Barcus (1969) noted a temporal decrease in nuclear γ-ray flux at the times of substorms which occur during an intense solar proton event. He interpreted the decrease to an increase of solar proton cut-off which, in turn, might result from the 'collapse' of the magnetotail field from a tail-like to a dipole-like configuration. Häusler *et al.* (1974) examined the distribution of solar proton fluxes with respect to the trapping boundary of energetic electrons during substorms. They found that as the electron distribution extends toward higher latitudes during a substorm, the penetration of protons toward lower latitudes appears to be suppressed as much as 2.5° in latitude. The authors suggest that this phenomenon is essentially due to an increase of the proton cut-off, observed by Barcus (1969) by balloon-borne detectors.

6.6. Plasma Sheet Expansion

After thinning, the plasma sheet expands, often very rapidly. The half-thickness of the expanding plasma sheet in the midnight sector may become as large as 6 R_E, but it becomes gradually thinner during a late epoch of a substorm. The temperature of the protons in the expanding plasma sheet is much higher ($\sim 10^8$ K) than that in a quiet time plasma sheet ($\sim 10^7$ K). There is also a much higher flux of energetic electrons (~ 40 keV) than in a quiet time plasma sheet. However, the number density becomes slightly less, so that the plasma pressures before and after a substorm are approximately the same. Figure 6.29 shows, from the top, variation of the density, temperature, pressure and flux of energetic electrons in the plasma sheet during two successive substorms. The expansion of the plasma sheet took place at about 1830, 2140 and 2220 UT, respectively. Note that energetic electrons reappear a little earlier than the main plasma.

When this phenomenon is observed in the high latitude lobe by an energetic electron detector, the electron flux increases suddenly and then slowly decays.

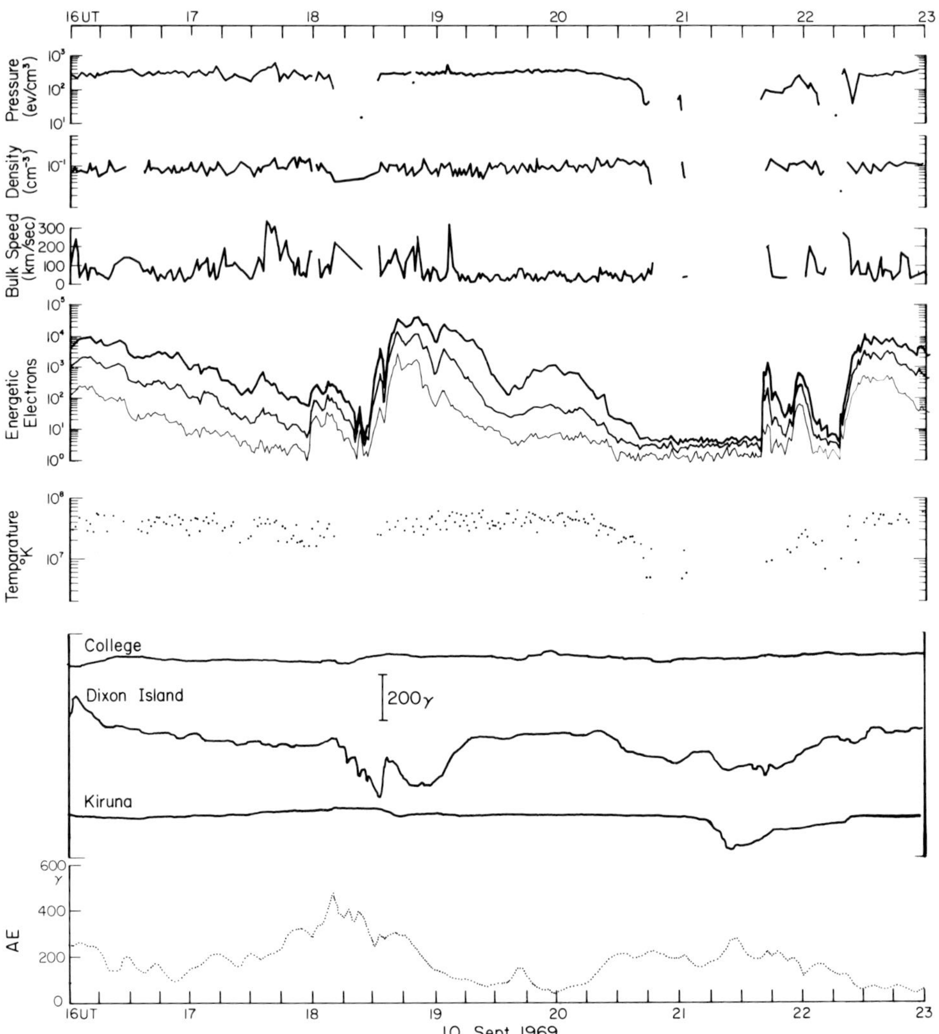

Fig. 6.29. Plasma sheet variations (from the top: pressure, density, bulk speed, energetic electron flux and temperature) during substorms on 1969, September 10, observed by the Vela satellite. The *H* component magnetic records from College, Dixon Island and Kiruna, as well as the AE index, are shown. (Courtesy of Lui, A. T. Y., Hones, E. W. Jr. and Akasofu, S.-I.)

This feature was first noted by Anderson (1965) who called high fluxes of electrons 'islands' (see also Rothwell and Wallington, 1968). However, Meng and Anderson (1971) later interpreted it as due to the fact that a satellite is suddenly submerged in the expanding plasma sheet.

The expansion of the plasma sheet has been studied by a number of workers (Hones *et al.*, 1967; Meng *et al.*, 1970; Akasofu *et al.*, 1970; Hones *et al.*, 1970; Hones *et al.*, 1971a; Aubry *et al.*, 1972). Figure 6.30 shows an example of the

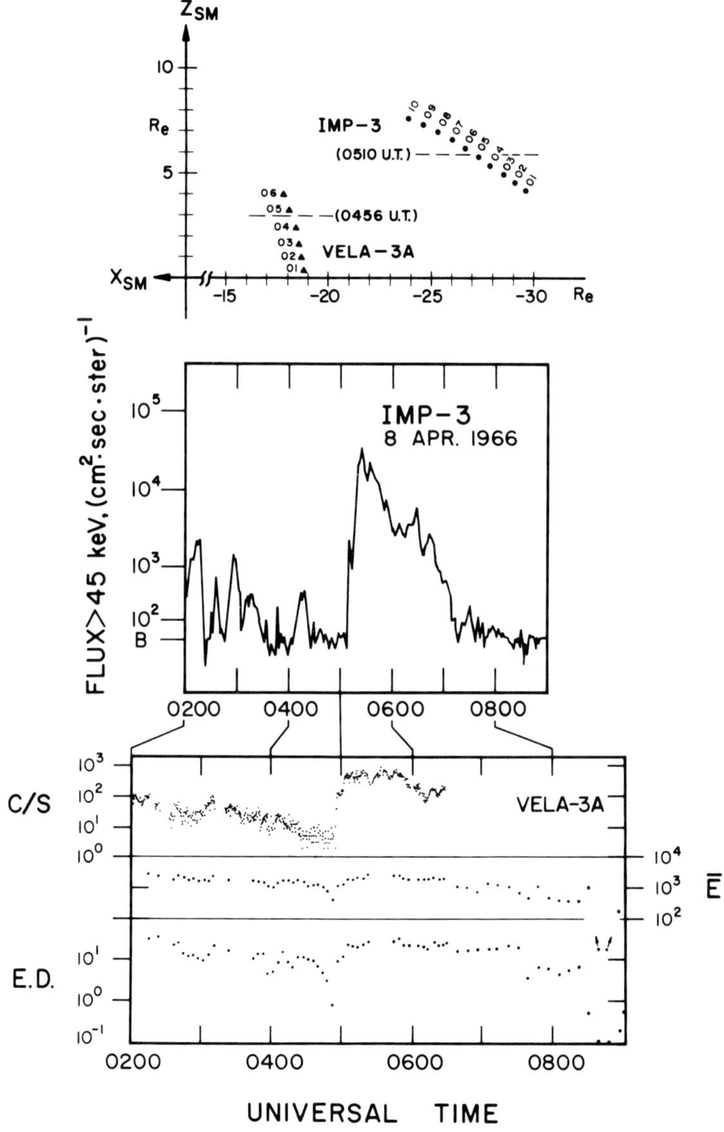

Fig. 6.30. Simultaneous observation of expansion of the plasma sheet by the IMP-3 and Vela-3A satellites (energetic electron fluxes). The location of the two satellites is shown at the top. (Meng, C.-I., Hones, E. W. Jr. and Akasofu, S.-I.: *J. Geophys. Res.* **75**, 7294, 1970.)

simultaneous observation of the expansion of the plasma sheet by two satellites (Vela-3A and IMP-3). Hones *et al.* (1970) showed that the velocity component of the expansion perpendicular to the plasma sheet is of order 20 km s^{-1}. Such a slow speed makes it very difficult to examine how the expansion proceeds as a function of geocentric distance, since there is no simple way to know accurately the precise distance between the midplane of the magnetotail and the satellites. During a substorm on 1968, August 15, Buck *et al.* (1973) observed the expansion speed of order 90 ± 30 km s^{-1} at a geocentric distance of 8 R_E.

Hones *et al.* (1970) showed further that at a geocentric distance of 18 R_E, the expansion of the plasma sheet is observed when the poleward advancing auroral (westward) electrojet reaches dp. lat. $\sim 75°$, namely at about the maximum epoch. Welcott *et al.* (1976) stressed the importance of their observation that the expansion of the plasma sheet at the Vela distance coincides with the 'poleward leap' of the auroral electrojet near its highest latitude. Therefore, at the Vela distance, the expansion of the plasma sheet may be regarded as a phenomenon which takes place at the very end of the expansive phase of magnetospheric substorms.

Oguti and Kokubun (1969) and Hones *et al.* (1973a) examined the time interval (ΔT_0) between the onset of a negative bay and the expansion. They noted that a short delay (0–10 min) is well confined within a geocentric distance of about 10 R_E, while beyond the distance of 20 R_E, ΔT_0 is greater than 30 minutes and does not depend clearly on the distance.

6.7. Plasma Flow

After the first identification by Hones *et al.* (1972b) of a large spin modulation (with one peak per rotation) of counting rates of protons in the plasma sheet as an indication of bulk motions of plasma, characteristics of plasma flow in the plasma sheet have been extensively studied by Hones and his colleagues (Hones, 1972a, b, 1973; Hones *et al.*, 1973; Hones *et al.*, 1974; Lui *et al.*, 1976). Since then an extensive amount of data has become available from Vela 5, 6, IMP-6 and other satellites. Here we examine plasma flow data from both the Vela and IMP-6 satellites. This is because the Vela satellites can observe flows only in a narrow cone angle which is perpendicular to the Earth-satellite line, so that they are not capable of observing a sunward or anti-sunward flow in the equatorward plane in the midnight sector, but are capable of observing an upward or downward flow, as well as a dawnward or duskward flow. On the other hand, the IMP-6 satellite is capable of observing flow of any direction in the equatorial plane, but not flows perpendicular to the equatorial plane. Therefore, it is necessary to combine both observations to obtain flow vectors in a three-dimensional space.

6.7.1. VELA SATELLITE OBSERVATIONS

Figure 6.31(a) shows five plasma parameters; pressure (ev cm^{-3}), density (cm^{-3}), bulk speed (km s^{-1}), energetic electrons (counts s^{-1}) and proton temperature (K), together with the AE index for the period between 11 and 18 UT, 1969, October

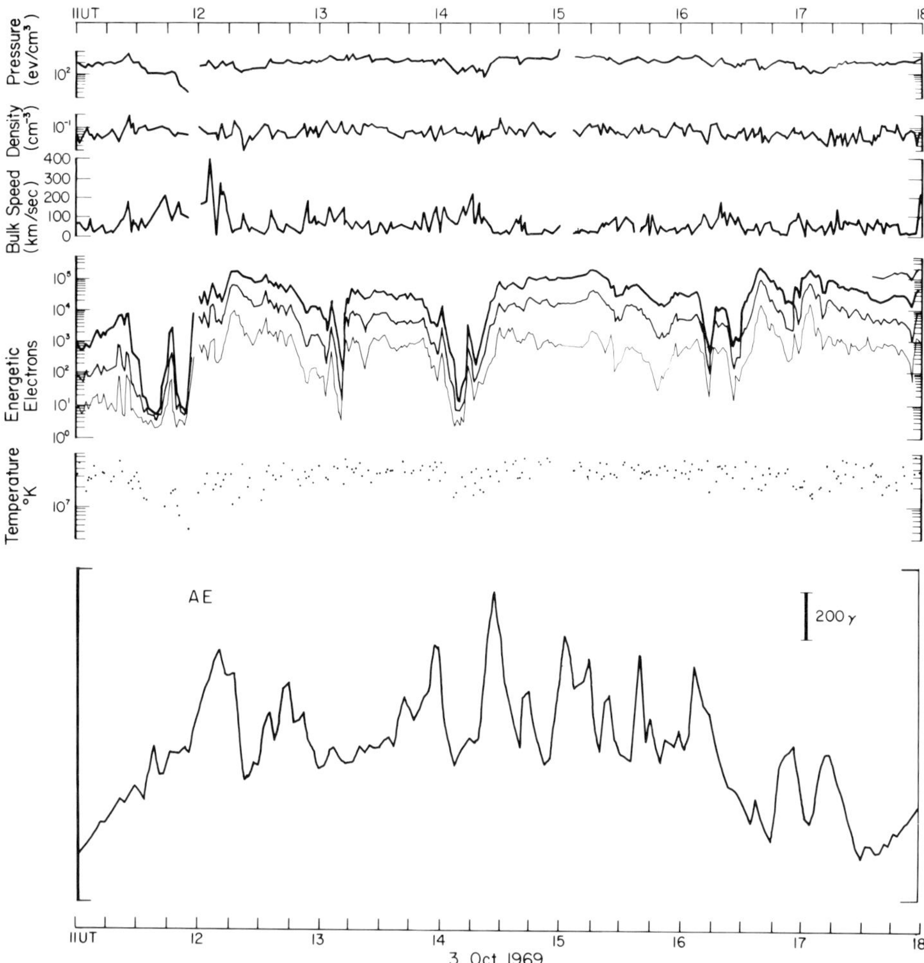

Fig. 6.31(a). Plasma sheet variations (from the top: pressure, density, bulk speed, energetic electron fluxes, temperature) during magnetospheric substorms on 1969, October 31, observed by the Vela satellite. The AE index is shown at the bottom; for the location of the satellite, see Figure 6.31(b).

3. The plasma flow speed during the corresponding period is also shown in the figure, together with the location of the satellite in the solar magnetospheric longitude ϕ-Z coordinates. The satellite was located in the plasma sheet in the evening sector, at about $Z \sim 2\,R_E$, throughout this period (Lui *et al.*, 1976).

The plasma sheet thinning and expansion can most easily be recognized by sharp decreases and increases in the energetic electron flux data. There were at least three major thinnings, beginning at about 1120, 1350 and 1610 UT, respectively. The corresponding decrease in plasma pressure by a factor of 2–3 can then be recognized. However, there is little corresponding change in density. Thus the thinnings were caused by a significant decrease in temperature, at least for the first two events.

There was a marked increase of bulk speed at about 1205 UT, during or soon after the sharp increase of energetic electron flux. In general, these two phenomena occur most commonly during the recovery phase of substorms. In this particular example, however, a number of substorms occurred in succession, and it is difficult to examine such a relationship. A similar but smaller increase of bulk speed was also observed at about 1410 UT, coinciding with the time of the increase of energetic electron flux. The third increase in bulk speed at about 1620 UT is also seen.

Figure 6.31(b) shows the 'direction' of bulk flow of plasma protons for the period in Figure 6.31(b). A flow vector which is consistent with a flow toward the Sun at the start of each hour interval is given by an arrow at that time and its orientation with respect to the vertical axis of the cross; the length of the arrow corresponds to a speed of 200 km s^{-1}. It can be seen that intense flows generally

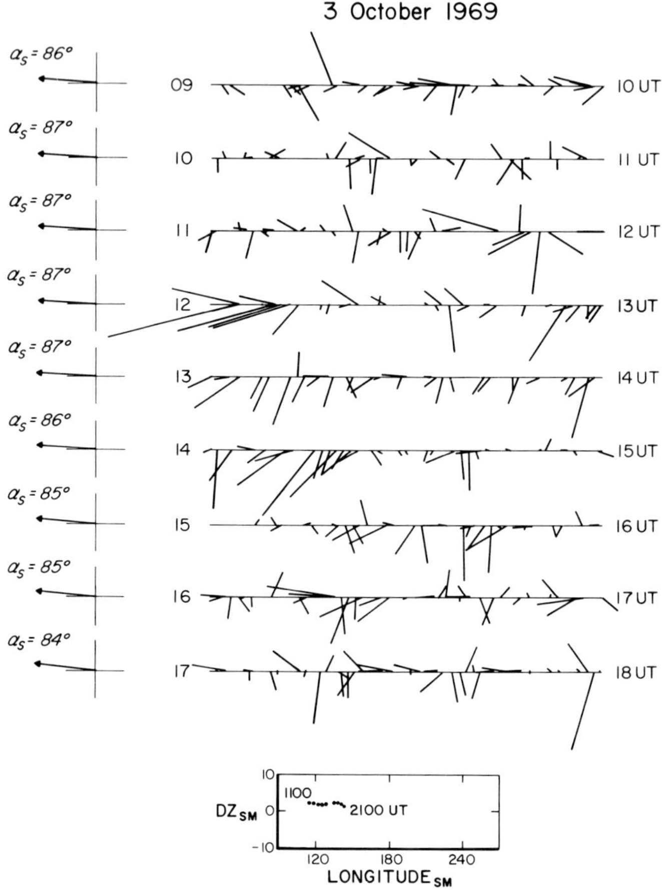

Fig. 6.31(b). 'Direction' and speed of plasma flow in the plasma sheet during the period in Figure 6.31(a). A flow vector which is consistent with a flow toward the Sun at the start of each hour is indicated by an arrow; the length of the arrow corresponds to a speed of 200 km s^{-1}. (Courtesy of Lui, A. T. Y., Hones, E. W. Jr., Yasuhara, F. and Akasofu, S.-I.)

directed toward the Sun (in the sense mentioned in the above) occurred during the period between 1205 and 1215 UT and during the period between 1412 and 1423 UT. However, the direction of the third flow was not clear.

Figure 6.32(a) shows the five plasma parameters, together with the *H* component magnetic record from Dixon (then a midnight station) and the AE index for the period between 1700 and 2030 UT, 1969, July 16. The plasma flow during the corresponding period is shown in Figure 6.32(b), together with the satellite location. The satellite was in the northern high latitude lobe at 1700 UT. A very rapid plasma sheet expansion was observed at about 1711 UT; the bulk speed reached as high as 350 km s^{-1} at about 1730 UT. The flow direction was approximately toward the Sun (again in the sense mentioned in the above) at that time. The second event began at about 1855 UT when the energetic electron flux and

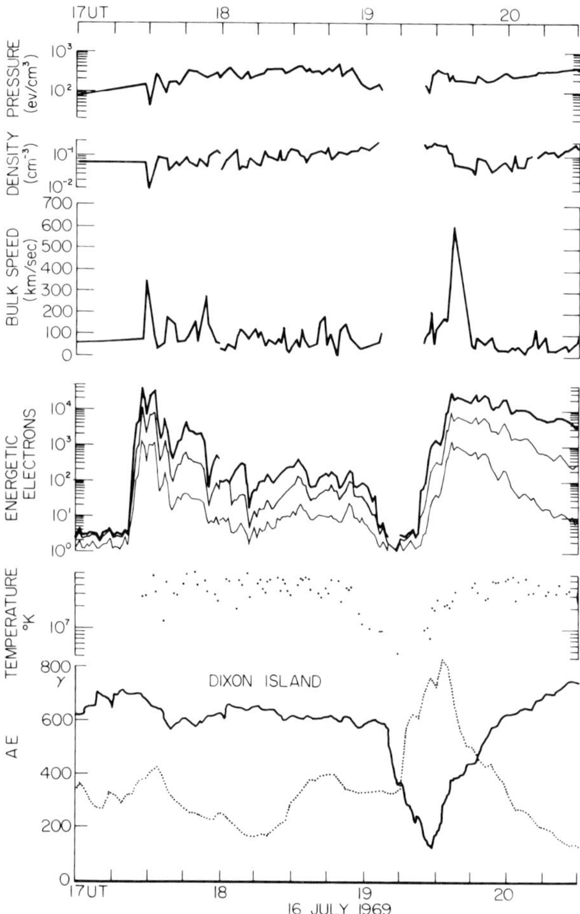

Fig. 6.32(a). Plasma sheet variations (from the top: pressure, density, bulk speed, energetic electron flux, temperature) during magnetospheric substorms on 1968, July 16. The AE index and the *H* component magnetic record from Dixon Island are also shown; for the satellite location, see Figure 6.32(b).

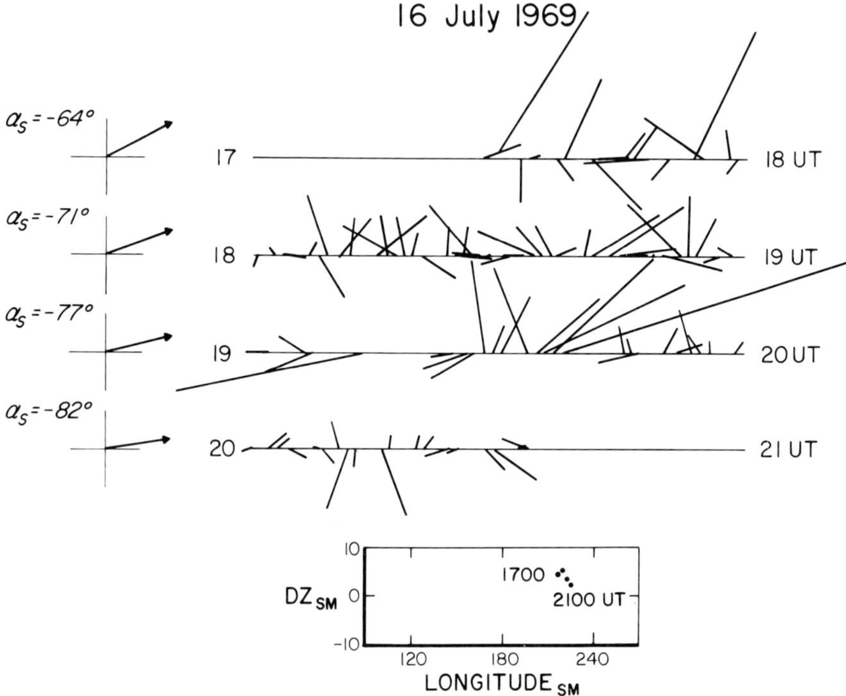

Fig. 6.32(b). 'Direction' and speed of plasma flow in the plasma sheet during the period in Figure 6.32(a). A flow vector which is consistent with a flow toward the Sun at the start of each hour is indicated by an arrow; the length of the arrows corresponds to a speed of 200 km s^{-1}. (Courtesy of Lui, A. T. Y., Hones, E. W. Jr., Yasuhara, F. and Akasofu, S.-I.)

plasma pressure began to decrease. Note that thinning was associated with a large decrease of proton temperature. The thinning lasted until about 1915 UT. A negative bay began at Dixon and the AE index began to increase at about 1907 UT, although the AE index was high ($\sim 300\ \gamma$) at that time.

As noted earlier, the Vela satellites cannot really observe the flow vector in a three-dimensional space. For this reason, Hones *et al.* (1972) devised a mercator map of the back of the Vela sphere (in the solar magnetospheric latitude and longitude) onto which the flow vector can be projected. Then, assuming that the flow is directed toward the Sun, they projected the expected flow direction onto the map. It is shown as an insert in Figure 6.33.

We have chosen 44 events during which the peak speed exceeded 100 km s^{-1}; the vectors are plotted in the longitude ϕ-Z coordinates in Figure 6.33. These events are selected by examining the accompanying large increase in the energetic electron flux data, together with the AE index. At the Vela distance ($\sim 18\ R_E$), the expansion of the plasma sheet is invariably accompanied by a sharp increase of the electron flux, so that it is reasonable to use such a signature in identifying the plasma sheet expansion. Several vectors are chosen for each event since it is generally difficult to choose a single vector as the representative one.

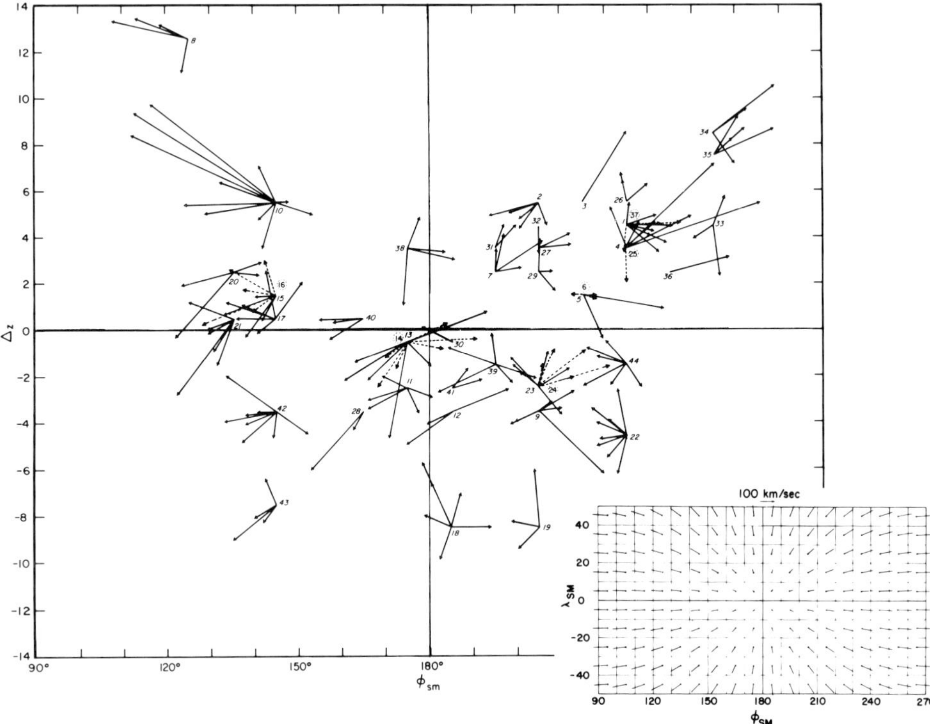

Fig. 6.33. Summary of plasma flow observations by the Vela satellites during the recovery phase of a number of substorms. The flow vectors are projected onto the Vela sphere in solar magnetospheric latitude-longitude coordinates. The insert shows the flow 'vectors' projected onto the Vela sphere when the flow is directed toward the Sun. (Courtesy of Lui, A. T. Y., Hones, E. W. Jr., Yasuhara, F. and Akasofu, S.-I.)

However, changes of the vectors did not exhibit any common systematic change, so their time variations are not indicated.

It can be seen from Figure 6.33 that except for the morning sector in the southern hemisphere, most of the vectors tend to point away from the center of the map; it is reasonable to conclude that the flow is, in general, directed toward the Sun during the expansion of the plasma sheet.

6.7.2. IMP-6 SATELLITE OBSERVATIONS

Figure 6.34(a) shows the plasma parameters (pressure, density, average proton energy, energetic electrons) and the AE index for the period between 07 UT on October 13 to 03 UT on 1972, October 14. The satellite was in the midnight sector and was inbound during this time. Substorm activity was frequent, and six thinnings and subsequent expansions of the plasma sheet were observed. Start of thinning was observed at 0824, 0928, 1239, 1842, 2200 and 2316 UT, and the corresponding full recovery of the plasma sheet was seen at 0900, 1100, 1448, 2031, 2235 and 0120 UT. Each of these recoveries was accompanied by the return

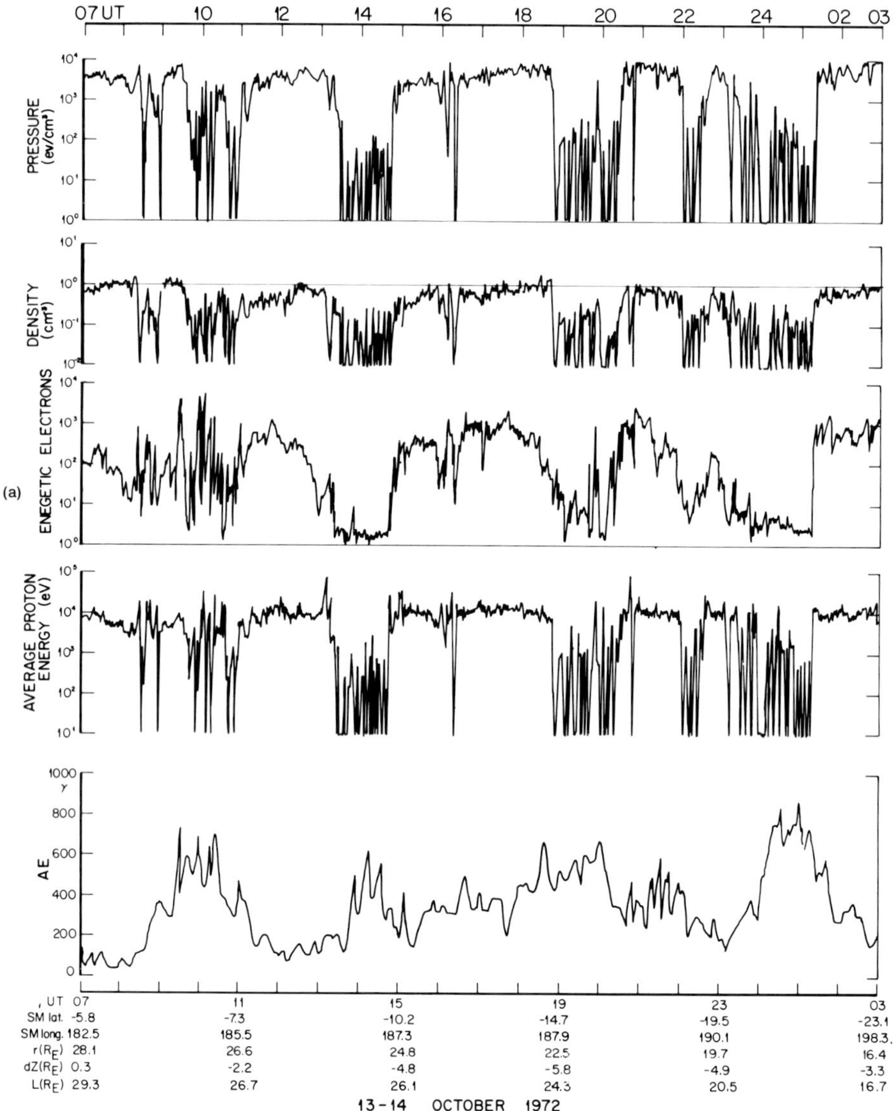

Fig. 6.34(a, b). Plasma sheet variations (from the top: pressure, density, energetic electron fluxes, average proton energy) and the AE index on 1972, October 13–14, and the direction of plasma flow in the *X*-*Y* plane. (Courtesy of Lui, A. T. Y., Hones, E. W. Jr., Yasuhara, F. and Akasofu, S.-I.)

of the energetic electron flux. Note that all the plasma parameters varied in unison during each thinning and recovery of the plasma sheet (Lui *et al.*, 1976).

The plasma flows observed during this period are given in Figure 6.34(b). Sunward flows were observed during the plasma sheet recovery of the second event at 1100 UT. The flow for this second event was observed to be small and fluctuating. Sunward flow was first seen for a brief interval of about 3 min. The flow changed to become duskward for about 10 min and finally became predomin-

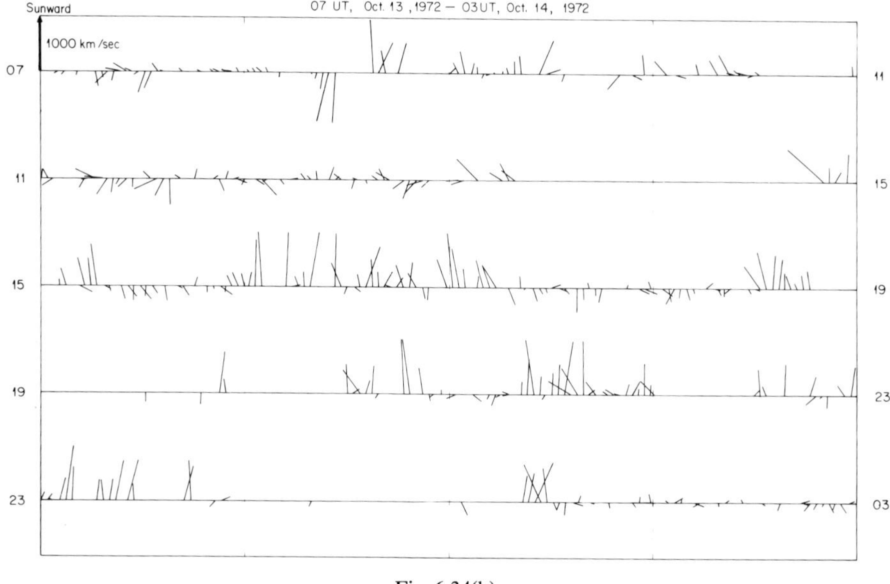

Fig. 6.34(b).

antly anti-sunward for the next 20 min. However, for all the other events plasma flows were usually large and directed generally sunward for at least 10 min. There were also two other intervals (1603–1715 UT and 2122–2140 UT) of strong sunward flows which were not associated with recovery of the plasma sheet at the location of the satellite. These periods were seen to correspond with continuous substorm activity as indicated by the AE index.

The observed plasma flows for 102 cases of plasma sheet expansion during the period from September 1971 to December 1972 were plotted on the solar magnetospheric equatorial plane in Figure 6.35. For all these events, the measured proton flux decreased by more than an order of magnitude during the thinning, and simultaneous substorm activity was indicated by the AE index. A dot without any flow vector indicates an event in which no significant flow ($> 100 \text{ km s}^{-1}$) was detectable. Note that the maximum flows were selected for cases in which the flows during the expansion of the plasma sheet were approximately in one consistent direction for at least more than 5 min. For cases in which the flow direction changed significantly ($> 30°$) during the plasma sheet recovery (e.g., the flows observed during the recovery at 1000 UT in Figure 6.34(b)), two or more arrows were displayed in the plot.

Figure 6.35 shows clearly that sunward flows were observed in an extensive region of the magnetosphere during the expansion of the plasma sheet. Further, the majority of the strong sunward flows were observed in the midnight sector where $|Y| \leq \sim 10\, R_E$, and weaker flows were usually observed at larger values of $|Y|$. Note that strong sunward flows usually lasted for only about 10–20 min.

In summary, the IMP-6 observations show clearly that the flow is generally directed toward the Sun in the equatorward plane. Thus it is reasonable to explain

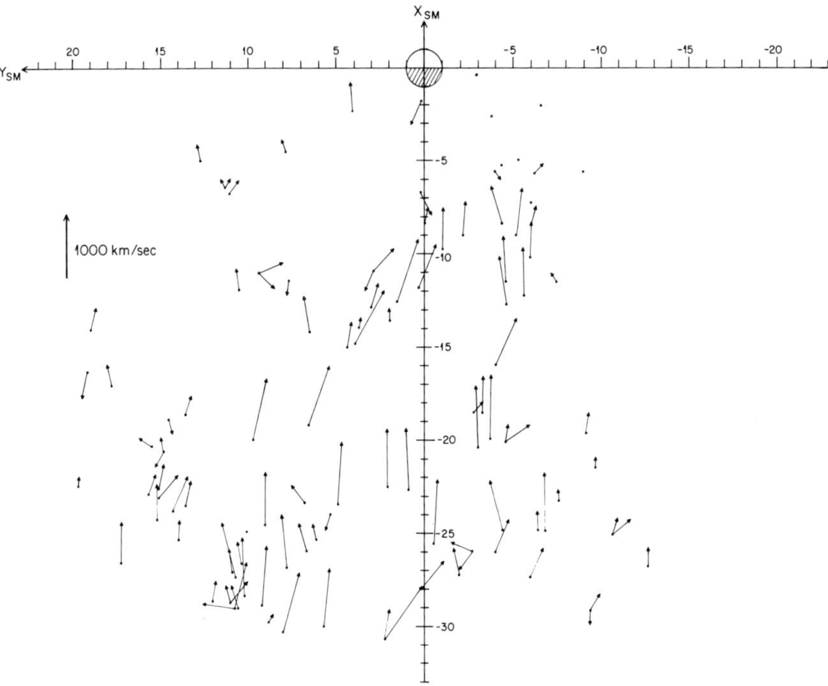

Fig. 6.35. Summary of plasma flow observations (the X-Y plane) during the recovery phase of a number of substorms, observed by the IMP-6 satellite. (Courtesy of Lui, A. T. Y., Hones, E. W. Jr., Yasuhara, F. and Akasofu, S.-I.)

the Vela observations near the center of the map in terms of slight fluctuations of the direction of the sunward flow. Note that the speed of the flow observed by the IMP-6 satellite is quite high ($\sim$1000 km s^{-1}). The Vela observations assure us also that there is no systematic high upward or downward flow.

In the dawn and dusk sector, both the IMP-6 and Vela observations indicate that the flow is directed toward the Sun. Note that in these sectors the Vela satellites can see a sunward or anti-sunward flow well.

The average duration of sunward flow is about 12 min, and the most enduring sunward flows are associated with continuous substorm activity rather than with the proximity to the neutral sheet. This result suggests that the sunward flow does not exist throughout the width of the plasma sheet.

6.7.3. FLOW REVERSAL

One of the critical tests of the reconnection theory will be to examine together both the B_z component and the plasma flow. In Figure 6.1, one would expect that when a satellite is located outside the neutral line, the B_z component reversal and an anti-solar plasma flow would be observed, while a positive B_z variation and a sunward plasma flow would occur within the distance of the neutral line. This test requires two satellites, one earthward of the proposed neutral line and the other anti-earthward. Unfortunately there is so far no report available on such a simultaneous observation. Thus, most of flow reversal observations have been interpreted in terms of hypothetical motions of the neutral line along the $-X$ axis in the vicinity of single satellites. Figure 6.36(a) shows schematically this situation.

Figure 6.36(b) shows an example presented by Hones *et al.* (1976c). It can be seen that a strong anti-sunward flow with the maximum speed of about 1000 km s^{-1} began at 0550 UT. The flow direction changed at about 0630 UT, and an intense sunward flow continued until about 10 UT or a little later.

Figure 6.36(c) shows details of the flow variations at about 0600 UT, together with the concurrent satellite magnetometer record. It can be seen that the B_z component became negative when the flow was directed in the anti-solar direction (0558–0603 UT). On the other hand, the magnitude of the field B increased rapidly at the time, indicating that the magnetic field vector dipped toward the equatorial plane (Section 6.2).

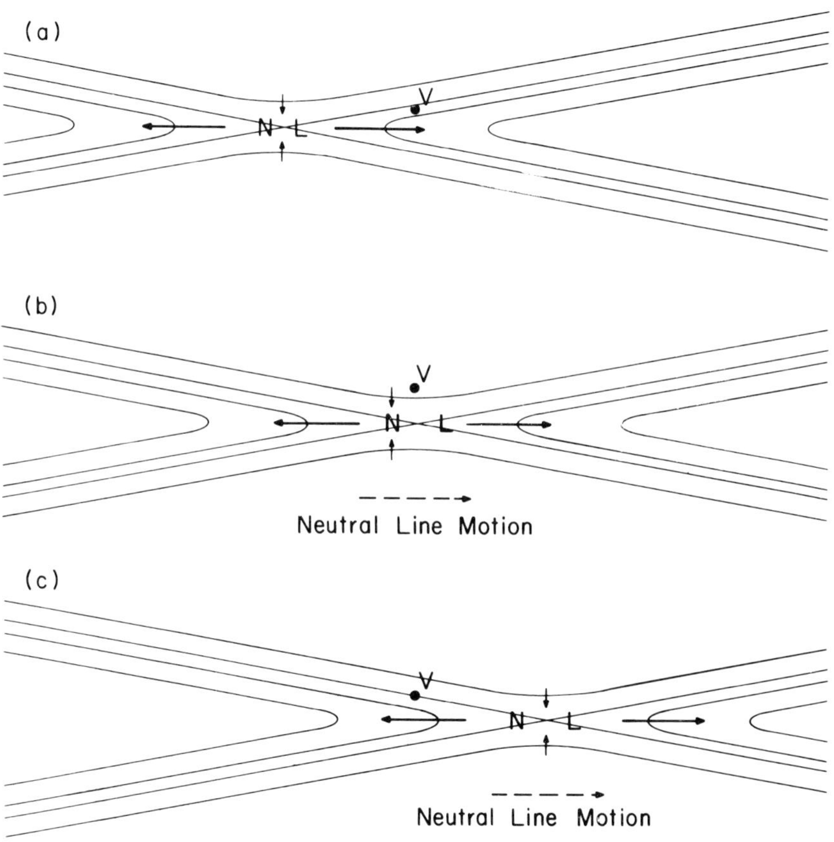

Fig. 6.36(a). Schematic diagram showing suggested changes of the magnetic field configuration (including the formation and shift of a hypothetical neutral line) during a magnetospheric substorm in the vicinity of the Vela satellite (indicated by a dot) at $X \simeq -18\ R_E$. (Hones, E. W. Jr., Lui, A. T. Y., Bame, S. J. and Singer, S.: *J. Geophys. Res.* **79**, 1385, 1974.)

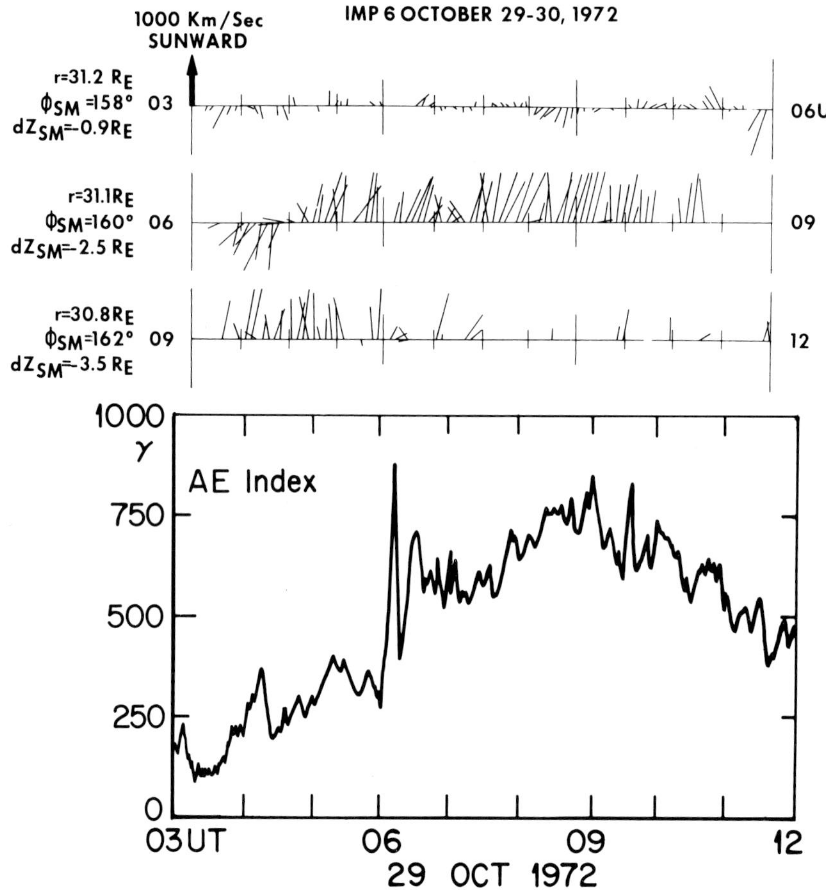

Fig. 6.36(b). Plasma flow observation by the IMP-6 satellite on 1972, October 6. The AE index is also shown. (Hones, E. W. Jr., Bame, S. J. and Asbridge, J. R.: *J. Geophys. Res.* **81**, 227, 1976.)

The unusually hot anti-sunward flow, with energetic electrons, lasted for about 25 min (until 0628 UT). The B_z component was slightly negative and oscillating during that period. When the flow direction reversed at 0628 UT, the B_z component became positive. This is a very unusual event in that an anti-solar flow of hot plasma appeared during an early epoch (0558–0628 UT) of the substorm and thus does not serve as a typical example.

Figure 6.37 shows another example of the simultaneous observations of both the magnetic field and plasma flows (Frank and Ackerson, 1976). Although the negative B_z component was observed simultaneously with an anti-solar flow of plasma, it was also associated with a large increase of the magnitude of B. Therefore, it is likely that the negative B_z component was caused by the dipping of $\boldsymbol{B}$ vector. As Frank and Ackerson (1976) pointed out, it is important to note that anti-sunward flows are associated with the positive B_z component, as well as the negative B_z component. Figure 6.38 shows an example of a long-lasting anti-solar flow near the midplane, which was associated with a large positive B_z component.

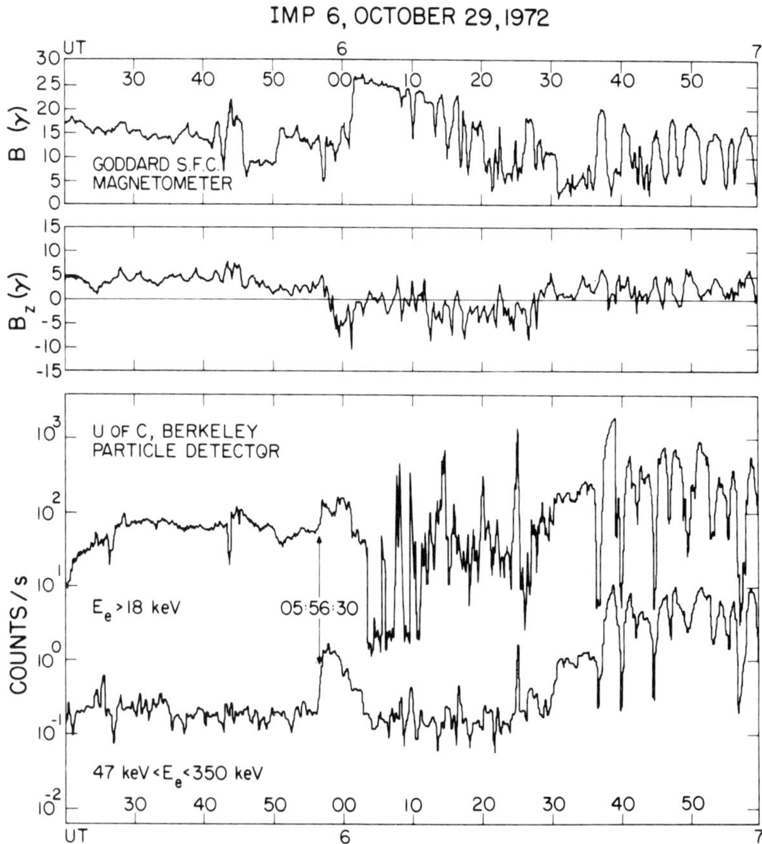

Fig. 6.36(c). Magnetic field (B, B_z) and energetic electron flux observations during the period shown in Figure 6.36(b).

Figure 6.39 shows another example of plasma sheet variations during a substorm, observed by IMP-7 in the dusk sector (Roelof *et al.*, 1976). In this particular event, an isotropic proton flux became suddenly anisotropic at about 0513 UT: the protons were flowing away in the anti-solar direction. About 5 min after this event, the flux in all sectors dropped suddenly for about 1 min below 1 count s^{-1}. Then a new burst appeared at about 0520 UT. Shortly after the onset of this new burst, the flux in the sectors between $\phi = 90°$ and 200° dropped below the threshold of the detector and remained depressed until about 0540 UT. This anti-sunward flow consisted of protons of energy of order 4.8 keV, flowing with a speed of about 800 km s^{-1}. At about 0540 UT, the plasma sheet returned to the satellite locations, and the protons (~5.9 keV) were flowing toward the Sun with a speed of 620 km s^{-1}.

Burke and Reasoner (1973) reported an anti-solar flow of plasma with a speed of order 250 km s^{-1} during the expansive phase at the lunar distance.

Bursts of anti-solar plasma flows were also detected by a superthermal ion detector located on the lunar surface during the magnetic storm of 1970, August

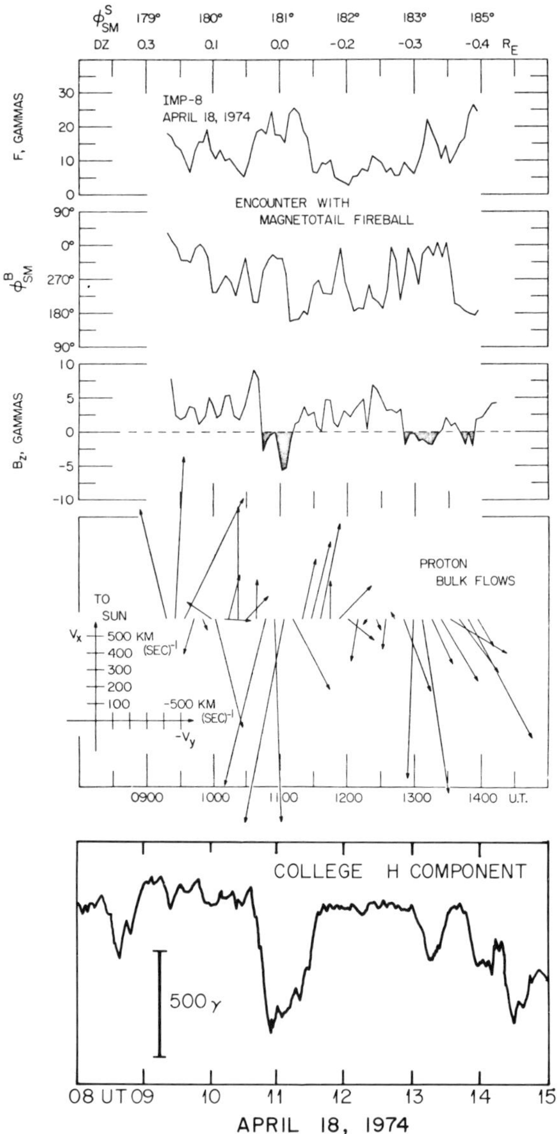

Fig. 6.37. Simultaneous observations of the magnetic field and plasma flow in the magnetotail. (Frank, L. A. and Ackerson, K. L.: *Dept. of Phys. Astronom. Rept.* 75–36, Univ. of Iowa, September 1975.)

17; Figure 6.40. Garrett *et al.* (1971) noted that two most intense bursts were each followed roughly 1/2 hour later by the sudden onset of an intense negative bay at the Earth's surface. They suggested that the bursts were caused by anti-sunward flows of plasma sheet particles during the growth phase. The only problem is the identification of two series of events during an intense geomagnetic storm, when intense substorms occur frequently.

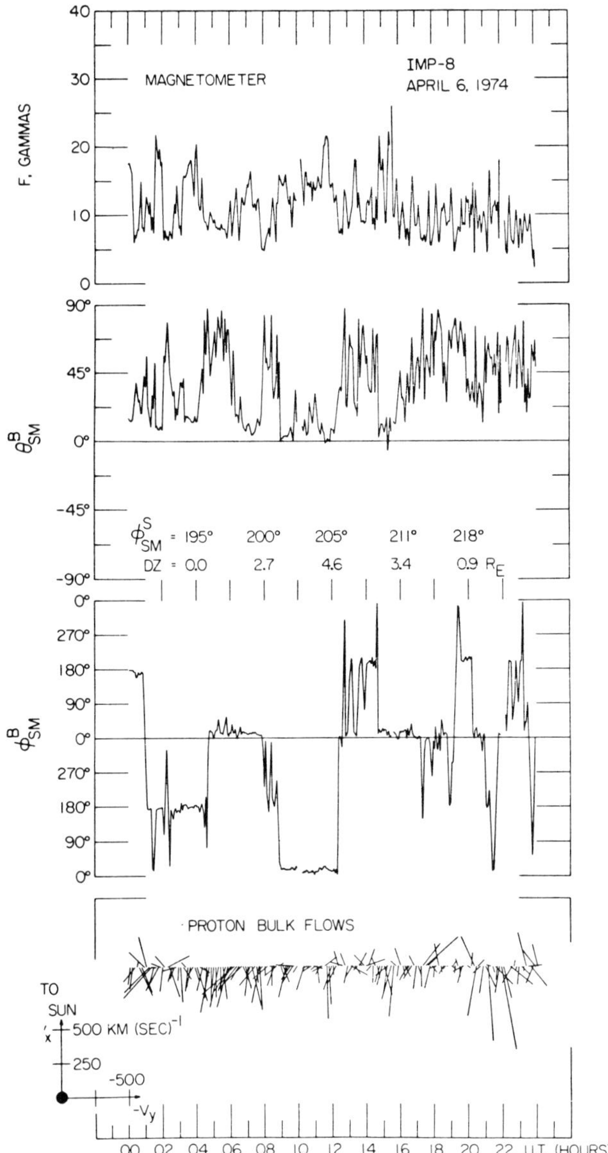

Fig. 6.38. Prolonged anti-solar flow of plasma in the plasma sheet (the bottom data) and the associated magnetic field variations ($|B| = F$, θ and ϕ). (Frank, L. A. and Ackerson, K. L.: *Dept of Phys. Astronom. Rept.* 76–1, Univ. of Iowa, January 1976.)

In summary, it appears that the flow pattern of plasma varies drastically during substorms, sometimes anti-sunward in an early epoch and invariably sunward when the plasma sheet reappears. It is too premature to conclude that the anti-sunward flow is closely associated with thinning of the plasma sheet. In fact, a recent extensive study of flow during thinning shows that within $X > -30\ R_E$

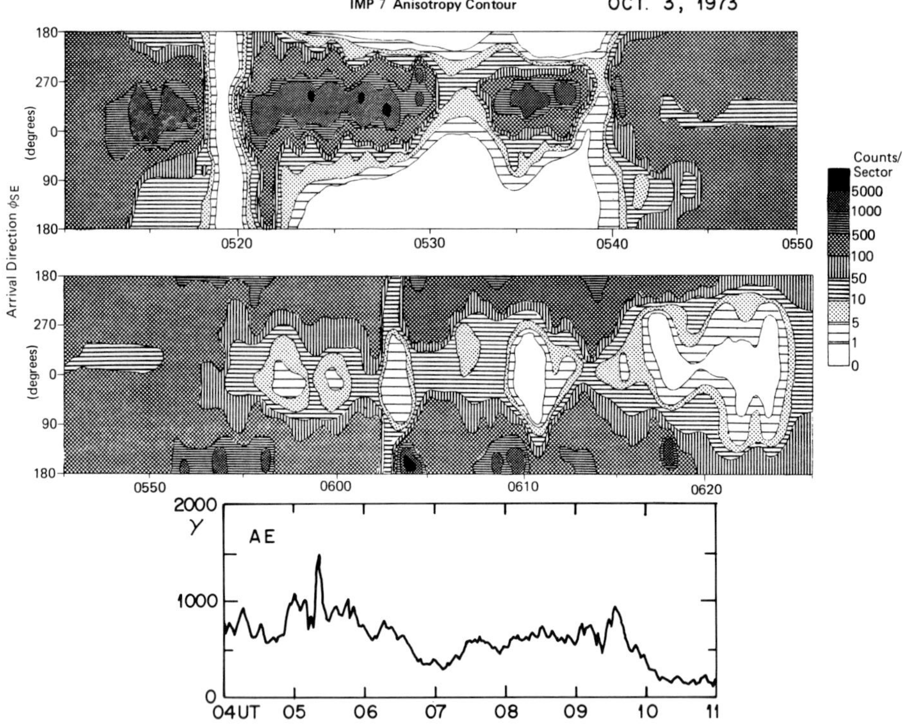

Fig. 6.39. Plasma flow observation by the IMP-7 satellite on 1973, October 1. The AE index is also shown. (Roelof, E. C., Keath, E. P., Boström, C. O. and Williams, D. J.: *J. Geophys. Res.* **81**, 8, 1976.)

the flow is directed toward the Sun during thinning. It may also be premature to conclude that the observed flow is a proof of the reconnection theory.

In fact, it is not known accurately whether the observed plasma flow is a field-aligned flow or an $(E \times B)$ drift, although it is likely that most of the observed flow is field-aligned. If so, is it in agreement with the reconnection theories? It should be noted that there is no indication of energization of plasma during thinning. The plasma in the anti-solar flow during thinning should, however, be hot, if the flow is caused by reconnection processes. These problems will be discussed in detail in Section 9.2.

6.8. Other Important Magnetotail Phenomena

6.8.1. BRIEF APPEARANCE OF ENERGETIC ELECTRONS IN THE PLASMA SHEET

Hones *et al.* (1971b) reported that a large impulsive flux of energetic electrons was observed at a geocentric distance of 18 R_E (Vela-4A) almost precisely ($\pm$20 s) at the onset of the expansive phase of a substorm that began at 0930 UT on 1968, September 14, (Figure 6.41(a)). An auroral arc located in the southern sky of College (Alaska) suddenly became bright and began to move poleward at that time

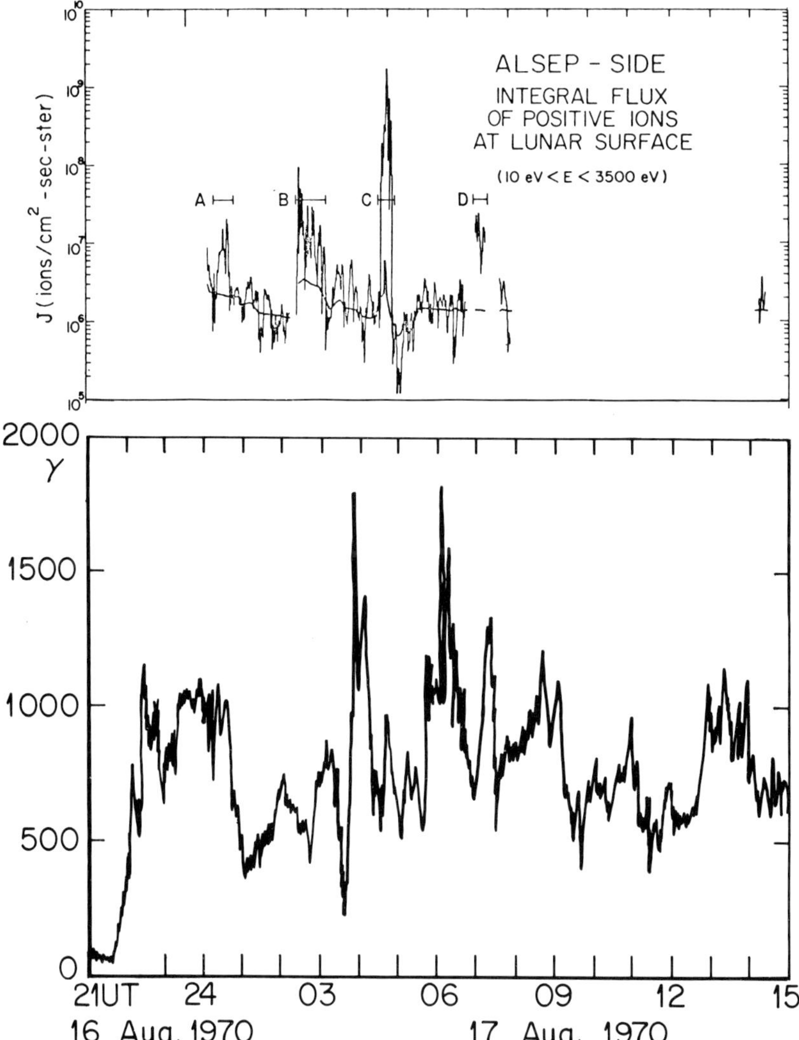

Fig. 6.40. Bursts of plasma observed at the lunar surface during the geomagnetic storm of 1970, August 16–17 (the AE index). (Garrett, H. B., Hill, T. W. and Fenner, M. A.: *Planet. Space Sci.* **19**, 1413, 1971.)

(Figure 6.41(b)). There seems little doubt that the sudden brightening of the arc was related to the burst of electrons observed by Vela-4A. Two important questions arise: (i) can the field lines originating near College reach as far as a geocentric distance of 18 R_E and (ii) why was the duration of the enhanced electron flux only about 15 min? Akasofu *et al.* (1971a) examined 29 substorms and found that 5 of them were associated with a similar impulsive flux; they also noted that a similar event was also observed at a geocentric distance of about 30 R_E (IMP-3).

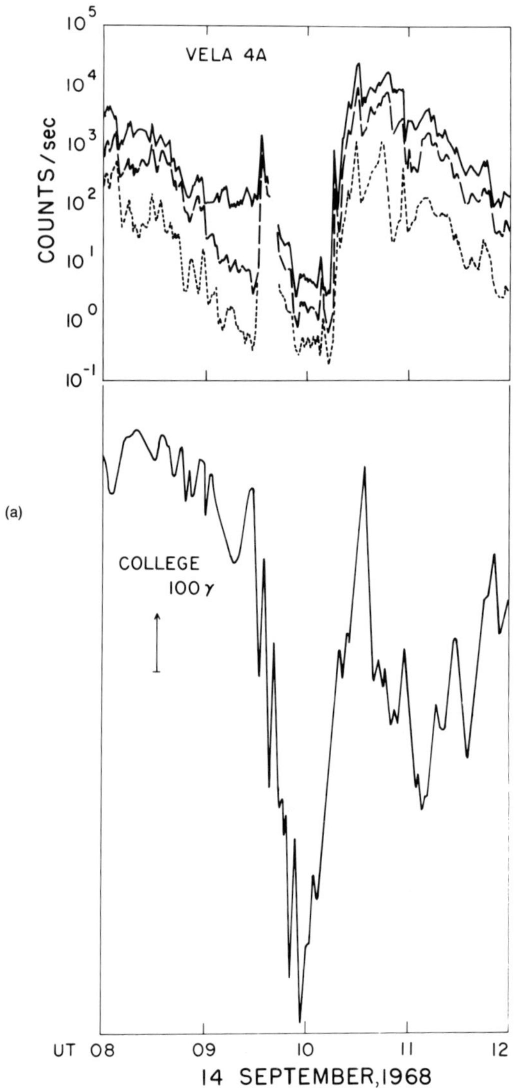

Fig. 6.41(a, b). Occurrence of an impulsive flux of energetic electrons observed by the Vela 4A satellite ($X \simeq -18\,R_E$) at the onset time of a magnetospheric substorm on 1968, September 14. The H component magnetic record and all-sky photographs taken from College are also shown. (Hones, E. W. Jr., Karas, R. H., Lanzerotti, L. J. and Akasofu, S.-I.: *J. Geophys. Res.* **76**, 6765, 1971.)

6.8.2. Leakage of Plasma Sheet Particles into the Magnetosheath

Vela satellites cross the magnetopause from outside to inside in the evening sector of the magnetotail and from inside to outside in the morning sector. A set of plasma particle detectors can, in general, clearly determine the moment of the crossing. Figure 6.42(a) shows the proton flux variations during an evening crossing. Here we have plotted the average counting rate of the 0.55–0.77 keV (the

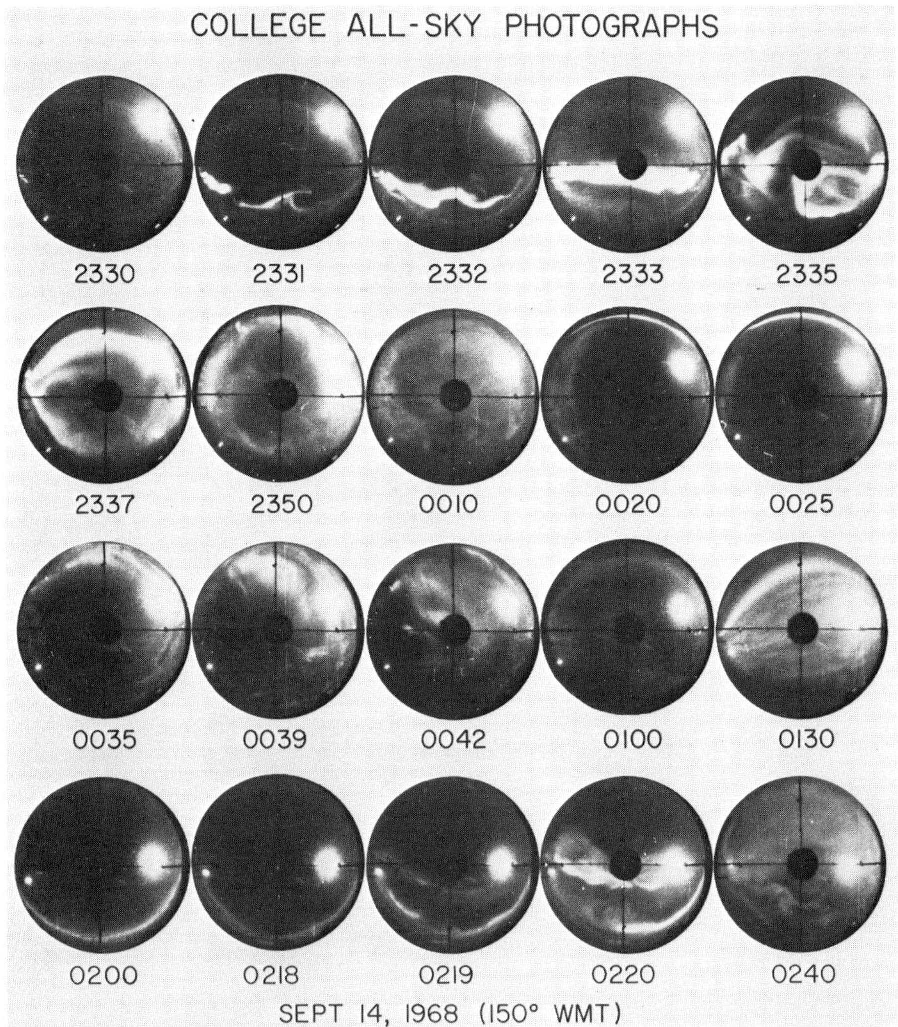

Fig. 6.41(b).

typical magnetosheath proton energy) and 19–28 keV channels (the typical plasma sheet proton energy at a geocentric distance of 18 R_E). The low energy (0.55–0.77 keV) flux disappeared at about 0755 UT, indicating that the satellite crossed the magnetopause into the magnetotail. After the crossing, the satellite encountered only the high energy plasma which is one of the characteristics of the magnetotail plasma sheet at about 18 R_E. One important feature to be noted here is that the high energy protons had been encountered, but at lower and less uniform intensities, *before* the magnetopause crossing (while the satellite was in the magnetosheath). The high energy (19–28 keV) protons began to increase rapidly after 0530 UT; note that owing to relative motions between the satellite and the magnetopause, the satellite crossed the magnetopause three times

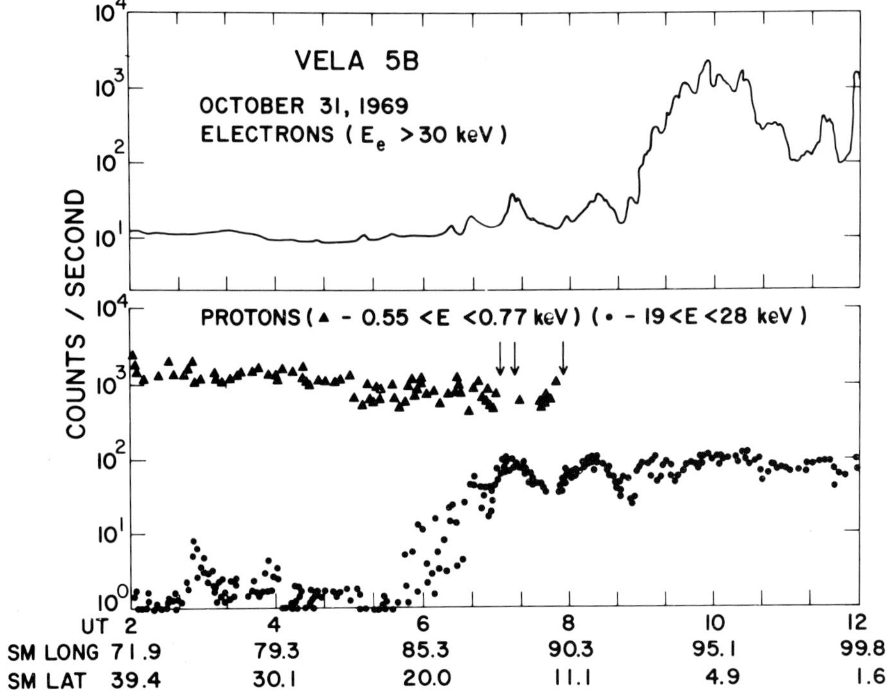

Fig. 6.42(a). Solar wind protons, energetic protons (of possible plasma sheet origin) and energetic electrons across the duskside magnetopause, observed by the Vela 5B satellite.

(indicated by arrows). This observation shows that protons in the energy range typical of the plasma sheet are seen well outside the magnetopause. Since the flux is generally highest in the plasma sheet and decreases from the magnetopause as the distance increases outward, it is not unreasonable to infer that the source of these protons lies within the magnetotail.

Hones *et al.* (1972a) noted that the variations of the proton flux are often more complicated than those shown in Figure 6.42(a) and that the high energy flux does not increase smoothly toward the magnetopause. There occur sporadic increases, superposed on the gradual changes described in the previous section. In many cases, it is found that those sporadic changes are associated with magnetospheric substorms. Figure 6.42(b) shows, from the top, energetic electron flux ($\varepsilon >$ 30 keV), the high energy proton flux and the AE index. Within about 5 to 10 min after an increase in the AE index, the proton flux began to increase. Another substorm began at about 0330 UT, and there were also the corresponding increases of the proton flux. The proton increases were associated with significant increases in the flux of the energetic electrons.

It is not difficult to show that these protons are not of solar origin. A suitable combination of two Vela satellites, one located well outside the bow shock and the other in the magnetosheath, is used to examine the flux of protons (9–13 keV) in the solar wind when an intense increase of such protons is observed near the magnetopause.

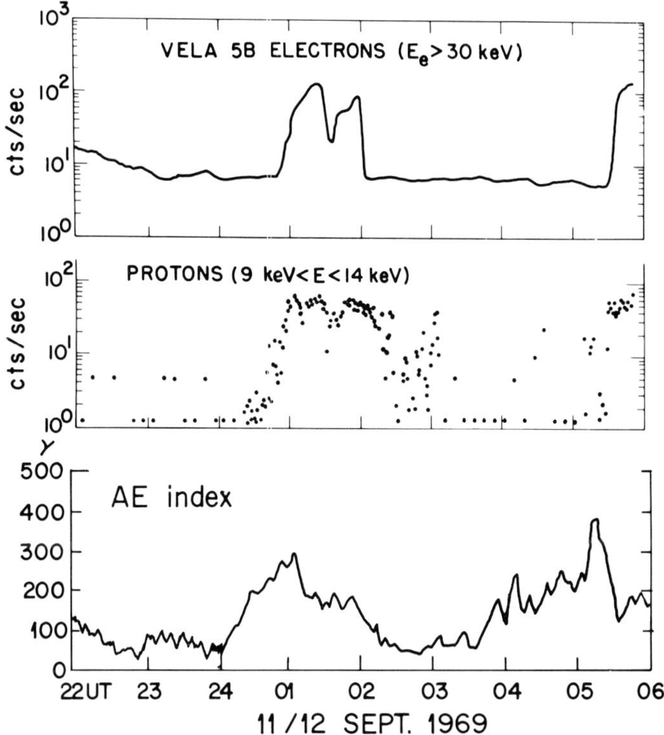

Fig. 6.42(b). Intense fluxes of both energetic electrons and protons just outside the magnetopause during a magnetospheric substorm (the AE index). (Hones, E. W. Jr., Akasofu, S.-I., Bame, S. J. and Singer, S.: *J. Geophys. Res.* 77, 6688, 1972.)

It should also be noted that Burke *et al.* (1973) reported that plasma sheet electrons leaked through the magnetosphere and became a component of the magnetosheath flow.

The examples of substorm-related appearances of the protons in the magnetosheath are about equally divided among the dawn and dusk sides of the magnetosheath. On one occasion, such protons were observed simultaneously in both the dawn and dusk sides by two Vela satellites. Therefore, there is a possibility that at least some of the protons in the plasma sheet leak out into the magnetosheath during the thinning period.

6.8.3. 'GEOMAGNETIC STORM PARTICLES'

Bame *et al.* (1971) found nearly monoenergetic energetic positive ions (0.3–3 keV) which flow in the anti-solar direction in the high latitude lobe of the magnetotail during geomagnetic storms. They are not present in the plasma sheet.

Figure 6.43(a) shows the *H* component magnetic records from several low latitude stations during the geomagnetic storm of 1970, April 20, and Figure 6.43(b) shows the trajectory of Vela 6A across the tail during the storm period.

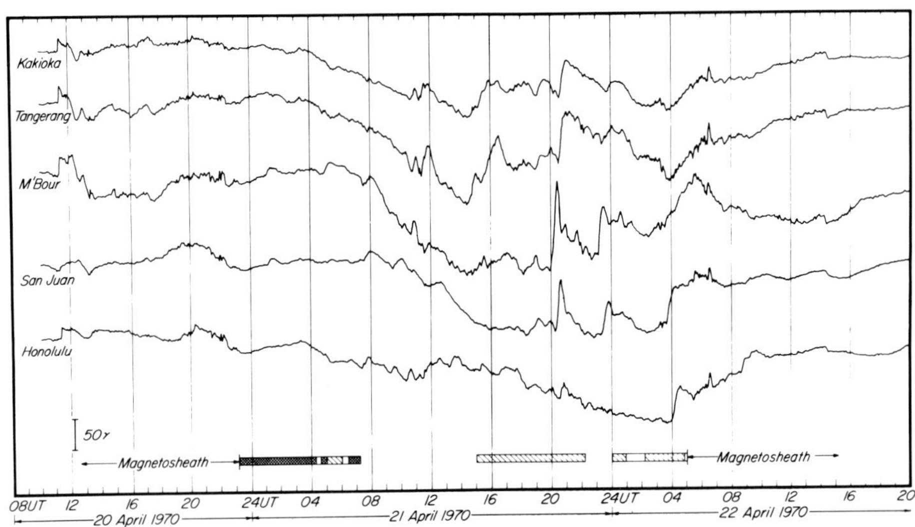

Fig. 6.43(a, b). Monoenergetic protons (0.3–3 keV) flowing in the anti-solar direction in the high latitude lobe in the magnetotail during the geomagnetic storm of 1970, April 20–22. The trajectory of the Vela 6A satellite is shown in (b). The *H* component magnetic records from five low latitude stations are shown to indicate the development of the storm. The periods when the streaming protons were observed are shown by hatched bars in (a) and thin lines along the trajectory in (b). Note that the observation was not continuous. (Bame, S. J., Hones, E. W. Jr., Akasofu, S.-I., Montgomery, M. D. and Asbridge, J. R.: *J. Geophys. Res.* **76**, 7566, 1971.)

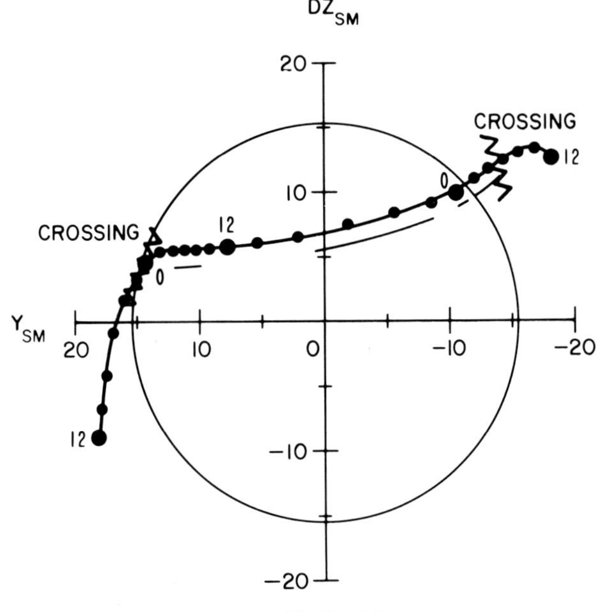

Fig. 6.43(b).

The satellite crossed the magnetopause at about 2300 UT, April 21, and about 0500 UT, April 22. Light lines drawn parallel to the trajectory show roughly where the storm particles were observed.

It appears that these storm particles are solar wind protons in the plasma mantle, which perhaps spread into the high latitude lobe even at as close as $X = -18\,R_E$ during a geomagnetic storm when intense substorms occur frequently.

6.8.4. DAWN-DUSK ASYMMETRY OF THE PROTON AND ELECTRON DISTRIBUTIONS

(a) *Production of Sub-Relativistic Protons and Electrons and Their Dawn-Dusk Asymmetry*

Fennell (1970) reported that bursts of energetic protons (> 0.32 MeV) occur at the lunar distance in the magnetotail during intense magnetospheric substorms. Further, these bursts tend to occur more frequently in the dusk half of the magnetotail than in the morning sector. A similar feature was also recently reported by Sarris *et al.* (1975) and Hones *et al.* (1976a). The energy range observed by the former workers was $0.29 < \varepsilon < 50$ MeV (Figure 6.44(a)).

A similar, but opposite, asymmetry in energetic electrons (~ 50 keV) has been reported by several workers (Montgomery, 1968; Meng, 1971) although relativistic electrons ($0.22 < \varepsilon < 2.5$ MeV) do not show any asymmetry (Sarris *et al.*, 1975).

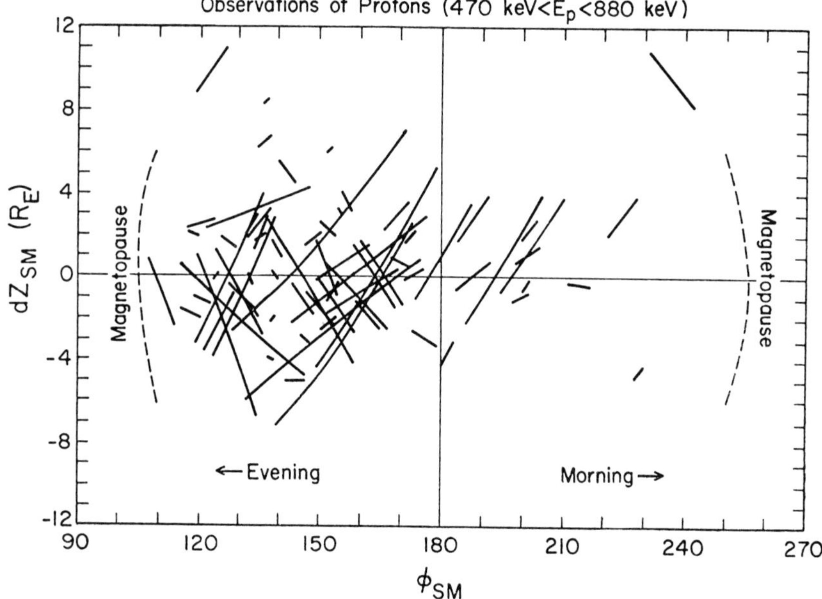

Fig. 6.44(a). Parts of Vela satellite trajectories along which intense fluxes of energetic protons ($470\,\text{keV} < \varepsilon < 880\,\text{keV}$) are observed. (Courtesy of Hones, E. W. Jr., Palmer, I. D. and Higbie, P. R.)

(b) *Dawn-Dusk Asymmetry of Precipitating Protons and Electrons*

Akasofu *et al.* (1969) noted that the hydrogen emission (Hβ) is absent in the westward traveling surge which is the brightest feature of the aurora in the evening sector. Their result was soon confirmed by Montbriand (1969). Frank and Ackerson (1971, 1972) found that there is no significant proton precipitation in the inverted V structure (Section 2.4.1).

The proton aurora in the morning sector has been extensively studied by Fukunishi and Tohmatsu (1973), Oguti (1973) and Fukunishi (1975). They showed that in the morning sector both the diffuse (electron) aurora and the proton aurora overlap during a quiet time, but the proton aurora appears poleward of the electron aurora during substorms. Figure 6.44(b) shows schematically the gross distribution of both types of auroras during a quiet and a substorm time.

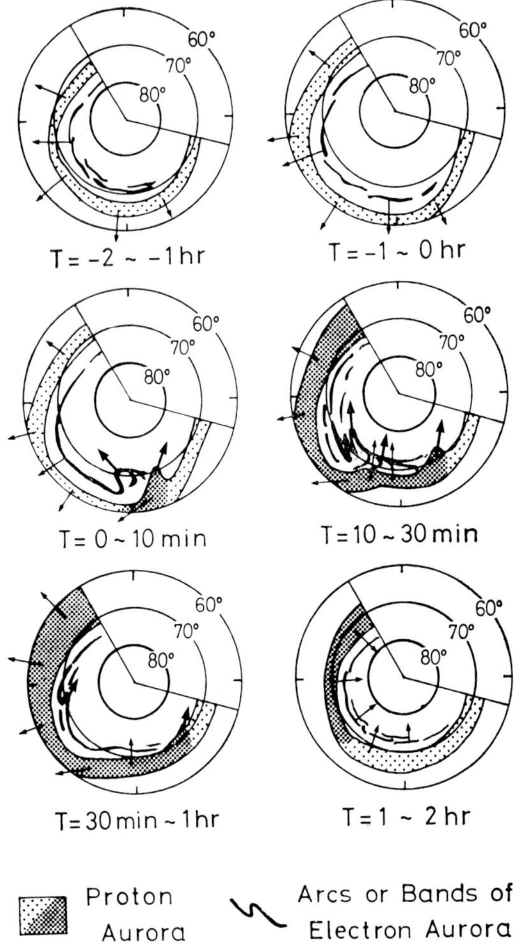

Fig. 6.44(b). Development of the proton and electron aurora substorms. (Fukunishi, H.: *J. Geophys. Res.* **80**, 553, 1975.)

Fukunishi (1975) pointed out: (i) during the period preceding the onset of a substorm expansive phase the proton aurora moves equatorward with a speed of 100–200 m s^{-1} (accompanying the development of the ring current); (ii) at the onset of the expansive phase the proton aurora rapidly expands poleward with a large increase in luminosity in the postmidnight sector; and (iii) the proton aurora is absent in the leading edge of the expanding electron auroral bulge, whereas breakup-type electron arcs and bands are not observed in the expanding proton auroral bulge. He suggested that there is a mechanism that accelerates electrons, in the premidnight region, and protons, in the postmidnight region, along the geomagnetic field lines down to the ionosphere.

(c) *Summary*

It is interesting to note that the asymmetry of the particle population is opposite in the ionosphere and in the magnetosphere. In the evening sector, electrons are accelerated downward, along the field lines, while protons are accelerated upward. In the morning sector, protons are accelerated toward the ionosphere, whereas electrons are accelerated upward. Akasofu (1970) suggested that this asymmetry is related to the primary field-aligned currents; the current flows into the morning half of the oval and out from the evening half of the oval (Section 1.3.2). Figure 6.45 illustrates this situation schematically.

6.8.5. SUMMARY OF THE MAGNETOTAIL OBSERVATIONS

The purpose of this section is to list the major features of the magnetospheric substorm in the magnetotail.

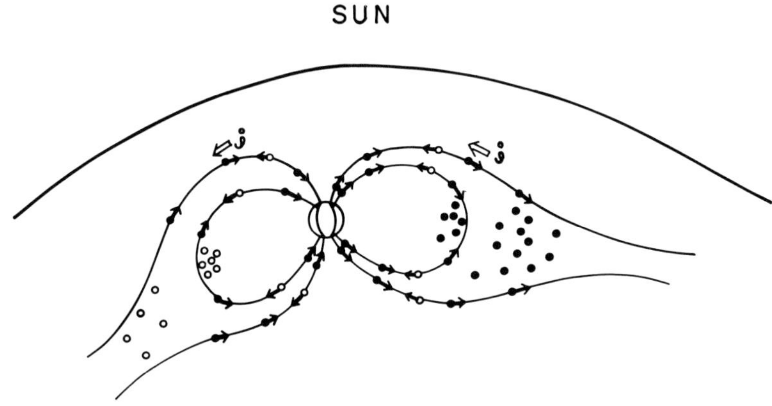

Fig. 6.45. Schematic diagram showing the direction of the primary field-aligned currents in the dawn and dusk meridians and the possible acceleration of protons associated with the currents. (Akasofu, S.-I.: *Solar-Terrestrial Physics/1970*, E. R. Dyer (ed.), p. 131, D. Reidel Publ. Co., 1972.)

(a) *Plasma Sheet Thinning*

(i) Plasma sheet thinning begins at about the onset time ($T = 0$) of the expansive phase, not $T = -30$ min.

(ii) At $X < -15\,R_E$, thinning proceeds until the poleward expanding auroral bulge (or the electrojet) reaches about dp. lat. 75°. Figure 6.46 illustrates schematically the configuration of the plasma sheet and of the magnetic field at about the maximum epoch of a substorm.

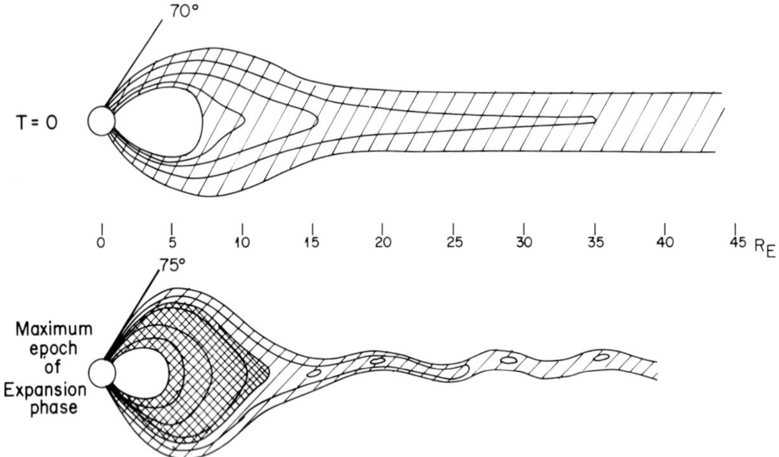

Fig. 6.46. Schematic diagram showing the configuration of the plasma sheet and of the magnetic field at $T = 0$ and at the maximum epoch of the magnetospheric substorm.

(iii) In general, plasma flow is weak during thinning. At $X > -15\,R_E$, when plasma flow is observable, it is directed sunward, while at $X \simeq -30\,R_E$, an anti-sunward flow is sometimes observed.

(iv) The plasma sheet becomes thinnest in the midnight sector. There is no indication of compression of the plasma sheet toward its midplane, since thinning occurs in the vicinity of the midplane (although it may not be a complete thinning or the so-called 'dropout').

(b) *Plasma Sheet Expansion*

(i) At $X \simeq -15\,R_E$, the plasma sheet begins to reappear or expand at about the maximum epoch of substorms.

(ii) Plasma particles in the expanding plasma sheet are much hotter than in a quiet time plasma sheet. Proton energy increases from about 1 keV (a quiet time value) to 20 keV.

(iii) However, plasma pressure does not show such a notable increase, indicating that the density is appreciably lower in the expanding plasma sheet than in the quiet time plasma sheet.

(iv) The expanding plasma sheet is almost always associated with a high speed (~ 500 km s^{-1}) sunward flow of plasma.

(c) *Magnetic Field Variations*

(i) There is no definite indication of the formation of a new magnetic neutral line in the near-Earth plasma sheet, which leads to a large-scale change of the magnetic field configuration suggested in Figure 6.1.

(ii) Most of the reported B_z component reversals are simply a slight dipping of the $\boldsymbol{B}$ vector.

(iii) The B_z component is predominantly positive in the midplane of the plasma sheet during substorms.

(iv) Small-scale loops appear to be common in the vicinity of the midplane during substorms.

(v) Anti-solar flows can be associated with both a negative B_z component (the dipping of the $\boldsymbol{B}$ vector) and a positive B_z component.

(vi) The enhanced B_z component is much greater than the dipole value.

(vii) The magnetic field configuration becomes tail-like again during a late epoch of the recovery phase.

Therefore, we are confronted with the problem of finding a new mechanism for the plasma sheet thinning. Further, since the plasma sheet thinning lasts until about the maximum epoch of a substorm at about $X = -18\ R_E$ (namely, until about the time when the auroral bulge reaches the highest latitude), we must find the auroral bulge formation mechanism which does not invoke an enhanced reconnection, as well as the plasma sheet thinning mechanism.

One possible mechanism for the plasma sheet thinning-auroral bulge formation which does not invoke an enhanced reconnection is the deflation of the plasma sheet, resulting in the inflation of the inner magnetosphere (namely, the ring current formation) and also auroral precipitation along the auroral oval. Figure 6.19 suggests that the plasma sheet thinning may be caused by a rarefaction wave which propagates in the anti-solar direction. Plasma behind the wave will flow toward the Earth, deflating the plasma sheet. Thus, the magnetic field lines (closed) *within* the plasma sheet would contract rapidly toward the Earth.

The contraction of the stretched field lines indicates also that the cross-tail current is disrupted in the near-Earth plasma sheet and is reduced in the distant plasma sheet. Therefore, it is possible that the sudden disruption of the current can trigger the deflation of the plasma sheet (Section 7.2). A large voltage could also develop across the tail (Section 8.2).

It is not implied here that enhanced reconnection does not occur in the magnetotail during the entire life of a substorm. In fact, near the maximum epoch of substorms, three well-established phenomena, the expansion of a hot plasma sheet, the associated large northward rotation of the tail field and the occurrence of intense sunward plasma flow, are all consistent with enhanced reconnection taking place at a location beyond the lunar distance.

In Section 9.2, an attempt will be made to combine these features in developing new ideas on the causes of magnetospheric substorms.

References

Akasofu, S.-I.: 1968, 'Auroral Observations by the Constant Local Time Flight', *Planet. Space Sci.* **16**, 1365.

Akasofu, S.-I.: 1970, 'A Model Current System for the Magnetospheric Substorm', *Particles and Fields in the Magnetosphere*, B. M. McCormac (ed.), p. 34, D. Reidel Publ. Co., Dordrecht-Holland.

Akasofu, S.-I.: 1972a, 'Photographs of the Front of the Expanding Auroral Bulge during an Auroral Substorm', *Planet. Space Sci.* **20**, 821.

Akasofu, S.-I.: 1972b, 'Midday Auroras at the South Pole during Magnetospheric Substorms', *J. Geophys. Res.* **77**, 2303.

Akasofu, S.-I.: 1972c, 'Midday Auroras and Magnetospheric Substorms', *J. Geophys. Res.* **77**, 244.

Akasofu, S.-I.: 1974a, 'The Midday Red Aurora Observed at the South Pole on August 5, 1972', *J. Geophys. Res.* **79**, 2904.

Akasofu, S.-I.: 1974b, 'The Aurora and the Magnetosphere: The Chapman Memorial Lecture', *Planet. Space Sci.* **22**, 885.

Akasofu, S.-I.: 1974c, 'A Study of Auroral Displays Photographed from the DMSP-2 Satellite and from the Alaska Meridian Chain of Stations', *Space Sci. Rev.* **16**, 617.

Akasofu, S.-I. and Kimball, D. S.: 1973, 'The Behavior of Midday Auroras during Substorms', *Planet. Space Sci.* **28**, 696.

Akasofu, S.-I., Eather, R. H. and Bradbury, J. N.: 1969, 'The Absence of the Hydrogen Emission Hβ in the Westward Traveling Surge', *Planet. Space Sci.* **17**, 1409.

Akasofu, S.-I., Hones, E. W. Jr. and Meng, C.-I.: 1970, 'Simultaneous Observations of an Energetic Electron Event in the Magnetotail by the Vela 3A and IMP 3 Satellites, 2', *J. Geophys. Res.* **75**, 7296.

Akasofu, S.-I., Hones, E. W. Jr. and Singer, S.: 1971a, 'Impulsive Energetic Electron Fluxes in the Distant Magnetotail Associated with the Onset of Magnetospheric Substorms', *J. Geophys. Res.* **76**, 6976.

Akasofu, S.-I., Hones, E. W. Jr., Montgomery, M. D., Bame, S. J. and Singer, S.: 1971b, 'Association of Magnetotail Phenomena with Visible Auroral Features', *J. Geophys. Res.* **76**, 5985.

Anderson, K. A.: 1965, 'Energetic Electron Fluxes in the Tail of the Geomagnetic Field', *J. Geophys. Res.* **70**, 4741.

Anger, D. C., Lui, A. T. Y. and Akasofu, S.-I.: 1973, 'Observations of the Auroral Oval and a Westward Traveling Surge from the ISIS 2 Satellite and the Alaska Meridian All-Sky Cameras', *J. Geophys. Res.* **78**, 3020.

Armstrong, T. P. and Krimigis, S. M.: 1968, 'Observations of Protons in the Magnetosphere and Magnetotail with Explorer 33', *J. Geophys. Res.* **73**, 143.

Aubry, M. P.: 1972, 'A Short Review of Magnetospheric Substorms', *Earth's Magnetospheric Processes*, B. M. McCormac (ed.), p. 357, D. Reidel Publ. Co., Dordrecht-Holland.

Aubry, M. P. and McPherron, R. L.: 1971, 'Magnetotail Changes in Relation to the Solar Wind Magnetic Field and Magnetospheric Substorms', *J. Geophys. Res.* **75**, 4381.

Aubry, M. P., Kivelson, M. G., McPherron, R. L., Russell, C. T. and Colburn, D. S.: 1972, 'Outer Magnetosphere near Midnight at Quiet and Disturbed Times', *J. Geophys. Res.* **77**, 5487.

Axford, W. I.: 1969, 'Magnetospheric Convection', *Rev. Geophys.* **7**, 421.

Bame, S. J., Hones, E. W. Jr., Akasofu, S.-I., Montgomery, M. D. and Asbridge, J. R.: 1971, 'Geomagnetic Storm Particles in the High-Latitude Magnetotail', *J. Geophys. Res.* **76**, 7566.

Barcus, J. R.: 1969, 'Increase in Cosmic-Ray Cutoffs at High Latitudes during Magnetospheric Substorms', *J. Geophys. Res.* **74**, 4694.

Berko, F. W.: 1974, 'Features of Polar Cusp Electron Precipitation Associated with a Large Magnetic Storm', Goddard Space Flight Center, *X-626-74-326, Nov.*

Bowling, S. B.: 1975, 'Transient Occurrence of Magnetic Loops in the Magnetotail', *J. Geophys. Res.* **80**, 4741.

Brown, J. W. and Stone, E. C.: 1972, 'High-Energy Electron Spikes at High Latitudes', *J. Geophys. Res.* **77**, 3384.

Buck, R. M., West, H. I. Jr. and D'Arcy, R. G. Jr.: 1973, 'Satellite Studies of Magnetospheric Substorms on August 15, 1968, 7. Ogo 5 Energetic Proton Observations – Spatial Boundaries', *J. Geophys. Res.* **78**, 3103.

Burke, W. J. and Reasoner, D. L.: 1973, 'Observation of Plasma Flow in the Neutral Sheet at Lunar Distance during Two Magnetic Bays', *J. Geophys. Res.* **78**, 6790.

Burke, W. J., Rich, F. J., Reasoner, D. L., Colburn, D. S. and Goldstein, B. E.: 1973, 'Effects on the Geomagnetic Tail at 60 R_E of the Geomagnetic Storm of April 9, 1971', *J. Geophys. Res.* **78**, 5477.

Burrows, J. R. and McDiarmid, I. B.: 1972, 'Trapped Particle Boundary Regions', *Critical Problems of Magnetospheric Physics*, E. R. Dyer (ed.), p. 83, National Academy of Sciences, Washington, D.C.

Caan, M. N., McPherron, R. L. and Russell, C. T.: 1975, 'Substorm and Interplanetary Magnetic Field Effects on the Geomagnetic Tail Lobes', *J. Geophys. Res.* **80**, 191.

Camidge, F. P. and Rostoker, G.: 1970, 'Magnetic Field Perturbations in the Magnetotail Associated with Polar Magnetic Substorms', *Can. J. Physics*, **48**, 2002.

Chubb, T. A. and Hicks, G. T.: 1970, 'Observations of the Aurora in the Far-Ultraviolet from OGO-4', *J. Geophys. Res.* **75**, 1290.

Coroniti, F. V. and Kennel, C. F.: 1972, 'Magnetospheric Substorms', *Cosmic Plasma Physics*, K. Schindler (ed.), p. 15, Plenum Press, New York.

Crooker, N. U.: 'A Comparison between Magnetospheric Energy and Solar Wind Kinetic Energy', Technical Report, M.I.T.

Cummings, W. D. and Coleman, P. J. Jr.: 1968, 'Simultaneous Magnetic Field Variations at the Earth's Surface and at Synchronous Equatorial Distance', *Radio Sci.* **3**, 758.

Currie, R. G.: 1972, 'A Note on the Geomagnetic Spectrum', *J. Geophys. Res.* **77**, 6893.

Davis, T. N.: 1962, 'The Morphology of the Auroral Displays of 1957–58, 2. Detailed Analyses of Alaska Data and Analyses of High Latitude Data', *J. Geophys. Res.* **67**, 75.

Davis, T. N.: 1972, 'Auroral Morphology', *Planet. Space Sci.* **20**, 1369.

Deehr, C. S., Egeland, A., Aarsnes, K., Amundsen, R., Lindalen, H. R., Söraas, F., Dalziel, R., Smith, P. A., Thomas, G. R., Stauning, P., Borg, H., Gustafsson, G., Holmgren, L. A., Riedler, W., Raitt, J., Skovli, G., Wedde, T. and Jaeschke, R.: 1973, 'Particle and Auroral Observations from the ESRO 1/AURORAE Satellite', *J. Atmosph. Terr. Phys.* **35**, 1979.

DeForest, S. E. and McIlwain, C. E.: 1971, 'Plasma Clouds in the Magnetosphere', *J. Geophys. Res.* **76**, 3587.

Domingo, V. and Page, D. E.: 1972, 'Two-Satellite Observation of Spatial and Temporal Particle Flux Variations over the Polar Caps', *J. Geophys. Res.* **77**, 1971.

Eather, R. H. and Mende, S. B.: 1973, 'Substorm Effects in Auroral Spectra', *J. Geophys. Res.* **78**, 7515.

Evlashin, L. S., Truttse, Yu. L. and Feldstein, Ya. I.: 1972, 'The Behavior of Low-Energy Particles during Substorms', *J. Atmosph. Terr. Phys.* **34**, 859.

Fairfield, D. H.: 1968: 'Average Magnetic Field Configuration of the Outer Magnetosphere', *J. Geophys. Res.* **73**, 7329.

Fairfield, D. H.: 1972, 'Magnetic Field Fluctuations during Substorms', *Planet. Space Sci.* **20**, 1455.

Fairfield, D. H.: 1973, 'Magnetic Field Signatures of Substorms on High-Latitude Field Lines in the Nighttime Magnetosphere', *J. Geophys. Res.* **78**, 1553.

Fairfield, D. H. and Mead, G. D.: 1975, 'Magnetospheric Mapping with a Quantitative Geomagnetic Field Model', *J. Geophys. Res.* **80**, 535.

Fairfield, D. H. and Ness, N. F.: 1970, 'Configuration of the Geomagnetic Tail during Substorms', *J. Geophys. Res.*, **75**, 7032.

Fairfield, D. H. and Ness, N. F.: 1972, 'Imp 5 Magnetic-Field Measurements in the High-Latitude Outer Magnetosphere near the Noon Meridian', *J. Geophys. Res.* **77**, 611.

Feldstein, Y. I.: 1969, 'Polar Auroras, Polar Substorms, and Their Relationships with the Dynamics of the Magnetosphere', *Rev. Geophys.* **7**, 179.

Feldstein, Y. I.: 1972, 'Auroras and Associated Phenomena', *Solar-Terrestrial Physics*, 1970, Part III, E. R. Dyer (ed.), p. 152, D. Reidel Publ. Co., Dordrecht-Holland.

Feldstein, Y.-I. and Starkov, G. V.: 1967, 'Dynamics of Auroral Belt and Polar Geomagnetic Disturbances', *Planet. Space Sci.* **15**, 209.

Fennell, J. F.: 1970, 'Observations of Proton Bursts in the Magnetotail with Explorer 35', *J. Geophys. Res.* **75**, 7048.

Frank, L. A. and Ackerson, K. L.: 1971, 'Observations of Charged Particle Precipitation into the Auroral Zone', *J. Geophys. Res.* **76**, 3612.

Frank, L. A. and Ackerson, K. L.: 1972, 'Local-Time Survey of Plasma at Low Altitudes over the Auroral Zones', *J. Geophys. Res.* **77**, 4116.

Frank, L. A. and Ackerson, K. L.: 1975, 'Examples of Plasma Flows within the Earth's Magnetosphere', *Dept. of Phys. Astronom. Rep.* 75-36, Univ. Iowa, Sept.

Frank, L. A. and Ackerson, K. L.: 1976, 'On Hot Tenuous Plasmas, Fireballs, and Boundary Layers in the Earth's Magnetotail', *Dept. of Phys. Astronom. Rep.* 76-1, Univ. of Iowa, Jan.

Fukunishi, H.: 1975, 'Dynamic Relationship between Proton and Electron Auroral Substorms', *J. Geophys. Res.* **80**, 553.

Fukunishi, H. and Tohmatsu, T.: 1973, 'Constitution of Proton and Electron Aurora Substorms, Part I, Meridian-Scanning Photometric System for Proton Auroras', *JARE Sci. Rep.* Series A No. 11, Polar Research Center, National Science Museum, Tokyo, March.

Garrett, H. B., Hill, T. W. and Fenner, M. A.: 1971, 'Plasma-Sheet Ions at Lunar Distance Preceding Substorm Onset', *Planet. Space Sci.* **19**, 1413.

Häusler, B., Scholer, M. and Hovestadt, D.: 1974, 'Variations of the Trapped Particle Population and the Cutoff and Pitch-Angle Distribution of Simultaneously Observed Magnetospheric Solar Protons during Substorm Activity', *J. Geophys. Res.* **79**, 1819.

Hoffman, R. A.: 1971, 'Properties of Low Energy Particle Impacts in the Polar Domain in the Dawn and Dayside Hours', Goddard Space Flight Center, *Rep. X-646-71-199.*

Hoffman, R. A. and Burch, J. L.: 1972, 'Electron Precipitation Patterns and Substorm Morphology', Goddard Space Flight Center, *X-646-72-370,* Oct.

Hoffman, R. A. and Burch, J. L.: 1973, 'Electron Precipitation Patterns and Substorm Morphology', *J. Geophys. Res.* **78**, 2867.

Hones, E. W. Jr.: 1972a, 'Plasma Sheet Variations during Substorms', *Planet. Space Sci.* **20**, 1409.

Hones, E. W. Jr.: 1972b, 'Substorm Behavior of Plasma Sheet Particles', *Earth's Magnetospheric Processes*, B. M. McCormac (ed.), p. 365, D. Reidel Publ. Co., Dordrecht-Holland.

Hones, E. W. Jr.: 1973, 'Plasma Flow in the Plasma Sheet and its Relation to Substorms', *Radio Sci.* **8**, 979.

Hones, E. W. Jr.: 1975, 'Recent Observations Relating to the Dynamics and Origin of the Magnetotail Plasma Sheet', *Physics of the Hot Plasma in the Magnetosphere,* B. Hultqvist and I. Stenflo (eds.), p. 137, Plenum Press, New York.

Hones, E. W. Jr., Asbridge, J. R., Bame, S. J. and Strong, I. B.: 1967, 'Outward Flow of Plasma in the Magnetotail Following Geomagnetic Bays', *J. Geophys. Res.* **72**, 5879.

Hones, E. W. Jr., Akasofu, S.-I., Perreault, P., Bame, S. J. and Singer, S.: 1970, 'Poleward Expansion of the Auroral Oval and Associated Phenomena in the Magnetotail during Auroral Substorms, 1', *J. Geophys. Res.* **75**, 7060.

Hones, E. W. Jr., Akasofu, S.-I., Bame, S. J. and Singer, S.: 1971a, 'Poleward Expansion of the Auroral Oval and Associated Phenomena in the Magnetotail during Auroral Substorms, 2', *J. Geophys. Res.* **76**, 8241.

Hones, E. W. Jr., Karas, R. H., Lanzerotti, L. J. and Akasofu, S.-I.: 1971b, 'Magnetospheric Substorms on September 14, 1968, *J. Geophys. Res.* **76**, 6765.

Hones, E. W. Jr., Asbridge, J. R. and Bame, S. J.: 1971c, 'Time Variations of the Magnetotail Plasma Sheet at 18 R_E Determined from Concurrent Observations by a Pair of Vela Satellites', *J. Geophys. Res.* **76**, 4402.

Hones, E. W. Jr., Akasofu, S.-I., Bame, S. J. and Singer, S.: 1972a, 'Outflow of Plasma from the Magnetotail into the Magnetosheath', *J. Geophys. Res.* **77**, 6688.

Hones, E. W. Jr., Asbridge, J. R., Bame, S. J., Montgomery, M. D., Singer, S. and Akasofu, S.-I.: 1972b, 'Measurements of Magnetotail Plasma Flow Made with Vela 4B', *J. Geophys. Res.* **77**, 5503.

Hones, E. W. Jr., Asbridge, J. R., Bame, S. J. and Singer, S.: 1973a, 'Substorm Variations of the Magnetotail Plasma Sheet from $X_{SM} \simeq -6\,R_E$ to $X_{SM} \simeq -60\,R_E$', *J. Geophys. Res.* **78**, 109.

Hones, E. W. Jr., Asbridge, J. R., Bame, S. J. and Singer, S.: 1973b, 'Magnetotail Plasma Flow Measured by Vela 4A', *J. Geophys. Res.* **78**, 5463.

Hones, E. W. Jr., Lui, A. T. Y., Bame, S. J. and Singer, S.: 1974, 'Prolonged Tailward Flow of Plasma in the Thinned Plasma Sheet Observed at $r \approx 18\,R_E$ during Substorms', *J. Geophys. Res.* **79**, 1385.

Hones, E. W. Jr., Palmer, I. D. and Higbie, P. R.: 1976a, 'Energetic Protons of Magnetospheric Origin in the Plasma Sheet Associated with Substorms', *J. Geophys. Res.* **81**, 2.

Hones, E. W. Jr., Akasofu, S.-I. and Perreault, P.: 1976b, 'Association of IMF Polarity, Plasma Sheet Thinning and Substorm Occurrence on March 6, 1970', *J. Geophys. Res.* **81** (in press), 1976.

Hones, E. W. Jr., Bame, S. J. and Asbridge, J. R.: 1976c, 'Proton Flow Measurements in the Magnetotail Plasma Sheet Made with IMP 6', *J. Geophys. Res.* **81**, 227.

Hruska, A.: 1973, 'Structure of High-Latitude Irregular Electron Fluxes and Acceleration of Particles in the Magnetotail', *J. Geophys. Res.* **78**, 7509.

Hruska, A. and Hruskova, J.: 1970, 'Transverse Structure of the Earth's Magnetotail and Fluctuations of the Tail Magnetic Field', *J. Geophys. Res.* **75**, 2449.

Hruska, A., Burrows, J. R. and McDiarmid, I. B.: 1972, 'High-Latitude Low-Energy Electron Fluxes and Variation of the Magnetospheric Structure with the Direction of the Interplanetary Magnetic Field and with the Geomagnetic Activity', *J. Geophys. Res.* **77**, 2770.

Iijima, T.: 1972, 'Relationship of Magnetospheric Substorms on the Ground and in the Distant Magnetotail', *Rep. Ionosph. Space Res. Japan*, **26**, 149.

Imhof, W. L., Gaines, E. E. and Reagan, J. B.: 1970, 'High-Resolution Studies of Electrons 25 kev to 2.7 Mev Precipitating at High Latitudes', *J. Geophys. Res.* **75**, 776.

Ivanov, K. G.: 1971, 'Space-Time Correlations during the Advance of Discontinuities on the Magnetosphere', *Geomag. Aeronom.* **11**, 863.

Ivanov, K. G. and Mikerina, N. V.: 1973, 'Interplanetary Magnetic Field, Reconnection, and the Structure of Magnetospheric Storms', *Geomag. Aeronom.* **13**, 754.

Ivanov, V. E. and Starkov, G. V.: 1972, 'Variations of the Auroral Electron Energy Spectra during Substorms', *Ann. Geophys.* **28**, 427.

Kamide, Y. and McIlwain, C. E.: 1974, 'The Onset Time of Magnetospheric Substorms Determined from Ground and Synchronous Satellite Records', *J. Geophys. Res.* **79**, 4787.

Kamide, Y., Burch, J. L., Winningham, J. D. and Akasofu, S.-I.: 1976, 'Dependence of the Latitude of the Cleft on the Interplanetary Magnetic Field and Substorm Activity', *J. Geophys. Res.* **81**, 698.

Kane, R. P.: 1972, 'Relationship between the Various Indices of Geomagnetic Activity and the Interplanetary Plasma Parameters', *J. Atmosph. Terrs. Phys.* **34**, 1941.

Kane, R. P.: 1974, 'Relationship between Interplanetary Plasma Parameters and Geomagnetic Dst', *J. Geophys. Res.* **79**, 64.

Kivelson, M. G., Farley, T. A. and Aubry, M. P.: 1973, 'Satellite Studies of Magnetospheric Substorms on August 15, 1968; 5. Energetic Electrons, Spatial Boundaries, and Wave-Particle Interactions at OGO 5', *J. Geophys. Res*, **78**, 3079.

Knox, F. B. and Andrews, M. K.: 1974, 'Vorticity in the Magnetosphere', *Planet. Space Sci.* **22**, 1703.

Kokubun, S.: 1972, 'Relationship of Interplanetary Magnetic Field Structure with Development of Substorm and Storm Main Phase', *Planet. Space Sci.* **20**, 1033.

Kovalevsky, J. V.: 1971, 'The Interplanetary Medium and Its Interaction with the Earth's Magnetosphere', *Space Sci. Rev.* **12**, 187.

Krassovsky, V. I.: 1972, 'Energy Releases in the Upper Atmosphere during Geomagnetic Disturbances', *Planet. Space Sci.* **20**, 1363.

Laird, M. J.: 1969, 'Structure of the Neutral Sheet in the Geomagnetic Tail', *J. Geophys. Res.* **74**, 133.

Lazarus, A. J., Siscoe, G. L. and Ness, N. F.: 1968, 'Plasma and Magnetic Field Observations during the Magnetosphere Passage of Pioneer 7', *J. Geophys. Res.* **73**, 2399.

Lui, A. T. Y. and Akasofu, S.-I.: 1976, 'Distribution of Plasma Sheet Thinning and Recovery Events in the Magnetotail Survey by the IMP-6 Satellite', *J. Geophys. Res.* (submitted).

Lui, A. T. Y., Hones, E. W. Jr., Venkatesan, D., Akasofu, S.-I. and Bame, S. J.: 1975a, 'Thinnings of the Plasma Sheet Observed by Vela Satellites and the Interplanetary Magnetic Field', *J. Geophys. Res.* **80**, 4649.

Lui, A. T. Y., Anger, C. D. and Akasofu, S.-I.: 1975b, 'The Equatorward Boundary of the Diffuse Aurora and Auroral Substorms as Seen by the ISIS 2 Auroral Scanning Photometer', *J. Geophys. Res.* **80**, 3603.

Lui, A. T. Y., Meng, C.-I. and Akasofu, S.-I.: 1976a, 'Search for the Magnetic Neutral Line in the Near-Earth Plasma Sheet: 1. Critical Re-Examination of Earlier Studies on Magnetic Field Observations', *J. Geophys. Res.* (in press).

Lui, A. T. Y., Meng, C.-I. and Akasofu, S.-I.: 1976b, 'Search for the Magnetic Neutral Line in the Near-Earth Plasma Sheet: 2. Systematic Study on IMP-6 Magnetic Field Observations', *J. Geophys. Res.* (in press).

Lui, A. T. Y., Hones, E. W. Jr., Yasuhara, F., Akasofu, S.-I. and Bame, S. J.: 1976c, 'Magnetotail Plasma Flow during Plasma Sheet Expansions: Vela-5, -6 and IMP-6 Observations', *J. Geophys. Res.* **81** (in press).

Lyatskaya, A. M. and Lyatskiy, V. B.: 1973, 'Motion of Auroras in the Course of a Substorm', *Geomag. Aeronom.* **13**, 324.

Maezawa, K.: 1975, 'Magnetotail Boundary Motion Associated with Geomagnetic Substorms', *J. Geophys. Res.* **80**, 3543.

McDiarmid, I. B. and Hruska, A.: 1972, 'Anistropy of High-Latitude Electron Fluxes during Substorms and Structure of the Magnetotail', *J. Geophys. Res.* **77**, 3377.

McDiarmid, I. B., Burrows, J. R. and Wilson, M. D.: 1972, 'Solar Particles and the Dayside Limit of Closed Field Lines', *J. Geophys. Res.* **77**, 1103.

McPherron, R. L., Parks, G. K., Colburn, D. S. and Montgomery, M. D.: 1973a, 'Satellite Studies of Magnetospheric Substorms on August 15, 1968, 2. Solar Wind and Outer Magnetosphere', *J. Geophys. Res.* **78**, 3054.

McPherron, R. L., Russell, C. T. and Aubry, M. P.: 1973b, 'Satellite Studies of Magnetospheric Substorms on August 16, 1968, 9. Phenomenological Model for Substorms', *J. Geophys. Res.* **78**, 3131.

Mead, G. D. and Fairfield, D. H.: 1975, 'A Quantitative Magnetospheric Model Derived from Spacecraft Magnetometer Data', *J. Geophys. Res.* **80**, 523.

Meng, C.-I.: 1970, 'Variation of the Magnetopause Position with Substorm Activity', *J. Geophys. Res.* **75**, 3252.

Meng, C.-I.: 1971, 'Energetic Electrons in the Magnetotail at 60 R_E', *J. Geophys. Res.* **76**, 862.

Meng, C.-I. and Akasofu, S.-I.: 1971, 'Magnetospheric Substorm Observations near the Neutral Sheet', *J. Geophys. Res.* **76**, 4679.

Meng, C.-I. and Anderson, K. A.: 1971, 'Energetic Electrons in the Plasma Sheet out to 40 R_E', *J. Geophys. Res.* **76**, 873.
Meng, C.-I. and Anderson, K. A.: 1974, 'Magnetic Field Configuration in the Magnetotail near 60 R_E', *J. Geophys. Res.* **79**, 5143.
Meng, C.-I., Hones, E. W. Jr. and Akasofu, S.-I.: 1970, 'Simultaneous Observations of an Energetic Electron Event in the Magnetotail by the Vela 3A and IMP 3 Satellites', *J. Geophys. Res.* **75**, 7294.
Meng, C.-I., Akasofu, S.-I., Hones, E. W. Jr. and Kawasaki, K.: 1971, 'Magnetospheric Substorms in the Distant Magnetotail Observed by IMP 3', *J. Geophys. Res.* **76**, 7584.
Meng, C.-I., Kawasaki, K. and Akasofu, S.-I.: 1974, 'Magnetotail Variations near 60 R_E Associated with Geomagnetic Activity along the Auroral Zone', *Space Sci. Lab.* June.
Mihalov, J. D.: 1970, 'On Geomagnetic Tail Structure near the Null Sheet', *Planet. Space Sci.* **18**, 1845.
Mihalov, J. D., Colburn, D. S., Currie, R. G. and Sonett, C. P.: 1968, 'Configuration and Reconnection of the Geomagnetic Tail', *J. Geophys. Res.* **73**, 943.
Mihalov, J. D., Sonett, C. P. and Colburn, D. S.: 1970, 'Reconnection and Noise in the Geomagnetic Tail', *Cosmic Electrodyn.* **1**, 178.
Montbriand, L. E.: 1969, 'Morphology of Auroral Hydrogen Emissions during Auroral Substorms', Ph.D. Thesis, Univ. of Saskatchewan.
Montgomery, M. D.: 1968, 'Observations of Electrons in the Earth's Magnetotail by Vela Launch 2 Satellites', *J. Geophys. Res.* **73**, 871.
Murayama, T.: 1970, 'Correlated Occurrence of Energetic Electron Bursts in the Magnetotail and Geomagnetic Impulsive Micropulsation', *Rep. Ionosph. Space Res. Japan* **24**, 135.
Ness, N. F.: 1972, 'Review of Magnetic Field Observations', *Earth's Magnetospheric Processes*, B. M. McCormac (ed.), p. 189, D. Reidel Publ. Co., Dordrecht-Holland.
Nishida, A.: 1972, 'Excitation of Polar Substorms by Northward Interplanetary Magnetic Field', *Earth's Magnetospheric Processes*, B. M. McCormac (ed.), p. 400, D. Reidel Publ. Co., Dordrecht-Holland.
Nishida, A. 1973, 'Neutral Line in the Magnetotail', Inst. of Space and Aeronautical Sci., Univ. of Tokyo, Report No. 497, July.
Nishida, A. and Hones, E. W. Jr.: 1974, 'Association of Plasma Sheet Thinning with Neutral Line Formation in the Magnetotail', *J. Geophys. Res.* **79**, 535.
Nishida, A. and Lyon, E. F.: 1972, 'Plasma Sheet at Lunar Distance: Structure and Solar-Wind Dependence', *J. Geophys. Res.* **77**, 4086.
Nishida, A. and Nagayama, N.: 1973, 'Synoptic Survey for the Neutral Line in the Magnetotail during the Substorm Expansion Phase', *J. Geophys. Res.* **78**, 3782.
Nishida, A. and Nagayama, N.: 1975, 'Magnetic-Field Observations in the Low-Latitude Magnetotail during Substorms', *Planet. Space Sci.* **23**, 1119.
Oguti, T.: 1973, 'Hydrogen Emission and Electron Aurora at the Onset of the Auroral Breakup', *J. Geophys. Res.* **78**, 7543.
Oguti, T. and Kokubun, S.: 1969, 'Polar Magnetic Disturbances and Associated Increase in Flux of Energetic Electrons in the Magnetosphere', *Rep. Ionosph. Space Res. Japan* **23**, 151.
Olson, W. P.: 1974a, 'Forces and Torques on the Earth Produced by Magnetospheric Currents', *J. Geophys. Res.* **79**, 1128.
Olson, W. P.: 1974b, 'A Model of the Distributed Magnetospheric Currents', *J. Geophys. Res.* **79**, 3731.
Olson, W. P. and Pfitzer, K. A.: 1974, 'A Quantitative Model of the Magnetospheric Magnetic Field', *J. Geophys. Res.* **79**, 3739.
Parks, G. K., Laval, G. and Pellat, R.: 1972, 'Behavior of Outer Radiation Zone and a New Model of Magnetospheric Substorm', *Planet. Space Sci.* **20**, 1391.
Pike, C. P.: 1972, 'Equatorward Shift of the Polar F Layer Irregularity Zone as a Function of the Kp Index', *J. Geophys. Res.* **77**, 6911.
Pike, C. P. and Whalen, J. A.: 1974, 'Satellite Observations of Auroral Substorms', *J. Geophys. Res.* **79**, 985.
Pudovkin, M. I. and Barsukov, V. M.: 1971, 'The Zones of Corpuscular Precipitations and the Structure of the Magnetosphere', *Planet. Space Sci.* **19**, 525.
Pudovkin, M. I., Shumilov, O. I. and Zaitzeva, S. A.: 1968, 'Polar Storms and Development of the DR-Currents, *Planet. Space Sci.* **16**, 891.
Roelof, E. C., Keath, E. P., Bostrom, C. O. and Williams, D. J.: 1976, 'Fluxes of $\geq$50 keV Protons and $\geq$30 keV Electrons at 35 R_E. 1. Velocity Anisotropies and Plasma Flow in the Magnetotail', *J. Geophys. Res.* **81**, 2304.
Rossberg, L.: 1971, 'Newly Observed Premidnight Asymmetry in the 40 keV Electron Flux Profiles', *J. Geophys. Res.* **76**, 6980.
Rossberg, L.: 1972, 'The Pre-Midnight Asymmetry in the 40 keV Electron Flux Profiles and Its

Relation to Magnetospheric Substorms', *Earth's Magnetospheric Processes*, B. M. McCormac (ed.), p. 147, D. Reidel Publ. Co., Dordrecht-Holland.

Rossberg, L: 1974, 'Remarks on the Growth Phase of Substorms', *Magnetospheric Physics*, B. M. McCormac (ed.), p. 367, D. Reidel Publ. Co., Dordrecht-Holland.

Rossberg, L., Lammers, E., Riedler, W., Skovli, G., Søraas, F., Stauning, P., Theile, B. and Thomas, G. R.: 1973, 'A Multi-satellite Study of Auroral-Zone Phenomena 1: Observations during a Moderately Disturbed Period on 11 November, 1969', ESRO SR-23, May.

Rossberg, L., Riedler, W., Skovli, G., Stauning, P. and Thomas, G. R.: 1974, 'Interplanetary Magnetic Field Direction and Auroral Zone Disturbances', *Correlated Interplanetary and Magnetospheric Observations*, D. E. Page (ed.), p. 85, D. Reidel Publ. Co., Dordrecht-Holland.

Rostoker, G. and Camidge, F. P.: 1971, 'Localized Character of Magnetotail Magnetic Fluctuations during Polar Magnetic Substorms', *J. Geophys. Res.* **76**, 6944.

Rothwell, P. and Wallington, V.: 1968, 'The Polar Substorm and Electron "Islands" in the Earth's Magnetic Tail', *Planet. Space Sci.* **16**, 1441.

Russell, C. T.: 1973, 'Comments on the Paper "The Internal Structure of the Geomagnetic Neutral Sheet" by K. Schindler and N. F. Ness', *J. Geophys. Res.* **78**, 7576.

Russell, C. T., McPherron, R. L. and Coleman, P. J. Jr.: 1971, 'Magnetic Field Variations in the Near Geomagnetic Tail Associated with Weak Substorm Activity,' *J. Geophys. Res.* **76**, 1823.

Russell, C. T., McPherron, R. L. and Burton, R. K.: 1974, 'On the Cause of Geomagnetic Storms', *J. Geophys. Res.* **79**, 1105.

Sarris, E. T., Krimigis, S. M. and Armstrong, T. P.: 1975, 'Observations of Magnetospheric Bursts of High Energy Protons and Electrons at $\sim 35\ R_E$ with IMP-7', Preprint Series, Applied Physics Lab. Johns Hopkins Univ., Silver Springs, Md., March.

Schindler, K.: 1974, 'A Theory of the Substorm Mechanism', *J. Geophys. Res.* **79**, 2803.

Schindler, K.: 1975, 'Plasma and Fields in the Magnetospheric Tail', *Space Sci. Rev.* **17**, 589.

Schindler, K. and Ness, N. F.: 1972, 'Internal Structure of the Geomagnetic Neutral Sheet', *J. Geophys. Res.* **77**, 91.

Schindler, K. and Ness, N. F.: 1974, 'Neutral Point Detection by Satellites', *J. Geophys. Res.* **79**, 5297.

Sergeev, V. A.: 1974, 'On the Longitudinal Localization of the Substorm Active Region and Its Changes during the Substorm', *Planet. Space Sci.* **22**, 1341.

Sharp, R. D. and Johnson, R. G.: 1972, 'The Behavior of Low-Energy Particles during Substorms', *Planet. Space Sci.* **20**, 1433.

Siscoe, G. and Crooker, N.: 1974, 'A Theoretical Relation between DST and the Solar Wind Merging Electric Field', *Geophys. Res. Letters* **1**, 17.

Smith, E. J. and Davis, L. Jr.: 1970, 'Magnetic Measurements in the Earth's Magnetosphere and Magnetosheath: Mariner 5', *J. Geophys. Res.* **75**, 1233.

Snyder, A. L. and Akasofu, S.-I.: 1972, 'Observations of the Auroral Oval by the Alaskan Meridian Chain of Stations', *J. Geophys. Res.* **77**, 3419.

Snyder, A. L. and Akasofu, S.-I.: 1974, 'Major Auroral Substorm Features in the Dark Sector by a USAF DMSP Satellite', *Planet. Space Sci.* **22**, 1511.

Snyder, A. L., Akasofu, S.-I. and Davis, T. N.: 1974, 'Auroral Substorms Observed from Above the North Polar Region by a Satellite', *J. Geophys. Res.* **79**, 1393.

Sonnerup, B. U. Ö.: 1971, 'Magnetopause Structure during the Magnetic Storm of September 24, 1961', *J. Geophys. Res.* **76**, 6717.

Sonnerup, B. U. Ö. and Ledley, B. G.: 1974, 'Magnetopause Rotational Forms', *J. Geophys. Res.* **79**, 4309.

Speiser, T. W.: 1970, 'Conductivity without Collisions or Noise', *Planet. Space Sci.* **18**, 613.

Speiser, T. W.: 1973, 'Magnetospheric Current Sheets', *Radio Sci.* **8**, 973.

Starkov, G. V. and Fel'dshteyn, Ya. I.: 1971, 'Substorm in Auroras', *Geomag. Aeronom.* **11**, 478.

Starkov, G. V., Fel'dshteyn, Ya. I. and Shevnina, N. F.: 1973, 'Auroras on the Dayside of the Oval during Substorms', *Geomag. Aeronom.* **13**, 72.

Turbin, Yu. G., Vakulin, Yu. I. and Urbanovich, V. D.: 1972, 'Some Statistical Characteristics of the Spectra of Polar Magnetic Substorms', *Geomag. Aeronom.* **12**, 928.

Vasyliunas, V. M.: 1968, 'A Survey of Low-Energy Electrons in the Evening Sector of the Magnetosphere with OGO 1 and OGO 3', *J. Geophys. Res.* **73**, 2839.

Vasyliunas, V. M.: 1972a, 'Non-uniqueness of Magnetic Field Line Motion', *J. Geophys. Res.* **77**, 6271.

Vasyliunas, V. M.: 1972b, 'The Interrelationship of Magnetospheric Processes', *Earth's Magnetospheric Processes*, B. M. McCormac (ed.), p. 29, D. Reidel Publ. Co., Dordrecht-Holland.

Vorobjev, V. G., Gustafsson, G., Starkov, G. V., Feldstein, Y. I. and Shevnina, N. F.: 1975, 'Dynamics of Day and Night Aurora during Substorms', *Planet. Space Sci.* **23**, 269.

Welcott, J. H., Pongratz, M. B., Hones, E. W. Jr. and Peterson, R. W.: 1976, 'Correlated Observations of Two Auroral Substorms from an Aircraft and from a Vela Satellite', *J. Geophys. Res.* **81**, (in press).

Wescott, E. M., Peek, H. M., Stenbaek-Nielsen, H. C., Murcray, W. B., Jensen, R. J. and Davis, T. N.: 1972, 'Two Successful Geomagnetic Field Line Tracing Experiments', *J. Geophys. Res.* **77**, 2982.

Wescott, E. M., Rieger, E. P., Stenbaek-Nielsen, H. C., Davis, T. N., Peek, H. M. and Bottoms, P. J.: 1974, '$L = 1.24$ Conjugate Magnetic Field Line Tracing Experiments with Barium Shaped Charges', *J. Geophys. Res.* **79**, 159.

Whalen, B. A. and McDiarmid, I. B.: 1970, 'Temporal Behavior of Energetic Particle Precipitation during an Auroral Substorm', *J. Geophys. Res.* **75**, 123.

Willis, D. M. and Pratt, R. J.: 1972, 'A Quantitative Model of the Geomagnetic Tail', *J. Atmosph. Terr. Phys.* **34**, 1955.

Winningham, J. D.: 1972, 'Characteristics of Magnetosheath Plasma Observed at Low Altitudes in the Dayside Magnetospheric Cusps', *Earth's Magnetospheric Processes*, B. M. McCormac (ed.), p. 68, D. Reidel Publ. Co., Dordrecht-Holland.

Winningham, J. D., Yasuhara, F., Akasofu, S.-I. and Heikkila, W. J.: 1975, 'The Latitudinal Morphology of 10-eV to 10-keV Electron Fluxes during Magnetically Quiet and Disturbed Times in the 2100–0300 MLT Sector', *J. Geophys. Res.* **80**, 3148.

Wolfe, J. H. and Intriligator, D. S.: 1970, 'The Solar Wind Interaction with the Geomagnetic Field', *Space Sci. Rev.* **10**, 511.

Yasuhara, F., Akasofu, S.-I., Winningham, J. D. and Heikkila, W. J.: 1973, 'Equatorward Shift of the Cleft during Magnetospheric Substorms as Observed by Isis I', *J. Geophys. Res.* **78**, 7286.

Yasuhara, F., Kamide, Y. and Akasofu, S.-I.: 1975, 'A Modeling of the Magnetospheric Substorms', *Planet. Space Sci.* **23**, 575.

CHAPTER 7

MAGNETOSPHERIC CURRENTS DURING SUBSTORMS

7.1. Introduction

One of the most important aspects of the magnetospheric substorm is that during it there occurs a large-scale change of the distribution of electric currents in the magnetosphere and the ionosphere. In Sections 5.5 and 6.5, it was suggested that the main cause of this change is the disruption of the cross-tail current in a limited region in the magnetotail and its diversion to the polar ionosphere. The diverted current should flow into the poleward half of the auroral oval in the morning and forenoon sectors, and out from the oval in the afternoon and evening sectors; see Figure 7.1. In this chapter, this current system is referred to as the substorm current system. It should be noted that the quiet time current system (S_q^p) has a similar distribution (Sections 1.3.2, 1.3.6 and 1.3.7) and thus the substorm current system may appear as an intensification of the S_q^p current system. In Section 7.2 of this chapter, we examine carefully whether the observed magnetic field disturbances $\Delta\boldsymbol{B}(X, Y, Z)$ at various points in the magnetosphere during substorms are consistent with those which we expected from the diverted cross-tail current.

The substorm current system generates complex current systems in the ionosphere, including concentrated currents along the auroral oval called the auroral electrojets. In Section 7.3, the geometrical relationship between the field-aligned currents and the electrojets will be investigated, both observationally and theoretically. Various aspects of the auroral electrojets, in particular their relationship to large-scale auroral activity, radar auroras and the IMF B_z component, will be studied in Sections 7.4, 7.5 and 7.6.

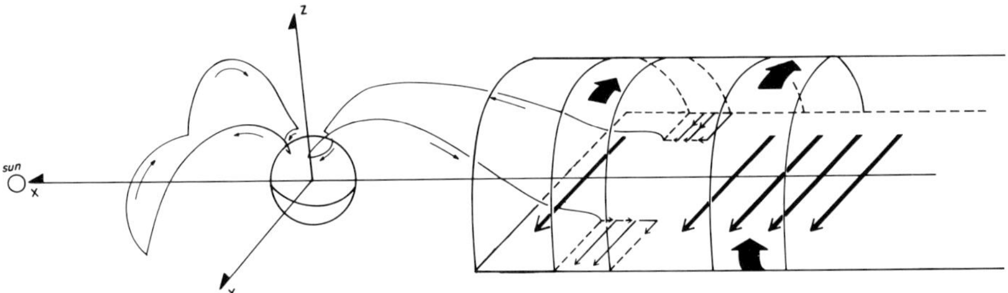

Fig. 7.1. Schematic diagram showing how the disrupted cross-tail current is diverted to the polar upper atmosphere during a substorm. Note that the location of the disruption is not known. The dayside current system is added to explain the equatorward motion of the cusp. (Yasuhara, F., Kamide, Y. and Akasofu, S.-I.: *Planet. Space Sci.* **23**, 575, 1975.)

In Section 7.7, we shall study the relationship between the ionospheric currents and the electric fields, both theoretically and observationally, during magnetospheric substorms. Incoherent scatter radar, balloon (probes), rocket (probes), rocket (barium releases) and satellite observations of the electric fields are briefly reviewed. The growth of the electrojet produces joule heat in the ionosphere. This heating process, together with the Lorentz force associated with the electrojets, causes a worldwide scale disturbance of the ionosphere by generating gravity waves, ion-drag horizontal and vertical winds, etc. These are the subject of Section 7.8. For earlier studies of polar magnetic substorms, see Akasofu (1968, Chapter 3).

7.2. Field-Aligned Currents

Since there is so far no simple method to measure directly the current distribution, we infer the current distribution $\boldsymbol{j}(X, Y, Z)$ on the basis of magnetic perturbation $\Delta \boldsymbol{B}(X, Y, Z)$, which is caused by the current system. In fact, this was the earliest subject of substorm studies and was pioneered by K. Birkeland. The problem associated with this particular study is, however, that it is not possible to uniquely determine a current system even if $\Delta \boldsymbol{B}$ is known at limited points. This problem was particularly serious in the past when $\Delta \boldsymbol{B}$ was measured only on the Earth's surface (Fukushima, 1969a, b, 1974a, b). However, recent satellite measurements of $\Delta \boldsymbol{B}$ have considerably reduced the number of possible current systems.

7.2.1. OBSERVATIONS IN THE MAGNETOTAIL

In Section 1.3.2, it was noted that the field-aligned currents flow along the upper (in the northern hemisphere) and lower (in the southern hemisphere) surfaces of the plasma sheet. They are directed toward the Earth in the morning sector and away from the Earth in the evening sector. When these currents are intensified, one would expect an eastward deflection of the magnetic field in the morning sector of the high latitude lobe of the magnetotail and a westward deflection in the evening sector. Fairfield (1973) showed that this is indeed the case. Figures 7.2(a) and 7.2(b) show examples of such changes in the morning and evening sectors, respectively.

However, it should be noted that these changes are associated with a sharp decrease in the magnitude of the field, indicating that the satellite was engulfed in the expanding plasma sheet (Section 6.4). Indeed, the simultaneous energetic electron data (the lower part of the figures) show a considerable increase at the time of the deflection. Therefore, the deflection occurred only near the surface of the plasma sheet, not in the high latitude lobe. Further, note that in both cases, the corresponding substorms were in progress at least 30 min prior to the deflection time, and the satellite was in the high latitude lobe during an early epoch of the substorms. Therefore, these observations suggest that the field-aligned current in each sector consisted of a double sheet of oppositely flowing currents; the magnetic field deflections are mostly limited to the region between the double

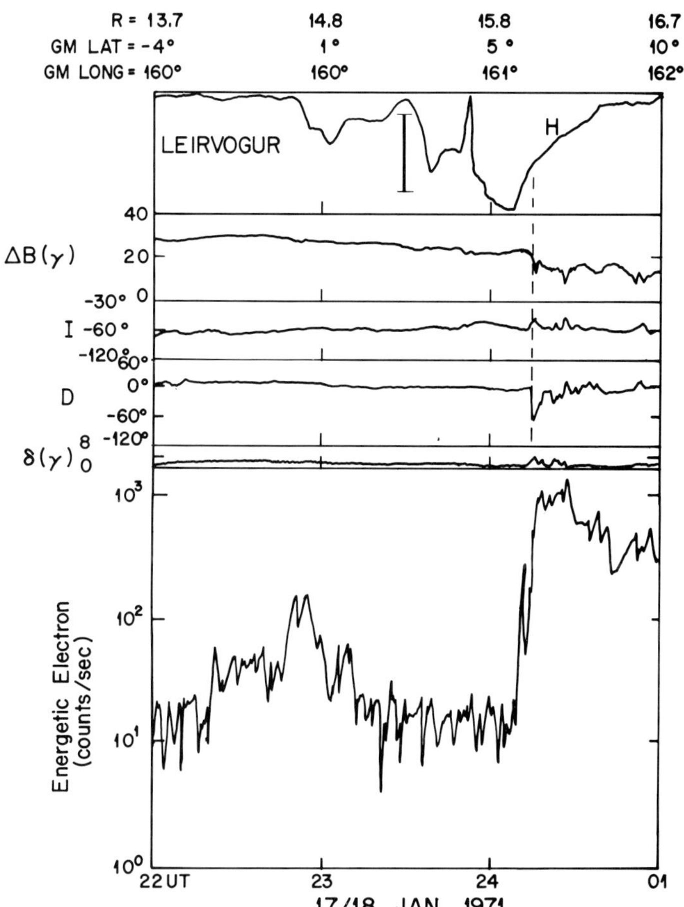

Fig. 7.2(a, b). Magnetic field and energetic electron flux observations in the magnetotail during substorms: (a) the dusk sector and (b) the dawn sector. The *H* component magnetic records from the midnight auroral zone stations are also shown. Note that a large deflection of the magnetic vector occurs at the moment when the satellite is submerged in the expanding plasma sheet. (Fairfield, D. H.: *J. Geophys. Res.* **78**, 1553, 1973; the energetic electron data, courtesy of Meng, C.-I.)

current sheet. On the other hand, Haerendel *et al.* (1971) reported that a significant deflection was observed in the high latitude lobe. Sugiura (1975) provided a few examples of the magnetic field variations across the plasma sheet boundary during disturbed periods, and his data (Figure 7.3) do not exhibit the very sharp changes seen in Figure 7.2.

7.2.2. OBSERVATIONS ABOVE THE IONOSPHERE

Yasuhara *et al.* (1975b) showed that the field-aligned currents, observed above the ionosphere by the TRIAD satellite, are considerably intensified during disturbed periods. In Figure 7.4, the intensities of the poleward current (I_1) and of the equatorward current (I_2) in the evening sector are plotted as a function of the Kp

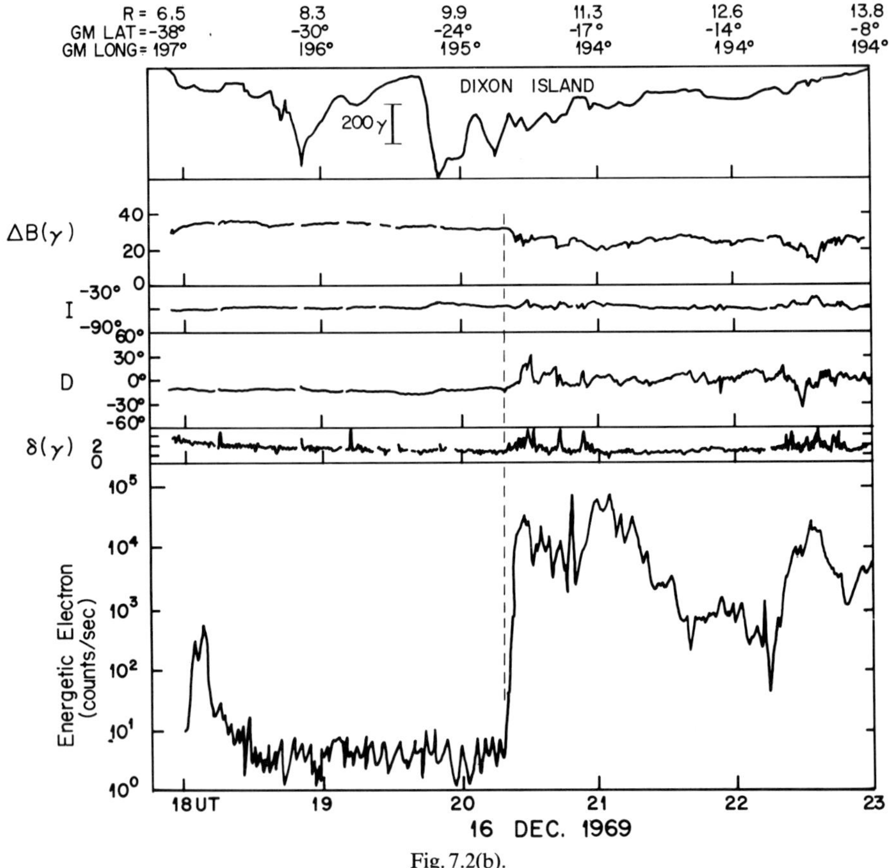

Fig. 7.2(b).

index. Note that the rate of increase with Kp is different for the poleward and for the equatorward current. Within the Kp range examined, the intensity of the poleward current sheet increases by an order of magnitude, while that of the equatorward one increases by a factor of only about two. This difference is important in interpreting ground and high latitude lobe magnetic observations.

7.2.3. GROUND OBSERVATIONS

A great advantage of ground observations compared with satellite observations is that the magnetic field produced by field-aligned currents can be monitored at a large number of fixed points on a continuous basis.

Figure 7.5 shows an example of the distribution of disturbance vectors during the maximum epoch of a substorm which occurred on 1964, December 16. There are several aspects to be discussed in the figure, but we concentrate first only on the distribution of the vectors in the late evening and morning sectors. There, the disturbance vectors are mostly directed in the east-west direction, eastward ($\Delta D > 0$) in the late evening sector and westward ($\Delta D < 0$) in the morning sector.

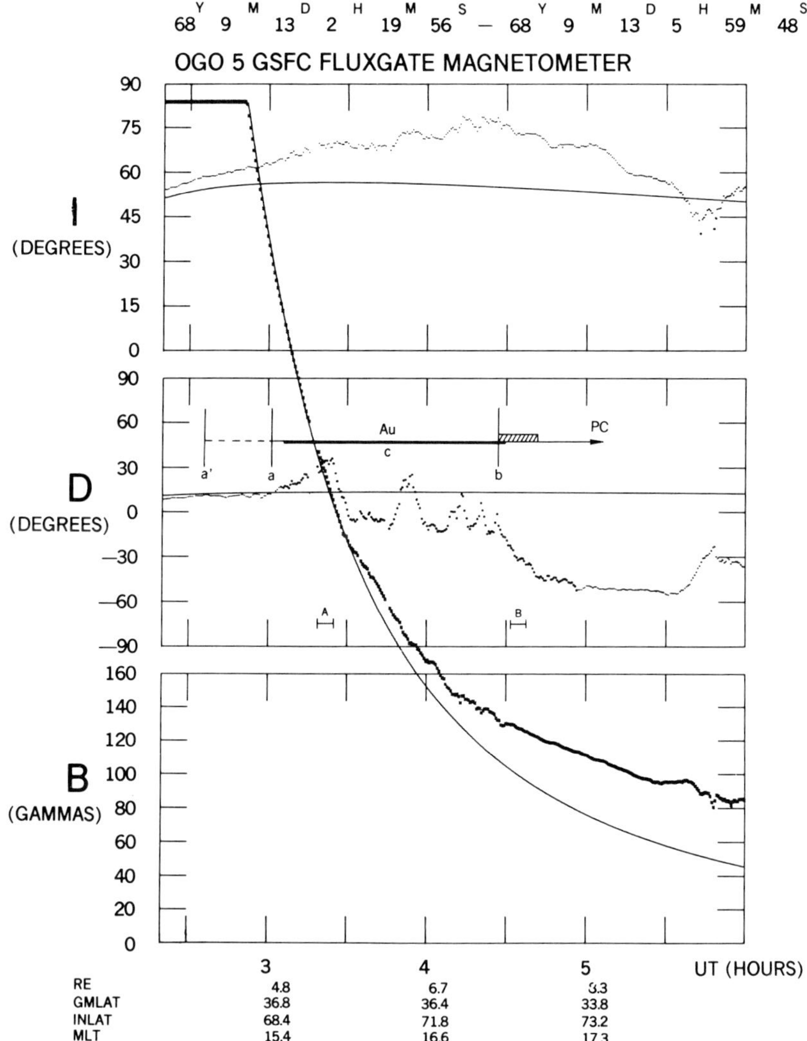

Fig. 7.3. Magnetic field observation across the 'horn' of the plasma sheet. If the observed D component variations are spatial ones, there are several current sheets embedded in the horn of the plasma sheet. (Sugiura, M.: *J. Geophys. Res.* **80**, 2057, 1975.)

In both sectors, the magnitude of the vectors increases toward higher latitudes. Akasofu and Meng (1969) and Meng and Akasofu (1969) showed that this particular distribution of the vectors ($\Delta \boldsymbol{B}$) can be explained in terms of an inward field-aligned current in the morning sector and an outward field-aligned current in the evening sector. This observed distribution may be compared with the computed distribution in Figure 7.6(a) for a simple three-dimensional current system shown in Figure 7.6(b) (see Kisabeth and Rostoker, 1971).

Since the TRIAD observations show that the field-aligned currents in each sector appear as a pair of oppositely directed currents and since the poleward

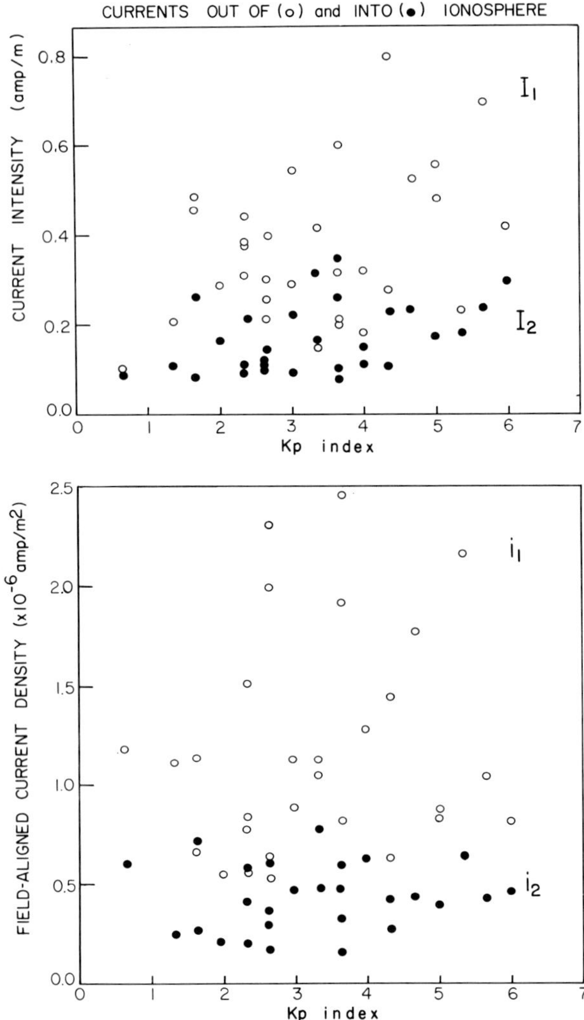

Fig. 7.4. Total current intensities (I) and densities (i) of the outward flowing (I_1, i_1) and inward flowing (I_2, i_2) field-aligned currents as a function of the Kp index in the event sector. (Yasuhara, F., Kamide, Y. and Akasofu, S.-I.: *Planet. Space Sci.* **23**, 575, 1975.)

field-aligned current (I_1) increases much more than the equatorward field-aligned current (I_2), it is reasonable to infer that the ΔD ground vectors in the evening or morning sector are produced by the net increase in the double sheet field-aligned current.

Now, we examine how the net current intensity is distributed in longitude and how it varies during substorms. Figure 7.7 shows both the H and the D component changes observed at a number of middle latitude stations at different dp. longitudes, ranging from 58.6° to 208.4°, during a substorm which occurred on 1965, February 8. It can be clearly seen that the D component changes its sign at about longitude 110°, positive at stations of longitude less than 110° and negative

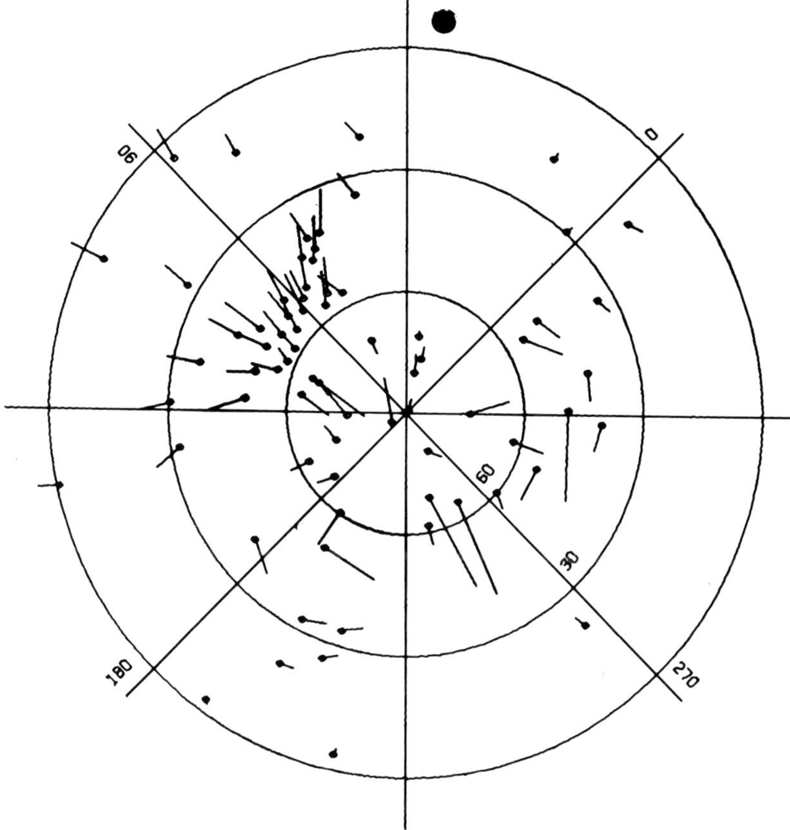

Fig. 7.5. Distribution of the magnetic disturbance vectors at 1400 UT on December 16, 1964. (Akasofu, S.-I. and Meng, C.-I.: *J. Geophys. Res.* **74**, 293, 1969.)

at stations of longitude greater than 110°. Here, we examine the growth and decay of the field-aligned (*net*) currents at three epochs during the substorm. In order to show the distribution and intensity of the field-aligned currents, the auroral electrojet and the ring current, a particular method of presentation is devised by using three concentric circles for each epoch. In Figure 7.8, the magnitude of the *D* component change is plotted as a radial arrow for each station in the middle circle. Its location is now given with respect to the Sun, in terms of MLT, and magnetic noon is shown by a circled dot at the top of each group of circles. A westward deflection is given by an inward pointing arrow (representing an inward field-aligned current) and an eastward deflection is given by an outward pointing arrow (representing an outward field-aligned current). The intensity of a negative bay at stations along the auroral zone is indicated by a westward directed arrow, representing the westward electrojet, along the inner circle in each epoch.

It can be seen that the dividing meridian between the inward and outward field-aligned currents lies approximately in the midnight sector and that the most intense part of the westward electrojet is located between the inward and outward

(a)

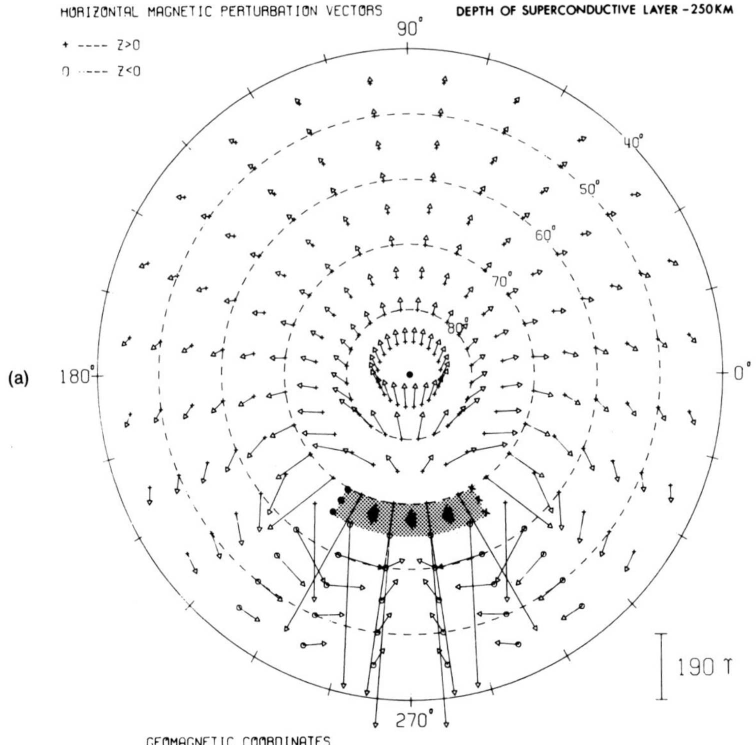

(b)

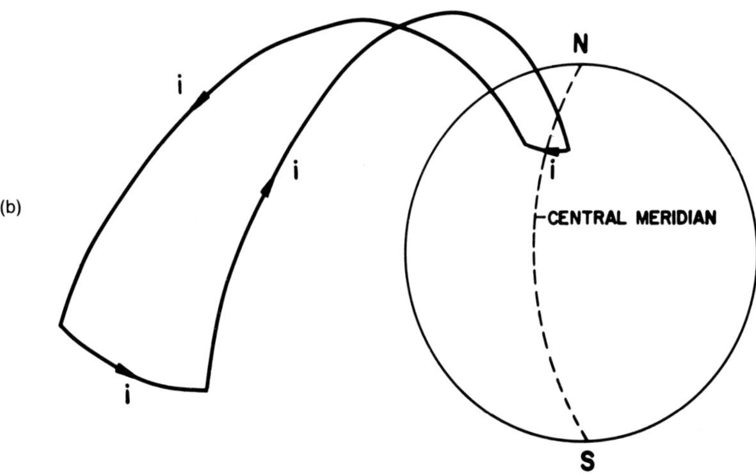

Fig. 7.6(a, b). Distribution of the magnetic field vectors for the current system illustrated in the lower part. (Kisabeth, J. L.: Ph.D. Thesis, Univ. of Alberta, 1972.)

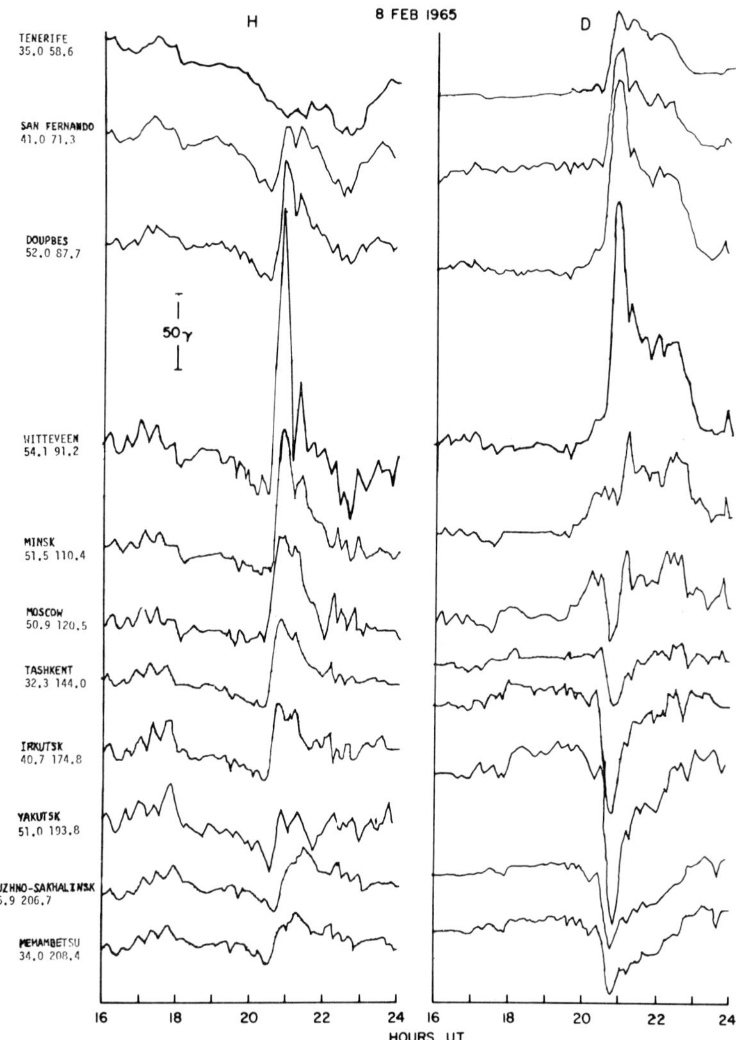

Fig. 7.7. *H* and *D* component records from mid-latitude stations during a substorm on 1965, February 8. (Meng, C.-I. and Akasofu, S.-I.: *J. Geophys. Res.* **74**, 4035, 1969.)

field-aligned currents. For bars along the outer circle in each epoch, see the next subsection.

Another interesting way to prove that the substorm current system appears as an intensification of the S_q^p current system is to obtain the distribution of the disturbance vectors at a number of high latitude stations on a disturbed day, when the AE index is high but fairly steady. Figure 7.9 shows the distribution of disturbance vectors on such a day, constructed by Iijima (1973). There is a considerable similarity between Iijima's figure and the S_q^p vector distribution (Figure 1.29). The same tendency was noted in Section 1.3.7.

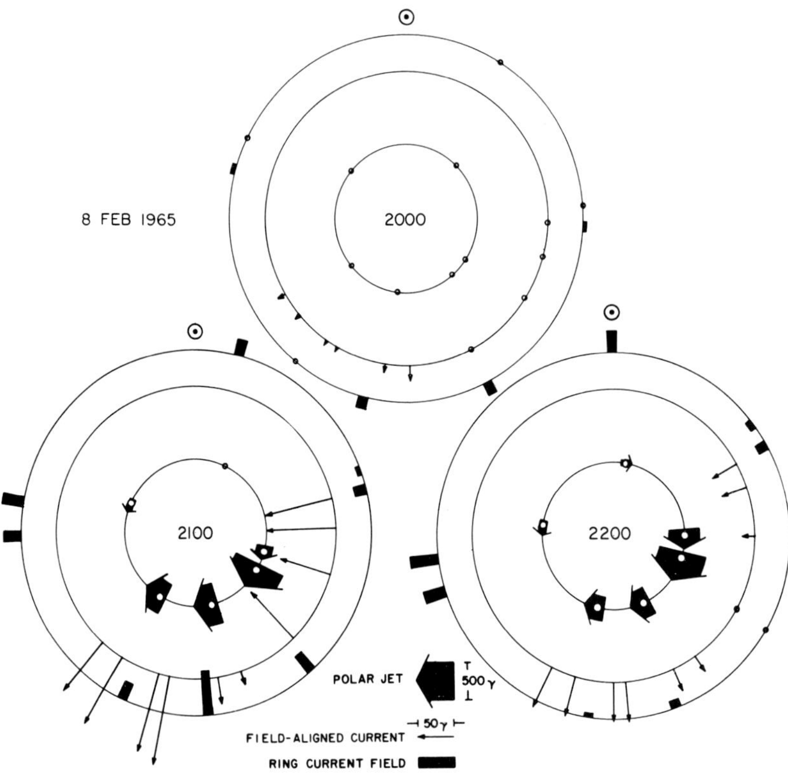

Fig. 7.8. Growth and decay of the field-aligned currents, the westward auroral electrojet and the partial ring current, deduced from ground magnetic data, during a substorm on 1965, February 8; for the format, see the text. (Meng, C.-I. and Akasofu, S.-I.: *J. Geophys. Res.* **74**, 4035, 1969.)

7.2.4. POSITIVE BAYS IN LOW LATITUDES AND POSITIVE B_z VARIATIONS AT THE SYNCHRONOUS DISTANCE AND IN THE MAGNETOTAIL

Geomagnetic variations in low latitudes during a polar magnetic substorm are most clearly manifested as a positive change in the H component, namely the so-called 'positive bay' in the night-morning sector. In Figure 7.5 such a positive change in low latitudes is indicated by a poleward pointing vector in the midnight sector. In Figure 7.8 inward directed bars along the outer circle in each epoch indicate also the magnitude and location (in MLT) of the positive changes. Figure 7.10 shows the H component records from the ATS-1 satellite, Honolulu and other low latitude stations; note a striking similarity between the first two records.

At least three source currents have been suggested for the positive bay. They are: (a) the return current from the westward auroral electrojet (Vestine and Chapman, 1938), (b) a reduction (or removal) of the westward ring current (Cummings *et al.*, 1968) and (c) field-aligned currents (Kamide and Fukushima, 1972; Fukushima and Kamide 1973b, c; Crooker and McPherron, 1972; McPherron *et al.*, 1973). This variance is due partly to the fact that it is not possible to

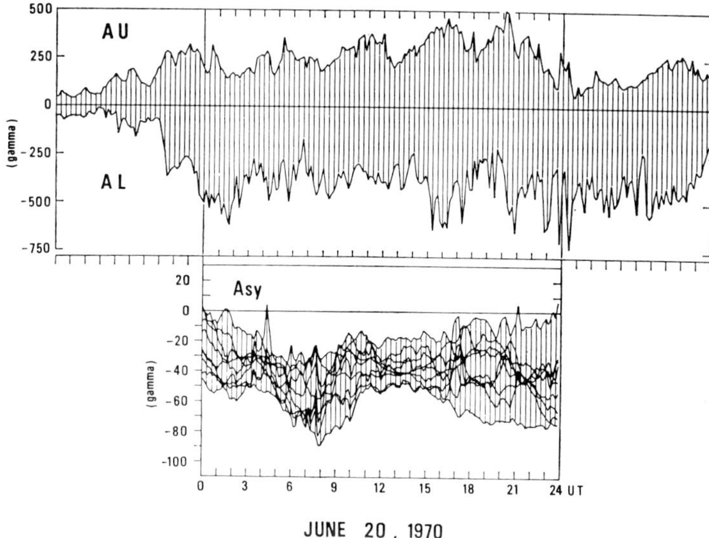

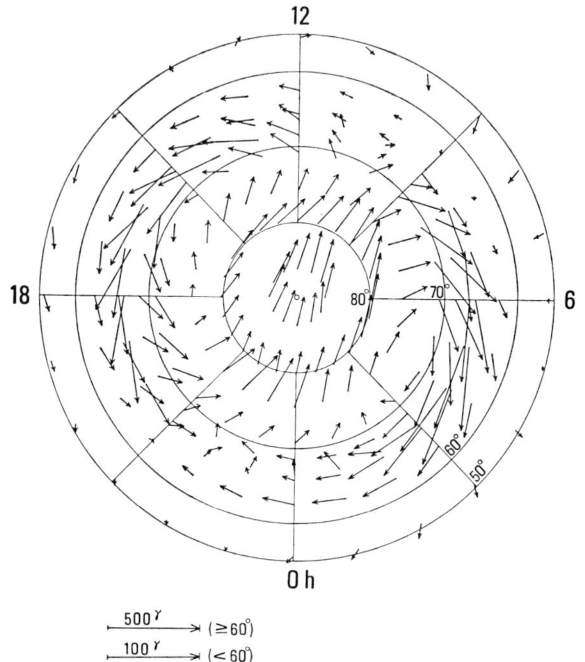

Fig. 7.9. Distribution of the equivalent current vectors in high latitudes. In constructing the figure, a highly disturbed day (of a constant AE index level) is chosen as seen in the upper two panels. (Iijima, T.: *Rep. Ionosph. Space Res. Japan* **27**, 199, 1973.)

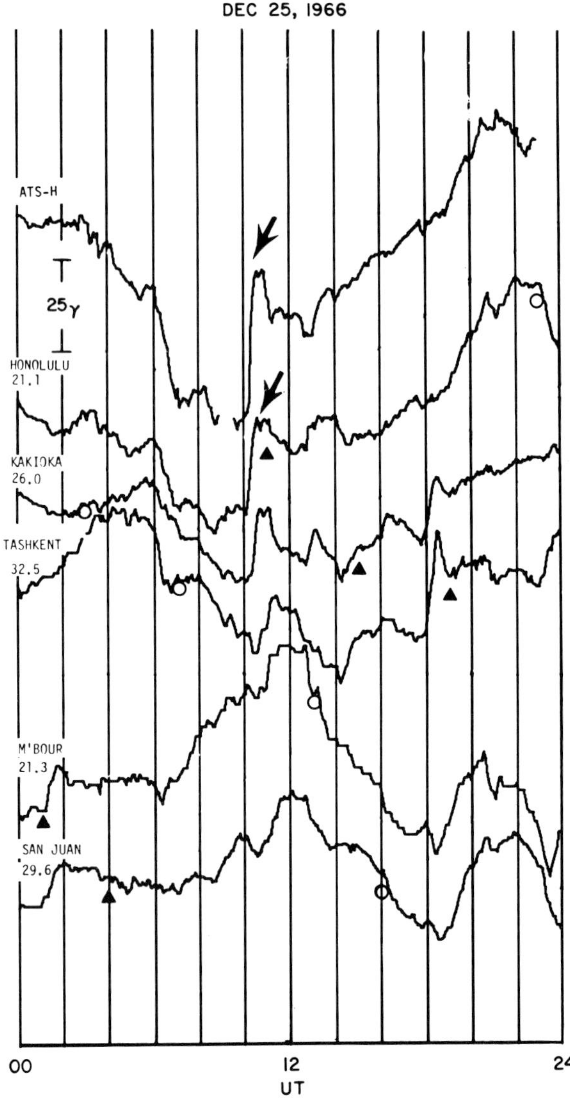

Fig. 7.10. *H* component magnetic variation at the ATS-1 satellite, at Honolulu (the sub-satellite point) and at several other low latitude stations. Note the similarity of the sharp increase at the ATS-1 satellite and at Honolulu. (Courtesy of W. D. Cummings and P. J. Coleman Jr., 1968.)

uniquely construct the actual three-dimensional current system on the basis of the magnetic vector distribution on the Earth's surface alone.

However, during the last decade numerous satellite magnetic observations have become available. These observations indicate that similar positive changes occur in the component perpendicular to the equatorial plane (the B_z component) at the synchronous distance and in the magnetotail (Cummings and Coleman, 1968; Fairfield and Ness, 1970; Meng *et al.*, 1971; Aubry *et al.*, 1972); see also Section 6.3.2. In the next subsection, we shall see that these positive B_z changes can be produced by the substorm current system.

7.2.5. MODELING THE MAGNETOSPHERIC SUBSTORM IN TERMS OF THE SUBSTORM CURRENT SYSTEM

(a) *Model*

In this subsection, we assume that the magnetotail current is disrupted in a narrow strip and is diverted to the polar ionosphere. Then we examine how the interval structure of the magnetosphere will be distorted by such a change of the current distribution.

We adopt here a model of quiet time magnetosphere, which is a modified version of Hones' model (1963). For the nightside current system, the cross-tail current is assumed to be disrupted in a narrow belt of width 7.7 R_E, centered at a geocentric distance of $X = -22.4\ R_E$; the dawn-dusk ends of the belt are located at $Y = \pm 7.2\ R_E$. This situation is shown schematically in Figure 7.1. The total current flowing along the circuit is 10^6 A for each hemisphere. It should be noted that the two field-aligned currents are connected by the westward current in the polar ionosphere. This connection will be discussed extensively in the next section (see also Section 1.3.3).

In the dayside magnetosphere, we assume a three-dimensional current system which is shown schematically in Figure 7.1. The two field lines are separated by 60° in longitude, and their feet are located at 79.5° in latitude. The total current flowing along the circuit is also 10^6 A. This portion of the circuit will be discussed later.

Figure 7.11(a) shows the magnetic field lines in the noon-midnight meridian for both the quiet time magnetosphere (dashed lines) and the substorm-time mag-

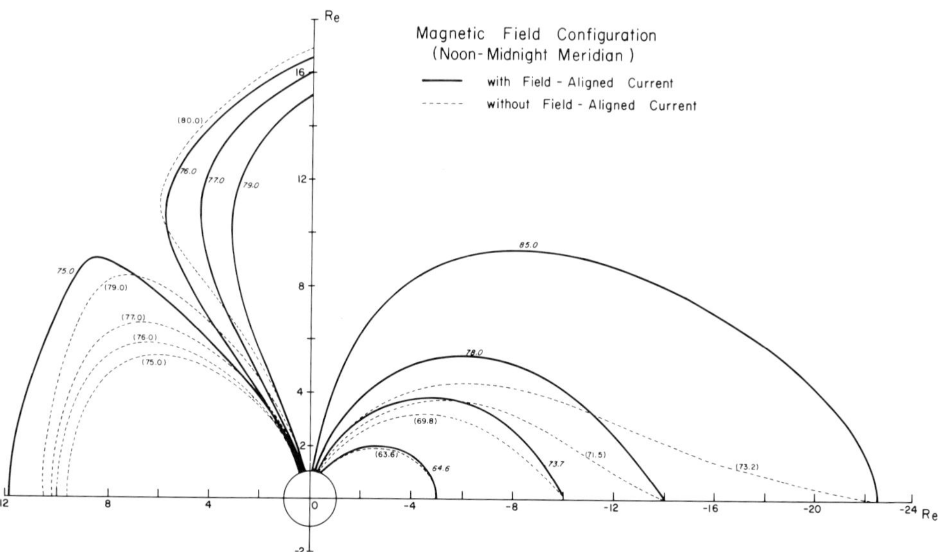

Fig. 7.11(a). Changes of the magnetic field configuration, which is caused by the diversion of the cross-tail current to the polar ionosphere, as illustrated in Figure 7.1. (Yasuhara, F., Kamide, Y. and Akasofu, S.-I.: *Planet. Space Sci.* **23**, 575, 1975.)

netosphere (solid lines). The number given for each field line is the latitude of its foot.

(b) *Positive B_z Changes*

In the midnight meridian, it can be seen that tail-like field lines in the northern and southern hemispheres become dipole-like. This tendency has been interpreted to indicate that the magnetosphere tends to regain its *original* Earth's dipolar configuration after the substorm. As noted in Section 6.5, however, the actual changes are very complicated. First of all, the B_z values at $X = -10.0$, -14.0 and -22.4 R_E, are 50, 34 and 14 γ, respectively, which may be compared with the corresponding dipole field values 31, 11 and 3 γ, respectively. In fact, the B_z increases observed by satellites are of order 5 γ at a geocentric distance of about 30 R_E (Section 6.5.3) where the dipole field is only about 1 γ. Secondly, the B_z component becomes small again during a late epoch of substorms. For these reasons, the observed changes of the B_z component can be understood better in terms of the growth and decay of the current system illustrated in Figure 7.1 than in terms of the tendency to regain the original Earth's dipole field.

Kamide *et al.* (1974) demonstrated also that the three-dimensional current system described here can reproduce the major parts of the well-known 'positive bay' in low latitudes on the Earth's surface, the positive B_z (or H) variation at the synchronous distance, as well as the positive B_z variation along the magnetotail. Figure 7.11(b) shows an example of their computed B_z values as a function of X. The parameter α is the angle between the incoming and outgoing field-aligned currents.

(c) *Poleward Shift of the Feet of the Geomagnetic Field Lines*

When we compare the latitude of the field line feet for a given equatorial crossing distance during quiet and substorm times, it can be seen that the latitude of feet is several degrees higher during the substorm than during the quiet time. Thus, auroral particles from the distant magnetotail region can reach a higher latitude during the substorm than the quiet time. This feature may explain the poleward motion of auroras in the midnight sector during substorms.

(d) *Equatorward Shift of the Cusp*

The proposed current system on the dayside (Figure 7.1) can also explain the equatorward shift of the cusp. This equatorward shift of several degrees in latitude is difficult to explain by a reasonable range of enhancement of the magnetotail current alone (see also Nishida, 1975). A strong local current is needed to shift the cusp location equatorward as much as several degrees. Kawasaki *et al.* (1974a,b) extended the above numerical study to explain details of the development of positive and negative bays in low latitudes and at the geosynchronous distance by varying the angle α as a function of the substorm time.

(e) *Summary*

A simple numerical study in the above shows that at least the two major features of the substorm can be explained reasonably by assuming the disruption and the

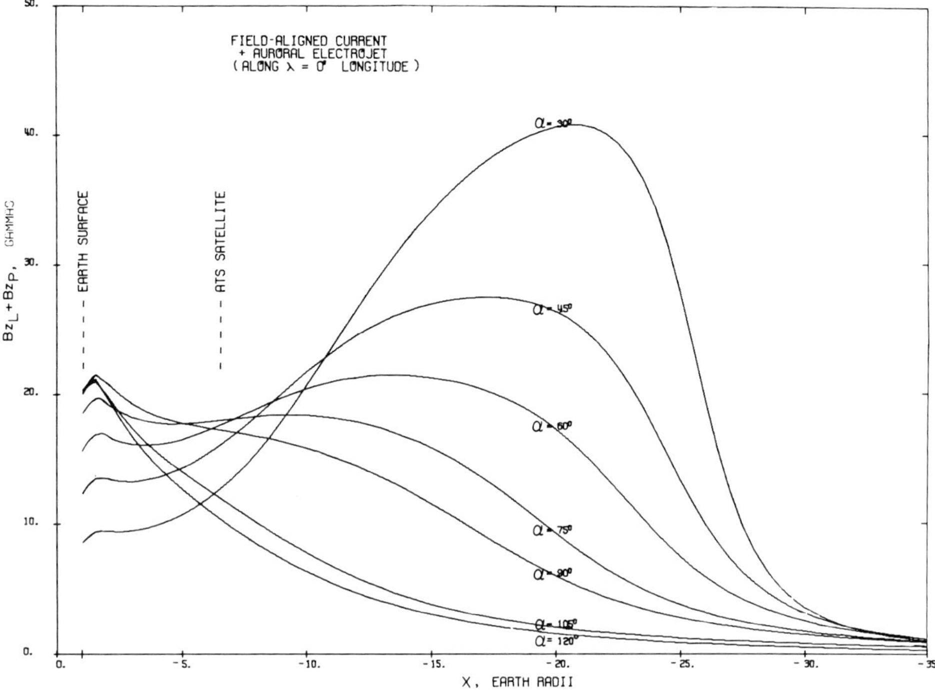

Fig. 7.11(b). B_z component as a function of X, produced by the diverted cross-tail current to the polar ionosphere. The parameter α is the 'wedge angle' between the inflow in the morning sector and the outflow in the afternoon sector. (Kamide, Y., Yasuhara, F. and Akasofu, S.-I.: *Planet. Space Sci.* **22**, 1219, 1974.)

subsequent diversion of the magnetotail current, namely, the B_z increase in the equatorial plane and the poleward expanding bulge. We have assumed so far that the disruption of the magnetotail current is the sole cause of the enhancement of the S_q^p current system.

It should also be noted that if the disruption of the cross-tail current takes place in a narrow strip of 5–10 R_E in length along the magnetotail, there will be large magnetic field variations of order 20–50 γ in the vicinity of the disruption region. Kamide *et al.* (1974) noted that the absence of such a large B_z variation within the lunar distance indicates that the disruption occurs at distances greater than 60 R_E.

One possibility is that the complete disruption occurs only near the inner edge of the plasma sheet, say $X = -6 \sim -10\ R_E$. At the geosynchronous distance ($X = -6.6\ R_E$), the B_z (or H) component increases by 20–50 γ, at about the onset time of a substorm (Figures 7.10, 8.9, 8.10 and 8.16). Since the plasma pressure increases at the same time (Section 8.3) and since γ is known to be of order unity, the diamagnetic effect of the plasma should be added to the observed value of B_z in estimating the total effect of the disruption.

It should be noted that the B_z component in the magnetotail is greatly increased in the recovering plasma sheet (Section 6.5.3). Since plasma parameters associated with the recovering plasma sheet are consistent qualitatively with an enhanced reconnection, a part of the B_z increase may be due to an enhanced reconnection during the recovery phase (Section 9.2).

7.3. Field-Aligned Currents and the Auroral Electrojets

7.3.1. INTRODUCTION

One of the most important problems associated with the field-aligned currents is how they are connected to ionospheric currents, in particular to the auroral electrojets. There had been no direct observation which indicates the link between the field-aligned currents and the electrojets in the past, although there have been some analytical and theoretical studies on the subject. Based primarily on magnetic variations on the Earth's surface, several three-dimensional current systems for the magnetospheric substorms have been inferred (Boström, 1964; Atkinson, 1967; Akasofu and Meng, 1969; Bonnevier *et al.*, 1970; Kisabeth and Rostoker, 1971; Fukushima, 1971; Kamide and Fukushima, 1972; Fukushima and Kamide, 1973a, b, c; Aubry *et al.*, 1972; Chen and Rostoker, 1974). They suggested that there is an inflow of current into the morning half of the auroral oval and an outflow from the evening half of the oval, and that these are connected through the westward electrojet.

The TRIAD satellite is the first polar orbiting satellite that carries a tri-axial magnetometer, providing vital information on the distribution of field-aligned currents just above the ionosphere. Therefore, it is now possible to make a direct comparison of the location of the field-aligned currents with that of the auroral electrojets in a wide latitudinal and longitudinal range. In this subsection, we examine the geometrical relationship between the field-aligned currents and the auroral electrojets on the basis of magnetic records from the TRIAD satellite and from the Alaska meridian chain of stations.

As has already been pointed out by Armstrong and Zmuda (1973) and Zmuda and Armstrong (1974a, b), the total perturbation field vector at auroral latitudes is mostly transverse to the main field to within experimental sensitivity, thus confirming that the magnetic perturbations result from field-aligned currents. Thus, the TRIAD magnetometer records are shown as $\Delta\boldsymbol{B}(=\boldsymbol{B}-\boldsymbol{B}_{\text{theoretical}})$ versus time (or latitude) only for the A(EW) and B(NS) axes. $\boldsymbol{B}_{\text{theoretical}}$ is calculated by using the International Geographic Reference Field (IGRF) 1965.0 model updated to the corresponding days of 1973.

7.3.2. EXAMPLES

(a) *1973, March 6*

The satellite passed over Alaska around the maximum epoch of an intense, isolated substorm; there was no appreciable substorm activity from the beginning of this day to the onset of this substorm; the substorm was characterized by an intense eastward auroral electrojet in the evening sector that is indicated by an abnormally large deviation in AU with a magnitude ($\sim 600\ \gamma$) comparable to the simultaneous negative deviation (AL).

Figure 7.12(a) shows the TRIAD magnetic field data as functions of both the universal time and the invariant latitude Λ, and the latitudinal profile of the ground H and Z perturbations at 0702 UT, together with the location of auroral arcs (determined from the all-sky camera data assuming the height of the auroras to be 110 km). The average magnetic local time of the satellite is also indicated. The ground magnetic perturbations ΔH and ΔZ are measured from the quietest level of the month (March 5 for this example). It should be noted that these data are plotted linearly in the satellite traverse time, with invariant latitude as noted.

Several points of interest are evident from the TRIAD data. First, an eastward magnetic perturbation occurred in the range between $\Lambda = 73.7°$ and 57.7° accompanied by a northward deviation, indicating the existence of field-aligned currents in this region. Second, the flow direction was upward on the poleward side and downward on the equatorward side; it is one of the typical features of the

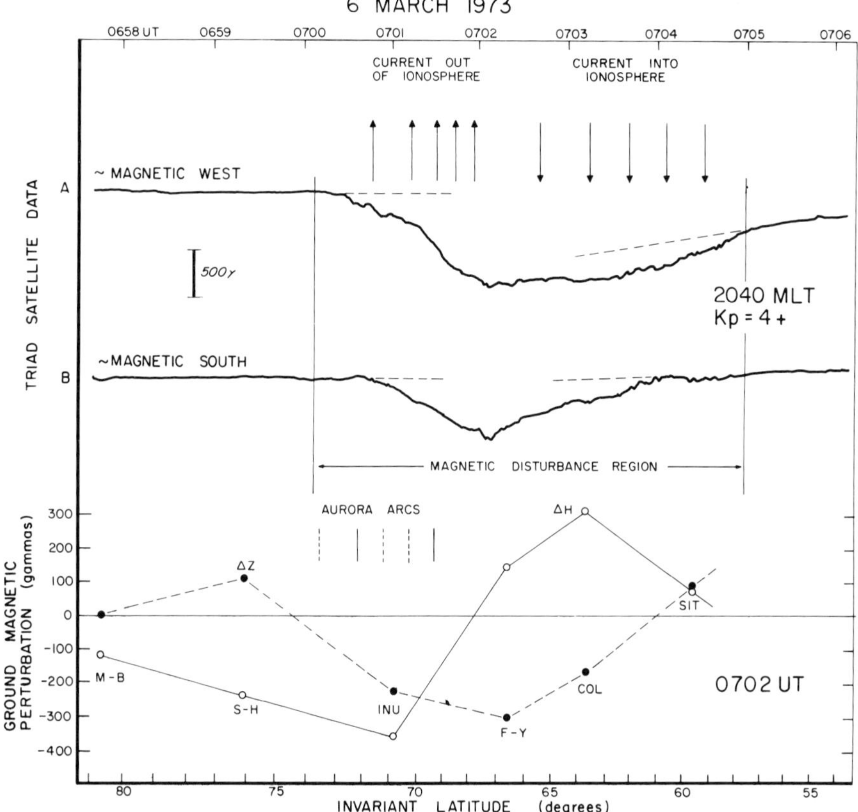

Fig. 7.12(a). Relationship between the field-aligned currents and the electrojets. The upper two curves show the east-west and north-south components of the magnetic field observed by the TRIAD satellite; the directions of the field-aligned currents deduced from the east-west component are shown by upward and downward directed arrows. Both ΔH and ΔZ changes at the Alaska meridian chain of stations are shown in the lower part. Note that the boundary of the upward and downward currents coincides approximately with the boundary of the region of $\Delta H > 0$ and $\Delta H < 0$. (Kamide, Y. and Akasofu, S.-I.: *Planet. Space Sci.* **24**, 203, 1976.)

field-aligned currents in the evening sector (Zmuda and Armstrong, 1974b). Third, for the east-west component, the extrapolation lines from the poleward and the equatorward traces (shown by dashed lines), did not match. This can be explained by supposing that the intensities of the upward and downward field-aligned currents are not equal (Yasuhara *et al.*, 1975b). Finally, if we assume that field-aligned current sheets had an infinite east-west extent, the average intensities of the upward and downward currents can be estimated from the slope of the east-west magnetic perturbation; they are 1.39×10^{-6} A m^{-2} (0.80 A m^{-1}) and 0.42×10^{-6} A m^{-2} (0.23 A m^{-1}), respectively.

The latitudinal profile of the auroral electrojets is deduced from the latitudinal profile of the magnetic H and Z perturbations along the meridian. It should be noted, however, that ground-based observations of the magnetic vector distribution alone cannot uniquely provide the latitudinal dependence of the ionospheric current without knowing the values of a number of variables that can affect ground magnetic perturbations (the location of the center and width of the auroral electrojets, geometry of the field-aligned currents and their intensities, height of the current, depth of the induced current inside the Earth, those parameters for the return currents in the ionosphere, etc.). However, we can infer the position of the center of the electrojet by knowing the latitudinal profile of both the H and Z components, in particular the locations of the maximum negative H deviation and of $\Delta Z = 0$, if we assume that these components are caused mainly by the east-west ionospheric current (i.e., the electrojets). Poleward and equatorward limits of the electrojets may be deduced from the locations of the Z component extrema in the profile. The basis of this technique is discussed in detail by Kisabeth (1972) and Wallis *et al.* (1976).

(b) *1973, March 9*

The Kp index for the first four 3-hour intervals of this day was 3_0, $3+$, 3_0 and 3_0, indicating that there was successive substorm activity during this period. The TRIAD passed over Alaska again during the maximum epoch of the substorm, which started at about 0600 UT. The satellite passed a little east of Fort Yukon. Although there was some enhancement in the AU index during the period 0600–0700 UT, to which the College H component mainly contributed, the Fort Yukon magnetogram (not shown here) showed a typical positive H bay at that time.

In Figure 7.12(b), we compare the location of the substorm region along the Alaska meridian relative to the field-aligned current region. The field-aligned currents were located in the latitudinal range between $\Lambda = 74.2°$ and 63.4°, inferred from the slope of the satellite magnetic field in the east-west component. Discrete auroral arcs were embedded in the upward field-aligned current region. Again, it was found that the extrapolated line from the poleward trace did not match with the line extrapolated from the equatorward trace, suggesting inequality of the current intensities between the upward and downward currents. By using the poleward extrapolation line for the upward field-aligned current and the equatorward line for the downward current, the intensities of the upward and downward currents are inferred to be 1.13×10^{-6} A m^{-2} (0.54 A m^{-1}) and 0.47×10^{-6} A m^{-2}

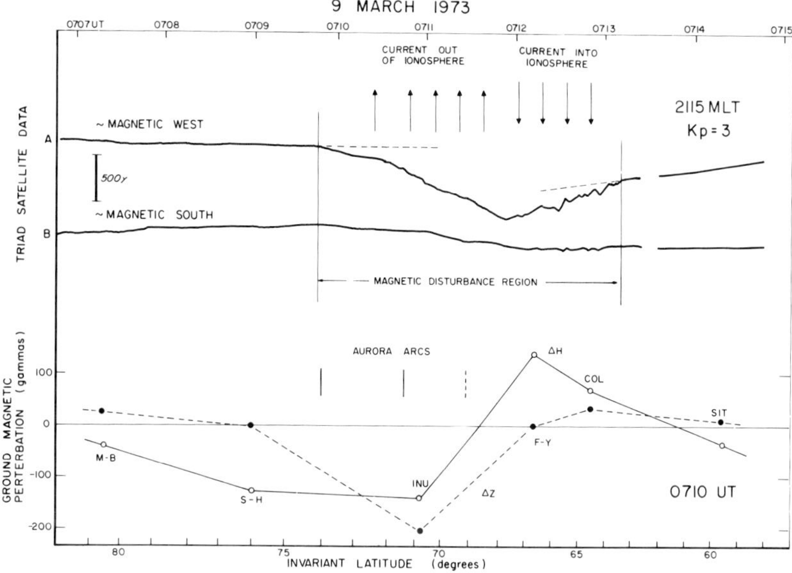

Fig. 7.12(b). Same as Figure 7.12(a).

($0.22\ \text{A m}^{-1}$), respectively; the ratio of the upward current density to the downward current density was 0.41 in this case.

The boundary between the upward and downward currents was located at about 68.0° invariant latitude at the time of the satellite passage. Along the Alaska meridian chain of stations, the sign of the *H* component perturbation changed between Inuvik and Fort Yukon. Connecting these two data points by a line, as indicated in Figure 7.12(b), the cross-over point is found to be located at 68.6°; the boundary of the oppositely directed field-aligned currents can be inferred to be located in this vicinity and is located near the boundary of the upward and downward field-aligned current regions with accuracy of about 1°. Figure 7.13 shows an example of the simultaneous observation of the field-aligned currents and the westward electrojet in the morning sector.

7.3.3. RELATIVE LOCATION OF THE AURORAL ELECTROJETS WITH RESPECT TO FIELD-ALIGNED CURRENTS IN THE EVENING SECTOR: A STATISTICAL RESULT AND SUMMARY

Figure 7.14 shows the locations (represented by circles) of the boundary between the upward and the downward field-aligned currents in the invariant latitude (Λ) and magnetic local time (MLT) coordinates determined from the 35 TRIAD passes in the evening sector. On the poleward side of this boundary the upward current prevails, whereas the downward current prevails on the equatorward side. The scattered distribution of the points is possibly caused by a wide range of magnetic disturbances (Kp maximum = 6). It is evident that the average location

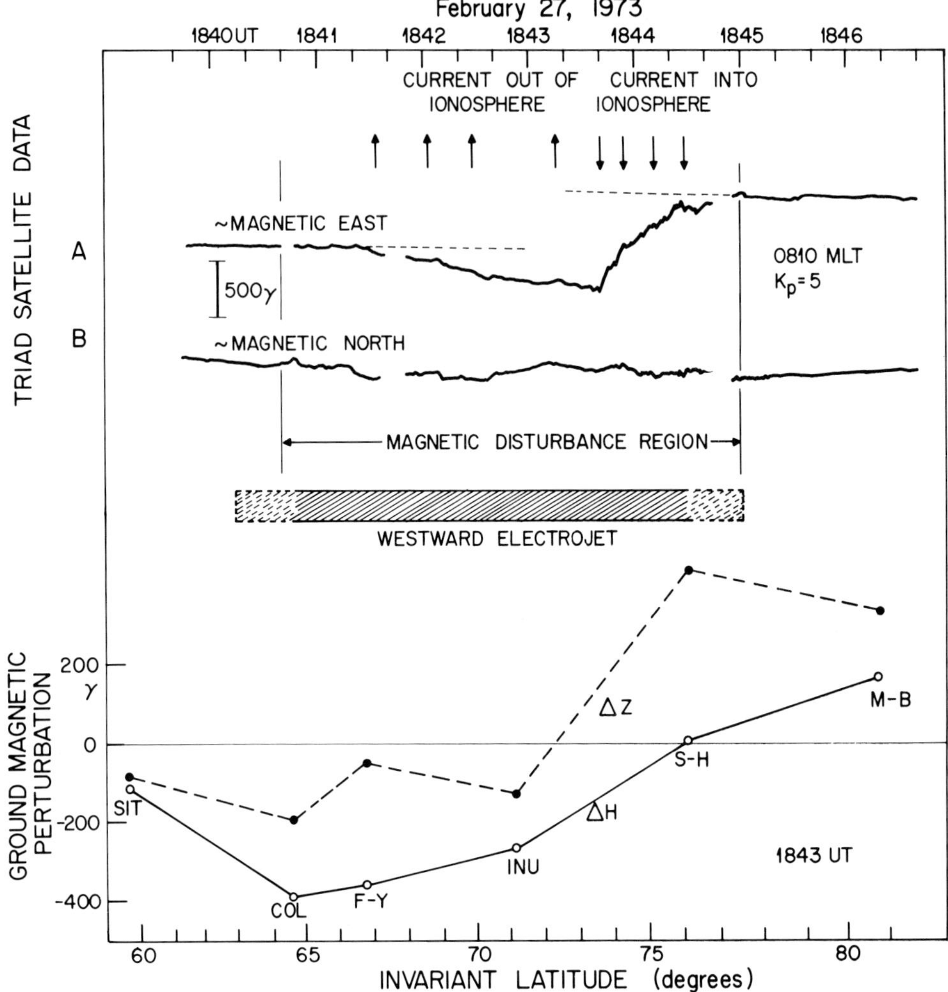

Fig. 7.13. Relationship between the field-aligned currents and the electrojet in the morning sector. (Courtesy of Kamide, Y., Akasofu, S.-I. and Rostoker, G.)

of the boundary, shown by a dashed line, is located near the 70° invariant latitude and approximately follows the auroral oval.

The relationship between the field-aligned currents and the auroral electrojets can be summarized as follows:

(1) The boundary between the upward and downward currents coincides in most cases with the boundary between the westward and eastward electrojets. Thus, the Harang discontinuity, defined by the line separating the negative and the positive H perturbations on the Earth's surface, corresponds to a narrow region in which no field-aligned current is present.

(2) As will be discussed in the next subsection, this finding suggests a three-dimensional current system which is quite different from what has been

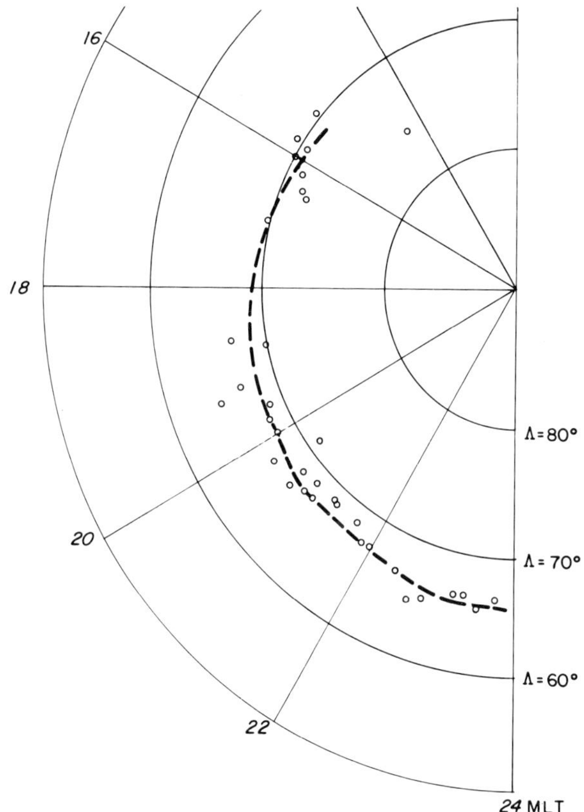

Fig. 7.14. Locations of the boundary of the outward and inward field-aligned current regions in the afternoon-evening sector in invariant latitude-MLT coordinates. (Kamide, Y. and Akasofu, S.-I.: *Planet. Space Sci.* **24**, 203, 1976.)

suggested in the earlier studies. Cummings *et al.* (1968), Kamide and Fukushima (1972) and Crooker and McPherron (1972) have put forward a model current system in which there is an upward field-aligned current at the eastern end of the eastward electrojet. Heppner *et al.* (1971) proposed that a sheet of the upward current originates from the Harang discontinuity.

(3) Although the equatorward boundary of the westward electrojet coincides with that of the upward current, the westward electrojet flows beyond the latitude of the upward current. The poleward edge of the upward current seems to be marked by the northernmost discrete auroral arc. There is an indication in the latitudinal profile of the H and Z perturbations which suggests that the upward current flows out only from the equatorward half of the westward electrojet. This makes an apparent similarity between the latitudinal profile of the magnetic east-west perturbation measured by the TRIAD satellite and that in the Z component perturbation on the ground.

7.3.4. MODEL CALCULATION

It is instructive to examine here theoretically how the field-aligned currents and the ionospheric currents are connected and how the observations in the previous sections can be explained. In Section 1.3.3, we made a model calculation in an attempt to examine the relationships between the two currents for a very simple ionospheric model. The model was intended to simulate a quiet time situation by having a simple annular belt of a high conductivity. The model provides most of the basic features of the relationship but needs some changes in simulating magnetospheric substorm conditions, in particular the conductivity distribution of the ionosphere.

In the following new model, the polar cap is bounded by the latitude circle of $\lambda = 72°$. The auroral oval is bounded by two latitude circles of $\lambda = 72°$ and $\lambda = 60°$. The oval is then divided into two belts by the latitude circle of $\lambda = 67°$. This is to simulate the two regions of auroras, the oval of discrete auroras and the oval of diffuse auroras (Section 2.2.1). The conductivity in the two belts is different (the conductivity in the oval of discrete auroras is assumed to be twice that in the oval of diffuse auroras) but is higher than that in the polar cap and the middle and low latitude belts (Table 7.1).

TABLE 7.1

The Pedersen and Hall conductivities

Polar cap (Region I)		Auroral oval (Region II)				Middle and low latitude (Region III)	
Region I		Region II		Region III		Region IV	
Σ_P	Σ_H	Σ_P	Σ_H	Σ_P	Σ_H	Σ_P	Σ_H
1.0	2.0	10.0	40.0	5.0	20.0	1.0	2.0

The ratio of the Hall to Pedersen conductivities is 4.0 in the auroral oval and 2.0 in the polar cap and the middle and low latitude regions (see Section 7.7.2). The field-aligned currents are located along two latitude circles of $\lambda = 70°$ and $\lambda = 63°$. The longitudinal dependence of the field-aligned current intensity is the same as that of the quiet time model for both the poleward and equatorward currents. The results are shown in Figures 7.15(a) and (b); for the format of the figures, see Section 1.3.3.

As we saw in Section 1.3.3, there is no N-S closure in the same meridian. All the currents that contribute to this closure are greatly deflected westward or eastward along the oval. Thus, the N-S closure current contributes significantly to the westward electrojet in the early morning sector.

One of the most important differences between the model discussed in Section 1.3.3 and the present one is the appearance of the eastward current in the equatorward half (the diffuse auroral region) of the oval in the late evening sector. This is due to the eastward deflection of the N-S closure. It is important to note that this eastward current is deflected at the poleward boundary of the diffuse auroral oval and becomes a westward current. Thus, it is not necessary to invoke

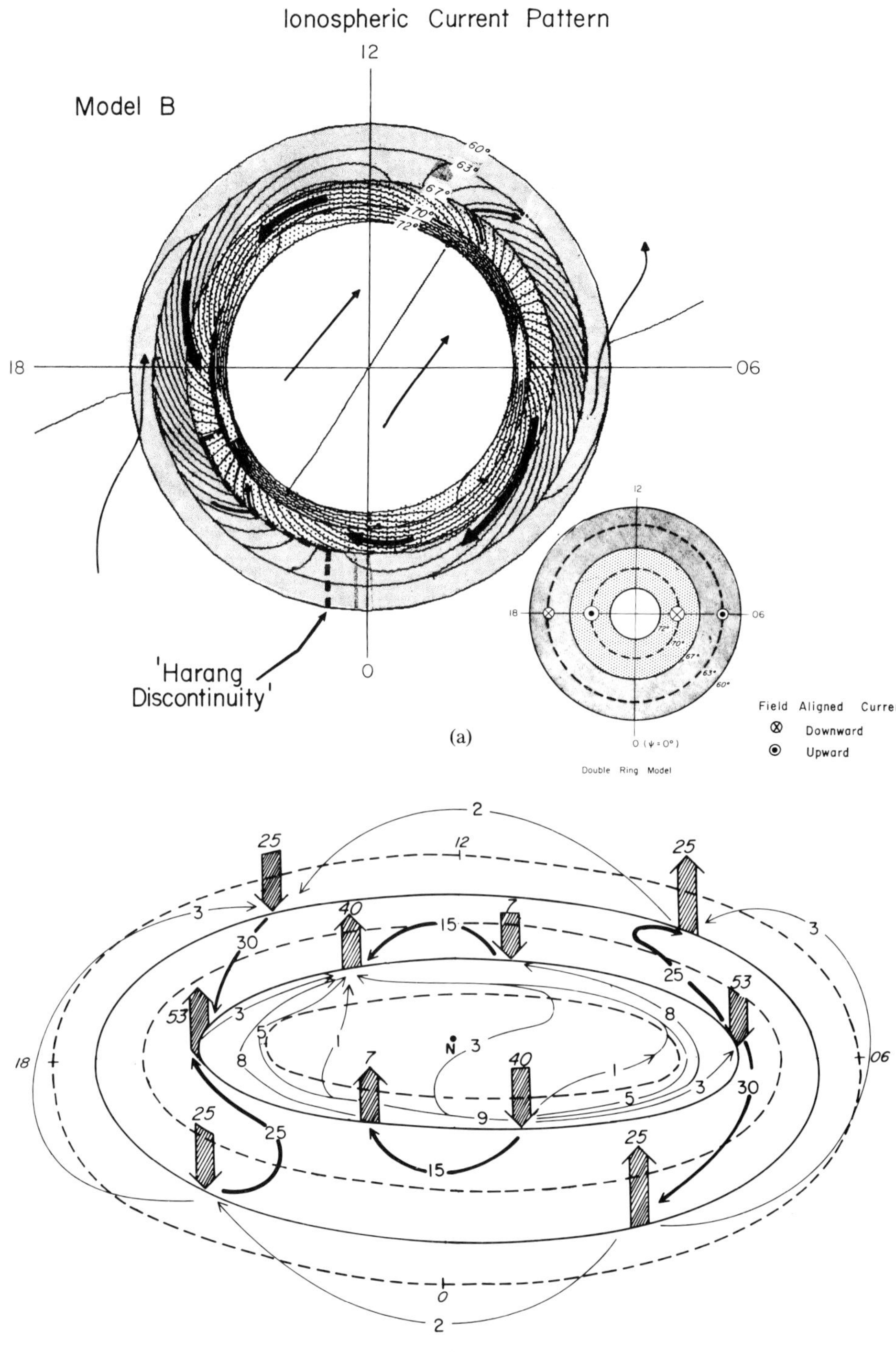

(a)

(b)

Fig. 7.15(a, b). Computed ionospheric current pattern for a double high conductive belt (see the insert). The thick dash line indicates the Harang discontinuity. The current pattern and its relation to the field-aligned currents are schematically shown in (b). (Courtesy of Yasuhara, F. and Akasofu, S.-I.)

the source of the eastward electrojet in the partial ring current. Further, the model successfully reproduces a concentrated westward current flow along the oval of discrete auroras in the evening sector, extending from the westward current in the early morning sector.

It is also possible to delineate the boundary between the regions of the eastward and westward currents in the evening sector. The boundary is shown by a dashed line. This boundary may be identified as the Harang discontinuity (Heppner, 1972b).

Another important feature in the model is that the equatorward field-aligned currents in the evening and the morning sectors do not connect with each other. All the equatorward field-aligned currents and one half of the poleward field-aligned currents are connected. This does not seem to depend on models of the ionosphere (for the model examined so far) or even the relative location of the peaks of the field-aligned currents.

The closure of the remaining half of the poleward field-aligned currents greatly depends on ionospheric conductivity. The direct E-W closure of this current increases when the conductivity is high in the oval compared with that in the other regions. More currents flow in the polar cap and the mid-low latitude region when the auroral oval conductivity is reduced. The deflection of the N-S closure in the westward or eastward direction also increases with an enhancement of the Hall conductivity in the oval.

The amount of ionospheric currents equatorward of the oval is surprisingly small for any of our ionospheric conductivity models. If this is indeed the case, the ground magnetic perturbations that have been ascribed to the two-dimensional currents might be mainly due to the magnetic fields of the field-aligned currents (Fukushima and Kamide, 1973a, b; Leont'yev *et al.*, 1974). On the other hand, it may be noted that Rostoker and Hron (1975) reported the presence of an eastward electrojet a little equatorward of the westward jet in the morning sector, suggesting the presence of the so-called 'return current'.

Here, it is instructive to compare the model study presented above with those studied by Iwasaki and Nishida (1967), Mal'tsev *et al.* (1973), Van'yan *et al.* (1973), Maeda and Maekawa (1973), Lyatsky *et al.* (1974) and Pudovkin (1974), who assumed a particular charge distribution or potential distribution either along specific latitude circles or at points, and then solved the Laplace equation for the potential function Φ for a given ionospheric model. Pudovkin (1974) also extensively examined the relationship between the auroral electrojet polarization field and the field-aligned currents.

One of the most important results obtained by Mal'tsev *et al.* (1973) is that they reproduced successfully the Harang discontinuity. Thus, the result presented above is in qualitative agreement with theirs. However, the present model reproduces a more realistic current system than theirs, providing the Harang discontinuity, a wide region of the eastward current in the diffuse auroral region in the evening sector and an intense westward current near the poleward boundary of the oval (which is the extension of the westward current in the morning sector). Lyatsky *et al.* (1974) also examined the potential and current patterns for different phases of the substorm.

In this section, we have so far dealt only with a steady state. One of the important problems in the magnetosphere-ionosphere is the time constant involved in the growth of a polarization electric field in the ionosphere. The magnetosphere plays the role of a capacitor in an electrical circuit. It is continuously charged by the polarization electric charges which leak away, along the geomagnetic field lines, into the magnetosphere. Meanwhile, it is simultaneously discharged by the Pedersen current in the ionosphere. DeWitt (1968) showed that this time constant is very short, of order of a few seconds, so that the leakage is not serious in the growth of the polarization field.

7.4. Auroral Electrojets

7.4.1. AURORAL ELECTROJET AND GLOBAL AURORAL FEATURES

A close relationship between polar geomagnetic disturbances and auroral activity was found in pioneering studies by Birkeland and Chapman. Owing to a lack of simultaneous auroral records, however, most pre-IGY studies of polar geomagnetic disturbances were conducted largely independently of auroral morphology, although Harang (1946), Heppner (1954) and Meek (1953) examined the relationship of auroras with geomagnetic activity on a local scale.

The IGY all-sky camera program provided the first opportunity to observe the distribution of the aurora over the entire polar region. Emerging from this was the concept of the auroral substorm, which is complementary to Birkeland's original idea of the polar elementary storm. However, even the IGY all-sky camera and magnetic data were not extensive enough to simultaneously study auroral and polar magnetic substorms over the entire polar region. It was only on a statistical basis that Akasofu *et al.* (1966) were able to indicate areas of negative and positive bays on maps of auroral forms associated with the development patterns of the auroral substorm.

With photographs from a scanning camera aboard the ISIS-2 and DMSP satellites, it is thus of great interest to examine the relationship between the large-scale auroral distribution and the auroral electrojets. In this section, we shall examine DMSP-2 satellite auroral photographs and the simultaneous ground magnetic data from a number of polar stations. It should be noted, however, that because the satellite photograph is available only once each orbital period, and this time is of the order of the lifetime of one substorm, the dynamic relationship of current systems to the auroral configuration throughout one substorm must await the availability of a series of photographs from a satellite with a very eccentric orbit, capable of viewing the polar region continuously for several hours.

(a) *Examples*

The auroral features on 1973, January 25–26 were described extensively by Snyder *et al.* (1974), based on the available DMSP photographs. Several isolated substorms occurred on these days, as can be seen in the AE and Kp indices (see Figure 7.16). Here, we have chosen six passes to demonstrate the relationship between auroras and equivalent currents (Figure 7.17).

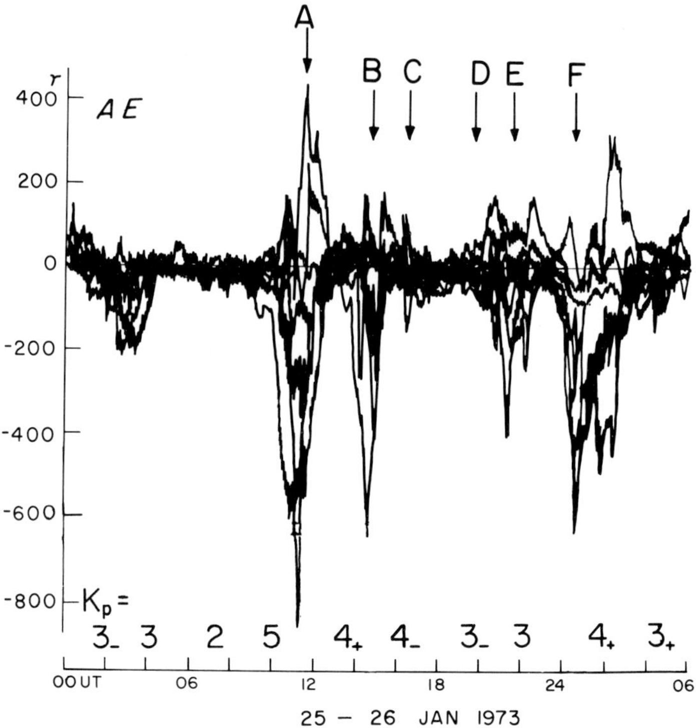

Fig. 7.16. AE and Kp indices on 1973, January 25–26. The letters A, B, C, D, E and F indicate the times when the DMSP-2 photographs, Figures 7.17a–f, were taken. (Kamide, Y. and Akasofu, S.-I.: *J. Geophys. Res.* **80**, 3585, 1975.)

The pass of 1122 UT (A): Figure 7.17(a). This pass took place during the maximum epoch of an intense substorm. The AL index was about $-900\ \gamma$ (College) and the AU index was of order $400\ \gamma$ (Tixie Bay) at that time. Indeed, the DMSP photograph shows very disturbed auroral features. The discrete auroras were considerably separated from the diffuse aurora, indicating their poleward displacement. A large-scale loop is seen in the evening sector.

The difference in the Z sign at Inuvik and Fort Yukon indicates that in the midnight sector the electrojet was centered between these two stations, near the poleward boundary of the diffuse aurora that showed a well-defined torch-like structure. Although the photograph did not cover the whole auroral oval, the magnetograms along the auroral region indicate that stations even in the noon sector (Leirvogur and Narssarssuaq) observed the westward electrojet. We can see the reversal in the current direction in the evening sector across active auroras (see the vectors at Cape Chelyuskin and Tixie Bay); note, however, that the eastward current was observed well inside the large auroral loop. Davis (1962a) found a similar case in the IGY all-sky camera data.

The pass of 1445 UT (B): Figure 7.17(b). This pass also took place during the maximum epoch of a new substorm which followed the previous one. The DMSP

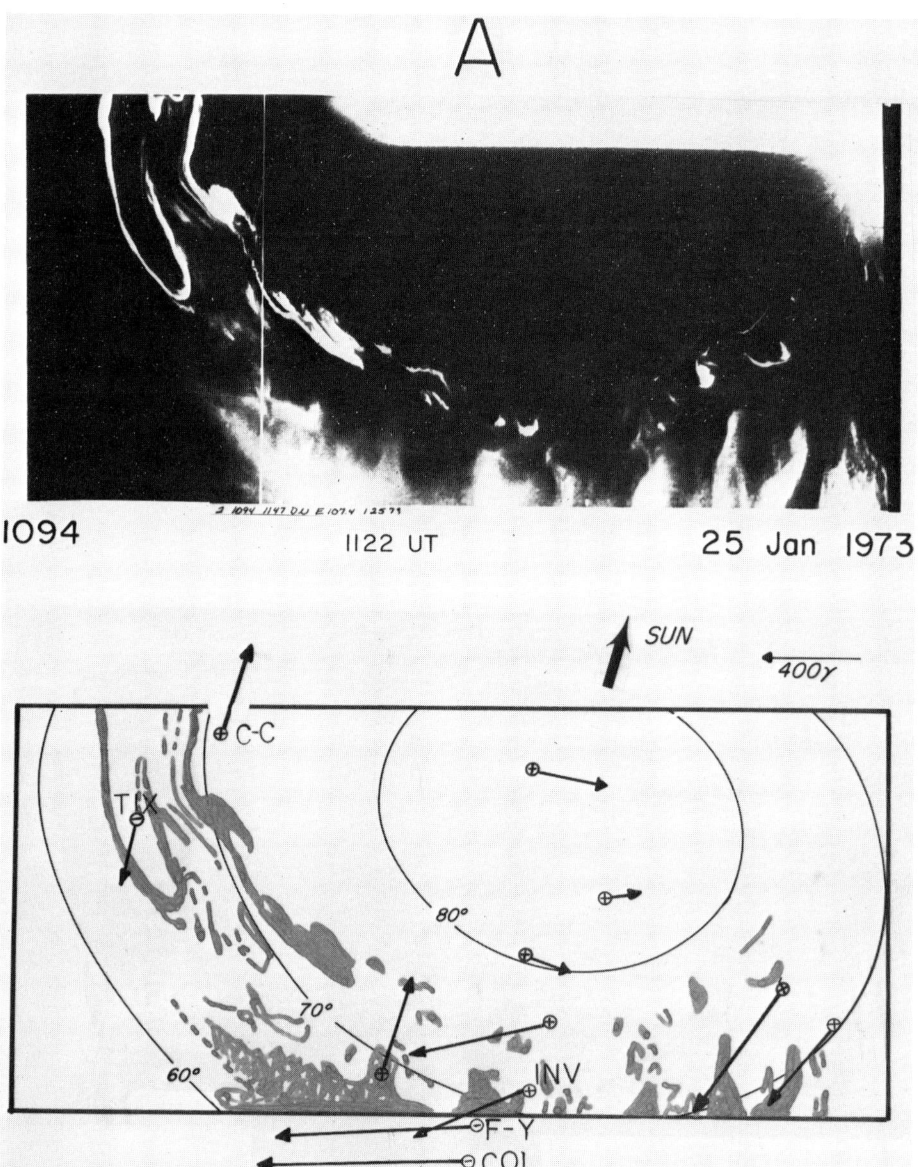

Fig. 7.17(a–f). DMSP-2 auroral photographs and the concurrent equivalent current vectors at the geomagnetic observatories in the field of view of the photographs. The signs of the Z component (+ positive and − negative) are also indicated. (Kamide, Y. and Akasofu, S.-I.: *J. Geophys. Res.* **80**, 3585, 1975.)

photograph shows that the diffuse aurora spread below invariant latitude of 60° in the early morning hours.

Comparing the equivalent current and auroral distributions, we notice that an intense electrojet appears to flow well poleward of the optical aurora; see the current vectors at Cape Chelyuskin and Point Barrow. However, the sign of the Z

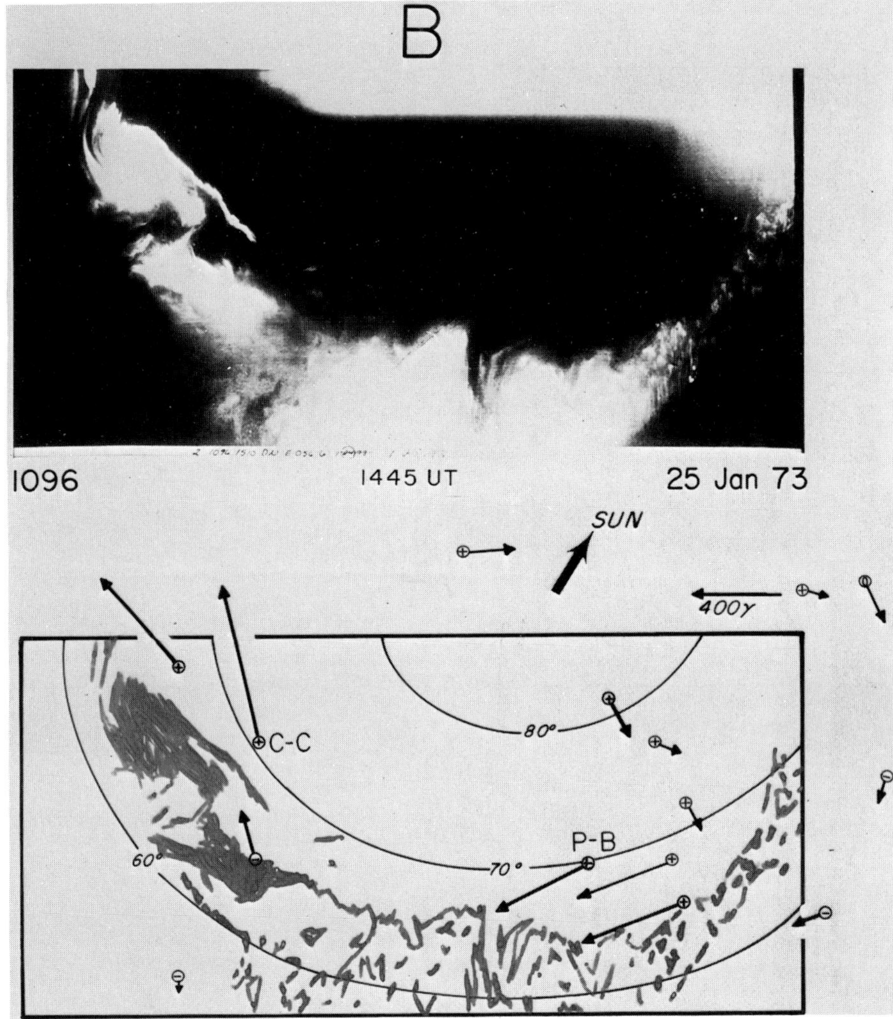

Fig. 7.17(b).

component indicates that the jet was flowing equatorward of the stations, and thus in the vicinity of auroras.

The pass of 1627 UT (C): Figure 7.17(c). This pass took place when weak disturbances (negative H bays of about 150 γ) were observed at Point Barrow and Tixie Bay; both stations were in the morning sector. The auroras were clearly visible only in the morning sector. Despite weakness of the disturbance as a whole, the current pattern was similar to that of an intense substorm.

The pass of 1951 UT (D): Figure 7.17(d). The pass marked D took place at about the onset time of an isolated substorm which is discussed in the following pass E. The satellite photograph shows faint auroras along the oval and a small

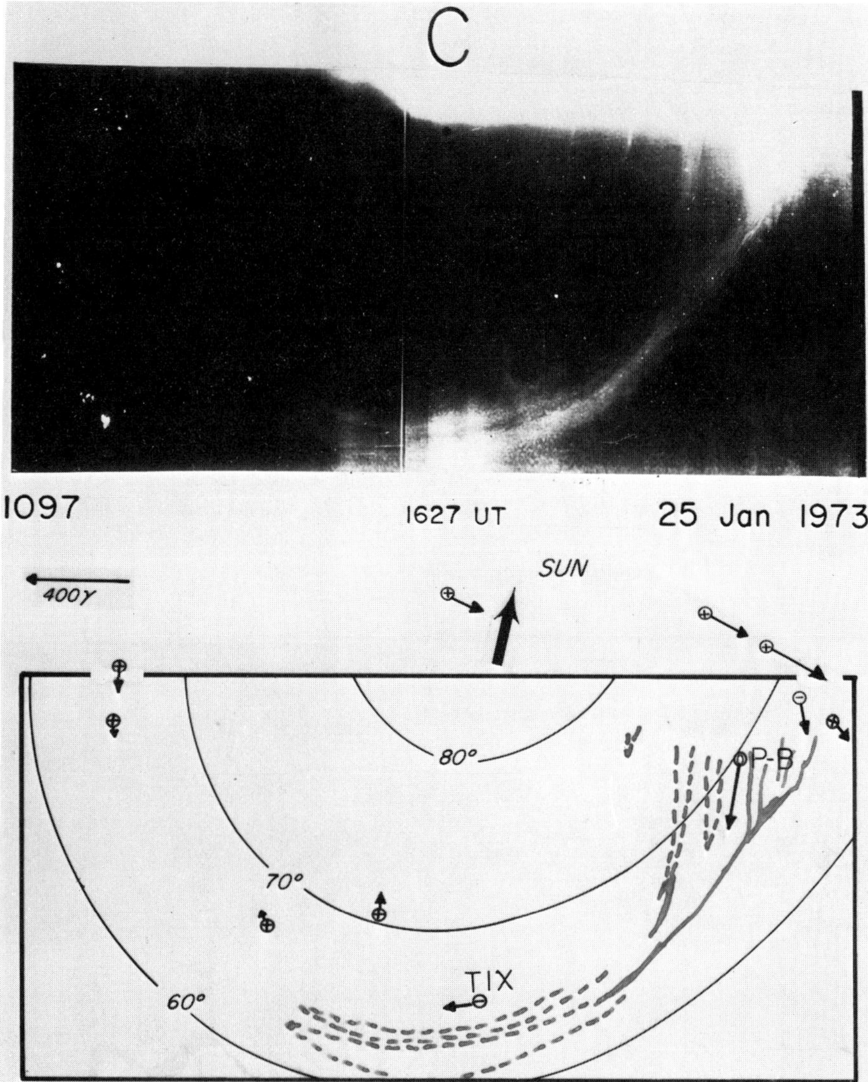

Fig. 7.17(c).

bright portion in the late evening sector. The equivalent currents along the auroras were directed eastward and westward in the evening and morning sectors, respectively.

The pass of 2132 UT (E): Figure 7.17(e). This particular pass took place during the maximum epoch of a substorm of moderate intensity (AL $\simeq -400\ \gamma$). In the midnight sector, a large dark area developed between the diffuse aurora and the poleward advancing arc. The general features of the current distribution are in good agreement with that of a typical substorm. An intense westward electrojet

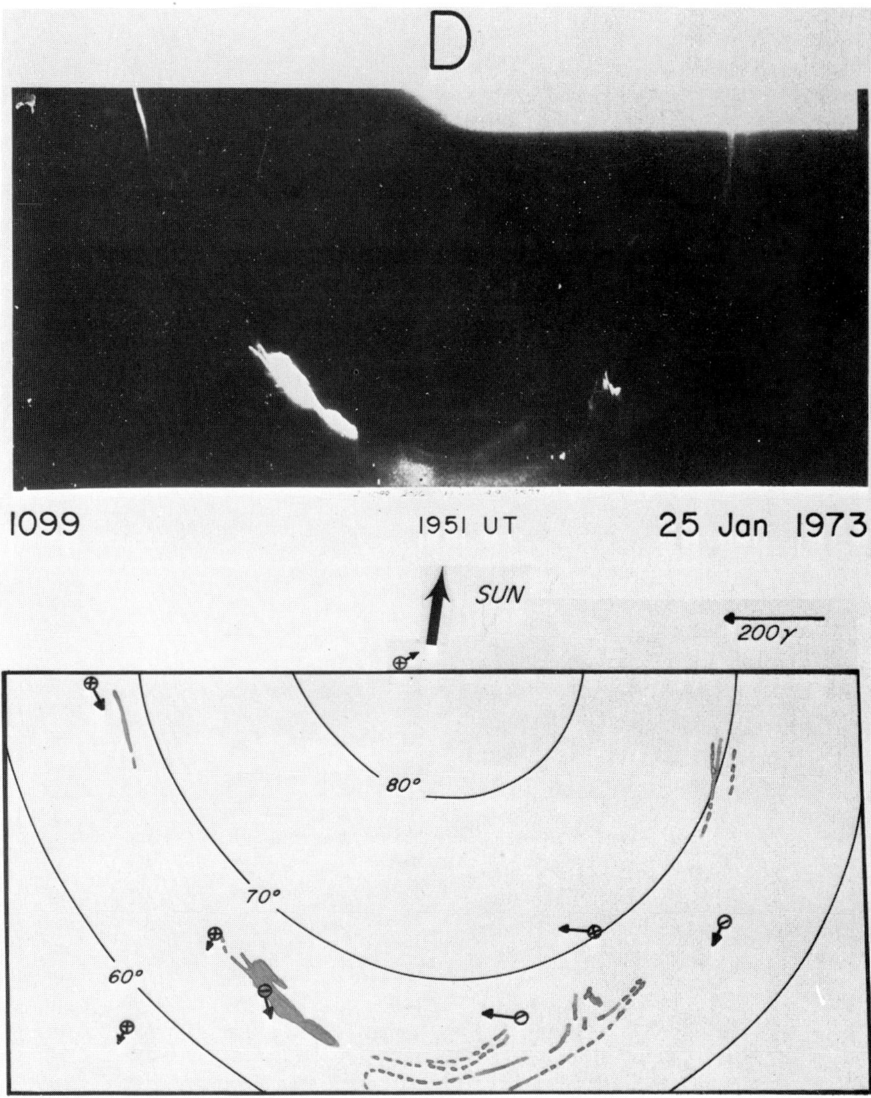

Fig. 7.17(d).

was observed in the diffuse aurora, but there was, unfortunately, no magnetic station in the dark area and along the poleward advancing arc.

The pass of 0046 UT (F): Figure 7.17(f). This example shows a well-developed auroral substorm. The DMSP photograph has a reasonable resemblance to the auroral pattern at $T = 30$ min of the auroral substorm development proposed by Akasofu (1968). The negative sign of the Z component perturbation at Leirvogur indicates that the electrojet center was located poleward of this station, perhaps along an arc.

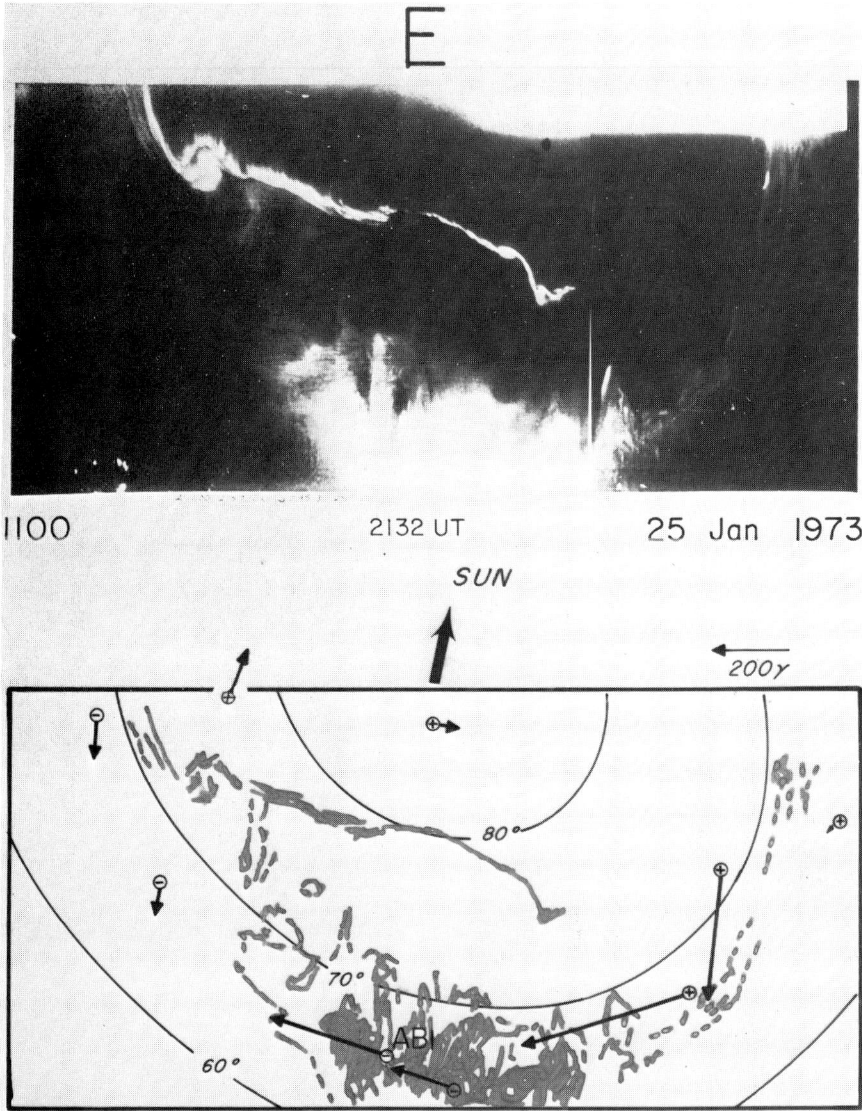

Fig. 7.17(e).

(b) *Summary*

Figure 7.18(a) shows a schematic diagram of characteristic features of an auroral substorm: this diagram is a composite of the major features appearing on a large number of DMSP photographs (Akasofu, 1974). Structured discrete auroras are active near the poleward boundary of the auroral oval, especially in the evening sector. On the other hand, the diffuse aurora delineates a fairly stable equatorward boundary of the auroral oval, showing structural asymmetry with respect to the midnight meridian.

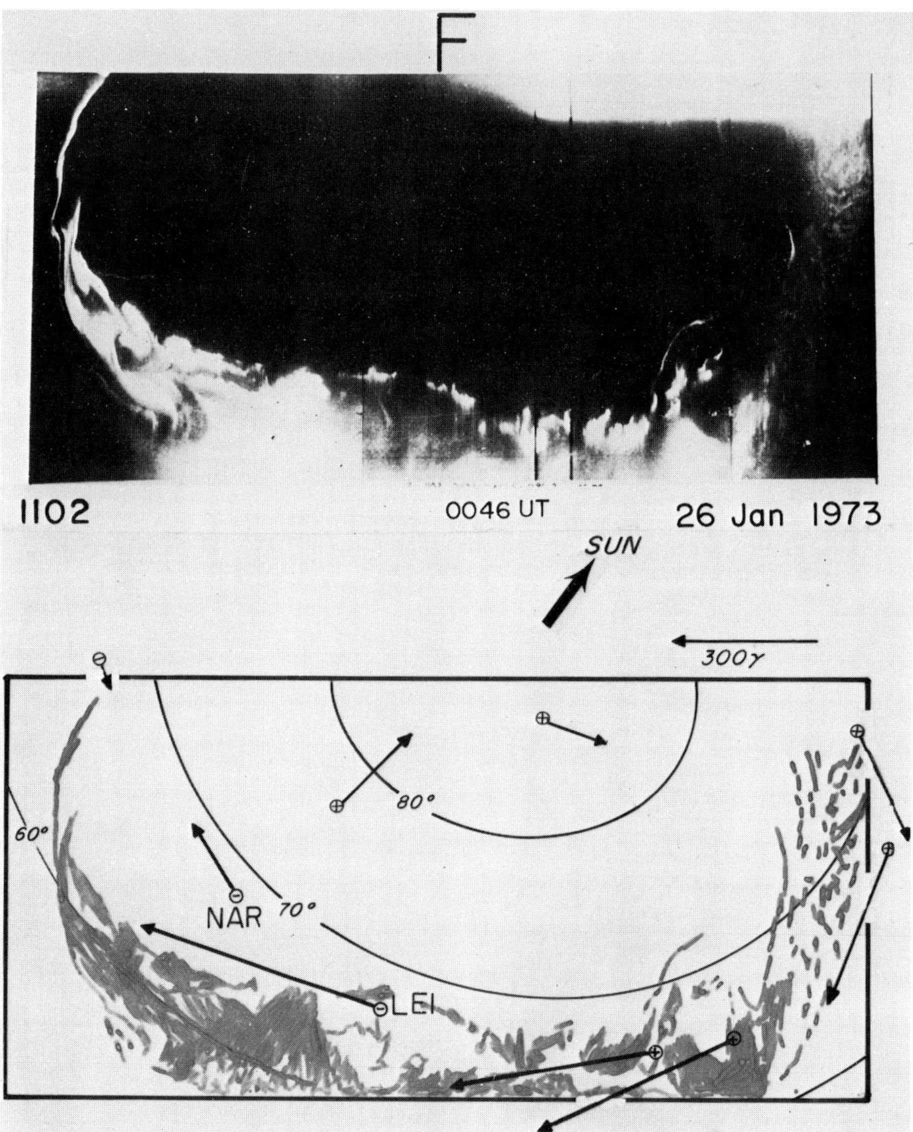

Fig. 7.17(f).

In order to identify global auroral behavior with the auroral electrojet and other features of polar magnetic substorms, the distribution of the equivalent current vectors are plotted in Figure 7.18(b), by using equivalent current vectors for 19 satellite passes which took place near the maximum epoch of substorms; eastward currents are represented by thick lines and westward currents by thin lines. The size of the Z component perturbations is also indicated by different arrow heads.

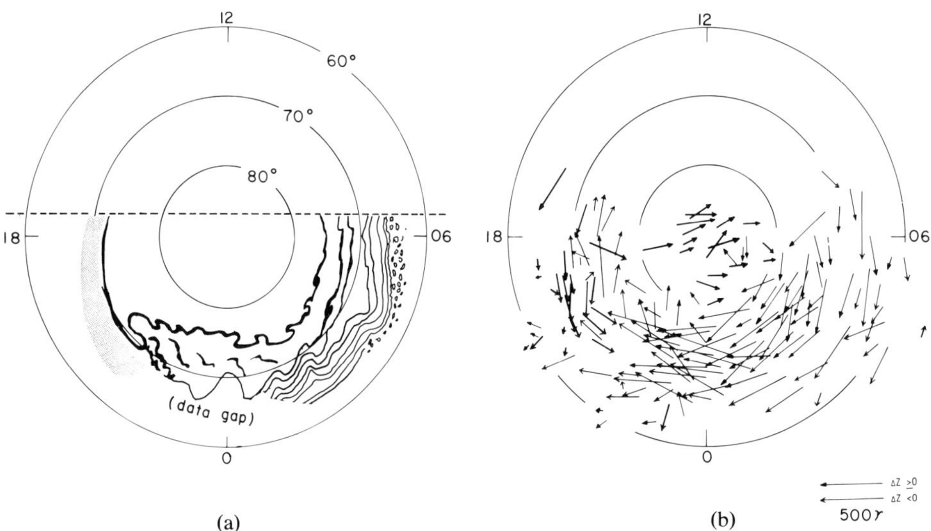

Fig. 7.18. Schematic diagram of auroral features during a typical auroral substorm and the distribution of equivalent current vectors with reference to the auroral features. The magnitude of each current vector is normalized for the maximum magnitude of the magnetic perturbation of 500 γ. (Kamide, Y. and Akasofu, S.-I.: *J. Geophys. Res.* **80**, 3585, 1975.)

Owing to the time interval between successive satellite passes (~ 102 min), the DMSP data examined here provide little opportunity to determine the progressive change of auroral features and of the corresponding ground magnetic response, although the detailed auroral features are quite variable from time to time, even in the same phase of substorms. Indeed, the size of the auroral oval has been found to vary considerably (Section 4.4.6(b)). Thus, in order to determine the current distribution for different ovals, it is necessary to use the *normalized auroral oval.* Here we have chosen the auroral oval shown in Figure 7.18(a) as a reference. The equivalent current vectors are plotted through the following processes: (i) The magnitude of each vector is normalized by defining the maximum intensity of the westward electrojet as $-500\ \gamma$ for each substorm. (ii) The location of each vector is adjusted in terms of the distance with respect to the observed location of the oval.

Figure 7.18(b) confirms that the electrojet is most intense in the midnight and early morning sectors. Although there is often a data gap near the equatorward edge around the midnight sector in the DMSP photographs, the latitudinal width of the area where the westward current is observed in the morning sector is much larger than that in the evening sector. It flows along both the discrete and diffuse auroras in the midnight and morning sectors, but only along the vicinity of discrete arcs in the evening sector.

One of the main westward electrojet features is that it does not end in the midnight meridian, but extends into the evening sector along the auroral oval (Harang, 1946; Akasofu *et al.*, 1965).

In the diffuse auroral region, the eastward currents (the positive bays in H) are common features in the evening sector, whereas the westward currents (the negative bays) are usually observed in the morning sector. Their magnitudes are, in general, less than the maximum magnitude of the sharp negative bay which is observed near the midnight meridian during substorms. Figure 7.18(b) indicates also that the reversal in the sign of the *H* component perturbation occurs at about 2100–2200 MLT, and its latitude varies as a function of local time. This locus may be identified as the Harang discontinuity. Perhaps the discontinuity can be best identified by finding the demarcation line between the regions of eastward auroral motions (observed most often in association with the westward electrojet) and of westward auroral motions (Heppner, 1954; Davis, 1962b); unfortunately, DMSP photographs cannot distinguish these two regions.

Figure 7.18(b) shows also that some disturbance vectors have a large east-west (*D*) component in the evening sector. This tendency can often be seen in individual events (such as the pass at 1122 UT on 1973, January 25). Further, the *D* sign is reversed between the poleward and equatorward stations of the surge, negative on the poleward side and positive on the equatorward side. This reversal was noted by Harang (1946) and recently by Kisabeth and Rostoker (1973). The latter authors suggested that the positive *D* variation can be explained in terms of a step-like equatorward displacement of the westward electrojet. However, Brekke *et al.* (1974b) showed that the ionospheric north-south current deduced from incoherent scatter radar measurements cannot explain the *D* perturbation at the ground, especially prior to midnight (Section 7.7.2). Therefore, it is quite likely that there is a concentrated upward field-aligned current from the surge which is the western end of the westward electrojet (Akasofu *et al.*, 1969).

Wallis *et al.* (1975) have also recently made a similar study on the relationship between a large-scale auroral display and geomagnetic disturbance vectors by using ISIS-2 photographs and the University of Alberta meridian chain magnetometer data. They found that the main westward current flows in the diffuse aurora and that discrete arcs are at times observed poleward of the electrojets.

7.5. Cross-Section of the Electrojets

7.5.1. LATITUDINAL PROFILE OF THE THREE-COMPONENT CHANGES

Until several years ago, there had been no way to infer the current density of the electrojets and their changes during substorms. Thus, the electrojets were assumed to be a line current or a narrowly concentrated current of width of order 500 km. The availability of magnetic records from Canadian and Alaskan meridian chains of stations as well as rocket magnetometer observations has considerably improved our knowledge on the electrojet during the last several years (Kisabeth and Rostoker, 1971, 1974; Akasofu *et al.*, 1971; Potter and Cahill, 1969; Cahill *et al.*, 1974). In this section, we review briefly studies made by Rostoker and his colleagues on the basis of their meridian chain of stations (Figure 7.19).

Before examining the magnetic records from the Canadian meridian chain of stations, it is instructive to study first the expected three-component changes as a

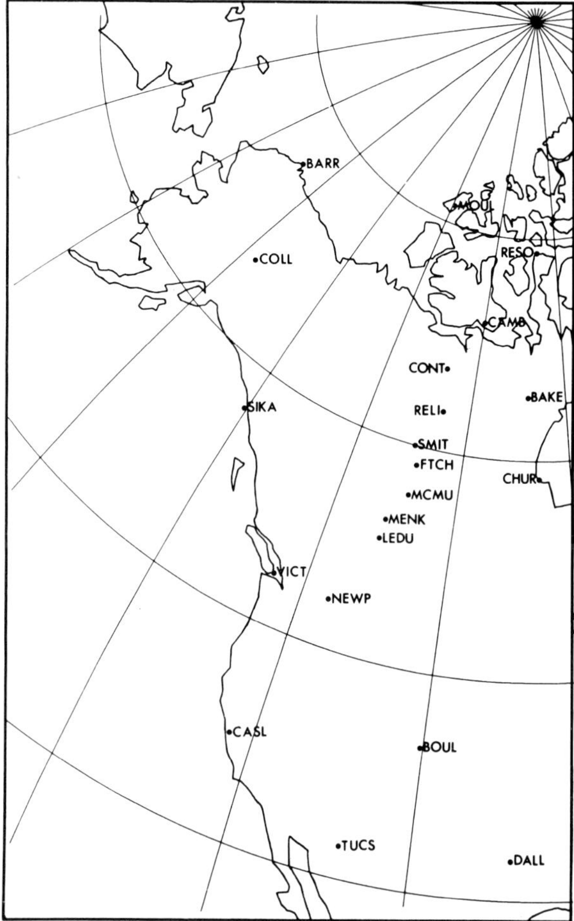

Fig. 7.19. Locations of the University of Alberta meridian chain of magnetometer stations; see Figures 7.22, 7.23, 7.26 and 7.27.

function of latitude, namely, the latitudinal profiles for a known current system. Figure 7.20 shows the latitudinal distribution of the H, D and Z components for a three-dimensional current system shown in Figure 7.6(b); note that the width and east-west length of the electrojet are assumed to be 5° and 20°, respectively. The total current in the system is 10^6 A, and the calculation is made along a meridian 4° east of the central meridian. Cahill *et al.* (1974) showed, by their rocket magnetometer observations, that the westward electrojet had a thickness of 20 km, a north-south width of 1000 km and a total current of 9×10^5 A during an intense substorm (a negative bay of about $-300\ \gamma$). On the basis of measured ion flow velocity V_i and inferred electron velocity V_e (from the probe measurement of electric fields), Bering and Mozer (1975) obtained a maximum current density of order 5×10^{-6} A m^{-2} at an altitude of 110 km, the height integrated current being 0.11 A m^{-1}.

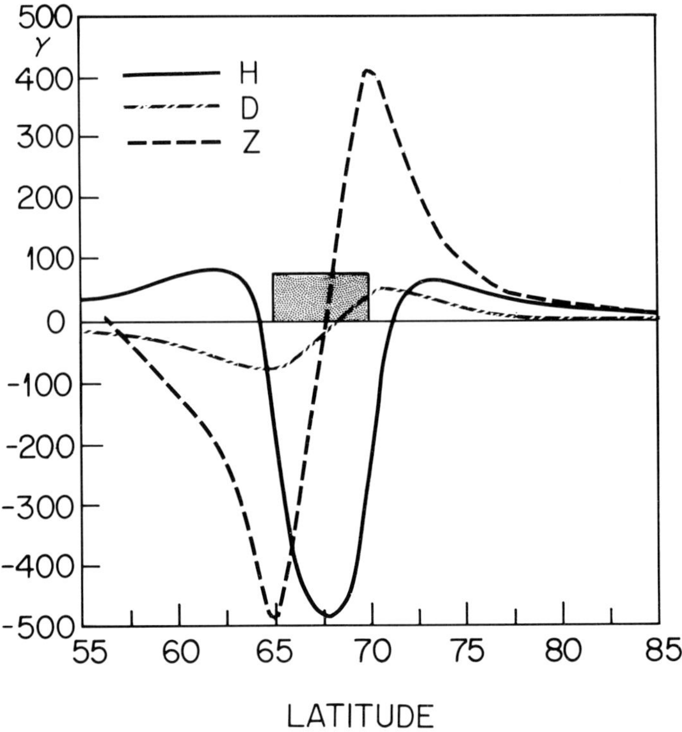

Fig. 7.20. Latitudinal profile of the three components, *H*, *D* and *Z*, for the three-dimensional current system shown in Figure 7.6(b). The calculation is made along a meridian 4° east of the central meridian. (Kisabeth, J. L.: Ph.D Thesis, University of Alberta, 1972.)

In Figure 7.20, the negative *H* deviation is considerably greater than the positive *H* deviation which is located equatorward of the former. However, the observed equatorward positive deviation is often greater than what is estimated from this simple three-dimensional current system, particularly in the evening sector. This is because there is an eastward ionospheric current a little equatorward of the westward electrojet. It is called the eastward electrojet. In order to simulate this situation, Kisabeth (1972) estimated the distribution of magnetic vectors and the latitudinal profile for a combination of two three-dimensional current systems, one with a westward current and the other with an eastward current. They are separated by 30° in longitude (see Figure 7.21(a)), and the current in each system is assumed to be 10^6 A.

Figure 7.21(b–d) shows a set of the estimated profiles. The profiles are estimated along the center of each system, as well as along the center of both systems. For details of the modeling of current systems and the associated geomagnetic disturbance fields, see Bonnevier *et al.* (1970), Kisabeth (1972), Kisabeth and Rostoker (1973) and Horning *et al.* (1974).

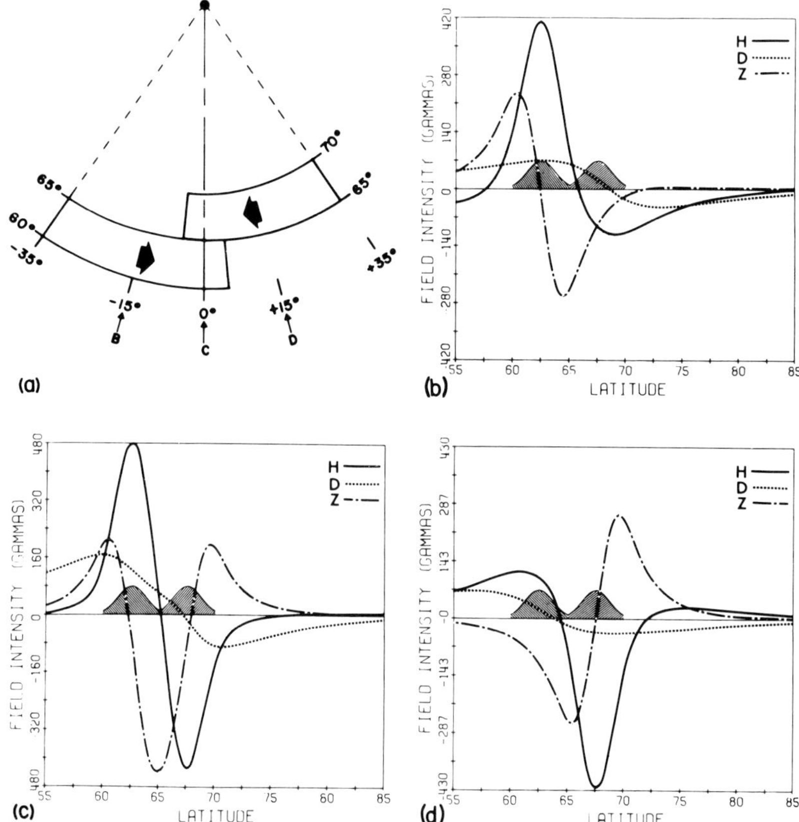

Fig. 7.21. Latitudinal profile of the three components, *H*, *D* and *Z*, for current system which consists of two three-dimensional current systems. Both are similar to that shown in Figure 7.6(b), but the current flows in the opposite direction; one of them is associated with the westward electrojet and the other the eastward electrojet. The geometry of both jets is shown in the upper part on the left-hand side (a). The profiles (b), (c) and (d) are computed along the meridians indicated in (a). (Kisabeth, J. L.: Ph.D. Thesis, University of Alberta, 1972.)

7.5.2. DEVELOPMENT OF THE ELECTROJETS

(a) *A Substorm on 1970, June 15*

The *H*, *D* and *Z* component magnetic records for this event are shown in Figure 7.22(a). The onset time of the substorm was about 0702 UT. The latitudinal profiles for the three components are shown in Figures 7.22(b), (c) and (d). About 4 min after the onset, a concentrated current was established, causing a maximum *H* component perturbation of $-140\ \gamma$. The center of the current was located at about 64.0°. The current grew steadily over the next few minutes without changing the configuration since the latitude profile for the three components did not change significantly, except for a slight poleward shift of the location of the maximum *H* perturbation. The estimated width of the electrojet was about 3.5° in latitude at that time.

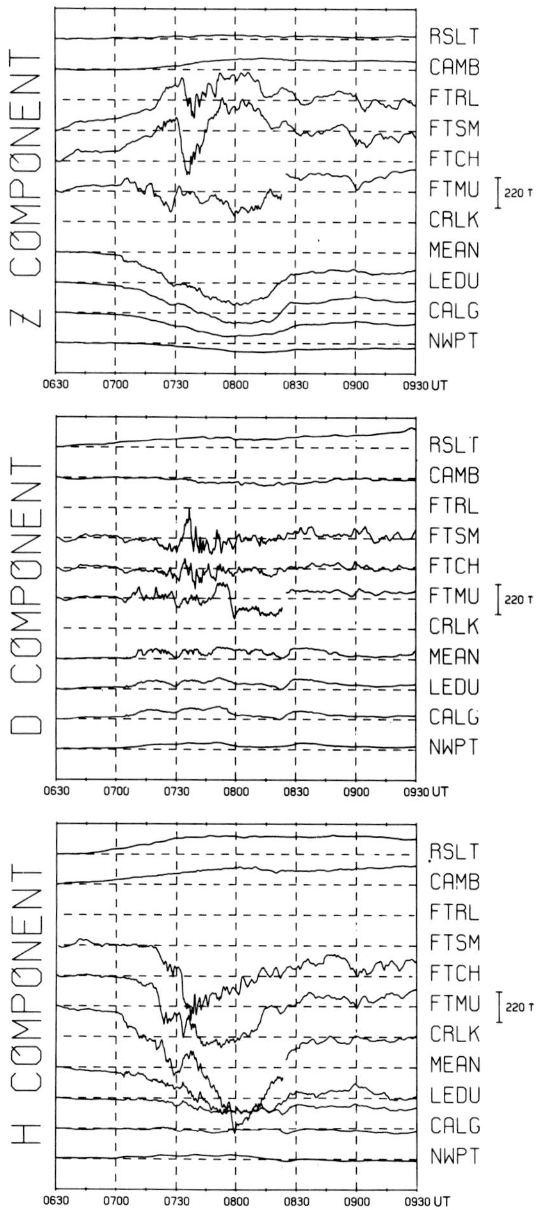

Fig. 7.22(a). Three-component magnetic records from the University of Alberta meridian chain of magnetic stations during a substorm on 1970, June 15.

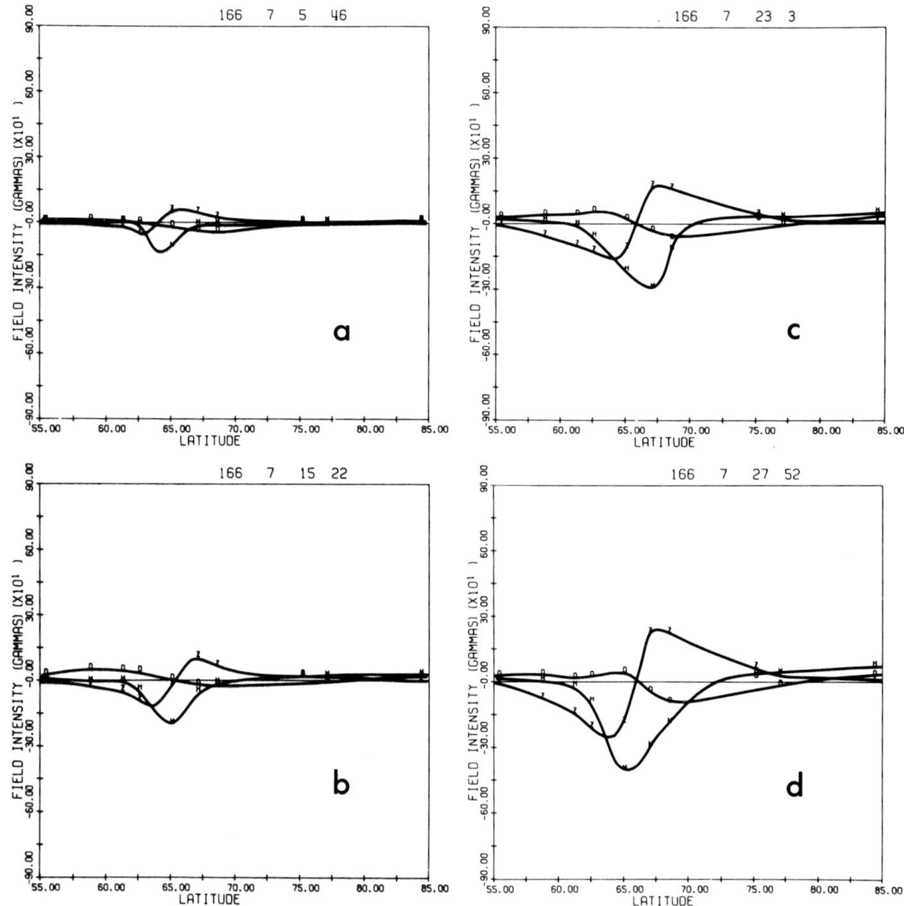

Fig. 7.22(b–d). Latitudinal profiles of the three components during the substorm indicated in Figure 7.22(a). (Kisabeth, J. L. and Rostoker, G.: *J. Geophys. Res.* **76**, 6815, 1971.)

At 0719 UT, the westward current intensified, particularly near the poleward boundary of the electrojet which had been steadily growing. This intensification can be seen in either the magnetic records from FTCH and FTMU or the latitude profile (c). By 0727 UT the amplitude of the H component perturbation under the current was about $-450\ \gamma$ and the width of the jet was 4.5° in latitude (d). It can also be inferred from the D component profile that the central line of the three-dimensional current system was east of the meridian chain of stations (compare Figure 7.20 and Figure 7.22(b)).

At 0732 UT, an additional new current began to grow near the poleward edge of the pre-existing current; the centers of the two currents were separated by about 5°. Note that the Z component perturbations from the two currents were oppositely directed in the region between them. As a consequence, they tended to cancel each other. By 0743 UT, the two currents appeared to merge together, and the total width became as wide as 7°. It is interesting to note that Rostoker

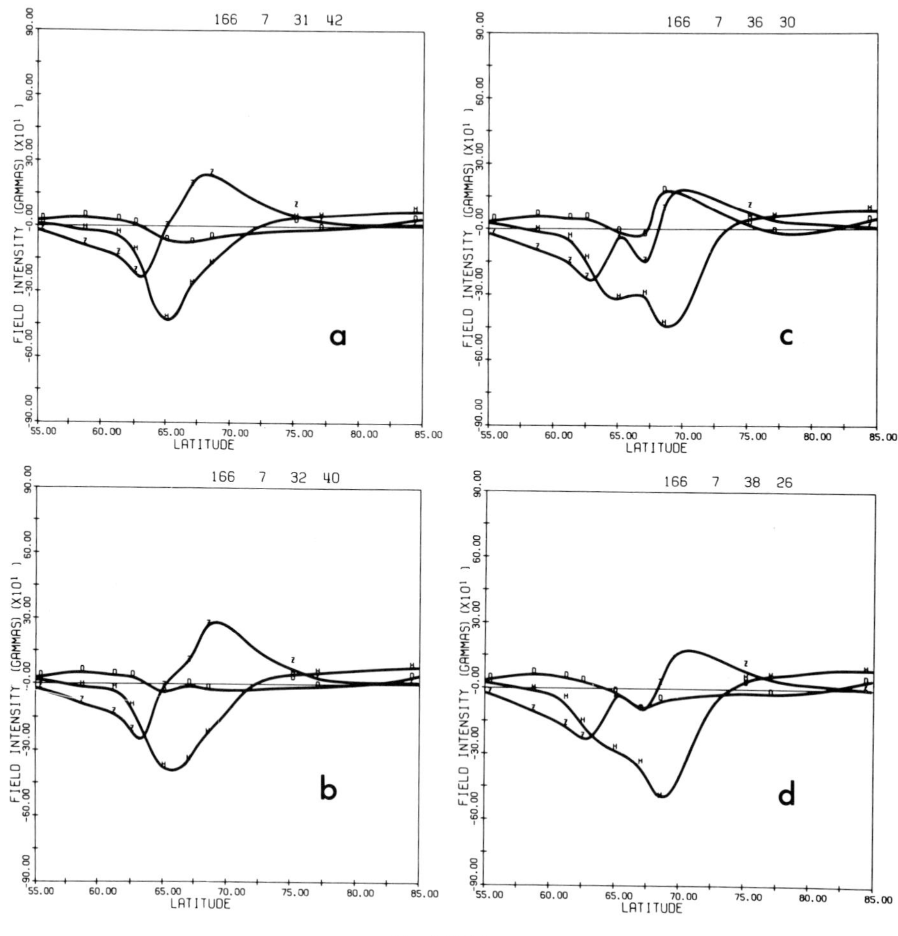

Fig. 7.22(c).

et al. (1975) showed that fluctuations of the westward electrojet arise primarily from changes of the conductivities.

(b) *A Substorm on 1970, July 14*

The three-component magnetic records and the associated micropulsation records (Pi 2) are shown in Figures 7.23(a) and (b), respectively. A substorm began at about 0730 UT. Figure 7.24 shows the latitudinal profile at 0733 UT, 3 min after the onset. It has most of the features which were studied in terms of the model calculation in Figure 7.20. Thus, a well-defined electrojet system was formed by then. The current decayed temporarily at about 0745 UT. During the next 15 min, the electrojet moved slowly equatorward, reaching the latitude of 68° by 0800 UT.

At 0818:50 UT, the electrojet was enhanced near its equatorward border. During the next two minutes the electrojet developed rapidly near the poleward border, resulting in a rapid poleward expansion of the width of the electrojet. Then there occurred a further enhancement of the electrojet near the poleward

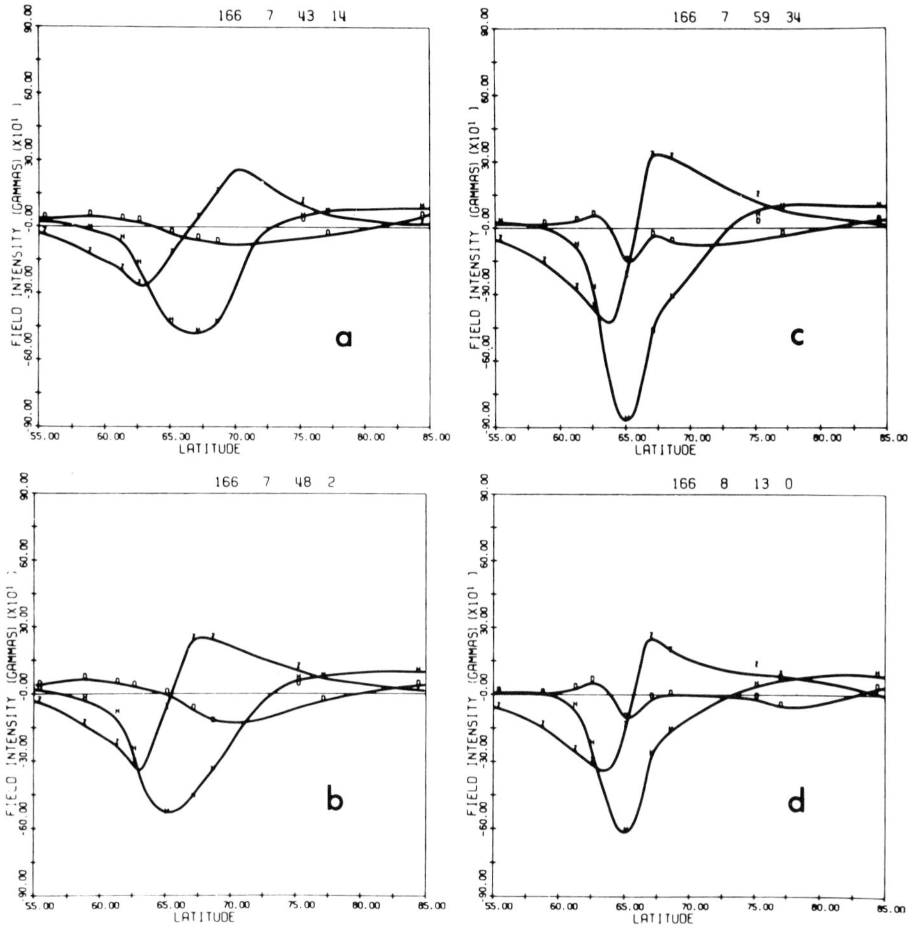

Fig. 7.22(d).

border. Figure 7.25 shows a three-dimensional display of the *H* component variation during the substorm (Kisabeth, 1972).

(c) *A Substorm on 1971, December 23*

Wiens and Rostoker (1975) found that the westward electrojet often develops in succession as a series of discrete steps or jumps in the evening sector. The advancing front of the new electrojet appears to the northwest (in the northern hemisphere) of the previous one. Figure 7.26 shows the magnetic records from their meridian chain of stations. In the *H* component records, this particular feature is clearly seen as impulsive changes, appearing first at FTCH and then at higher latitude stations as the time progresses. Wiens and Rostoker illustrated this feature pictorially on a geographic map, reproduced here as Figure 7.27.

It is known that a large westward traveling surge consists of a number of smaller surges (Akasofu, 1974). The small surges grow one upon another near the

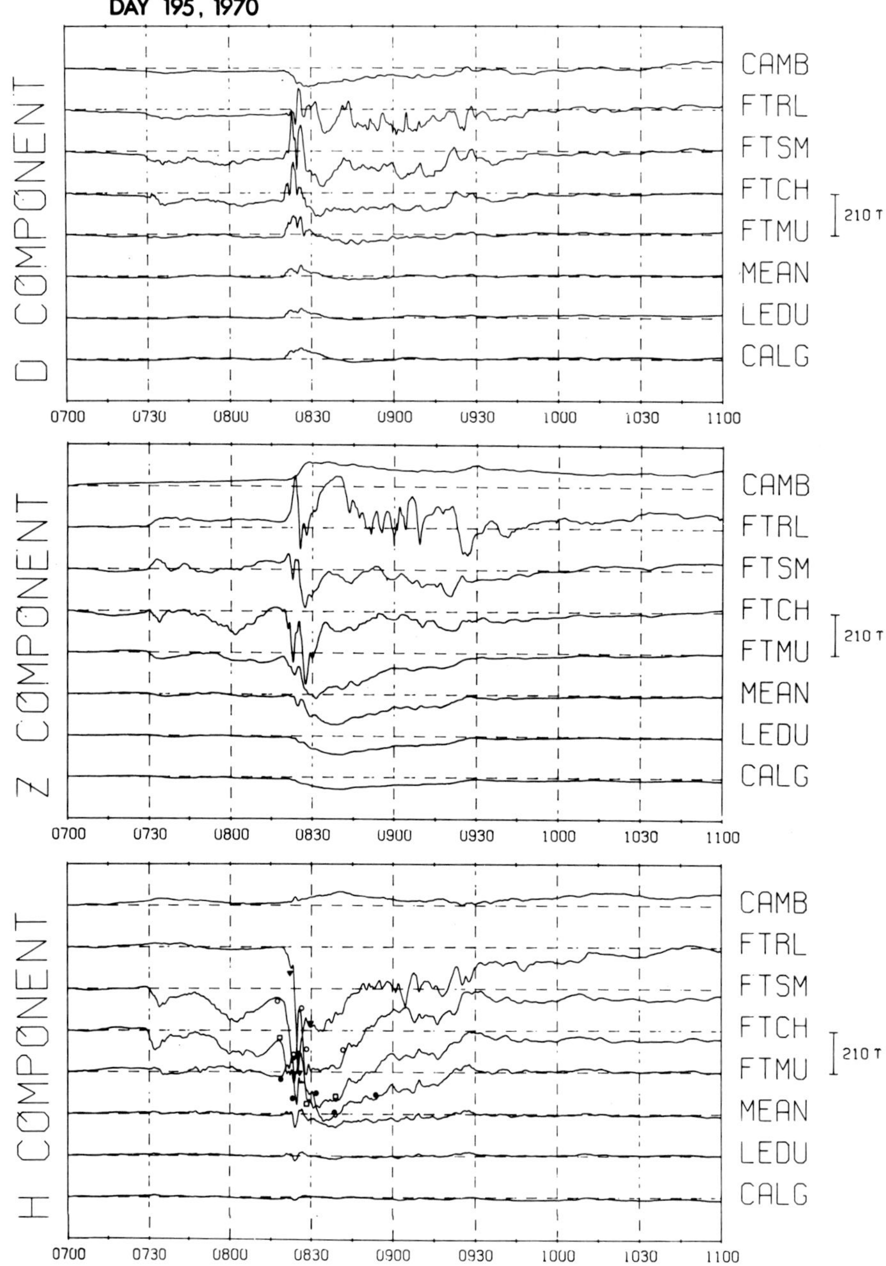

Fig. 7.23(a). Three-component magnetic records from the University of Alberta meridian chain of magnetic stations during a substorm on 1970, July 14.

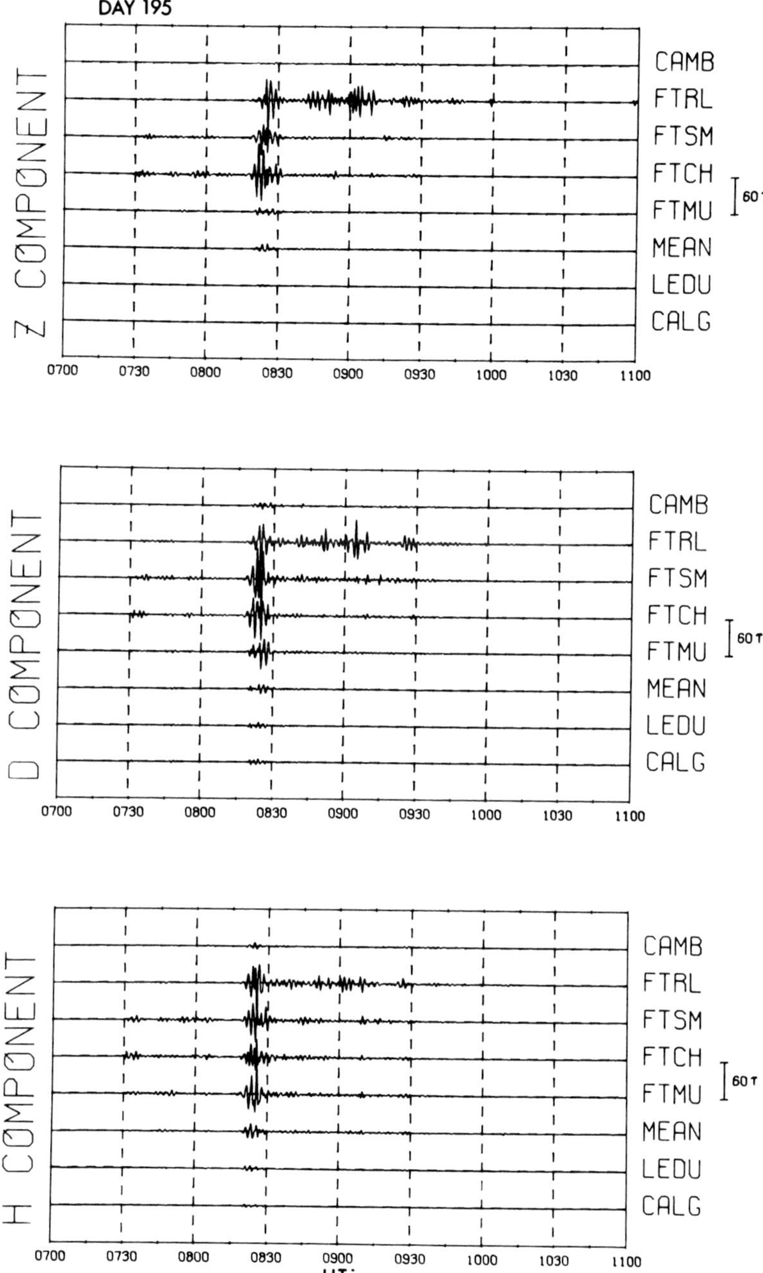

Fig. 7.23(b). Three-component micropulsation records from the University of Alberta meridian chain of magnetic stations during the substorm indicated in Figure 7.23(a).

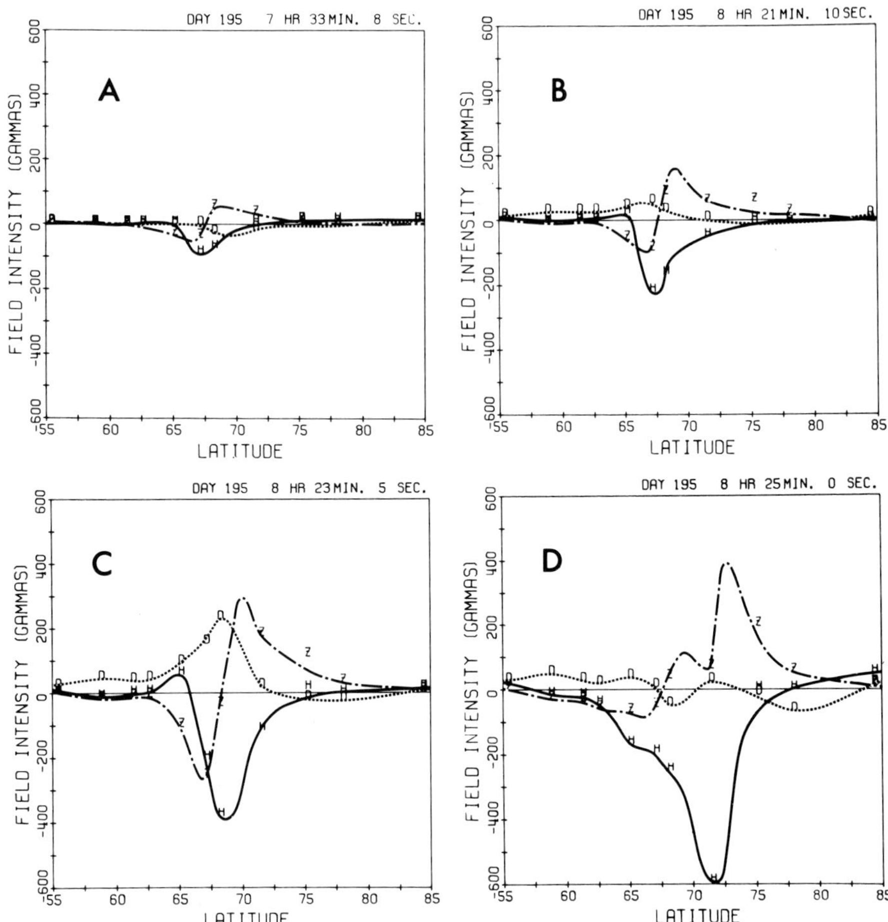

Fig. 7.24. Latitudinal profiles of the three components during the substorm indicated in Figure 7.23(a). (Kisabeth, J. L. and Rostoker, G.: *J. Geophys. Res.* **76**, 6815, 1971.)

poleward front. Thus, one can expect that the westward electrojet in the evening sector develops in the manner explained by Wiens and Rostoker (1975) and earlier by Rostoker (1968). Rostoker *et al.* (1970) examined the development of substorm magnetic fields and traced the propagation of westward traveling surges on the basis of a close magnetometer network in the subauroral zone and in middle latitudes.

7.5.3. RADAR AURORAS

Radar or radio auroras have been studied extensively in the past, but we shall only briefly review some of the recent studies (*cf.* McNamara, 1972 and Unwin, 1966a, b; Lange-Hesse, 1967, 1969). There are at least two types of radar echoes, discrete and diffuse. It is generally believed that radar auroral irregularities are

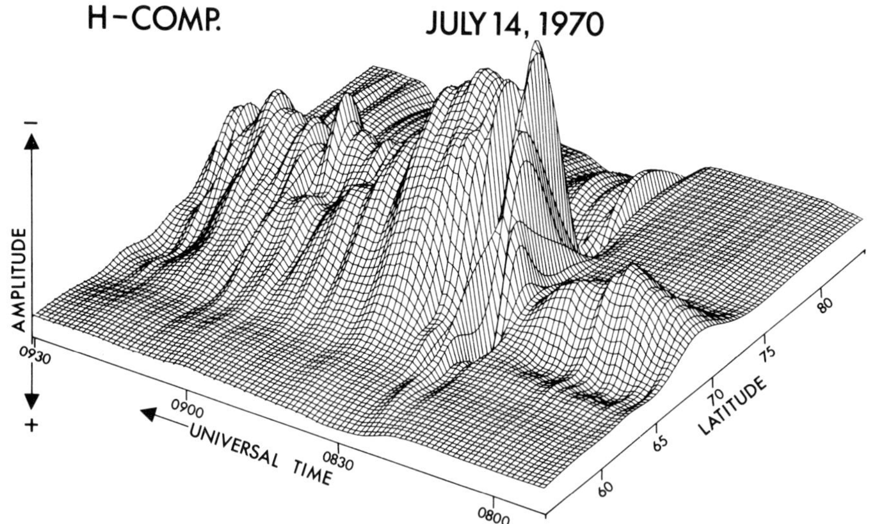

Fig. 7.25. Perspective view plot of the *H* component showing the dynamic development of a substorm on 1970, July 14. Note that negative changes are plotted upward. (Kisabeth, J. L.: Ph.D. Thesis, University of Alberta, 1972.)

generated by various current-associated instabilities. Once the exact nature of the instabilities can be identified, a radar will become a powerful tool in studying the auroral electrojet. It has been suggested that both the Buneman two-stream instability and the gradient drift instability are primarily responsible for the generation of ionospheric irregularities which scatter radio waves. The two-stream instability is excited when the electron drift velocity relative to ion-drift velocity exceeds the ion-acoustic velocity, while the gradient drift instability may be excited for a much lower threshold speed (Moorcroft, 1972; Greenwald, 1974; Wang and Tsunoda, 1975; Greenwald *et al.*, 1975a; Ossakow *et al.*, 1975). Kelley and Mozer (1973) reported evidence for the presence of the two-stream instability in the aurora by a rocket observation. On the other hand, Tsunoda (1975) showed, on the basis of the simultaneous UHF and incoherent scatter radar observations, that the diffuse auroral echoes do not seem to be caused by the Buneman-Farley instability.

Greenwald *et al.* (1973, 1975a) found that there is sometimes a good linear relationship between the total horizontal disturbance observed at College and the square root of the range-integrated backscattered 50 MHz power, indicating that the intensity of ionospheric currents is linearly related to the range-integrated irregularity amplitude. Figure 7.28 shows an example of the linear relationship. They noted also that at some other times such a good correlation cannot be found. They suggested that one possible explanation for the lack of consistency may be that the linear relationship can be found only when current variations are proportional to electric field variations; current variations associated with conductivity variations should not cause such a relationship.

Figure 7.29 shows an example of radar observation of a substorm conducted by the Anchorage auroral radar at 50 MHz (Greenwald *et al.*, 1975a). The middle

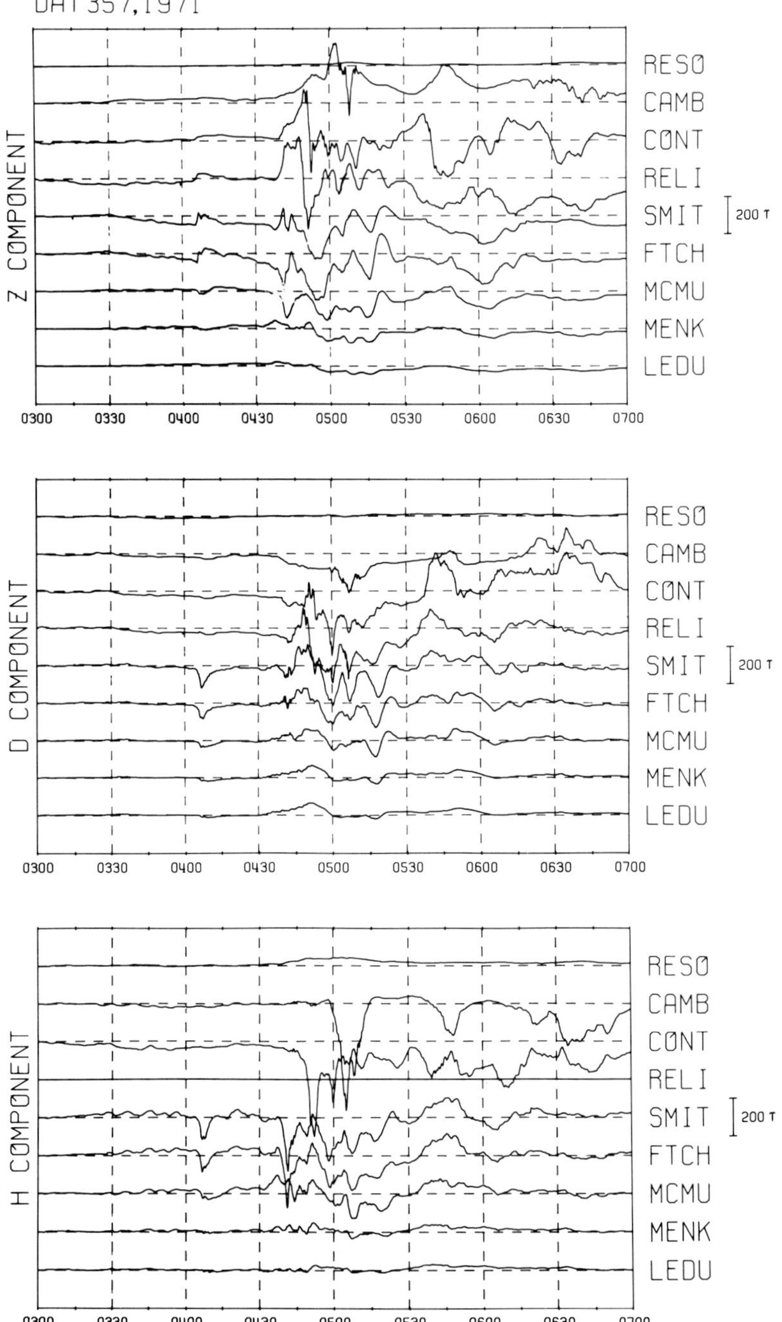

Fig. 7.26. Three-component magnetic records from the University of Alberta meridian chain of magnetic stations during a substorm of 1971, December 23.

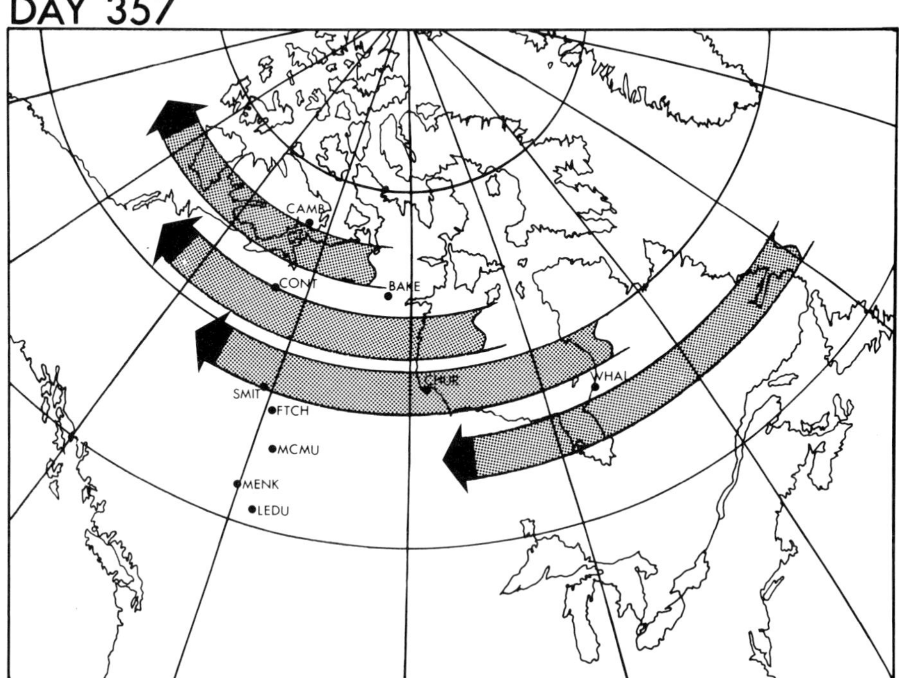

Fig. 7.27. Discrete northwestward shifts of the westward auroral electrojet during the substorm shown in Figure 7.26. (Wiens, R. G. and Rostoker, G.: *J. Geophys. Res.* **80**, 2109, 1975.)

portion of the figure shows range-time display of the echo activity, indicated by different shades. It can be seen that the diffuse auroral echo returned from a little poleward of College between 0800 and 1000 UT. At that time, the H and Z components at College showed weak positive changes, indicating that an eastward electrojet was located slightly poleward of College.

Discrete auroral echoes began to appear at about 1000 UT. They were spatially structured and short-lived echoes. Unwin and Keys (1975) and Ecklund *et al.* (1974) noted that such a feature appears simultaneously at geomagnetically conjugate areas. Figure 7.30 shows another example of a substorm observed by the Anchorage radar. The substorm occurred along a contracted oval. It can be seen from both the radar and magnetometer data that the entire activity was north of College. The westward current spread equatorward to the latitude of College and then receded poleward.

Auroral displays have also been studied by a UHF radar (the 398-MHz phased-array radar located at Homer, Alaska) by Tsunoda *et al.* (1974) and Tsunoda and Fremouw (1975). They showed that auroral substorm activity can be well monitored by the radar; UHF radar auroral features are found to move in concert with the visual aurora, and all major features of the visual auroral substorm have the corresponding radar auroral signatures. For example, a sudden brightening of an arc is often accompanied by a bifurcation of a diffuse band (DB) radar aurora. Figure 7.31 shows an example of the bifurcation. The poleward

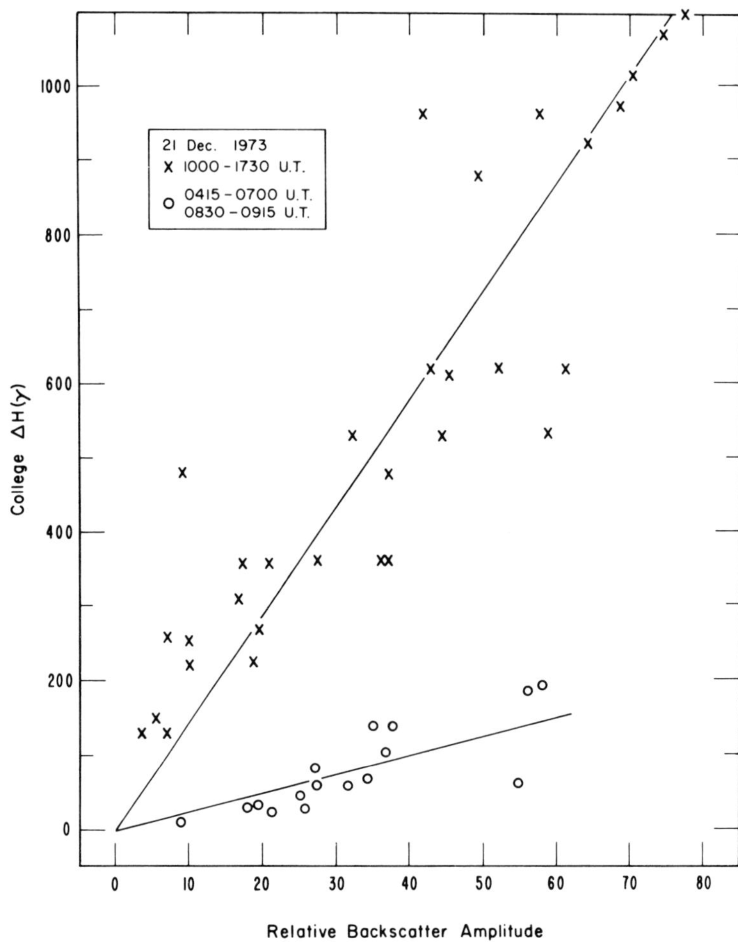

Fig. 7.28. Correlation between 15-min backscatter amplitude values and ΔH from the College magnetogram. (Gray, A. M. and Ecklund, W. L.: NOAA Tech. Rep. ERL 306-AL 9, Sept. 1974.)

expansion of auroras is seen as a poleward motion of radar auroral forms. The authors noted that the bifurcation is related to an increase in electron density and a corresponding decrease in the electric field intensity (Section 7.7.2(c)), which in turn prevents the growth of plasma waves responsible for auroral backscatter. A westward traveling surge often appears as a hook-shaped echo which coincides in space with or trails behind the visual surge (Figure 7.31).

7.6. Latitudinal Cross-Section of the Auroral Electrojet and its Relation to the Interplanetary Magnetic Field Polarity

Kamide and Akasofu (1974) examined the total current and width of the westward electrojet, across the Alaska meridian, as a function of the IMF B_z component. The

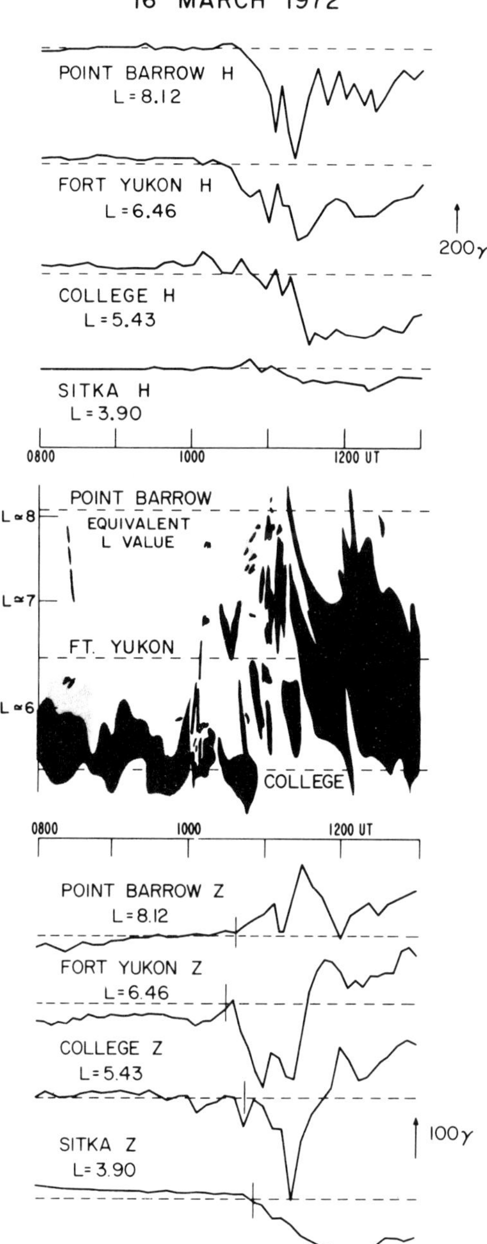

Fig. 7.29. 50 MHz auroral radar backscattered power and magnetic records (*H*, *Z*) from four Alaskan stations during a substorm on 1972, March 16. (Greenwald, R. A., Ecklund, W. L. and Balsley, B. B.: *J. Geophys. Res.* **80**, 3635, 1975a.)

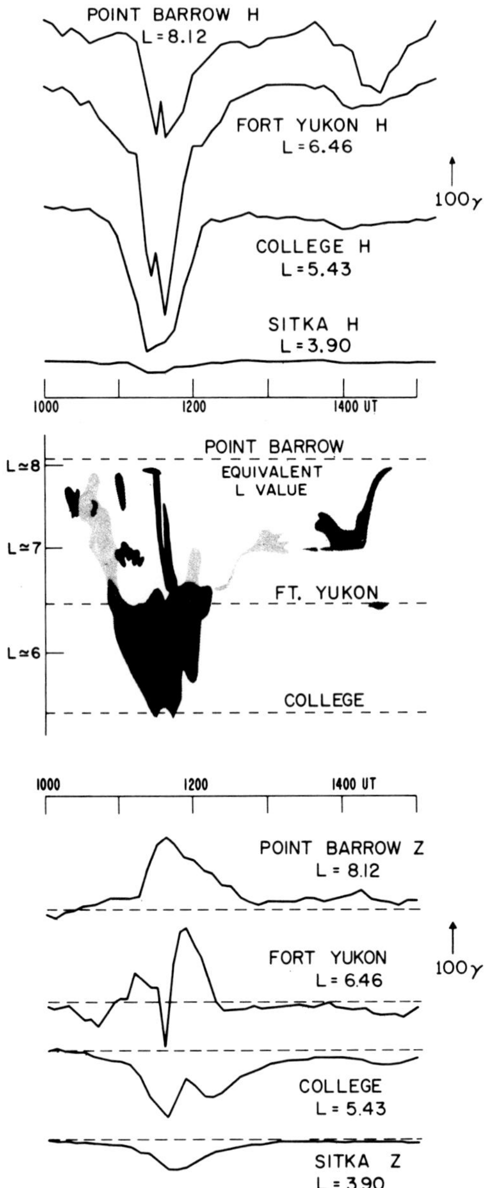

Fig. 7.30. 50 MHz auroral radar backscattered power and magnetic records (H, Z) from four Alaskan stations during a substorm on 1972, March 13. (Greenwald, R. A., Ecklund, W. L. and Balsley, B. B.: *J. Geophys. Res.* **80**, 3635, 1975a.)

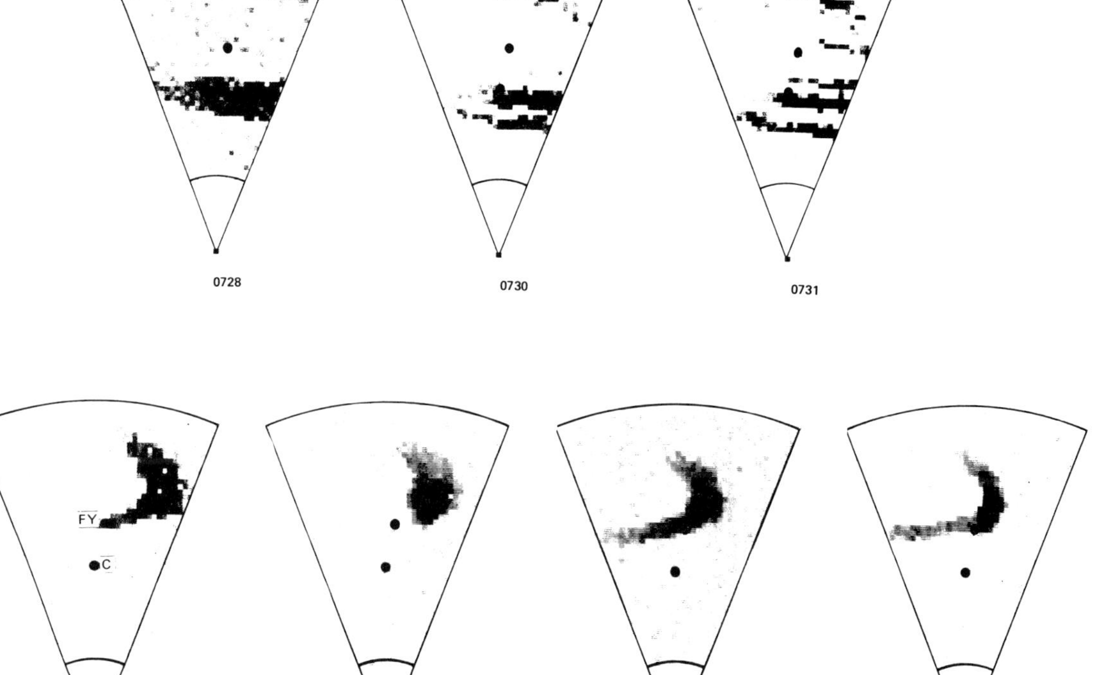

Fig. 7.31. 398 MHz UHF radar echoes. The upper part shows an example of the bifurcation associated with a sudden brightening of an arc and its subsequent poleward motion. The lower part shows a hook-shaped echo, associated with a westward traveling surge. (Tsunoda, R. T. and Fremouw, E. J.: Stanford Research Inst. Rept., May 1975.)

total current I was estimated by using the following equation:

$$I = \int_{\phi_1}^{\phi_2} j_y \mathrm{R}\, \mathrm{d}\phi \simeq \int_{\phi_1}^{\phi_2} \Delta X_m \mathrm{K}\, \mathrm{d}\phi \simeq \int_{\phi_1}^{\phi_2} \Delta H \mathrm{R}\, \mathrm{d}\phi$$

where j_y is the current density of the westward electrojet

$$j_y(\mathrm{A\ km^{-1}}) = \left(\frac{2}{3}\right)\left(\frac{10}{2\pi}\right) \Delta X_m$$

The relationship between the total current I and B_z is shown in Figure 7.32; different marks are for different ranges of the corresponding positive bays in low latitudes. Figure 7.33 shows the relationship between the total intensity and the central locations (and the half-widths) of the westward electrojet at the maximum epoch of substorms; different symbols are for different B_z values. Both figures show clearly that the total current, location and half-widths are related to the B_z component. For a larger negative B_z value, the total current and the half-widths

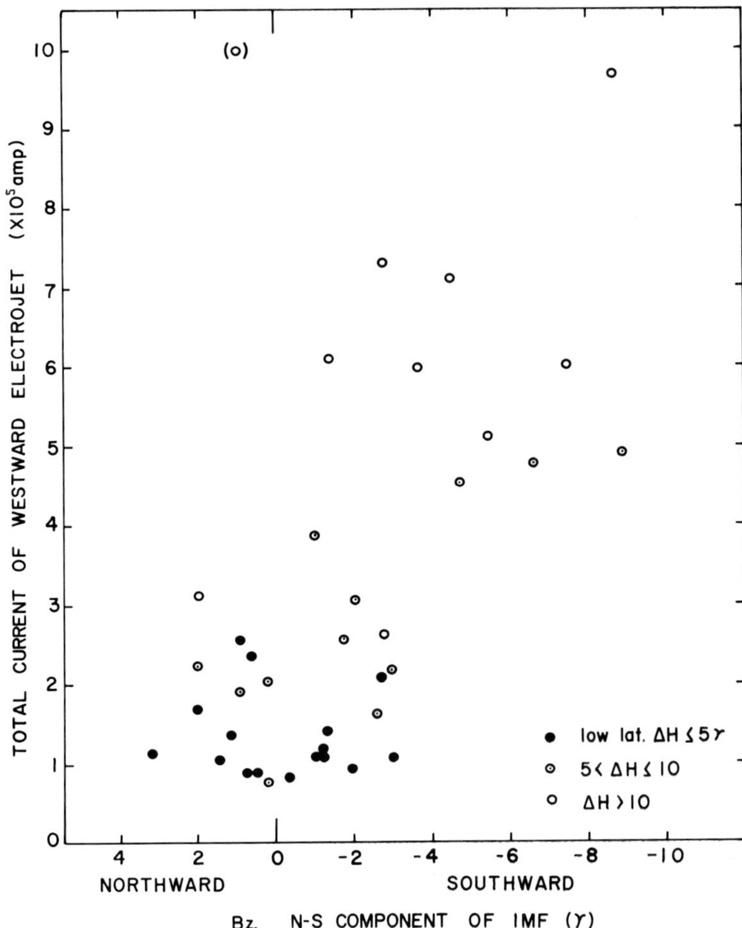

Fig. 7.32. Relationship between the total westward electrojet intensity and the IMF B_z component. (Kamide, Y. and Akasofu, S.-I.: *J. Geophys. Res.* **79**, 3755, 1974.)

are larger, and the center is located at a lower latitude. It should be noted, however, that both quantities are continuous functions of B_z, and there is no discontinuity at $B_z = 0$.

7.7. Ionospheric Currents and Electric Fields

During the last several years, a number of ways to measure, directly or indirectly, magnetospheric electric fields have become available. They are described in Section 1.3.4. In this section, we review the electric field observations in the ionospheric level during magnetospheric substorms.

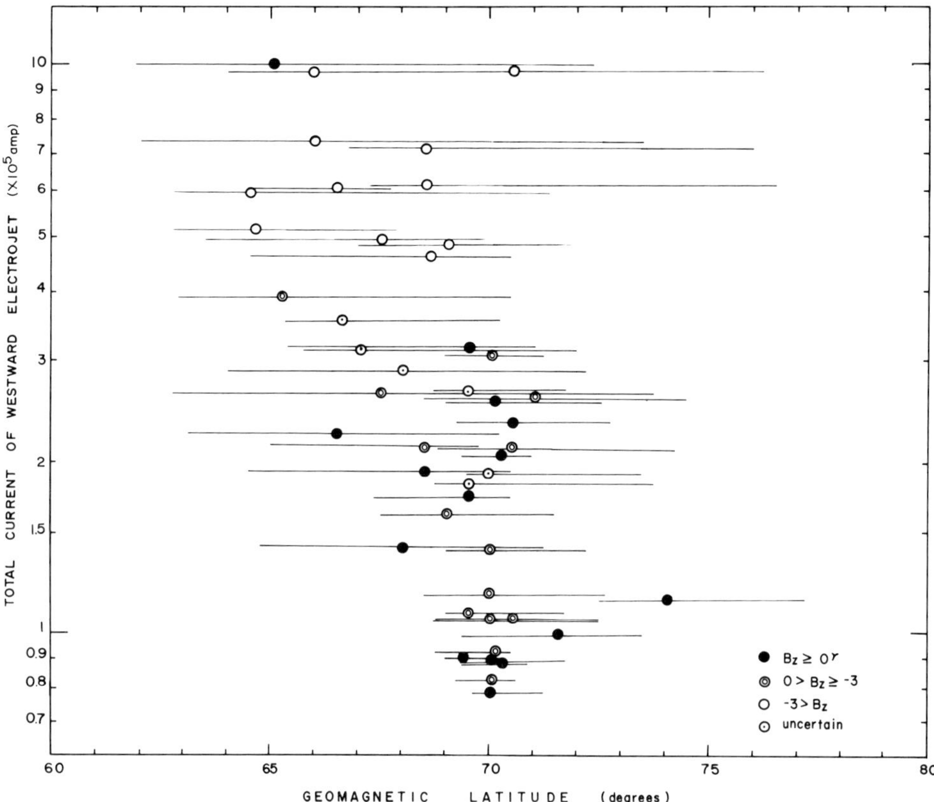

Fig. 7.33. Center locations of the westward electrojet and its half-widths (different to the north and south of the center) and the total intensity. The IMF B_z component at the times of the observation is also indicated. (Kamide, Y. and Akasofu, S.-I.: *J. Geophys. Res.* **79**, 3755, 1974.)

7.7.1. MODEL STUDY

As an introduction to the review of various electric field observations, it is instructive to study the ionospheric electric field for a model case. For this purpose, we examine here the electric field for the model which was examined in Section 7.3.4.

Figure 7.34 shows the potential distribution for the model. It can be seen that the equipotential contour lines are considerably distorted in the most conductive belt, namely the poleward half of the oval. This is caused by a large accumulation of the space charges at the boundary between the discrete and diffuse auroral regions due to the conductivity discontinuity there. Since these accumulated space charges cannot be removed by the field-aligned current in this particular model, they modify the electric field imposed on the ionosphere from outside.

Figure 7.34 may be compared with Figure 1.18(b). The latter represents a quiet condition and the former, the substorm condition. The main difference is the presence of the second conductive belt (the diffuse auroral belt); note that the

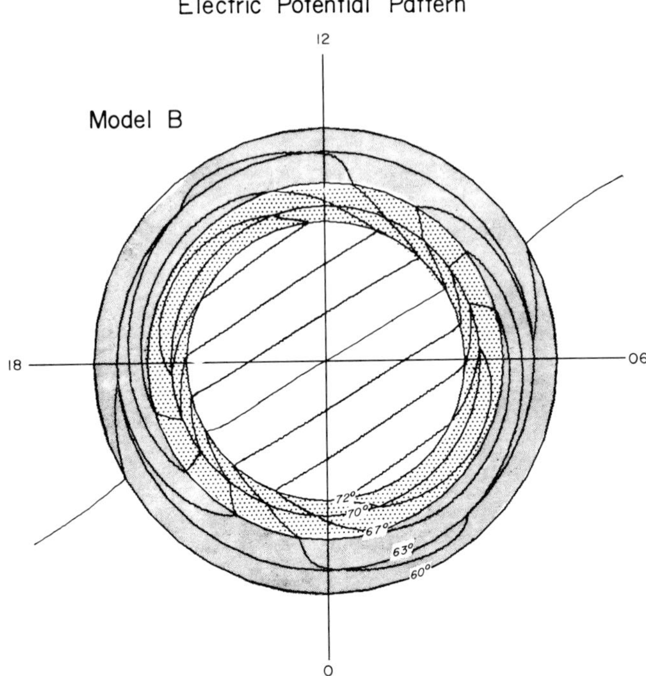

Fig. 7.34. Computed pattern of electric equipotential for the model in Figure 7.15(a). (Courtesy of Yasuhara, F. and Akasofu, S.-I.)

distribution of the field-aligned currents is the same. Thus, the formation of a new high conductive belt can modify the quiet time pattern into a substorm-like pattern. In particular, a significant part of the eastward electrojet arises as a result of the conductivity change.

Figure 7.35(a) shows the electric field profile along the dawn-dusk meridian. It may be compared with an observed profile obtained by Heppner (1972); see Figure 7.35(b). It is quite likely that the large variation of the field in the evening sector was associated with a westward traveling surge. The model reproduces this feature well.

In order to compare the computed electric field with Chatanika (near Fairbanks, Alaska) radar data (i.e., data from a fixed point on the Earth, which rotates under the electric field pattern once a day), the electric field vectors along the constant latitudes (74°, 71°, 68°, 66°, 64°, 62°, 58°) are plotted in Figure 7.36. In the auroral oval (62°–68°), the electric field vector rotates from north to south and counterclockwise. Indeed, Banks and Doupnik (1975) demonstrated that the Chatanika radar observation shows this tendency. Maynard (1974) found also that the electric field rotates, counterclockwise, from the northward to the southward direction across the Harang discontinuity.

In the previous section, it was shown that it is possible to infer approximately the distribution of the electrojet current intensity as a function of latitude on the basis of magnetic records from a meridian chain of stations. However, it is

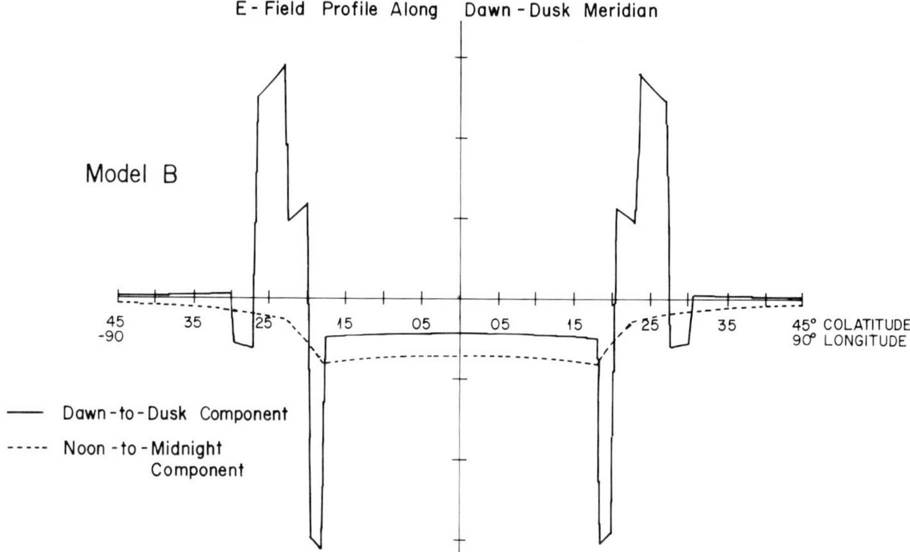

Fig. 7.35(a). Computed dawn-to-dusk and noon-to-midnight components of the electric field along the dawn-dusk meridian for the model in Figure 7.15(a). (Courtesy of Yasuhara, F. and Akasofu, S.-I.)

important to find a more direct way to measure the current density. Fortunately, the incoherent scatter radar at Chatanika has made it possible to estimate the ionospheric current density from simultaneous observations of the ionospheric electric field and conductivity, even though both quantities cannot be measured directly. In this subsection, we review mainly Chatanika radar results on this particular subject and compare them with results from the Alaska meridian chain of magnetic stations.

7.7.2. CHATANIKA RADAR OBSERVATIONS

(a) *Electric Field, Current and Conductivity*

(i) *1972, May 25.* Figure 7.37(a) shows the height-integrated Hall and Pedersen conductivities, the north-south and east-west components of the electrostatic field ($\boldsymbol{E}$) and the north-south and east-west components of the dynamo field ($\boldsymbol{V} \times \boldsymbol{B}$) induced by the neutral wind. The method of estimating these quantities from incoherent radar data is described in detail by Brekke *et al.* (1974b). This particular day was a quiet day. Thus, it is reasonable to infer that the auroral oval was located poleward of the radar even in the midnight sector. A large daily variation of the two conductivities resulted from the daily variation of the solar zenith angle. It should be noted that the ratio Σ_H/Σ_P was about 2 throughout the day.

There was, however, a minor increase of the conductivities between 08 and 10 UT. There was also a large increase of the northward electric field from about 0630 to 1100 UT, which was associated with a weak substorm along the contracted oval. This is an important example to infer variations of the electric field near the

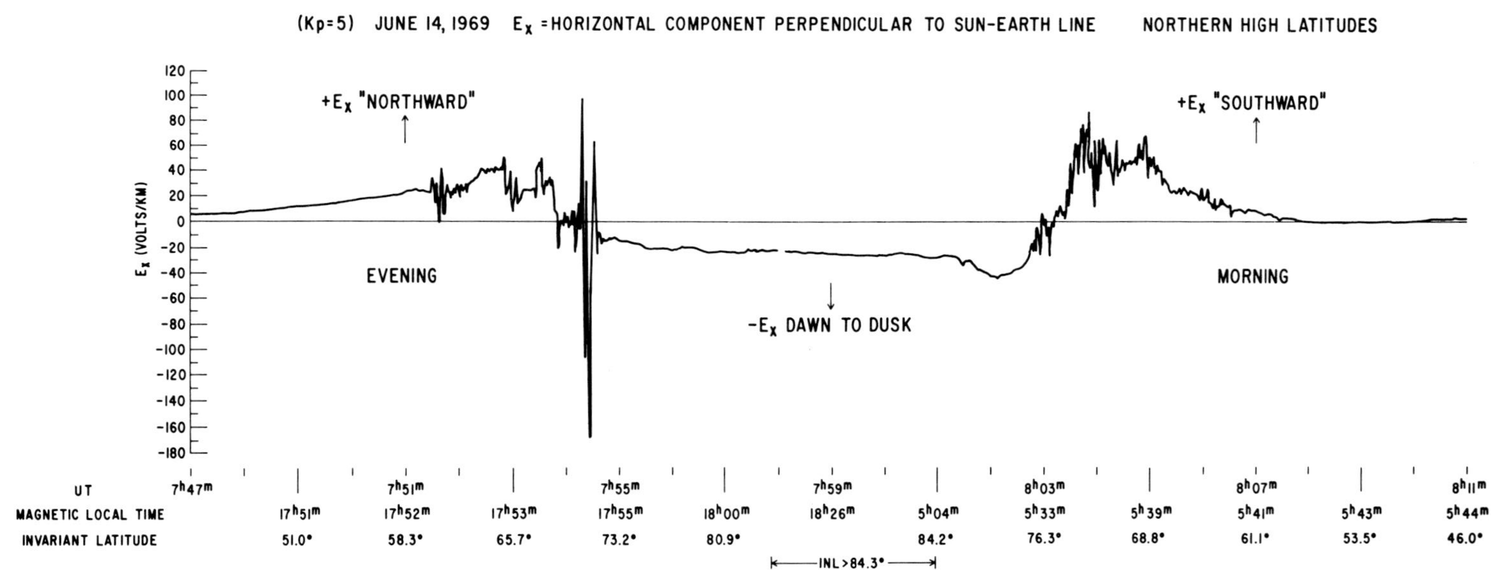

Fig. 7.35(b). Observed dawn-dusk component of the electric field during a substorm. (Heppner, J. P.: *Planet. Space Sci.* **20**, 1475, 1972.)

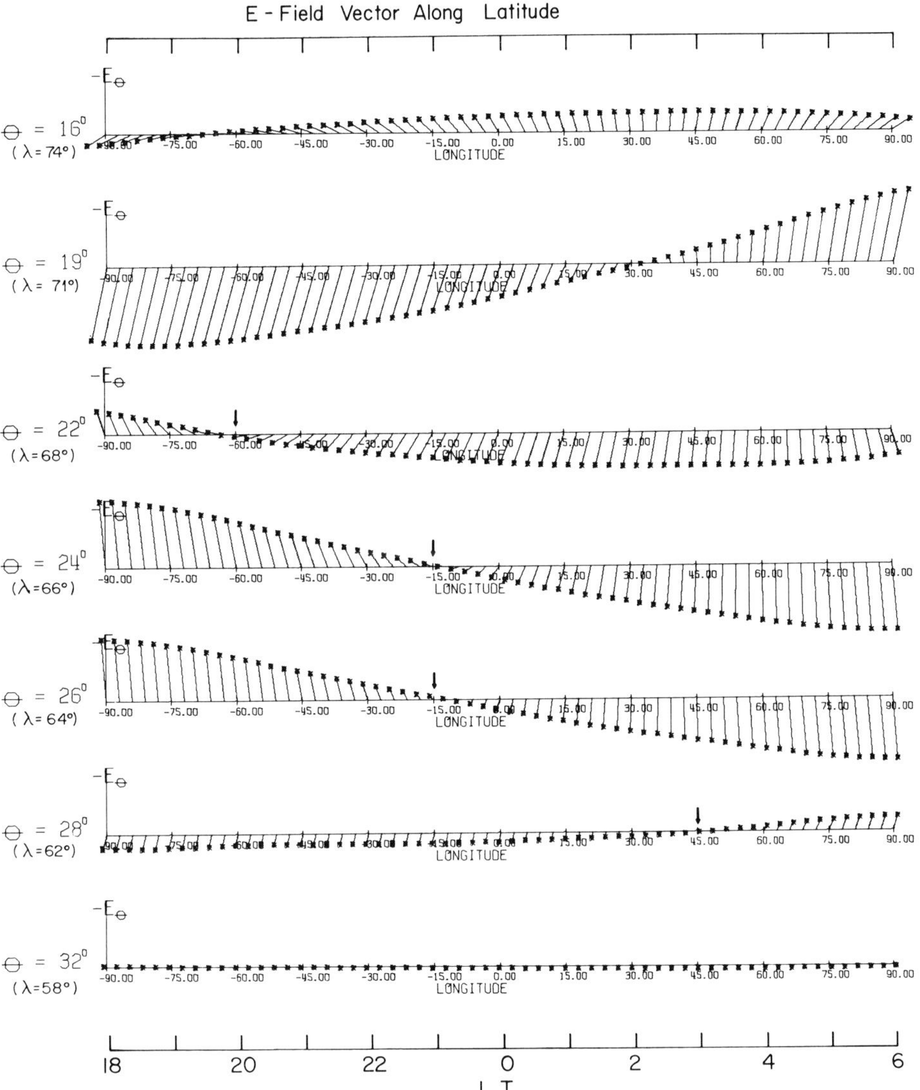

Fig. 7.36. Computed electric field vectors along several latitudinal circles as a function of LT for the model in Figure 7.15(a). (Courtesy of Yasuhara, F. and Akasofu, S.-I.)

equatorward boundary of the oval. Brekke *et al.* (1974b) also pointed out that the dynamo-induced electric field is more intense than the electrostatic field in afternoon hours, but that they are comparable during the rest of the day.

Figure 7.37(b) shows the north-south and east-west components of the ionospheric current, estimated on the basis of the inferred conductivities and both the electrostatic and dynamo-induced electric fields. They are compared with the east-west and north-south components of the magnetic record from College on that day. All the quantities concerned here were too small to compare, but it is

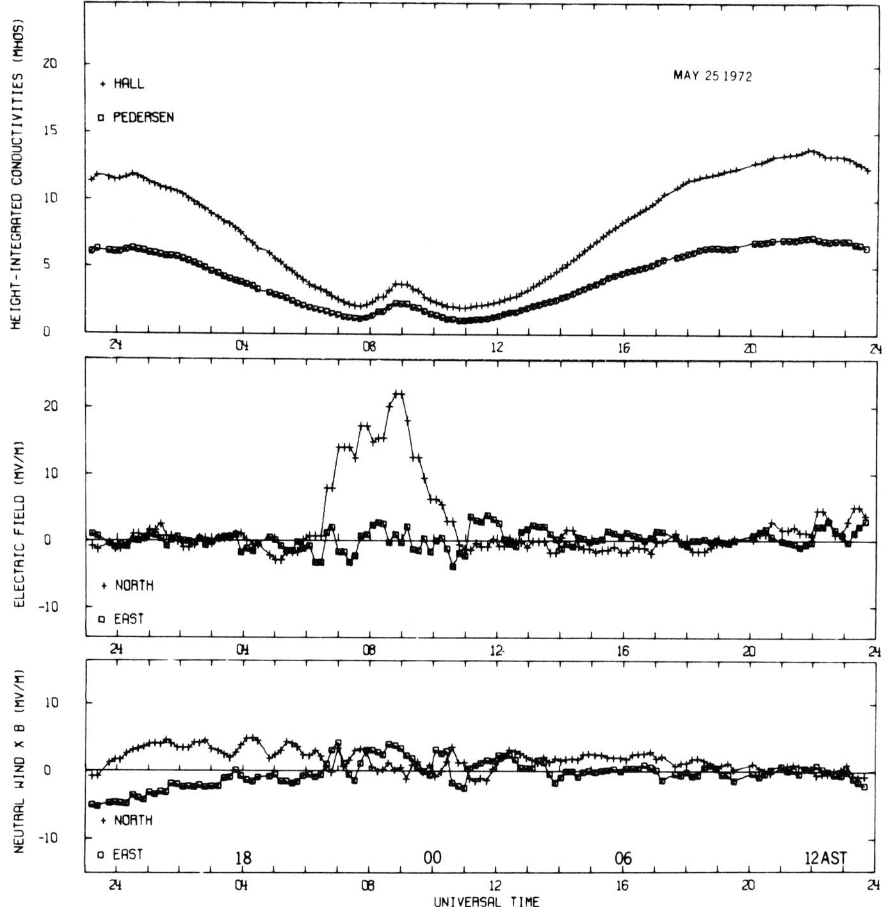

Fig. 7.37(a). Estimated Hall and Pedersen conductivities, observed north-south and east-west components of the neutral wind by the Chatanika incoherent scatter radar on 1972, May 25.

significant that the daily variation of the D component cannot be explained in terms of the overhead ionospheric current.

(ii) *1972, March 13–14.* This was a fairly quiet day, but there was a weak isolated substorm around local midnight. In Figure 7.38(a), substorm effects can be seen as a large increase of the two conductivities and of the ratio Σ_H/Σ_P. A large increase of the north-south and east-west components of the electric field also occurred, particularly the former. Figure 7.38(b) shows the inferred north-south and the east-west current intensity, together with the corresponding H and D component magnetic records from College. The inferred north-south current density variation agrees well with the corresponding H component variation at College. However, there is no obvious relation between the east-west current density and the D component. This problem will be discussed shortly in great detail.

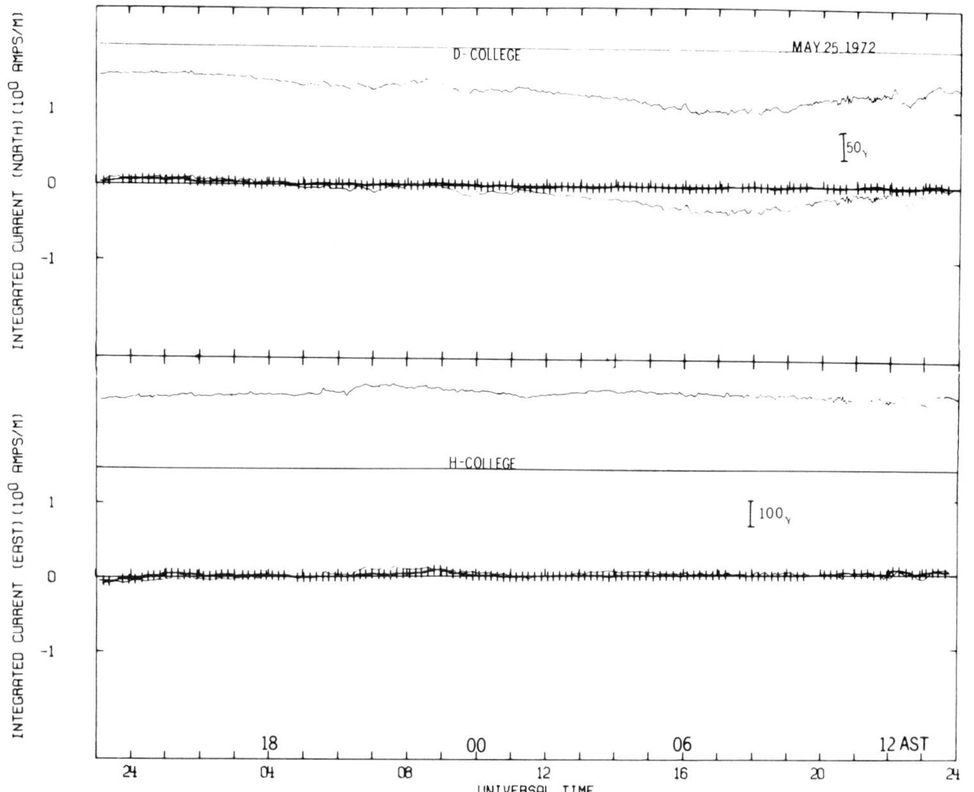

Fig. 7.37(b). Estimated north-south and east-west current densities deduced from the radar data shown in Figure 7.37(a), together with the D and H component magnetic records from College. (Brekke, A., Doupnik, J. R. and Banks, P. M.: *J. Geophys. Res.* **79**, 3773, 1974.)

(iii) *1972, October 13–14.* This day was more disturbed than March 13–14, 1972, and it is instructive to compare the Chatanika radar data on both days. One of the major differences between the two days is the appearance of three large increases in the northward electric field in afternoon-evening hours on October 13–14; see Figure 7.39(a). They were associated with intense substorm activities. Another important feature, which was absent in the previous example, is a rapid change in the sign of the north-south component of the electric field, from northward to southward, at about 10 UT. In Section 7.7.3, we shall examine in detail auroral conditions at the time of this change. The inferred change of the north-south and east-west current density is shown in Figure 7.39(b), together with the corresponding magnetic record. Again, there is a reasonable agreement between the east-west component of the current density and the H component, but no obvious relation between the north-south component of the current density and the D component.

It is interesting to compare the conductivity ratio Σ_H/Σ_P on 1972, October 13–14, and on 1972, May 25 (the first example of this subsection). The ratio was nearly constant and was about 1.9–2.0 on the latter day, but was significantly

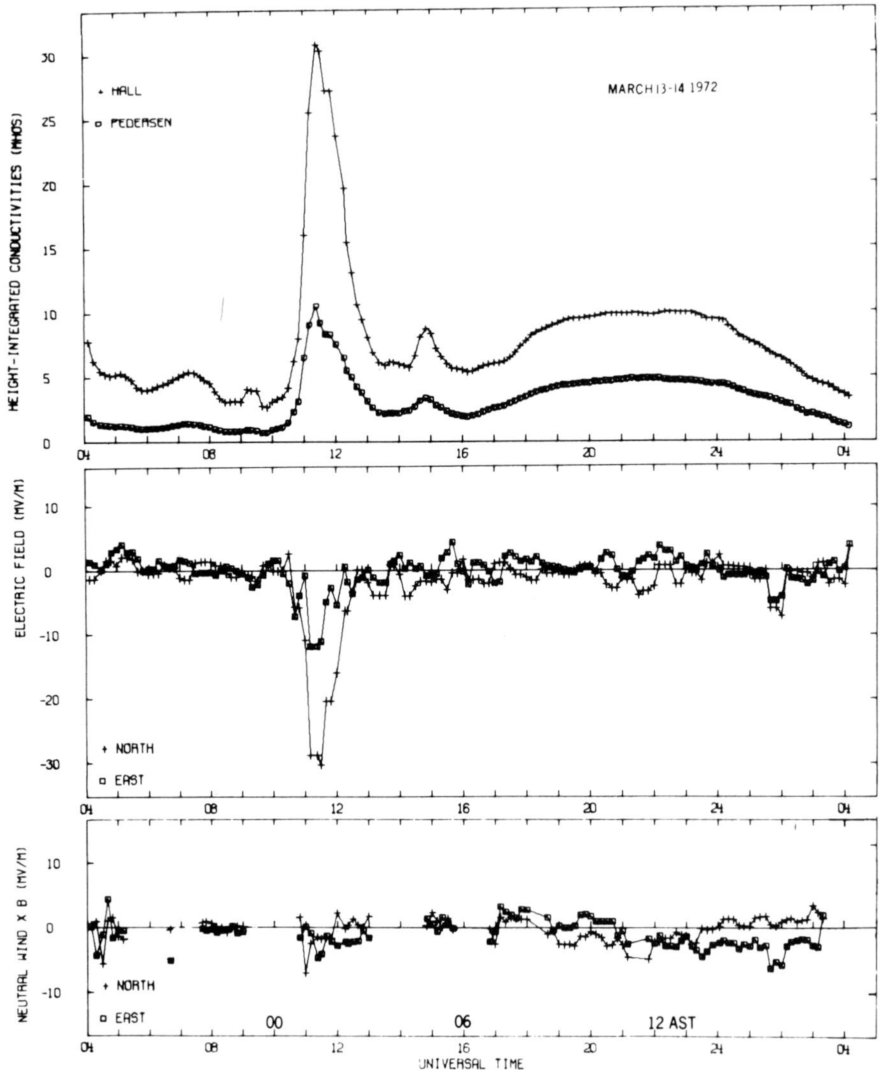

Fig. 7.38(a). Estimated Hall and Pedersen conductivities, observed north-south and east-west components of the electric field and the estimated north-south and east-west components of the neutral wind by the Chatanika incoherent scatter radar on 1972, March 13–14.

larger (2.5–4) during midnight and morning hours on the former day. The ionization at about a 100-km level contributes most to the large ratio of Σ_H/Σ_P.

Kamide and Brekke (1975) examined the use of the equivalent overhead current approximation for the H component disturbances associated with the electrojets, namely

$$i(\mathrm{A\,km^{-1}}) \simeq H(\gamma)$$

by using the height-integrated ionospheric current density deduced by the

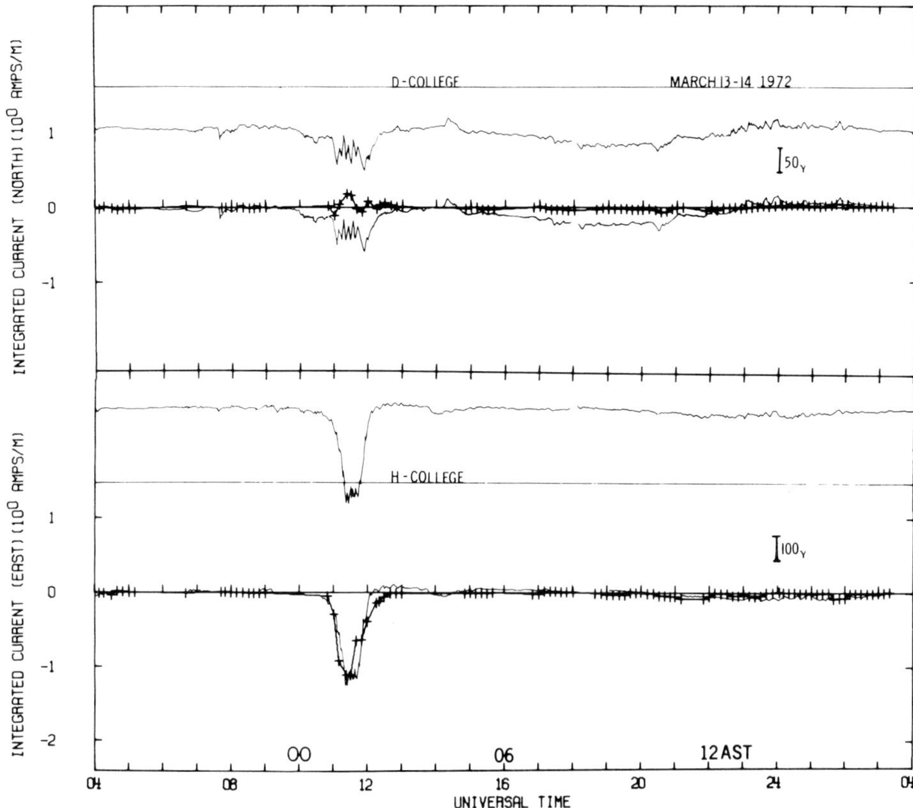

Fig. 7.38(b). Estimated north-south and east-west current densities deduced from the radar data in Figure 7.38(a), together with the *D* and *H* component magnetic records from College.

Chatanika radar. They found that by using the conventional infinite overhead current approximation, the height-integrated current in the ionosphere is, on the average, underestimated by a factor of 2 or more.

Another important parameter obtained by the Chatanika radar is the recombination coefficient in the E region of the auroral ionosphere. Baron (1974) and Wickwar *et al.* (1975) estimated it to be $3–6 \times 10^{-7}$ cm^3 s.

(iv) *1974, May 17.* Figure 7.40 shows an example of the '6 position experiment' which can determine ion drifts at several L shells (Banks and Doupnik, 1975). It shows the electric field vector at five points at a given time. This result may be compared with the theoretical one in Figure 7.36.

(b) *North-South Current and the D Component*

It was shown in the previous section that there is a serious disagreement between changes of the north-south component of ionospheric current and those of the *D* component, although the east-west component of ionospheric current and the *H*

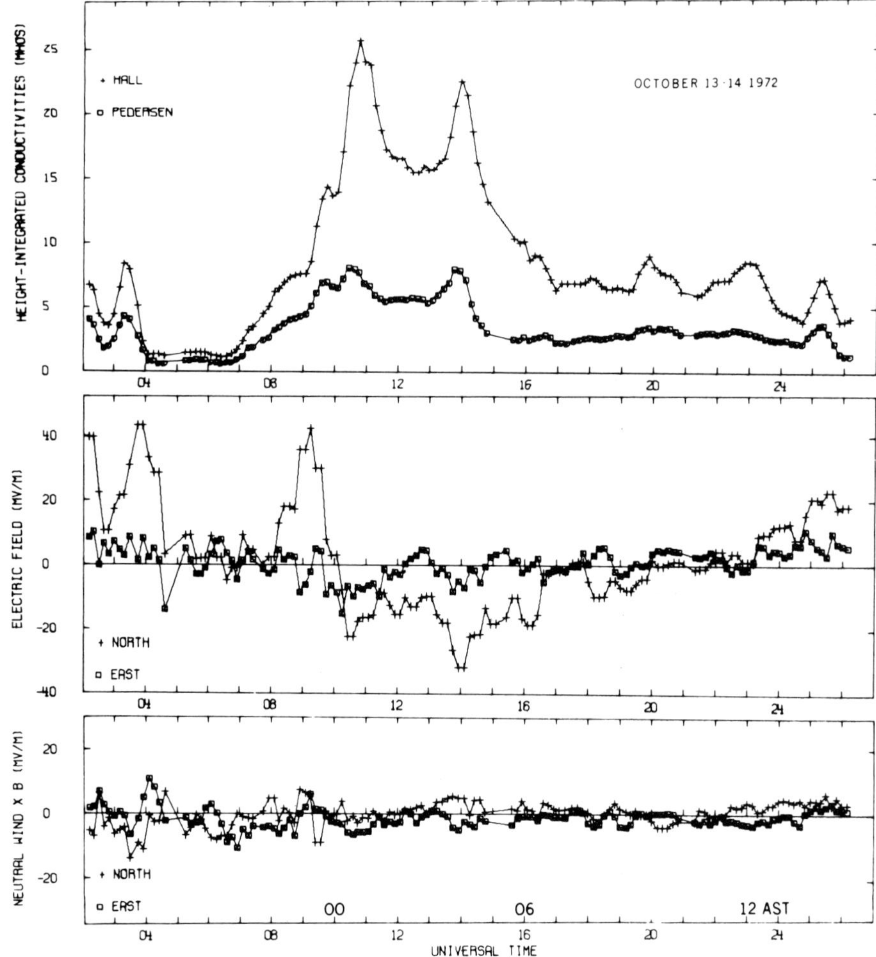

Fig. 7.39(a). Estimated Hall and Pedersen conductivities, and the observed north-south and east-west components of the neutral wind by the Chatanika incoherent scatter radar on 1972, May 25.

component show similar variations. In this subsection, we examine reasons for this disagreement.

Figure 7.41 shows an example of the deduced ionospheric current densities in the eastward and northward components at Chatanika and the corresponding magnetic traces of the H and D components observed at Poker Flat, which is only 3 km north of the radar site, on 1973, October 20. Moderate magnetic disturbances were observed all day. In the figure the ordinate scales are given in such a way that 1 A in the current density and 400 γ in the magnetic perturbation are represented by the same scale length. Time variations of the east-west current density are quite similar to the corresponding variations in the H component of the geomagnetic field near the radar site. There are some disagreements, in magnitude, between the two during the period 0400–0703 UT, but they can be

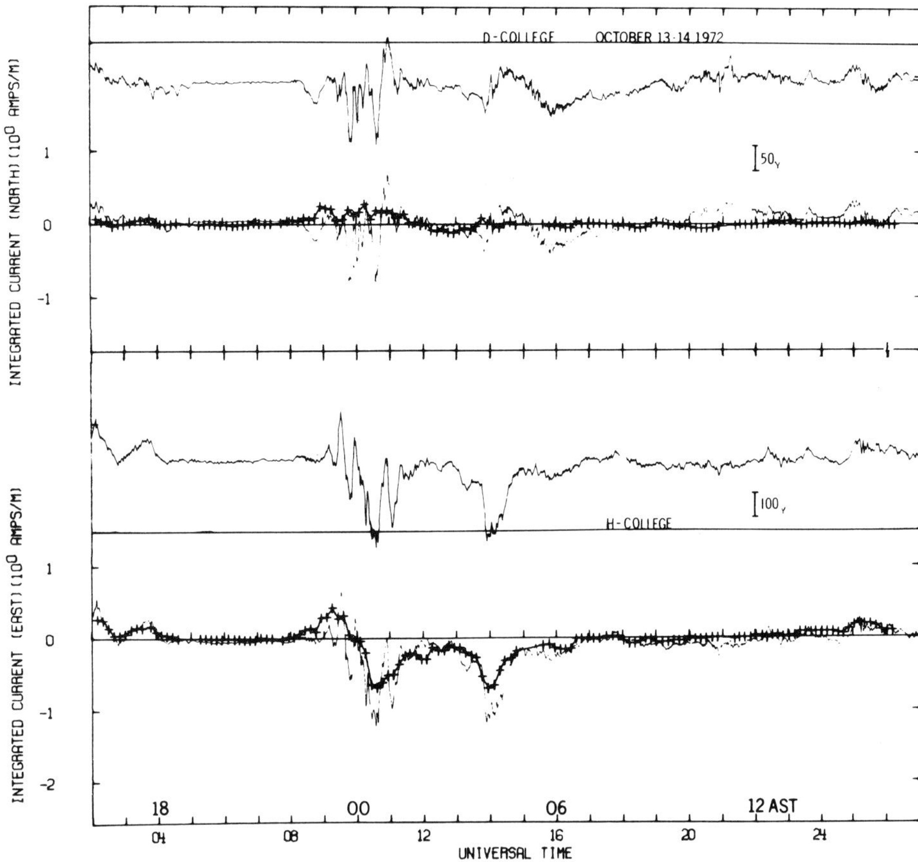

Fig. 7.39(b). Estimated north-south and east-west current densities deduced from the radar data shown in Figure 7.39(a).

reasonably explained by the fact that the eastward electrojet had a small latitudinal width during that period.

The relationships between the north-south current density and the ground D component appear, however, to be more complicated, although it has been the common practice to assume that changes of the D component arise from the north-south component of ionospheric currents (*cf.* Rostoker and Kisabeth, 1973). From about 0500 UT, the Chatanika radar observed a gradual increase of the northward current followed by its sudden intensification at about 0703 UT. Until about 1130 UT, the ionospheric current had essentially a northward component, regardless of the change of the sign in the D component. The disagreement between the north-south ionospheric current and the ground D is most serious prior to local midnight, except for the short time intervals 0750–0850 UT and 0950–1020 UT. During such periods, the Chatanika radar data give results completely contrary to the observed D perturbations on the Earth's surface. This makes it difficult to explain changes of the D component entirely in terms of the

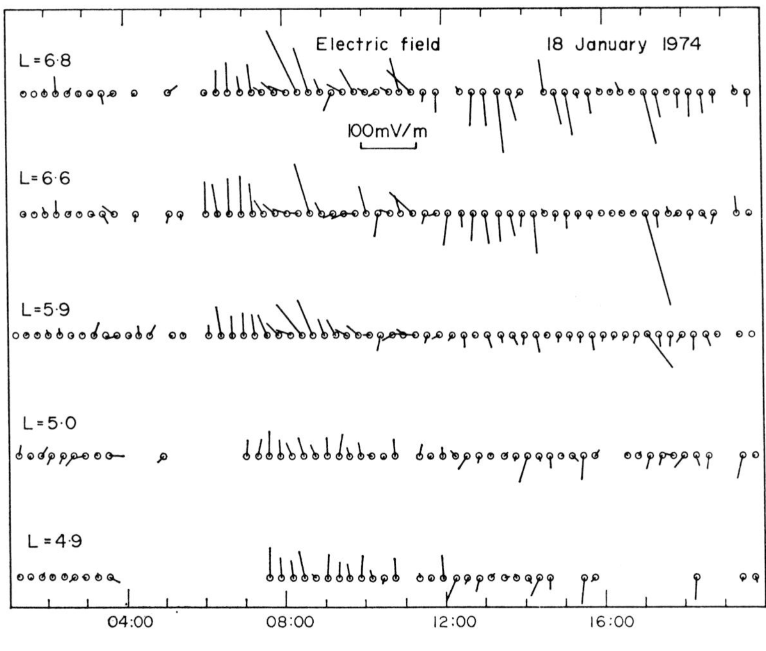

Fig. 7.40. Electric fields observed in the vicinity of the Harang discontinuity by the Chatanika incoherent scatter radar. (Banks, P. M. and Doupnik, J. R.: *J. Atmosph. Terr. Phys.* **37**, 951, 1975.)

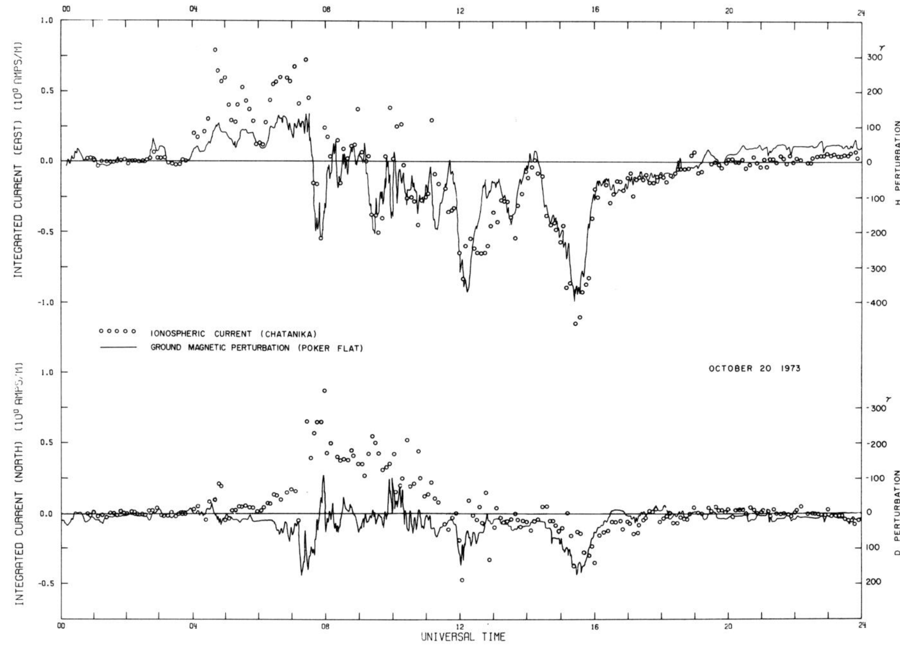

Fig. 7.41. Upper diagram: The comparison of the estimated east-west current intensity and the H component magnetic record. The lower diagram: the comparison of the estimated north-south current intensity and the D component magnetic records. The ionospheric currents are deduced from Chatanika radar data, and the magnetic records are from the College observatory. (Kamide, Y., Akasofu, S.-I. and Brekke, A.: *Planet. Space Sci.* **24**, 193, 1976.)

north-south component of ionospheric current. On the other hand, the disagreement becomes less serious in the morning sector. There are at least two eastward deviations ($\Delta D > 0$) at Poker Flat, which occurred about 1145 and 1420 UT, associated with intense negative H bays; these eastward perturbations were observed at all the Alaskan observatories except Sitka along the meridian. They were caused by the southward ionospheric current actually observed at Chatanika.

These characteristics may be more clearly seen in the vector representation at four instances (two of them were observed in the evening sector and the other two in the morning sector). In Figure 7.42, we show the vector $\boldsymbol{J}$ of the deduced ionospheric current at Chatanika and the magnetic perturbation vector $\boldsymbol{\Delta F}$ in the

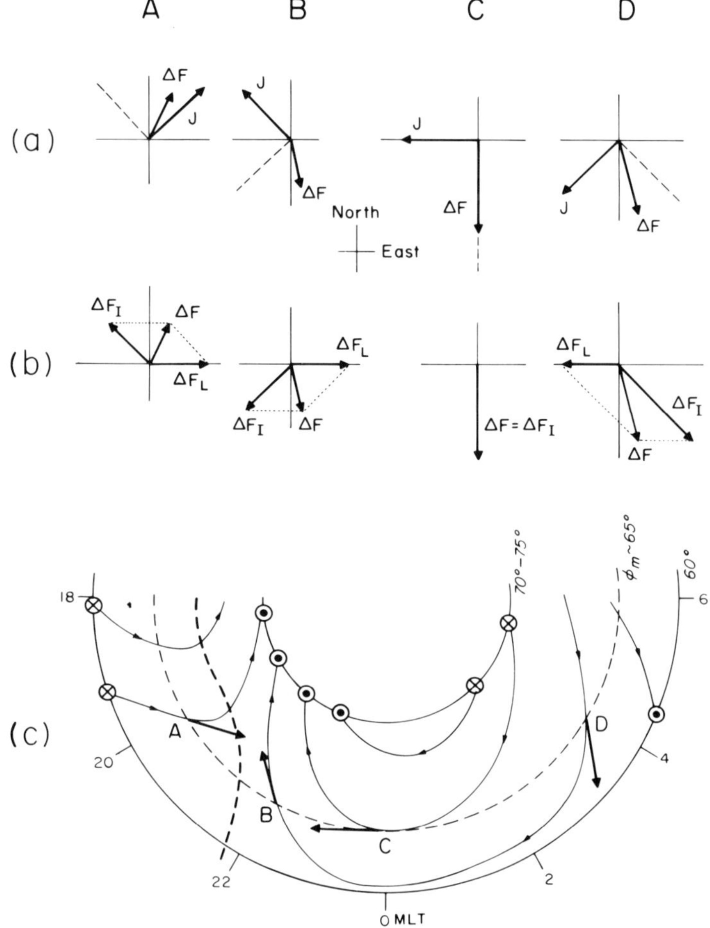

Fig. 7.42. Comparison of the observed ionospheric current intensity ($\boldsymbol{J}$) and the observed magnetic disturbance vector ($\boldsymbol{\Delta F}$) at four points, A, B, C and D. The two vectors are not at right angles to each other in the evening sector (a). In (b), it is assumed that the observed disturbance vector ($\boldsymbol{\Delta F}$) arises from the ionospheric current ($\boldsymbol{\Delta F}_{\mathrm{I}}$) and field-aligned currents ($\boldsymbol{\Delta F}_{\mathrm{L}}$). The direction of the field-aligned currents and the ionospheric current stream lines are shown, together with the Harang discontinuity (dashed line). (Kamide, Y., Akasofu, S.-I. and Brekke, A.: *Planet. Space Sci.* **24**, 193, 1976.)

horizontal plane ($|\Delta F| = \sqrt{\Delta H^2 + \Delta D^2}$) observed at Poker Flat, along with the predicted direction of $\Delta \boldsymbol{F}$ (assuming that it results purely from the observed overhead ionospheric current $\boldsymbol{J}$). Disturbance vectors at Point Barrow and Fort Yukon are also studied for comparison. At 0730 UT, when a typical positive H bay was in progress at Poker Flat, the strong westward electrojet was flowing in higher latitudes, producing negative H bays at both Point Barrow and Fort Yukon. The radar measurement indicates that the ionospheric current was directed northeastward; thus the expected $\Delta \boldsymbol{F}$ on the Earth's surface should be directed northwestward ($\Delta H > 0$, $\Delta D < 0$) as illustrated by the dashed line. However, the actually observed vector $\Delta \boldsymbol{F}$ was directed northeastward ($\Delta H > 0, \Delta D > 0$).

At 0745 UT, the ground perturbations in the H component at all the three stations were negative, whereas the D perturbation at Poker Flat was positive, and was negative at higher latitudes. Again there is a large discrepancy between the predicted and the observed $\Delta \boldsymbol{F}$. The angle between these two vectors is almost 90° for both cases.

In morning hours, however, the direction of $\Delta \boldsymbol{F}$, predicted on the basis of the deduced ionospheric current vector, agrees reasonably well with that observed at Poker Flat for both cases at 1205 and 1522 UT, during the maximum phases of the two substorms.

Therefore, the problem is why the D component shows an eastward deviation when the overhead ionospheric current has a northward component in the evening sector. This contradiction may be at least partially solved by noting that there is a pair of field-aligned sheet currents, outward along the poleward boundary of the oval and inward along the equatorward boundary of the oval, and that the intensity of these currents is, in general, not equal (Section 1.3.2); Yasuhara *et al.* (1975b) have demonstrated that the upward field-aligned current is more intense than the downward current, a typical ratio being 2:1. In this situation, all the equatorward field-aligned currents (downward in the evening sector and upward in the morning sector) can be discharged through one half of the poleward currents (downward in the morning sector and upward in the evening sector) in the way suggested by Zmuda and Armstrong (1974a, b). However, the other half of the poleward field-aligned currents must find their connecting currents in the ionosphere. Now, in Figure 7.42 consider a downward field-aligned current along the equatorward boundary of the oval and an upward field-aligned current along the poleward boundary in the evening sector: Points A and B. Since the upward field-aligned current is twice as intense as the downward current, the net east-west perturbation on the Earth's surface should be directed eastward. If we assume that the intensities of the upward current and the downward current are 1 A m^{-1} and 0.5 A m^{-1} (which are commonly observed values during magnetospheric substorms), respectively, the eastward perturbation $\Delta \boldsymbol{F}_L$ (due to the sheet currents with infinite east-west extent) is 313 γ on the ground. When it is combined with the magnetic field of the ionospheric current $\Delta \boldsymbol{F}_I$, (which is the same order of magnitude as illustrated in Figure 7.42 for A and B), the total $\Delta \boldsymbol{F}$ will be directed in the direction that is observed by the ground magnetometers.

At Point C (in the midnight sector), the field-aligned currents cannot greatly affect the ground magnetic field perturbation under average conditions, and the westward electrojet is the main source; this is because the magnetic fields caused by the field-aligned currents tend to cancel each other.

On the other hand, the disagreement between the time variations of the southward ionospheric current and of the ground D perturbation is less serious in the morning sector. This suggests that in the morning sector the effect of the ionospheric current is much more dominant than that of the field-aligned current. The typical relationship between $\boldsymbol{J}$ and $\Delta\boldsymbol{F}$ is shown schematically in Figure 7.42 for Point D. If we assume that the downward field-aligned current is more intense than the upward current, the net ground perturbation $\Delta\boldsymbol{F}_{\mathrm{L}}$ should be directed westward. Since the ionospheric current is directed southwestward, $\Delta\boldsymbol{F}_{\mathrm{I}}$ should be directed southeastward. Thus, the total field $\Delta\boldsymbol{F}$ will be directed also southeastward if $|\Delta\boldsymbol{F}_{\mathrm{I}}| > |\Delta\boldsymbol{F}_{\mathrm{L}}|$. It should be noted that the current pattern in the figure is well reproduced in the model calculation in Section 7.3.4 (Figure 7.15).

(c) *Electric Field and Auroral Activity*

In the previous subsection, we examined in detail how the electric field varies during substorms and how the current density variations are related to magnetic perturbations on the ground. Here we examine how the electric field variations in the ionosphere are related to auroral activity (Banks and Doupnik, 1975). Doupnik *et al.* (1972), Banks *et al.* (1973, 1974) and Rino *et al.* (1974) studied this problem on the basis of data taken from the Chatanika radar and the Alaska meridian chain of all-sky cameras and magnetometers. Banks *et al.* (1973) showed that a large northward electric field is observed in the evening sector when a substorm is in progress in the midnight sector, and that a large increase of the westward electric field is associated with the passage of a westward traveling surge in the poleward sky of the radar.

The east-west component of the electric field during substorms was studied by Rino *et al.* (1974). Figure 7.43(a) shows the College magnetometer record and the EW component of the electric field observed on 1973, February 24. In the following, we examine in detail the electric field and concurrent auroral variations on this day. In Figure 7.43(b), it can be seen that the magnitude of the EW component electric field was considerably reduced for a brief period around 0720 UT; see the all-sky photograph taken at that time. Note that auroras within the measured magnetic flux tube were bright and very active. Again, from 0756 through 0808 UT, the electric field was considerably reduced and there were again bright active auroras within the measured magnetic flux tube. This phenomenon may be interpreted as a shorting effect of the field caused by local conductivity enhancements. The same effect has been observed by Mozer *et al.* (1973) as an anti-correlation between balloon-borne probe measurements and simultaneous X-ray data, and more directly with satellite particle detectors (Bogott and Mozer, 1971); see also Maynard *et al.* (1973).

The midnight sector enhancement coincided with a period of accelerated equatorward motion of the auroras; see Figure 7.43(c). This activity commenced at ~ 0940 UT. A negative bay began to develop at about that time. However, the westward electric field had subsided almost to its previous mean level of $\sim -15\ \mathrm{mV\ m^{-1}}$ by 1010 UT.

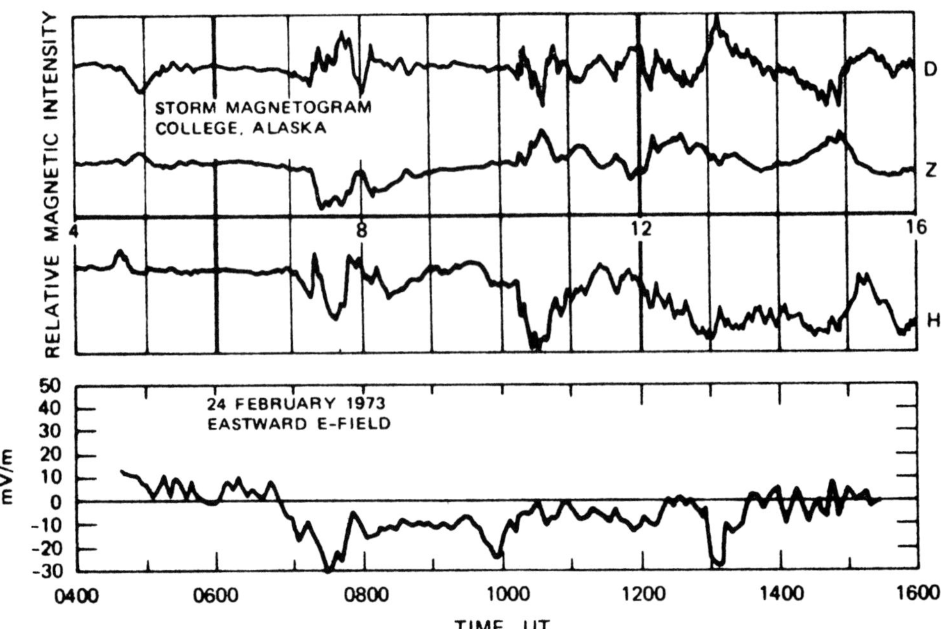

Fig. 7.43(a). East-west component of the ionospheric electric field observed by the Chatanika incoherent scatter radar and the three-component magnetic records from College. (Rino, C. L., Wickwar, V. B., Banks, P. M., Akasofu, S.-I. and Rieger, E.: *J. Geophys. Res.* **79**, 4669, 1974.)

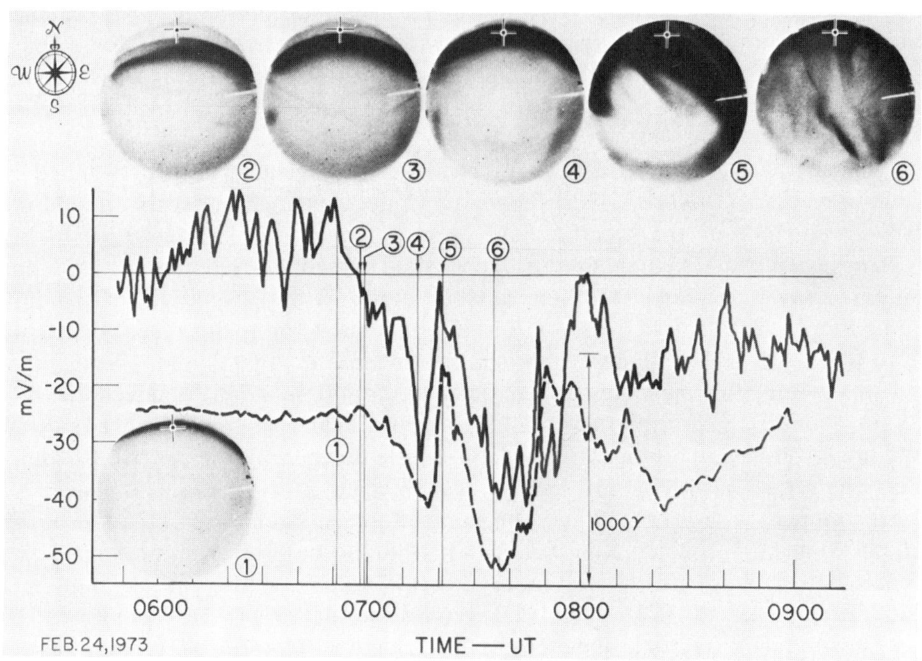

Fig. 7.43(b).

Fig. 7.43(b–e). High time resolution electric field and the *H* component data in (a), together with several all-sky photographs. The point where the electric field was measured is indicated by a cross in the photograph.

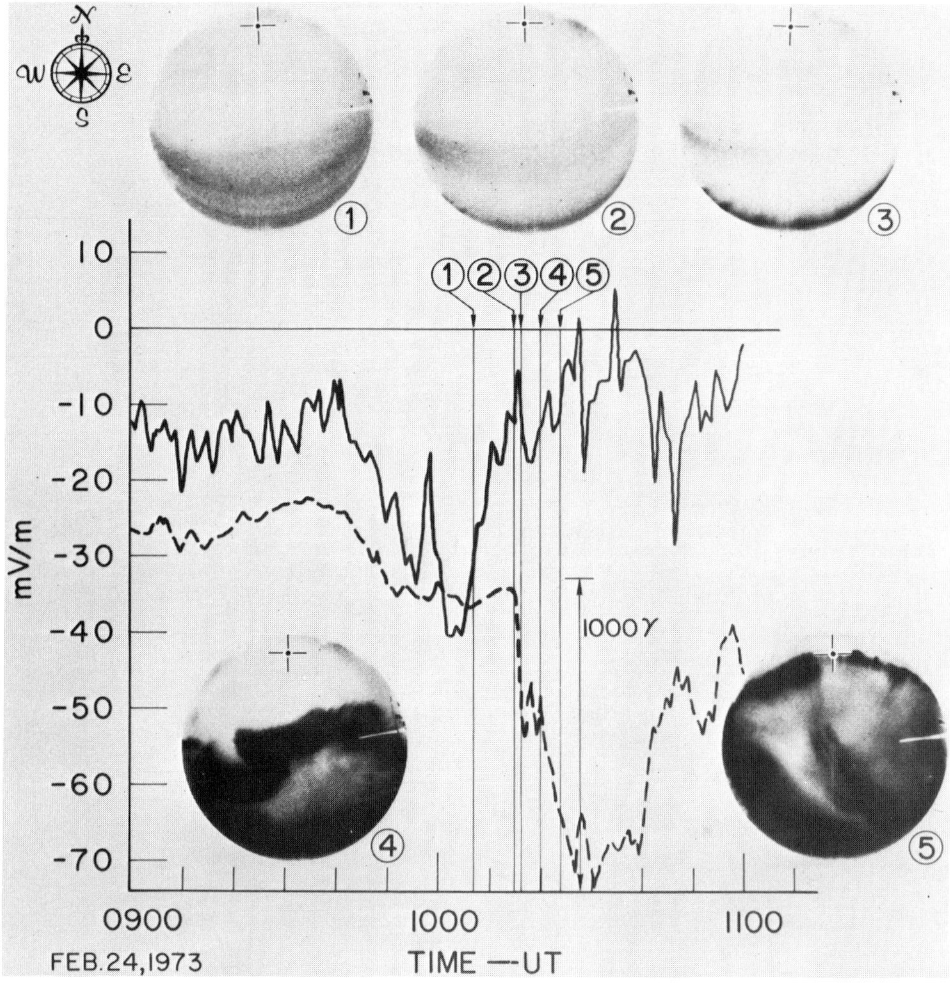

Fig. 7.43(c).

At 1016 UT the second major substorm occurred, as indicated by the brightening of an arc on the southern horizon at Chatanika and its subsequent violent poleward expansion. Evidently the enhanced westward component field subsided just before the substorm onset. The shortening of the electric field occurred again when the expanding bulge reached the measured magnetic flux tube at approximately 1025 UT. From 1130 UT until dawn, active auroras were visible over the entire sky at College; Figures 7.43(d) and (e).

7.7.3. BALLOON OBSERVATIONS

Extensive observations of the electric field have also been conducted by Mozer and his associates on the basis of balloon-borne electric probes (Mozer and Serlin, 1969; Mozer, 1971; Mozer and Manka, 1971; Mozer, 1972, 1973a, b, c; Mozer and

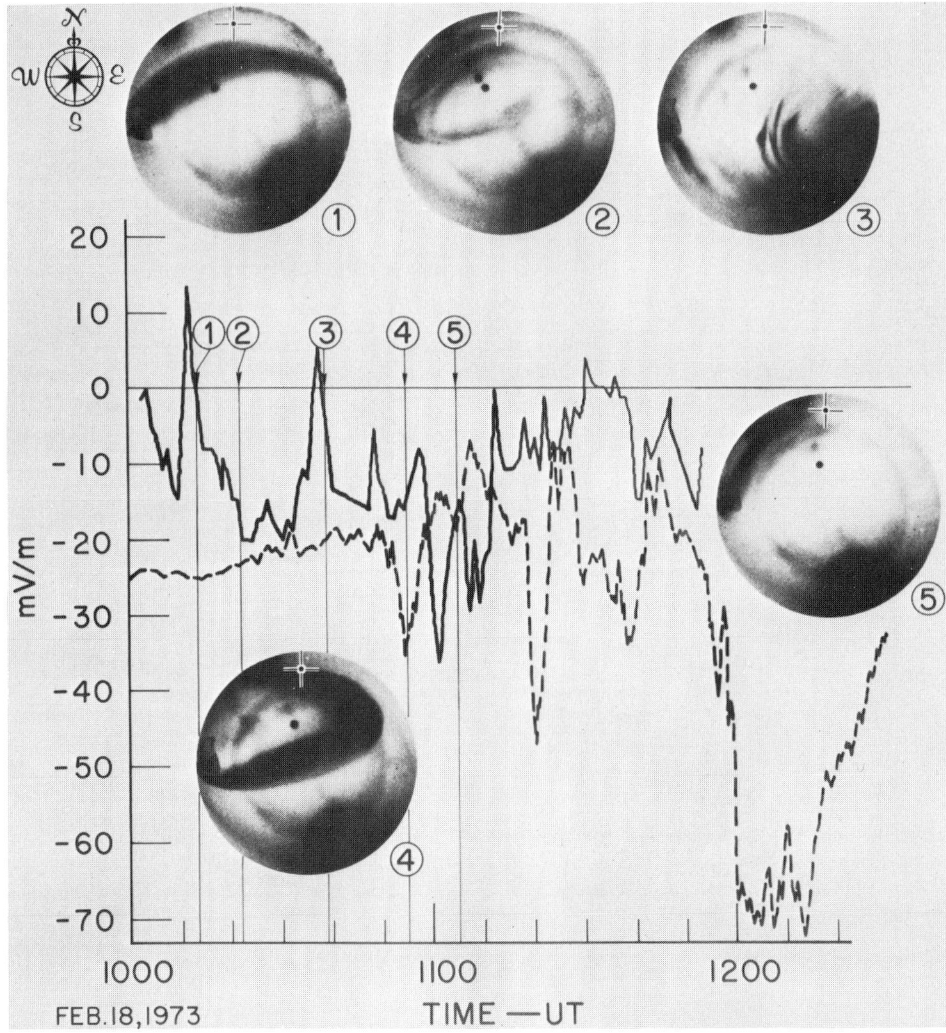

Fig. 7.43(d).

Lucht, 1974; Mozer *et al.*, 1974; Kelley and Mozer, 1975; Kelley *et al.*, 1975). Their study of the large-scale polar cap electric field (Mozer and Serlin, 1969; Mozer and Lucht, 1974), was summarized in Section 1.3.4. Thus, we shall be concerned here with electric field observations during magnetospheric substorms.

Mozer and Manka (1971) and Mozer (1971, 1973c) showed that a westward component of the electric field E_w develops about one hour prior to the onset of the expansive phase of isolated substorms, while an equatorward component grows at the expansive onset. Figure 7.44 shows 15-min averages of balloon data for 19 events; from the top, the westward, southward components and the total intensity. Mozer (1973c) associated the initial growth of E_w with a growth phase

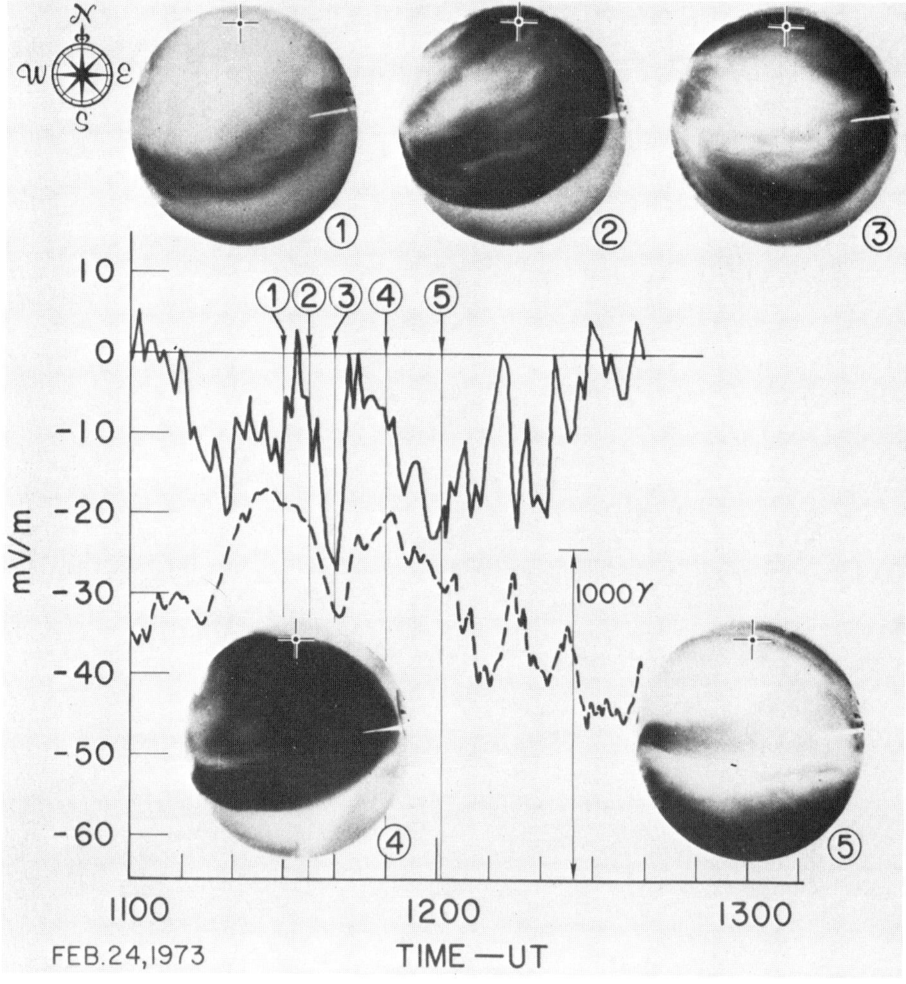

Fig. 7.43(e).

feature. Kelley *et al.* (1971) showed that the growth of E_w is well correlated with an equatorward motion of auroral forms and that the drift speed of auroras can be estimated from E_w/B. On the other hand, poleward motions of auroras during the expansive phase are not related to the eastward electric field, so that the poleward motion of auroras cannot be due to the $(E \times B)$ drift motion. The growth of the equatorward directed electric field during the expansive phase is in agreement with the radar observation in the previous subsection and other electric field observations which will be described in the following subsection.

The fact that the westward electric field is well correlated with an equatorward motion of auroras suggests that it is associated with the expansion of the auroral oval. As mentioned in Section 4.4.6(b), the IMF B_z component controls the size of the oval, so that the initial growth of the westward electric field can be due to the southward turning of the IMF vector.

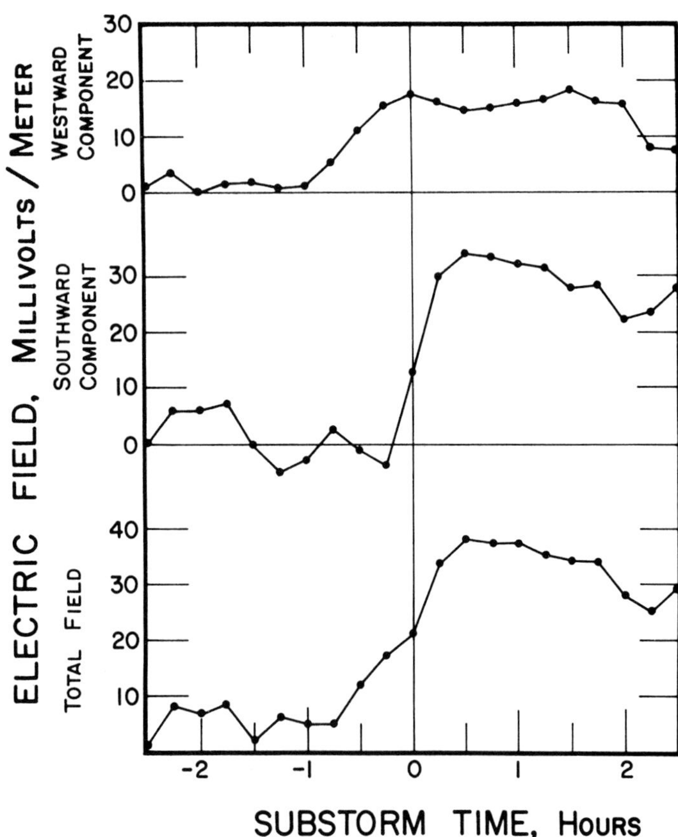

Fig. 7.44. Fifteen-minute averages of the east-west and north-south components, of the electric field and of the total field obtained during 19 balloon flights. (Mozer, F. S.: *J. Geophys. Res.* **76**, 7595, 1971.)

7.7.4. BARIUM CLOUD OBSERVATIONS

(a) *Auroral Oval*

Early studies of ionospheric electric fields by using drift motions of barium clouds, released from rockets, have been summarized by Wescott *et al.* (1969, 1970), Haerendel and Lüst (1970) and Haerendel (1972). These studies show that the westward electrojet is associated with a southward electric field (10–130 mV m^{-1}), while the eastward electrojet is associated with a northward electric field (50 mV m^{-1}), indicating that the electrojets are essentially Hall currents (see also Section 1.3.4). Wescott *et al.* (1969) also noted that the electric field is rather weak within an auroral arc.

(b) *Polar Cap*

There are so far only three series of barium cloud observations in the polar cap. Heppner *et al.* (1971) showed that the electric field is fairly uniform in the polar cap and is directed roughly from dawn to dusk; their data are included in Figure 1.19. They also found that the ground magnetic disturbance vectors expected from the electric field observations differ considerably from the observed ones. Thus, the ground observations should not be interpreted to be caused solely by the ionospheric Hall current in the form of an infinite sheet current across the polar cap. There is little doubt that field-aligned currents which flow into or out of the auroral oval have a significant contribution to the ground magnetic field in the polar cap. This does not mean, however, that there is no current in the polar cap (Primdahl *et al.*, 1974; Olesen *et al.*, 1975). Mikkelsen *et al.* (1975) observed drift motions of barium clouds in the polar cap during two magnetospheric substorms. On one occasion near the end of a substorm, the cloud drifted with a speed of about 400 m s^{-1} in the anti-solar direction, with a dawn-dusk component. During the second release, the cloud first moved slightly poleward (sunward) and then equatorward. The simultaneous auroral observations indicated that the aurora moved in a similar way. The authors suggested that the poleward motion of the cloud was caused by an eastward electric field which was induced by a rapid change of the B_z component ($\partial B_z/\partial t$) in the vicinity of the newly formed neutral line in the magnetotail.

Jeffries *et al.* (1975) showed that one of the barium clouds, released a little poleward of the midday part of the oval, drifted across the entire polar cap along the noon-midnight meridian with a speed of order 1400 m s^{-1}, implying an electric field of 82 mV m^{-1} directed dawn-to-dusk. Figure 7.45 shows the trajectory of the barium cloud. This release took place during a highly disturbed period but a little prior to an intense substorm. By the time the cloud reached the midnight part of the oval (near Godhavn, Greenland), the intense substorm began. The cloud turned eastward with a high speed near the front of the expanding auroral bulge.

7.7.5. ROCKET OBSERVATIONS

Rocket observations of electric fields have been conducted by several groups. Since the electric field component parallel to the geomagnetic field was discussed in Section 3.9, we shall be concerned here only with the perpendicular component. Most of the rocket observations show that the electric field in the vicinity of active auroras is directed southward and has a magnitude of order 50–100 mV m^{-1} (Mozer and Fahleson, 1970; Potter, 1970; Kelley *et al.*, 1971a; Cahill *et al.*, 1974; Kelley *et al.*, 1975). However, there is some disagreement as to the electric field within auroral forms. Kelley *et al.* (1975) found that the electric field is not always weakest within auroral arcs, while Whalen *et al.* (1975) found that the ion drift speed in an active arc was small, implying that the electric field is weak there. It is also interesting to note that Whalen *et al.* (1975) found that ions drift poleward in the region slightly poleward of the expanding auroral bulge.

7.8. Thermospheric and Ionospheric Disturbances

The polar ionosphere is greatly disturbed by particle bombardment, electric currents and electric fields during the magnetospheric substorm. The ionospheric

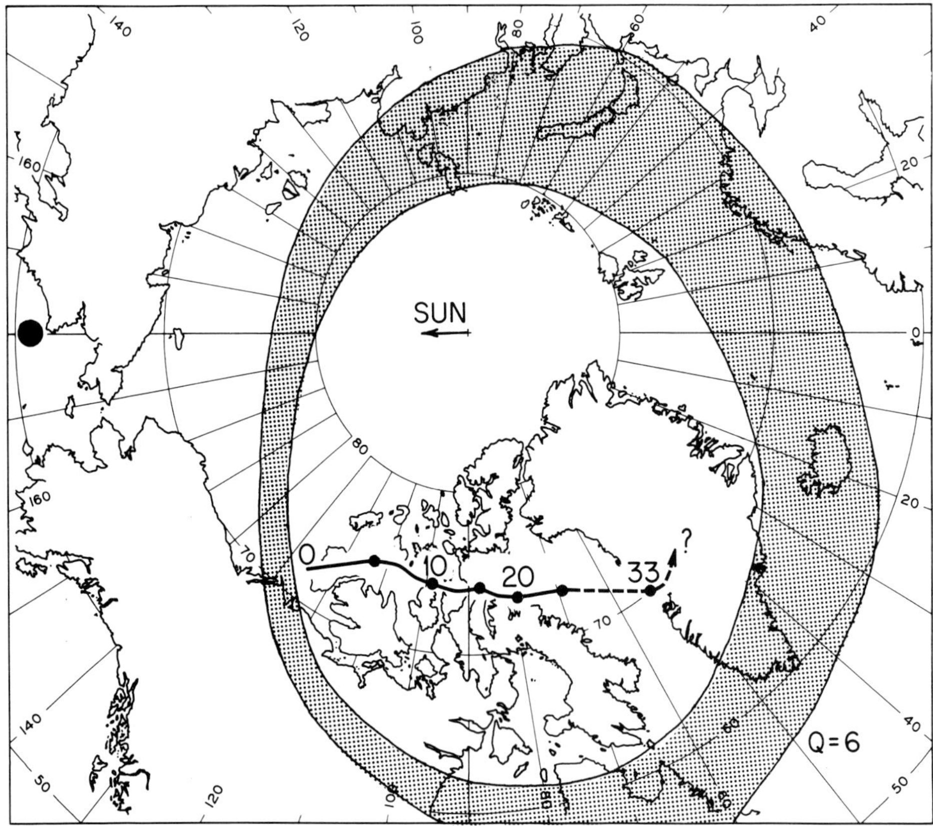

Fig. 7.45. Drift motion of a barium cloud injected into the cusp region during a fairly disturbed condition. The cloud drifted across the polar cap along the noon-midnight meridian. (Jeffries, R. A., Roach, W. H., Hones, E. W. Jr., Wescott, E. M., Stenbaek-Nielsen, H. C., Davis, T. N. and Winningham, J. D.: *Geophys. Res. Lett.* **2**, 285, 1975.)

disturbances thus manifested are as a whole called the ionospheric substorm (Akasofu, 1968, Chapter 4). One of the reasons for studying the ionospheric substorm is that one can learn different modes by which the substorm energy is dissipated. Eventually most of the substorm energy is converted into thermal energy in the thermosphere. However, the various effects of the particle bombardment, electric currents and electric fields on thermospheric and ionospheric disturbances are not independent, but couple with each other in complicated ways. Figure 7.46 illustrates some of the ionospheric processes associated with the magnetospheric substorm (Rees, 1975). In this section, we review only briefly a few aspects of ionospheric substorms.

A typical maximum energy deposition rate to the ionosphere by auroral particles is known to be of order of 10^{-7}–10^{-6} erg cm^{-3} s^{-1}. Rees (1975) estimates that at least 60% of the energy deposited by typical auroral particle bombardment heats the neutral atmosphere, 11% maintains the ionization, and 10% is radiated as auroral light.

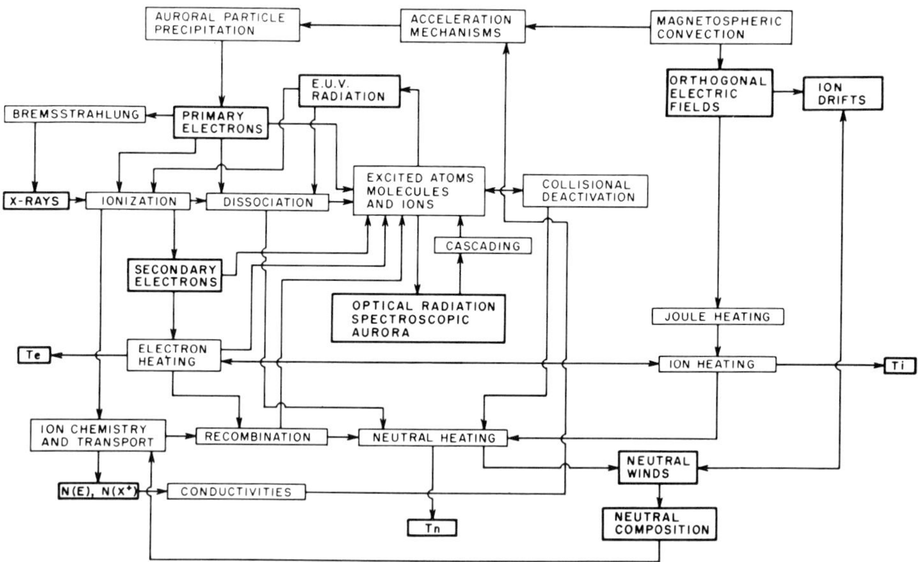

Fig. 7.46. Flow chart of auroral processes and effects due to electron bombardment. (Rees, M. H.: *Atmosphere of Earth and the Planets*, B. M. McCormac (ed.), p. 323, D. Reidel Publ. Co., 1975.)

The importance of joule heating of the ionosphere by the electrojet was pointed out first by Cole (1962, 1969, 1971a). Recent theoretical studies by Hays *et al.* (1973) and Rees (1975) suggest that joule heating is comparable to or a little greater than the heating by the bombardment of auroral particles. Wickwar *et al.* (1975) estimated the joule heat input to be 30 erg cm^{-2} s^{-1} on the basis of their incoherent scatter radar observations. There is also little doubt that the composition of the upper atmosphere is considerably affected by both the bombardment of auroral particles and the electric field (Donahue *et al.*, 1970; Prasad and Furman, 1975).

Both the particle bombardment and joule dissipation cause heating of the ionosphere and the subsequent neutral winds. Neutral winds are also generated by the $(E \times B)$ convection of ionospheric plasma, because ions, drifting with a speed of (E/B), impart their momentum to neutral particles by colliding with them (Akasofu and DeWitt, 1965; Fedder and Banks, 1972). In terms of the equation of motion of the ionospheric gas, this effect is expressed by the Lorentz force acceleration term and is often called the ion drag effect.

Neutral winds in the polar region have been deduced from the motion of barium clouds (the neutral component) by Rieger (1974) and Meriwether *et al.* (1973); from the motion of meteor trails by Hook (1970); and on the basis of the incoherent scatter radar observations by Brekke *et al.* (1973, 1974a). These observations indicate that the $(E \times B)$ convection is an important cause of the neutral winds in the polar ionosphere. Figure 7.47(a) shows the neutral wind pattern observed by Meriwether *et al.* (1973). Note the anti-solar flow in the polar cap, the duskward component in the evening sector and the dawn component in

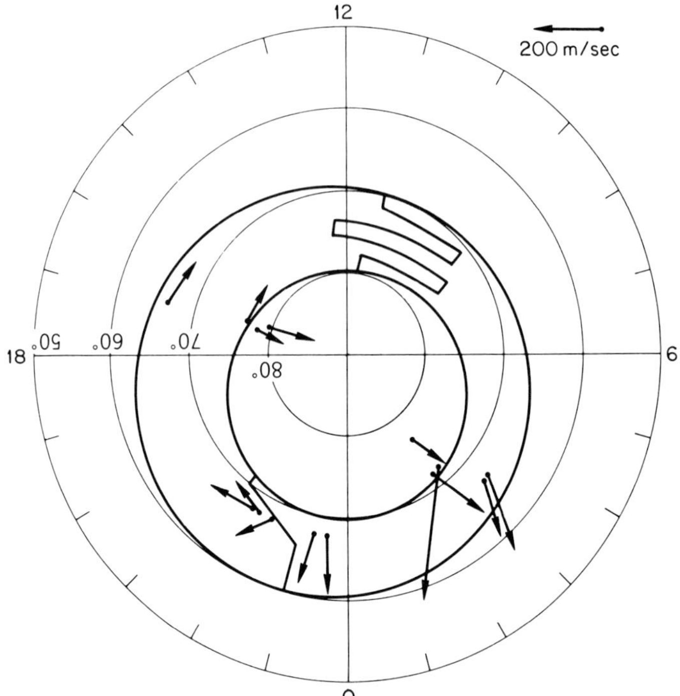

Fig. 7.47(a). Neutral wind vectors deduced from the drift motion of barium clouds in invariant latitude-MLT coordinates. (Meriwether, J. W., Heppner, J. P., Stolarik, J. D. and Wescott, E. M.: *J. Geophys. Res.* **78**, 6643, 1973.)

the morning sector. Figure 7.47(b) shows the neutral wind pattern deduced from Chatanika radar data (Brekke *et al.*, 1973, 1974a).

The generation of the neutral wind by joule heating of the westward electrojet has recently been studied by a number of workers, including Cole (1971b), Bates (1973, 1974a, b), Heaps and Megill (1975) and Heaps (1974). Since the heating takes place along a rather narrow belt, namely along the auroral oval, an intense upward flow of the atmosphere above the oval is expected. Figure 7.48 shows an example of model calculation showing the upward flow (Heaps and Megill, 1975). Bates (1974a, b) showed, on the basis of the Chatanika radar observations, that such a flow indeed occurs.

Traveling ionospheric disturbances (TID) are also known to be generated in the auroral region and to propagate to great distances (Hines, 1960; Davis, 1971; Harper, 1972; Rao, 1975; Testud, 1970, 1972; Vasseur *et al.*, 1972; Malingre, 1973; Hunsucker and Tveten, 1967). This subject was extensively reviewed by Francis (1975). Auroral infrasonic waves have been observed by Wilson (1969, 1972, 1973a, b, 1974, 1975), Wilson and Hargreaves (1974) and Johnson (1972). Their generation mechanisms have been considered by Chimonas and Hines (1970), Chimonas (1970), Chimonas and Peltier (1970), Fedder and Banks (1972), Wilson (1973) and Swift (1973). These mechanisms were critically discussed in the light of observations by Wilson (1975).

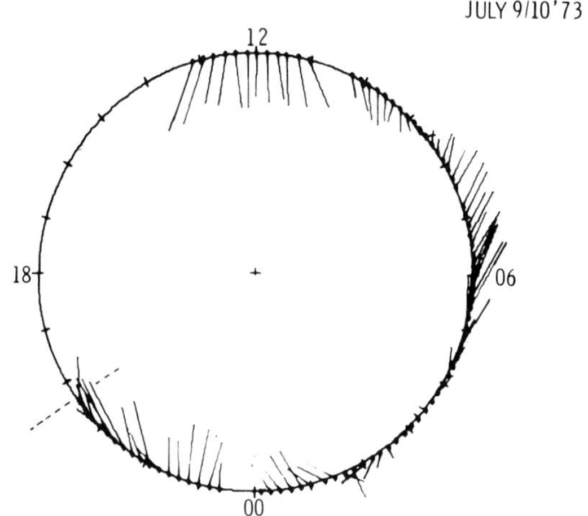

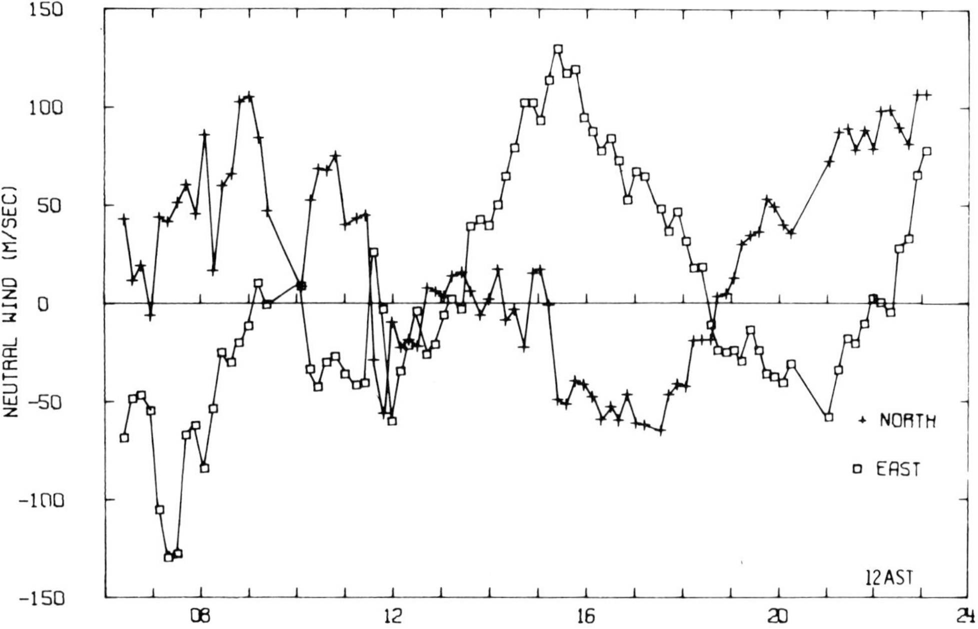

Fig. 7.47(b). Neutral wind vectors in the ionosphere, deduced from the Chatanika incoherent radar observations in geographic latitude-LT coordinates. The east-west and north-south components of the wind are also shown as a function of LT (Alaska standard time = UT − 10 h). (Brekke, A., Doupnik, J. R. and Banks, P. M.: *J. Geophys. Res.* **78**, 8235, 1973.)

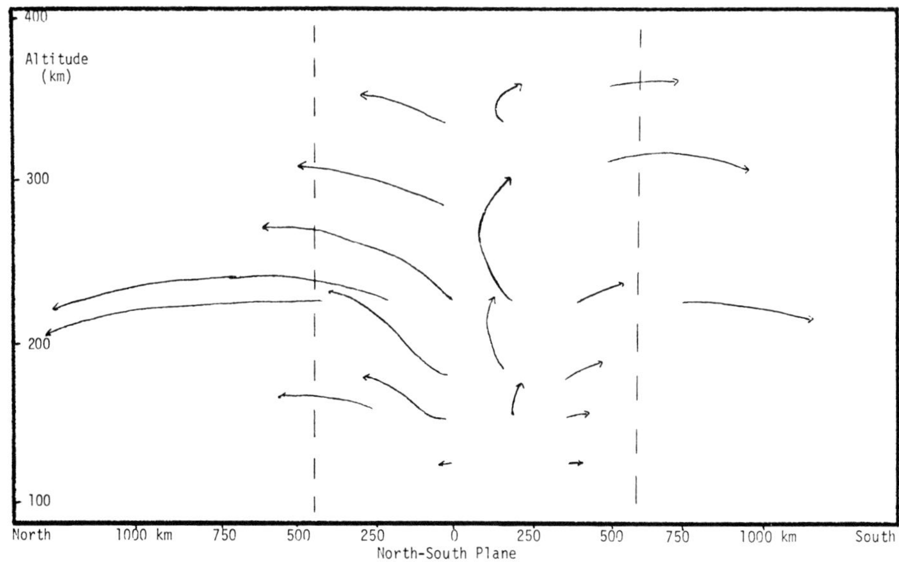

Fig. 7.48. Net displacement of air parcels in a meridian plane for a period of storms. The tails and heads of the arrows mark the beginning and end points of the displacement. (Heaps, M. G. and Megill, L. R.: *J. Geophys. Res.* **80**, 1829, 1975.)

There is little doubt that the intense heating of the lower thermosphere along the auroral oval causes changes of the ionosphere and the thermosphere on a worldwide scale during substorms, and particularly during storms (when intense substorms occur frequently; Section 5.6). A worldwide increase of temperature and winds in the thermosphere has been observed directly by a satellite-borne Fabry-Perot interferometer (Blamont and Luton, 1972); by a ground-based interferometer (Hays *et al.*, 1969; Truttse, 1968a, b, 1972; Truttse and Yurchenko, 1971; Truttse and Shefov, 1970); by a satellite-borne Langmuir probe (Raitt, 1974); indirectly by a mass spectrometer on the basis of a large increase of the density of N_2 and the concurrent decrease of He (Hedin and Reber, 1972; Reber and Hedin, 1974); by satellite drag effect and acceleration measurements (DeVries *et al.*, 1972; Jacchia and Slowey, 1964; Jacchia *et al.*, 1967; Jacchia, 1971; DeVries, 1972; Ching and Rugge, 1975); by a microphone density gauge (OGO-6) (Anderson, 1973); and by an incoherent scatter radar (Evans, 1970a, b; Reddy, 1974). Thermospheric responses to the heating and the ion drag have also been investigated theoretically by Volland and Mayr (1971), Mayr and Volland (1973) and Richmond and Matsushita (1975). The ionospheric ionization is also affected on a worldwide scale by such thermospheric changes, as well as by many other storm effects. This phenomenon is called the ionospheric storm (*cf.* S.T.P., Section 8.10) and has been extensively studied by a number of workers (Matsushita, 1959; Bauer and Krishnamurthy, 1968; Jani and Kotadia, 1969; Arendt, 1969; Thomas, 1970; Obayashi, 1972; Matuura, 1972; Lakshmi and Reddy, 1970; Davies, 1974a, b; Chandra and Herman, 1969; Chandra and Stubbe, 1971; Evans, 1970a, b, 1972,

1973; Ben'kova and Zevakina, 1974; Papagiannis and Mendillo, 1971; Mendillo *et al.*, 1974; Mendillo *et al.*, 1974; Jones, 1973; Spurling and Jones, 1973; Bates and Hunsucker, 1974; Carpenter *et al.*, 1975).

A full review of this subject is obviously beyond the scope of this book. It may be suggested, however, that one of the ways to study the ionospheric storm is to examine in detail ionospheric phenomena which occur during individual magnetospheric substorms, namely the ionospheric substorm. Since we have begun to understand a geomagnetic storm in terms of a successive occurrence of intense substorms, a significant part of the ionospheric storm may be understood in terms of a successive occurrence of ionospheric substorms. Fortunately, there has already been a considerable effort on studies of ionospheric substorms along this line of thought. In the following, we shall review briefly some of the recent studies of ionospheric substorms.

Richmond and Matsushita (1975) simulated, by an extensive computer calculation, the winds and temperature variations in the thermosphere which result from the growth of electric currents along the auroral oval during a large isolated substorm. The current intensity is assumed to vary as follows:

$$A(t) = t/30\text{ min}, \qquad t < 30\text{ min}$$
$$A(t) = (120\text{ min} - t)/90\text{ min}, \qquad 30\text{ min} < t < 120\text{ min}$$

The equation of motion of the atmosphere is solved by including the heat source (the joule dissipation) and the Lorentz force (the ion drag effect). The results are discussed in terms of gravity waves, ion drag winds and vertical wind (Figure 7.49).

One of the prominent features in Figure 7.49(a) is the generation of disturbances propagating both equatorward and poleward. The front of the disturbance moves at a nearly constant speed of 750 m s^{-1}, which is considerably greater than the north-south air speed ($\simeq$200 m s^{-1}). The disturbance reaches the equator at about $t = 175$ min and is reflected back toward higher latitudes (note that all changes in the thermosphere are assumed to occur symmetrically with respect to the equator). Therefore, this phenomenon may be identified as gravity waves (Chimonas and Hines, 1970; Testud, 1970; Volland and Mayr, 1971). The next interesting feature is the generation of the eastward wind which arises from the ion drag effect (Figure 7.49(b)). This wind continues to blow well after $t =$ 120 min, namely after the electric field is 'switched off', and thus the ion drag ceases. In fact, after $t = 120$ min, this neutral wind drags the ionospheric plasma, generating the dynamo effect, which drives the poleward current. In turn, this current produces space charges near the poleward and equatorward boundaries of the oval. The resulting polarization electric field is directed equatorward and lasts as long as the eastward motion of the neutral air continues. The authors believe, however, that their model seems to overestimate the ion drag effect.

The heating causes also a vertical motion of the atmosphere above the oval. This arises from two causes: one is due to the atmospheric expansion by the increased temperature and the other is due to the buoyancy. In Figure 7.49(c) it can be seen that the time behavior of the vertical velocities roughly follows that of

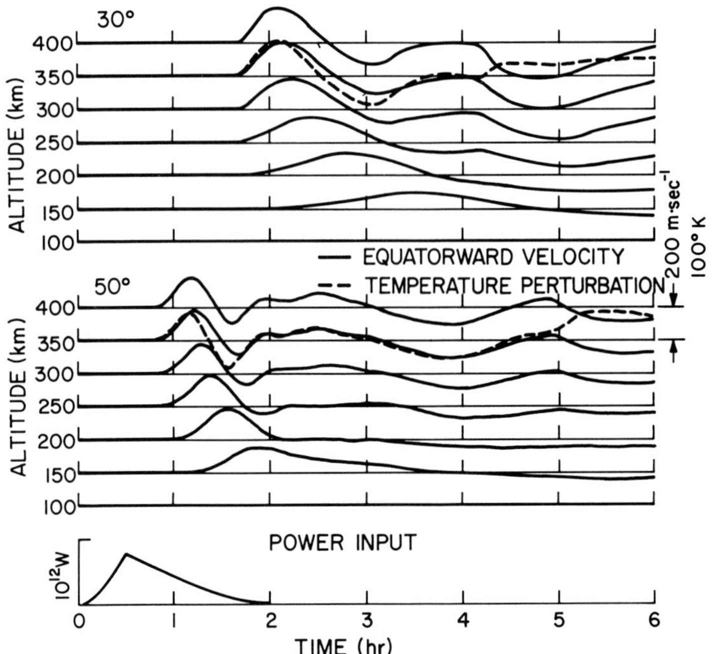

Fig. 7.49(a). Time histories of the equatorward wind velocity at 30° and 50° latitude at various altitudes. The temperature perturbations at a 350-km level are also indicated. The total power input into the auroral oval, by the combined effect of joule heating and particle bombardment, is shown as a function of substorm time.

the energy input. The authors pointed out that the accompanying horizontal divergent flow will supply N_2 and O_2 in the upper F regions in the middle latitude, causing a profound change in ion composition and electron density.

The ionosphere in the middle latitude is greatly disturbed during magnetospheric substorms. One of the causes for the disturbance arises from the $(E \times B)$ drift motion of ionospheric plasma. Therefore, this particular aspect of the ionospheric substorm is of great value, not only from the point of view of a study of the distortion of the ionosphere by the electric field, but also from the point of view of the origin of the electric field. As we shall see in Section 8.2.2, during an early epoch of substorms, the cross-tail electric field and also the polar cap electric field penetrate into the region between the Earth and the inner edge of the magnetosphere (and also below the latitude of the equatorward boundary of the oval) causing a sudden earthward advance of the plasma sheet, namely the plasma injection. The penetrating electric field affects also the plasmasphere (Section 8.6). In fact, the ionospheric substorm in the middle latitude cannot be discussed without understanding the distortion of the plasmasphere (Park, 1971, 1974; Testud *et al.*, 1975; Park and Banks, 1974). Therefore, the ionospheric substorm in the middle latitude belt will be discussed in Section 8.6, after studying the penetration of the electric field into the middle latitude. There is so far no definitive evidence that substorm activity affects the mesosphere. Maeda and Aikin (1968) discussed theoretically auroral effects on mesospheric ozone. Hays

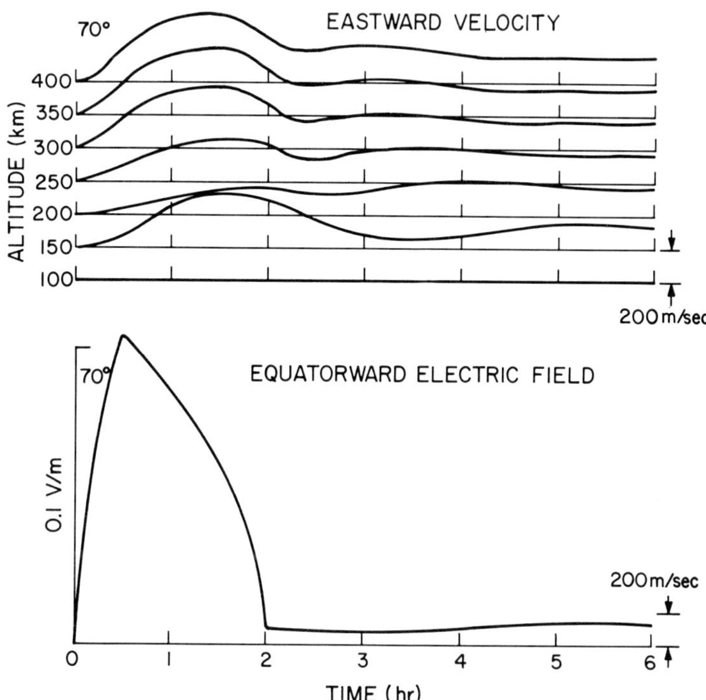

Fig. 7.49(b). Time histories of the eastward wind velocity at 70° latitude at various altitudes. The growth and decay of the equatorward electric field are also given as a function of substorm time at the bottom.

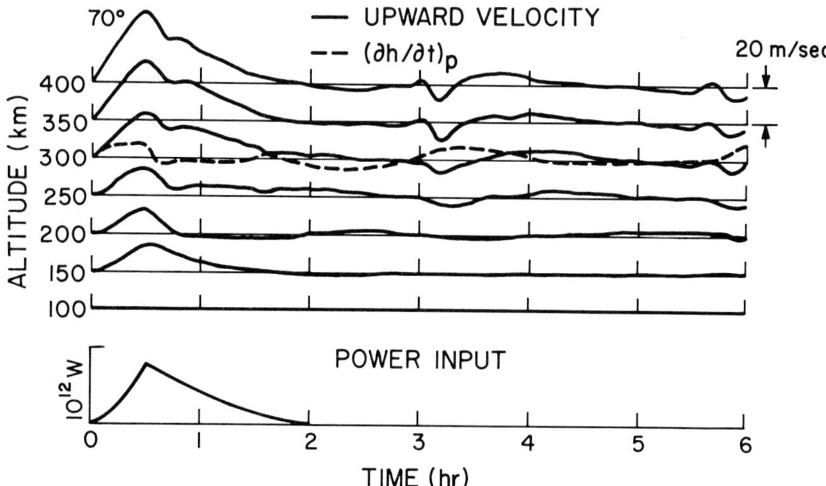

Fig. 7.49(c). Time histories of the upward neutral wind velocity at 70° latitude at various altitudes. The vertical velocity of a constant pressure surface at a 300-km level is also shown. (Richmond, A. D. and Matsushita, S.: *J. Geophys. Res.* **80**, 2839, 1975.)

and Roble (1973) found no change in the ozone layer in low latitudes during magnetic storms. On the other hand, it is interesting to note that there are a few observations which report a reduction of the brightness or fading of noctilucent clouds after intense auroral activity, suggesting a possible auroral-induced heating at or below the mesopause (Fogle, 1966).

References

Afonia, R. G. and Fel'dshteyn, Ya. I.: 1971a, 'Relationship between Geomagnetic Field Variations in the Polar Region in the Morning-Night Sectors of the Auroral Oval', *Geomag. Aeronom.* **11**, 485.

Afonina, R. G. and Fel'dshteyn, Ya. I.: 1971b, 'Intensity Variations of the Easterly and Westerly Polar Electrojets in Universal Time', *Geomag. Aeronom.* **11**, 947.

Akasofu, S.-I.: 1960, 'Large-Scale Auroral Motions and Polar Magnetic Disturbances – I: A Polar Disturbance at about 1100 hours on 23 September, 1957', *J. Atmosph. Terr. Phys.*, **19**, 10.

Akasofu, S.-I.: 1968, *Polar and Magnetospheric Substorms*, D. Reidel Publ. Co., Dordrecht-Holland.

Akasofu, S.-I.: 1974, 'A Study of Auroral Displays Photographed from the DMSP-2 Satellite and from the Alaska Meridian Chain of Stations', *Space Sci. Rev.* **16**, 617.

Akasofu, S.-I. and DeWitt, R. N.: 1965, 'Dynamo Action in the Ionosphere and Motions of the Magnetospheric Plasma, IV The Pedersen Conductivity Generalized to Take Account of Acceleration of the Neutral Gas', *Planet. Space Sci.* **13**, 737.

Akasofu, S.-I. and Meng, C.-I.: 1969, 'A Study of Polar Magnetic Substorms', *J. Geophys. Res.* **74**, 293.

Akasofu, S.-I., Chapman, S. and Meng, C.-I.: 1965, 'Equivalent Current System for Intense Polar Magnetic Substorm', *J. Atmosph. Terr. Phys.* **27**, 1275.

Akasofu, S.-I., Meng, C.-I. and Kimball, D.S.: 1966, 'Dynamics of the Aurora, 4, Polar Magnetic Substorms and Westward Traveling Surges', *J. Atmosph. Terr. Phys.* **28**, 489.

Akasofu, S.-I., Eather, R. H. and Bradbury, J. N.: 1969, 'The Absence of the Hydrogen Emission (Hβ) in the Westward Traveling Surge', *Planet. Space Sci.* **17**, 1409.

Akasofu, S.-I., Wilson, C. R., Synder, L. and Perreault, P.: 1971, 'Results from a Meridian Chain of Observatories in the Alaskan Sector (I)', *Planet. Space Sci.* **19**, 477.

Anderson, A. D.: 1973, 'The Relation between Low-Latitude Neutral Density Variations near 400 km and Magnetic Activity Indices', *Planet. Space Sci.* **21**, 2049.

Arendt, P. R.: 1969, 'Comparison of the Topside Ionosphere during Magnetic Storms of Various Types', *Planet. Space Sci.* **17**, 1993.

Armstrong, J. C.: 1974, 'Field-Aligned Currents in the Magnetosphere', *Magnetospheric Physics*, B. M. McCormac (ed.), p. 155, D. Reidel Publ. Co., Dordrecht-Holland.

Armstrong, J. C. and Zmuda, A. J.: 1973, 'Triaxial Magnetic Measurements of Field-Aligned Currents at 800 Kilometers in the Auroral Region: Initial Results', *J. Geophys. Res.* **78**, 6802.

Atkinson, G.: 1967, 'The Current System of Geomagnetic Bays', *J. Geophys. Res.* **72**, 6063.

Aubry, M. P., Kivelson, M. G., McPherron, R. L., Russell, C. T. and Colburn, D. S.: 1972, 'Outer Magnetosphere near Midnight at Quiet and Disturbed Times', *J. Geophys. Res.* **77**, 5487.

Balsley, B. B. and Ecklund, W. L.: 1972: 'VHF Power Spectra of the Radar Aurora', *J. Geophys. Res.* **77**, 4746.

Balsley, B. B., Ecklund, W. L. and Greenwald, R. A.: 1973, 'VHF Doppler Spectra of Radar Echoes Associated with a Visual Auroral Form: Observations and Implications', *J. Geophys. Res.* **78**, 1681.

Banks, P. M. and Doupnik, J. R.: 1975, 'A Review of Auroral Zone Electrodynamics Deduced from Incoherent Scatter Radar Observations', *J. Atmosph. Terr. Phys.* **37**, 951.

Banks, P. M., Doupnik, J. R. and Akasofu, S.-I.: 1973, 'Electric Field Observations by Incoherent Scatter Radar in the Auroral Zone', *J. Geophys. Res.* **78**, 6607.

Banks, P. M., Rino, C. L. and Wickwar, V. B.: 1974, 'Incoherent Scatter Radar Observations of Westward Electric Fields and Plasma Densities in the Auroral Ionosphere, 1', *J. Geophys. Res.* **79**, 187.

Baron, M. J.: 1974, 'Electron Densities within Aurorae and Other Auroral E Region Characteristics', *Radio Sci.* **9**, 341.

Bates, H. F.: 1973, 'Atmospheric Expansion through Joule Heating by Horizontal Electric Fields', *Planet. Space Sci.* **21**, 2073.

Bates, H. F.: 1974a, 'Atmospheric Expansion from Joule Heating', *Planet. Space Sci.* **22**, 925.

Bates, H. F.: 1974b, 'Thermospheric Changes Shortly after the Onset of Daytime Joule Heating', *Planet. Space Sci.* **22**, 1625.

Bates, H. F. and Hunsucker, R. D.: 1974, 'Quiet and Disturbed Electron Density Profiles in the Auroral Zone Ionosphere,' *Radio Sci.* **9**, 455.

Bates, H. F., Sharp, R. D., Belon, A. E. and Boyd, J. S.: 1969, 'Spatial Relationships between HF Radar Aurora, Optical Aurora and Electron Precipitation', *Planet. Space Sci.* **17**, 83.

Bates, H. F., Akasofu, S.-I., Kimball, D. S. and Hodges, J. C.: 1973, 'First Results from the North Polar Auroral Radar', *J. Geophys. Res.* **78**, 3857.

Bauer, S. J. and Krishnamurthy, B. V.: 1968, 'Behaviour of the Topside Ionosphere during a Great Magnetic Storm', *Planet. Space Sci.* **16**, 653.

Ben'kova, N. P. and Zevakina, R. A.: 1974, 'Ionospheric Disturbance of November 5–12, 1970', *Geomag. Aeronom.* **14**, 392.

Bering, E. A. and Mozer, F. S.: 1975, 'A Measurement of Perpendicular Current Density in an Aurora', *J. Geophys. Res.* **80**, 3961.

Blamont, J. E. and Luton, J. M.: 1972, 'Geomagnetic Effect on the Neutral Temperature of the F Region during the Magnetic Storm of September 1969', *J. Geophys. Res.* **77**, 3534.

Bogott, F. J. and Mozer, F. S.: 1971, 'Equatorial Proton and Electron Angular Distributions in the Loss Cone and at Large Angles', *J. Geophys. Res.* **76**, 6790.

Bonnevier, B., Boström, R. and Rostoker, G.: 1970, 'A Three-Dimensional Model Current System for Polar Magnetic Substorms', *J. Geophys. Res.* **75**, 107.

Boström, R.: 1964, 'A Model of the Auroral Electrojet', *J. Geophys. Res.* **69**, 4983.

Boström, R.: 1969, 'Auroral Current Systems', *Atmospheric Emissions*, B. M. McCormac and A. Omhold (eds.), p. 277, Van Nostrand Reinhold Co., New York.

Brekke, A., Doupnik, J. R. and Banks, P. M.: 1973, 'A Preliminary Study of the Neutral Wind in the Auroral E Region', *J. Geophys. Res.* **78**, 8235.

Brekke, A., Doupnik, J. R. and Banks, P. M.: 1974a, 'Observations of Neutral Winds in the Auroral E Region during the Magnetospheric Storm of August 3–9, 1972', *J. Geophys. Res.* **79**, 2448.

Brekke, A., Doupnik, J. R. and Banks, P. M.: 1974b, 'Incoherent Scatter Measurements of E Region Conductivities and Currents in the Auroral Zone', *J. Geophys. Res.* **79**, 3773.

Cahill, L. J., Potter, W. E., Kintner, P. M., Arnoldy, R. L. and Choy, L. W.: 1974, 'Field-Aligned Currents and the Auroral Electrojet', *J. Geophys. Res.* **79**, 3147.

Carpenter, D. L., Foster, J. C., Rosenberg, T. J. and Lanzerotti, L. J.: 1975, 'A Sub-Auroral and Mid-Latitude View of Substorm Activity', *J. Geophys. Res.* **80**, 4279.

Chandra, S. and Herman, J. R.: 1969, 'F Region Ionization and Heating during Magnetic Storms', *Planet. Space Sci.* **17**, 841.

Chandra, S. and Stubbe, P.: 1971, 'Ion and Neutral Composition Changes in the Thermospheric Region during Magnetospheric Storms', *Planet. Space Sci.* **19**, 491.

Chen, A. J. and Rostoker, G.: 1974, 'Auroral-Polar Currents during Periods of Moderate Magnetospheric Activity', *Planet. Space Sci.* **22**, 1101.

Chimonas, G.: 1970, 'Infrasonic Waves Generated by Auroral Currents', *Planet. Space Sci.* **18**, 591.

Chimonas, G. and Hines, C. O.: 1970, 'Atmospheric Gravity Waves Launched by Auroral Currents', *Planet. Space Sci.* **18**, 565.

Chimonas, G. and Peltier, W. R.: 1970, 'The Bow Wave Generated by an Auroral Arc in Supersonic Motion', *Planet. Space Sci.* **18**, 599.

Ching, B. K. and Rugge, H. R.: 1975, 'Atmospheric Density at 169 km from Accelerometer Measurements and Orbital Decay of a Low Altitude Satellite', *Planet. Space Sci.* **23**, 1301.

Clauer, C. R. and McPherron, R. L.: 1974a, 'Mapping the Local Time-Universal Time Development of Magnetospheric Substorms Using Mid-Latitude Magnetic Observations', *J. Geophys. Res.* **79**, 2811.

Clauer, C. R. and McPherron, R. L.: 1974b, 'Variability of Mid-Latitude Magnetic Parameters Used to Characterize Magnetospheric Substorms', *J. Geophys. Res.* **79**, 2898.

Cole, K. D.: 1962, 'Joule Heating of the Upper Atmosphere', *Aust. J. Phys.* **15**, 223.

Cole, K. D.: 1968, 'Particle Heating Effects on the Upper F Region', *Planet. Space Sci.* **16**, 525.

Cole, K. D.: 1969, 'Theory of Electric Currents in Ionospheric E Layers', *Planet. Space Sci.* **17**, 1977.

Cole, K. D.: 1971a, 'Electrodynamic Heating and Movement of the Thermosphere', *Planet. Space Sci.* **19**, 59.

Cole, K. D.: 1971b, 'Thermospheric Winds Induced by Auroral Electrojet Heating', *Planet. Space Sci.* **19**, 1010.

Coroniti, F. V. and Kennel, C. F.: 1972, 'Polarization of the Auroral Electrojet', *J. Geophys. Res.* **77**, 2835.

Crooker, N. U. and McPherron, R. L.: 1972, 'On the Distinction between the Auroral Electrojet and Partial Ring Current Systems', *J. Geophys. Res.* **77**, 6886.

Cummings, W. D. and Coleman, P. J. Jr.: 1968, 'Simultaneous Magnetic Field Variations at the Earth's Surface and at Synchronous, Equatorial Distance, 1. Bay-Associated Events', *Radio Sci.* **3**, 758.

Cummings, W. D., Barfield, J. N. and Coleman, P. J.: 1968, 'Magnetospheric Substorms Observed at the Synchronous Orbit', *J. Geophys. Res.* **73**, 6687.

D'Angelo, N.: 1973, 'Type III Spectra of the Radar Aurora', *J. Geophys. Res.* **78**, 3987.

Davies, K.: 1974a, 'A Model of Ionospheric F-2 Region Storms in Middle Latitudes', *Planet. Space Sci.* **22**, 237.
Davies, K.: 1974b, 'Studies of Ionospheric Storms Using a Simple Model', *J. Geophys. Res.* **79**, 605.
Davis, M. J.: 1971, 'On Polar Substorms as the Source of Large-Scale Traveling Ionospheric Disturbances', *J. Geophys. Res.* **76**, 4525.
Davis, M. J. and da Rosa, A. V.: 1969, 'Traveling Ionospheric Disturbances Originating in the Auroral Oval during Polar Substorms', *J. Geophys. Res.* **74**, 5721.
Davis, T. N.: 1962a, 'The Morphology of the Auroral Displays of 1957–1958, 1, Statistical Analysis of Alaska Data', *J. Geophys. Res.* **67**, 59.
Davis, T. N.: 1962b, 'The Morphology of the Auroral Displays of 1957–1958, 2, Detailed Analysis of Alaskan Data and Analysis of High-Latitude Data', *J. Geophys. Res.* **67**, 75.
Davydov, V. M.: 1971, 'Characteristics of the Ionospheric Hall Current and of Its Magnetic Field', *Geomag. Aeronom.* **11**, 198.
DeVries, L. L.: 1972, 'Structure and Motion of the Thermosphere Shown by Density Data from the Low-G Accelerometer Calibration System (LOGALS)', *Space Research XIII*, S. A. Bowhill, L. D. Jaffe and M. J. Rycroft (eds.), p. 867, Akademie-Verlag, Berlin.
DeVries, L. L., Schusterman, L. and Bruce, R. W.: 1972, 'Atmospheric Density Variations at 140 Kilometers Reduced from Precise Satellite Radar Tracking Data', *J. Geophys. Res.* **77**, 1905.
DeWitt, R. N.: 1968, 'Polarization of the Auroral Electrojet', *J. Geophys. Res.* **73**, 6307.
Donahue, T. M., Zipf, E. C. and Parkinson, T. D.: 1970, 'Ion Composition and Ion Chemistry in Aurora', *Planet. Space Sci.* **18**, 171.
Doupnik, J. R., Banks, P. M., Baron, M. J., Rino, C. L. and Petriceks, J.: 1972, 'Direct Measurements of Plasma Drift Velocities at High Magnetic Latitudes', *J. Geophys. Res.* **77**, 4268.
Ecklund, W. L., Balsley, B. B. and Greenwald, R. A.: 1973, 'Doppler Spectra of Diffuse Radar Auroras', *J. Geophys. Res.* **78**, 4797.
Ecklund, W. L., Carter, D. A., Keys, J. G. and Unwin, R. S.: 1974, 'Conjugate Auroral Radar Observations of a Substorm', *J. Geophys. Res.* **79**, 3211.
Ecklund, W. L., Balsley, B. B. and Greenwald, R. A.: 1975, 'Crossed Beam Measurements of the Diffuse Radar Aurora', *J. Geophys. Res.* **80**, 1805.
Evans, J. V.: 1970a, 'Mid-Latitude Ionospheric Temperatures during Three Magnetic Storms in 1965', *J. Geophys. Res.* **75**, 4803.
Evans, J. V.: 1970b, 'F Region Heating Observed during the Main Phase of Magnetic Storms', *J. Geophys. Res.* **75**, 4815.
Evans, J. V.: 1972, 'Measurements of Horizontal Drifts in the E and F Regions at Millstone Hill', *J. Geophys. Res.* **77**, 2341.
Evans, J. V.: 1973, 'The Causes of Storm-Time Increases of the F Layer at Mid-Latitudes', *J. Atmosph. Terr. Phys.* **35**, 593.
Fairfield, D. H.: 1973, 'Magnetic Field Signatures of Substorms on High-Latitude Field Lines in the Nighttime Magnetosphere', *J. Geophys. Res.* **78**, 1553.
Fairfield, D. H. and Ness, N. F.: 1970, 'Configuration of the Geomagnetic Tail during Substorms', *J. Geophys. Res.* **75**, 7032.
Fedder, J. A. and Banks, P. M.: 1972, 'Convection Electric Fields and Polar Thermospheric Winds', *J. Geophys. Res.* **77**, 2328.
Feldstein, Y. I. and Zaitzev, A. N.: 1968, 'Quiet and Disturbed Solar-Daily Variations of Magnetic Field at High Latitudes during the IGY', *Tellus* **20**, 338.
Fogle, B. T.: 1966, 'Noctilucent Clouds', Ph. D. Thesis, University of Alaska, May.
Forbes, J. M. and Marcos, F. A.: 1973, 'Thermospheric Density Variations Associated with Auroral Electrojet Activity', *J. Geophys. Res.* **78**, 3841.
Francis, S. H.: 1974, 'A Theory of Medium-Scale Traveling Ionospheric Disturbances, *J. Geophys. Res.* **79**, 5245.
Francis, S. H.: 1975, 'Global Propagation of Atmospheric Gravity Waves: A Review', *J. Atmosph. Terr. Phys.* **37**, 1011.
Fukushima, N.: 1953, 'Polar Magnetic Storms and Geomagnetic Bays', *J. Fac. Sci. Univ. Tokyo, Sect.* **2**, **8**, 292.
Fukushima, N.: 1968, 'Three-Dimensional Electric Current and Toroidal Magnetic Field in the Ionosphere', *Rep. Ionosph. Space Res. Japan* **22**, 173.
Fukushima, N.: 1969a, 'Spatial Extent of the Return Current of the Auroral-Zone Electrojet, Part 1', *Rep. Ionosph. Space Res. Japan* **23**, 209.
Fukushima, N.: 1969b, 'Equivalence in Ground Geomagnetic Effect of Chapman-Vestine's and Birkeland-Alfvén's Electric Current-Systems for Polar Magnetic Storms', *Rep. Ionosph. Space Res. Japan* **23**, 219.

Fukushima, N.: 1972, 'Polar Magnetic Substorms', *Planet. Space Sci.* **20**, 1443.
Fukushima, N.: 1971, 'Electric Current Systems for Polar Substance and Their Magnetic Effect below and above the Ionosphere', *Radio Sci.* **6**, 289.
Fukushima, N.: 1974a, 'Equivalent Current Pattern for Ground Geomagnetic Effect when the Ionospheric Conductivity is Discontinuous at the Foot of a Field-Aligned Current', *Rep. Ionosph. Space Res. Japan* **28**, 139.
Fukushima, N.: 1974b, 'Equivalent Current Pattern for Ground Geomagnetic Effect when the Ionospheric Conductivity is Discontinuous at the Foot of Field-Aligned Current Sheet', *Rep. Ionosph. Space Res. Japan* **28**, 147.
Fukushima, N. and Kamide, Y.: 1973a, 'Partial Ring Current Models for World-Wide Geomagnetic Disturbances', *Rev. Geophys. Space Phys.* **11**, 795.
Fukushima, N. and Kamide, Y.: 1973b, 'Contribution to Low-Latitude Geomagnetic DS(H) from Field-Aligned Currents in the Magnetosphere', *Rep. Ionosph. Space Res. Japan* **27**, 57.
Fukushima, N. and Kamide, Y.: 1973c, 'Contributions of Magnetospheric Field-Aligned Current to Geomagnetic Bays and S_q^p Fields: A Comment on Partial Ring-Current Models', *Radio Sci.* **8**, 1013.
Galperin, Y. I., Ponomarev, V. N. and Zosimova, A. G.: 1974, 'Plasma Convection in Polar Ionosphere', *Ann. de Geophys.* **30**, 1.
Greenwald, R. A.: 1974, 'Diffuse Radar Aurora and the Gradient Drift Instability', *J. Geophys. Res.* **79**, 4807.
Greenwald, R. A. and Ecklund, W. L.: 1975, 'A New Look at Radar Auroral Motions', *J. Geophys. Res.* **80**, 3642.
Greenwald, R. A., Ecklund, W. L. and Balsley, B. B.: 1973, 'Auroral Currents, Irregularities and Luminosity', *J. Geophys. Res.* **78**, 8193.
Greenwald, R. A., Ecklund, W. L. and Balsley, B. B.: 1975a, 'Radar Observations of Auroral Electrojet Currents', *J. Geophys. Res.* **80**, 3635.
Greenwald, R. A., Ecklund, W. L. and Balsley, B. B.: 1975b, 'Diffuse Radar Aurora: Spectral Observations of Non-Two-Stream Irregularities', *J. Geophys. Res.* **80**, 131.
Gurnett, D. A.: 1970, 'Satellite Measurements of DC Electric Fields in the Ionosphere', *Particles and Fields in the Magnetosphere*, B. M. McCormac (ed.), p. 239, D. Reidel Publ. Co., Dordrecht-Holland.
Gurnett, D. A.: 1972a, 'Electric Field and Plasma Observations in the Magnetosphere', *Critical Problems of Magnetospheric Physics*, Proceedings of the Joint COSPAR/IAGA/URSI Symposium, Madrid, Spain, 11–13 May, 1972, E. R. Dyer (ed.), p. 123, National Academy of Sciences, Washington, D.C.
Gurnett, D. A.: 1972b, 'INJUN-5 Observations of Magnetospheric Electric Fields and Plasma Convection', *Earth's Magnetospheric Processes*, B. M. McCormac (ed.), p. 233, D. Reidel Publ. Co., Dordrecht-Holland.
Haerendel, G.: 1972, 'Plasma Drifts in the Auroral Ionosphere Derived from Barium Releases', *Earth's Magnetospheric Processes*, B. M. McCormac (ed.), p. 246, D. Reidel Publ. Co., Dordrecht-Holland.
Haerendel, G. and Lüst, R.: 1970, 'Electric Fields in the Ionosphere and Magnetosphere', *Particles and Fields in the Magnetosphere*, B. M. McCormac (ed.), p. 213, D. Reidel Publ. Co., Dordrecht-Holland.
Haerendel, G., Hedgecock, P. C. and Akasofu, S.-I.: 1971, 'Evidence of Magnetic Field Aligned Currents during the Substorms on March 18, 1969', *J. Geophys. Res.* **76**, 2382.
Hagfors, T., Johnson, R. G. and Power, R. A.: 1971, 'Simultaneous Observation of Proton Precipitation and Auroral Radar Echoes', *J. Geophys. Res.* **76**, 6093.
Harang, L.: 1946, 'The Mean Field Disturbance of Polar Geomagnetic Storms', *Terr. Mag. Atmosph. Elec.* **51**, 353.
Harper, R. M.: 1972, 'Observation of a Large Nightime Gravity Wave at Arecibo', *J. Geophys. Res.* **77**, 1311.
Hays, P. B. and Roble, R. G.: 1971, 'Direct Observations of Thermospheric Winds during Geomagnetic Storms', *J. Geophys. Res.* **76**, 5316.
Hays, P. B. and Roble, R. G.: 1973, 'Observation of Mesospheric Ozone at Low Latitudes', *Planet. Space Sci.* **21**, 273.
Hays, P. B., Nagy, A. F. and Roble, R. G.: 1969, 'Interferometric Measurements of the 6300 A Doppler Temperature during a Magnetic Storm', *J. Geophys. Res.* **74**, 4162.
Hays, P. B., Jones, R. A. and Rees, M. H.: 1973, 'Auroral Heating and the Composition of the Neutral Atmosphere', *Planet. Space Sci.* **21**, 559.
Heaps, M. G.: 1974, 'The Effects of Including the Coriolis Force of Joule Dissipation in the Upper Atmosphere', *Planet. Space Sci.* **22**, 1031.
Heaps, M. G. and Megill, L. R.: 1975, 'Circulation in the High-Latitude Thermosphere Due to Electric Fields and Joule Heating', *J. Geophys. Res.* **80**, 1829.

Hedin, A. E. and Reber, C. A.: 1972, 'Longitudinal Variations of Thermospheric Composition Indicating Magnetic Control of Polar Heat Input', *J. Geophys. Res.* **77**, 2871.

Heppner, J. P.: 1954, 'Time Sequences and Spatial Relations in Auroral Activity during Magnetic Bays at College, Alaska', *J. Geophys. Res.* **59**, 329.

Heppner, J. P.: 1969, 'Magnetospheric Convection Patterns Inferred from High-Latitude Activity', *Atmospheric Emissions*, B. M. McCormac and A. Omholt (eds.), p. 251, Van Nostrand Reinhold, New York.

Heppner, J. P.: 1972a, 'Electric Field Variations during Substorms: OGO-6 Measurements', *Planet. Space Sci.* **20**, 1475.

Heppner, J. P.: 1972b, 'The Harang Discontinuity in Auroral Belt Ionospheric Currents', *Geofys. Pub.* **29**, 105.

Heppner, J. P.: 1972c, 'Electric Fields in the Magnetosphere', Goddard Space Flight Center, *X-645-72-154*, May.

Heppner, J. P., Stolarik, J. D. and Wescott, E. M.: 1971, 'Electric-Field Measurements and the Identification of Currents Causing Magnetic Disturbances in the Polar Cap', *J. Geophys. Res.* **76**, 6028.

Hines, C. O.: 1960, 'Internal Atmospheric Gravity Waves at Ionospheric Heights', *Canad. J. Phys.* **38**, 1441.

Hones, E. W. Jr.: 1963, 'Motions of Charged Particles Trapped in the Earth's Magnetosphere', *J. Geophys. Res.* **68**, 1209.

Hook, J. L.: 1970, 'Winds at the 75–110 km Level at College, Alaska', *Planet. Space Sci.* **18**, 1623.

Horning, B. L., McPherron, R. L. and Jackson, D. D.: 1974, 'Application of Linear Inverse Theory to a Line Current Model of Substorm Current Systems', *J. Geophys. Res.* **79**, 5202.

Hunsucker, R. D. and Tveten, L. H.: 1967, 'Large Traveling-Ionospheric Disturbances Observed at Midlatitudes Utilizing the High Resolution HF Backscatter Technique', *J. Atmosph. Terr. Phys.* **29**, 909.

Hunsucker, R. D., Romick, G. R., Ecklund, W. L., Greenwald, R. A., Balsley, B. B. and Tsunoda, T. R.: 1975, 'Structure and Dynamics of Ionization and Auroral Luminosity during the Auroral Events of March 16, 1972 near Chatanika, Alaska', *Radio Sci.* **10**, 813.

Iijima, T.: 1973, 'Enhancement of the S_q^p Field as the Basic Component of Polar Magnetic Disturbances', *Rep. Ionosph. Space Res. Japan* **27**, 199.

Iwasaki, N. and Nishida, A.: 1967, 'Ionospheric Current System Produced by an External Electric Field in the Polar Cap', *Ionosph. Space Res. Japan* **21**, 17.

Jacchia, L. G.: 1971, 'Revised Statistical Models of the Thermosphere and Exosphere with Empirical Temperature Profiles', Spec. Rep. 332, Smithsonian Astrophys. Obs. Cambridge, Mass.

Jacchia, L. G. and Slowey, J.: 1964, 'Atmospheric Heating in the Auroral Zones: A Preliminary Analysis of the Atmospheric Drag of the INJUN-3 Satellite', *J. Geophys. Res.* **69**, 905.

Jacchia, L. G., Slowey, J. and Verniani, F.: 1967, 'Geomagnetic Perturbation and Upper-Atmosphere Heating', *J. Geophys. Res.* **72**, 1423.

Jani, K. G. and Kotadia, K. M.: 1969, 'Changes in the Ionospheric F2-Layer at a Place near the S_q^p Current Focus during Geomagnetic Storms', *Planet. Space Sci.* **17**, 1949.

Jeffries, R. A., Roach, W. H., Hones, E. W. Jr., Wescott, E. M., Stenbaek-Nielsen, H. C., Davis, T. N. and Winningham, J. D.: 1975, 'Two Barium Plasma Injections into the Northern Magnetospheric Cleft', *Geophys. Res. Lett.* **2**, 285.

Johnson, R. E.: 1972, 'An Infrasonic Pressure Disturbance Study of Two Polar Substorms', *Planet. Space Sci.* **20**, 313.

Jones, K. L.: 1973, 'Wind, Electric Field and Composition Perturbations of the Mid-Latitude F-Region during Magnetic Storms', *J. Atmos. Terr. Phys.* **35**, 1515.

Kamide, Y.: 1970, 'Spatial Extent of the Return Current of the Auroral-Zone Electrojet, Part III', *Rep. Ionosph. Space Res. Japan* **24**, 347.

Kamide, Y. and Akasofu, S.-I.: 1974, 'Latitudinal Cross Section of the Auroral Electrojet and Its Relation to the Interplanetary Magnetic Field Polarity', *J. Geophys. Res.* **79**, 3755.

Kamide, Y. and Akasofu, S.-I.: 1975, 'The Auroral Electrojet and Global Auroral Features', *J. Geophys. Res.* **80**, 3585.

Kamide, Y. and Akasofu, S.-I.: 1976, 'The Auroral Electrojet and Field-Aligned Current', *Planet. Space Sci.* **24**, 203.

Kamide, Y. and Brekke, A.: 1975, 'Auroral Electrojet Current Density Deduced from the Chatanika Radar and from the Alaska Meridian Chain of Magnetic Observatories', *J. Geophys. Res.* **80**, 587.

Kamide, Y. and Fukunishi, H.: 1973, 'Simultaneous Growth of High-Latitude Positive Bay and DR-Field in the Course of Proton Aurora Substorm', *J. Atmos. Terr. Phys.* **35**, 2085.

Kamide, Y. and Fukushima, N.: 1970, 'Spatial Extent of the Return Current of the Auroral-Zone Electrojet, Part II', *Rep. Ionosph. Res. Japan* **24**, 115.

Kamide, Y. and Fukushima, N.: 1971, 'Analysis of Magnetic Storms with DR-Indices for Equatorial Ring Current Field', *Rep. Ionosph. Space Res. Japan* **25**, 125.

Kamide, Y. and Fukushima, N.: 1972, 'Positive Geomagnetic Bays in Evening High-Latitudes and Their Possible Connection with Partial Ring Current', *Rep. Ionosph. Space Res. Japan* **26**, 79.

Kamide, Y., Yasuhara, F. and Akasofu, S.-I.: 1974, 'On the Cause of Northward Magnetic Field along the Negative X Axis during Magnetospheric Substorms', *Planet. Space Sci.* **22**, 1219.

Kamide, Y., Akasofu, S.-I. and Rostoker, G.: 1976a, 'Field-Aligned Currents and the Auroral Electrojet in the Morning Sector', *J. Geophys. Res.* **81** (in press).

Kamide, Y., Akasofu, S.-I. and Brekke, A.: 1976b, 'Ionospheric Currents Obtained from the Chatanika Radar and Ground Magnetic Perturbations at the Auroral Latitude', *Planet. Space Sci.* **24**, 193.

Kawasaki, K. and Akasofu, S.-I.: 1971, 'The Computed Distribution of Geomagnetic Disturbance Vectors for Model Three-Dimensional Current Systems', *Planet. Space Sci.* **19**, 543.

Kawasaki, K., Fukushima, N. and Kamide, Y.: 1974a, 'Time Variation of Low Latitude Geomagnetic Bay with Three-Dimensional Substorm Current Development', *Rep. Ionosph. Space Res. Japan* **28**, 77.

Kawasaki, K., Meng, C.-I. and Kamide, Y.: 1974b, 'The Development of the Three-Dimensional Current System during a Magnetospheric Substorm', *Planet. Space Sci.* **22**, 1471.

Kelley, M. C. and Mozer, F. S.: 1973, 'Electric Field and Plasma Density Oscillations Due to the High-Frequency Hall Current Two-Stream Instability in the Auroral E Region', *J. Geophys. Res.* **78**, 2214.

Kelley, M. C. and Mozer, F. S.: 1975, 'Simultaneous Measurement of the Horizontal Components of the Earth's Electric Field in the Atmosphere and in the Ionosphere', *J. Geophys. Res.* **80**, 3275.

Kelley, M. C., Mozer, F. S. and Fahleson, U. V.: 1971a, 'Electric Fields in the Night-Time and Daytime Auroral Zone', *J. Geophys. Res.* **76**, 6054.

Kelley, M. C., Starr, J. A. and Mozer, F. S.: 1971b, 'Relationship between Magnetospheric Electric Fields and the Motion of Auroral Forms', *J. Geophys. Res.* **76**, 5269.

Kelley, M. C., Haerendel, G., Kappler, H., Mozer, F. S. and Fahleson, U. V.: 1975, 'Electric Field Measurements in a Major Magnetospheric Substorm', *J. Geophys. Res.* **80**, 3181.

Kisabeth, J. L.: 1972, 'The Dynamical Development of the Polar Electrojets', Ph.D. Thesis, Univ. of Alberta.

Kisabeth, J. L. and Rostoker, G.: 1971, 'Development of the Polar Electrojet during Polar Magnetic Substorms', *J. Geophys. Res.* **76**, 6815.

Kisabeth, J. L. and Rostoker, G.: 1973, 'Current Flow in Auroral Loops and Surges Inferred from Ground-Based Magnetic Observations', *J. Geophys. Res.* **78**, 5573.

Kisabeth, J. L. and Rostoker, G.: 1974, 'The Expansive Phase of Magnetospheric Substorms, 1, Development of the Auroral Electrojets and Auroral Arc Configuration during a Substorm', *J. Geophys. Res.* **79**, 972.

Lakshmi, D. R. and Reddy, B. M.: 1970, 'Nighttime Response of the Topside Ionosphere to Magnetic Storms', *J. Geophys. Res.* **75**, 4335.

Lange-Hesse, G.: 1967, 'Radio Aurora, I. Observations, II. Comparison of the Observations with a Theoretical Model', *Aurora and Airglow*, B. M. McCormac (ed.), p. 519, Reinhold Publishing Co., New York.

Lange-Hesse, G.: 1969, 'Radio Observations of the Aurora by Continuous Wave Transmissions', *Atmospheric Emissions*, B. M. McCormac and A. Omholt (ed.), p. 201, Van Nostrand Reinhold Co., New York.

Lange-Hesse, G.: 'Analogies between Substorm Phenomena in Visual and Radio Aurora', *Kleinbeubacher Berichte* **16**, 141–152, issued by Fernmeldetechnisches Zentralamt, D 33a Darmstadt, W. Germany.

Lennartsson, O. W.: 1973, 'Ionospheric Electric Field and Current Distribution Associated with Altitude Electric Field Inhomogeneities', *Planet. Space Sci.* **21**, 2089.

Leont'yev, S. V., Lyatskiy, V. B. and Mal'tsev, Yu. P.: 1974, 'DP2 Current System and Its Place in the Development of a Substorm', *Geomag. Aeronom.* **14**, 91.

Lyatskiy, V. B. and Mal'tsev, Yu. P.: 1971, 'Model of a Magnetospheric Substorm', *Geomag. Aeronom.* **11**, 711.

Lyatskiy, V. B., Maltsev, Yu. P. and Leont'yev, S. V.: 1974, 'Three-Dimensional Current System in Different Phases of a Substorm', *Planet. Space Sci.* **22**, 1231.

Maeda, K. and Aikin, A. C.: 1968, 'Variations of Polar Mesospheric Oxygen and Ozone during Auroral Events', *Planet. Space Sci.* **16**, 71.

Maeda, H. and Maekawa, K.: 1973, 'A Numerical Study of Polar Ionospheric Currents', *Planet. Space Sci.* **21**, 1287.

Malingre, M.: 1973, 'De Focalisation des Ondes TBF dans l'Ionosphere Superieure an Voisinage de la de'Pression de Moyenne Latitude', *Ann. Geophys.* **29**, 239.

Mal'tsev, Yu. P. and Lyatsky, W. B.: 1975, 'Field-Aligned Currents and Erosion of the Dayside Magnetosphere', *Planet. Space Sci.* **23**, 1257.

Mal'tsev, Yu. P., Leont'yev, S. V. and Lyatsky, V. B.: 1973, 'Electric Fields and Currents in the Boundary Region of Electrojets', *Geomag. Aeronom.* **13**, 910.

Matsushita, S.: 1959, 'A Study of the Morphology of Ionospheric Storms', *J. Geophys. Res.* **64**, 305.

Matuura, N.: 1972, 'Theoretical Models of Ionospheric Storms', *Space Sci. Rev.* **13**, 124.

Maynard, N. C.: 1974, 'Electric Field Measurements across the Harang Discontinuity', *J. Geophys. Res.* **79**, 4620.

Maynard, N. C. and Heppner, J. P.: 1970, 'Variations in Electric Fields from Polar Orbiting Satellites', *Particles and Fields in the Magnetosphere*, B. M. McCormac (ed.), p. 247, D. Reidel Publ. Co., Dordrecht-Holland.

Maynard, N. C., Bahnsen, A., Christophersen, P., Egeland, A. and Lundin, R.: 1973, 'An Example of Anticorrelation of Auroral Particles and Electric Fields', *J. Geophys. Res.* **78**, 3976.

Mayr, H. G. and Volland, H.: 1972, 'Magnetic Storm Effects in the Neutral Composition', *Planet. Space Sci.* **20**, 379.

Mayr, H. G. and Volland, H.: 1973, 'Magnetic Storm Characteristics of the Thermosphere', *J. Geophys. Res.* **78**, 2251.

McDiarmid, D. R. and McNamara, A. G.: 1972, 'Periodically Varying Radio Aurora', *Ann. Geophys.* **28**, 433.

McNamara, A. G.: 1972, 'The Occurrence of Radio Aurora at High Latitudes: The IGY Period, 1957–1959', *Geofysiske Publik.* **29**, 135.

McPherron, R. L., Russell, C. T., Kivelson, M. G. and Coleman, P. J. Jr.: 1973, 'Substorms in Space: The Correlation between Ground and Satellite Observations of the Magnetic Field', *Radio Sci.* **8**, 1059.

Meek, J. H.: 1953, 'Correlation of Magnetic Auroral and Ionospheric Variations at Saskatoon', *J. Geophys. Res.* **58**, 445.

Mendillo, M.: 1971, 'Ionospheric Total Electron Content Behavior during Geomagnetic Storms', *Nature* **234**, 23.

Mendillo, M., Papagiannis, M. D. and Klobuchar, J. A.: 1972, 'Average Behavior of the Mid-Latitude F-Region Parameters N_T, N_{max} and z during Geomagnetic Storms', *J. Geophys. Res.* **77**, 4891.

Mendillo, M., Klobuchar, J. A. and Hajeb-Hosseinieh, H.: 1974, 'Ionospheric Disturbances: Evidence for the Contraction of the Plasmapause during Severe Geomagnetic Storms', *Planet. Space Sci.* **22**, 223.

Meng, C.-I. and Akasofu, S.-I.: 1969, 'A Study of Polar Magnetic Substorms, 2. Three-Dimensional Current System', *J. Geophys. Res.* **74**, 4035.

Meng, C.-I., Akasofu, S.-I., Hones, E. W. Jr. and Kawasaki, K.: 1971, 'Magnetospheric Substorms in the Distant Magnetotail Observed by IMP-3', *J. Geophys. Res.* **76**, 7584.

Meriwether, J. W., Heppner, J. P., Stolarik, J. D. and Wescott, E. M.: 1973, 'Neutral Winds above 200 km at High Latitudes', *J. Geophys. Res.* **78**, 6643.

Mikkelsen, I. S., Jørgensen, T. S. and Kelley, M. C.: 1975, 'Observation and Interpretation of Plasma Motions in the Polar Cap Ionosphere during Magnetic Substorms', *J. Geophys. Res.* **80**, 3197.

Moorcroft, D. R.: 1972, 'Vertical Gradients in the Theory of Radio Aurora', *J. Geophys. Res.* **77**, 769.

Mozer, F. S.: 1971, 'Origin and Effects of Electric Fields during Isolated Magnetospheric Substorms', *J. Geophys. Res.* **76**, 7595.

Mozer, F. S.: 1972, 'Simultaneous Electric-Field Measurements on Nearby Balloons', *J. Geophys. Res.* **77**, 6129.

Mozer, F. S.: 1973a, 'Analysis of Techniques for Measuring DC and AC Electric Fields in the Magnetosphere', *Space Sci. Rev.* **14**, 272.

Mozer, F. S.: 1973b, 'On the Relationship between the Growth and Expansion Phases of Substorms and Magnetospheric Convection', *J. Geophys. Res.* **78**, 1719.

Mozer, F. S.: 1973c, 'Electric Fields and Plasma Convection in the Plasmasphere', *Rev. Geophys. Space Phys.* **11**, 755.

Mozer, F. S. and Fahleson, U. V.: 1970, 'Parallel and Perpendicular Electric Fields in an Aurora', *Planet. Space Sci.* **18**, 1563.

Mozer, F. S. and Lucht, P.: 1974, 'The Average Auroral Zone Electric Field', *J. Geophys. Res.* **79**, 1001.

Mozer, F. S. and Manka, R. H.: 1971, 'Magnetospheric Electric Field Properties Deduced from Simultaneous Balloon Flights', *J. Geophys. Res.* **76**, 1697.

Mozer, F. S. and Serlin, R.: 1969, 'Magnetospheric Electric Field Measurements with Balloons', *J. Geophys. Res.* **74**, 4739.

Mozer, F. S., Bogott, F. H. and Tsurutani, B.: 1973, 'Relations between Ionospheric Electric Fields and Energetic Trapped and Precipitated Electrons', *J. Geophys. Res.* **78**, 630.
Mozer, F. S., Serlin, R., Carpenter, D. L. and Siren, J.: 1974, 'Simultaneous Electric Field Measurements near $L = 4$ from Conjugate Balloons and Whistlers', *J. Geophys. Res.* **79**, 3215.
Murphy, C. H., Wang, C. S. and Kim, J. S.: 1974, 'Inductive Electric Field of a Field-Aligned Current System', *J. Geophys. Res.* **79**, 2901.
Nishida, A.: 1975, 'Interplanetary Field Effect on the Magnetosphere', *Space Sci. Rev.* **17**, 353.
Nopper, R. W. Jr. and Hermance, J. F.: 1974, 'Phase Relations between Polar Magnetic Substorm Fields at the Surface of a Finitely Conducting Earth', *J. Geophys. Res.* **79**, 4799.
Obayashi, T.: 1972, 'World-Wide Electron Density Changes and Associated Thermospheric Winds during an Ionospheric Storm', *Planet. Space Sci.* **20**, 511.
Olesen, J. K., Primdahl, F., Spangslev, F. and D'Angelo, N.: 1975, 'On the Farley Instability in the Polar Cap E Region', *J. Geophys. Res.* **80**, 696.
Osipov, N. K., Pivovarov, V. G. and Pivovarova, N. B.: 1971, 'Magnetic Disturbances and Auroral Luminosity', *Geomag. Aeronom.* **11**, 487.
Ossakow, S. L., Papadopoulos, K., Orens, J. and Coffey, T.: 1975, 'Parallel Propagation Effects on the Type I Electrojet Instability', *J. Geophys. Res.* **80**, 141.
Papagiannis, M. D. and Mendillo, M.: 1971, 'Simultaneous Storm-Time Increases of the Ionospheric Total Electric Content and the Geomagnetic Field in the Dusk Sector', *Planet. Space Sci.* **19**, 503.
Park, C. G.: 1971, 'Westward Electric Fields as the Cause of Nighttime Enhancements in Electron Concentrations in Midlatitude F-Region', *J. Geophys. Res.* **76**, 4560.
Park, C. G.: 1974, 'A Morphological Study of Substorm-Associated Disturbances in the Ionosphere', *J. Geophys. Res.* **79**, 2821.
Park, C. G. and Banks, P. M.: 1974, 'Influence of Thermal Plasma Flow on the Mid-Latitude Nighttime F2 Layer: Effects of Electric Fields and Neutral Winds Inside the Plasmasphere', *J. Geophys. Res.* **79**, 4661.
Park, D.: 1974, 'Magnetic Field at the Earth's Surface Produced by a Horizontal Line Current', *J. Geophys. Res.* **79**, 4802.
Potter, W. E.: 1970, 'Rocket Measurements of Auroral Electric and Magnetic Fields', *J. Geophys. Res.* **75**, 5415.
Potter, W. E. and Cahill, L. J. Jr.: 1969, 'Electric and Magnetic Field Measurements near an Auroral Electrojet', *J. Geophys. Res.* **74**, 5159.
Prasad, S. A. and Furman, D. R.: 1975, 'Auroral Electric Field Heating: Its Implications in the Altitude Profile of Ionic Composition and Emissions', *J. Geophys. Res.* **80**, 3701.
Primdahl, F., Olesen, J. K. and Spangslev, F.: 1974, 'Backscatter from a Postulated Plasma Instability in the Polar Cap Ionosphere and the Direct Measurement of a Horizontal E Region Current', *J. Geophys. Res.* **79**, 4262.
Prölss, G. W. and von Zahn, V.: 1974, 'Esro 4 Gas Analyzer Results, 2. Direct Measurements of Changes in the Neutral Composition during an Ionospheric Storm', *J. Geophys. Res.* **79**, 2535.
Pudovkin, M. I.: 1974, 'Electric Fields and Currents in the Ionosphere', *Space Sci. Rev.* **16**, 727.
Raitt, W. J.: 1974, 'The Temporal and Spatial Development of Mid-Latitude Thermospheric Electron Temperature Enhancements during a Geomagnetic Storm', *J. Geophys. Res.* **79**, 4703.
Raitt, W. J., Dorling, E. B., Sheather, P. H. and Blades, J.: 1975, 'Ionospheric Measurements from the ESRO-4 Satellite', *Planet. Space Sci.* **23**, 1085.
Rao, N. N.: 1975, 'A Large-Scale Traveling Ionospheric Disturbance of Polar Origin', *Planet. Space Sci.* **23**, 381.
Raspopov, O. M. and Roldugin, V. K.: 1972, 'Regular Auroral and Magnetic Field Intensity Fluctuations', *Geomag, Aeronom.* **12**, 502.
Reber, C. A. and Hedin, A. E.: 1974, 'Heating of the High-Latitude Thermosphere during Magnetically Quiet Periods', *J. Geophys. Res.* **79**, 2457.
Reddy, C. A.: 1974, 'Evidence of a Meridional Circulation Cell in the Lower-Thermosphere during a Magnetic Storm', *Atmosph. Terr. Phys.* **36**, 156.
Rees, D.: 1972, 'Winds and Temperatures in the Auroral Zone and Their Relationship with Geomagnetic Activity', *Phil. Trans. Roy. Soc. London, Ser. A.* **271**, 563.
Rees, D.: 1973, 'Neutral Wind Structure in the Thermosphere during Quiet and Disturbed Geomagnetic Periods', *Physics and Chemistry of Upper Atmospheres*, B. M. McCormac (ed.), p. 11, D. Reidel Publ. Co., Dordrecht-Holland.
Rees, M. H.: 1975a, 'Magnetospheric Substorm Energy Dissipation in the Atmosphere', *Planet. Space Sci.* **23**, 1589.
Rees, M. H.: 1975b, 'Processes and Emissions Associated with Electron Precipation', *Atmosphere of Earth and Planets*, B. M. McCormac (ed.), p. 323, D. Reidel Publ. Co. Dordrecht-Holland.

Richmond, A. D. and Matsushita, S.: 1975, 'Thermospheric Response to a Magnetic Substorm', *J. Geophys. Res.* **80**, 2839.

Rieger, E.: 1974, 'Neutral Air Motions Deduced from Barium Release Experiments, I. Vertical Winds', *J. Atmosph. Terr. Phys.* **36**, 1377.

Rino, C. L., Wickwar, V. B., Banks, P. M., Akasofu, S.-I. and Rieger, E.: 1974, 'Incoherent Scatter Radar Observations of Westward Electric Fields, 2', *J. Geophys. Res.* **79**, 4669.

Roemer, M.: 1971, 'Geomagnetic Activity Effect on Atmospheric Density in the 250 to 800 km Altitude Region', *Space Research XII*, S. A. Bowhill, L. D. Jaffe and M. J. Rycroft (eds.), p. 965, Akademie-Verlag, Berlin.

Romick, G. J., Ecklund, W. L., Greenwald, R. A., Balsley, B. B. and Imhof, W. L.: 1974, 'The Interrelationship between the > 130 keV Electron Trapping Boundary, the VHF Radar Backscatter, and the Visual Aurora', *J. Geophys. Res.* **79**, 2439.

Rostoker, G.: 1968, 'Macrostructure of Geomagnetic Bays', *J. Geophys. Res.* **73**, 4217.

Rostoker, G.: 1972, 'Interpretation of Magnetic Field Variations during Substorms', *Earth's Magnetospheric Processes*, B. M. McCormac (ed.), p. 379, D. Reidel Publ. Co., Dordrecht-Holland.

Rostoker, G. and Hron, M.: 1975, 'The Eastward Electrojet in the Dawn Sector', *Planet. Space Sci.* **23**, 1377.

Rostoker, G. and Kisabeth, J. L.: 1973, 'Responses of the Polar Electrojets in the Evening Sector to Polar Magnetic Substorms', *J. Geophys. Res.* **78**, 5559.

Rostoker, G., Anderson, C. W. III, Oldenburg, D. W., Camfield, P. A., Gough, D. I. and Porath, H.: 1970, 'Development of a Polar Magnetic Substorm Current System', *J. Geophys. Res.* **75**, 6318.

Rostoker, G. Kisabeth, J. L., Sharp, R. D. and Shelley, E. G.: 1975, 'The Expansive Phase of Magnetospheric Substorms, 2, The Response at Synchronous Altitude of Particles of Different Energy Range', *J. Geophys. Res.* **80**, 3557.

Rothwell, P., Mountford, R. and Martelli, G.: 1974, 'Neutral Wind Modifications above 150 km Altitude Associated with the Polar Substorm', *J. Atmosph. Terr. Phys.* **36**, 1915.

Shaftan, V. A. and Voshchina, L. K.: 1974, 'Irregular Geomagnetic Pulsations and Radar Auroras as a Result of the Turbulence of the Polar Electrojet', *Geomag. Aeronom.* **14**, 262.

Shipstone, D. M.: 1969, 'Morphology of the Southern Radio Auroral Zone', *Planet. Space Sci.* **17**, 403.

Silsbee, H. C. and Vestine, E. H.: 1942, 'Geomagnetic Bays, Their Frequency and Current Systems', *J. Geophys. Res. Terr. Mag. Atmosph. Elec.* **47**, 195.

Skadron, G. and Weinstock, J.: 1969, 'Nonlinear Stabilization of Two-Stream Plasma Instability in the Ionosphere', *J. Geophys. Res.* **74**, 5113.

Snyder, A. L., Akasofu, S.-I. and Davis, T. N.: 1974, 'Auroral Substorms Observed from above the North Polar Region by a Satellite', *J. Geophys. Res.* **79**, 1393.

Sofko, G. J. and Kavadas, A.: 1971, 'Polarization Characteristics of 42-MHz Auroral Backscatter', *J. Geophys. Res.* **76**, 1778.

Spurling, P. H. and Jones, K. L.: 1973, 'The Nature of Seasonal Changes in the Effects of Magnetic Storms on Mid-Latitude F-Layer Electron Concentration', *J. Atmosph. Terr. Phys.* **35**, 921.

Sugiura, M.: 1975, 'Identifications of the Polar Cap Boundary and the Auroral Belt in the High-Altitude Magnetosphere: A Model for Field-Aligned Currents', *J. Geophys. Res.* **80**, 2057.

Swift, D. W.: 1973 'The Generation of Infrasonic Waves by Auroral Electrojets', *J. Geophys. Res.* **78**, 8205.

Taeusch, D. R., Carigann, G. R. and Reber, C. A.: 1971, 'Neutral Composition Variation above 400 Kilometers during a Magnetic Storm', *J. Geophys. Res.* **76**, 8318.

Tashikinova, L. G. and Tverskoy, B. A.: 1973, 'Effect of Ionospheric Conductivity on the Oscillation of DPI and DP2 Current Systems', *Geomag. Aeronom.* **13**, 621.

Testud, J.: 1970, 'Gravity Waves Generated during Magnetic Substorms', *J. Atmosph. Terr. Phys.* **32**, 1793.

Testud, J.: 1972, 'Interaction between Gravity Waves and Ionization in the Ionospheric F-Region', *Space Research XII*, S. A. Bowhill, L. D. Jaffe and M. J. Rycroft (eds.), p. 1163, Akademie-Verlag, Berlin.

Testud, J., Amayenc, P. and Blanc, M.: 1975, 'Middle and Low Latitude Effects of Auroral Disturbances from Incoherent Scatter', *J. Atmosph. Terr. Phys.* **37**, 989.

Thomas, L.: 1970, 'F2-Region Disturbances Associated with Major Magnetic Storms', *Planet. Space Sci.* **18**, 917.

Truttse, Yu. L.: 1968a, 'Oxygen Emission at λ 6300 Å and Geomagnetic Activity', *Planet. Space Sci.* **16**, 140.

Truttse, Yu. L.: 1968b, 'Upper Atmosphere during Geomagnetic Disturbances – 1. Some Regular Features of Low-Latitude Auroral Emissions', *Planet. Space Sci.* **16**, 981.

Truttse, Yu. L.: 1972, 'Oxygen Emission at 6300 Å', *Ann. Geophys.* **28**, 169.

Truttse, Yu. L. and Shefov, N. N.: 1970, 'On Middle-Latitude Aurorae at the Current Solar Maximum', *Planet. Space Sci.* **18**, 1850.

Truttse, Yu. L. and Yurchenko, O. T.: 1971, 'Temperature of the Upper Atmosphere from the 6300 Å Emission Data', *Planet. Space Sci.* **19**, 545.

Tsunoda, R. T.: 1975, 'Electric Field Measurements above a Radar Scattering Volume Producing "Diffuse" Auroral Echoes', *J. Geophys. Res.* **80**, 4297.

Tsunoda, R. T. and Fremouw, E. J.: 1975, 'Radar Studies of Fast Auroral Forms', Stanford Research Institute Report, Menlo Park, California, May.

Tsunoda, R. T., Presnell, R. I. and Leadabrand, R. L.: 1973, 'First Results of Radar Auroral Substorm Characteristics as Seen by a 398-MHz Phased-Array Radar Operated at Homer, Alaska', Presented at the Second General Science Assembly of the International Association of Geomagnetism and Aeronomy, (IAGA), Kyoto, Japan, September 9–21.

Tsunoda, R. T., Presnell, R. I. and Leadabrand, R. L.: 1974, 'Radar Auroral Echo Characteristics as Seen by a 938-MHz Phased Array Radar Operated at Homer, Alaska', *J. Geophys. Res.* **79**, 4709.

Unwin, R. S.: 1966a, 'The Morphology of the VHF Radio Aurora at Sunspot Maximum, 1. Diurnal and Seasonal Variations', *J. Atmosph. Terr. Phys.* **28**, 1167.

Unwin, R. S.: 1966b, 'The Morphology of VHF Radio Aurora at Sunspot Maximum, 2. The Behavior of Different Echo Types', *J. Atmosph. Terr. Phys.* **28**, 1183.

Unwin, R. S. and Baggaley, W. J.: 1972, 'The Radio Aurora', *Ann. de Geophys.* **28**, 111.

Unwin, R. S. and Keys, J. G.: 1975, 'Characteristics of the Radio Aurora during the Expansive Phase of Polar Substorms', *J. Atmosph. Terr. Phys.* **37**, 55.

Van'yan, L. L. and Debabov, A. S.: 1973, 'Three-Dimensional Current System of a Substorm Allowing for the Mutual Influence of Magnetically Conjugated Regions of the Ionosphere', *Geomag. Aeronom.* **13**, 810.

Van'yan, L. L., Debabov, A. S. and Osipova, I. L.: 1973, 'Nature of the Auroral Electrojet', *Geomag. Aeronom.* **13**, 278.

Vasseur, G., Reddy, C. A. and Testud, J.: 1972, 'Observations of Waves and Travelling Disturbances', *Space Research XII*, S. A. Bowhill, L. D. Jaffe and M. J. Rycroft (eds.), p. 1109, Akademie-Verlag, Berlin.

Vestine, E. H. and Chapman, S.: 1938, 'The Electric Current-System of Geomagnetic Disturbance', *Terr. Mag. Atmosph. Elec.* **43**, 351.

Volland, H. and Mayr, H. G.: 1971, 'Response of the Thermospheric Density to Auroral Heating during Geomagnetic Disturbances', *J. Geophys. Res.* **76**, 3764.

Wagner, C.-U.: 1971, 'Electric Currents and Polarization Fields at the Base of the Magnetosphere', *J. Atmosph. Terr. Phys.* **33**, 751.

Wallis, D. D., Anger, C. D. and Rostoker, G.: 1976, 'The Spatial Relationship of the Auroral Electrojets and Visible Aurora in the Evening Sector', *J. Geophys. Res.* **81**, 2857.

Wang, T. N. C. and Tsunoda, R. T.: 1975, 'On a Crossed Field Two-Stream Plasma Instability on the Auroral Plasma', *J. Geophys. Res.* **80**, 2172.

Wescott, E. M., Stolarik, J. D. and Heppner, J. P.: 1969, 'Electric Fields in the Vicinity of Auroral Forms from Motions of Barium Vapor Releases', *J. Geophys. Res.* **74**, 3469.

Wescott, E. M., Stolarik, J. D. and Heppner, J. P.: 1970, 'Auroral and Polar Cap Electric Fields from Barium Releases', *Particles and Fields in the Magnetosphere*, B. M. McCormac (ed.), p. 229, D. Reidel Publ. Co., Dordrecht-Holland.

Wescott, E. M., Stenbaek-Nielsen, H. C., Davis, T. N., Murcray, W. B., Peek, H. M. and Bottoms, P. J.: 1975, 'The $L = 6.6$ Oosik Barium Plasma Injection Experiment and Magnetic Storm on March 7, 1972', *J. Geophys. Res.* **80**, 951.

Whalen, B. A., Verschell, H. J. and McDiarmid, I. B.: 1975, 'Correlations of Ionospheric Electric Fields and Energetic Particle Precipitation', *J. Geophys. Res.* **80**, 2137.

Wickwar, V. B., Baron, M. J. and Sears, R. D.: 1975, 'Auroral Energy Input from Energetic Electrons and Joule Heating at Chatanika', *J. Geophys. Res.* **80**, 4364.

Wiens, R. G. and Rostoker, G.: 1975, 'Characteristics of the Development of the Westward Electrojet during the Expansive Phase of Magnetospheric Substorms', *J. Geophys. Res.* **80**, 2109.

Wilson, C. R.: 1969, 'Auroral Infrasonic Waves', *J. Geophys. Res.* **74**, 1812.

Wilson, C. R.: 1972, 'Auroral Infrasonic Wave-Generation Mechanism', *J. Geophys. Res.* **77**, 1820.

Wilson, C. R.: 1973a, 'Seasonal Variation of Auroral Infrasonic Wave Activity', *J. Geophys. Res.* **78**, 4801.

Wilson, C. R.: 1973b, 'Comments on a Paper by J. A. Fedder and P. M. Banks, "Convection Electric Fields and Polar Thermospheric Winds",' *J. Geophys. Res.* **78**, 778.

Wilson, C. R.: 1974, 'Trans-Auroral Zone Auroral Infrasonic Wave Observations', *Planet. Space Sci.* **22**, 151.

Wilson, C. R.: 1975, 'Infrasonic Wave Generation by Aurora', *J. Atmosph. Terr. Phys.* **37**, 973.
Wilson, C. R. and Hargreaves, J. K.: 1974, 'The Motions of Peaks in Ionospheric Auroral Absorption and Auroral Infrasonic Waves', *J. Atmosph. Terr. Phys.* **36**, 1555.
Yasuhara, F. and Akasofu, S.-I.: 1976, 'Field-Aligned Currents and Ionospheric Electric Fields', *J. Geophys. Res.* **81** (in press).
Yasuhara, F., Kamide, Y. and Akasofu, S.-I.: 1975a, 'A Modeling of Magnetospheric Substorms', *Planet. Space Sci.* **23**, 575.
Yasuhara, F., Kamide, Y. and Akasofu, S.-I.: 1975b, 'Field-Aligned and Ionospheric Currents', *Planet. Space Sci.* **23**, 1355.
Zaitzev, A. N. and Boström, R.: 1971, 'On Methods of Graphical Displaying of Polar Magnetic Disturbances', *Planet. Space Sci.* **19**, 643.
Zmuda, A. J. and Armstrong, J. C.: 1974a, 'The Diurnal Variation of the Region with Vector Magnetic Field Changes Associated with Field-Aligned Currents', *J. Geophys. Res.* **79**, 2501.
Zmuda, A. J. and Armstrong, J. C.: 1974b, 'The Diurnal Flow Pattern of Field-Aligned Currents', *J. Geophys. Res.* **79**, 4611.

CHAPTER 8

PENETRATING CONVECTION ELECTRIC FIELD, PLASMA INJECTION AND PLASMASPHERE DISTURBANCES

8.1. Introduction

One of the most important processes which takes place in the magnetosphere during the magnetospheric substorm is the 'injection' of energized plasma particles from the plasma sheet into the Van Allen belt, the trapping region. It has generally been believed that the plasma 'injection' arises simply from an enhanced plasma flow caused by the southward turning of the interplanetary magnetic field (IMF) vector. However, it was shown in Section 4.4.5 that the plasma injection does not occur simply as a result of the southward turning of the IMF vector and that other processes are involved in causing it.

In Section 1.3.5(b), it was noted that the geocentric distance L_G of the inner edge of the plasma sheet for a given potential Φ across the magnetotail is given (Jaggi and Wolf, 1973) by

$$L_G \simeq 7.93\left(\frac{31\ \gamma\eta\mu}{\Phi\langle\sigma_p\rangle}\right)^{1/3}$$

The reason for such a limited penetration of the plasma sheet is that space charges near the advancing front of the plasma sheet tend to cancel the convection electric field, shielding the inner magnetosphere from the convection electric field.

During the last few years, several important attempts have been made to interpret precisely the observed particle behavior in the inner magnetosphere in terms of the 'injection' (McIlwain, 1972, 1974; Williams *et al.*, 1974; Smith and Hoffman, 1974; Konradi *et al.*, 1975; Kivelson and Southwood, 1975). There seems to be some disagreement among the above workers on the definition of the term 'injection', particularly whether the injection is simply a sudden earthward advance of the plasma sheet *as a whole* or an onrush of *isolated clouds*. However, since they agree that the injection takes place simultaneously from a wide longitudinal (or local time) range of the inner part of the plasma sheet, the crucial problem is how to interpret, without contradiction, the apparent dispersion effect for the entire range of observed energies of both protons and electrons.

Newly injected plasma particles drift in the trapping region under the influence of both the magnetic and electric fields. This is the subject of Section 8.3. The drifting plasma particles undergo a variety of wave-particle interactions. It has been suggested that the injected protons are precipitated into the upper atmosphere, after interacting with 'cold' plasma in the plasmasphere, and that this loss process becomes the energy source for the mid-latitude red arc. This subject has recently been reviewed by Rees and Roble (1975) Hultqvist (1975a, b) and Gendrin (1975) and thus will be mentioned only very briefly here.

In addition to its interaction with the ring current protons, the plasmasphere undergoes a considerable change during substorms, caused by the penetrating electric field. This subject will be discussed in Section 8.4. Such disturbances in the plasmasphere also affect the underlying ionosphere, causing some aspects of the ionospheric substorm. Section 8.5 will be concerned with this subject.

8.2. Penetration of the Convection Electric Field into the Inner Magnetosphere and the Resulting Plasma Injection

8.2.1. OBSERVATIONS

(a) *Plasma Injection at the Geosynchronous Distance*

In Section 4.4.5, it was shown that the plasma injection observed at the geosynchronous distance is almost simultaneous with the onset of substorms, but there is no such close relation between the plasma injection and the southward turning of the IMF vector. Here we shall examine two more examples to confirm this conclusion.

(i) *1969, November 27 (Figures 8.1(a) and (b)).* A well-defined southward turning was estimated to arrive at the magnetosphere at 0329 UT. The AE index shows that weak magnetic activity was present between 01 and 04 UT, consisting of at least three increases, at 0130–0200, 0230–0300 and 0330–0400 UT, respectively. These changes were actually due to an increase of the AU index, so that they are likely to be caused by the expansion of the S_q^p current system. Indeed, there were three sharp decreases of the B_z component corresponding to the three AE increases, including the southward turning at 0329 UT. It should be noted that there was no plasma injection between 0130 and 0400 UT, although there were drifting protons and electrons in the vicinity of the satellite. At about 0400 UT, the AE index began to increase and reached a 600 γ level in about one hour. At the location of the ATS-5 satellite, 20 keV protons began to arrive at 0358 UT, and the injection time is estimated to be about 0356 UT. At about 0410 UT the satellite encountered electrons of energies ranging from 0 to ~1 keV.

Therefore, the plasma injection and the onset of the AE increase were almost simultaneous, but there was a time lag of 50 min after the southward turning before the plasma injection.

(ii) *1969, September 26 (Figures 8.2(a) and (b)).* The IMF southward turning at 0553 UT was not sharp. The AE index was very low between 0415 and 0600 UT. It began to increase slowly at about 0600 UT, arising from the expanding oval (AU $\gg$ AL). The AE index began to increase rather suddenly at about 0640 UT, and an increase in the flux of 20 keV protons was observed at the ATS-5 at 0647 UT. The estimated injection time was 0645 $\pm$ 5 min UT. Subsequently, the

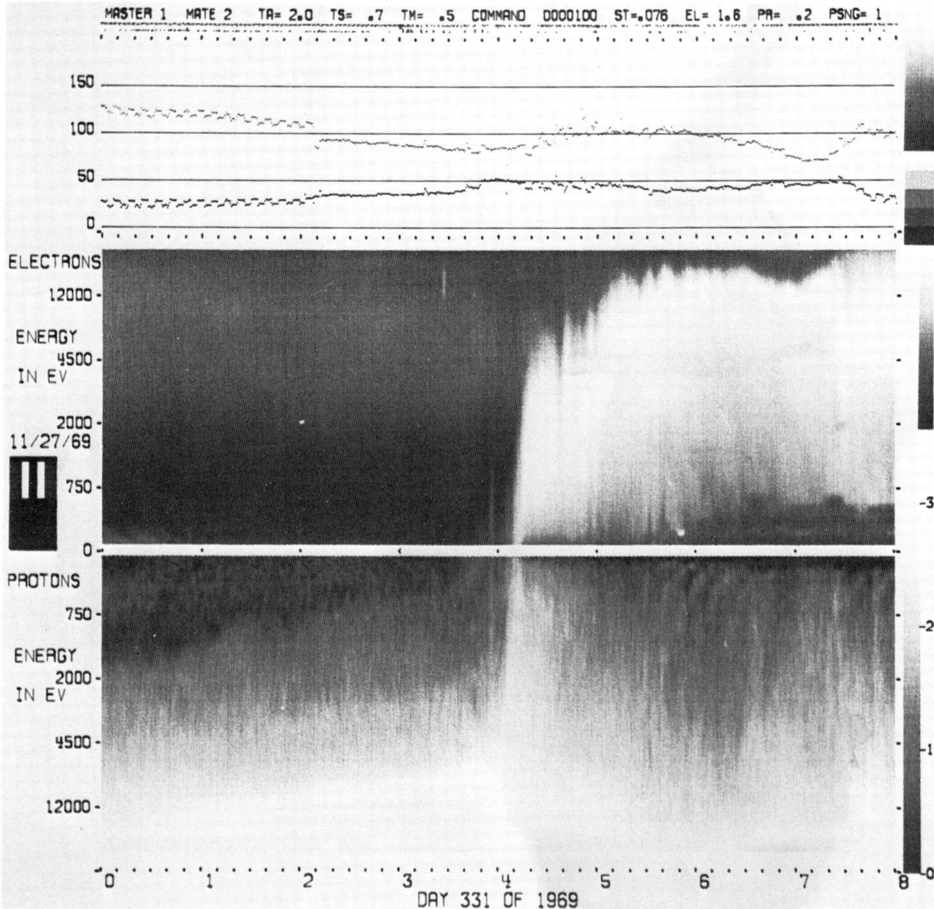

Fig. 8.1(a, b). Injection of plasma observed at the synchronous distance (the ATS-5 satellite) and its relation to substorm activity and the IMF B_z component. (Courtesy of Yasuhara, F., McIlwain, C. E. and Akasofu, S.-I.)

corresponding electrons were observed at about 0700 UT, at 0659 UT up to 0.1 keV and 0700 UT up to 5 keV. Again, the plasma injection at 0645 UT took place about 5 ± 5 minutes after the AE increase, but about 1 hour after the southward turning.

(iii) *1970, March 13 (Figures 8.3(a) and (b)).* There were two IMF southward turnings, after a prolonged period of a large B_z value ($\sim +4\ \gamma$). The first turning was complex and the changes were small (the average B_z component after the turning was greater than $-2\ \gamma$), but it is instructive to include it in the description. The B_z component began to decrease at about 0410 UT and turned southward at about 0430 UT, after a few oscillatory changes. Then the B_z component turned northward sometime between 0600 and 0630 UT (note the data gap in this interval). At about 0830 UT it again turned southward very suddenly and remained southward for at least 1.5 h.

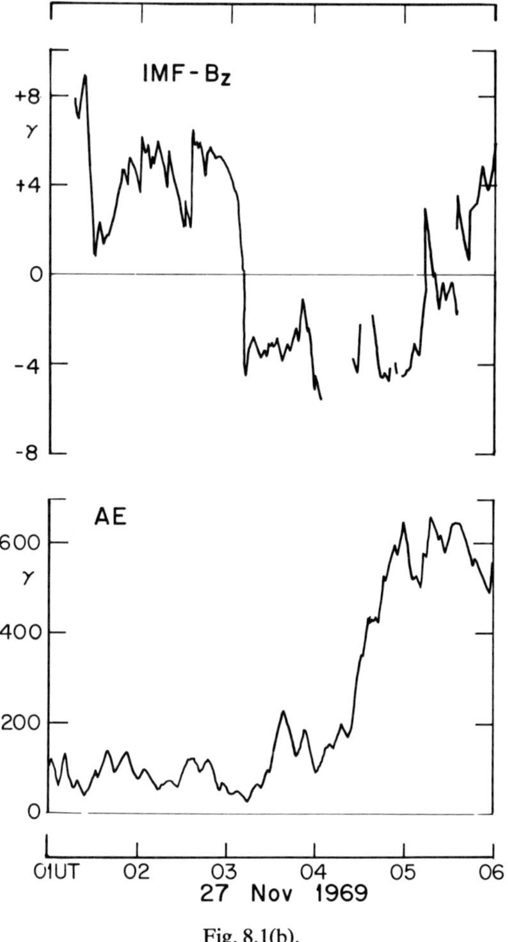

Fig. 8.1(b).

The AE index was quite low until about 0630 UT, when it suddenly increased, although the magnetic records from Great Whale River and Churchill showed the first sign of the activity at about 0613 UT. After this increase, there were at least two major increases. The corresponding studies indicate that the plasma injection and the enhancement of the westward electric field in the plasmasphere took place 'simultaneously' with the substorm onset, within accuracy of the determinations of the substorm onset time and the injection time. DeForest and McIlwain (1971), Kamide and McIlwain (1974) and Akasofu *et al.* (1974) also showed that there is a very close time relation between the plasma injection and the substorm onset.

However, there was no such close relation between the IMF southward turning and the plasma injection. In many cases, the plasma injections occurred about 1 to 2 h after the southward turning. There are at least two possibilities in interpreting the observed delay. The first is that the time constant for the enhanced electric field to penetrate deep into the inner magnetosphere is of order of

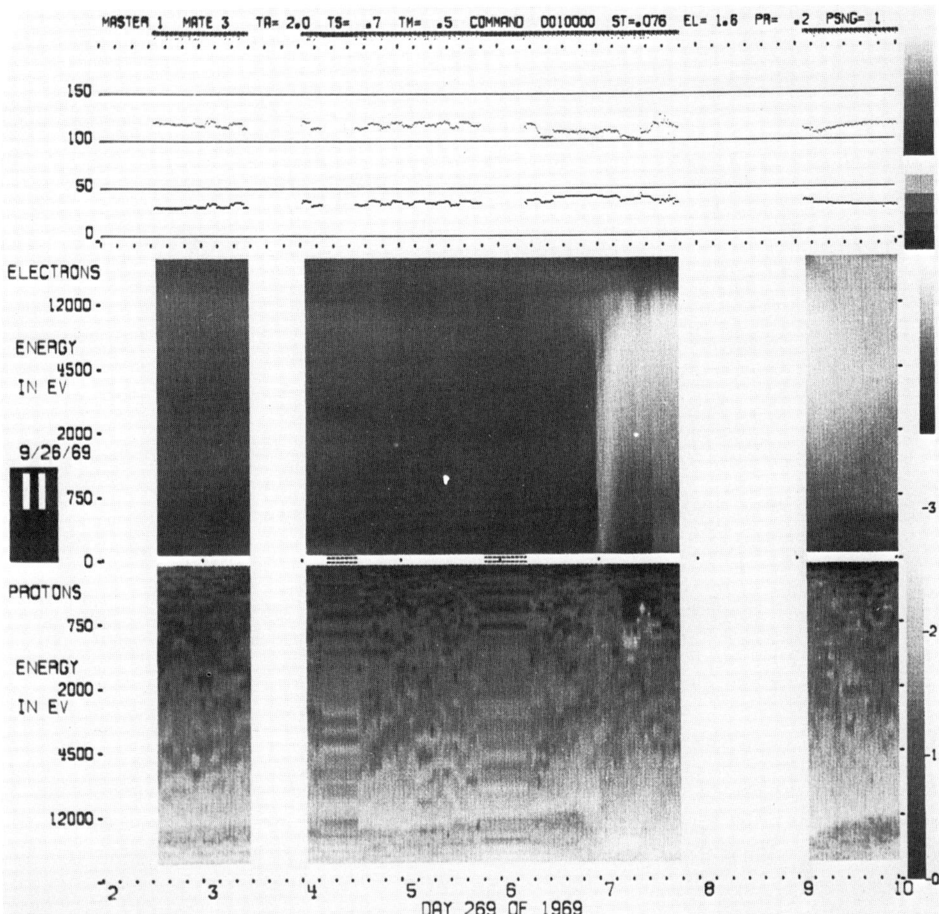

Fig. 8.2(a, b). Injection of plasma observed at the synchronous distance (the ATS-5 satellite) and its relation to substorm activity and the IMF B_z component. (Courtesy of Yasuhara, F., McIlwain, C. E. and Akasofu, S.-I.)

one or two hours or that it takes about one or two hours for the inner edge of the plasma sheet to reach the synchronous distance from its quiet time location (say, $\sim 10\ R_E$); see Shelley *et al.* (1971). The second possibility is that there is no *direct* relationship between the southward turning and the plasma injection. In fact, some plasma injections took place even after northward turnings.

The 'simultaneity' of the plasma injection and substorm onset seems to rule out the first possibility, since there is no reason why the arrival time of the front of the advancing plasma sheet to the geosynchronous distance should almost always coincide exactly with the substorm onset.

The 'simultaneity' of the plasma injection at the substorm onset also suggests that a sudden change occurs in the distribution of the electric field at the onset time of substorms. More specifically, some substorm processes make it possible for the enhanced electric field to penetrate into the inner magnetosphere, allowing

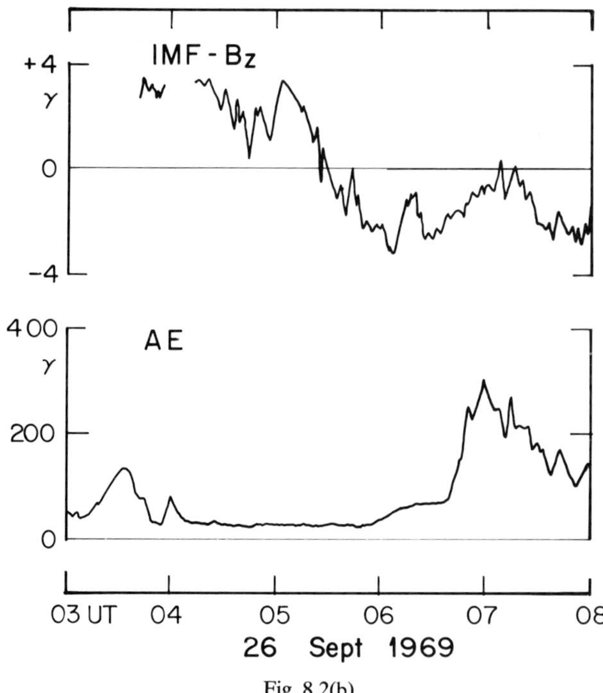

Fig. 8.2(b).

the plasma sheet to advance closer toward the Earth. Barfield *et al.* (1975) reported that following the onset of an 800 γ negative bay in the auroral zone, there was a sharp dropout of Van Allen belt electrons and of plasma in the plasmasphere and also a sudden appearance of plasma sheet protons. They suggested that there is a common cause for these phenomena, a sudden enhancement of the convection electric field at $L \simeq 5$ in the dusk sector.

(b) *Enhancement of the Westward Electric Field in the Plasmasphere*

Carpenter and Stone (1967) showed that the east-west component of electric fields in the plasmasphere can be inferred from radial motions of whistler ducts. Carpenter and Akasofu (1972) showed that the westward electric field is considerably intensified during substorms. Figure 8.4 shows an example of their data; from the top, all-sky photographs from Byrd (Antarctic), ULF pulsations recorded at Eights Station (Antarctic), the west-east component of the electric field deduced from the whistler duct data, the H and D component magnetic records from Fredericksburg and the IMF B_z component. It can be seen that the westward electric field began to grow at about 0520–25 UT, rather than 0430 UT (when the IMF B_z component became negative).

Carpenter and Kirchhoff (1975) made an interesting comparison of the electric fields observed by incoherent scatter radars at Chatanika, Alaska (65.1° N, 147.5° W) and Millstone Hill, Massachusetts (42.6° N, 71.5° W) and found that the

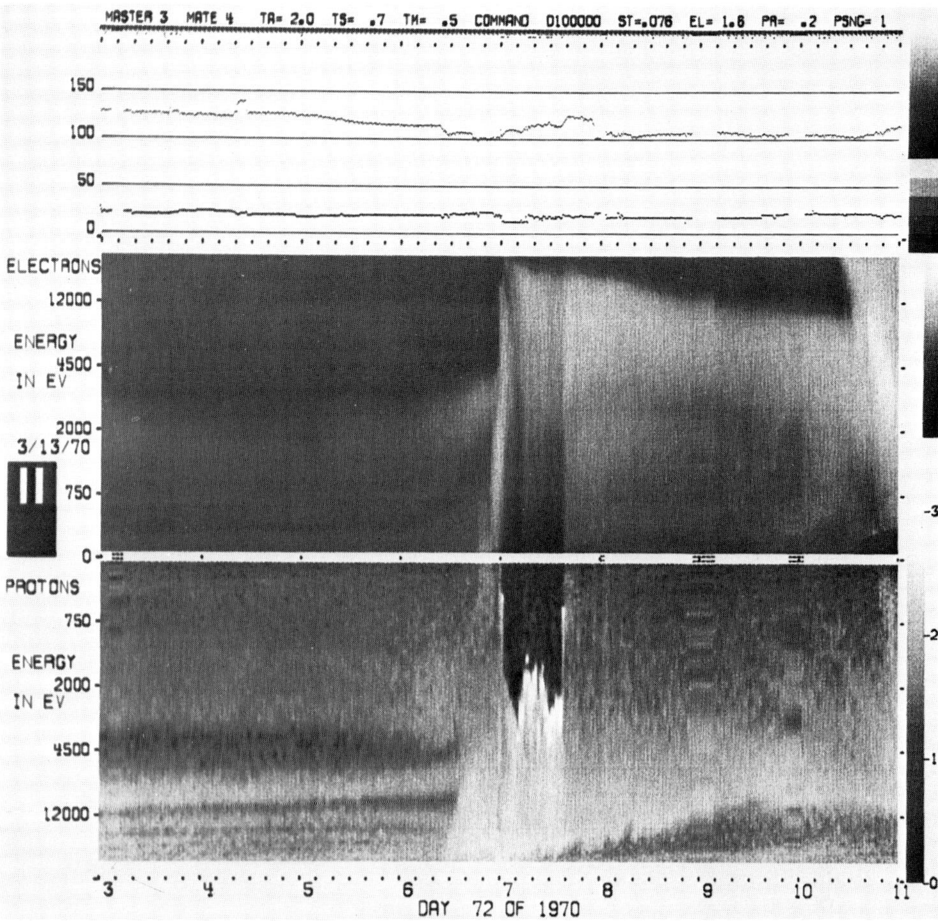

Fig. 8.3(a, b). Injection of plasma observed at the synchronous distance (the ATS-5 satellite) and its relation to substorm activity and the IMF B_z component. (Courtesy of Yasuhara, F., McIlwain, C. E. and Akasofu, S.-I.)

daily variations at the two stations are quite similar. Figure 8.5 shows an example of this observation. They suggested that the convection electric field has an appreciable effect to as low as at least $L = 3.2$.

Testud *et al.* (1975) reported an enhancement of the westward component of the electric field observed by an incoherent scatter radar at St. Santin–Nancay An example of their observation is shown in Figure 8.6. They noted that not all substorms appeared to cause the enhancement. It is quite likely that the enhancement is observed only when the radar is located in the dark sector.

8.2.2. THEORETICAL ESTIMATES

One of the important quantities associated with the electric field in the inner magnetosphere, and also in the middle and low latitudes, is its magnitude relative

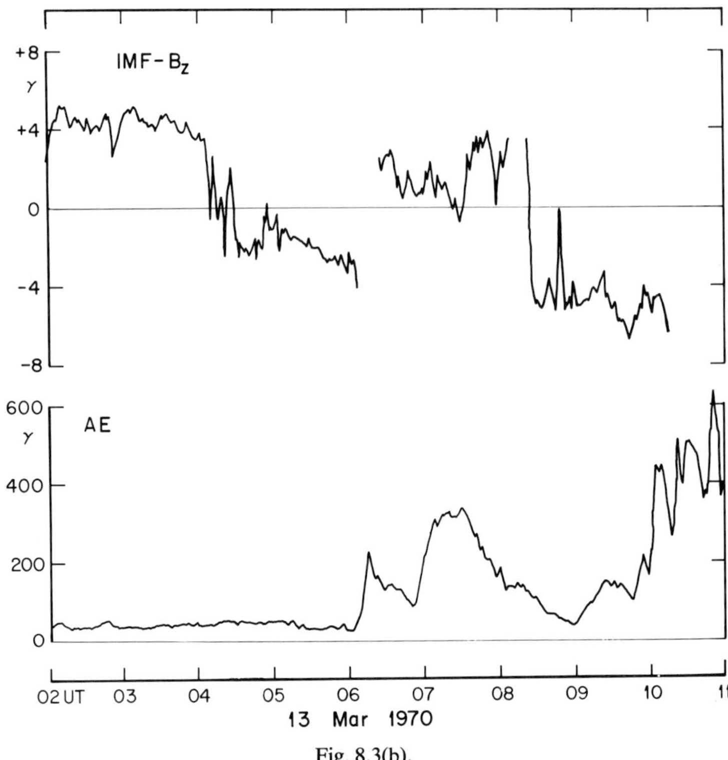

Fig. 8.3(b).

to that of the high latitude convection field. This relative magnitude or the ratio ($\Delta\phi_S/\Delta\phi_N$) – the magnitude of the potential along the equatorward boundary ($\Delta\phi_S$) to that along the poleward boundary ($\Delta\phi_N$) of the oval – represents a parameter which indicates the degree of the penetration of the high latitude convection electric field to the low latitude ionosphere. The magnitude of the potential is defined as the sum of the magnitudes of the positive and negative peak values along the latitude circles. The problem has been studied extensively by many workers. Particularly, Swift (1971), Vasyliunas (1972) and Jaggi and Wolf (1973) solved a set of equations which allows the coupling between the ionosphere and the magnetosphere through the field-aligned currents, under the self-consistent electric field for particular models of the ionospheric conductivity (Section 1.3.5). In the following, we describe briefly some of the results obtained by Yasuhara (1975).

In Figure 8.7, the magnitude of the potential along different constant latitude circles ($\Delta\phi(\lambda)$) is plotted for the different models, (1) to (5), in Table 8.1; note that $\Delta\phi(\lambda)$ is normalized by $\Delta\phi_N$. These models differ from each other only in the conductivity values in the auroral oval, $\Sigma^A/\Sigma^P = 1.0$, 1.5, 2.0, 5.0 and 10.0 for the models (1), (2), (3), (4) and (5), respectively. The ratio of the total intensity of the equatorward field-aligned current to that of the poleward one is fixed to be 0.5.

In this figure, it can be seen that when the conductivity in the auroral oval

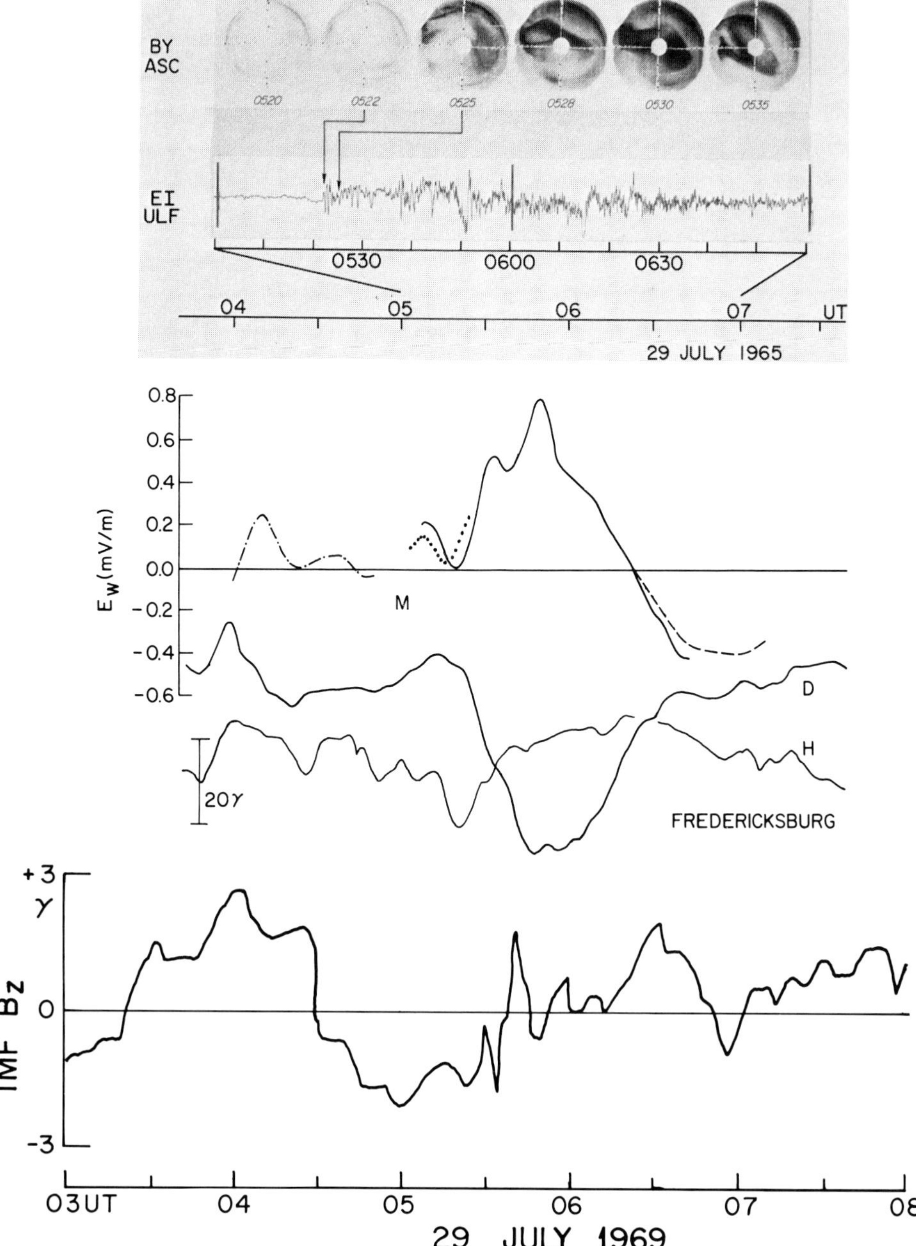

Fig. 8.4. Penetration of the westward electric field into the plasmasphere and its relation to auroral and geomagnetic activities and the IMF B_z component. From the top: selected all-sky photographs from Byrd station, Antarctica, ULF noise from Eights station, Antarctica, the electric field, the *D* and *H* component magnetic records from Fredericksburg and the IMF B_z component. (Carpenter, D. L. and Akasofu, S.-I.: *J. Geophys. Res.* **77**, 6854, 1972.)

TABLE 8.1

List of the parameters of various models used in the computation. The values of the Pedersen and Hall conductivities in the polar cap and middle and low latitudes are fixed to be $\Sigma_P^P = \Sigma_P^M = 1.0$ mho and $\Sigma_H^P = \Sigma_H^M = 2.0$ mho, respectively.

	Σ_P^A/Σ_P^P	Σ_H^A/Σ_P^A	I_S^T/I_N^T	Location of peaks I_N		Location of peaks I_S	
[1]	1.0	2.0	0.5	90°	270°	45°	315°
[2]	1.5	2.0	0.5	90°	270°	45°	315°
[3]	2.0	2.0	0.5	90°	270°	45°	315°
[4]	5.0	2.0	0.5	90°	270°	45°	315°
[5]	10.0	2.0	0.5	90°	270°	45°	315°

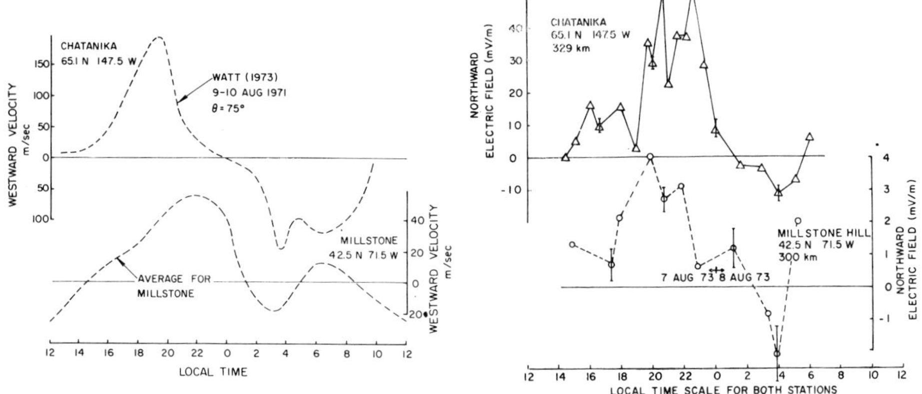

Fig. 8.5. Similarity of the electric field variations as a function of LT in the auroral zone (the Chatanika radar) and in the middle latitude (the Millstone Hill radar). (Carpenter, L. A. and Kirchhoff, V. W. J. H.: *J. Geophys. Res.* **80**, 1810, 1975.)

increases, the slope of the curves decreases, and thus the penetration of the high-latitude electric field to the latitudes lower than the equatorward field-aligned current increases.

This result differs considerably from that obtained by Vasyliunas (1972) and Swift (1971), who showed that the potential variation of the latitude just below the latitude of the ring current (or the inner boundary of the plasma sheet) decreases abruptly and becomes much less than $\Delta\phi_N$ by one to three orders of magnitude. However, their ionospheric models lack the highly conductive belt, and therefore are unable to sufficiently discharge the space charges developed near the inner edge of the plasma sheet.

In the Jaggi and Wolf model (1973), a highly conductive belt along the auroral region is included. However, since their field-aligned currents are infinitesimally thin, while the conductivity variation is continuous, the gradient of the conductivity cannot contribute significantly to the field-aligned currents. Further, their field-aligned currents appear to differ from the observed ones. In particular, for

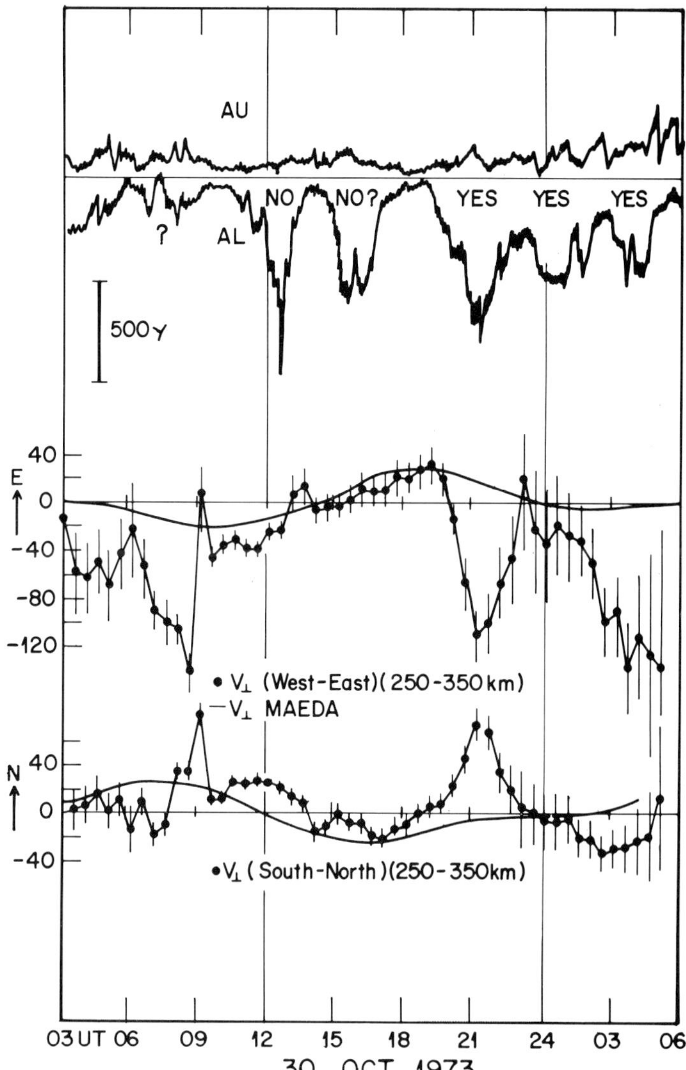

Fig. 8.6. Penetration of the westward and northward electric fields into mid-latitudes, observed at St. Santin-Nancay in terms of the drift speed of plasma (m s^{-1}). The AU and AL indices are also shown at the top. Note that the substorm effects are absent in daylight hours. (Testud, J., Amayenc, P. and Blanc, M.: *J. Atmosph. Terr. Phys.* **37**, 989, 1975.)

the ionosphere with a uniform conductivity, the ratio of the intensity of the equatorward field-aligned currents (I_S^T) to that of the poleward currents (I_N^T) is quite sensitive to the penetration of the high-latitude convection electric field to lower latitudes.

In order to examine how the ratio I_S^T/I_N^T is related to the degree of the penetration of the electric field below the equatorward boundary of the oval, the magnitude ratio ($\Delta\phi_S/\Delta\phi_N$) is plotted as a function of I_S^T/I_N^T in Figure 8.8.

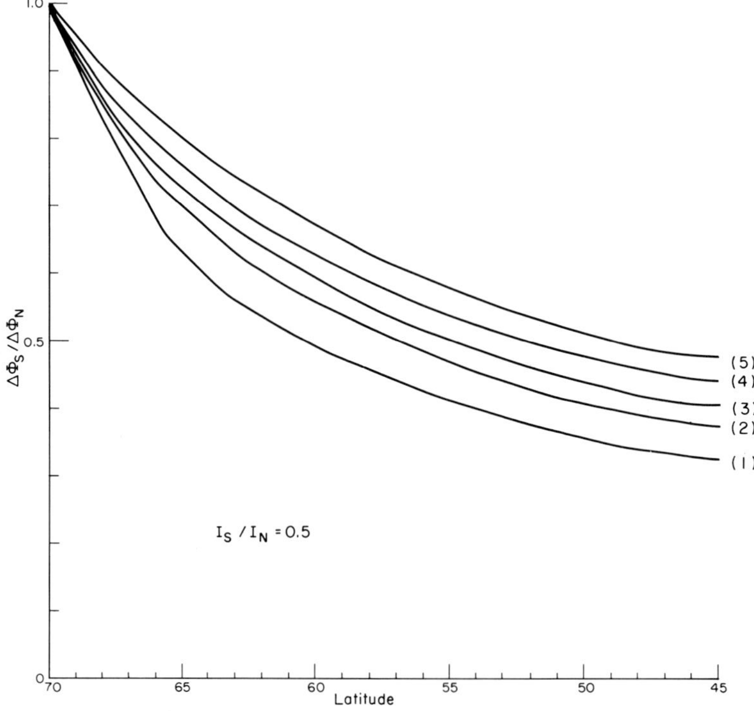

Fig. 8.7. Normalized magnitude of the potential ($\Delta\phi_S/\Delta\phi_N$) as a function of latitude for the five different models of the ionosphere. The figure shows that when the conductivity in the auroral oval increases, the penetration of the polar cap electric field to lower latitudes is improved. (Yasuhara, F.: Ph.D. Thesis, University of Alaska, 1975.)

In this calculation, the oval is bounded by two latitude circles $\lambda = 70°$ and $\lambda = 65°$. The intensity of the field-aligned currents is given by $I_\parallel = I_N \sin\psi$ along the $\lambda = 70°$ circle and $I_\parallel = I_S \sin\psi$ along the $\lambda = 65°$ circle ($I_N, I_S > 0$). The Pedersen and Hall conductivities in both the polar cap and the middle and low latitudes are 1 and 2 (in arbitrary units), respectively. The Pedersen and Hall conductivities Σ_P^A and Σ_H^A in the oval are assumed to be

$\Sigma_P^A = 1$	$\Sigma_H^A = 2$	for curve (1)
1	4	(2)
2	4	(3)

and

$$\Sigma_P^A = 10 \qquad \Sigma_H^A = 40 \qquad (4)$$

The curve (1) represents the case of a uniform conductivity over the entire ionosphere. Therefore, this is the only case by which the results of Swift (1971) and Vasyliunas (1972) and the results shown in Figure 8.8 can be compared. Note that the ratio $\Delta\phi_S/\Delta\phi_N$ decreases sharply between $I_S^T/I_N^T = 0.7$ and 0.8. This indicates that when the conductivity of the ionosphere is uniform, the penetration of the electric field decreases drastically for the ratio $I_S^T/I_N^T > 0.7$. The same

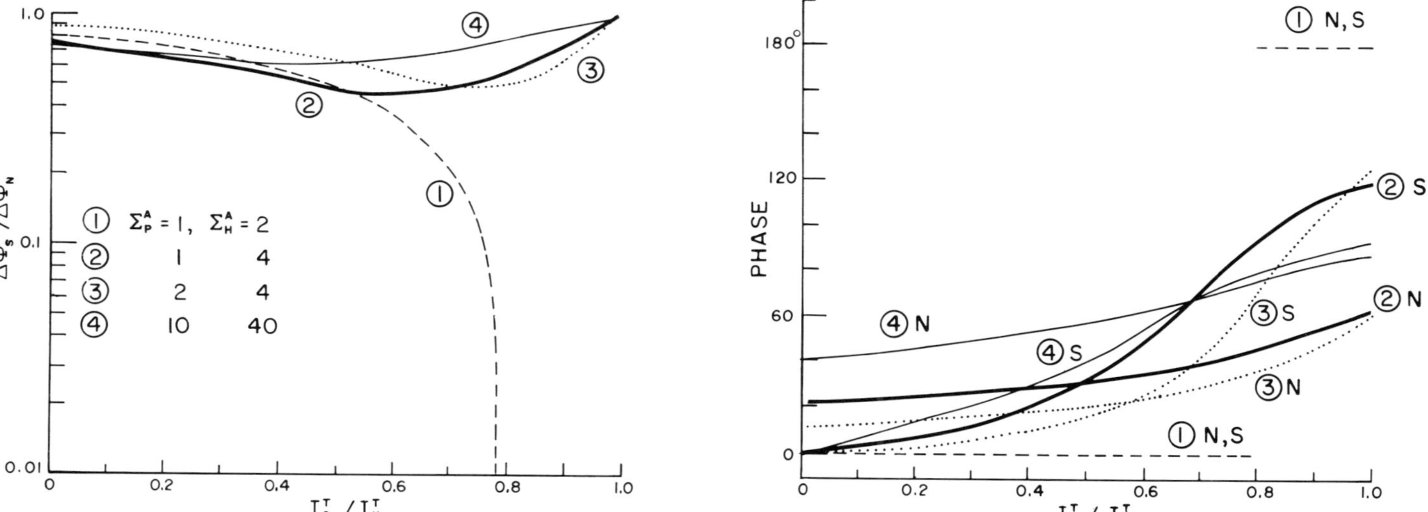

Fig. 8.8. Ratio of the magnitude and the phase angle of the potential ($\Delta\phi_S$) along the equatorward boundary of the oval to the potential ($\Delta\phi_N$) along the poleward boundary as a function of the ratio I_S^T/I_N^T (the intensity of the equatorward field-aligned current to that of the poleward current) for different models of the ionosphere. (Yasuhara, F.: Ph.D. Thesis, University of Alaska, 1975.)

tendency is obtained for the model in which the equatorward field-aligned currents are located outside the highly conductive belt; this model simulates the Jaggi and Wolf model.

Since Swift (1971), Vasyliunas (1972) and Jaggi and Wolf (1973) showed that their models do not result in a deep penetration of the convection electric field, the above calculation suggested that the ratio I_S^T/I_N^T in their models is of order 0.7 to 0.8.

When a highly conductive belt appears, the penetration of the electric field is drastically increased (curves (2), (3) and (4) in Figure 8.7). It is suggested that this increase is partly responsible for a sudden earthward plasma injection.

Here, the ionosphere is assumed to play only a passive role in injecting plasma from the plasma sheet. However, as mentioned in Section 5.4.5, the ionosphere may play an active role in the injection process. This is because the observed westward electric fields are much more intense ($E_{wi} \sim 30$ mV m^{-1} in the ionosphere and $E_{we} \sim 2$ mV m^{-1} in the equatorial plane) than the expected cross-tail electric field ($E_{wi} \sim 4.4$ mV m^{-1} and $E_{we} \sim 0.3$ mV m^{-1}). This suggestion is discussed in Section 9.2.

8.3. Relationship between Particles at the Geosynchronous Distance and Auroral Activity near the Geomagnetically Conjugate Point

Since the geosynchronous distance is ideal for observing the injection of plasma from the plasma sheet, it is of great interest to identify specific particle features with specific auroral displays (Akasofu *et al.*, 1974).

(i) *1969, December 16: Figures 8.9(a) and (b).* Auroras were fairly active during the first 3 h on 1969, December 16 (Figure 8.9(a)); note that the all-sky data presented here are shown in negative. The activity was associated with a substorm centered in the midnight sector (see the AE index). A traveling surge was passing near the zenith of Great Whale River (GWR) at about 0140 UT. An arc-like luminosity (possibly the diffuse aurora) was seen in the southern sky between 0150 and 0200 UT. A weak negative bay was also observed at GWR at that time. Between 0220 and 0300 UT, auroras were very quiet. At about 0300 UT, the diffuse aurora in the southern sky brightened. A westward traveling surge was seen near the northern horizon between 0320 and 0450 UT; an arc near the northern boundary of the diffuse aurora increased considerably in brightness between 0410 and 0435 UT.

Between 0440 and 0600 UT, auroral activity was very low. At 0610 UT, the arc at the northern boundary increased in brightness, but it was after 0650 UT that a violent poleward motion and other complicated displays followed. The corresponding magnetogram from GWR shows a rather ‘isolated bay’ which began at about 0650 UT; it recovered at about 0830 UT. Large-scale folds drifted rapidly eastward after 0750 UT, and the patchy structure covered the whole sky between 0850 and 1020 UT. A new and weaker negative bay began at about 0835 UT, lasting until about 1030 UT. After 1040 UT, however, auroras were seen only near the northern horizon; the twilight obscured the display at about 1150 UT. A new substorm began at College at about 1100 UT.

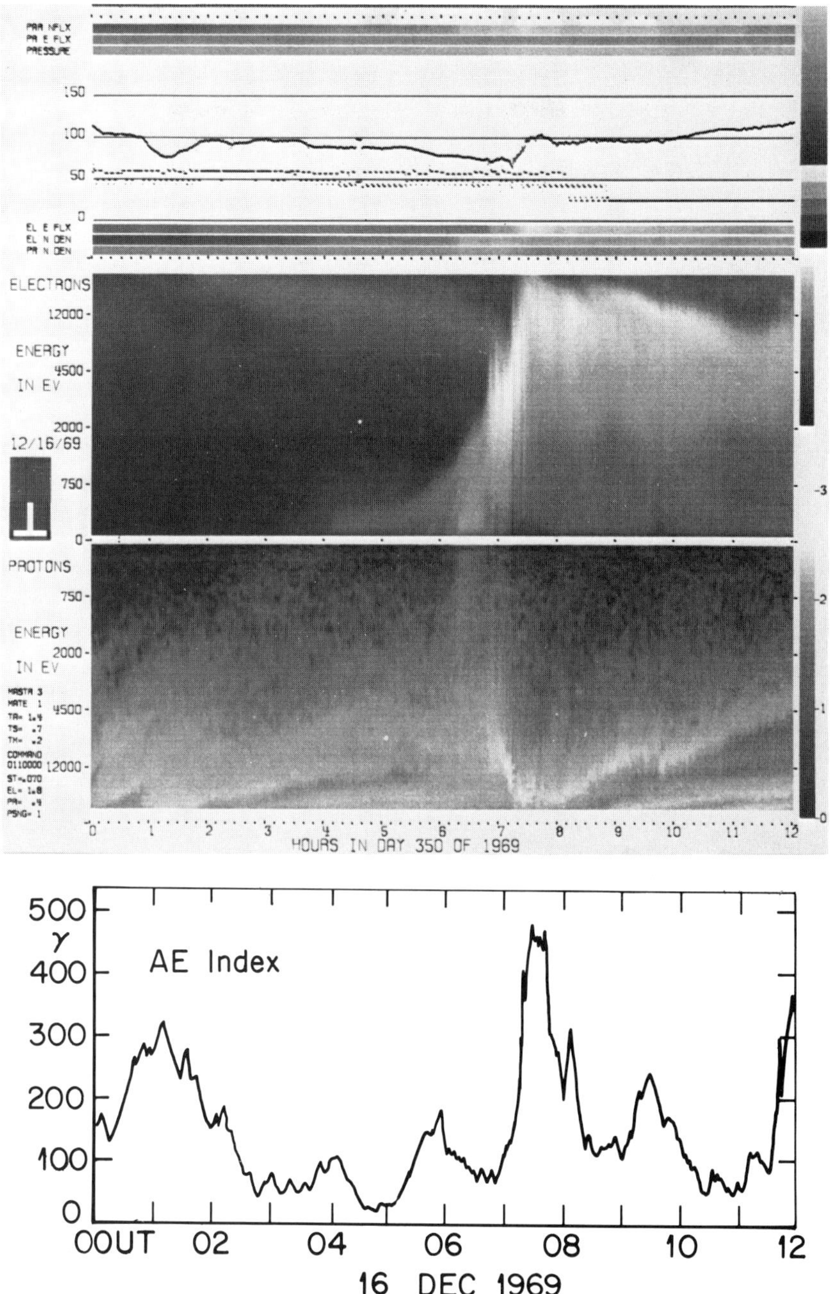

Fig. 8.9(a, b). Hot plasma particles observed at the ATS-5 satellite and selected all-sky photographs from Great Whale River, Canada. The AE index and the satellite magnetic record are also shown. (Akasofu, S.-I., DeForest, S. and McIlwain, C.: *Planet. Space Sci.* **22**, 25, 1974.)

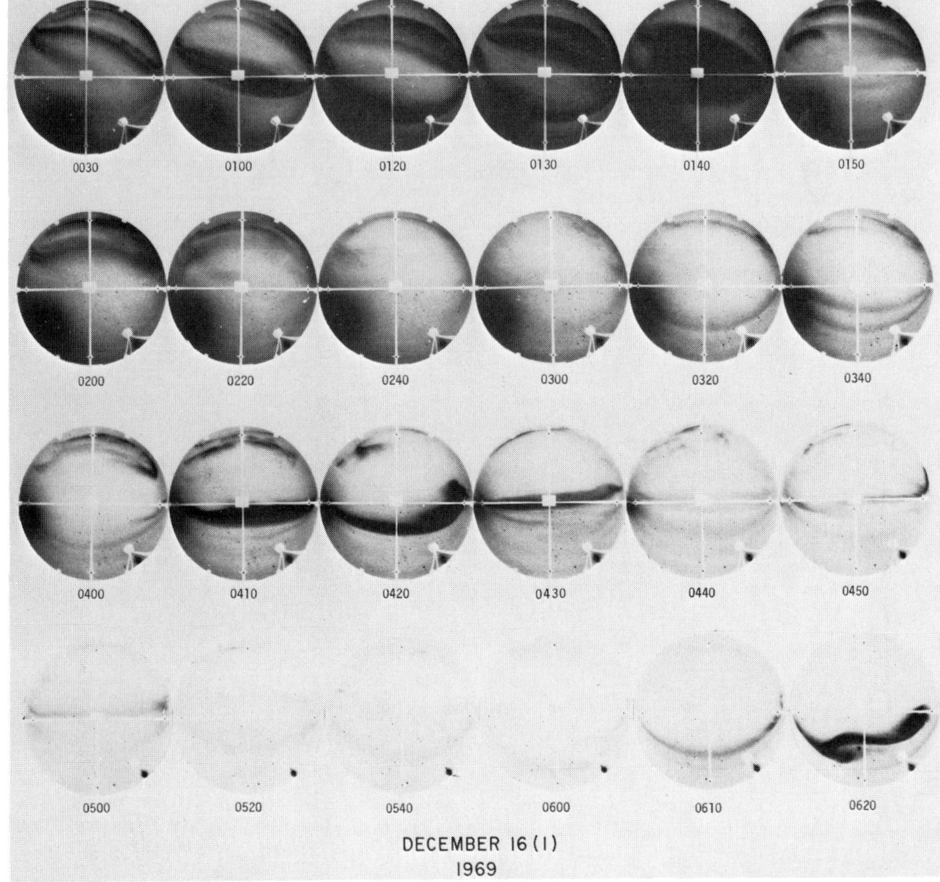

Fig. 8.9(b.1).

The injection of both ring current protons and Van Allen electrons associated with the first substorm and their subsequent drift motions are clearly seen as 'dispersion curves' in the spectrograms in Figure 8.9(b) (DeForest and McIlwain, 1971; McIlwain, 1972). The highest energy protons began to arrive at ATS-5 at about 0050 UT, and subsequently, lower energy protons reached there at later times. The magnetic field decreased at the satellite during the substorm. However, since there were no detectable low energy (~ 1 keV) electrons, one can infer that the inner boundary of the plasma sheet was located beyond the distance of ATS-5. It was only after 0300 UT that the satellite encountered the low energy plasma sheet particles which probably advanced earthward during the substorm that occurred near the end of December 15; see the injected protons and their dispersion curve, which appeared at the beginning of December 16. At about 0550 and 0620 UT, the satellite encountered high density plasmas, and at 0650 UT it encountered a hotter plasma with a more intense flux. At 0715 UT, the satellite was embedded in an even hotter plasma. Both the energy flux and pressure of

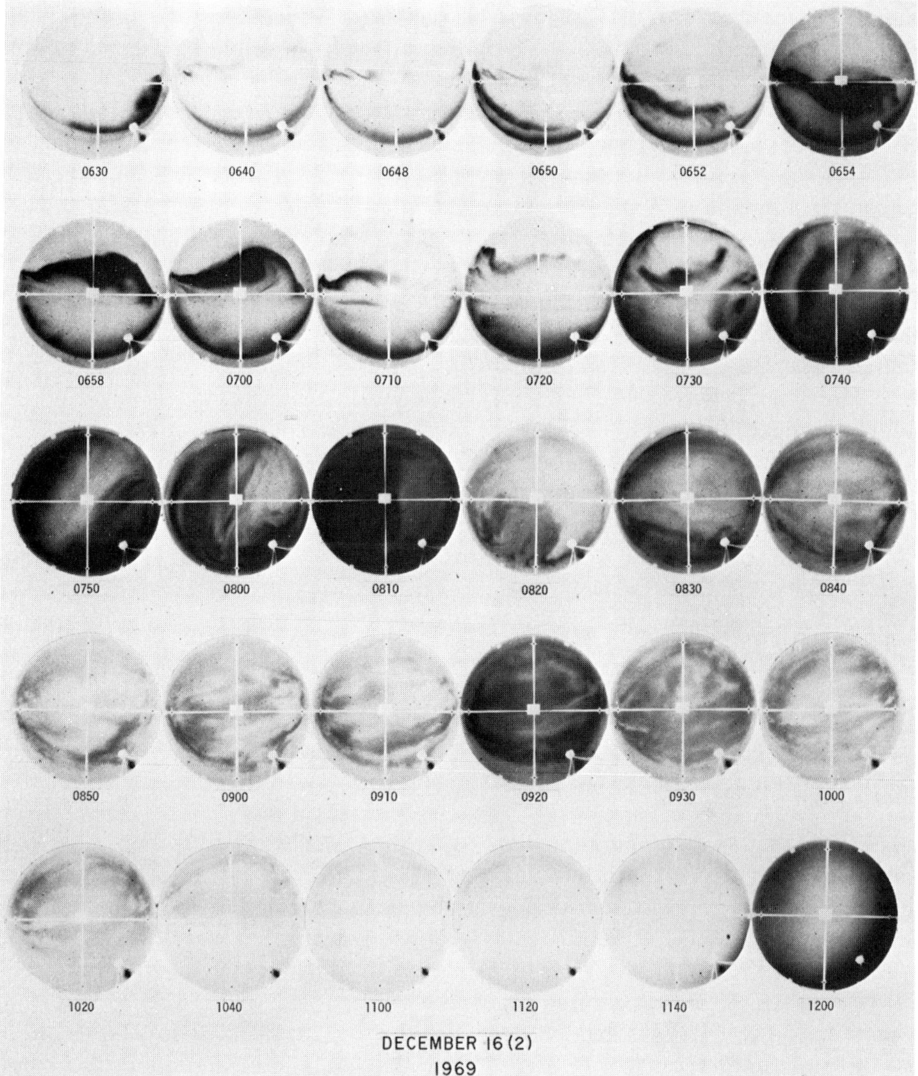

Fig. 8.9(b.2).

electrons increased appreciably at that time, while the associated change of the number density was not significant. Note that the magnetic field increased at that time. The injection of ring current protons was not clear until about 0715 UT, while the 'dispersion curves' for the Van Allen electrons are evident. After 0715 UT, electrons over a wide range of energies were present at the location of the satellite, indicating that the satellite was in the plasma sheet.

The complete lack of low energy electron response to the first auroral substorm activity at the location of the satellite indicates that the satellite was well on the inner side of the earthward boundary of the plasma sheet. On the other hand, the injected protons drifted westward and reached the location of the satellite. A part of the magnetic field depression at the satellite is likely to be due to the growth of an asymmetric proton belt.

It appears that the minor auroral activity between 0340 and 0430 UT advanced the front of the plasma sheet to the synchronous distance. In particular, at 0410 UT when ATS-5 encountered an intense flux of plasma, a bright arc was extending along the E-W direction; one can infer that the western 'end' of the arc seen in the GWR photograph was located near the 'foot' of ATS-5 at that time.

The brightening of the arc at 0610 UT in the southern sky at GWR appeared to be associated with an earthward advance of a hotter plasma to the synchronous distance or with an enhancement of fluxes well above the detectors' threshold values. Again, the western 'end' of the arc was near the 'foot' of ATS-5 at that time. However, it was as late as 0650 UT when a very intense plasma engulfed the satellite; indeed, the southern-most arc brightened considerably at that time. The auroral activity further intensified at 0720 UT, and the satellite encountered a plasma of even more intense fluxes.

Particular features of eastward drifting patches and irregular folds are not easily associated with any specific features in the spectrograms. However, it is quite likely that these morning features of auroras are associated with the disturbed plasma sheet, since ATS-5 was embedded in it during the period when such features were dominant.

An important point to be noted is that when the magnetic field increased at 0715 UT, the kinetic pressure of the plasma also increased, so that this field increase cannot be due to the removal of the diamagnetic effect of plasma at the satellite location, as suggested by Atkinson (1971) and others.

(ii) *1970, January 29: Figures 8.10(a) and (b).* A westward traveling surge was seen very near the northern horizon at GWR at 0030 UT; it was associated with the substorm which was recovering at the beginning of the day. The activity subsided after 0100 UT, and the oval contracted poleward; it was beyond the field of view at GWR. At 0330 UT, the oval began to expand as a substorm grew. A diffuse aurora (with some internal structures) crossed the zenith of GWR at 0400 UT. The activity was suddenly enhanced at 0452 UT and continued throughout the night until about 1000 UT. The oval receded poleward at about that time. In particular, a reasonably well-defined substorm occurred at 0545 UT when an arc in the southern sky at GWR brightened and moved violently poleward. The corresponding magnetic record from GWR showed an enhancement of the negative bay at that time. After 0650 UT, the auroral activity decreased; this is also reflected in the magnetogram from GWR. However, there was no definite sign of new substorm activity at about 0715 UT when the negative bay was enhanced, although eastward drifting patches began to appear at about 0720 UT.

The arrival of the protons and electrons injected in the midnight region during the recovery phase of the first substorm is evident as a well-defined dispersion curve in early UT hours on 1970, January 29. However, ATS-5 had not been immersed in the plasma when it encountered the plasma sheet. The flux was very weak until about 0550 UT. Between 0620 and 0745 UT, ATS-5 encountered a

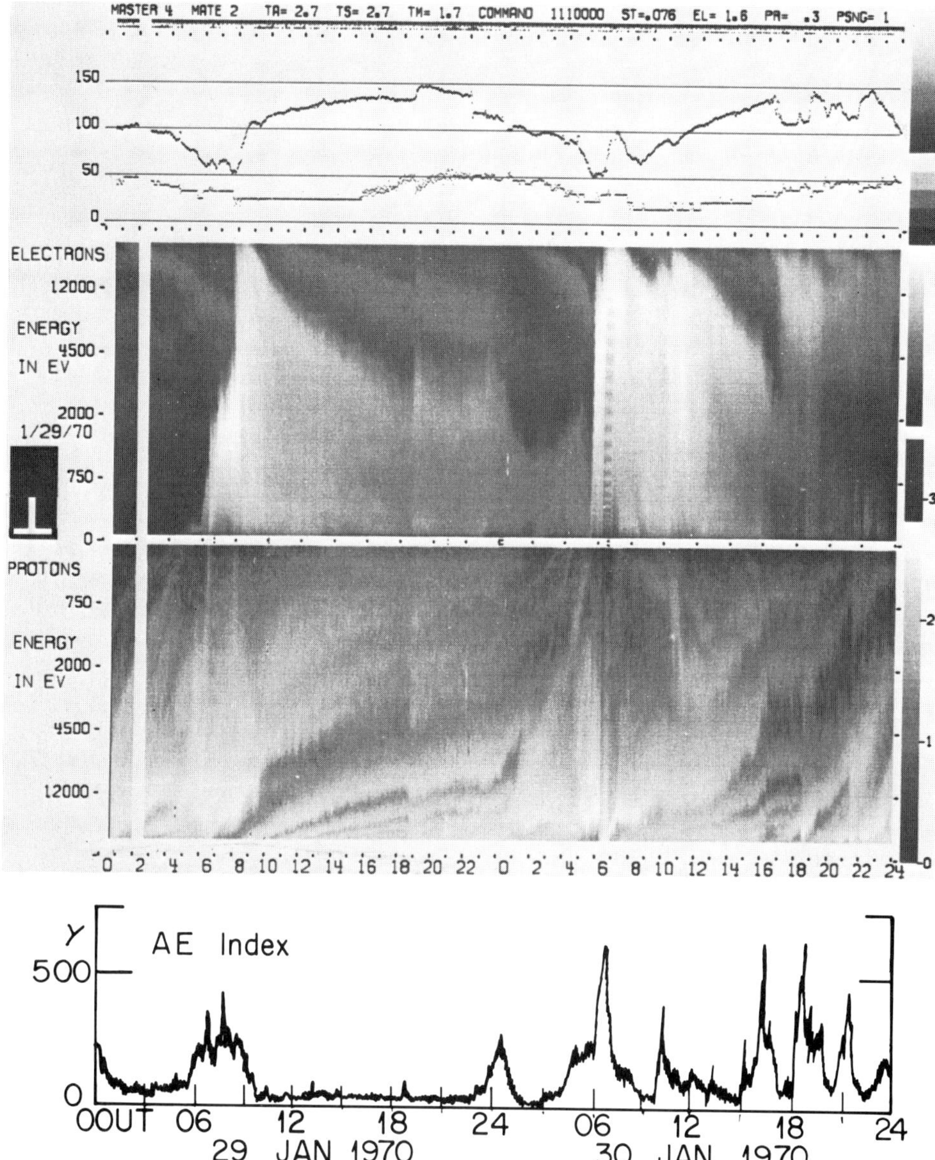

Fig. 8.10(a, b). Hot plasma particles observed at the ATS-5 satellite and selected all-sky photographs from Great Whale River, Canada. The AE index and the satellite magnetic record are also shown. (Akasofu, S.-I., DeForest, S. and McIlwain, C.: *Planet. Space Sci.* **22**, 25, 1974.)

plasma of a much higher flux, but the electron energy was not more than 5 keV. (Unfortunately, there was some bad telemetry at this time, causing the peculiar stripes.) At about 0735 UT, electrons of much higher energies arrived at ATS-5. The magnetic field was continuously decreasing from about 0400 UT until about 0600 UT when it recovered slightly; a significant recovery began at about 0750 UT.

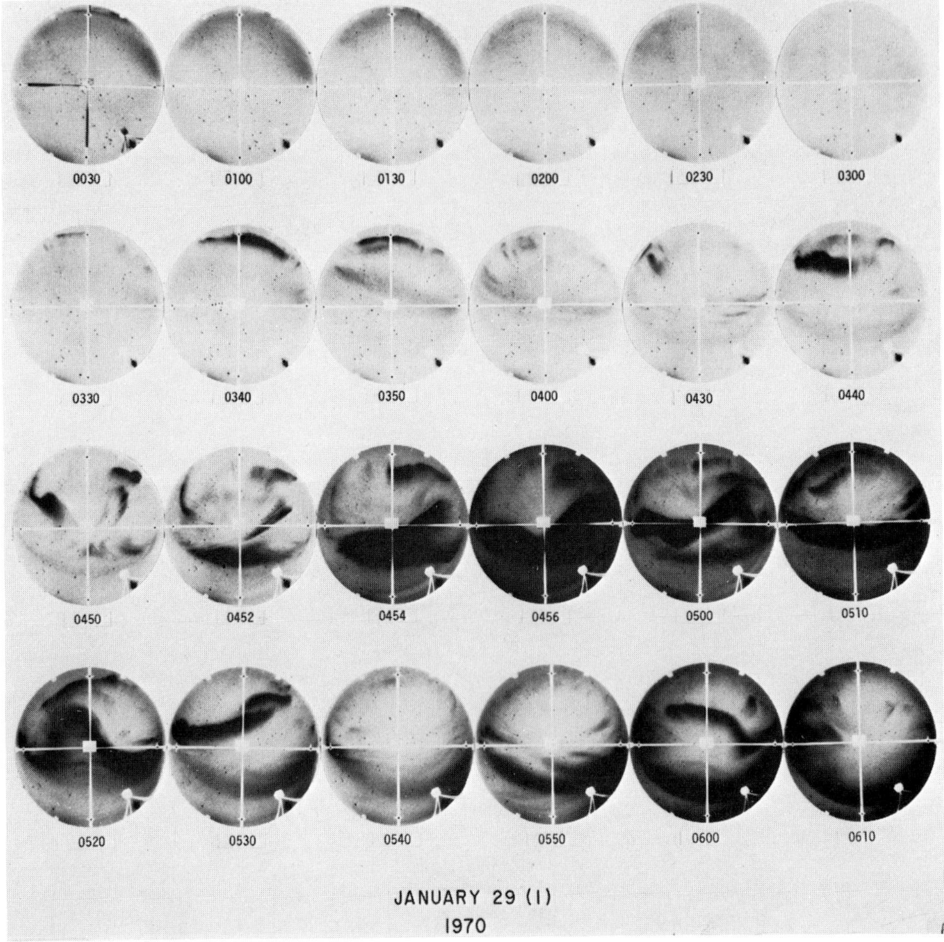

Fig. 8.10(b.1).

This is another example in which most evening substorm phenomena in the plasma sheet occurred beyond the synchronous distance. ATS-5 detected low energy electrons at about the time when an arc near its 'foot' brightened (0452 UT).

At about 0610 UT, an arc near the 'foot' of ATS-5 brightened, and the satellite encountered an intense flux of plasma, although the maximum energy of electrons was less than 5 keV. This particular type of plasma was observed at ATS-5 at least until 0735 UT, although the violent auroral activity subsided at GWR by 0700 UT.

There was no drastic change of auroral displays at the time of the arrival of the hotter plasma, at about 0735 UT, but it may well be that the arrival at about that time was associated with the appearance of patchy auroras in the southern sky which became more evident after 0804 UT (particularly after 0844 UT). No distinct auroral display was associated with the increase of the magnetic field at

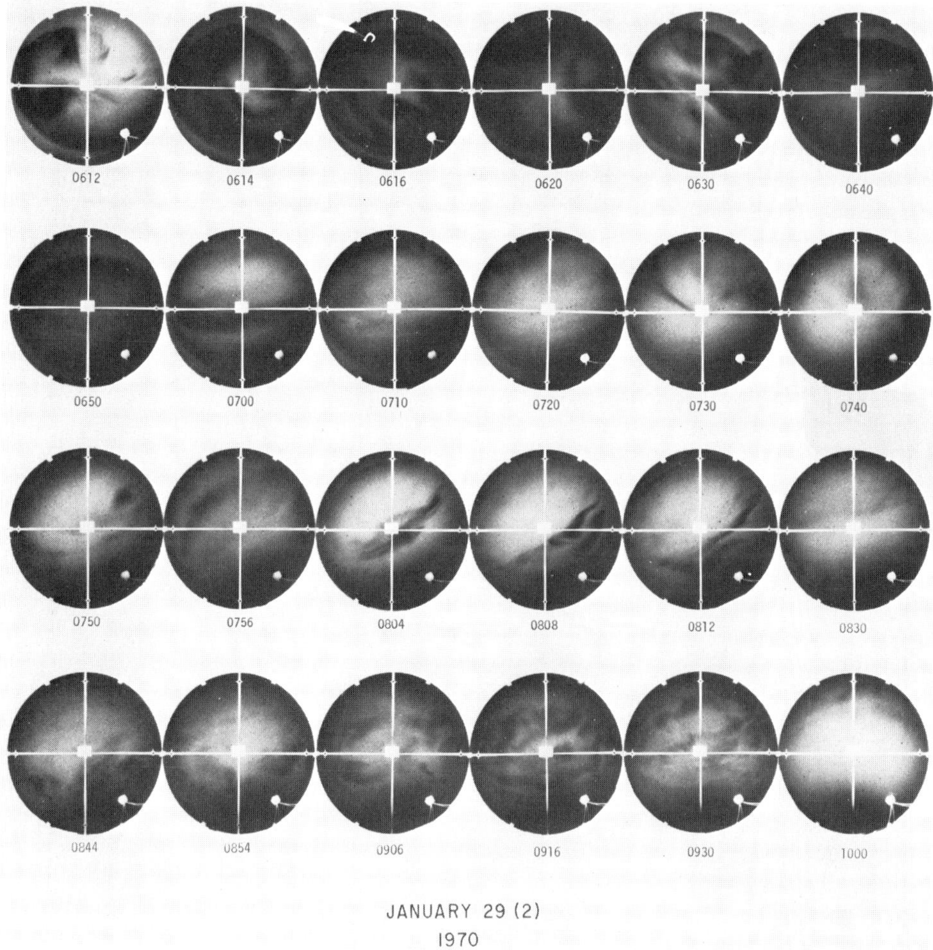

Fig. 8.10(b.2).

0750 UT. Again, it is not possible to find any specific features in the spectrogram that can be associated with individual auroral features after 0735 UT, although one can infer that the drifting patches are associated with the disturbed plasma sheet and the Van Allen belt.

(iii) *1970, January 30: Figures 8.10(a) and 8.11.* Auroral activity on this day was similar to that on 1969, December 16. A substorm of moderate intensity was in progress at the beginning of 1970, January 30. The associated auroral activity is seen as a westward traveling surge near the northern horizon of GWR (0030 and 0100 UT). After 0330 UT, a diffuse aurora began to shift equatorward. An arc near the poleward boundary of the diffuse aurora showed an indication of brightening at about 0416 UT. It brightened considerably at 0436 UT, and a westward traveling surge arrived over GWR at 0432 UT. Since the surge activity was mostly confined

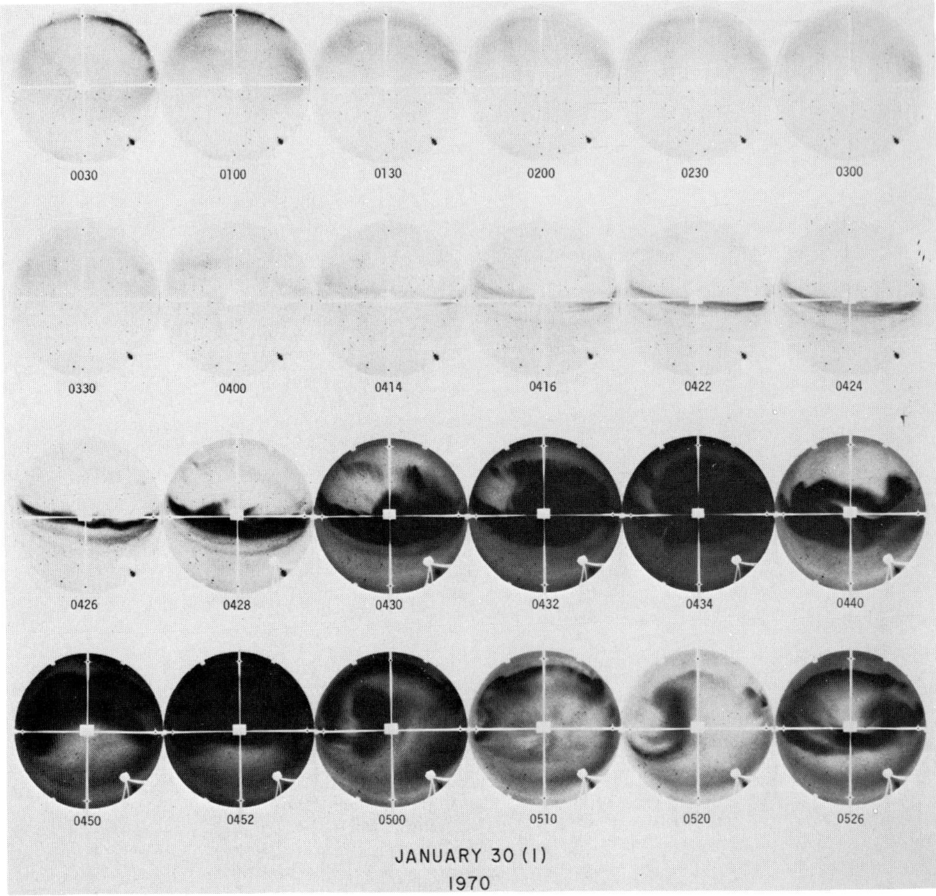

Fig. 8.11(a, b). All-sky photographs taken from Great Whale River on 1970, January 30. See Figure 8.10(a).

to the northern sky, the corresponding magnetic variations were not well defined. After the passage of the surge, auroral activity subsided considerably (0500–0520 UT); note that the diffuse aurora was located near the foot of ATS-5 after 0440 UT. New activity began at 0526 UT, when an arc near the northern boundary of the diffuse aurora suddenly became active, but it subsided by 0620 UT; a weak negative bay was observed at GWR during this period. Further activity at 0624 UT near the eastern horizon spread over the whole sky in 12 min (0636 UT). An intense negative bay developed at GWR. Thereafter there was an eastward drift motion of patches until the dawn twilight obscured the display.

This day was different from the two previous days in that the satellite had been embedded in a plasma of low energies (electrons < 2 keV; protons < 10 keV) from the beginning of the day; note also that injected ring current protons and Van Allen electrons were also present in the vicinity of ATS-5. The nature of the low energy plasma can be examined from the spectrograms on both January 29 and 30,

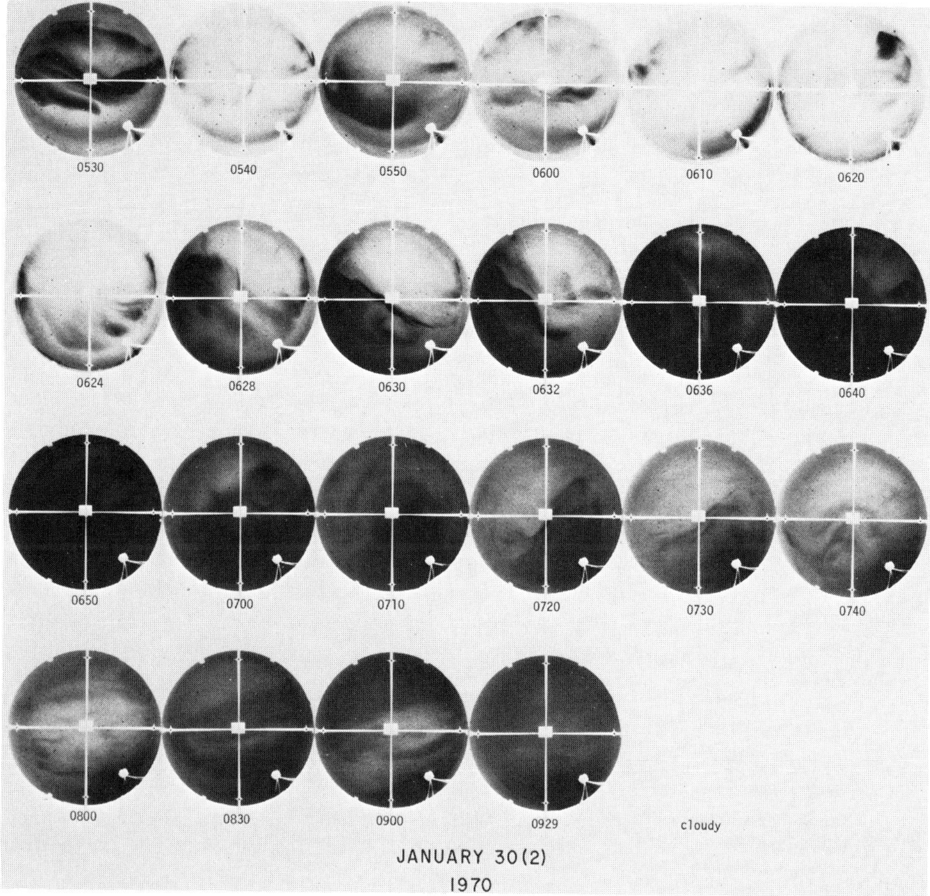

Fig. 8.11(b).

in which one can see clearly that the particular low energy plasma observed during the early UT hours of January 30 was a component of the plasma injected around 0600 UT on January 29. The low energy plasma was corotating with the Earth and ATS-5 at that time (McIlwain, 1972). Thus, strictly speaking, this particular plasma does not belong to either the plasma sheet or the plasmasphere.

A fairly intense flux of protons, which were injected during the substorm in progress at the beginning of this day, arrived at the location of the satellite, the 1700 MLT sector, during the first hour of this day. The Van Allen belt electrons began to arrive at the ATS location at about 0130 UT. No significant change of the spectrograms occurred until about 0430 UT when the energies of the trapped component of the Van Allen electrons and their fluxes began to decrease. The magnetic field began to decrease, while the number density and pressure of protons (both the parallel and perpendicular components) increased.

The satellite began to encounter the plasma injected during the first substorm at about 0515 UT. It encountered a hot plasma with an intense flux at about

0535 UT, an even hotter plasma at 0620 UT and finally another plasma at about 0617 UT. During the last encounter, the proton density and pressure decreased momentarily, and the magnetic field component suddenly increased. The satellite was embedded in the hot plasma at least until 1200 UT.

There was no obvious detectable luminosity near the 'foot' of ATS-5 until about 0428 UT, although ATS-5 observed low energy plasma from the beginning of the day. A low level (< 1 kR) uniform glow is difficult to identify in all-sky photographs.

During the first 6 hours of 1970, January 30, ATS-5 was located in the plasma region where the average energy of plasma was significantly lower than that in the deep plasma sheet. There the satellite observed neither plasma nor magnetic field variations, which might have been associated both with the auroral activity at the very beginning of the day and the intense westward traveling surge at 0432 UT. In the midnight sector, however, the situation appears to be different. Hot plasmas are often seen at the time of auroral activation.

Another important feature to be noted is an enhanced decrease of the magnetic field at ATS-5 which began at about 0400 UT. An intense westward traveling surge was observed during this period. The integral data showed a slight increase of both proton and electron pressure, and it is likely that the depression of the magnetic field was partly caused by drifting protons which were injected a little beyond the distance of ATS-5. The sudden increase of the magnetic field at 0617 UT was not associated with any significant change of plasma pressure except for a momentary proton and density decrease.

(iv) *Summary*

Evening sector. On relatively quiet days, ATS-5 is located between the earthward boundary of the plasma sheet and the Earth. Plasma sheet disturbances associated with westward traveling surges occur beyond the synchronous distance. On the other hand, the injected protons can reach the satellite after drifting westward, and the injected electrons can reach the satellite after drifting eastward. GWR is located well equatorward of the oval on such days. Further, on such days the first encounter with a hot plasma is often, but not always, associated with the equatorward shifts of the diffuse aurora.

On a moderately disturbed day, ATS-5 is often surrounded by a plasma cloud which had been injected on the previous day and which corotates with it. This plasma cloud is located in the vicinity of the earthward boundary of the plasma sheet. The average energy of the electrons has been observed to decrease monotonically as a satellite traverses this region toward the Earth (Section 3.8.1).

On a moderately disturbed day, in spite of the fact that ATS-5 is embedded in this particular region of the plasma sheet, no appreciable change occurs in the condition of the plasma when an intense surge is observed near the 'foot' of ATS-5. Thus in the evening sector on a moderately disturbed day, the inner boundary region of the plasma sheet is almost free of disturbance during substorms. Also, it appears that surge-associated disturbances occur in the upper (in the northern hemisphere) and the lower (in the southern hemisphere) boundary layers of the plasma sheet. Particle observations from the polar orbiting satellites support this conclusion. On a very disturbed day, ATS-5 tends to encounter freshly injected plasmas early in the evening hours. This indicates that ATS-5 is located deeper in the plasma sheet on such a day.

Midnight sector. In the midnight sector, the correlation between the initial brightening of an auroral arc and an increase of plasma flux is good at ATS-5, in particular where an arc is located near the foot of ATS-5. However, this does not mean that the whole plasma sheet in the midnight sector is seriously disturbed during substorms. The plasma near the inner boundary region (which is located inside the orbit of a synchronous satellite) appears to be relatively free of disturbances.

ATS-5 is often located outside the plasma sheet in early evening hours, but it is a daily event for the satellite to encounter the plasma sheet in midnight hours. Thus, the plasma sheet is considerably asymmetric with respect to the dipole axis (Vasyliunas, 1968; McIlwain, 1972); its inner boundary is much closer to the Earth in the midnight sector than in the evening sector.

The initial brightening and subsequent poleward expansion of auroras are often, but not always, associated with a sharp increase of the magnetic field (namely, the B_z component) at ATS-5. In general the plasma pressure increases at such times. Thus, one must be cautious in interpreting the magnetic field changes as changes in diamagnetism at the location of the satellite (DeForest and McIlwain, 1971). In fact, as shown in Section 7.2.5(b), it is reasonable to interpret the B_z increase in terms of the magnetic field which is caused by the disruption and diversion of the cross-tail current to the ionosphere.

Morning sector. Even on relatively quiet days, a single substorm can bring the plasma sheet to the synchronous distance and even closer to the Earth than during the midnight sector. Then, it remains there during the rest of the night, even after substorm activity subsides. At the 'foot' of ATS-5, the most common forms of auroras are eastward drifting patches or a multiple arc system (the structured diffuse aurora). Except for a few special occasions, it is not possible to associate specific features of such types of auroras to those of either the electron or proton spectrograms.

However, since these auroral features cover the entire sky of GWR and since there is little day-to-day change of characteristics of electron spectrograms in the morning plasma sheet, one can infer that the morning auroras originate in the plasma sheet which is relatively uniform as a function of radial distance, at least in the vicinity of the synchronous distance.

Rostoker *et al.* (1975) examined relationships between the westward electrojet and particles observed at the ATS-5 satellite. They showed that when marked increases in the fluxes of energetic particles over a wide range of energies occur simultaneously (namely, when there is no observable dispersion), it is the time when the satellite is on field lines which map to the poleward border of the substorm-intensified westward electrojet. When the satellite is on field lines which penetrate the central part of the westward electrojet, there are only steady high fluxes of energetic particles and there are no well-defined changes in fluxes associated with impulsive intensifications of the electrojet (Section 7.5.2(d)).

8.4. 'The Fault Line'

One of the interesting features of the magnetospheric substorm in the vicinity of the inner edge of the plasma sheet is that both plasma and magnetic field behavior are quite different across a certain local time meridian which is usually located in the evening sector. This particular aspect was most comprehensively studied by Lezniak and Winckler (1970) and Winckler (1970), who called the demarcation meridian line 'the fault line'.

8.4.1. DAWNSIDE OF THE FAULT LINE

On the dawnside of the fault line, typical substorm features are:

(1) The increase in the geomagnetic field intensity is of order 20 γ.

(2) The field increase is associated with a marked decrease in Θ, the angle between the magnetic field vector and the geographic north-pointing vector.

(3) The rate of increase of the field intensity is approximately 40 γ min^{-1} or greater.

(4) The 50–150 keV electron flux increases by at least a factor of 10 or the flux increases by at least a factor of 5 and the final (maximum) flux becomes greater than 4.5×10^5 cm^{-2} s^{-1} ster^{-1} keV^{-1}.

The authors found that this particular type of change tends to occur in the local time interval 2340–0220 LT, and that it can be explained in terms of the 'collapse' of the tail-like magnetic field structure to a more or less dipole-like structure. (Figures 8.12(a) and (b) show this feature schematically.) They suggested also that the collapse and the associated betatron and Fermi acceleration processes are responsible for the production of a high flux of electrons of 50–150 keV. They estimated the convection speed of plasma to be of order 0.5 R_E/min. The field intensity in the equatorial plane may increase from about 10 γ to over 100 γ; the field line may shorten to 1/3 of its pre-substorm length. In this regard, it should be noted that Kivelson *et al.* (1973) carefully examined changes of energetic electron fluxes (50 keV), both parallel and transverse to the magnetic field vector, during the period of the 'collapse' which was observed by the OGO-5 satellite at about $X = -7.7\,R_E$ on 1968, August 15. The transverse flux was increased by a factor of 3 or more, while the parallel flux showed little change (Figure 8.13), and suggested that the betatron acceleration is the major process for the acceleration. Kaufmann (1974) examined particle characteristics in low altitudes for a simple model of the collapse, by assuming that the first two invariants are conserved. He showed that there is a general tendency for electron fluxes to become slightly peaked at 90°. As stressed in the previous section, the 'collapse' is not caused by the removal of diamagnetism of local plasma. Actually, plasma pressure increases in the 'collapsed' region.

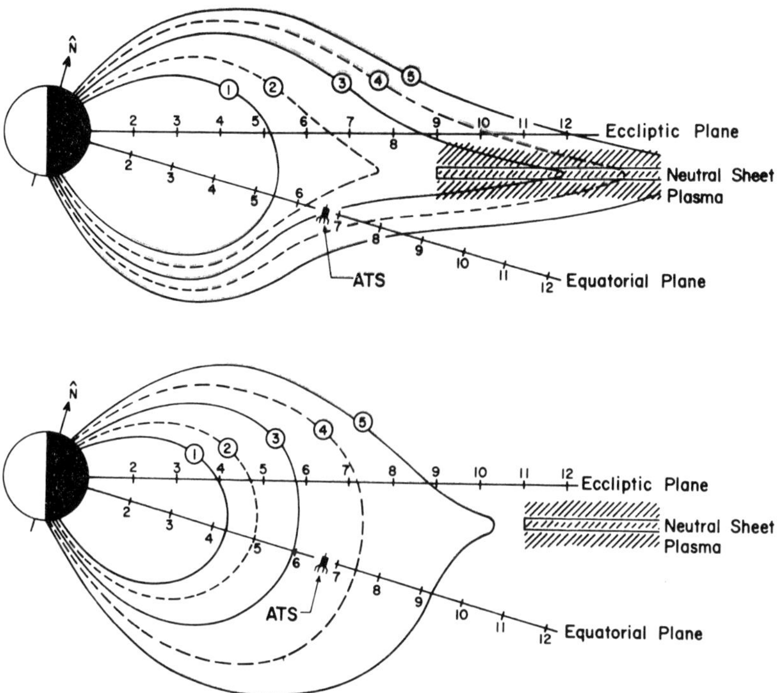

Fig. 8.12(a). 'Collapse' of the configuration of the magnetic field in the vicinity of the inner edge of the plasma sheet just east of the fault line before (top) and after (bottom) the substorm. The location of the ATS-1 satellite is indicated.

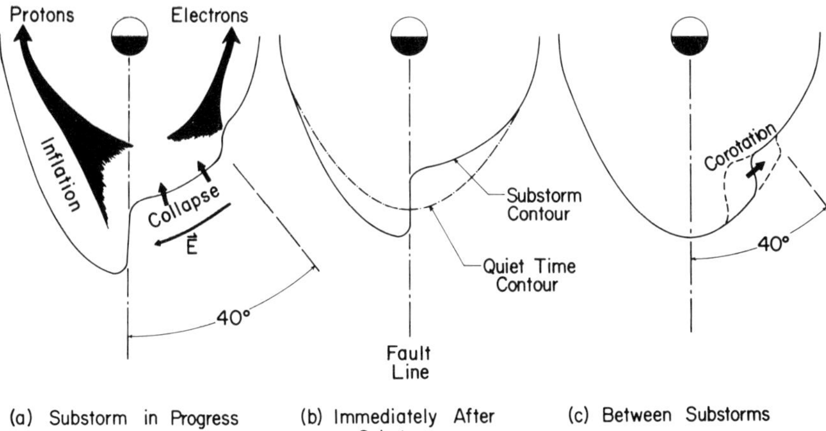

Fig. 8.12(b). Changes of the magnetic field configuration near the equatorial plane during and after a substorm. The solid curves indicate the equatorial loci of field lines intercepting the Earth at constant latitude. Inflated field lines in a 40° sector just east of the fault line 'collapse' inward producing (1) energetic protons that drift westward and inflate the evening sector of the magnetosphere, (2) energetic electrons that drift eastward and (3) a westward electric field in the 'collapsed' sector. Immediately after the substorm, field lines west of the fault line remain collapsed. (Lezniak, T. W. and Winckler, J. R.: *J. Geophys. Res.* **75**, 7075, 1970.)

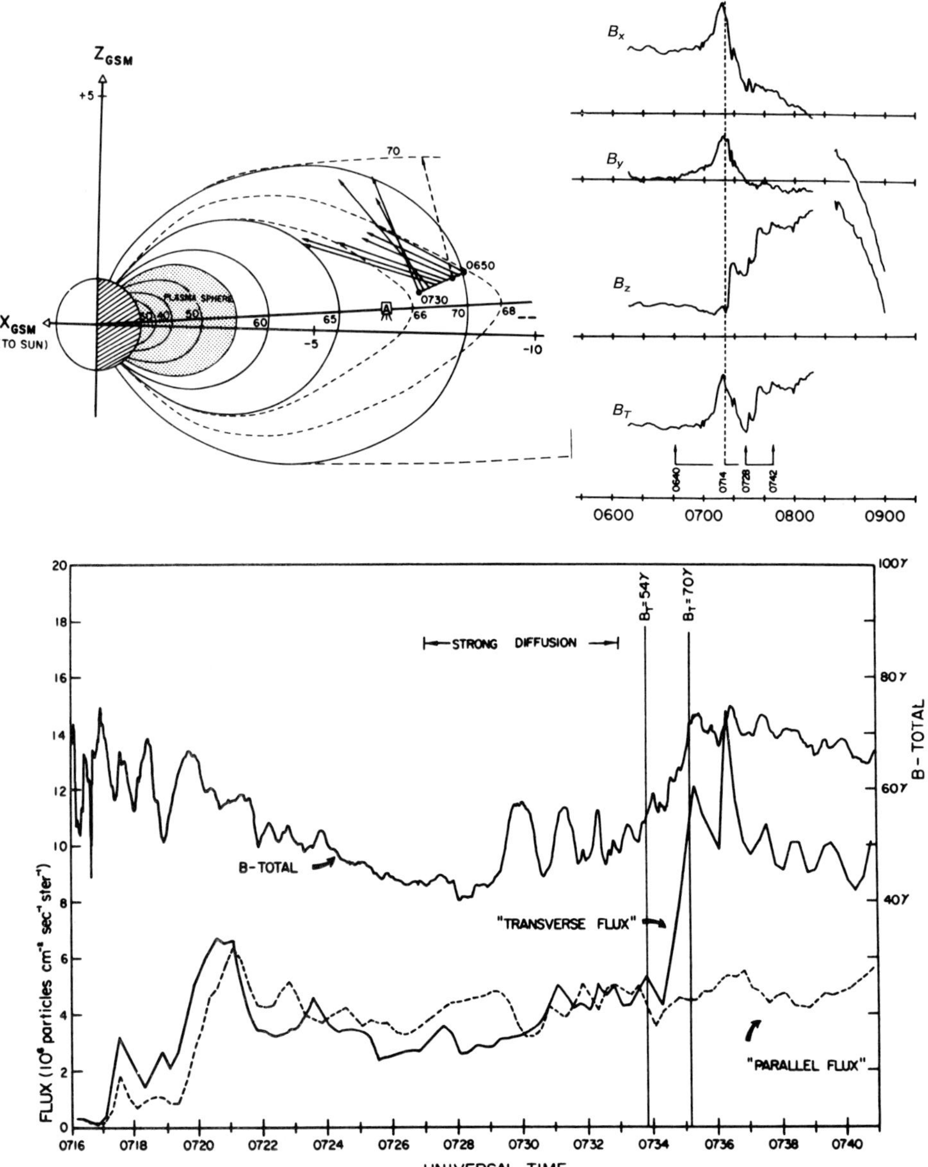

Fig. 8.13. Changes of integral directional flux (15-second averages) of energetic electrons (>50 keV) with pitch-angle transverse (90° ± 30°) and parallel (within 35°) to the magnetic field line, observed at the ATS-1 satellite, during a substorm on 1968, August 15. The magnetic field vectors projected onto the X-Y plane and the components B_x, B_y and B_z and B_T during the event are also shown. (Courtesy of McPherron, R. L., Kivelson, M. G., Farley, T. A. and Aubry, M. P.)

8.4.2. DUSKSIDE OF THE FAULT LINE

On the duskside of the fault line, typical substorm features are:

(1) Electron fluxes, particularly high energy fluxes (500–1000 keV), decrease.

(2) The angle Θ increases.

(3) The magnetic field intensity increases.

Lezniak and Winckler (1970) noted that this particular type of change tends to occur in the evening sector and that it must arise from the inflation of the geomagnetic field and the resulting distention of the field lines.

It appears that the inflation is caused by westward drifting protons from the midnight sector, which form a partial ring current.

Bogott and Mozer (1973b) also showed that sharp decreases of proton and electron fluxes are observed before the onset of the expansive phase for some substorms at the synchronous distance. They suggested that the decreases arise from the earthward motion of trapped particles, caused by a westward electric field of order 0.4 mV m^{-1} (corresponding to an inward speed of 1 R_E/45 min, since it takes typically 45 min for the fluxes to drop to their lowest level). They showed that this particular phenomenon is observed between 1700 and 0800 LT. Figures 8.14(a) and (b) show two typical examples of this feature. Recently, Barfield *et al.* (1975) reported that following the onset of an 800 γ negative bay in the auroral zone, a sharp drop of Van Allen belt electrons (35–360 keV) was observed by the S^3 satellite while plasma sheet protons (3.5–138.5 keV) increased simultaneously. Kaufmann *et al.* (1972) observed rapid changes of particle fluxes near the trapping boundary and interpreted them in terms of wave-like deformations of the outer boundary, which propagate toward the dayside (see also Su *et al.*, 1976).

8.5. Drift Motions

8.5.1. PROTONS

(a) *Satellite Observations of Drifting Protons*

(i) *ATS-5 observations.* Motions of injected protons and electrons at the synchronous distance during magnetospheric substorms have been extensively studied by Freeman *et al.* (1968), DeForest and McIlwain (1971), Bogott and Mozer (1973a, 1974) and McIlwain (1974).

We have already examined several examples of plasma data obtained by the ATS-5 satellite. When such drift motions are observed at a point in the trapping region (at the ATS-5 satellite in this case), particles of different energies (ε) arrive at different times (t), indicating the dispersion.

McIlwain (1972) constructed theoretical dispersion curves for both protons and electrons which drift under the influence of an assumed set of time-independent electric and magnetic fields, after being injected at a geocentric distance of 6.6 R_E at 2300 LT. In Figure 8.15, each curve indicates the expected dispersion effect at the ATS-5 satellite for protons and electrons of a wide range of energies at different UT. Such expected dispersion curves may be compared with

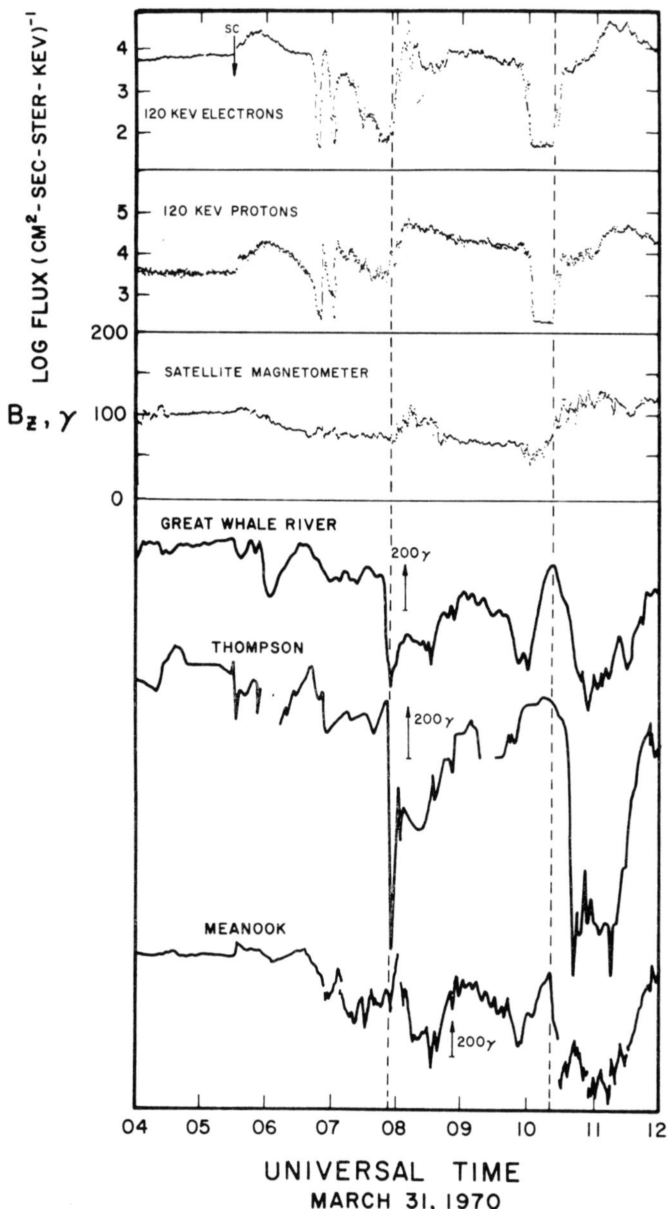

Fig. 8.14(a, b). Sharp decreases of proton and electron fluxes, observed by the ATS-5 satellite, just before the onset of negative bays at several auroral zone stations. (Bogott, F. H. and Mozer, F. S.: *J. Geophys. Res.* **78**, 8119, 1973.)

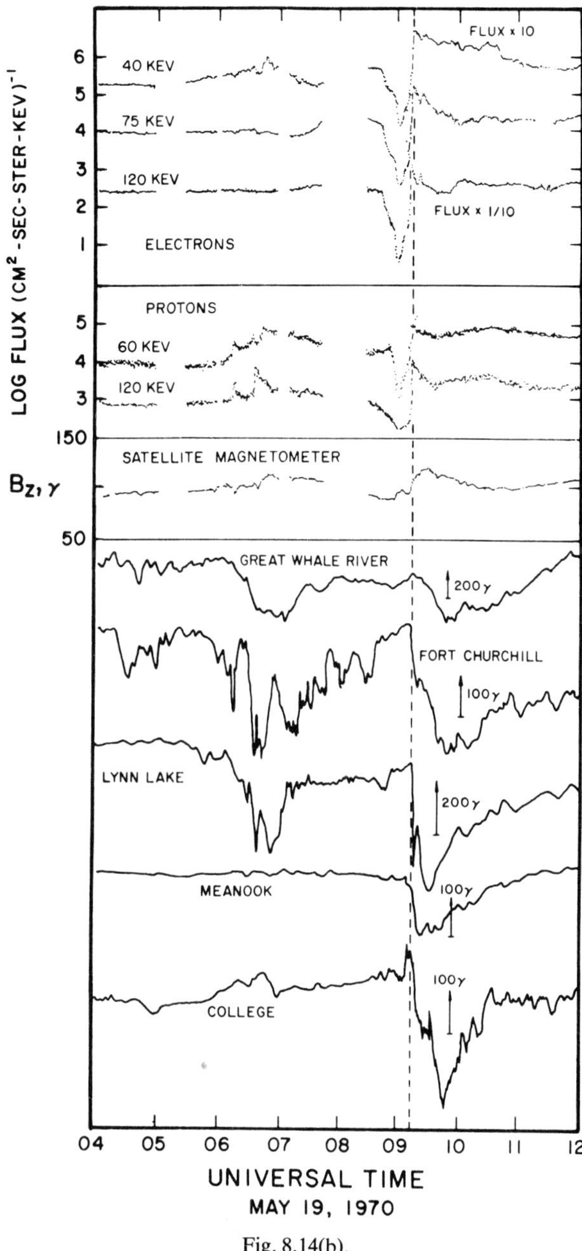

Fig. 8.14(b).

an example of the observed dispersion curves (Figure 8.16(a)). Figure 8.16(b) indicates the major features in Figure 8.16(a). There is a reasonable similarity between the expected dispersion curves and some of the observed ones. Figure 8.16(a) also shows the electron pressure perpendicular to $\boldsymbol{B}$. It can be seen that it increased by two orders of magnitude during an intense injection at about

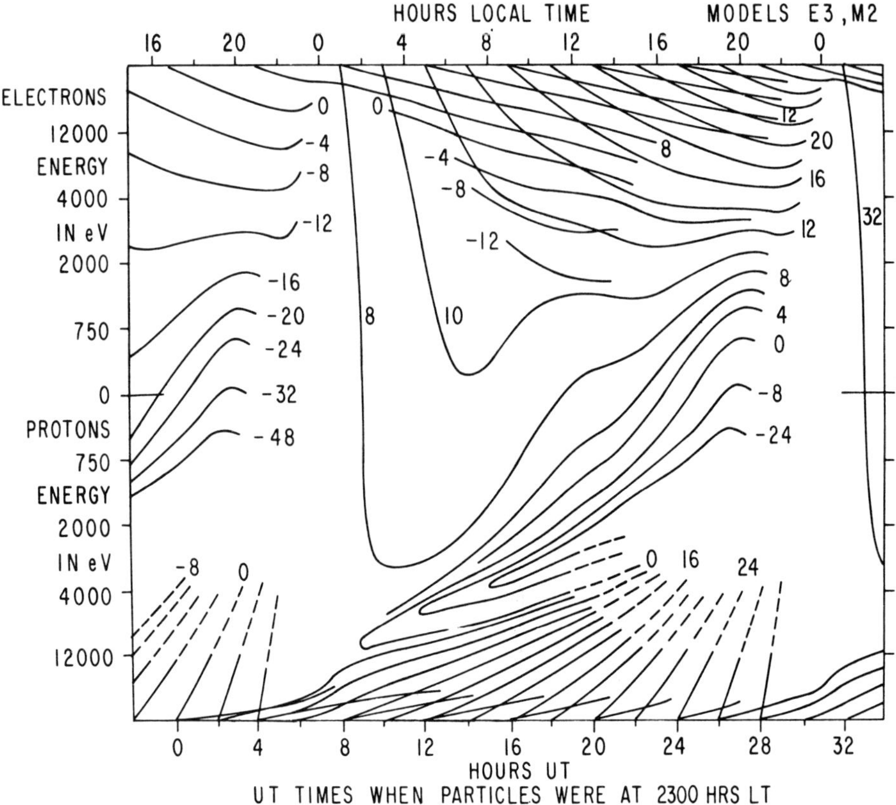

Fig. 8.15. Computed dispersion curves for the injected electrons and protons. The universal times are plotted at which the particles encountered by the ATS-5 satellite were at 23 LT. The dashed portions correspond to trajectories which do not meet the 23 LT meridian, so that neighboring starting locations had to be used. The curves for the energetic trapped particles making between one and two orbits around the Earth are also included. (McIlwain, C. E.: *Earth's Magnetospheric Processes*, B. M. McCormac (ed.), p. 268, D. Reidel Pub., Co., 1972.)

0300 UT on January 2. At the same time, the magnetic field component perpendicular to the equatorial plane increased from about 50 to 120 γ. Therefore, the magnetic field increase is not caused by the removal of diamagnetism so one must be cautious in the use of the term 'collapse'.

In a later paper, McIlwain (1974) formulated the problem differently. The electric and magnetic fields can be thought of as an optical system which images particle trajectories from the injection boundary onto the satellite path. The problem then is to deduce the electric field (the optical system) and the geometry of the injection boundary (the objects) on the basis of the observed energy-time spectrograms which correspond to the images in the optical analogy. However, this problem cannot be solved on the basis of the spectrogram obtained from a single satellite. McIlwain attempted to determine the injection boundary in such a way that the results are at least consistent for both protons and electrons. It is also assumed that both the magnetic and electric fields remain unchanged during their

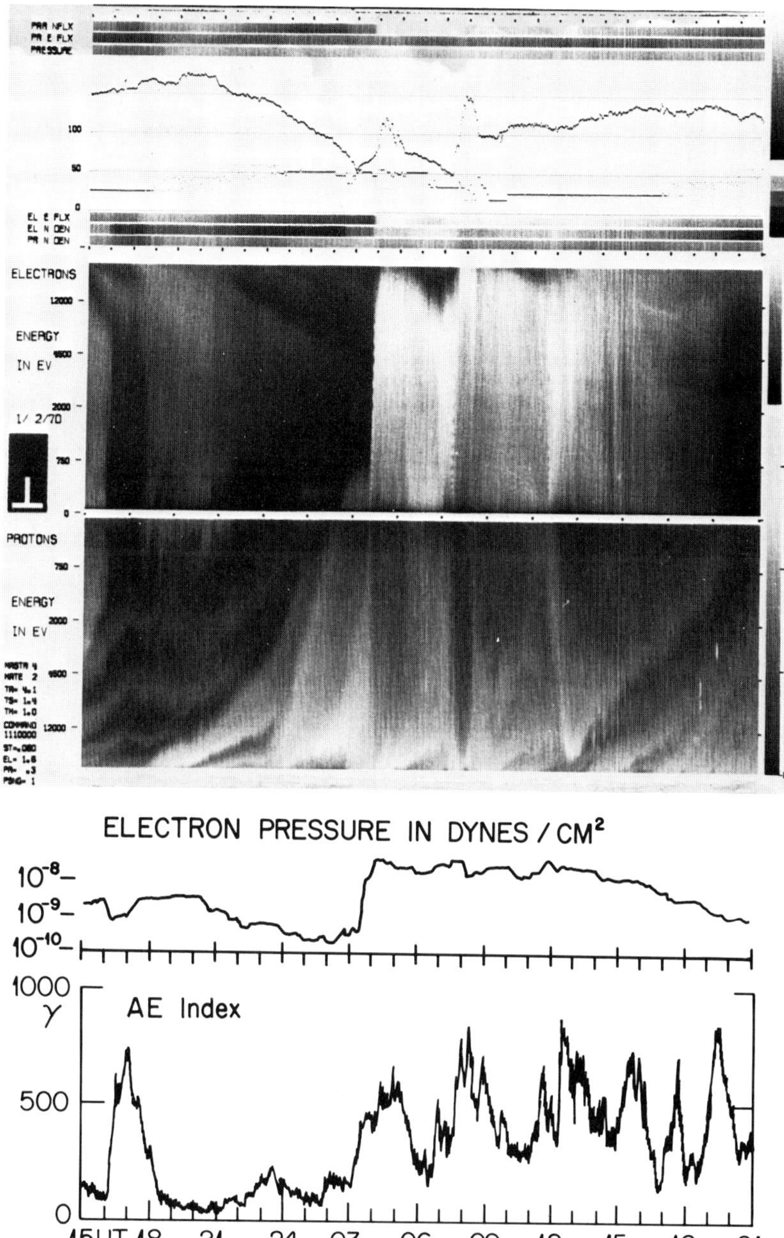

Fig. 8.16(a, b). Injection of electrons and protons, observed at the ATS-5 satellite on 1970, January 2. From the top: the *H* component magnetic record from the satellite, the electron and proton spectrograms, plasma pressure and the AE index. In (b), some of the major features in the spectrograms are reproduced. (DeForest, S. E. and McIlwain, C. E.: *J. Geophys. Res.* **76**, 3587, 1971.)

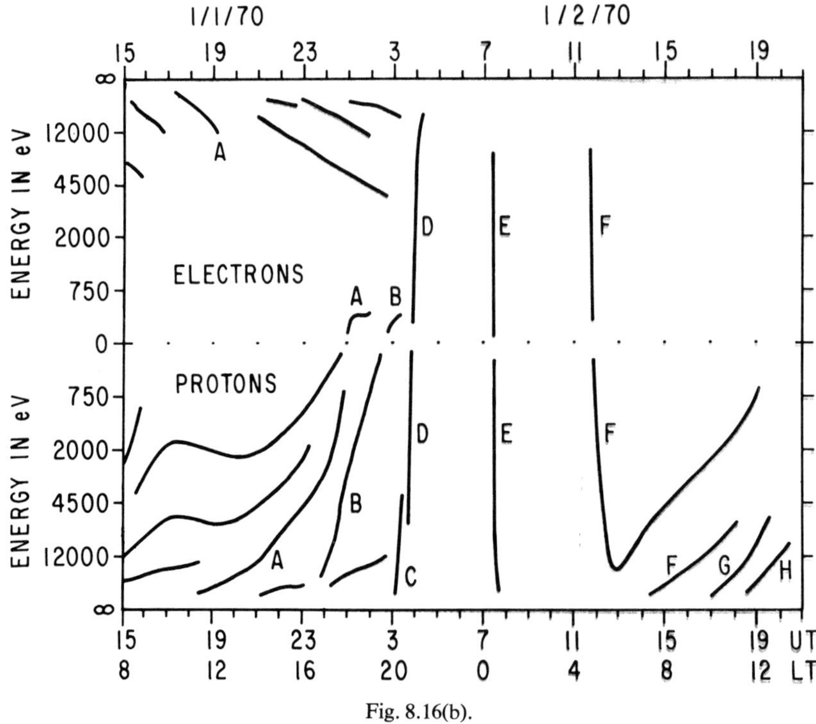

Fig. 8.16(b).

drift motions. The procedure of this study is then to 'backtrack' particle trajectories for a given combination of particular electric and magnetic fields and determine the injection point; his model electric field (E3) is shown in Figure 1.21. Figure 8.17(a) shows an example of his results. The location of the arrow head indicates the point of encounter of the satellite with particles. The circle with a dot indicates the computed location of those particles at 0030 UT, the injection time. The inferred injection boundary based on an independent study (Mauk and McIlwain, 1974; Figure 8.17(b)) is shown by a dashed line with circles. Both curves show a spiral feature and agree reasonably well. They may be identified as the inner edge of the plasma sheet. Thus, this study suggests that the plasma injection takes place from the inner boundary of the plasma sheet over a wide local time range. Mauk and McIlwain (1974) showed that the location of the injection boundary differs for different Kp values, although the spiral geometry is well maintained. It is closer to the Earth for greater Kp indices. Recently, Freeman (1974) also obtained an empirical expression for the encounter time (LT) of a geosynchronous satellite with the plasma sheet, LT = 27.6–1.6 Kp.

Bogott and Mozer (1974) also made an extensive study of plasma particles, particularly the dispersion characteristics at the synchronous distance. They found that particles with energies between 30 and 200 keV are shown to be produced simultaneously (within tens of minutes) in the local time range between 0000 and 0400 LT and that particles with energies greater than 75 keV drift in

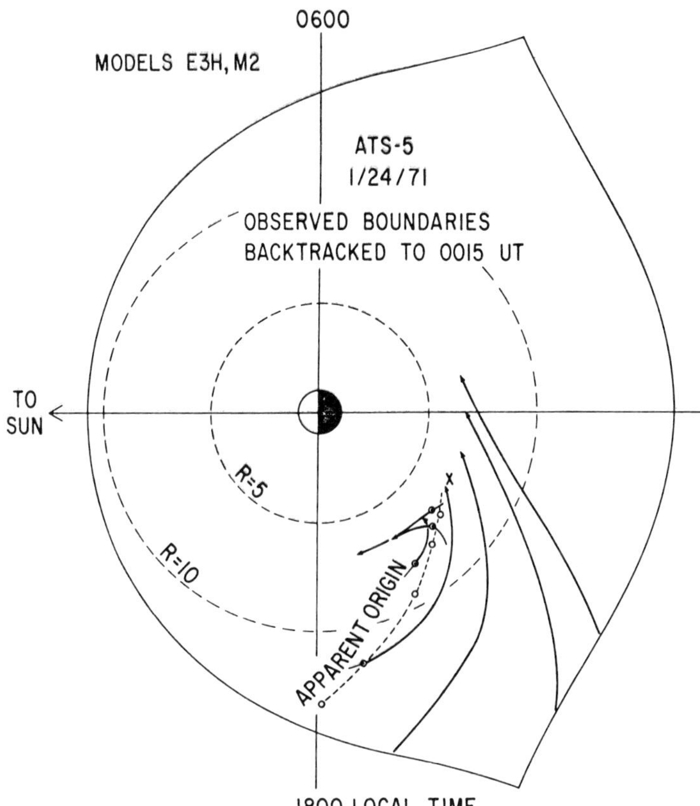

Fig. 8.17(a). Computed trajectories of protons. The arrows indicate the satellite (ATS-5) encounter, and the circle with a dot indicates the computed location at 0030 UT, the onset time of a substorm on 1971, February 6. The cross indicates the encounter with the inner boundary of plasma sheet electrons which is presumably the same as the outer boundary of the low energy (substorm injected) plasma boundary. (McIlwain, C. E.: *Magnetospheric Physics*, B. M. McCormac (ed.), p. 143, D. Reidel Publ. Co., 1974.)

longitude in the direction and with the speed expected from the gradient B drift. On the other hand, the arrival times of the lower energy particles occur much earlier than they would if the particles would have drifted from the midnight sector. Roederer and Hones (1974) examined drift motions of protons and electrons in a time-dependent electric field (a fast rise, lasting about 10 min and slow decay, lasting about 2 h) and showed that the observed dispersion of the protons at the ATS-5 satellite can be reasonably reproduced.

(ii) *Explorer 45 (S^3) observations.* The formation of the ring current belt has been extensively studied by Smith and Hoffman (1973, 1974), Konradi *et al.* (1973), Konradi *et al.* (1975) and others in terms of a combined effect of the inward connection, gradient drift and corotation.

One of the most interesting features observed by the S^3 satellite is the 'nose' structure in the proton energy spectrograms, as the satellite traverses outward

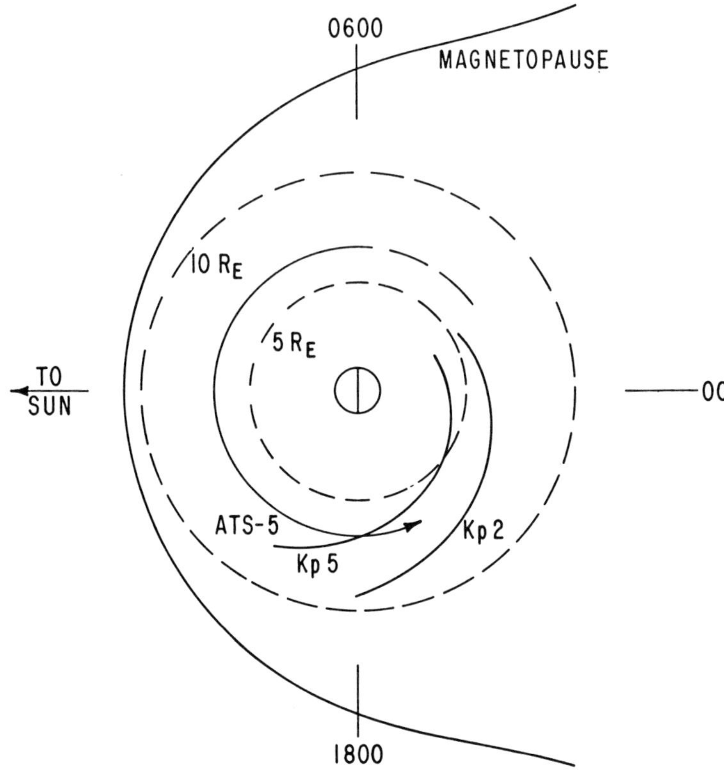

Fig. 8.17(b). Position of very low energy (injected) plasma boundaries for two values of Kp as derived from the expression $L_b = (122 - 10\ \mathrm{Kp})/(\mathrm{LT} - 7.3)$. (Mauk, B. H. and McIlwain, C. E.: *J. Geophys. Res.* **79**, 3193, 1974.)

through plasma in the evening sector. A few examples of the nose structure are shown in Figure 8.18. The satellite first encounters protons of energies 15–20 keV; this occurs inside the plasmapause. Then the flux increase spreads toward higher and lower energies, as the satellite moves away from the Earth. Subsequently, there occurs a sudden increase of electron fluxes of energies ranging from 1 to 50 keV and proton fluxes of energies 1–2 keV.

As plasma particles are convected toward the Earth, their drift paths deviate from each other, depending critically on their magnetic moment μ. The electrons and low energy protons are carried away eastward by the corotation effect. The higher energy protons become dominated by the gradient drift and follow westward trajectories. Smith and Hoffman (1974) called attention to the theoretical estimates by Chen (1970) and Jaggi and Wolf (1973) that the minimum geocentric distance to which the convection field can bring protons to the dusk meridian is roughly proportional to $\mu^{1/3}$. Thus, protons of lower magnetic moments can penetrate deeper into the Van Allen belt region. However, there is a critical value of μ; protons with μ smaller than the critical value would drift eastward. This critical value of μ is of order 40 V/γ, which corresponds to an energy of about 45 keV at $L = 5$.

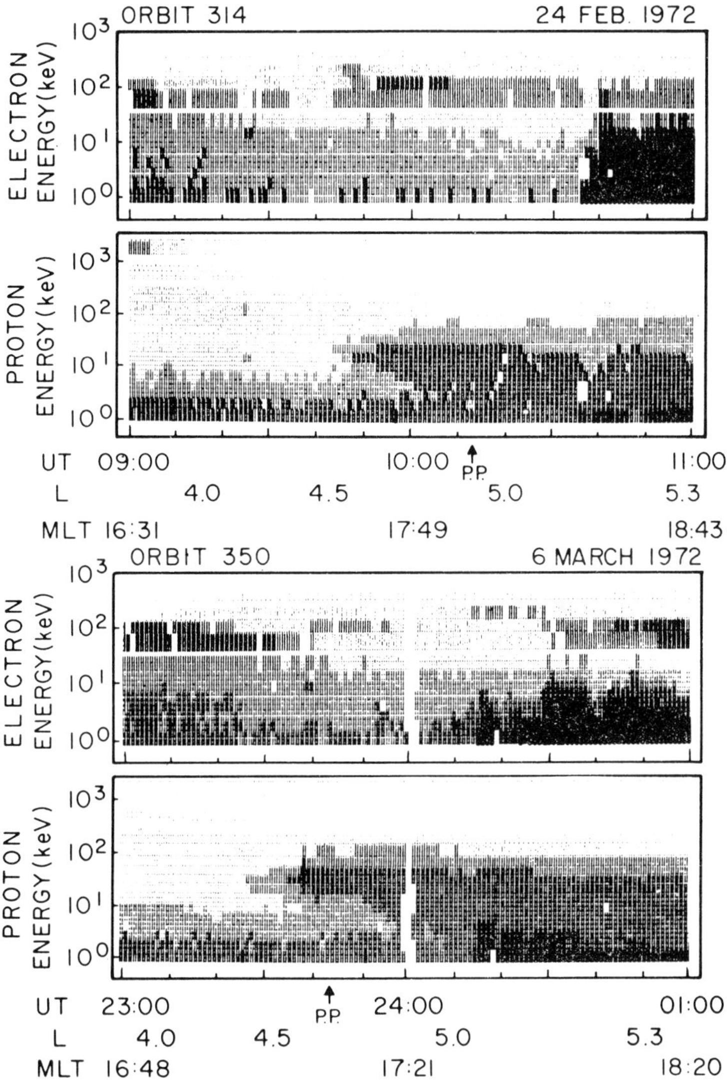

Fig. 8.18. 'Nose' structure in the proton and electron spectrograms obtained by the S^3 satellite during the development of the main phase of two magnetic storms. The gray shading is a measure of the flux (particles $cm^{-2}\,s^{-1}\,ster^{-1}\,keV^{-1}$) of the equatorially mirroring particles. Black represents the most intense flux. (Smith, P. H. and Hoffman, R. A.: *J. Geophys. Res.* **79**, 966, 1974.)

Smith and Hoffman (1974) suggested that the observed nose structure arises from the fact that the S^3 satellite travels through this spatial structure in the region of the *continuous plasma flow* which appears to be fairly stationary during an early epoch of magnetic storms. Their view differs from that of DeForest and McIlwain (1971) who attempted to explain the behavior of protons and electrons in the trapping region in terms of an injected plasma cloud. However, Williams *et al.* (1974) indicated that substorm-associated changes in the electric and magnetic fields make it difficult to assume either of the two simplified views.

Konradi *et al.* (1975) also made a detailed study of the motions of protons and electrons on the basis of their S^3 observation during a substorm on 1972, February 13. They showed they could reproduce the observation reasonably well by assuming that both protons and electrons were injected at the onset of the substorm instantaneously (less than 5 min) into the magnetosphere along and beyond (several Earth radii) the injection boundary and that they drift under the influence of the model electric (E3) and magnetic fields proposed by McIlwain (1972). Figure 8.19 shows sample trajectories of protons of energies 2.7, 6.0, 13.5, 30.0 and 43.0 keV from the injection point to the point where they were detected by the satellite; actually, the proton trajectories are traced backward in time. The period during which each energy is observed corresponds to the arc of the satellite orbit forming the terminus of the drift trajectory. It can be seen that all protons originate at or beyond the injection boundary for Kp = 5.9 (dashed line). Further, the figure suggests that protons of energies between 10 and 35 keV can be convected in from the magnetotail and that the injection boundary and the plasmapause coincide in the evening sector.

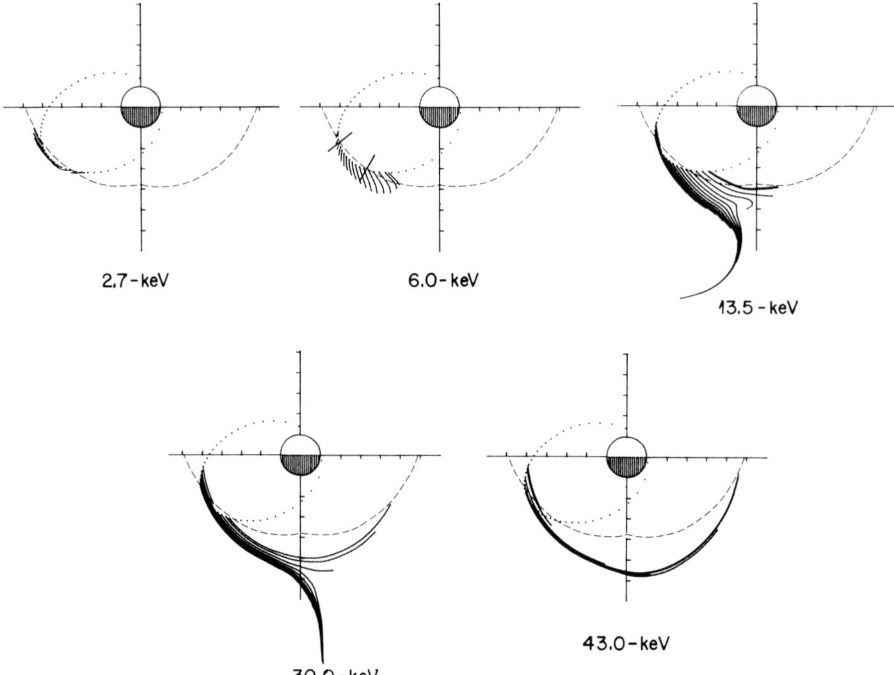

Fig. 8.19. Sample trajectories of protons drifting in a model magnetic and electric field. The protons are assumed to be injected into the trapping region at 1116 UT (1972, February 13) and drift to the S^3 satellite orbit, where they are detected with the indicated energy. The period during which each energy level is observed corresponds to the arc of the orbit forming the terminus of the drift trajectories; the injection boundary for Kp = 5.9 is indicated by the dashed curve. (Konradi, A., Semar, C. L. and Fritz, T. A.: *J. Geophys. Res.* **80**, 543, 1975.)

(iii) *OGO-3 observations*. The observation of ring current protons by the OGO-3 satellite is of particular interest because it shows clearly how rapidly and deeply plasma can penetrate into the trapping region (to about $L = 4$ or less) even during a weak storm. Figure 8.20 shows variations in the flux of ring current protons as a function of L between 1966, June 9 and July 23. There were three periods when the protons penetrated to a distance less than $L = 4$. None of them was associated with intense storms. Even during the most intense disturbance, the maximum Dst deviation was only of order 50 γ, but the protons reached a geocentric distance of as low as $L = 3$.

(iv) *Magnetic field distortion*. The injected protons form the proton belt which seriously distorts the geomagnetic field. Since the proton belt reduces the horizontal component of the geomagnetic field on a worldwide scale, particularly in middle and low latitudes, the magnetic field associated with the proton belt has been intensively investigated in the past as the cause of the main phase of geomagnetic storms; for recent studies on this subject, see Hoffman and Cahill (1968), Hoffman (1973), Davis (1969), Cahill (1970, 1973), Parady and Cahill (1972),

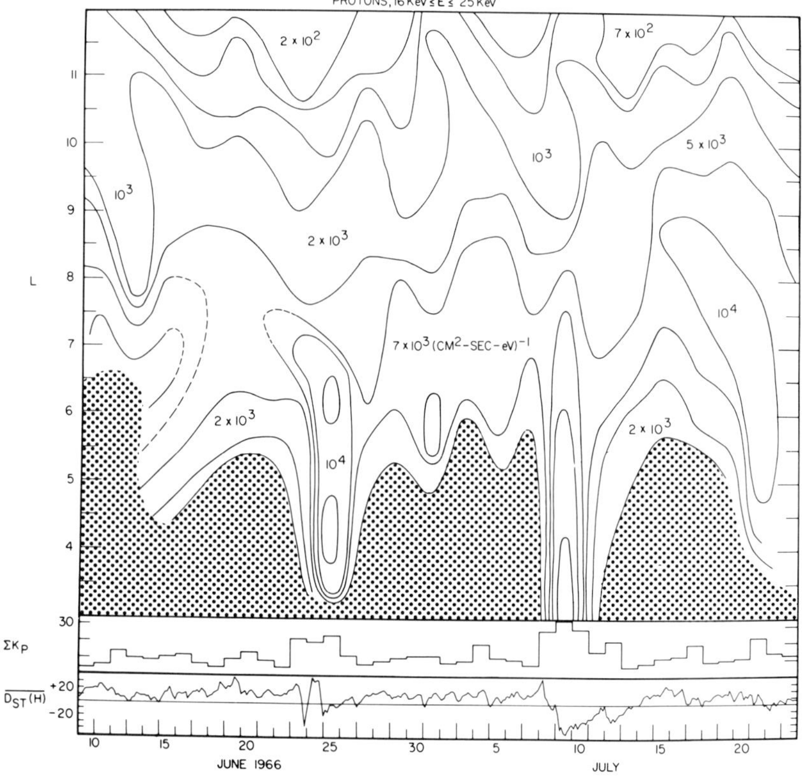

Fig. 8.20. Proton ($0.5 < \varepsilon < 50$ keV) fluxes at the magnetic equator at different L values during the period 1966, June 9–July 23. Both the Kp (daily sum) and Dst indices are shown at the bottom. (Frank, L. A. and Owens, H. D.: *J. Geophys. Res.* **75**, 1269, 1970.)

Coleman and Cummings (1971), Dolginov *et al.* (1972), Crooker and Siscoe (1974), Grafe (1974), Kawasaki and Akasofu (1971a), Langel and Sweeney (1971), Shabanskiy (1971), Shevnin (1971, 1973a, b), Sozou and Windle (1970), Zaytseva and Glazhevska (1972); see also S.T.P., Section 8.8. More recently, Berko *et al.* (1975) showed that the observed proton flux can account for the observed magnetic field of the ring current within experimental uncertainties.

(b) *Precipitating Protons*

Precipitating protons have also been extensively studied on the basis of records obtained from polar orbiting satellites (Hultqvist *et al.*, 1971; Riedler and Borg, 1972; Lindalen *et al.*, 1971; Aarsnes *et al.*, 1970; Hultqvist *et al.*, 1974; Mizera, 1974; Bernstein *et al.*, 1974). Most of the results and related issues are well summarized by Hultqvist (1975a, b).

The location of ring current proton precipitation with respect to the plasmapause has been one of the important issues in terms of precipitation processes. It appears that the following results have been confirmed:

(i) The peak of the precipitation region is located well outside the plasmapause (Figure 8.21).

(ii) A typical flux of protons of energies 5–10 keV above the atmosphere (>600 km in altitude) is of order $10^6\ \text{cm}^{-2}\ \text{s}^{-1}\ \text{ster}^{-1}\ \text{keV}^{-1}$.

(iii) A typical flux of protons of energies 5–10 keV in the equatorial plane is also of order $10^6\ \text{cm}^{-2}\ \text{s}^{-1}\ \text{ster}^{-1}\ \text{keV}^{-1}$.

These observations have an important implication in terms of precipitation processes of the protons. It has been proposed by Cornwall *et al.* (1970, 1971a),

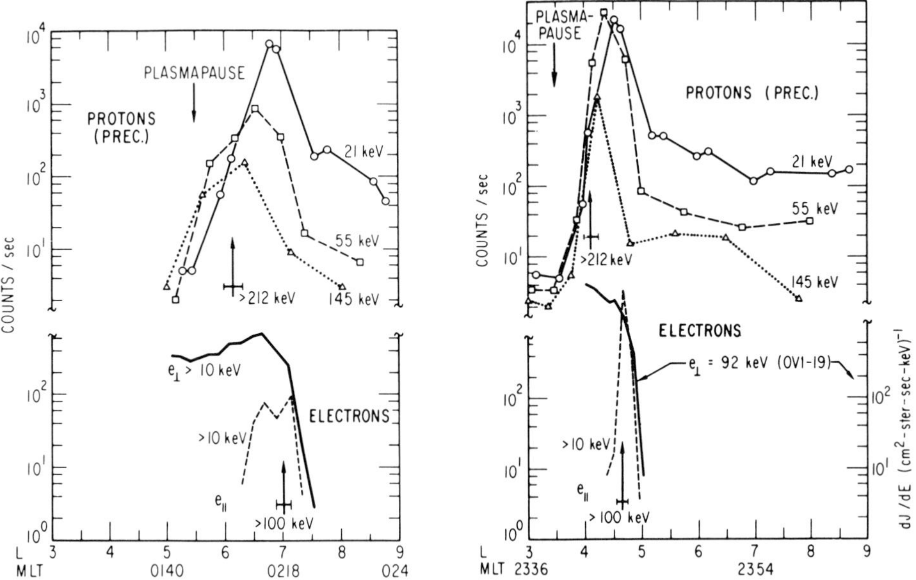

Fig. 8.21. Location of the peak of the proton precipitation with respect to the plasmapause and the trapping boundary of electrons (Mizera, P. F.: *J. Geophys. Res.* **79**, 581, 1974.)

Eather and Carovillano (1971), Coroniti *et al.* (1972) and Brice and Lucas (1975) that proton-cyclotron wave instabilities and the associated strong pitch-angle diffusion processes play the major role in precipitating the protons. The instabilities have been predicted to occur at or just inside the plasmapause, where ion-cyclotron waves are amplified by the interaction of the cold plasma in the plasmasphere with the hot ring current plasma. Mizera (1974) noted that his observation is in disagreement with the predicted dominance of the proton-cyclotron processes (Figure 8.21). Hultqvist (1975a, b) also noted that the ring current plasma is always in a turbulent state outside the plasmapause, in a wide range of distance at all local times and at all disturbance levels, and that the turbulence is always so intense that it gives rise to strong pitch-angle diffusion. The turbulence affecting keV protons is not the ion-cyclotron instability, because turbulence is both found where this particular instability is not expected to operate and not found where it is expected to be present (namely, just inside the plasmapause). He also noted that the turbulence should contain a complex set of wave-particle interactions, because it acts on protons and electrons of the same energies simultaneously in the same region of the magnetopause.

Williams and Lyons (1974a, b) and Williams (1975) examined in great detail the pitch-angle distribution of ring current protons in the vicinity of the plasmapause, on the basis of S^3 satellite data, during the recovery phase of the 1971, December 17 magnetic storm. They found that outside the plasmapause the ring current protons exhibit a pitch-angle distribution that is nearly isotropic except for an empty loss cone, but that just outside the plasmapause this distribution switches rapidly to one that is peaked around $\pi/2$, suggesting that the ring current protons are stably trapped there, with negligible losses due to pitch-angle scattering. On the other hand, within the plasmasphere a moderate pitch-angle diffusion process is present, and the measured quantities, including the cold plasma density, can be interpreted in terms of scattering by ion-cyclotron waves.

In a recent review, Gendrin (1975) summarized these observations as follows:

(i) Well outside the plasmapause (at high L values), the protons interact with electrostatic waves, resulting in strong diffusion; this explains an isotropic pitch-angle distribution of low-energy protons (>6 keV).

(ii) Proceeding inward, but still outside the plasmapause, the electrostatic instabilities are quenched as the density of cold plasma increases. Protons of high energy may interact with electromagnetic cyclotron waves.

(iii) In the vicinity of the plasmapause, the critical energy for cyclotron interaction is high, preventing the precipitation of the protons; this explains the flat distribution with the empty loss cone and the absence of precipitating protons at low altitudes.

(iv) Proceeding still further inward, the ion-cyclotron electromagnetic interactions may start again, the high energy protons precipitating first and their equatorial pitch-angle distribution being rounded with a peak at 90°, indicating the presence of a weak diffusion process.

It should also be noted that the energy lost from the ring current belt has been suggested as the source of energy for the mid-latitude subvisual red arc (Cole, 1965; Cornwall *et al.*, 1971a). The energy associated with the amplified ion-cyclotron waves is eventually converted into heat energy of the cold electrons in the ionosphere. Recently, Williams (1975) showed that the energy lost by the pitch-angle diffusion (though not a strong one) was more than the energy needed to excite the mid-latitude red arc, which was observed from Fritz Peak Observatory during the 1971, December 17 storm. For a recent review on the subject of the mid-latitude red arc, see Rees and Roble (1975). Substorm-associated changes of the precipitating protons have been studied by direct observations of the protons by polar orbiting satellites and the Hβ emission observed by ground-based photometers. The latter was discussed in Section 6.4.4(b). Lindalen *et al.* (1971) and Hauge and Søraas (1975) showed that (i) during quiet geomagnetic conditions there is a single zone of precipitating protons (100–200 keV) located at an invariant latitude of 67° on the nightside and two zones located at invariant latitudes of 70° and 77° (anisotropic pitch-angle) on the dayside; (ii) closely correlated with substorms, the proton trapping boundary moves poleward, the two zones of proton precipitation on the dayside merging into one zone; (iii) preceding the poleward movement of the trapping boundary, there is an intensity increase in the already existing proton precipitation on the nightside; (iv) the proton precipitation depends on substorm activity, as well as the ring current intensity; (v) the equatorward boundary of the main precipitation is well related to the Dst index, in the evening sector, and shifts equatorward with increasing ring current intensity (Figure 8.22); and (vi) the poleward precipitation boundary moves equatorward as the ring current intensity reaches about $-100\ \gamma$.

Prölss (1973) and Smith *et al.* (1975) noted that the charge exchange process can also account for the decay of the storm-time proton ring current. They demonstrated that the observed decay rate agrees well with the estimated one by Liemohn (1961).

(c) *IPDP Pulsations*

It has been known that a particular type of geomagnetic field pulsations, IPDP (intervals of pulsations of diminishing periods), is observed in the evening sector during magnetospheric substorms (Troitskaya, 1961; Fukunishi, 1969; Saito, 1969; Jacobs, 1970; Gendrin, 1970). One of their most characteristic features is that the mid-frequency of the pulsations rises from about 0.1 to 0.5 Hz s^{-1} over 1 h (Figure 8.23a). It has been suggested that IPDP pulsations originate from the proton-cyclotron resonance instability which is expected to cause pitch-angle diffusion and precipitation of protons. The presence of cold plasma is important in this process (Cornwall *et al.*, 1970, 1971), and so it has been suggested that the instability is expected to occur where ring current protons interact with the plasmasphere (see 8.5.1 (b)); see Figure 8.23(b). Fukunishi (1973) showed that IPDP pulsations are associated with the proton (Hβ) aurora and also with a weak cosmic noise absorption (Figure 8.24).

Heacock *et al.* (1976) note that there are at least two possibilities as the cause for the rising frequency of IPDP pulsations. The first is that the rising frequency is the manifestation of the proton dispersion (higher energy protons arriving first and preferentially generating lower frequencies), and the second is that it is produced as the IPDP source region is convected inward (the proton gyro-frequency

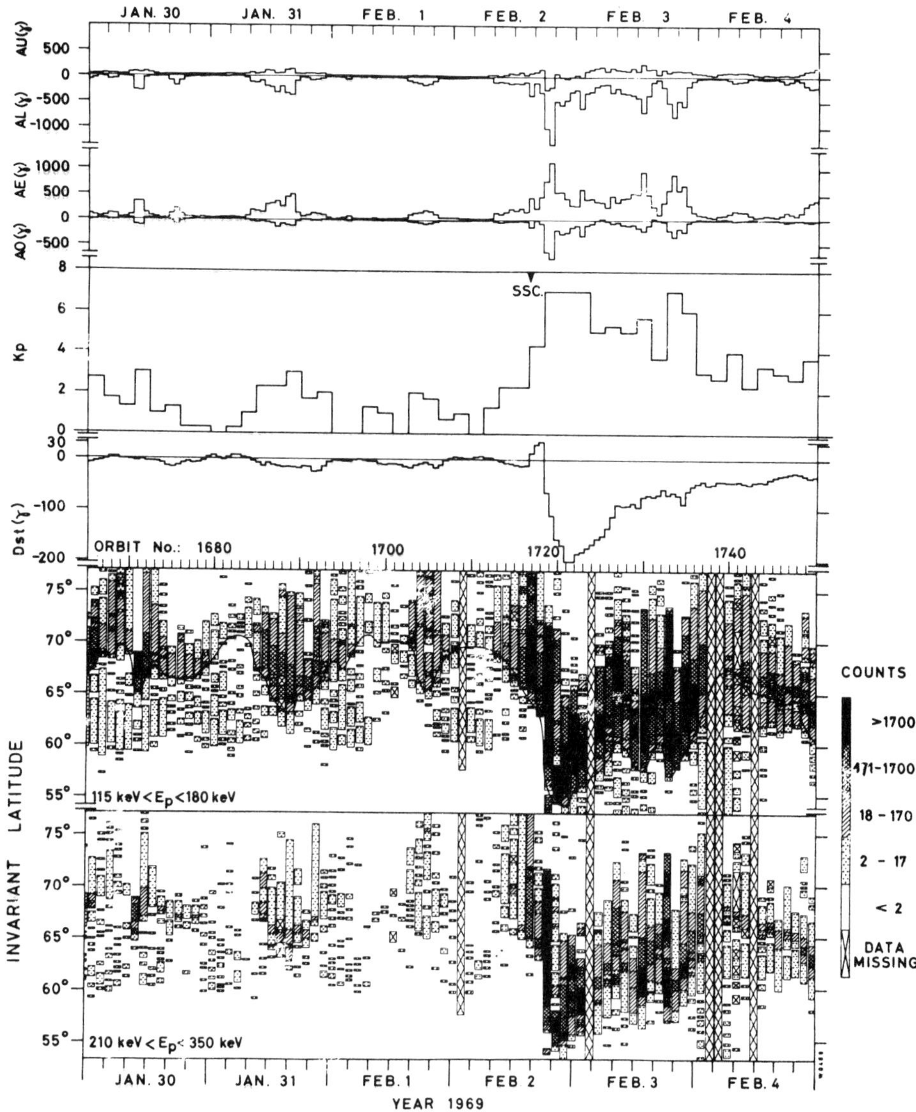

Fig. 8.22. Latitudinal range of precipitating energetic protons (115 keV < ε < 180 keV; 210 keV < ε < 350 keV) in the midnight sector before, during and after the geomagnetic storm of 1969, February 2–3. (Courtesy of Søraas, F., Aarsnes, K., Lindalen, H. R. and Madsen, M. M.)

increases, causing the rising frequency). Heacock *et al.* (1976) showed that in order to account for the frequency shift of about 0.9 Hz h^{-1}, the required convection electric field is of order 0.1–2 mV m^{-1}. They also noted that the rate of the frequency rise is linearly related to the maximum AE attained during the IPDP event.

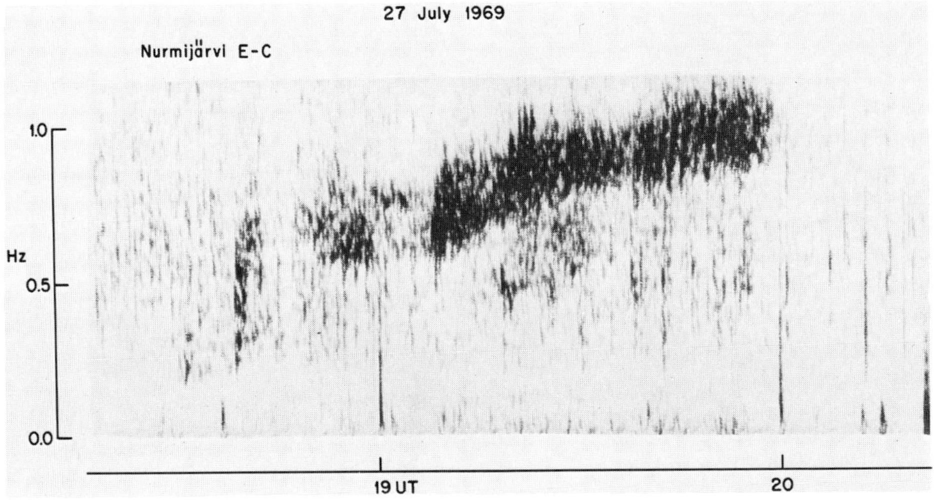

Fig. 8.23(a). Example of IPDP pulsations during a substorm of 1969, July 27 observed at Nurmijarvi. (Heacock, R. R., Henderson, D. J., Reid, J. S. and Kivinen, M.: *J. Geophys. Res.* **81**, 273, 1976.)

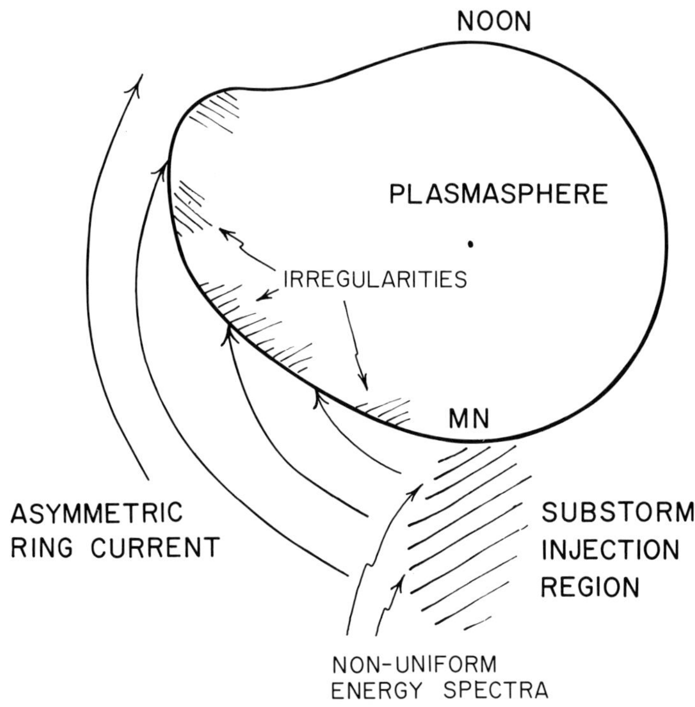

Fig. 8.23(b). Schematic diagram showing the trajectories of injected protons which interact with the bulge region of the plasmasphere in the dusk sector. (Heacock, R. R.: *Nature* **246**, 93, 1973.)

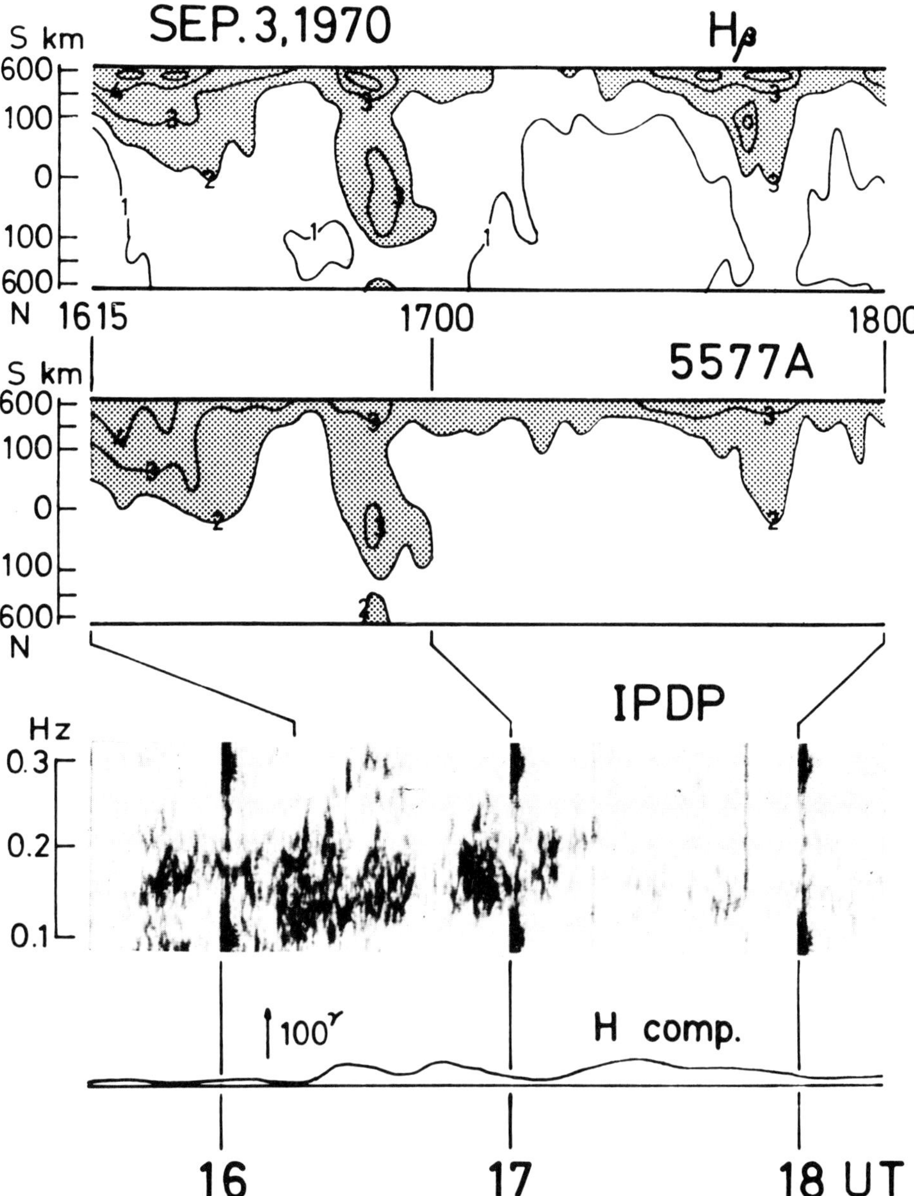

Fig. 8.24. Relationship between the proton aurora and IPDP pulsations. From the top: the distribution of the Hβ emission (with respect to the station), the 5577 Å emission, the IPDP pulsations and the *H* component magnetic record (showing positive bays) observed at Syowa station, Antarctica. Note that the van Phijn effect and the atmospheric extinction are not corrected. (Fukunishi, H.: *J. Geophys. Res.* **78**, 3981, 1973.)

8.5.2. ENERGETIC ELECTRONS

(a) *Satellite Observations of Drifting Electrons*

It has been well established that magnetospheric substorms are the major source process in feeding energetic electrons (~45 keV) into the Van Allen belt (Parks *et al.*, 1968; Parks and Winckler, 1968, 1969; Lezniak *et al.*, 1968; Pfitzer and Winckler, 1969; Arnoldy and Chan, 1969). Figure 8.25 shows an interesting observation of drifting electron clouds by two satellites (ATS-1 and OGO-3), which were widely separated in longitude. A substorm began at about 1255 UT on 1967, January 11. The ATS-1 satellite was located at about 03 LT sector and observed an increase of electron flux at about the same time the substorm began. The same electron cloud was observed later at OGO-3, which was located in the 11 LT sector; note that electrons of energies 150–500 keV arrived at the satellite a little earlier than those of energies 50–150 keV. Pfitzer and Winckler (1969) noted that the observed delay at OGO-3 can be ascribed to the eastward drift motion of electrons after being injected into the midnight sector. The insert in the figure will be discussed shortly.

Hoffman (1970) and Hoffman and Burch (1973) also extensively studied the drift motion of electrons after their injection into the midnight sector. In Figure 8.26 a negative bay grew at College at about 1100 UT on 1967, October 23 and lasted until about 1400 UT. About 6 h later, the OGO-4 satellite began to detect an intense flux of 7.3 keV electrons at latitudes from 65° to 75°; it takes about 6 h for the 7 keV electrons to drift from the midnight sector to the noon sector. Figure 8.27 shows the average precipitation area for 7.3 keV electrons, observed by the OGO-4 satellite. The evening precipitation is associated with the inverted V precipitation (Section 2.4.1), but the extensive morning precipitation arises from the drifting electrons.

Williams *et al.* (1974) also examined the arrival time of electrons at the S^3 satellite after the onset of a substorm. They showed that the observed arrival times for high energy electrons (30–400 keV) are in a reasonable agreement with what would be expected from a simple magnetic field gradient drift. This tendency is shown in the upper part of Figure 8.28. However, they noted also that electrons of energies of 2–3 keV arrived at the satellite almost simultaneously with the high energy electrons. Further, in this low energy range, lower energy electrons arrive earlier than higher energy ones (see the lower part of Figure 8.28). The authors suggested that this feature could be explained by the drift motion of electrons under the influence of the magnetic and electric fields, but that McIlwain's electric field (Section 8.5.1) would not be strong enough to inject them to the distance where such electrons were detected. Bondareva and Trerskaya (1973) and Kasymov and Shabanskiy (1974) theoretically examined drift trajectories of charged particles in a dipole field and in a nonstationary electric field that varies sinusoidally in time; see also Gurevich and Tsedilina (1969a, b).

Kivelson and Southwood (1975) and Walker and Kivelson (1975) have recently made an extensive study of time variations of the flux of energetic electrons arriving at the synchronous orbit after a sudden enhancement of a uniform cross-tail (convection) electric field (2 kV/R_E). On the basis of adiabatic

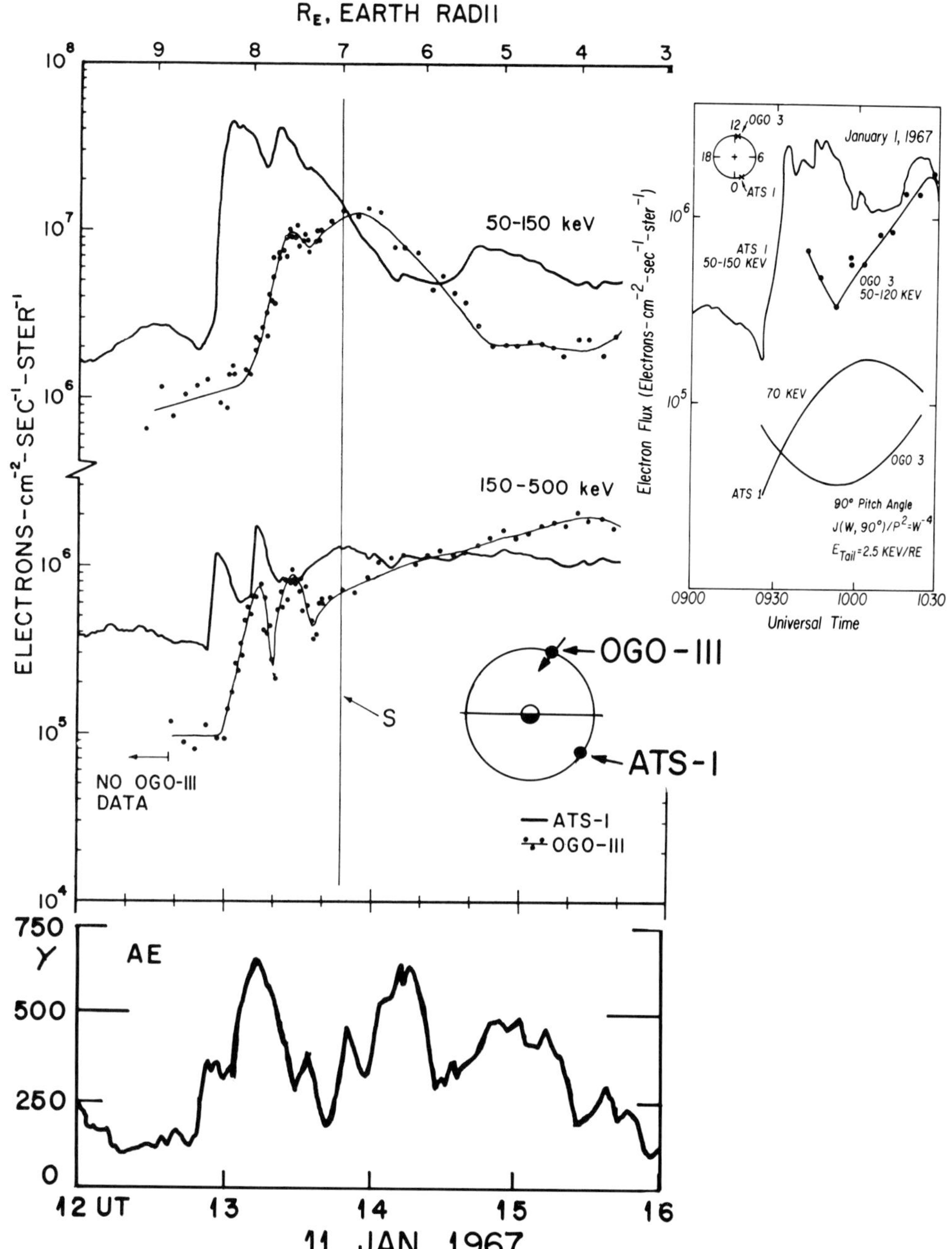

Fig. 8.25. Drifting electrons in the morning sector, observed by two satellites, ATS-1 (solid lines) and OGO-3 (dots). (Pfitzer, K. A. and Winckler, J. R.: *J. Geophys. Res.* **74**, 5005, 1969.) In the inserted panel, the observations are theoretically reproduced. (Walker, R. J. and Kivelson, M. G.: *J. Geophys. Res.* **80**, 2074, 1975.)

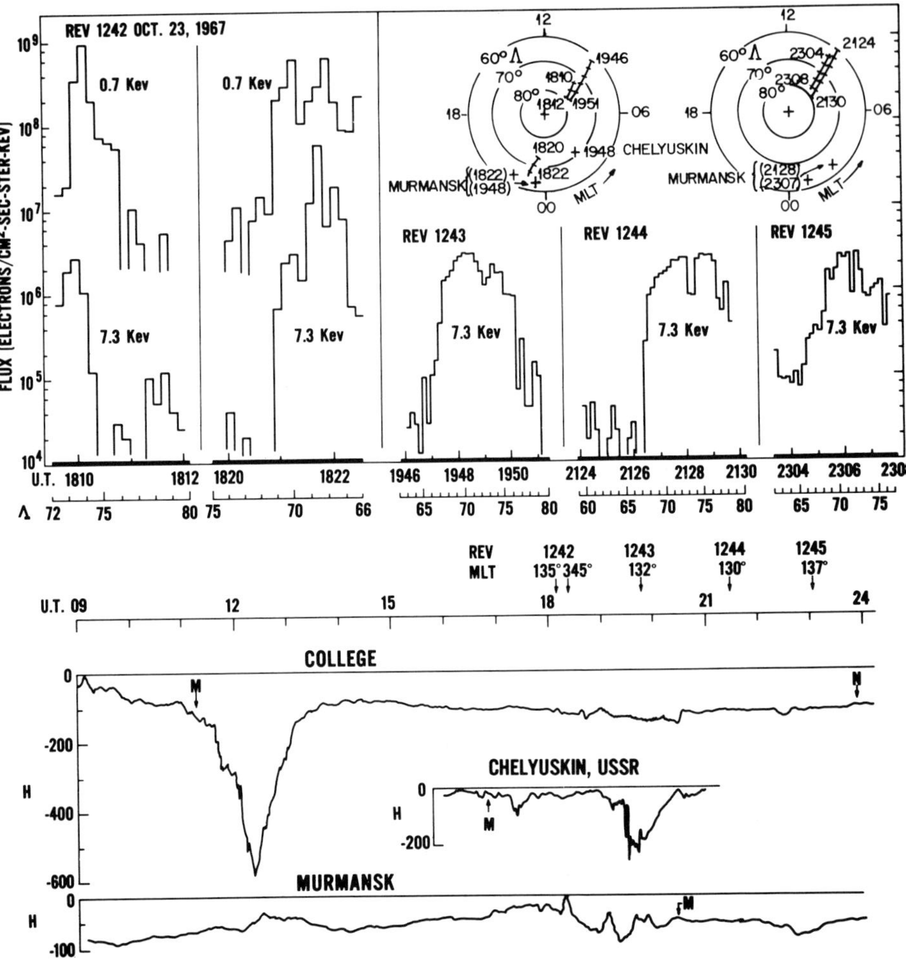

Fig. 8.26. Drifting electrons (~ 7.3 keV) in the morning sector, which are injected into the trapping region during the substorm that began at about 11 UT on 1967, October 23 (see the H component magnetic record from College). Parts of the satellite trajectories along which the electrons were observed are also indicated in the two inserted diagrams (in invariant latitude-MLT coordinates). (Hoffman, R. A.: *Goddard Space Flight Center Rep. X-646-70-205*, June, 1970.)

theory for motions of electrons, the rate of change of kinetic energy W of a particle of charge e along its trajectory is given by

$$\frac{\mathrm{d}W}{\mathrm{d}t} = e\boldsymbol{E}_0(\boldsymbol{R}, t) \cdot \langle \boldsymbol{V}_0 \rangle$$

$$\langle \boldsymbol{V}_0 \rangle = \nabla_0 J \times \boldsymbol{e}_0 / e\tau_b B_0$$

where $\boldsymbol{E}_0(\boldsymbol{R}, t)$ is the total electric field evaluated at the equatorial crossing of the bounce path, $\langle \boldsymbol{V}_0 \rangle$ the bounce-averaged velocity of the equatorial crossing point of the bounce path, J the second adiabatic invariant, $\boldsymbol{B}_0 = \boldsymbol{e}_0 B_0$ at the equator and τ_b

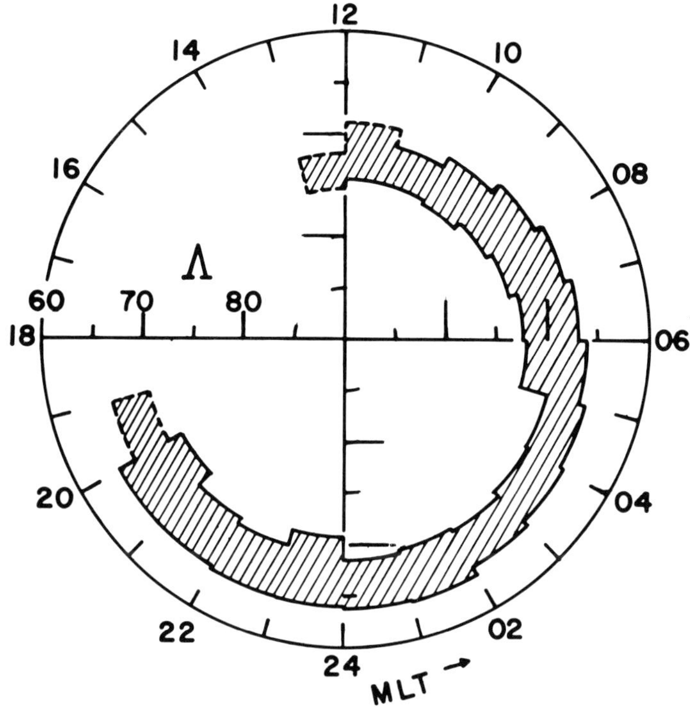

Fig. 8.27. Precipitation region of 7.3 keV electrons in invariant latitude-MLT coordinates. (Hoffman, R. A. and Burch, J. L.: *J. Geophys. Res.* **78**, 2867, 1973.)

the bounce period. The particle flux $j(t)$ at $\boldsymbol{R}$ as a function of time is obtained by invoking the Liouville theorem,

$$\mathrm{d}/\mathrm{d}t\ [j(W, \boldsymbol{R}, \alpha_0)/p^2] = 0$$

and thus

$$j(W, \boldsymbol{R}, \alpha_0)|_t = (p^2/p'^2)j(W', \boldsymbol{R}', \alpha_0)$$

where p denotes the particle momentum and α_0 the pitch-angle.

As the initial condition, the particle distribution was assumed to be spatially uniform (within the region from which particles are convected to the synchronous orbit in one hour). The initial energy spectrum was described by a power law for high energies (>50 keV) and by a Maxwellian distribution with a temperature of 1 keV for lower energies. In this circumstance, electrons gain energy by crossing electric potential lines, particularly in the morning sector.

They showed that their model can successfully reproduce many substorm associated variations of energetic electron fluxes at the synchronous distance, such as a rapid increase of the flux in the midnight to dawn sector and a marked softening of energy spectrum. The observed energy-dependent delay which increases with local time, namely the 'dispersion', is also reproduced. One of their results is shown in the insert of Figure 8.25. They have also succeeded in

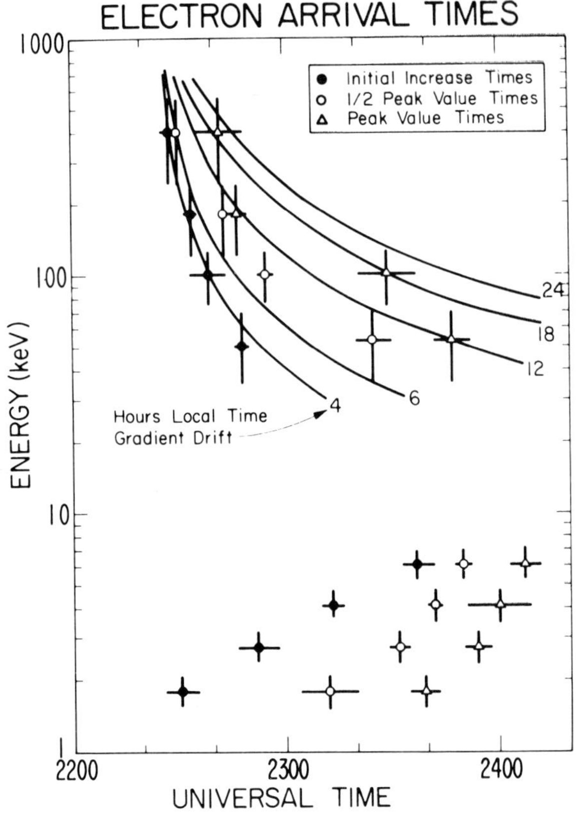

Fig. 8.28. Electron arrival times for energy versus T_i (the initial increase time), $T_{m/2}$ (one-half maximum value time) and T_m (the maximum value time). (Williams, D. J., Barfield, J. N. and Fritz, T. A.: *J. Geophys. Res.* **79**, 554, 1974.)

reproducing some aspects of McIlwain's spectrogram, in particular the intense electron features in the late evening and midnight sectors (Figure 8.29).

It is not an easy task to study how the radial distribution of electrons in the Van Allen belt is affected by a single substorm, although there have been a large number of studies which deal with storm-time variations (Craven, 1966; Lanzerotti, 1968; Brown *et al.*, 1968; Williams and Ness, 1966; Williams *et al.*, 1968; Rothwell *et al.*, 1970; Vampola, 1971; Häusler and Sckopke, 1974; Kirsch *et al.*, 1975). Most recently, Lyons and Williams (1975b) studied in detail the radial and pitch-angle distribution of energetic electron (35–560 keV) during the geomagnetic storms of 1971, December 17 and of 1972, June 17. Figure 8.30 shows the radial distribution of the perpendicular (90° measured local pitch-angle) electron flux near the equatorial plane for the former storm. A large increase of the flux extended to as far as $L \simeq 2.5$; the storm-time profiles slowly return to their pre-storm equilibrium structure over a period of a few weeks. They noted that the increase occurs even at $L \simeq 2.0$ for more intense storms and that the pitch-angle distribution is greatly disturbed during the storm.

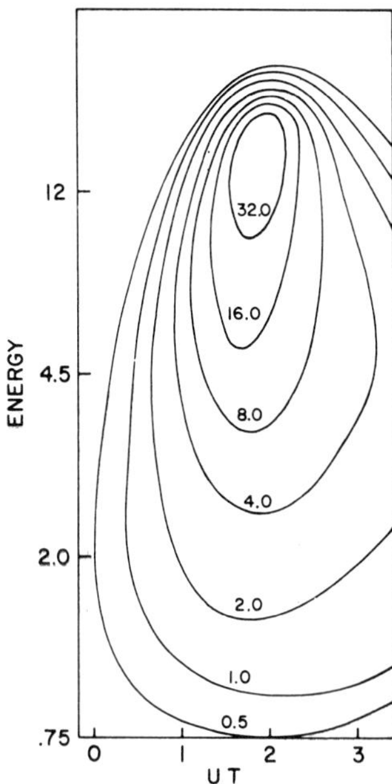

Fig. 8.29. Theoretically reproduced spectrogram (contours of constant energy flux) at the ATS-5 satellite for a midnight event. (Walker, R. J. and Kivelson, M. G.: *J. Geophys. Res.* **80**, 2074, 1975.)

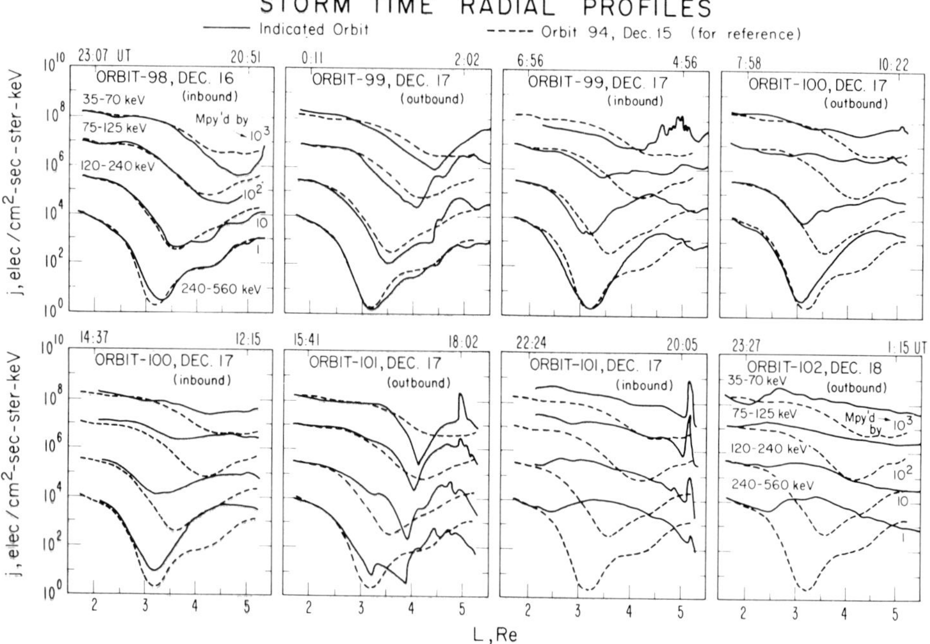

Fig. 8.30. Changes of the radial distribution (L) of electrons during and after the geomagnetic storm of 1971, December 19. The electron distribution during orbit 94 (December 15) is given by the dashed curve for reference. (Lyons, L. R. and Williams, D. J.: *J. Geophys. Res.* **80**, 3985, 1975.)

(b) *Ground-Based Observations of Drifting Electrons*

The drifting electrons are subject to various wave-particle interactions. Consequently, some of them are precipitated into the upper atmosphere. This subject has been extensively studied by a large number of workers after the pioneering work by Brice (1964) and Kennel and Petschek (1966); for recent reviews, see Ashour-Abdalla and Cowley (1974), Kimura (1974), Nambu (1974), Fredricks (1975), Gendrin (1975) and Scarf (1975).

It is generally agreed that VLF electrostatic waves are responsible for precipitating the drifting electrons in the Van Allen belt. Young *et al.* (1973) showed that in order for VLF electrostatic waves to grow in magnetospheric plasma it is necessary to have a cold and a warm species of electrons such that (i) the warm component has an anomalous velocity distribution function that is nonmonotonic in $f(v)$ and is the source of free energy driving the instabilities, (ii) the density ratio of the cold component to the hot component is greater than 10^{-2}, and (iii) the temperature ratio of the two components for the case of high density is no less than 0.1. The growth rate of the waves is maximized at 1.4 Ω_{ce} where Ω_{ce} denotes the gyro-frequency of electrons (see also Gendrin, 1975). Lyons (1974) computed the diffusion coefficient for electrons which are in resonance with intense electrostatic waves of frequency 1.5 Ω_{ce} and showed that the wave can cause strong pitch-angle diffusion within and near the loss cone, as well as significant energy diffusion for electrons of energies between a few tenths and a few keV. He found also that the most intense observed waves can cause strong diffusion of electrons of energies up to 100 keV.

It is also interesting to note that a recent study of pulsating features of auroras by Royrvik (1976) shows that most auroras tend to pulsate and that the observed features appear to agree with theoretical studies by Coroniti and Kennel (1970a, b) who demonstrated that micropulsations modulate the particle distribution which affects the growth of chorus emissions.

(i) *Riometer observations.* The eastward drift motion of energetic electrons has also been investigated by using riometer records from a number of stations along the auroral zone (Driatskiy, 1968, 1969; Lichtenstein, 1970; Rosen and Winckler, 1970; Driatskiy and Shumilov, 1970, 1972; Driatskiy *et al.*, 1972; Gustafsson, 1969; Hargreaves, 1969, 1970, 1971, 1974; Hargreaves *et al.*, 1975; Jelly, 1970; and most extensively by Berkey *et al.*, 1974). Figure 8.31 shows an example of the development of the absorption event. It began in the early morning sector and rapidly spread toward the forenoon sector.

A typical absorption event grows as follows (Berkey *et al.*, 1974):

T = 0

The absorption starts around 65° corr. geomag. lat. on the nightside, and the absorption area expands in all directions.

T = 0–1 h 15 min

In this period the eastward expansion dominates, and at the end of this period the absorption pattern forms a continuous belt of a mean width of 10°. A secondary maximum develops, and the strongest absorption is in the late morning sector. This phase may be called the expansion phase.

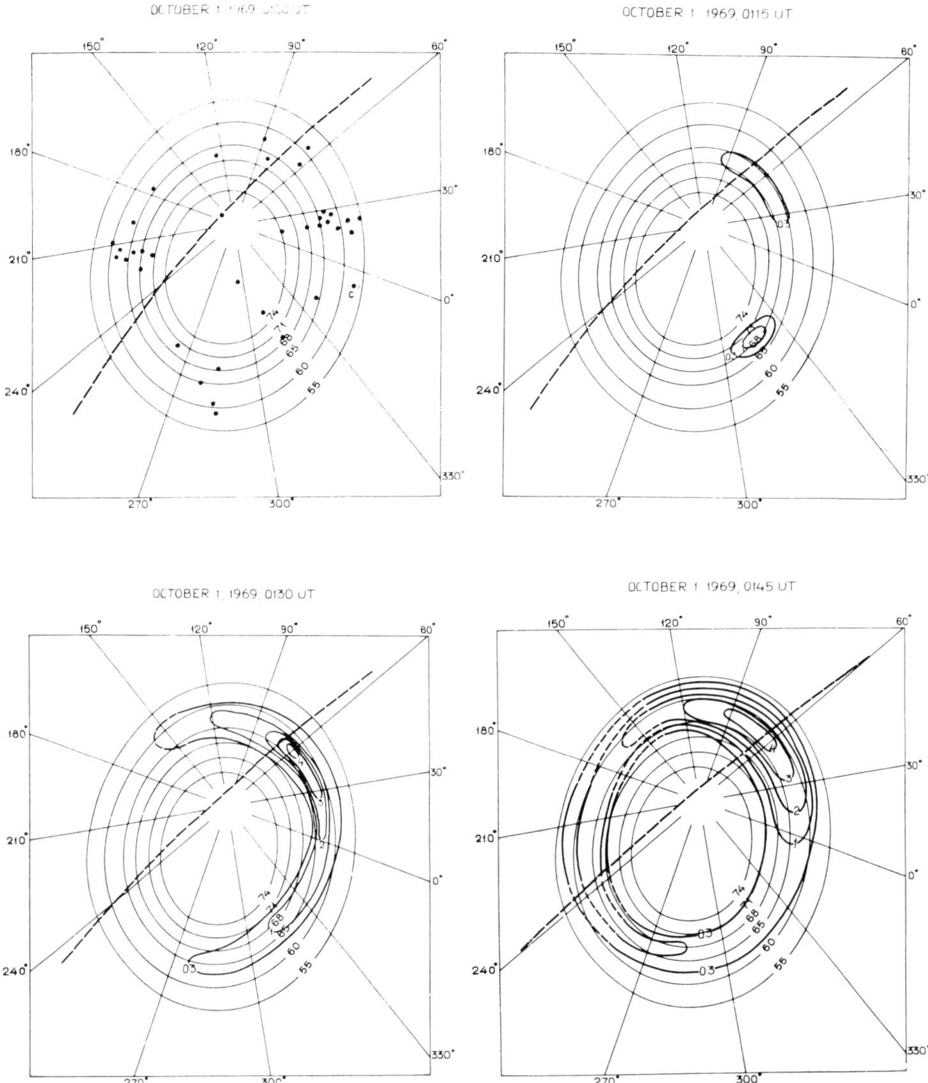

Fig. 8.31. Development of the region of cosmic radio noise absorption during a substorm of 1965, October 1. The locations of riometers whose data were used in the study are shown in the first diagram. (Berkey, F. T., Driatskiy, V. M., Henriksen, K., Hultqvist, B., Jelly, D., Shchuka, T.I., Theander, A. and Ylinieni, J.: *Planet. Space Sci.* **22**, 255, 1974.)

T = 1 h 15 min–2 h 45 min

This interval is characterized by the decrease of absorption both in magnitude and extent. The pattern splits into patches, and the absorption remains longest on the dayside.

Berkey *et al.* (1974) obtained the eastward expansion speeds of the absorption region (Table 8.2).

TABLE 8.2
The average eastward expansion velocity ($km\ s^{-1}$) of the absorption area as function of time after substorm onset for IQSY and IASY.

Time passed since substorm onset min	1964–65	1969
15	3.2 + 0.9	3.4 + 0.8
30	2.9 + 1.0	3.0 + 1.1
35	2.3 + 1.1	2.0 + 0.8

TABLE 8.3
Drift velocity (at the Earth's surface) of electrons on the magnetic shell of $L = 5.6$ for two different equatorial pitch-angles ($\alpha = 90°$ and $10°$).

Energy keV	Velocity ($km\ s^{-1}$) Electrons	
	$\alpha = 90°$	$\alpha = 10°$
50	1.49	1.12
100	2.82	2.12
200	5.41	4.06
300	7.67	5.76
500	11.5	8.66
1000	20.1	15.2

They compared the observed expansion speeds of the area of cosmic noise absorption with the computed drift speeds of electrons of different energies and of different pitch-angles. Table 8.3 gives the computed speeds. Hargreaves (1969, 1970, 1971, 1974) and Hargreaves *et al.* (1975) examined extensively time structures within the region where absorption develops. The last authors noted that the movement of the region of the maximum absorption differs significantly from the movement of the boundary of the absorption region. Hones *et al.* (1971) examined in detail the drift motion of energetic electrons and the development of the absorption.

(ii) *X-ray observations.* Bremsstrahlung X-ray observations by balloon-borne detectors have been very useful in studying motions of energetic electrons in the morning and the noon sectors. During recent years, an intensive X-ray observation program was conducted by European groups (Christensen and Karas, 1970; Maral, 1970; Bjordal *et al.*, 1971; Jentsch and Kremser, 1972; Sørensen *et al.*, 1973; Kremser *et al.*, 1973), as well as by Rosenberg *et al.* (1971) and also by Parks (1970) in conjunction with the simultaneous ATS-1 satellite observations. Parks (1970), Pilkington *et al.* (1968), Pilkington (1972), and Sletten *et al.* (1971) established that X-ray events observed in the auroral zone are always associated with substorm activity.

X-ray observations were also conducted at geomagnetically conjugate points by Barcus *et al.* (1973).

Parks *et al.* (1968), Parks and Winckler (1969) and Parks (1970) showed that time variations of precipitated auroral energetic electrons observed by a balloon-borne X-ray detector are very well correlated with the corresponding electron flux variations in the equatorial plane. They suggested that the precipitation process is intimately associated with mechanisms which are responsible for accelerating fresh particles in the magnetosphere. Figure 8.32 shows an example of such a correlation; such good correlation is observable in the local time range 00–12 LT. Parks (1970) suggested thus that the acceleration and precipitation of electrons take place locally.

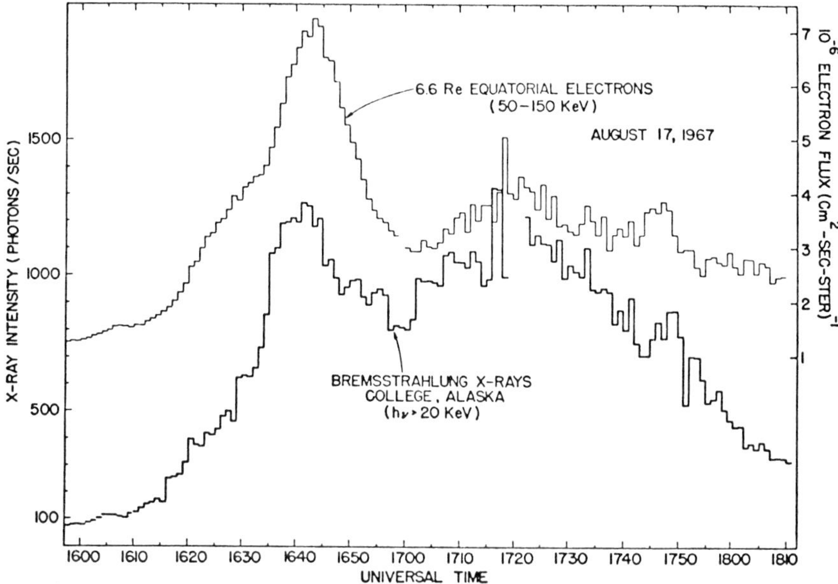

Fig. 8.32. Comparison of energetic electron fluxes observed at the ATS-1 satellite and X-ray intensity observed at the 'foot' (College, Alaska) of the field line which intersects with the satellite in the equatorial plane. (Parks, G. K. and Winckler, J. R.: *J. Geophys. Res.* **74**, 4003, 1969.)

Mozer *et al.* (1973) also made an extensive simultaneous observation of X-rays and electric fields by balloon-borne detectors and of particles in the equatorial plane by their ATS-5 detectors. They noted that X-ray events are often ($\sim 70\%$) preceded by an enhancement of the westward electric field.

Kremser *et al.* (1973) classified X-ray events into two groups – direct and drift precipitations. The latter are caused by energetic electrons drifting from the midnight sector toward the forenoon sector. In the early morning sector, both types of precipitation can occur simultaneously. This is because the auroral oval and the drift shells (along which the electrons drift) are lying closely together. However, the distance between them increases progressively toward the forenoon sector. Then it is possible, in many cases (but not always), to separate both events. Figure 8.33 shows an example of simultaneous X-ray observations at Andenes, Kiruna and Sodankyla, together with the supporting riometer and magnetometer observations in Scandinavia. Kremser *et al.* concluded that the intense X-ray event observed at Andenes (corr. geomag. lat. 66.6° N) was associated with the direct precipitation which followed along the oval and also along the auroral electrojet; see the iso-intensity contour lines of ΔH and $\Delta Z = 0$. At Kiruna (corr. geomag. lat. 64.8° N) the precipitation event between 01 and 02 UT was the direct precipitation, and the precipitation event after 02 UT was the drift precipitation. At Sodankyla (corr. geomag. lat. 63.9°N), the entire precipitation resulted from drift precipitation.

Maral *et al.* (1973) examined the east-west extent of both types of precipitation by using a network of balloon-borne detectors. They found that the direct

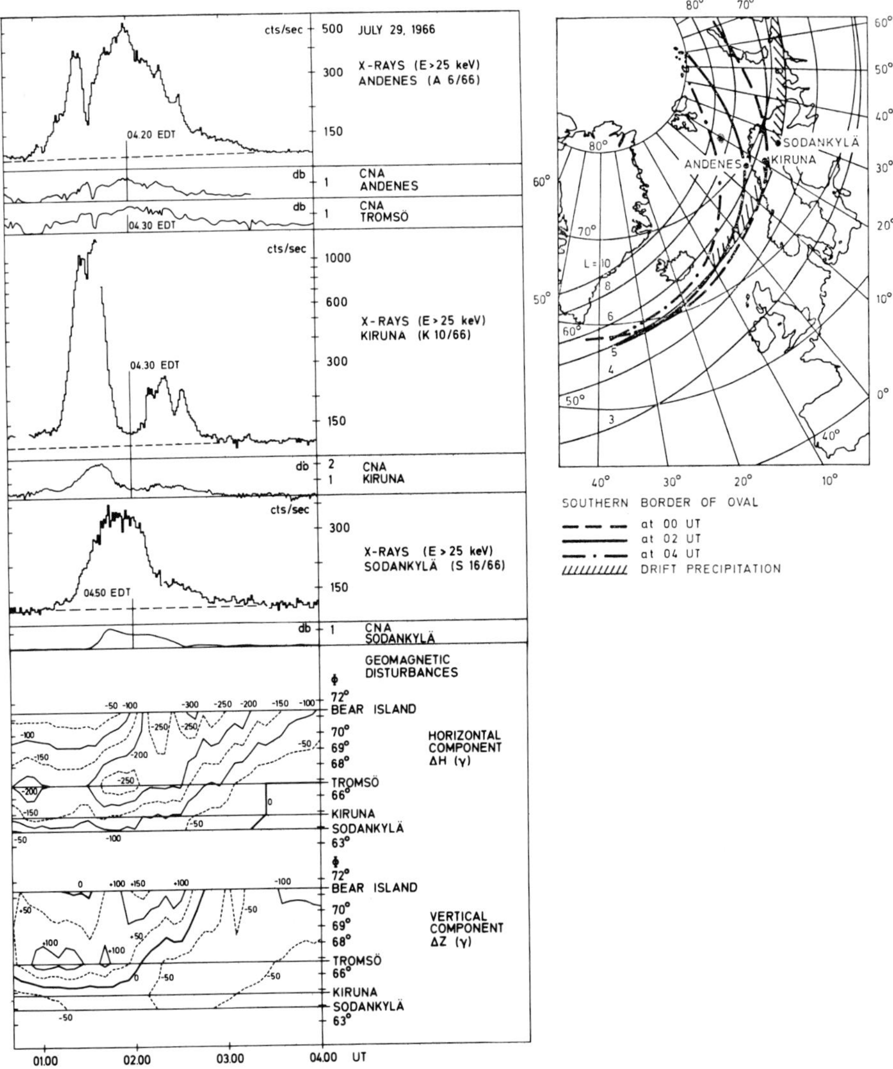

Fig. 8.33. Simultaneous observations of X-rays by balloon-borne instruments at Andenes, Kiruna and Sodankyla. The corresponding riometer records and the iso-intensity contours of ΔH and ΔZ are also given. The location of the balloons, the drift precipitation zone and the southern border of the auroral oval (assumed to coincide with the direct precipitation region) are given in the inserted panel. (Kremser, G., Wilhelm, K., Riedler, W., Brønstad, K., Trefall, H., Ullaland, S. L., Legrand, J. P., Kangas, J. and Tanskanen, P.: *J. Atmosph. Terr. Phys.* **35**, 713, 1973.)

precipitation event occurs almost simultaneously over a distance of 2000 km and is located near the auroral electrojet. On the other hand, the drift precipitation starts progressively later toward the forenoon sector from the midnight sector. Figure 8.34 shows an example of their simultaneous observation. Here, we shall be concerned only with a few major features:

(1) The onset of an X-ray event was recorded at Vik, Iceland, at 0356 UT and

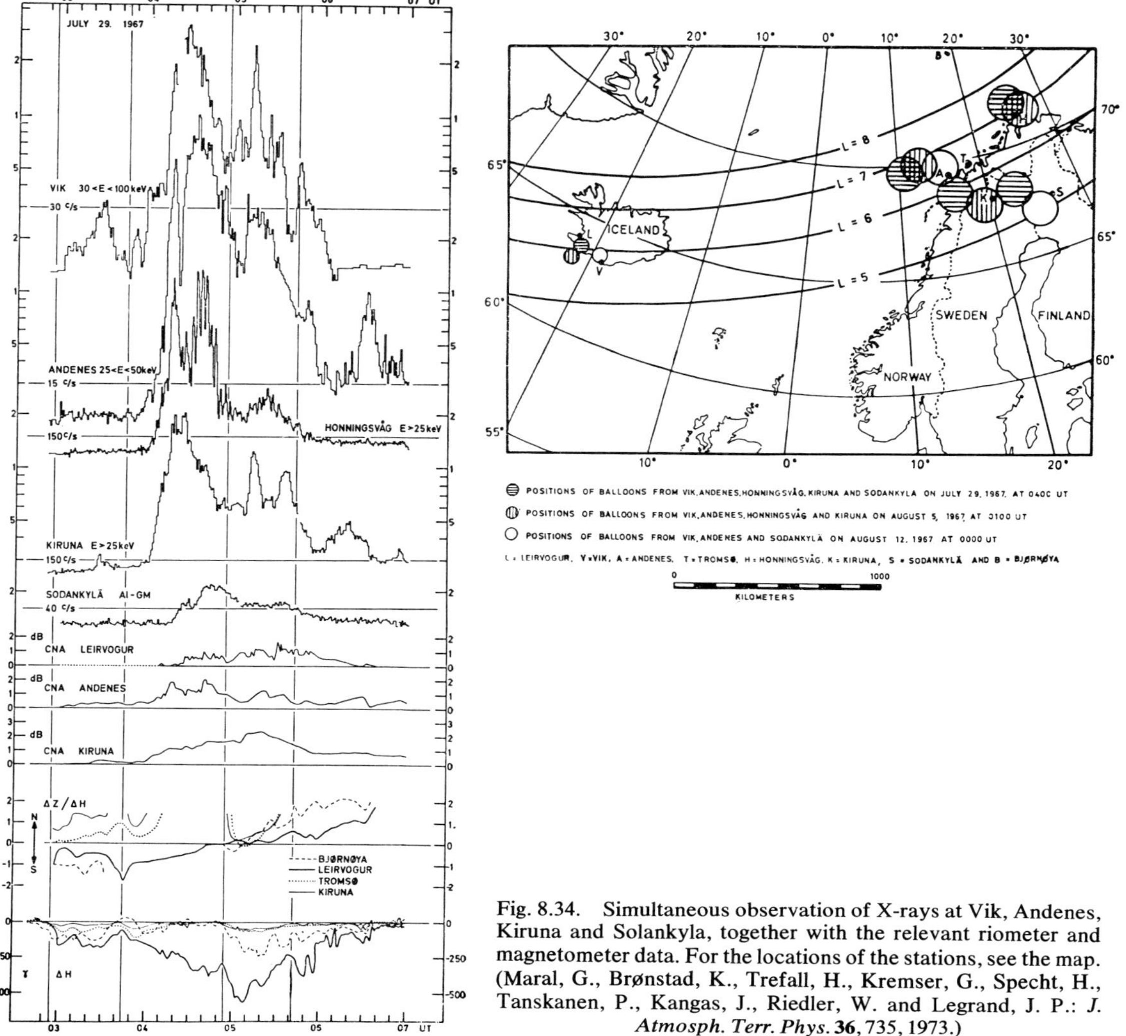

Fig. 8.34. Simultaneous observation of X-rays at Vik, Andenes, Kiruna and Solankyla, together with the relevant riometer and magnetometer data. For the locations of the stations, see the map. (Maral, G., Brønstad, K., Trefall, H., Kremser, G., Specht, H., Tanskanen, P., Kangas, J., Riedler, W. and Legrand, J. P.: *J. Atmosph. Terr. Phys.* **36**, 735, 1973.)

the possibly corresponding event at Andenes at 0402 UT, a delay of about 6 minutes. Then the corresponding event began at Honningsvag a few minutes later. At Kiruna, however, the corresponding event began at 0400 UT, 2 min earlier than that at Andenes. The authors pointed out that the electrojet was located well south of Vik when the event began and moved rapidly poleward. Thus, the electrons precipitating over Kiruna may have started to drift earlier than those precipitating over Andenes. The Sodankyla record shows that the corresponding event began there at 0414 UT, indicating the trend toward a later appearance of drift precipitation as one goes farther east and south. This delay can partly be explained as an effect of the L-dependence of drift speeds.

(2) The onset of an X-ray event was recorded at Vik at about 0450 UT. The corresponding events began at 0550, 0507 and 0506 UT at Andenes, Honningsvag and Kiruna, respectively. There is some indication for the arrival of additional drift precipitation at Sodankyla. It appears to be caused by the drifting electrons, because the electrojet was located far north of these stations at that time.

Barcus *et al.* (1973) examined X-ray events at geomagnetically conjugate points – Reykjavik, Iceland and Syowa, Antarctica. They noted that dissimilarities exist between the conjugate events, especially for midnight events, but that a

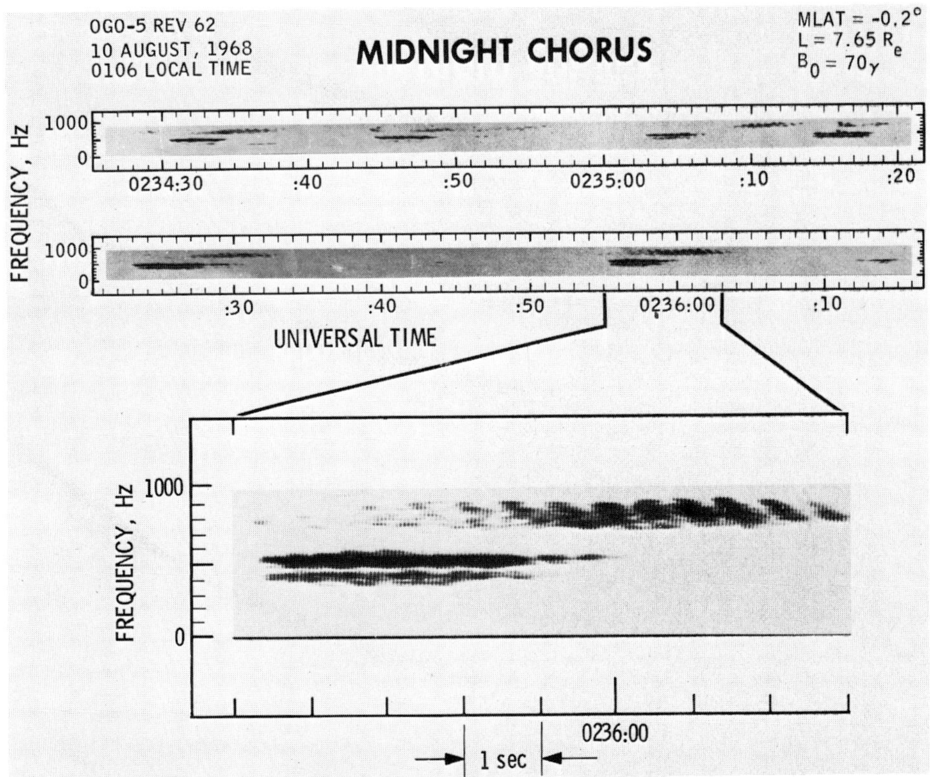

Fig. 8.35(a). Example of chorus, observed by the OGO-5 satellite, near the equatorial plane in the midnight sector. (Tsurutani, B. T. and Smith, E. J.: *J. Geophys. Res.* **79**, 118, 1974.)

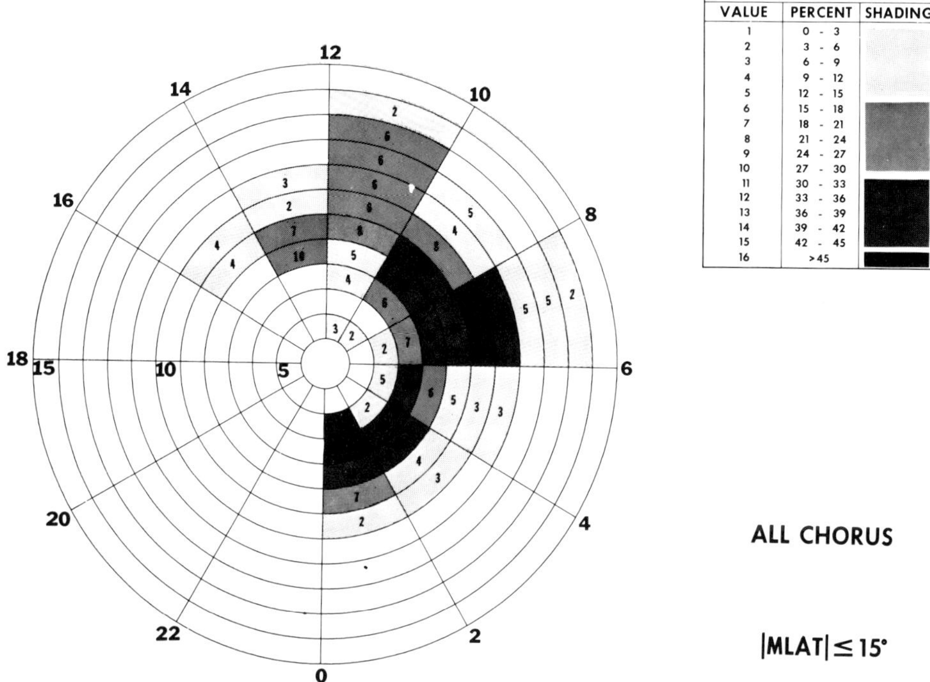

Fig. 8.35(b). Occurrence frequency of chorus near the equatorial plane (magnetic latitude ≤ 15°). (Courtesy of B. T. Tsurutani.)

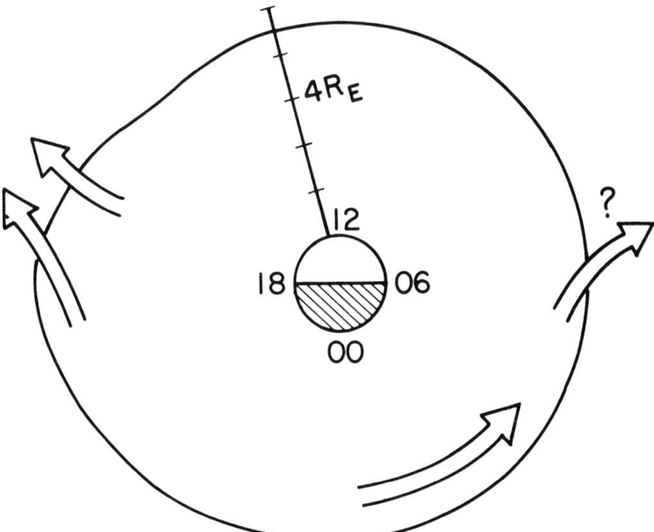

Fig. 8.36. Schematic diagram indicating the directions of the deformation of the plasmasphere during a substorm. (Carpenter, D. L.: *J. Geophys. Res.* **75**, 3837, 1970.)

close conjugate relation exists for day events (despite the fact that the two balloons were out of conjugacy by approximately 700 km).

It should be mentioned that Imhof *et al.* (1974) succeeded in observing auroral X-rays from a satellite, making it possible to observe a large-scale precipitation pattern of energetic electrons over the entire polar cap region.

(iii) *Chorus emissions.* The drifting electrons can also be detected by observing a particular type of VLF emission, called chorus, and other emissions. An example of midnight chorus is shown in Figure 8.35(a). Chorus frequencies vary from less than 0.25 Ω_{ce} to as high as 0.75 Ω_{ce}. The emission arises from a Doppler-shifted cyclotron resonance between the waves and the electrons. Tsurutani and Smith (1974) estimated that the energy range of electrons should be between 10 and 100 keV to satisfy the observed frequency of the waves. Tsurutani

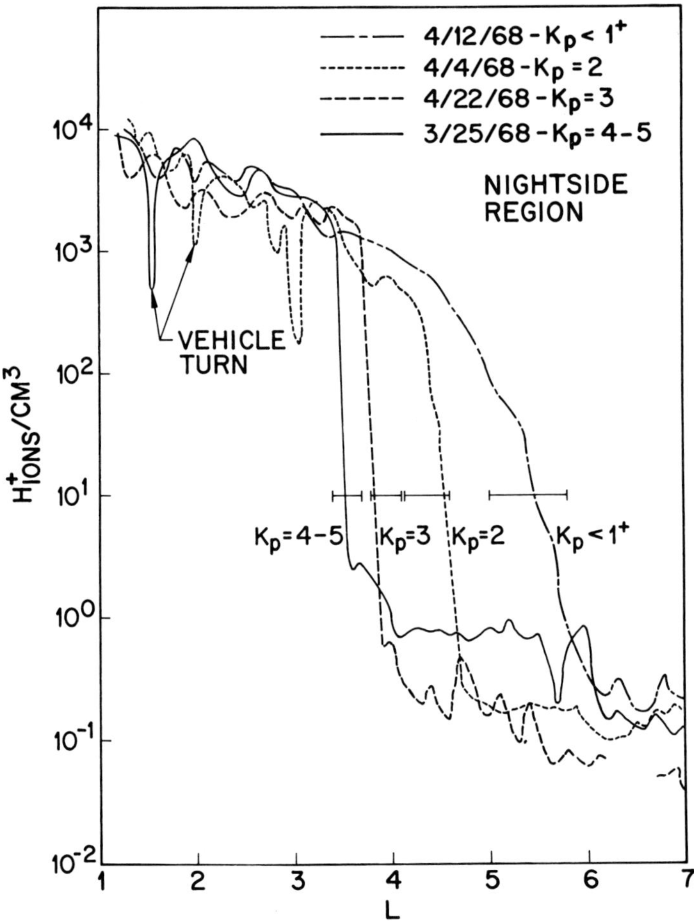

Fig. 8.37. Earthward shift of the plasmapause in the midnight sector for different Kp values. (Chappell, C. R., Harris, K. K. and Sharp, G. W.: *J. Geophys. Res.* **75**, 50, 1970.)

and Smith (1974) and Thorne *et al.* (1974) found that chorus was detected during the expansive phase of substorms. The distribution of chorus as a function of local time and L value is quite similar to that of the drifting electrons. Figure 8.35(b) shows the normalized occurrence of chorus near the equatorial plane (magnetic latitude $\leq 15°$); see also Kelley *et al.* (1975). At high magnetic latitudes ($\geq 15°$), chorus tends to occur in the midday sector.

8.6. Deformation of the Plasmasphere and Associated Ionospheric Disturbances

The plasmasphere deforms considerably during magnetosphere substorms, particularly during magnetospheric storms, when intense substorms occur very frequently (Carpenter and Stone, 1967; Taylor *et al.*, 1968). This phenomenon was discovered first by Carpenter (1970) who found that the bulge region of the plasmasphere swings toward the afternoon sector during substorms. Figure 8.36 shows schematically the direction of plasma flow during substorms. Chappell *et al.* (1970a, b), Harris *et al.* (1970) and Carpenter and Chappell (1973) showed also that the geocentric distance to the plasmapause decreases with increasing

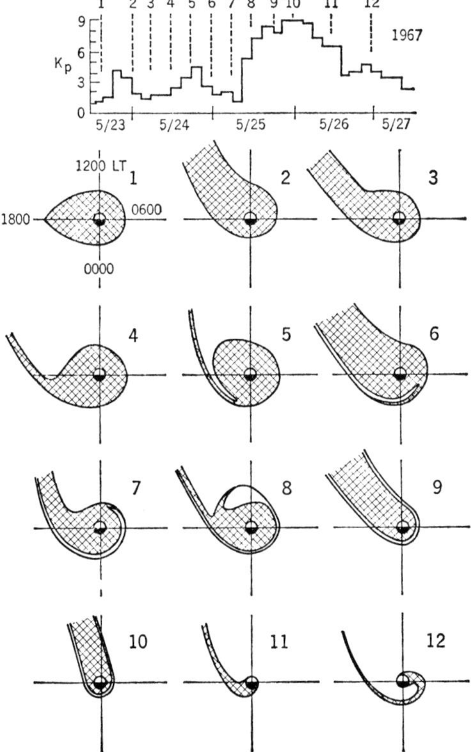

Fig. 8.38(a). Computed evolution of the boundary of the plasmasphere during the geomagnetic disturbance of 1967, May 23–27. The Kp index during the event is shown at the top, and the convection electric field E is assumed to be given by $E = 0.125(1 + \frac{2}{3}\,\mathrm{Kp})\ \mathrm{mV\ m^{-1}}$. (Grebowsky, J. M., Tulunay, Y. and Chen, A. J.: *Planet. Space Sci.* **22**, 1089, 1974.)

magnetic activity, and the decrease is accompanied by an increased steepness in the density gradient at the plasmapause, although the density inside and outside the plasmapause remains approximately the same (Figure 8.37).

Subsequently, Taylor *et al.* (1970), Chappell *et al.* (1970) and Chappell (1974) found 'patches' of cold plasma well outside the plasmasphere. Chappell *et al.* (1970a) suggested that such patches were detached from the bulge region of the plasmasphere by an enhanced convection. On the other hand, Grebowsky (1970, 1971) and Chen and Wolf (1972) showed that an enhanced convection does not result in detached clouds of plasma, but that the plasmasphere is distorted, forming strands of plasma which they propose to call 'plasma tails'.

The deformation of the plasmasphere by an enhanced cross-tail electric field has recently been studied by Chen *et al.* (1975), Grebowsky *et al.* (1974) and Grebowsky and Chen (1975). They examined how long a particular flux tube at a given UT and location in the equatorial plane had been closed on the dayside by tracking its motion backward in time. The primary source of ionization originates in the dayside ionosphere and it is believed to take about 6 days to fill a tube. Thus, the density in a particular tube depends on the duration of the supply of plasma from the dayside ionosphere. In tracking backward the motions of flux tubes, they

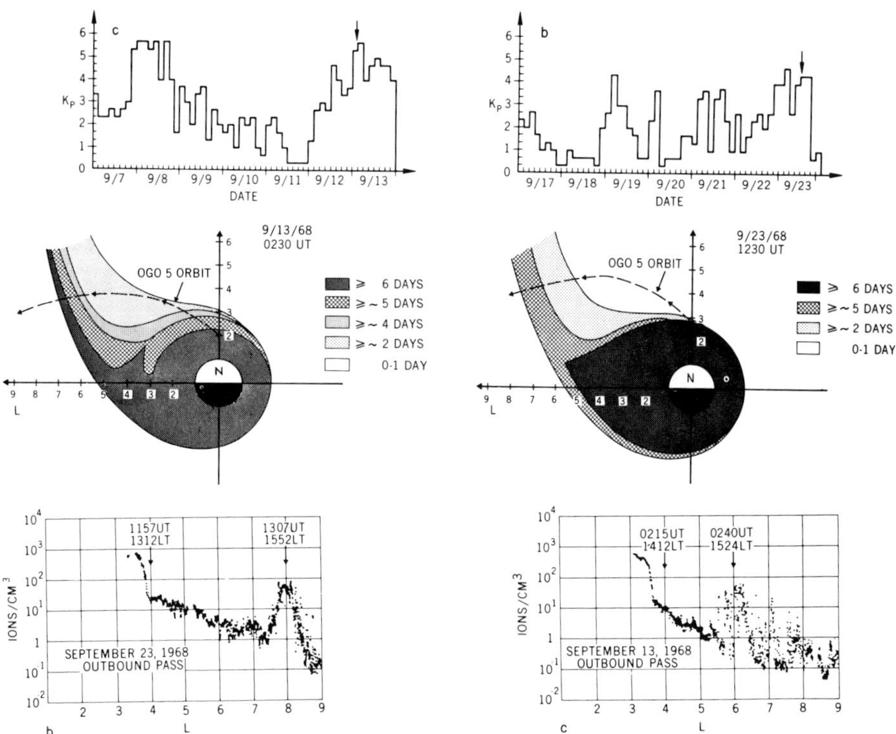

Fig. 8.38(b). Computed deformation of the boundary of the plasmasphere during two disturbed periods. Note in particular the development of the tail-like structure, the trajectory of the OGO-5 satellite and the actually observed distribution of plasma ('detached plasma'). (Chen, A. J. and Grebowsky, J. M.: *J. Geophys. Res.* **79**, 3851, 1974.)

used an empirical equation for the electric field E;

$$E = 0.125(1 + \tfrac{2}{3}\,\mathrm{Kp})\ \mathrm{mV\ m^{-1}}$$

Thus, by knowing the Kp index as a function of time, the time variation of the electric field can be inferred.

Figure 8.38(a) shows how the plasmasphere deforms from its normal (or quiet time) shape to different shapes as the electric field varies in the way indicated by the time variation of Kp. In particular, note the development of a tail-like structure, the plasma tail. Chen and Grebowsky (1974) proposed that the 'detached' plasma, observed by Chappell *et al.* (1970a), can be explained in terms of the traverse of a satellite across the plasma tail. Figure 8.38(b) shows a few more examples of their computed shape of the plasmasphere, the satellite trajectory (traversing the plasma tail) and the observed plasma 'cloud'.

On the basis of whistler data recorded in Antarctica (Eights) and North America, Park (1971, 1973) also showed that the total plasma content in magnetic tubes of force is drastically reduced during substorms. Figure 8.39 shows the tube content as a function of L just before and soon after an intense substorm activity on 1965, June 25–26. A large decrease in tube content can be seen beyond $L \simeq 2.7$. Park suggested that an enhanced westward electric field caused a downward motion of plasma, as well as an earthward motion. Indeed, he showed the critical

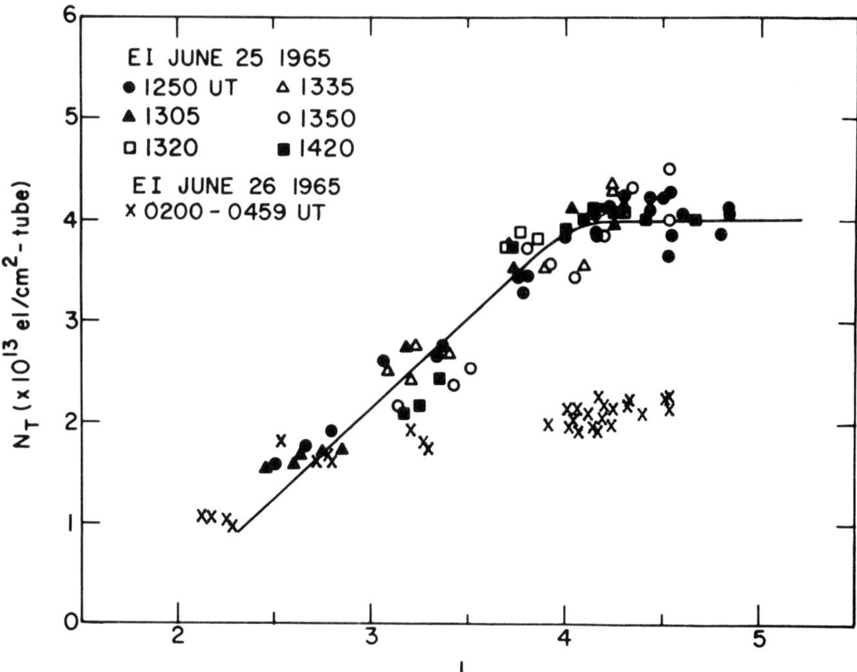

Fig. 8.39. Drastic reduction of the total electron content in magnetic tubes of force between 1420 UT on 1965, June 25 and 0459 UT on 1965, June 26, deduced on the basis of whistler data observed at Eights station, Antarctica. (Park, C. G.: *J. Geophys. Res.* **78**, 672, 1973.)

frequency of the F2 region (foF2) over Ottawa which was located near the meridian of Eights. Rüster (1971), VanZandt *et al.* (1971) and Park and Meng (1973) examined ionospheric data from widely spaced locations and found that the F layer is pushed upward (by an eastward electric field) in the premidnight sector and downward (by a westward electric field) in the postmidnight sector. Park (1974) also made a more detailed study of the ionosphere during the event on 1965, June 25–26 and found that the critical frequency foF2 was enhanced up to 25% above the monthly median value in the region where a large decrease of the tube content in the plasmasphere was observed.

It should be noted that both Pushkova *et al.* (1972) and Park and Meng (1973) noted that the magnitude of the electric field cannot be deduced from simultaneously observed geomagnetic variations since they may arise from non-ionospheric currents.

More recently, Park and Banks (1974, 1975) made a detailed theoretical study

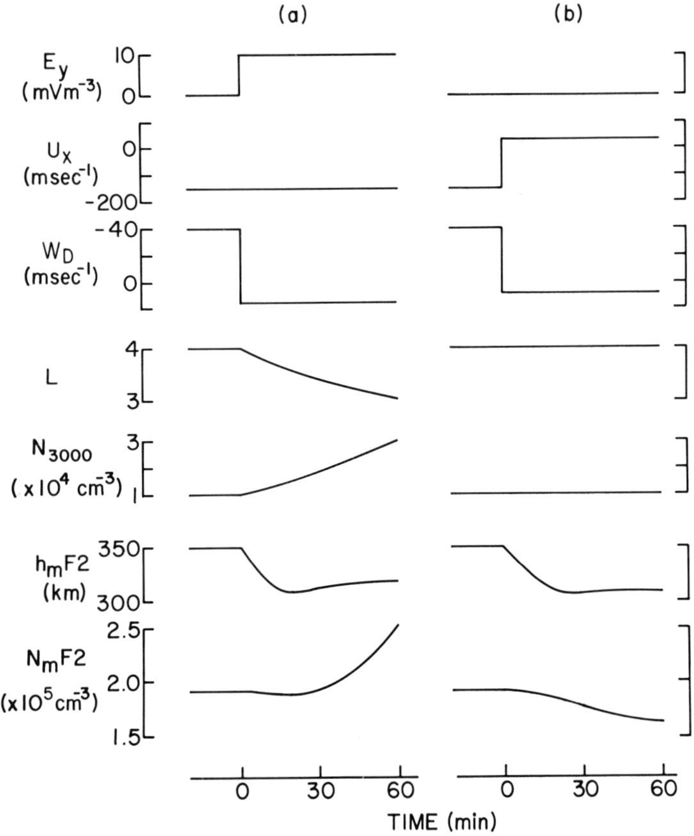

Fig. 8.40. Response of the ionosphere to a transient increase of the westward electric field (a) and the southerly thermospheric wind (b). W_D gives the applied vertical drift velocity, L indicates the location of the flux tube, N_{3000} the density of H^+ at a 3000 km level, h_mF2 the height of the maximum electron density in the F2 region and N_m the maximum electron density. (Park, C. G. and Banks, P. M.: *J. Geophys. Res.* **79**, 4661, 1974.)

of the coupling between the ionosphere and the plasmasphere. They found that a suddenly enhanced westward electric field squeezes plasma out of the plasmasphere along a contracting field tube. The induced downward flow causes an increase of foF2, even though the layer is strongly pushed into regions of rapid ionization loss. Figure 8.40 shows how the ionosphere-plasmasphere system responds to a step function-like change of the electric field. Brace *et al.* (1974) noted that in addition to the earthward shift of the plasmapause, a deep trough developed in the region $L = 1.3–1.8$ and it extended down to the F region.

As mentioned in Section 8.2.1, Barfield *et al.* (1975) also reported that the disappearance of plasma in the plasmasphere was simultaneously associated with the appearance of plasma sheet protons, magnetic field distortion and Van Allen belt electron dropout.

References

Aarsness, K., Amundsen, R., Lindalen, H. R. and Søraas, F.: 1970, 'Proton Precipitation in the Noon-Midnight Meridian and Its Relation to Geomagnetic Activity', *Physica Norvegica* **4**, 73.

Akasofu, S.-I. and Meng, C.-I.: 1969, 'Non-Uniform Growth of the Ring Current Belt', *Planet. Space Sci.* **17**, 707.

Akasofu, S.-I., DeForest, S. and McIlwain, C.: 1974, 'Auroral Displays near the "Foot" of the Field Line of the ATS-5 Satellite', *Planet. Space Sci.* **22**, 25.

Anderson, D. N. and Roble, R. G.: 1974, 'The Effect of Vertical $E \times B$ Ionospheric Drifts on F-Region Neutral Winds in the Low-Latitude Thermosphere', *J. Geophys. Res.* **79**, 5231.

Arnoldy, R. L. and Chan, K. W.: 1969, 'Particle Substorms Observed at the Geostationary Orbit', *J. Geophys. Res.* **74**, 5019.

Ashour-Abdalla, M. and Cowley, S. W. H.: 1974, 'Wave-Particle Interactions near the Geostationary Orbit', *Magnetospheric Physics*, B. M. McCormac (ed.), p. 241, D. Reidel Publ. Co., Dordrecht-Holland.

Atkinson, G. J.: 1971, 'Magnetospheric Flows and Substorms', Review paper presented at the Advanced Study Inst. on Magnetosphere–Ionosphere Interactions, Dalseter, Norway, April 14–23.

Barcus, J. R., Brown, R. R., Karas, R. H., Rosenberg, T. J., Trefall, H. and Brønstad, K.: 1971, Auroral X-Ray Pulsations in the 1.2 to 4-Second Period Range', *J. Geophys. Res.* **76**, 3811.

Barcus, J. R., Brown, R. R., Karas, R. H., Brønstad, K., Trefall, H., Kodama, M. and Rosenberg, T. J.: 1973, 'Balloon Observations of Auroral-Zone X-Rays in Conjugate Regions', *J. Atmosph. Terr. Phys.* **35**, 497.

Barfield, J. N., Burch, J. L. and Williams, D. J.: 1975, 'Substorm-Associated Reconfiguration of the Dusk Side Equatorial Magnetosphere: A Possible Source Mechanism for Isolated Plasma Regions', *J. Geophys. Res.* **80**, 47.

Basu, S.: 1974, 'VHF Ionospheric Scintillations at $L = 2.8$ and Formation of Stable Auroral Red Arcs by Magnetospheric Heat Conduction', *J. Geophys. Res.* **79**, 3155.

Bauer, S. J. and Krishnamurthy, B. V.: 1968, 'Possible Relationship between the Storm-Time Whistler Cutoff and the Magnetospheric Ring Current', *J. Geophys. Res.* **73**, 1853.

Berkey, F. T.: 1968, 'Coordinated Measurements of Auroral Absorption and Luminosity Using the Narrow Beam Technique', *J. Geophys. Res.* **73**, 319.

Berkey, F. T., Driatskiy, V. M., Henriksen, K., Hultqvist, B., Jelly, D., Shchuka, T. I., Theander, A. and Yliniemi, J.: 1974, 'A Synoptic Investigation of Particle Precipitation Dynamics for 60 Substorms in IQSY (1964–65) and IASY (1969)', *Planet. Space Sci.* **22**, 255.

Berko, F. W., Cahill, L. J., Jr. and Fritz, T. A.: 1975, 'Protons as the Prime Contributors to Storm Ring Current', *J. Geophys. Res.* **80**, 3549.

Bernstein, W., Inouye, G. T., Sanders, N. L. and Wax, R. L.: 1969, 'Measurements of Precipitated 1–20 keV Protons and Electrons during a Breakup Aurora', *J. Geophys. Res.* **74**, 3601.

Bernstein, W., Hultqvist, B. and Borg, H.: 1974, 'Some Implications of Low Altitude Observations of Isotropic Precipitation of Ring Current Protons beyond the Plasmapause', *Planet. Space Sci.* **22**, 767.

Bjordal, H., Trefall, H., Ullaland, S., Bewersdorff, A., Kangas, J., Tanskanen, P., Kremser, G., Saeger, K. H. and Specht, H.: 1971, 'On the Morphology of Auroral-Zone X-Ray Events – 1, Dynamics of Midnight Events', *J. Atmosph. Terr. Phys.* **33**, 605.

Bogott, F. H. and Mozer, F. S.: 1971, 'Equatorial Proton and Electron Angular Distributions in the Loss Cone and at Large Angles', *J. Geophys. Res.* **76**, 6790.

Bogott, F. H. and Mozer, F. S.: 1973a, 'ATS-5 Observations of Energetic Proton Injection', *J. Geophys. Res.* **78**, 8113.

Bogott, F. H. and Mozer, F. S.: 1973b, 'Nightside Energetic Particle Decreases at the Synchronous Orbit', *J. Geophys. Res.* **78**, 8119.

Bogott, F. H. and Mozer, F. S.: 1974, 'Drifting Energetic Particle Bunches Observed on ATS 5, *J. Geophys. Res.* **79**, 1825.

Bondareva, T. V. and Trerskaya, L. V.: 1973, 'Drift of Radiation-Belt Particles during a Substorm', *Geomag. Aeronom.* **13**, 612.

Brace, L. H. and Theis, R. F.: 1974, 'The Behavior of the Plasmapause at Mid-Latitudes: Isis 1 Langmuir Probe Measurements', *J. Geophys. Res.* **79**, 1871.

Brace, L. H., Maier, E. J., Hoffman, J. H., Whitteker, J. and Shepherd, G. G.: 1974, 'Deformation of the Night Side Plasmasphere and Ionosphere during the August 1972 Geomagnetic Storm', *J. Geophys. Res.* **79**, 5211.

Bradbury, J. N., Evans, J. E., Joki, E. G., Moe, C. R. and Hook, J. L.: 1968, 'Simultaneous Measurements of Electron Content and Charged-Particle Precipitation in a Region of Auroral Activity', *J. Geophys. Res.* **73**, 2363.

Brekke, A.: 1971, 'On the Correlation between Pulsating Aurora and Cosmic Radio Noise Absorption', *Planet. Space Sci.* **19**, 891.

Brice, N.: 1964, 'Fundamentals of Very Low Frequency Emission Generation Mechanisms', *J. Geophys. Res.* **69**, 4515.

Brice, N. and Lucas, C.: 1975, 'Interaction between Heavier Ions and Ring Current Protons', *J. Geophys. Res.* **80**, 936.

Brown, W. L., Cahill, L. J., Davis, L. R., McIlwain, C. E. and Roberts, C. S.: 1968, 'Acceleration of Trapped Particles during Magnetic Storm on April 18, 1965', *J. Geophys. Res.* **73**, 153.

Burrows, J. R., McDiarmid, I. B. and Wilson, M. D.: 1972, 'Pitch Angles and Spectra of Particles in the Outer Zone near Noon', *Earth's Magnetospheric Processes*, B. M. McCormac (ed.), p. 153, D. Reidel Publ. Co., Dordrecht-Holland.

Cahill, L. J. Jr.: 1970, 'Magnetosphere Inflation during Four Magnetic Storms in 1965', *J. Geophys. Res.* **75**, 3778.

Cahill, L. J. Jr.: 1973, 'Magnetic Storm Inflation in the Evening Sector', *J. Geophys. Res.* **78**, 4724.

Carman, E. H., Heeran, M. P. and Stevenson, R. W. H.: 1973, 'Observation of Stable Auroral Red Arcs from Southern Africa', *Planet. Space Sci.* **21**, 683.

Carpenter, D. L.: 1970, 'Whistler Evidence of the Dynamic Behavior of the Duskside Bulge in the Plasmasphere', *J. Geophys. Res.* **75**, 3837.

Carpenter, D. L.: 1971, 'Ogo 2 and 4 VLF Observations of the Asymmetric Plasmapause near the Time of SAR Arc Events', *J. Geophys. Res.* **76**, 3644.

Carpenter, D. L. and Akasofu, S.-I.: 1972, 'Two Substorm Studies of Relations between Westward Electric Fields in the Outer Plasmasphere, Auroral Activity and Geomagnetic Perturbations', *J. Geophys. Res.* **77**, 6854.

Carpenter, D. L. and Chappell, C. R.: 1973, 'Satellite Studies of Magnetospheric Substorms on August 15, 1968, 3. Some Features of Magnetospheric Convection', *J. Geophys. Res.* **78**, 3062.

Carpenter, D. L. and Stone, K.: 1967, 'Direct Detection by a Whistler Method of the Magnetospheric Electric Field Associated with a Polar Substorm', *Planet. Space Sci.* **15**, 395.

Carpenter, D. L., Park, C. G., Taylor, H. A. Jr. and Brinton, H. C.: 1969, 'Multi-Experiment Detection of the Plasmapause from EOGO Satellites and Antarctic Ground Stations', *J. Geophys. Res.* **74**, 1837.

Carpenter, D. L., Park, C. G., Arens, J. F. and Williams, D. J.: 1971a, 'Position of the Plasmapause during a Stormtime Increase in Trapped Energetic ($E > 280$ keV) Electrons', *J. Geophys. Res.* **76**, 4669.

Carpenter, D. L., Fraser-Smith, A. C., Unwin, R. S., Hones, E. W. Jr. and Heacock, R. R.: 1971b, 'Correlation between Convection Electric Fields in the Nightside Magnetosphere and Several Wave and Particle Phenomena during Two Isolated Substorms', *J. Geophys. Res.* **76**, 7778.

Carpenter, L. A. and Kirchhoff, V. W. J. H.: 1975, 'Comparison of High-Latitude and Mid-Latitude Ionosphere Electric Fields', *J. Geophys. Res.* **80**, 1810.

Chandra, S. and Krishnamurthy, B. V.: 1968, 'The Response of the Upper Atmospheric Temperature to Changes in Solar EUV Radiation and Geomagnetic Activity', *Planet. Space Sci.* **16**, 231.

Chandra, S., Maier, E. J., Troy, B. E. Jr. and Rao, B. C. N.: 1971, 'Subauroral Red Arcs and Associated Ionospheric Phenomena', *J. Geophys. Res.* **76**, 920.

Chandra, S., Maier, E. J. and Stubbe, P.: 1972, 'The Upper Atmosphere as a Regulator of Subauroral Red Arcs', *Planet. Space Sci.* **20**, 461.

Chappell, C. R.: 1974, 'Detached Plasma Regions in the Magnetosphere', *J. Geophys. Res.* **79**, 1861.

Chappell, C. R., Harris, K. K. and Sharp, G. W.: 1970a, 'The Reaction of the Plasmapause to Varying Magnetic Activity', *Particles and Fields in the Magnetosphere*, B. M. McCormac (ed.), p. 148, D. Reidel Publ. Co., Dordrecht-Holland.

Chappell, C. R., Harris, K. K. and Sharp, G. W.: 1970b, 'A Study of the Influence of Magnetic Activity on the Location of the Plasmapause as Measured by OGO 5', *J. Geophys. Res.* **75**, 50.

Chappell, C. R., Harris, K. K. and Sharp, G. W.: 1971, 'Ogo 5 Measurements of the Plasmasphere during Observations of Stable Auroral Red Arcs', *J. Geophys. Res.* **76**, 2357.

Chen, A. J.: 1970, 'Penetration of Low-Energy Protons Deep into the Magnetosphere', *J. Geophys. Res.* **75**, 2458.

Chen, A. J. and Grebowsky, J. M.: 1974, 'Plasma Tail Interpretations of Pronounced Detached Plasma Regions Measured by OGO 5', *J. Geophys. Res.* **79**, 3851.

Chen, A. J. and Wolf, R. A.: 1972, 'Effects on the Plasmasphere of a Time-Varying Convection Electric Field', *Planet. Space Sci.* **20**, 483.

Chen, A. J., Grebowsky, J. M. and Taylor, H. A. Jr.: 1975, 'Dynamics of Mid-Latitude Light Ion Trough and Plasma Tails', *J. Geophys. Res.* **80**, 968.

Christensen, A. B. and Karas, R.: 1970, 'Energy Spectra of Precipitating Electrons from Observations of Optical Aurora, Bremsstrahlung X-Rays, and Auroral Absorption', *J. Geophys. Res.* **75**, 4266.

Cladis, J. B.: 1971, 'Multiple Coupled Oscillations of Field Lines in the Magnetosphere: Modulation of Trapped Particles and Ionospheric Currents', *J. Geophys. Res.* **76**, 2345.

Cladis, J. B.: 1973, 'Effect of Magnetic Field Gradient on Motion of Ions Resonating with Ion Cyclotron Waves', *J. Geophys. Res.* **78**, 8129.

Cole, K. D.: 1965, 'Stable Auroral Red Arcs, Sinks for Energy of Dst Main Phase', *J. Geophys. Res.* **70**, 1689.

Cole, K. D.: 1970, 'Magnetospheric Processes Leading to Mid-Latitude Auroras', *Ann. de Geophys.* **26**, 187.

Coleman, P. J. Jr. and Cummings, W. D.: 1971, 'Stormtime Disturbance Field at ATS I', *J. Geophys. Res.* **76**, 51.

Cornwall, J. M.: 1970, 'Mutually Interacting Instabilities in the Magnetosphere', *Particles and Fields in the Magnetosphere*, B. M. McCormac (ed.), p. 266, D. Reidel Publ. Co., Dordrecht-Holland.

Cornwall, J. M.: 1974, 'Magnetosphere Dynamics with Artificial Plasma Clouds', *Space Sci. Rev.* **15**, 841.

Cornwall, J. M., Coroniti, F. V. and Thorne, R. M.: 1970: 'Turbulent Loss of Ring Current Protons', *J. Geophys. Res.* **75**, 4699.

Cornwall, J. M., Coroniti, F. V. and Thorne, R. M.: 1971a, 'Unified Theory of SAR Arc Formation at the Plasmapause', *J. Geophys. Res.* **76**, 4428.

Cornwall, J. M., Hilton, H. H. and Mizera, P. F.: 1971b, 'Observations of Precipitating Protons in the Energy Range 2.5 kev $\leq E \leq$ 200 kev', *J. Geophys. Res.* **76**, 5220.

Coroniti, F. V. and Kennel, C. F.: 1970a, 'Electron Precipitation Pulsations', *J. Geophys. Res.* **75**, 1279.

Coroniti, F. V. and Kennel, C. F.: 1970b, 'Auroral Micropulsation Instability', *J. Geophys. Res.* **75**, 1863.

Coroniti, F. V., Fredricks, R. W. and White, R.: 1972, 'Instability of Ring Current Protons beyond the Plasmapause during Injection Events', *J. Geophys. Res.* **77**, 6245.

Craven, J. D.: 1966, 'Temporal Variations of Electron Intensities at Low Altitudes in the Outer Radiation Zone as Observed with Injun 3', *J. Geophys. Res.* **71**, 5643.

Crooker, N. U. and Siscoe, G. L.: 1974, 'Model Geomagnetic Disturbance from Asymmetric Ring Current Particles', *J. Geophys. Res.* **79**, 589.

Cummings, W. D.: 1966, 'Asymmetric Ring Currents and the Low Latitude Disturbance Daily Variation', *J. Geophys. Res.* **71**, 4495.

Davis, T. N.: 1969, 'Temporal Behavior of Energy Injection into the Geomagnetic Ring Current', *J. Geophys. Res.* **74**, 6266.

DeForest, S. E. and McIlwain, C. E.: 1971, 'Plasma Clouds in the Magnetosphere', *J. Geophys. Res.* **76**, 3587.

Dolginov, Sh. Sh., Zhigalov, L. N., Strunnikova, L. V., Fel'dshteyn, Ya. I., Cherevko, T. N. and Sharova, V. A.: 1972, 'Magnetic Storm of March 8–10, 1970, According to Ground-Based and Kosmos-321 Observations', *Geomag. Aeronom.* **12**, 909.

Driatskiy, V. M.: 1968, 'Diurnal Pattern of Auroral Absorption in the Auroral Zone', *Geomag. Aeronom.* **8**, 33.

Driatskiy, V. M: 1969, 'Movement of Auroral Absorption along the Auroral Zone', *Geomag. Aeronom.* **9**, 398.

Driatskiy, V. M.: 1971, 'Auroral Absorption and Development of DR-Currents', *Geomag. Aeronom.* **11**, 306.

Driatskiy, V. M. and Shumilov, O. I.: 1970, 'Meridional Movement of Auroral Absorption Bays', *Geomag. Aeronom.* **10**, 235.
Driatskiy, V. M. and Shumilov, O. I.: 1972, 'Ionospheric Substorms', *Planet. Space Sci.* **20**, 1375.
Driatskiy, V. M., Shumilov, O. I. and Frank-Kamenetskiy, A. I.: 1972, 'Dynamics of Auroral Absorption', *Geomag. Aeronom.* **12**, 396.
Dubinin, E. M. and Podgorny, I. M.: 1974, 'Particle Precipitation and Radiation Belt in Laboratory Experiments', *J. Geophys. Res.* **79**, 1426.
Eather, R. H. and Carovillano, R. L.: 1971, 'The Ring Current as the Source Region for Proton Auroras', *Cosmic Electrodyn.* **2**, 105.
Fel'dshteyn, Ya. I.: 1972, 'Dynamics of the Inner Boundary of the Plasma Sheet in the Tail of the Magnetosphere during Substorms', *Geomag. Aeronom.* **12**, 321.
Ferraro, V. C. A. and Davies, C. M.: 1972, 'A Further Note on the Geomagnetic Ring-Current', *Geophys. J. R. Astr. Soc.* **29**, 241.
Findlay, J. A., Dyson, P. L., Brace, L. H., Zmuda, A. J. and Radford, W. E.: 1969, 'Ionospheric and Magnetic Observations at 1000 Kilometers during the Geomagnetic Storm and Aurora of May 25–26, 1967', *J. Geophys. Res.* **74**, 3705.
Frank, L. A.: 1970a, 'On the Presence of Low-Energy Protons ($5 \lesssim E \lesssim 50$ kev) in the Interplanetary Medium', *J. Geophys. Res.* **75**, 707.
Frank, L. A.: 1970b, 'Direct Detection of Asymmetric Increases of Extraterrestrial "Ring Current" Proton Intensities in the Outer Radiation Zone', *J. Geophys. Res.* **75**, 1263.
Frank, L. A. and Owens, H. D.: 1970, 'Omnidirectional Intensity Contours of Low-Energy Protons ($0.5 \leq E \leq 50$ keV) in the Earth's Outer Radiation Zone at the Magnetic Equator', *J. Geophys. Res.* **25**, 1269, 1970.
Frank, L. A., Saflekos, N. A. and Ackerson, K. L.: 1976, 'Electron Precipitation in the Postmidnight Sector of the Auroral Zones', *J. Geophys. Res.* **81**, 155.
Fredricks, R. W.: 1975, 'Wave-Particle Interactions and Their Relevance to Substorms', *Space Sci. Rev.* **17**, 449.
Freeman, J. W. Jr.: 1974, 'Kp Dependence of the Plasma Sheet Boundary', *J. Geophys. Res.* **79**, 4315.
Freeman, J. W. Jr., Warren, C. S. and Maguire, J. J.: 1968, 'Plasma Flow Directions at the Magnetopause on January 13 and 14, 1967', *J. Geophys. Res.* **73**, 5719.
Fukunishi, H.: 1969, 'Occurrences of Sweepers in the Evening Sector Following the Onset of Magnetospheric Substorms', *Rep. Ionosph. Space Res. Japan* **23**, 21.
Fukunishi, H.: 1973, 'Occurrence of IPDP Events Accompanied by Cosmic Noise Absorption in the Course of Proton Aurora Substorms', *J. Geophys. Res.* **78**, 3981.
Gendrin, R.: 1970, 'Substorm Aspects of Magnetic Pulsations', *Space Sci. Rev.* **11**, 54.
Gendrin, R.: 1972, 'Changes in the Distribution Functions of Magnetospheric Particles Associated with Gyroresonant Interactions', *Earth's Magnetospheric Processes*, B. M. McCormac (ed.), p. 311, D. Reidel Publ. Co., Dordrecht-Holland.
Gendrin, R.: 1975, 'Waves and Wave-Particle Interactions in the Magnetosphere: A Review', *Space Sci. Rev.* **18**, 145.
Glass, N. W., Wolcott, J. H., Miller, L. W. and Robertson, M. M.: 1970, 'Local Time Behavior of the Alignment and Position of a Stable Auroral Red Arc', *J. Geophys. Res.* **75**, 2579.
Grafe, A.: 1974, 'Anomalous DS-Variation in Equatorial Latitudes during Geomagnetic Storms', *Planet. Space Sci.* **22**, 991.
Grebowsky, J. M.: 1970, 'Model Study of Plasmapause Motion', *J. Geophys. Res.* **75**, 4329.
Grebowsky, J. M.: 1971, 'Time-Dependent Plasmapause Motion', *J. Geophys. Res.* **76**, 6193.
Grebowsky, J. M. and Chen, A. J.: 1975, 'Effects of Convection Electric Field on the Distribution of Ring Current Type Protons', *Planet. Space Sci.* **23**, 1045.
Grebowsky, J. M., Tulunay, Y. K. and Chen, A. J.: 1974, 'Temporal Variations in the Dawn and Dusk Mid-Latitude Trough and Plasmapause Position', *Planet. Space Sci.* **22**, 1089.
Gurevich, A. V. and Tsedilina, Ye. Ye.: 1969a, 'Dynamics of Fast Electron and Ion Inhomogeneities in the Earth's Magnetosphere, I', *Geomag. Aeronom.* **9**, 372.
Gurevich, A. V. and Tsedilina, Ye. Ye.: 1969b, 'Dynamics of Inhomogeneities of Fast Electrons and Ions in the Earth's Magnetosphere, II', *Geomag. Aeronom.* **9**, 519.
Gustafsson, G.: 1969, 'Spatial and Temporal Relations between Auroral Emission and Cosmic Noise Absorption', *Planet. Space Sci.* **17**, 1961.
Hall, W. N.: 1974, 'Mid-Latitude Pulsating Auroras', *Planet. Space Sci.* **22**, 1315.
Hargreaves, J. K.: 1969, 'Auroral Absorption of HF Radio Waves in the Ionosphere: A Review of Results from the First Decade of Riometry', *Proc. IEEE* **57**, 1348.
Hargreaves, J. K.: 1970, 'Conjugate and Closely-Spaced Observations of Auroral Radio Absorption – IV. The Movement of Simple Features', *Planet. Space Sci.* **18**, 1691.
Hargreaves, J. K.: 1971, 'Conjugate and Closely-Spaced Observations of Auroral Radio Absorption –

V. Alternation of Auroral Particle Precipitation between Hemispheres', *Planet. Space Sci.* **19**, 513.
Hargreaves, J. K.: 1974, 'Dynamics of Auroral Absorption in the Midnight Sector – The Movement of Absorption Peaks in Relation to the Substorm Onset', *Planet. Space Sci.* **22**, 1427.
Hargreaves, J. K., Chivers, H. J. A. and Axford, W. I.: 1975, 'The Development of the Substorm in Auroral Radio Absorption', *Planet. Space Sci.* **23**, 905.
Harris, K. K., Sharp, G. W. and Chappell, C. R.: 1970, 'Observations of the Plasmapause from OGO 5', *J. Geophys. Res.* **75**, 219.
Haskell, G. P. and Hynds, R. J.: 1972, 'Mechanisms for the Injection of Protons into the Magnetosphere', *Earth's Magnetospheric Processes*, B. M. McCormac (ed.), p. 81, D. Reidel Publ. Co., Dordrecht-Holland.
Hauge, R. and Söraas, F.: 1975, 'Precipitation of >115 keV Protons in the Evening and Forenoon Sectors in Relation to the Magnetic Activity', *Planet. Space Sci.* **23**, 1141.
Häusler, B. and Sckopke, N.: 1974, 'The Behavior of Outer Radiation Belt Electrons (>1.5 meV) during a Geomagnetically Disturbed Time on 8 March 1970 and Its Relationship to Magnetospheric Processes', *Planet. Space Sci.* **22**, 1249.
Heacock, R. R.: 1973, 'Type IPDP Magnetospheric Plasma Wave Events', *Nature* **246**, 93.
Heacock, R. R., Henderson, D. J., Reid, J. S. and Kivinen, M.: 1976, 'Type IPDP Pulsation Events in the Late Evening-Midnight Sector', *J. Geophys. Res.* **81**, 273.
Hoch, R. J. and Clark, K. C.: 1970, 'Recent Occurrences of Stable Auroral Red Arcs', *J. Geophys. Res.* **75**, 2511.
Hoch, R. J., Marovich, E. and Clark, K. C.: 1968, 'Reappearance of a Stable Auroral Red Arc at Midlatitudes', *J. Geophys. Res.* **73**, 4213.
Hoch, R. J., Carver, F. I., Smith, L. L. and Clark, K. C.: 1970, 'Auroral Activity near Banff, Alberta', *J. Geophys. Res.* **75**, 1935.
Hoch, R. J., Batishko, C. R. and Clark, K. C.: 1971, 'Simultaneous Spectrographic Triangulation of an SAR Arc and a Hydrogen Arc', *J. Geophys. Res.* **76**, 6185.
Hoffman, R. A.: 1970, 'Auroral Electron Drift and Precipitation: Cause of the Mantle Aurora', *Goddard Space Flight Center Rep. X-646-70-205.*
Hoffman, R. A.: 1973, 'Particle and Field Observations from Explorer 45 during the December 1971 Magnetic Storm Period', *J. Geophys. Res.* **78**, 4771.
Hoffman, R. A. and Burch, J. L.: 1973, 'Electron Precipitation Patterns and Substorm Morphology', *J. Geophys. Res.* **78**, 2867.
Hoffman, R. A. and Cahill, L. J. Jr.: 1968, 'Ring Current Particle Distributions Derived from Ring Current Magnetic Field Measurements', *J. Geophys. Res.* **73**, 6711.
Hoffman, R. A., Cahill, L. J. Jr., Anderson, R. R., Maynard, N. C., Smith, P. H., Fritz, T. A., Williams, D. J., Konradi, A. and Gurnett, D. A.: 1974, 'Explorer 45 (S^3-A) Observations of the Magnetosphere and Magnetopause during the August 4–5, 1972, Magnetic Storm Period', *Goddard Space Flight Center, Rep. X-621-74-114.*
Hofmann, D. J. and Greene, R. A.: 1972, 'Balloon Observations of Simultaneous Auroral X-Ray and Visible Bursts', *J. Geophys. Res.* **77**, 776.
Holzer, R. E., Farley, T. A., Burton, R. K. and Chapman, M. C.: 1974, 'A Correlated Study of ELF Waves and Electron Precipitation on Ogo 6', *J. Geophys. Res.* **79**, 1007.
Hones, E. W. Jr., Singer, S., Lanzerotti, L. J., Pierson, J. D. and Rosenberg, T. J.: 1971, 'Magnetospheric Substorm of August 25–26, 1967', *J. Geophys. Res.* **76**, 2977.
Hovestadt, D., Achtermann, E., Ebel, B., Häusler, B. and Paschmann, G.: 1972, 'New Observations of the Proton Population of the Radiation Belt between 1.5 and 104 MeV', *Earth's Magnetospheric Processes*, B. M. McCormac, (ed.), p. 115, D. Reidel Publ. Co., Dordrecht-Holland.
Hultqvist, B.: 1971, 'Temporal Development of the Geographical Distribution of Auroral Absorption for 30 Substorm Events in Each of IQSY (1964–65) and IASY', *World Data Center A, UAG.*
Hultqvist, B.: 1975a, 'Some Experimentally Determined Characteristics of the Turbulence in the Magnetosphere', *Physics of the Hot Plasma in the Magnetosphere*, B. Hultqvist and L. Stenflo (eds.), p. 291, Plenum Press, New York.
Hultqvist, B.: 1975b, 'The Ring Current and Particle Precipitation near the Plasmapause', *Ann. Geophys.* **31**, 111.
Hultqvist, B., Borg, H., Christophersen, P. and Riedler, W.: 1971, 'Observations of Magnetic Field Aligned Anisotropy for 1 and 6 keV Positive Ions in the Upper Ionosphere', *Planet. Space Sci.* **19**, 279.
Hultqvist, B., Borg, H., Christophersen, P., Riedler, W. and Bernstein, W.: 1974, 'Energetic Protons in the keV Energy Range and Associated keV Electrons Observed at Various Local Times and Disturbance Levels in the Upper Ionosphere', *NOAA, ERL, Space Environmental Laboratory, Boulder, Colorado, USA, Report.*

Ichikawa, T., Old, T. and Kim, J. S.: 1969, 'Relationship between a Monochromatic Auroral Arc of 6300 Å and a Visible Aurora', *J. Geophys. Res.* **74**, 5819.

Imholf, W. L., Nakano, G. H., Johnson, R. G. and Reagan, J. B.: 1974, 'Satellite Observations of Bremsstrahlung from Widespread Energetic Electron Precipitation Events', *J. Geophys. Res.* **79**, 565.

Jacchia, L. G. and Slowey, J.: 1964, 'Atmospheric Heating in the Auroral Zones: A Preliminary Analysis of the Atmospheric Drag of the Injun 3 Satellite', *J. Geophys. Res.* **69**, 905.

Jacchia, L. G., Slowey, J. and Verniani, F.: 1967, 'Geomagnetic Perturbations and Upper-Atmosphere Heating', *J. Geophys. Res.* **72**, 1423.

Jacobs, J. A.: 1970, *Geomagnetic Micropulsations*, Springer-Verlag.

Jaggi, R. K. and Wolf, R. A.: 1973, 'Self-Consistent Calculation of the Motion of a Sheet of Ions in the Magnetosphere', *J. Geophys. Res.* **78**, 2852.

Jani, K. G. and Kotadia, K. M.: 1969, 'Changes in the Ionospheric F2-Layer at a Place near the S_q-Current Focus during Geomagnetic Storms', *Planet. Space Sci.* **17**, 1949.

Jelly, D. H.: 1970, 'On the Morphology of Auroral Absorption during Substorms', *Canadian Journal of Physics* **48**, 335.

Jentsch, V. and Kremser, G.: 1972, 'On Observations of Auroral-Zone X-Rays with Energies up to and Greater than 200 keV', *J. Atmosph. Terr. Phys.* **34**, 499.

Kamide, Y. and McIlwain, C. E.: 1974, 'The Onset Time of Magnetospheric Substorms Determined from Ground and Synchronous Satellite Records', *J. Geophys. Res.* **79**, 4787.

Kane, R. P.: 1973, 'Global Evolution of the DS Component during Geomagnetic Storms', *J. Geophys. Res.* **78**, 5585.

Kangas, J., Lukkari, L. and Heacock, R. R.: 1974, 'On the Westward Expansion of Substorm-Correlated Particle Phenomena', *J. Geophys. Res.* **79**, 3207.

Kasymov, U. and Shabanskiy, V. P.: 1972, 'Computation of Angular Particle Distribution in the Magnetosphere', *Geomag. Aeronom.* **12**, 507.

Kasymov, U. and Shabanskiy, V. P.: 1974, 'Determination of the Parameters of Magnetic Drift Shells and of Particle Distribution in the Magnetosphere', *Geomag. Aeronom.* **14**, 77.

Kaufmann, R. L.: 1974, 'Electron Acceleration during Tail Collapse', *J. Geophys. Res.* **79**, 549.

Kaufmann, R. L., Horng, J.-T. and Konradi, A.: 1972, 'Trapping Boundary and Field-Line Motion during Geomagnetic Storms', *J. Geophys. Res.* **77**, 2780.

Kawasaki, K. and Akasofu, S.-I.: 1971a, 'Geomagnetic Storm Fields near a Synchronous Satellite', *Planet. Space Sci.* **19**, 1339.

Kawasaki, K. and Akasofu, S.-I.: 1971b, 'Low-Latitude DS Component of the Geomagnetic Storm Field', *J. Geophys. Res.* **76**, 2396.

Kelley, M. C., Tsurutani, B. T. and Mozer, F. S.: 1975, 'Properties of ELF Electromagnetic Waves in and above the Earth's Ionosphere Deduced from Plasma Wave Experiments on the OV1-17 and Ogo 6 Satellites', *J. Geophys. Res.* **80**, 4603.

Kennel, C. F. and Petschek, H. E.: 1966, 'Limit on Stably Trapped Particle Fluxes', *J. Geophys. Res.* **71**, 1.

Khocholava, G. M., Kadukhin, G. F. and Mebagishvili, N. N.: 1973, 'Ionospheric Effects of Geomagnetic Storms at Middle Latitudes', *Geomag. Aeronom.* **13**, 743.

Kimura, I.: 1974, 'Interrelation between VLF and ULF Emissions', *Space Sci. Rev.* **16**, 389.

Kirsch, E., Münch, J. W. and Scholer, M.: 1975, 'Double Structure in the Outer Electron Radiation Belt during Geomagnetic Activity', *Planet. Space Sci.* **23**, 913.

Kivelson, M. G. and Southwood, D. J.: 1975, 'Local Time Variations of Particle Flux Produced by an Electrostatic Field in the Magnetosphere', *J. Geophys. Res.* **80**, 56.

Kivelson, M. G., Farley, T. A. and Aubry, M. P.: 1973, 'Satellite Studies of Magnetospheric Substorms on Aug. 15, 1968; 5. Energetic Electrons, Spatial Boundaries, and Wave-Particle Interactions at Ogo 5', *J. Geophys. Res.* **78**, 3079.

Kleckner, E. W. and Hoch, R. J.: 1973, 'Simultaneous Occurrences of Hydrogen Arcs and Mid-Latitude Stable Auroral Red Arcs', *J. Geophys. Res.* **78**, 1187.

Konradi, A., Williams, D. J. and Fritz, T. A.: 1973, 'Energy Spectra and Pitch Angle Distributions of Storm-Time and Substorm Injected Protons', *J. Geophys. Res.* **78**, 4739.

Konradi, A., Semar, C. L. and Fritz, T. A.: 1975, 'Substorm-Injected Protons and the Injection Boundary Model', *J. Geophys. Res.* **80**, 543.

Kremser, G., Wilhelm, K., Riedler, W., Brønstad, K., Trefall, H., Ullaland, S. L., Legrand, J. P., Kangas, J. and Tanskanen, P.: 1973, 'On the Morphology of Auroral Zone X-Ray Events – II. Events during the Early Morning Hours', *J. Atmosph. Terr. Phys.* **35**, 713.

Lackner, K.: 1970, 'Deformation of a Magnetic Dipole Field by Trapped Particles', *J. Geophys. Res.* **75**, 3180.

Langel, R. A. and Sweeney, R. E.: 1971, 'Asymmetric Ring Current at Twilight Local Time', *J. Geophys. Res.* **76**, 4420.
Lanzerotti, L. J.: 1968, 'Outer-Zone Electrons and the Interplanetary Magnetic Fields during Two Geomagnetic Storms', *J. Geophys. Res.* **73**, 4388.
LaValle, S. R. and Elliott, D. D.: 1972, 'Observations of SAR Arcs from OV1-10', *J. Geophys. Res.* **77**, 1802.
Lezniak, T. W. and Winckler, J. R.: 1970, 'Experimental Study of Magnetospheric Motions and the Acceleration of Energetic Electrons during Substorms', *J. Geophys. Res.* **75**, 7075.
Lezniak, T. W., Arnoldy, R. L., Parks, G. K. and Winckler, J. R.: 1968, 'Measurement and Intensity of Energetic Electrons at the Equator at 6.6 R_e', *Radio Sci.* **3**, 710.
Lichtenstein, P. R.: 1970, 'Mid-Latitude Magnetic Bays and Auroral Absorption', *Planet. Space Sci.* **18**, 1301.
Liemohn, H.: 1961, 'The Lifetime of Radiation Belt Protons with Energies between 1 keV and 1 MeV', *J. Geophys. Res.* **66**, 3593.
Lin, C. S. and Parks, G. K.: 1974, 'Further Characteristics of the Evening Energetic Electron Decreases during Substorms', *J. Geophys. Res.* **79**, 3201.
Lindalen, H. R. and Egeland, A. L.: 1972, 'Observations of Trapped and Precipitated Protons on March 8, 1970', *Ann. de Geophys.* **28**, 129.
Lindalen, H. R., Sørass, F., Aarsnes, K. and Amundsen, R.: 1971, 'Variations in the High Latitude Proton Trapping Boundary Associated with Polar Magnetic Substorms', *Planet. Space Sci.* **19**, 1041.
Liu, C. S.: 1970, 'Low-Frequency Drift Instabilities of the Ring Current Belt', *J. Geophys. Res.* **75**, 3789.
Lyons, L. R.: 1974, 'Electron Diffusion Driven by Magnetospheric Electrostatic Waves', *J. Geophys. Res.* **79**, 575.
Lyons, L. R. and Williams, D. J.: 1975a, 'The Quiet Time Structure of Energetic (35–560 keV) Radiation Belt Electrons', *J. Geophys. Res.* **80**, 943.
Lyons, L. R. and Williams, D. J.: 1975b, 'The Storm and Poststorm Evolution of Energetic (35–560 keV) Radiation Belt Electron Distributions', *J. Geophys. Res.* **80**, 3985.
Mal'tseva, N. F., Fel'dshteyn, Ya. I. and Gul'yel'mi, A. V.: 1971, 'Intervals of Pulsations of Decreasing Period and Development of Asymmetry in the Ring Current', *Geomag. Aeronom.* **11**, 255.
Maral, G.: 1970, 'Motions Associated with Auroral Zone Electron Precipitation', *J. Geophys. Res.* **75**, 2601.
Maral, G., Brønstad, K., Trefall, H., Kremser, G., Specht, H., Tanskanen, P., Kangas, J., Riedler, W. and Legrand, J. P.: 1973, 'On the Morphology of Auroral-Zone X-Ray Events – III, Large-Scale Observations in the Midnight-to-Morning-Sector', *J. Atmosph. Terr. Phys.* **36**, 735.
Märk, E.: 1974, 'Growth Rates of the Ion Cyclotron Instability in the Magnetosphere', *J. Geophys. Res.* **79**, 3218.
Marovich, E.: 1970, 'Recent Occurrences of Stable Auroral Red Arcs', *J. Geophys. Res.* **75**, 4893.
Mauk, B. H. and McIlwain, C. E.: 1974, 'Correlation of K_p with the Substorm-Injected Plasma Boundary', *J. Geophys. Res.* **79**, 3193.
Maynard, N. C. and Chen, A. J.: 1974, 'Isolated Cold Plasma Regions: Observations and Their Relation to Possible Production Mechanisms', *Goddard Space Flight Center*, May.
McIlwain, C. E.: 1972, 'Plasma Convection in the Vicinity of the Geosynchronous Orbit', *Earth's Magnetospheric Processes*, B. M. McCormac (ed.), p. 268, D. Reidel Publ. Co., Dordrecht-Holland.
McIlwain, C. E.: 1974, 'Substorm Injection Boundaries', *Magnetospheric Physics*, B. M. McCormac (ed.), p. 143, D. Reidel Publ. Co., Dordrecht-Holland.
McPherron, R. L., Russell, C. T. and Coleman, P. J. Jr.: 1972, 'Fluctuating Magnetic Fields in the Magnetosphere, II. ULF Waves', *Space Sci. Rev.* **13**, 411.
Mizera, P. F.: 1974, 'Observations of Precipitating Protons with Ring Current Energies', *J. Geophys. Res.* **79**, 581.
Mizera, P. F. and Blake, J. B.: 1973, 'Observations of Ring Current Protons at Low Altitudes', *J. Geophys. Res.* **78**, 1058.
Mozer, F. S., Bogott, F. H. and Tsurutani, B.: 1973, 'Relations between Ionospheric Electric Fields and Energetic Trapped and Precipitating Electrons', *J. Geophys. Res.* **78**, 630.
Nagy, A. F., Hanson, W. B., Hoch, R. J. and Aggson, T. L.: 1972, 'Satellite and Ground-Based Observations of a Red Arc', *J. Geophys. Res.* **77**, 3613.
Nagy, A. F., Brace, L. H., Maynard, N. C. and Hanson, W. B.: 1974, 'Is the Red Arc a Good Indicator of Ionosphere-Magnetosphere Conditions?', *J. Geophys. Res.* **79**, 4331.
Nambu, M.: 1973, 'Electrostatic Turbulent Loss of Ring Current Protons', *J. Geophys. Res.* **78**, 1203.
Nambu, M.: 1974, 'Wave-Particle Interactions between the Ring Current Particles and Micropulsations Associated with the Plasmapause', *Space Sci. Rev.* **16**, 427.

Olbert, S., Siscoe, G. L. and Vasyliunas, V. M.: 1968, 'A Simple Derivation of the Dessler-Parker-Sckopke Relation', *J. Geophys. Res.* **73**, 1115.

Olson, W. P. and Cummings, W. D.: 1970, 'Comparison of the Predicted and Observed Magnetic Field at ATS 1,' *J. Geophys. Res.* **75**, 7117.

Page, D. E. and Shaw, M. L.: 1972, 'Some Parameters Affecting the Poleward Boundary of Trapped Electrons', *Earth's Magnetospheric Processes*, B. M. McCormac (ed.), p. 175, D. Reidel Publ. Co., Dordrecht-Holland.

Parady, B. and Cahill, L. J. Jr.: 1972, 'Comparison of Magnetosphere Models with Storm Observations from Explorer 26', *J. Geophys. Res.* **77**, 6235.

Park, C. G.: 1971, Westward Electric Fields as the Cause of Nighttime Enhancements in Electron Concentrations in the Mid-Latitude F-Region, *J. Geophys. Res.* **76**, 4560.

Park, C. G.: 1973, 'Whistler Observations of the Depletion of the Plasmasphere during a Magnetospheric Substorm', *J. Geophys. Res.* **78**, 672.

Park, C. G.: 1974, 'Some Features of Plasma Distribution in the Plasmasphere Deduced from Antarctic Whistlers', *J. Geophys. Res.* **79**, 169.

Park, C. G. and Banks, P. M.: 1974, 'Influence of Thermal Plasma Flow on the Mid-Latitude Nighttime F2 Layer: Effects of Electric Fields and Neutral Winds Inside the Plasmasphere', *J. Geophys. Res.* **79**, 4661.

Park, C. G. and Banks, P. M.: 1975, 'Influence of Thermal Plasma Flow on the Daytime F2 Layer', *J. Geophys. Res.* **80**, 2819.

Park, C. G. and Carpenter, D. L.: 1970, 'Whistler Evidence of Large-Scale Electron Density Irregularities in the Plasmasphere', *J. Geophys. Res.* **75**, 3852.

Park, C. G. and Meng, C.-I.: 1973, 'Distortions of the Nightside Ionosphere during Magnetospheric Substorms', *J. Geophys. Res.* **78**, 3828.

Parks, G. K.: 1970, 'The Acceleration and Precipitation of Van Allen Outer Zone Energetic Electrons', *J. Geophys. Res.* **75**, 3802.

Parks, G. K. and Winckler, J. R.: 1968, 'Acceleration of Energetic Electrons Observed at the Synchronous Altitude during Magnetospheric Substorms', *J. Geophys. Res.* **73**, 5786.

Parks, G. K. and Winckler, J. R.: 1969, 'Simultaneous Observations of 5- to 15-Second Period Modulated Energetic Electron Fluxes at the Synchronous Altitude and the Auroral Zone', *J. Geophys. Res.* **74**, 4003.

Parks, G. K., Arnoldy, R. L., Lezniak, T. W. and Winckler, J. R.: 1968, 'Correlated Effects of Energetic Electrons at the 6.6 R_e Equator and the Auroral Zone during Magnetospheric Substorms', *Radio Sci.* **3**, 715.

Paulikas, G. A., Blake, J. B., Freden, S. C. and Imamoto, S. S.: 1968a, 'Observations of Energetic Electrons at Synchronous Altitude, 1. General Features and Diurnal Variations', *J. Geophys. Res.* **73**, 4915.

Paulikas, G. A., Blake, J. B., Freden, S. C. and Imamoto, S. S.: 1968b, 'Boundary of Energetic Electrons during the January 13–14, 1967, Magnetic Storm', *J. Geophys. Res.* **73**, 5743.

Pfitzer, K. A. and Winckler, J. R.: 1969, 'Intensity Correlations and Substorm Electron Drift Effects in the Outer Radiation Belt Measured with the OGO 3 and ATS 1 Satellites', *J. Geophys. Res.* **74**, 5005.

Pfitzer, K. A., Lezniak, T. W. and Winckler, J. R.: 1969, 'Experimental Verification of Drift-Shell Splitting in the Distorted Magnetosphere', *J. Geophys. Res.* **74**, 4687.

Pilkington, G. R.: 1972, 'X-Ray Observations and Interpretations', *Earth's Magnetospheric Processes*, B. M. McCormac (ed.), p. 391, D. Reidel Publ. Co., Dordrecht-Holland.

Pilkington, G. R. and Anger, C. D.: 1971, 'A Monte Carlo Analysis of the Passage of Auroral X-Rays through the Atmosphere', *Planet. Space Sci.* **19**, 1069.

Pilkington, G. R., Anger, C. D. and Clark, T. A.: 1968, 'Auroral X-Rays and Their Association with Rapidly Changing Auroral Forms', *Planet. Space Sci.* **16**, 815.

Prölss, G. W.: 1973, 'Decay of the Magnetic Storm Ring Current by the Charge-Exchange Mechanism', *Planet. Space Sci.* **21**, 983.

Pudovkin, M. I., Isaev, S. I. and Zaitzeva, S. A.: 1970, 'Development of Magnetic Storms and the State of the Magnetosphere According to the Data of Ground-Based Observations', *Ann. Geophys.* **26**, 761.

Pushkova, G. N., Yudovich, L. A., Petviashvily, V. I. and Feldstein, Ya. I.: 1972, 'Magneto-Ionospheric Effect of the Substorm', *J. Atmosph. Terr. Phys.* **34**, 1097.

Rao, C. S. R.: 1969, 'Some Observations on Energetic Electrons in the Outer Van Allen Zone during Auroral Substorms in Relation to Open and Closed Field Lines', *J. Geophys. Res.* **74**, 6513.

Raspopov, O. M. and Koshelevskiy, V. K.: 1972, 'Quasi-Stationary Pinch Effect and the Structure of the Magnetosphere', *Geomag. Aeronom.* **12**, 323.

Reed, E. I. and Blamont, J. E.: 1974, 'Observations of the Conjugate SAR Areas of September 28–30, 1967', *J. Geophys. Res.* **79**, 2524.

Rees, M. H. and Roble, R. G.: 1975, 'Observations and Theory of the Formation of Stable Auroral Red

Arcs', *Rev. Geophys. Space Phys.* **13**, 201.

Riedler, W. and Borg, H.: 1972, 'High-Latitude Precipitation of Low Energy Particles as Observed by ESRO 1A', *Space Research XII*, S. A. Bowhill, L. D. Jaffe and M. J. Rycroft (eds.), p. 1397, Akademie-Verlag, Berlin.

Roble, R. G. and Dickinson, R. E.: 1970, 'Atmospheric Response to Heating within a Stable Auroral Red Arc', *Planet. Space Sci.* **18**, 1489.

Roble, R. G. and Dickinson, R. E.: 1972, 'Time-Dependent Behaviour of a Stable Auroral Red Arc Excited by an Electric Field', *Planet. Space Sci.* **20**, 591.

Roble, R. G., Hays, P. B. and Nagy, A. F.: 1970a, 'Photometric and Interferometric Observations of a Mid-Latitude Stable Auroral Red Arc', *Planet. Space Sci.* **18**, 431.

Roble, R. G., Hays, P. B. and Nagy, A. F.: 1970b, 'Comparison of Calculated and Observed Features of a Stable Midlatitude Red Arc', *J. Geophys. Res.* **75**, 4261.

Roble, R. G., Norton, R. B., Findlay, J. A. and Marovich, E.: 1971, 'Calculated and Observed Features of Stable Auroral Red Arcs during Three Geomagnetic Storms', *J. Geophys. Res.* **76**, 7648.

Roederer, J. G. and Schulz, M.: 1971, 'Splitting of Drift Shells by the Magnetospheric Electric Field', *J. Geophys. Res.* **76**, 1055.

Roederer, J. G. and Hones, E. W. Jr.: 1974, 'Motion of Magnetospheric Particle Clouds in a Time-Dependent Electric Field Model', *J. Geophys. Res.* **79**, 1432.

Rosen, L. H. and Winckler, J. R.: 1970, 'Evidence for the Large-Scale Azimuthal Drift of Electron Precipitation during Magnetospheric Substorms', *J. Geophys. Res.* **75**, 5576.

Rosenberg, T. J., Bjordal, J., Trefall, H., Kvifte, G. J., Omholt, A. and Egeland, A.: 1971, 'Correlation Study of Auroral Luminosity and X-Rays', *J. Geophys. Res.* **76**, 122.

Rostoker, G., Kisabeth, J. L., Sharp, R. D. and Shelley, E. G.: 1975, 'The Expansive Phase of Manetospheric Substorms, 2, The Response at Synchronous Altitude of Particles of Different Energy Ranges', *J. Geophys. Res.* **80**, 3557.

Rothwell, P. L. and Katz, L.: 1973, 'Enhancement of 0.24- to 0.96-Mev Trapped Protons during the May 25, 1967, Magnetic Storm', *J. Geophys. Res.* **78**, 5490.

Rothwell, P. L., Webb, V. H. and Katz, L.: 1970, 'Trapped and Polar Particles during the June 9, 1968, Magnetic Storm', *Particles and Fields in the Magnetosphere*, B. M. McCormac (ed.), p. 132, D. Reidel Publ. Co., Dordrecht-Holland.

Royrvik, O.: 1976, 'Pulsating Aurora: Local and Global Morphology', Ph.D. Thesis, Univ. of Alaska, May.

Russell, C. T., McPherron, R. L. and Coleman, P. J. Jr.: 1972, 'Fluctuating Magnetic Fields in the Magnetosphere, 1. ELF and VLF Fluctuations', *Space Sci. Rev.* **12**, 810..

Rüster, R.: 1971, 'The Relative Effects of Electric Fields and Atmospheric Composition Changes on the Electron Concentration in the Mid-Latitude F-Layer', *J. Atmosph. Terr. Phys.* **33**, 275.

Saeger, K. H., Kremser, G., Pfotzer, G., Specht, H., Riedler, W. and Geophy. Observatory: 1972, 'Auroral-Zone X-Ray Measurements at Kiruna in 1970', *SPARMO-Bulletin*, Vol. V, No. 1, Aug.

Saito, T.: 1969, 'Geomagnetic Pulsations', *Space Sci. Rev.* **10**, 319.

Saito, T., Takahashi, F., Morioka, A. and Kuwashima, M.: 1974, 'Fluctuations of Electrons Precipitation to the Dayside Auroral Zone Modulated by Compression and Expansion of the Magnetosphere', *Planet. Space Sci.* **22**, 939.

Sarma, S. B. S. S. and Sharma, M. C.: 1971, 'Latitudinal Variation of Cosmic Noise Absorption', *Planet. Space Sci.* **19**, 1579.

Scarf, F. L.: 1975, 'Characteristics of Instabilities in the Magnetosphere Deduced from Wave Observations', *Physics of the Hot Plasma in the Magnetosphere*, B. Hultqvist and L. Stenflo (eds.), p. 271, Plenum Press, New York.

Scholer, M., Morfill, G. and Hovestadt, D.: 1975, 'Weak Pitch Angle Scattering of Energetic Protons in the Magnetosphere', *J. Geophys. Res.* **80**, 2745.

Sergeyev, V. A. and Shumilov, O. I.: 1974a, 'Dynamics and Mechanism of Formation of the Morning Auroral Absorption Maximum', *Geomag. Aeronom.* **14**, 381.

Sergeyev, V. A. and Shumilov, O. I.: 1974b, 'Mechanism of Electron Precipitation in the Morning and Afternoon Magnetosphere', *Geomag. Aeronom.* **14**, 150.

Shabanskiy, V. P.: 1971, 'Depression of the Geomagnetic Field during a Magnetic Storm', *Geomag. Aeronom.* **11**, 148.

Sharp, R. D., Shelley, E. G. and Rostoker, G.: 1975, 'A Relationship between Synchronous Altitude Electron Fluxes and the Auroral Electrojet', *J. Geophys. Res.* **80**, 2319.

Shelley, E. G., Johnson, R. G. and Sharp, R. D.: 1971, 'Plasma Sheet Convection Velocities Inferred from Electron Flux Measurements at Synchronous Altitude', *Radio Sci.* **6**, 305.

Shepherd, G. G.: 1975, 'The Global Pattern of 6300Å Atomic Oxygen Emission as Seen from the ISIS-2 Spacecraft', *Atmospheres of Earth and Planets*, B. M. McCormac (ed.), p. 283, D. Reidel Publ. Co., Dordrecht-Holland.

Shevnin, A. D.: 1971, 'Asymmetry of the Field of a Partial Ring Current', *Geomag. Aeronom.* **11**, 778.
Shevnin, A. D.: 1973a, 'Rate of Decay of the Ring Current and of Polar Disturbances', *Geomag. Aeronom.* **13**, 100.
Shevin, A. D.: 1973b, 'Some Patterns in the Rate of Decay of the Ring Current', *Geomag. Aeronom.* **13**, 282.
Siscoe, G. L. and Crooker, N. U.: 1974, 'On the Partial Ring Current Contribution to Dst', *J. Geophys. Res.* **79**, 1110.
Sletten, A., Stadsnes, J. and Trefall, H.: 1971, 'Auroral-Zone X-Ray Events and Their Relation to Polar Magnetic Substorms', *J. Atmosph. Terr. Phys.* **33**, 589.
Smith, L. L., Hoch, R. J., Owen, R. W., Hernandez, G. and Marovich, E.: 1972, 'Altitudes of the λ6300-A, λ5577-A, and λ4278-A Emissions of Stable Auroral Red Arcs of March 8–9, 1970', *J. Geophys. Res.* **77**, 2987.
Smith, P. H. and Hoffman, R. A.: 1973, 'Ring Current Particle Distributions during the Magnetic Storms of December 16–18, 1971', *J. Geophys. Res.* **78**, 4731.
Smith, P. H. and Hoffman, R. A.: 1974, 'Direct Observations in the Dusk Hours of the Characteristics of the Storm Time Ring Current Particles during the Beginning of Magnetic Storms', *J. Geophys. Res.* **79**, 966.
Smith, P. H., Hoffman, R. A. and Fritz, T.: 1975, 'Ring Current Proton Decay by Charge Exchange', *Goddard Space Flight Center Rep. X-626-75-251*, Oct.
Søraas, F.: 1972, 'ESRO 1A/B Observations at High Latitudes of Trapped and Precipitating Protons with Energies above 100 keV', *Earth's Magnetospheric Processes*, B. M. McCormac (ed.), p. 120, D. Reidel Publ. Co., Dordrecht-Holland.
Søraas, F. and Berg, L. E.: 1974, 'Correlated Satellite Measurements of Proton Precipitation and Plasma Density', *Scientific/Technical Report No. 60*, Univ. of Bergen, Norway, March.
Søraas, F., Aarsnes, K., Lindalen, H. R. and Madsen, M. M.: 1970, 'A Satellite Instrument for Measuring Protons in the Energy Range 0.1 MeV to 6 MeV', Arbok for Univ. 1 Bergen, *Mat.-Naturv. Serie No. 6.*
Sørensen, J., Bjordal, J., Trefall, H., Kvifte, G. J. and Pettersen, H.: 1973, 'Correlation between Pulsations in Auroral Luminosity Variations and X-Rays', *J. Atmosph. Terr. Phys.* **35**, 961.
Southwood, D. J. and Kivelson, M. G.: 1975, 'An Approximate Analytic Description of Plasma Bulk Parameters and Pitch Angle Anisotropy under Adiabatic Flow in a Dipolar Magnetospheric Field', *J. Geophys. Res.* **80**, 2069.
Sozou, C. and Windle, D. W.: 1970, 'The Effect of a Large Ring Current on the Topology of the Magnetosphere', *Planet. Space Sci.* **18**, 699.
Stern, D. P.: 1971, 'Shell Splitting Due to Electric Fields', *J. Geophys. Res.* **76**, 7787.
Stern, D. P.: 1975, 'The Motion of a Proton in the Equatorial Magnetosphere', *J. Geophys. Res.* **80**, 595.
Stern, D. P. and Palmadesso, P.: 1975, 'Drift-Free Magnetic Geometries in Adiabatic Motion', *Goddard Space Flight Center, X-602-75-15*, Feb.
Su, S.-Y., Fritz, T. A. and Konradi, A.: 1976, 'Repeated Sharp Dropouts Observed at 6.6 R_E during a Geomagnetic Storm', *J. Geophys. Res.* **81**, 245.
Swift, D. W.: 1971, 'Possible Mechanisms for Formation of the Ring Current Belt', *J. Geophys. Res.* **76**, 2276.
Swisher, R. L. and Frank, L. A.: 1968, 'Lifetimes for Low-Energy Protons in the Outer Radiation Zone', *J. Geophys. Res.* **73**, 5665.
Taylor, H. A. Jr., Brinton, H. C. and Pharo, M. W. III: 1968, 'Contraction of the Plasmasphere, during Geomagnetically Disturbed Periods', *J. Geophys. Res.* **73**, 961.
Taylor, H. A. Jr., Brinton, H. C. and Deshmukh, A. R.: 1970, 'Observations of Irregular Structure in Thermal Ion Distributions in the Duskside Magnetosphere', *J. Geophys. Res.* **75**, 2481.
Testud, J., Amayenc, P. and Blanc, M.: 1975, 'Middle and Low Latitude Effects of Auroral Disturbances from Incoherent Scatter', *J. Atmosph. Terr. Phys.* **37**, 989.
Thorne, R. M. and Kennel, C. F.: 1971, 'Relativistic Electron Precipitation during Magnetic Storm Main Phase', *J. Geophys. Res.* **76**, 4446.
Thorne, R. M., Smith, E. J., Burton, R. K. and Holzer, R. E.: 1973, 'Plasmaspheric Hiss', *J. Geophys. Res.* **78**, 1581.
Thorne, R. M., Smith, E. J., Fiske, K. J. and Church, S. R.: 1974, 'Intensity Variation of ELF Hiss and Chorus during Isolated Substorms', *Geophys. Res. Lett.* **1**, 193.
Troitskaya, V. A.: 1961, 'Pulsations of the Earth's Electromagnetic Field with Periods of 1 to 15 Seconds and Their Connection with Phenomena in the High Atmosphere', *J. Geophys. Res.* **66**, 5.
Troshichev, O. A. and Feldstein, Ya. I.: 1972, 'The Ring Current in the Magnetosphere and the Polar Magnetic Substorms', *J. Atmosph. Terr. Phys.* **34**, 845.
Tsurutani, B. T. and Smith, E. J.: 1974, 'Postmidnight Chorus: a Substorm Phenomenon', *J. Geophys. Res.* **79**, 118.

Tsurutani, B. T., Smith, E. J. and Thorne, R. M.: 1975, 'Electromagnetic Hiss and Relativistic Electron Losses in the Inner Zone', *J. Geophys. Res.* **80**, 600.

Vampola, A. L.: 1971, 'Electron Pitch-Angle Scattering in the Outer Zone during Magnetically Disturbed Times', *J. Geophys. Res.* **76**, 4685.

VanZandt, T. E., Peterson, L. and Laird, A. R.: 1971, 'Electromagnetic Drift of the Mid-Latitude F2 Layer during a Storm', *J. Geophys. Res.* **76**, 278.

Vasyliunas, V. M.: 1968, 'A Survey of Low-Energy Electrons in the Evening Sector of the Magnetosphere with OGO 1 and OGO 3', *J. Geophys. Res.* **73**, 2839.

Vasyliunas, V. M.: 1972, 'The Interrelationship of Magnetospheric Processes', *Earth's Magnetospheric Processes*, B. M. McCormac (ed.), p. 29, D. Reidel Publ. Co., Dordrecht-Holland.

Walker, J. C. G. and Rees, M. H.: 1968, 'Excitation of Stable Auroral Red Arcs at Subauroral Latitudes', *Planet. Space Sci.* **16**, 915.

Walker, R. J. and Kivelson, M. G.: 1975, 'Energization of Electrons at Synchronous Orbit by Substorm-Associated Cross-Magnetosphere Electric Fields', *J. Geophys. Res.* **80**, 2074.

Walt, M.: 1970, 'Radial Diffusion of Trapped Particles', *Particles and Fields in the Magnetosphere*, B. M. McCormac (ed.), p. 410, D. Reidel Publ. Co., Dordrecht-Holland.

Wang, C. S. and Kim, J. S.: 1972, 'Electric Field Induced by a Slowly Decaying Ring Current', *J. Geophys. Res.* **77**, 4867.

Wilhelm, K., Kremser, G., Münch, J., Schnell, M., Legrand, J. P., Petrou, N. and Riedler, W.: 1972, 'Measurements of Energetic Particle Fluxes during a Slowly Varying Absorption Event by Two Co-ordinated Rocket Flights', *Space Research XII*, S. A. Bowhill, L. D. Jaffe and M. J. Rycroft (eds.), p. 1437, Akademie-Verlag, Berlin.

Williams, D. J.: 1970, 'Trapped Protons $\geq$ 100 keV and Possible Sources', *Particles and Fields in the Magnetosphere*, B. M. McCormac (ed.), p. 396, D. Reidel Publ. Co., Dordrecht-Holland.

Williams, D. J.: 1975, 'Hot Plasma Dynamics within Geostationary Altitudes', *Physics of the Hot Plasma in the Magnetosphere*, B. Hultqvist and L. Stenflo (eds.), p. 159, Plenum Press, New York.

Williams, D. J. and Lyons, L. R.: 1974a, 'The Proton Ring Current and Its Interaction with the Plasmapause: Storm Recovery Phase', *J. Geophys. Res.* **79**, 4195.

Williams, D. J. and Lyons, L. R.: 1974b, 'Further Aspects of the Proton Ring Current Interaction with the Plasmapause: Main and Recovery Phases', *J. Geophys. Res.* **79**, 4791.

Williams, D. J. and Ness, N. F.: 1966, 'Simultaneous Trapped Electron and Magnetic Tail Field Observations', *J. Geophys. Res.* **71**, 5117.

Williams, D. J., Arens, J. F. and Lanzerotti, J.: 1968, 'Observations of Trapped Electrons at Low and High Altitudes', *J. Geophys. Res.* **73**, 5673.

Williams, D. J., Fritz, T. A. and Konradi, A.: 1973, 'Observations of Proton Spectra ($1.0 \leq E_p \leq 300$ keV) and Fluxes at the Plasmapause', *J. Geophys. Res.* **78**, 4751.

Williams, D. J., Barfield, J. N. and Fritz, T. A.: 1974, 'Initial Explorer 45 Substorm Observations and Electric Field Considerations', *J. Geophys. Res.* **79**, 554.

Winckler, J. R.: 1970, 'The Origin and Distribution of Energetic Electrons in the Van Allen Radiation Belts', *Particles and Fields in the Magnetosphere*, B. M. McCormac (ed.), p. 332, D. Reidel Publ. Co., Dordrecht-Holland.

Winckler, J. R.: 1974, 'An Investigation of Wave-Particle Interactions and Particle Dynamics Using Electron Beams Injected from Sounding Rockets', *Space Sci. Rev.* **15**, 751.

Wolf, R. A.: 1970, 'Effects of Ionospheric Conductivity on Convective Flow of Plasma in the Magnetosphere', *J. Geophys. Res.* **75**, 4677.

Yasuhara, F.: 1975, 'Field-Aligned and Ionospheric Currents', Ph.D. Thesis, Univ. of Alaska, May.

Young, T. S. T., Callen, J. D. and McCune, J. E.: 1973, 'High-Frequency Electrostatic Waves in the Magnetosphere', *J. Geophys. Res.* **78**, 1082.

Zaytseva, S. A. and Glazhevska, A.: 1972, 'Formation of the Belt of DR-Currents', *Geomag. Aeronom.* **12**, 260.

Zaytseva, S. A., Pudovkin, M. I., Oryakhlov, V. V. and D'yachenko, V. N.: 1971, 'Dynamics of the belt of DR-Currents and Middle Latitude Red Arcs', *Geomag. Aeronom.* **11**, 721.

Zhulina, Ye. M., Driatskiy, V. M., Shchuka, T. I., Zhulin, I. A., Kambou, F. and Sait-Marc, A.: 1972, 'Dynamics of Auroral-Particle Injection during Substorms', *Geomag. Aeronom.* **12**, 212.

CHAPTER 9

SOLAR-TERRESTRIAL RELATIONS AND MAGNETOSPHERIC SUBSTORMS

9.1. Interplanetary Disturbances

9.1.1. BASIC SOLAR-INTERPLANETARY MAGNETIC FIELD STRUCTURE

In studying solar-terrestrial relationships, it is important to recognize first of all that there is the basic solar-interplanetary magnetic field structure which has been revealed by two important discoveries during the last two decades. The first is the sector structure of the magnetic field at the photospheric level and the second is the interplanetary magnetic field sector structure (Ness and Wilcox, 1967; Severny *et al.*, 1970).

Most recently, Svalgaard *et al.* (1974), Wilcox and Svalgaard (1974) and Svalgaard *et al.* (1975) have shown that the magnetic sector boundary at the photospheric level is associated with the magnetic field line arcade and helmet streamers at the coronal level and also with the interplanetary magnetic field sector structure. Figure 9.1 illustrates this geometrical relationship. The magnetic field line arcade structure confines hot coronal plasma in it, so that the coronal green line intensity and X-ray emission are high there. The outward flow of coronal plasma (namely, the solar wind) is considerably suppressed above the arcade structure in spite of the fact that the underlying corona is very hot. This problem has been theoretically studied by Pneuman (1973), Durney and Pneuman (1975) and others.

In the areas bounded by magnetic field line arcades, there are regions where the magnetic field lines are 'open'. Furthermore, Timothy *et al.* (1976) found that when this region is observed at soft X-ray wave lengths, it is seen as open features, devoid of X-ray emission. For this reason, such regions are called *coronal holes.* Figure 9.2 shows a photograph of the coronal holes. Further, Krieger *et al.* (1973) and Neupert and Pizzo (1974) found that a high speed solar wind stream originates from such a coronal hole.

The relationship between coronal holes and the corresponding high speed solar wind streams is most dramatically demonstrated by the figure constructed by Sheeley *et al.* (1976); it is reproduced here as Figure 9.3.

The figure shows also the corresponding geomagnetic activity during the same period. There is little doubt that 27-day recurrent disturbances are associated with the coronal holes and the corresponding high speed solar wind streams. Therefore, we have finally identified photographically the source region of 27-day recurrent disturbances, the coronal hole. This source region has long been a

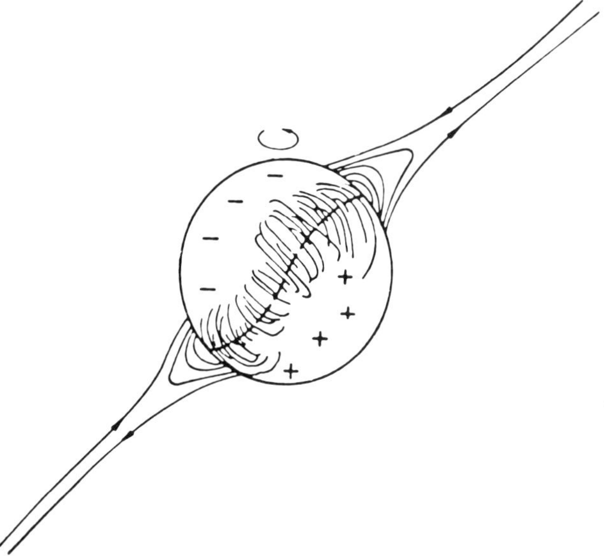

Fig. 9.1. Geometrical relationship between the solar sector boundary, the magnetic field line arcade and the helmet streamers. (Svalgaard, L., Wilcox, J. M. and Duvall, T. L.: *Solar Phys.* **37**, 157, 1974.)

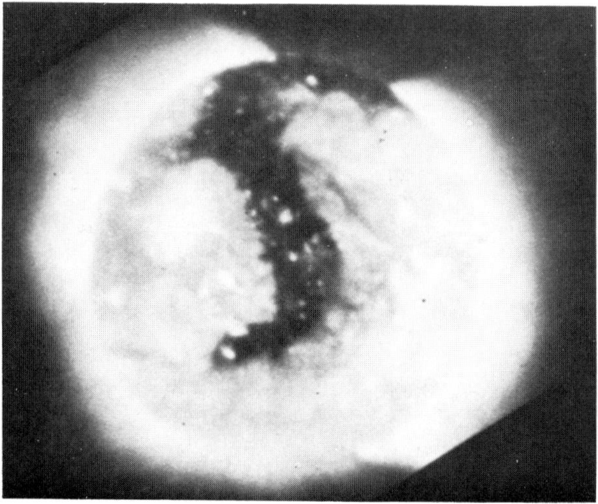

Fig. 9.2. Coronal holes observed on 1973, August 19 and 21. (Courtesy of Timothy, A. F., Krieger, A. S., and Vaiana, G. S.)

matter of controversy. Bartels (1932) once noted: "The recurrent storms tended to appear when the central portions of the sun were free of any optical sign of activity." Bartels named this region the "*M region*"; the letter M signifies 'magnetically active' and/or 'mysterious.' It may also be recalled that Pecker and Roberts (1955) proposed the concept of '*cone of avoidance*' for the source region,

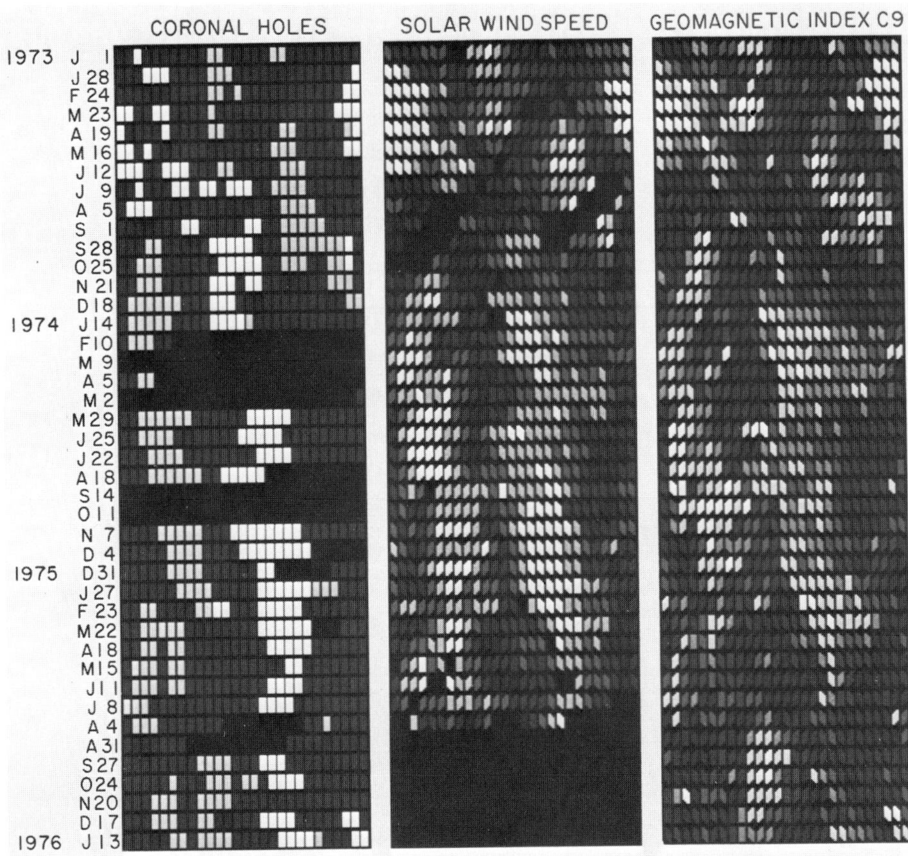

Fig. 9.3. Comparison between coronal central-meridian-passage dates (plus 3 days), solar wind speed and the C9 geomagnetic disturbance index for the interval 1973, January 1–1976, February 8. The dates are arranged in the Bartels' 27-day rotation sequence. (Courtesy of Sheeley, N. R., Jr., Harvey, J. W. and Feldman, W. C.)

whereas Mustel (1964) suggested that M regions can be identified with young active regions. There have also been a large number of papers which discuss the correlation or the anti-correlation between the coronal green line intensity and geomagnetic activity. Most recently, however, Gulbrandsen (1973, 1974, 1975) and Hansen *et al.* (1976) showed that recurrent geomagnetic disturbances are associated with regions of low coronal intensities. The correlation between the coronal green line intensity and solar wind speed has also been discussed by a number of workers. Roelof *et al.* (1975) showed that recurrent geomagnetic disturbances are associated with a low coronal intensity, while transient geomagnetic disturbances are associated with a high coronal intensity. However, a number of workers in geomagnetism and in cosmic ray physics had in the past believed that the region of a high coronal intensity is the base of a high speed solar wind stream.

One can summarize the basic solar-interplanetary magnetic structure by the following table.

TABLE 9.1
Basic solar-interplanetary magnetic structure

Photosphere	Coronal level	Interplanetary space
Sector boundary	Magnetic arcade Helmet streamer Intense coronal emission	Interplanetary sector structure
Open field line region	Coronal hole	High speed solar wind stream

9.1.2. HIGH SPEED SOLAR WIND STREAMS AND GEOMAGNETIC DISTURBANCES

It should be cautioned at this point that the speed of solar wind is not directly responsible for geomagnetic disturbances. Although it has been shown by Snyder, Neugebauer and Rao (1963) and Olbert (1968) that there is a simple linear relation between the daily sum of the Kp index (ΣKp) and the bulk speed of the solar wind (*cf.* S.T.P., pp. 546–547), Arnoldy (1971) pointed out that there is no such relation between the AE index (a substorm index) and the solar wind speed (Section 4.4.7). This means that geomagnetic disturbances tend to occur during the period when the magnetosphere is engulfed by a high speed solar wind stream, but that the speed of the solar wind is not the crucial parameter. There must be some crucial parameters which tend to be associated with a high speed stream. Indeed, Arnoldy (1971) and other workers have shown that the most crucial interplanetary parameter is the north-south component of the interplanetary magnetic field.

Therefore, magnetospheric physicists have just begun to understand the interaction between the interplanetary magnetic field and the magnetosphere and in particular how the north-south component of the IMF controls the efficiency of the solar wind-magnetosphere dynamo which generates the power needed for geomagnetic disturbances or, more specifically, magnetospheric substorms.

Then, what is the nature of fluctuations, in particular those which cause changes of the north-south component of the IMF? Burlaga (1975) recently pointed out that relatively little is known about magnetic fluctuations, although the fluctuations are generally high when the speed is high. He examined extensively the interaction region which consists of the forward shock, the stream interface (separating the denser region from the hotter region of the stream), the reverse shock and the rarefaction region and pointed out that the most intense fluctuations tend to occur in the interaction regions. Figure 9.4(a) illustrates these features. Figures 9.4(b), (c) and (d) show the solar wind parameters and the AE index across a few interaction regions. It is a future problem for interplanetary and magnetospheric physicists to find the nature of magnetic fluctuations which lead to magnetospheric substorms. The interaction between a high speed solar wind stream and the ambient solar wind has recently been studied by a number of workers. Coleman *et al.* (1966) attributed it to the Kelvin-Helmholz instability, while Belcher

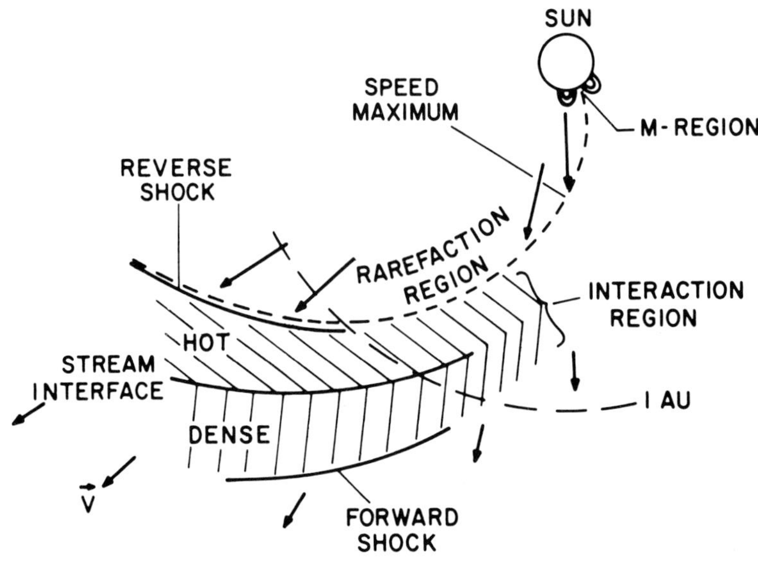

Fig. 9.4(a). Interaction of a high solar wind stream with the ambient solar wind. (Burlaga, L. F.: *Space Sci. Rev.* **17**, 327, 1975.)

et al. (1969) identified it as Alfvén waves. Belcher and Davis (1971) concluded that Alfvén waves are propagating away from the sun. Figure 9.5 shows a set of solar-wind and magnetic field data which indicate the presence of Alfvén waves. Burlaga and Ogilvie (1970) suggested that the fluctuations in the interaction region are due to an unidentified interplanetary process associated with the steepening of a stream. For recent reviews on the interaction region, see Hundhausen (1972) and Dryer (1975), and on waves and microstructures in the solar wind, see Volk (1975) and Hollweg (1975).

In summary, the open field region in the photospheric level appears to be related to geomagnetic disturbances in the following way:

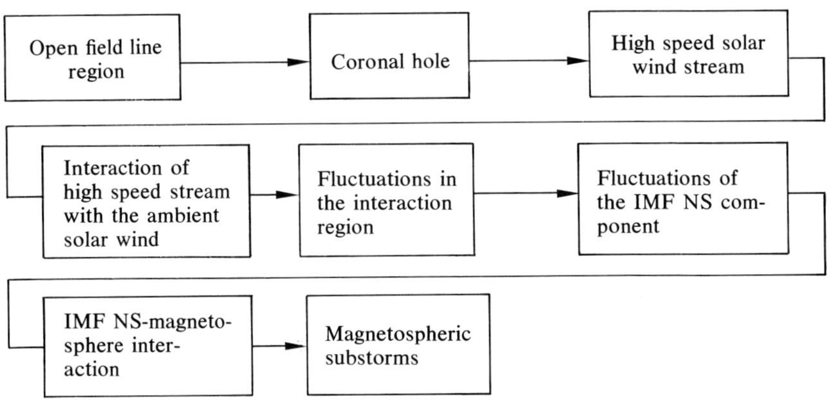

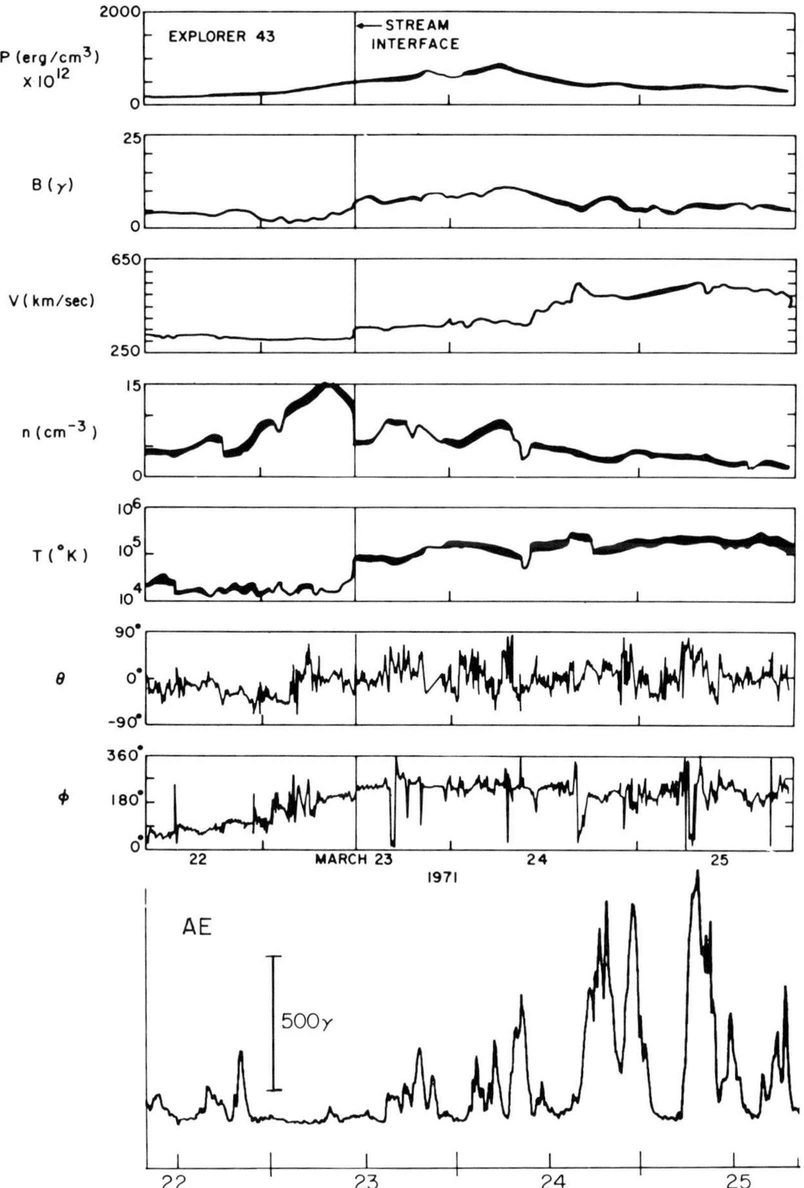

Fig. 9.4(b)–(d). Examples of the structure of the interaction region (Burlaga, L. F.: *J. Geophys. Res.* **79**, 3717, 1974) and the corresponding AE index.

Thus, the major geomagnetic disturbances, magnetospheric substorms, result as an end product of a long chain of processes. Actually, there are a number of processes involved even in the last step in the above chain process, and in fact the main purpose of this book is an attempt to clarify these processes.

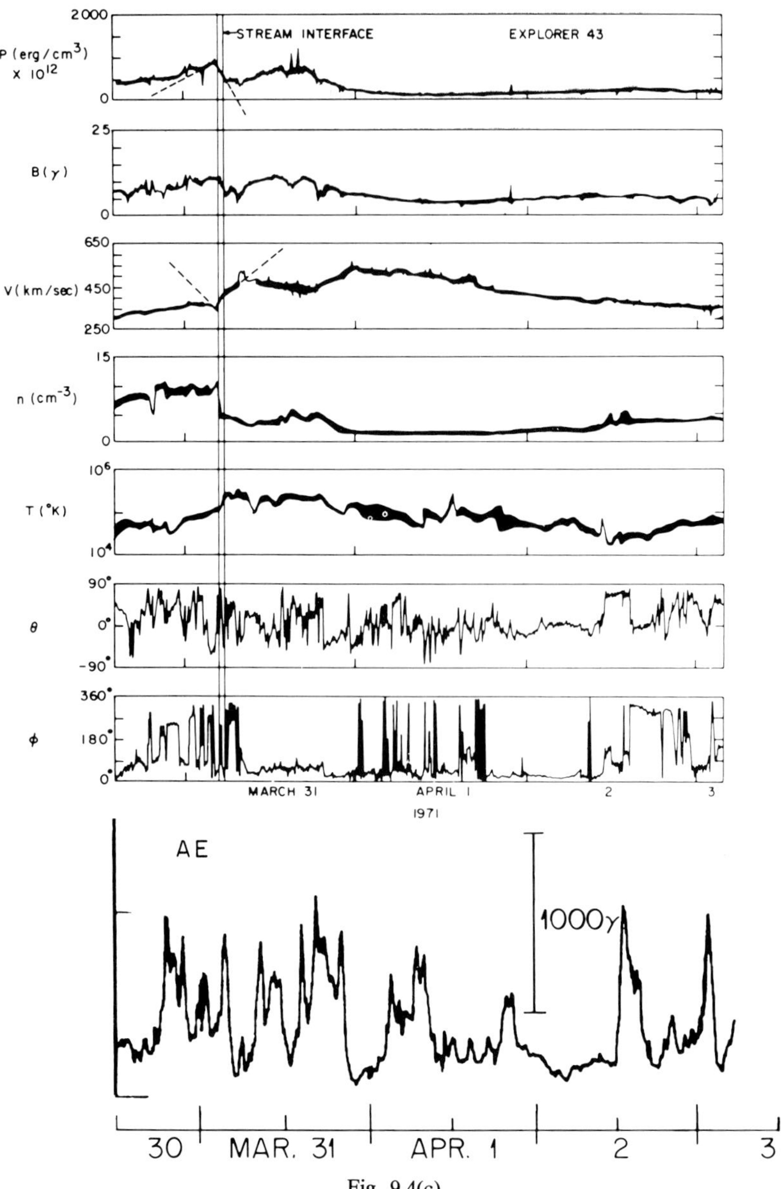

Fig. 9.4(c).

9.1.3. TRANSIENT SOLAR ACTIVITIES AND ASSOCIATED INTERPLANETARY DISTURBANCES

The most spectacular transient solar activity occurs in the vicinity of sunspots (namely, centers of activity) and is called a *solar storm* (*cf.* S.T.P., Chapter 7). Optically, it is manifested by a sudden increase in the brightness of the Hα

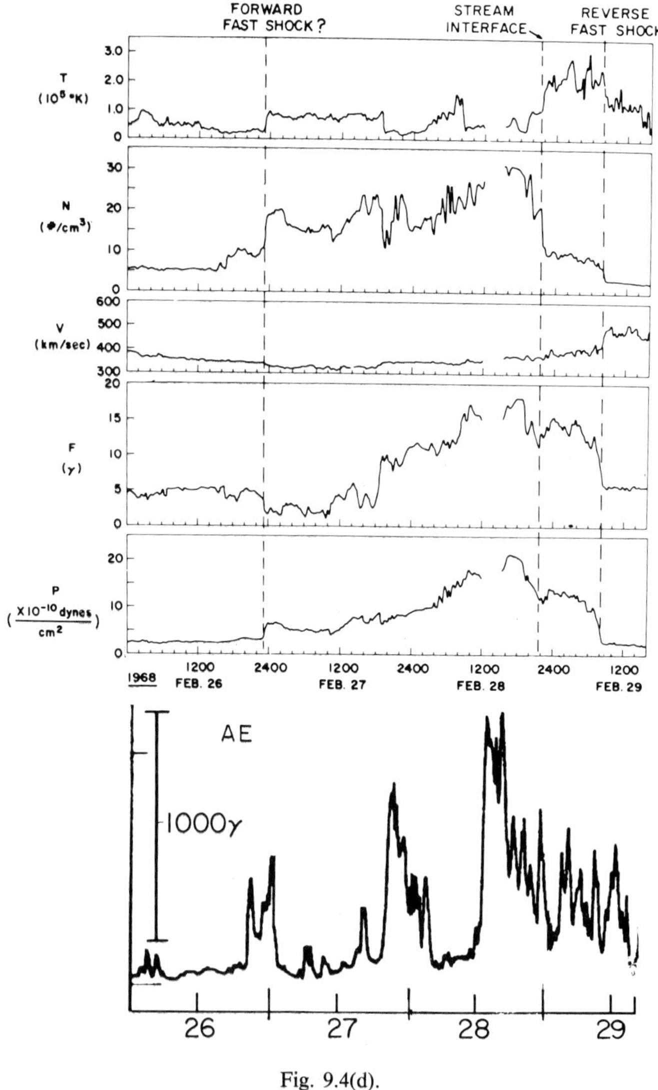

Fig. 9.4(d).

radiation, called a *solar flare.* Most recently, it has been revealed that the sun ejects material and magnetic fields in the form of 'bubbles' into interplanetary space (Gosling *et al.*, 1974, 1975). It is found also that the bubble ejection occurs far more frequently than solar flares and that the bubbles may play an important role in many interplanetary processes (Newkirk, 1975).

Solar storms and magnetic bubbles cause a transient perturbation in the basic solar-interplanetary magnetic field structure. After Parker's pioneering study, extensive theoretical and analytical studies of interplanetary disturbances as-

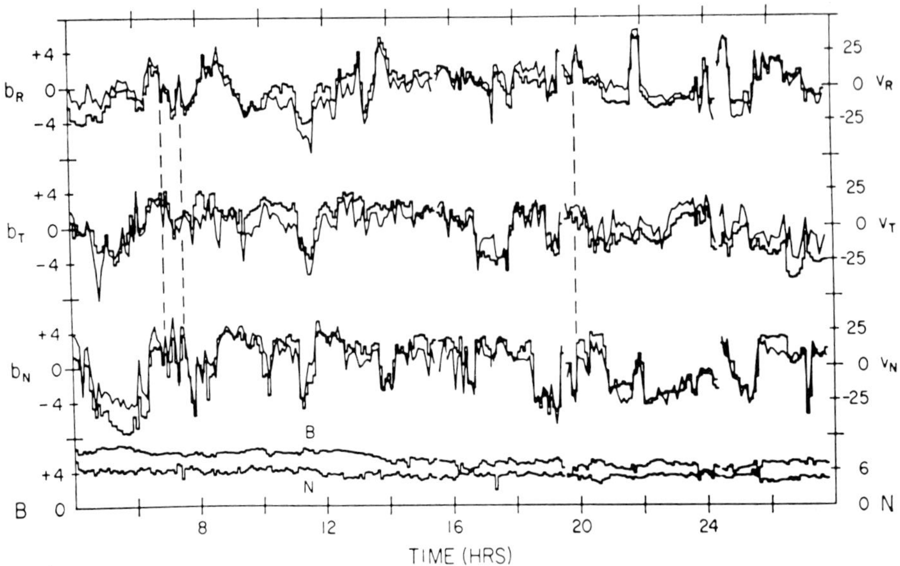

Fig. 9.5. An example of solar wind and the IMF data which indicate the presence of large amplitude Alfvén waves. (Belcher, J. W. and Davis, L. Jr.: *J. Geophys. Res.* **76**, 3534, 1971.)

sociated with solar storms have been conducted by a number of workers (cf. S.T.P., Chapter 7). It is of particular interest that the so-called 'piston gas' from storm regions has a helium-rich shell (Hirschberg *et al.*., 1972; Sakurai and Chao, 1973); see Figure 9.6. The interactions of the piston gas with the ambient solar wind and the resulting shock structure were a subject of another intensive study during the last decade. The figure also shows a schematic illustration of the piston gas-shock structure. For recent reviews on this subject, see Hundhausen (1972), Sakurai (1973) and Dryer (1975).

Most recently, Uchida *et al.* (1973) examined numerically how the wave front of MHD fast-mode waves, emanating from storm regions, propagates in the corona; Figure 9.7 shows an example of their illustrations. Hirshberg *et al.* (1974) examined how a sudden spherically symmetric disturbance, introduced at a distance of 30 solar radii, propagates through the solar wind which contains a high speed stream; see Figure 9.8. In the upper diagram, perspective representation of the magnitude of the velocity of the undisturbed steady state solar wind stream is given between 30 and 215 solar radii (1 AU). The lower diagram shows the velocity configuration 50 h after the disturbance is initiated, shortly before the disturbance reaches a distance of 1 AU. It should be noted that their study removed the spherical symmetry with respect to the solar center which was assumed in most early studies of the subject.

The interaction of the piston gas structure with the magnetosphere has also been a hotly debated topic during the last decade. A storm period is found to be a period during which intense substorms occur frequently. Therefore, one can infer that the development of a geomagnetic storm is controlled by the north-

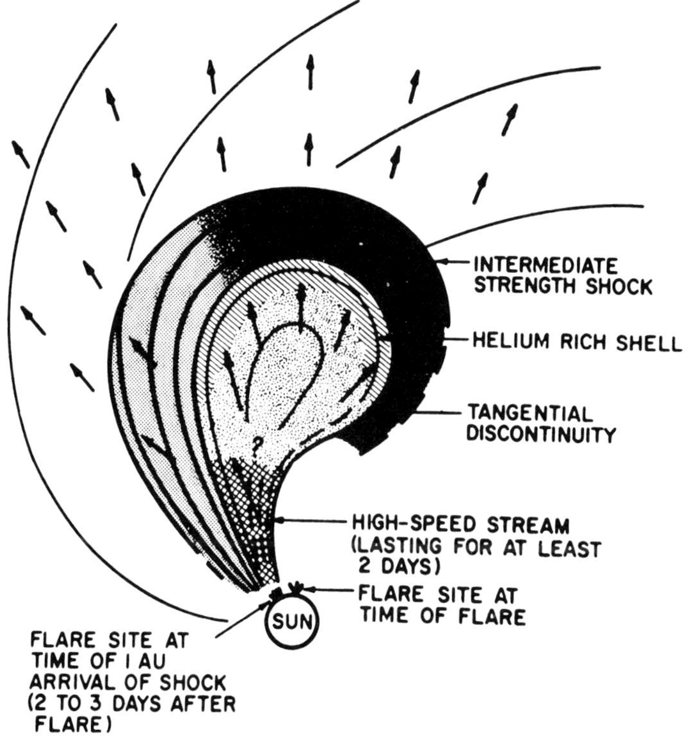

Fig. 9.6. Interaction of the piston-gas, with a helium rich shell, with the ambient solar wind. (Burlaga, L.F.: *Space Sci. Rev.* **17**, 327, 1975.)

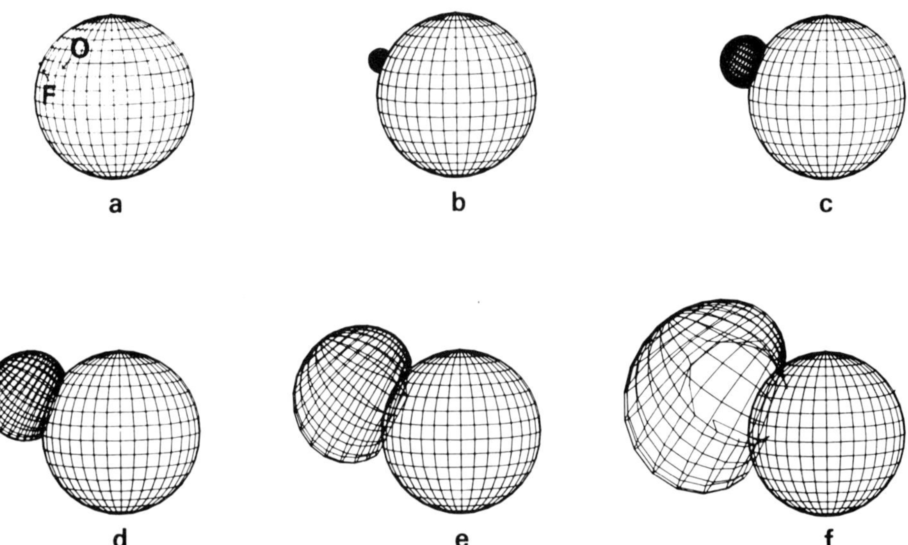

Fig. 9.7. Time development of the front of MHD-fast-mode wave front, emitted at the limb of the Sun. (Uchida, Y., Altschuler, M. D. and Newkirk, G., Jr.: *Solar Phys.* **28**, 495, 1973.)

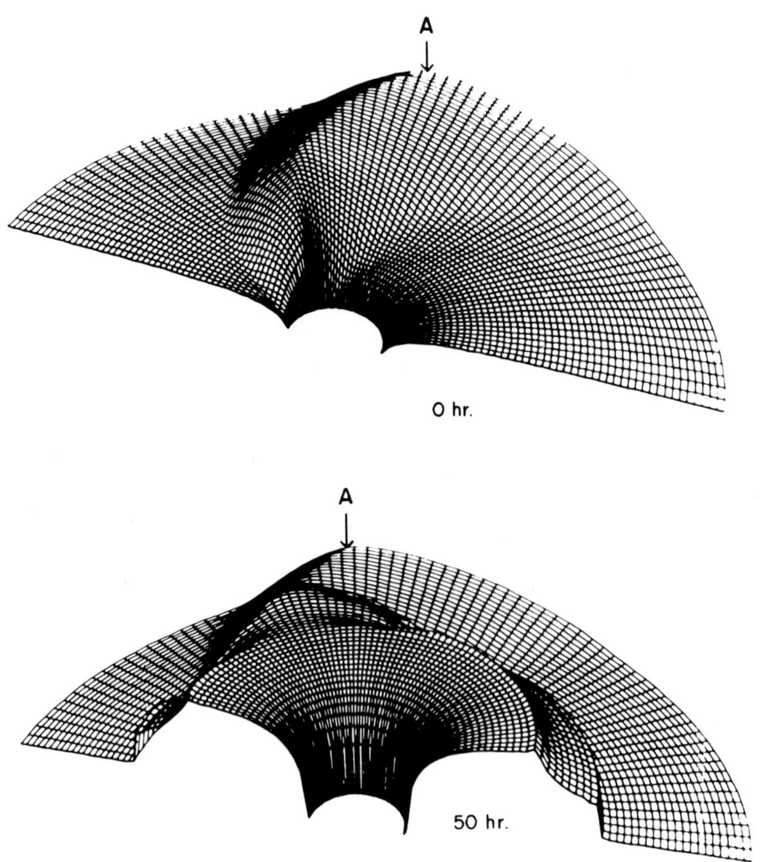

Fig. 9.8. Perspective representation of the magnitude of the velocity of the undisturbed steady state solar wind stream between 30 and 215 (1 AU) solar radii and the velocity configuration 50 hours after the disturbance is initiated. (Hirshberg, J., Nakagawa, Y. and Wellck, R. E.: *J. Geophys. Res.* **79**, 3726, 1974.)

south component of the IMF. In Section 5.6, we found that this is indeed the case. However, the problem is again the nature of the fluctuations. Are they 'turbulence' or Alfvén waves? This is another new problem which should be studied jointly by both interplanetary physicists and magnetospheric physicists.

9.2. Morphological Model of Magnetospheric Substorms

9.2.1. BASIC REQUIREMENTS FOR MODELS

(i) *Response of the Magnetosphere to the Southward Component of the IMF*

In studying substorm processes, it is important to identify various responses of the magnetosphere to the southward turning of the IMF (or more accurately to $\partial B_z/\partial t < 0$) and to distinguish them from substorm processes. Such a step will clarify some past confusion, in particular, the problem of the so-called 'growth phase'. In constructing a model of magnetospheric substorms, one must identify processes which are common to *most* substorms. Some workers have proposed that a growth phase is apparent mostly for 'isolated substorms', the first of a series of substorms and "... perhaps occasionally for specific ideal substorms in the middle of a sequence of substorms" (McPherron *et al.*, 1973; p. 3147). However, such a procedure excludes a large percentage of substorms and thus the concept of growth phase cannot be applicable. In fact, the first of a series of substorms occurs often after the southward turning of the IMF vector, and thus one is tempted to identify the responses of the magnetosphere to the IMF southward turning as growth phase features. However, as we noted in Section 4.4, a series of magnetospheric substorms result when the magnetosphere is pulsed once by a single southward turning of the IMF vector. Thus, every substorm is not preceded by an IMF southward turning. It is for this and many other reasons why the responses of the magnetosphere to the north-south component of the IMF should be identified and distinguished from substorm processes.

Figure 9.9 illustrates schematically the responses of the magnetosphere to a simple IMF B_z 'pulse'. As we found in Section 5.2, the area bounded by the auroral oval provides a measure of magnetic energy accumulated in the magnetotail. Thus, after it is elevated to an 'excited state' by a single IMF southward turning, the magnetosphere tends to produce successive substorms until it exhausts the excess energy in the magnetotail and eventually returns to the 'ground state'. In other words, the magnetosphere is capable of generating substorms and is thus always in the state of 'growth phase' so long as it has the excess energy. The excess energy was defined in Section 5.2. Changes of the

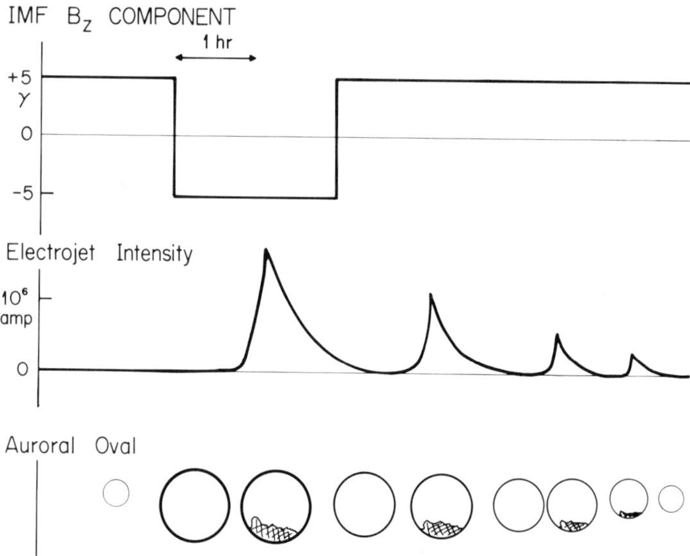

Fig. 9.9. Responses of the magnetosphere, in terms of the electrojet intensity, the size of the auroral oval, substorm activity, when the magnetosphere is pulsed by a single southward turning of the IMF.

magnetosphere caused by such an IMF 'pulse' are studied in detail in Section 4.4 and are indicated in Figure 9.10. The major responses are as follows:

(1) Slight earthward shift of the dayside magnetopause.
(2) Equatorward shift of the cusp.
(3) Expansion of the auroral oval.
(4) Enhancement of the cross-tail electric field.
(5) Enhancement of the S_q^p current.
(6) Absence of magnetopause motion in the magnetotail.
(7) Slight increase (~ 10%) of the magnetic field intensity (B_T).
(8) Absence of plasma sheet thinning.
(9) Absence of the plasma injection at the geosynchronous distance.

As mentioned in the Introduction, we have been accustomed to using such terms as 'erosion of the dayside magnetopause,' 'transfer of the merged field lines to the magnetotail' or 'reconnection' without seriously attempting to understand basic processes involved. As a result, effects of the IMF B_z component have been overemphasized. The reasons for (6)–(9) are given in Section 9.2.2(a).

(ii) *Basic Requirements for Models*

In the following, we shall describe two basic requirements for a model of magnetospheric substorms. Other requirements, conforming to the observed substorm features, will be listed in Section 9.2.2.

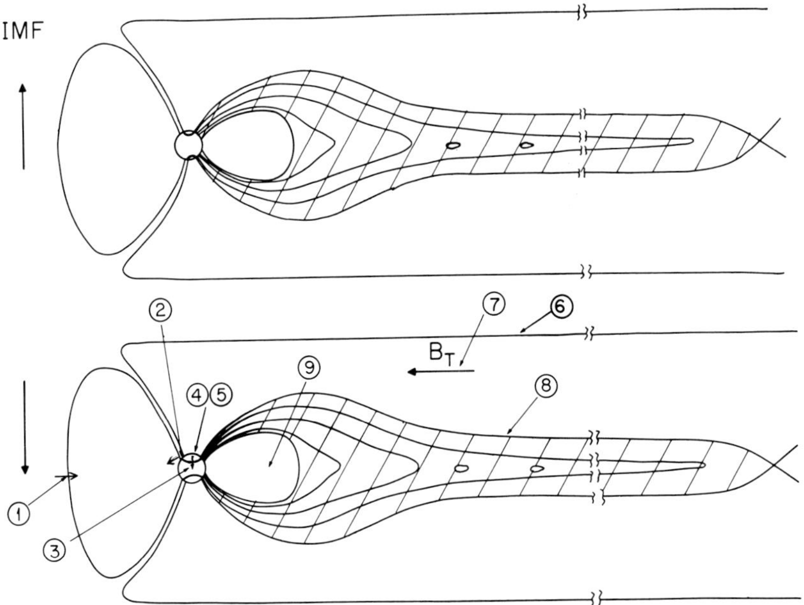

Fig. 9.10. Schematic presentation of major magnetospheric responses to a southward turning of the IMF.

(α) *Energy production rate and the total energy production.* In Section 5.3, it was found that the energy injected into the ring current belt and into the upper atmosphere amounts to 3×10^{18} erg s^{-1}; for a weak substorm it may be one order of magnitude less than the above value. Then, assuming that a substorm lasts for 2 h = 7200 s, the total energy associated with a single intense magnetospheric substorm is estimated to be 2.3×10^{22} erg; again, for a weak substorm it may be one order of magnitude less than the above value.

(β) *Two substorm phases.* A magnetospheric substorm has the two characteristic phases, the expansive phase and the recovery phase. Active auroras advance rapidly poleward in the midnight sector during the explosive first phase, while they fade out and recede equatorward during the second phase. An intense injection of plasma takes place from the plasma sheet into the Van Allen belt during the expansive phase.

On the other hand, in the magnetotail ($X \simeq -10 \sim -40\, R_E$), the plasma sheet thins during the first phase, and a high speed flow of hot plasma appears at about the maximum epoch of the substorm and remains during the second phase:

	Upper atmosphere and Van Allen belt	*Magnetotail*
Expansive phase	Auroral bulge formation Injection and the ring current formation	Thinning of plasma sheet
Recovery phase	Contraction of the bulge	Appearance of a high speed flow of hot plasma

Therefore, substorm features during each phase are contrasting in the upper atmosphere and in the distant magnetotail ($X \simeq -10 \sim -40\, R_E$) in the sense that when auroras are most active in the polar upper atmosphere, it is rather uneventful in the magnetotail; also, when auroral activity subsides during the recovery phase, a high speed flow of hot plasma appears rather suddenly.

Suitable theories of magnetospheric substorms must explain these contrasting features. Further, during the expansive phase the energy estimated in (α) appears mostly as the energy of ring current particles in the trapping region and also as auroral energy which is eventually converted into thermal energy in the upper atmosphere. It is only during the recovery phase when hot plasma particles appear in the distant magnetotail and thus the plasma sheet recovers. That is to say, the substorm energy is finally fed into the plasma sheet during the recovery phase.

9.2.2. DESCRIPTION OF A MODEL

In this subsection, we shall present a model of the magnetospheric substorm which satisfies the two basic requirements given in Section 9.2.1 and a number of observed substorm features (Chapters 5, 6, 7 and 8). However, the proposed model is simply meant to be an exercise in synthesizing a variety of magnetospheric and ground-based observations in terms of the concept of magnetospheric substorm. We are still far from constructing even a first approximation model which has a sound physical ground, so that the proposed model may not be entirely self-consistent.

Before describing details of the proposed model of magnetospheric substorms, it is useful to quote a few paragraphs from the last chapter of *Polar and Magnetospheric Substorms* (Akasofu, 1968).

"... suitable theory of magnetospheric substorms must answer at least the following questions (Figure 9.11(a) taken from p. 224).

(1) What is the original form of the solar wind energy for the magnetospheric substorm?
(2) How is it converted into energy suitable for storage in the magnetosphere?
(3) What is the form of this energy during the storage?
(4) How is it converted into energy for the magnetospheric substorm?"

We have a reasonable first approximation answer to each of the first three questions. They were discussed in terms of the solar wind-magnetosphere dynamo in Chapters 1–5. In particular, we examined the questions (2) and (3) in terms of the excess energy in Section 5.2.

Unfortunately, however, we are far from providing a satisfactory answer to the last question (4). The processes involved in (4) may generally be described in terms of the following block diagram (Figure 9.11(b)). In constructing such a block diagram, it is important to keep in mind that the magnetospheric substorm begins with an explosive phase.

First of all, the presence of the excess (stored) energy in the magnetotail (defined in Section 5.2) may be manifested in an enhanced convection of magnetospheric plasma or in an enhanced cross-tail current (1). Many substorm models suggest that either an enhanced convection or an enhanced cross-tail current initiates the growth of perturbation in the magnetotail or in the ionosphere (2). However, such a perturbation cannot grow in the presence of the tight magnetosphere-ionosphere coupling. A magnetospheric perturbation, such as the interchange or flute instability, tends to be suppressed by the ionosphere which discharges the resulting space charges needed for its positive feedback process. Similarly, an ionospheric perturbation may be suppressed by the magnetosphere which can discharge the resulting ionospheric space charges.

Thus, in order for the initial perturbation to grow, the magnetosphere and the ionosphere must be decoupled. The development of an anomalous resistivity along the field lines, produced by microscopic plasma processes, such as current-driven instabilities, can achieve the decoupling, since the magnetosphere

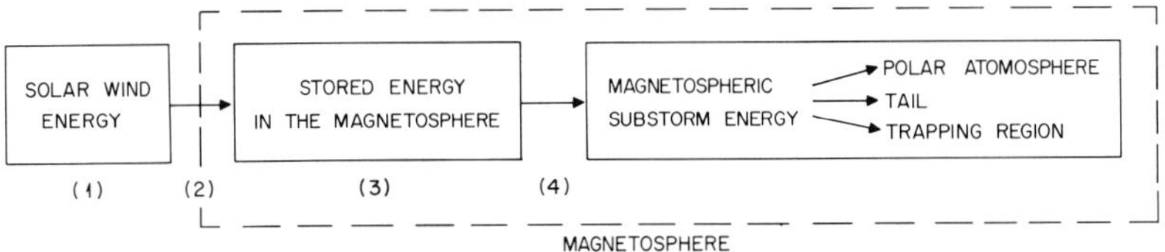

Fig. 9.11(a). Block diagram showing the energy flow for the magnetospheric substorm.

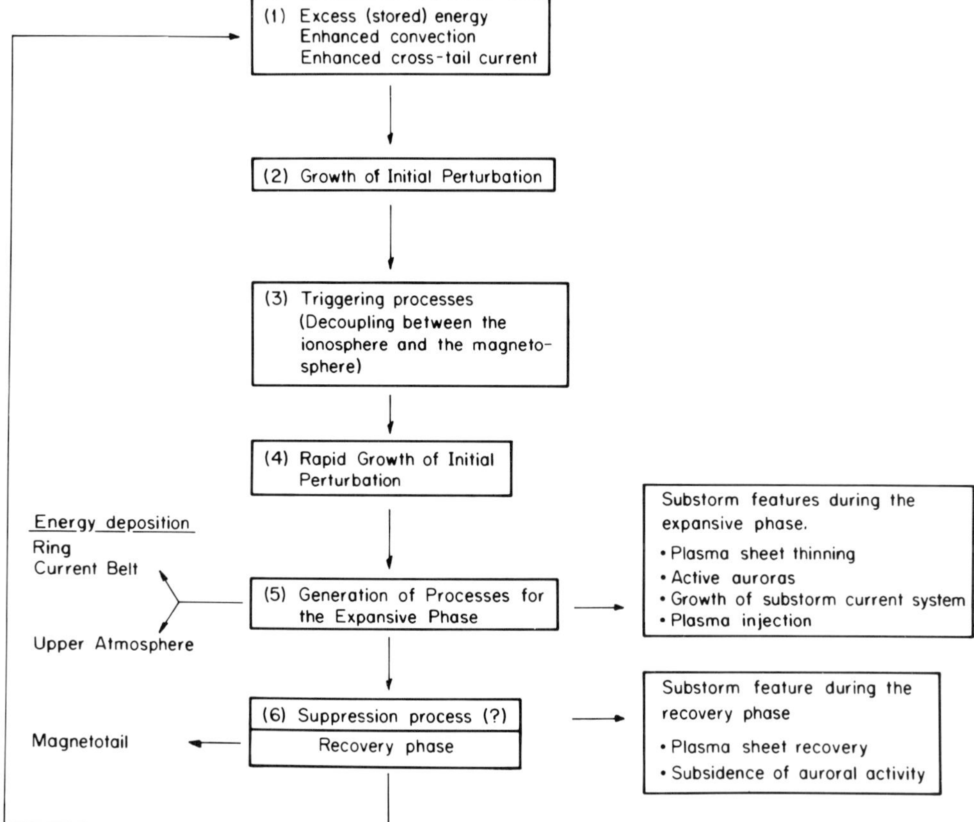

Fig. 9.11(b). Block diagram showing the chain of processes for the magnetospheric substorm.

and the ionosphere will no longer be connected by highly conductive field lines (3). For this reason, it has been proposed by some workers that the current-driven instabilities are generated by the discharge process itself, when the resulting current intensity reaches a certain threshold value. It is in this situation that the initial perturbation may grow rapidly (4). The rapidly growing initial perturbation itself may become the cause of the expansive phase when it grows beyond a certain limit (4). Another possibility is that a rapid growth of the initial perturbation generates a new process which becomes responsible for the expansive phase (5).

A successful model should be able to identify the specific perturbations which satisfy the two basic requirements in Section 9.2.1, and should explain all major substorm features (5). Some of the major substorm features during the expansive phase are

(i) Plasma injection into the trapping region
(ii) Plasma sheet thinning
(iii) Development of an auroral substorm

(iv) Growth of the substorm current system

At present, it is not known why a series of substorms is generated after the magnetosphere is pulsed once. More specifically, why does a substorm (or the initial perturbation) tend to subside before exhausting all available excess energy at once? It may well be that some suppression processes appear when the substorm grows to a certain limit (6). In any case, the initial explosive phase comes to an end in about 30 min, and the magnetosphere begins to recover from the perturbation. The plasma sheet, deflated during the expansive phase, is inflated again, although the new plasma is hotter than before the substorm onset. Auroral activity during the expansive phase begins to subside, and the expanded auroral bulge contracts. It is important to keep in mind the timing of the recovery of the plasma sheet at $X \simeq -15\, R_{\mathrm{E}}$, which coincides approximately with the maximum epoch of the substorm. Two important recovery features are:

(i) Recovery of the plasma sheet.

(ii) Subsidence of auroral activity.

Naturally, a number of models can be considered even for a given set of the observed substorm features, depending on how one interprets them (in terms of a cause or an effect, etc.). Further, one has a formidable task of synthesizing satellite data observed at different locations and at different times for different substorms and analyzed by different workers. As emphasized in Section 4.4.1, it is important to keep in mind that magnetospheric phenomena are not a controlled experiment and that most magnetospheric quantities are a function of several parameters. For example, the magnetic field intensity B_T in the high latitude lobe of the magnetotail is a function of solar wind pressure, the production rate of open field lines, the radius of the magnetotail R_T, etc. Thus, it is not possible to obtain a functional relationship between B_T and the production rate of open field lines (say), unless the other parameters can remain constant. Unfortunately, there are a number of papers which discuss changes of the magnetic energy $\int B_T^2/8\pi \, \mathrm{d}V$ in the magnetotail just on the basis of changes of B_T^2. We shall see later that such a procedure is incorrect, simply because even if B_T is reduced initially by some substorm process, the magnetotail tends to adjust itself immediately in keeping with the requirement of pressure balance with the solar wind (R_T will be reduced until B_T is increased again to maintain the pressure balance).

In the following, an attempt will be made to construct a model of the magnetospheric substorm. In the model, it is proposed that the magnetosphere achieves the energy conversion in a two-stage process, first by deflating the plasma sheet and then converting the excess magnetic energy in the magnetotail. In terms of the block diagram given earlier, we have

(1) Same as the block diagram.

(2) Growth of the polarization electric field in the ionosphere.

(3) Growth of current-driven instabilities and decoupling.

(4) Rapid growth of the polarization field.

(5a) Sudden earthward displacement of plasma near the inner edge.

(5b) Generation of a rarefaction wave.

(5c) Deflation of the plasma sheet (= plasma sheet thinning) and the resulting plasma flow which causes active auroral features.

(6) Completion of the deflation by the arrival of the rarefaction wave to the anti-solar end of the X-line, the end of the expansive phase. Onset of an enhanced reconnection (the recovery of the plasma sheet).

(a) *Deflation of the Plasma Sheet*

It is suggested that the deflation is initiated by a magnetosphere-ionosphere decoupling process which suddenly displaces the plasma near the inner boundary of the plasma sheet toward the Earth ($T = 0$–2 min). At the geosynchronous distance ($r = 6.6\ R_E$), we identify this process as the injection of plasma into the trapping region. This sudden earthward displacement of the plasma generates a rarefaction wave which propagates in the anti-solar direction with almost the speed of sound ($\sim 1000\ \mathrm{km\ s^{-1}}$). Behind the rarefaction wave, the plasma flows toward the Earth, causing deflation – or thinning – of the plasma sheet.

As mentioned in Section 5.4.5, such an idea is not new. Coroniti and Kennel (1972) assumed that electrons precipitating into the equatorward boundary of the auroral oval are more energetic than the rest, so that a peak of conductivity ratio Σ_H/Σ_P results there; Figure 9.12. Such a non-uniformity of the conductivity tends to cause an equatorward polarization electric field as the Hall current (arising from an enhanced plasma convection) flows across it. However, so long as the conductivity along the field lines is high, the space charge accumulation will be small, since plasma sheet electrons and ionospheric electrons will discharge the space charges.

It is expected that when the magnetosphere contains excess energy, the electric

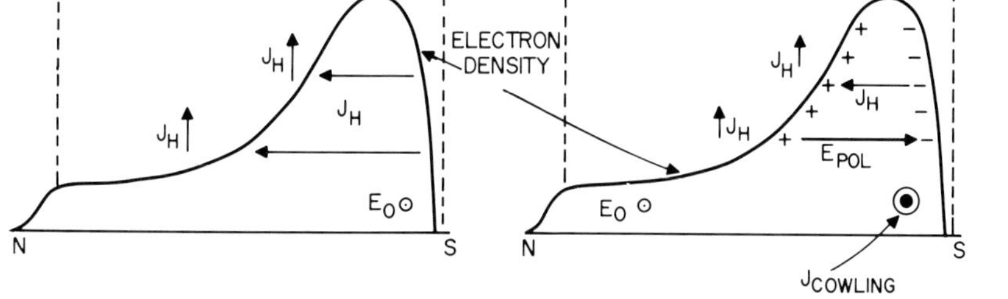

Fig. 9.12. Schematic diagram showing how the equatorward polarization electric field develops when the ratio Σ_H/Σ_P has a non-uniform distribution and when current-driven instabilities take place. (Coroniti, F. V. and Kennel, C. F.: *Cosmic Plasma Physics*, K. Schindler (ed.), p. 15, Plenum Press, New York, 1972.)

potential, the electric field and the Hall current in the polar cap are large and thus that the discharge process is also efficient. (As noted earlier, the magnetosphere is always in the state of 'growth phase' so long as it has the excess energy.) In such a situation, if the intensity of the discharge current exceeds a certain limit or if local plasma parameters allow, current-driven instabilities may grow (Section 3.9.4). When such an anomalous feature takes place along the discharge circuit, it tends to reduce the discharge rate, resulting in the accumulation of the space charges and increasing the equatorward electric field which drives an eastward motion of plasma and the westward auroral electrojet. Coroniti and Kennel (1972) suggested that this plasma flow resembles a hydromagnetic piston, which launches a rarefaction wave into the magnetotail.

In Chapter 8, we examined the injection process of plasma from the plasma sheet into the Van Allen belt in terms of an increase of the electric conductivity along the auroral oval, resulting in short-circuiting process of the space charges near the inner edge of the plasma sheet and the subsequent penetration of the convection electric field into the inner magnetosphere. In this process, the ionosphere plays only a *passive* role, while the idea suggested by Coroniti and Kennel (1972) assumes that the ionosphere plays an *active* role in the triggering processes of magnetospheric substorms.

Perhaps, two major changes are required in Coroniti and Kennel's process. The first would be that the non-uniformity of the conductivity occurs along the boundary between the oval of discrete auroras and the oval of the diffuse aurora rather than near the equatorward boundary of the diffuse aurora where the precipitating electrons are much less energetic than the rest (Section 2.4). Indeed, one of the first indications of the magnetospheric substorm is a sudden brightening of an auroral arc which is located near the equatorward boundary of the oval of discrete auroras and thus near the poleward boundary of the diffuse aurora. The second change would be that in order for the rarefaction wave to propagate into the anti-solar direction, the sudden plasma motion near the inner edge of the plasma would have a significant radially inward component (toward the Earth), as well as toward the dawn sector. Thus, an intense westward electric field must develop to cause the earthward displacement of plasma near the boundary of the plasma sheet. In Section 7.7, it was shown that an intense westward electric field does develop in the ionosphere; it is also interesting to note in this connection that auroral arcs *within* the expanding auroral bulge drift rapidly equatorward (Section 6.5).

It is possible to estimate the energy flux F_T associated with the deflation of the plasma sheet. It is given (Rossi and Olbert, 1970; p. 293) by

$$F_T = V_T(\tfrac{1}{2}\rho V_T^2 + \tfrac{5}{2}n\kappa T) \simeq 6\times 10^{-2}\ \text{erg cm}^{-2}\ \text{s}^{-1}$$

where

$V_T =$ the flow speed of plasma $\simeq 100$ km s^{-1}

$\rho =$ the mass $= mn = 1.55\times 10^{-24}$ g $\times\, 0.3$ cm^{-3}

$T =$ the temperature of plasma protons $\simeq 5$ keV

Assuming then the thickness and width of the plasma sheet to be 2 R_E and 30 R_E, respectively, the total energy flux $\bar{F}_T$ becomes 1.4×10^{18} erg s^{-1} which is comparable to the required total energy flux (Section 9.2.1).

As noted in Sections 5.3 and 9.2.1, a part of the energy thus estimated is fed into the ring current belt and the rest into the polar upper atmosphere. The former results from the injection process of plasma particles into the Van Allen belt. The plasma thus injected into the trapping region becomes the ring current particles. Since the formation of the ring current belt is often referred to as the 'inflation' of the inner magnetosphere and since the energy associated with the ring current belt is greater than the energy brought into the polar ionosphere by auroral particles and the auroral electrojet, the initial stage of the magnetospheric substorm can, as a first approximation, be considered to be displacement of plasma within the magnetosphere, from the plasma sheet to the inner magnetosphere.

On the other hand, in order for the flow energy of plasma to appear as auroral energy, there must be a process by which the flow energy is converted into the kinetic energy of auroral particles. In Section 3.9.3 we examined such energy conversion processes. It was suggested that the flow creates a shock wave. The electric polarization current arises from the plasma flow in the vicinity of the shock wave and/or of the converging magnetic field geometry, and the resulting upward current along an auroral arc is in part carried by precipitating auroral electrons.

The sudden earthward displacement of the plasma near the inner boundary of the plasma sheet appears to initiate the disruption and the subsequent diversion of the cross-tail current into the polar atmosphere, although the present model cannot provide a specific process for the disruption. As a result, a part of the disrupted cross-tail current flows into the morning half of the oval and out of the evening half, after flowing in the polar ionosphere.

It is suggested that the magnetic field associated with this current circuit is responsible for a large-scale deformation of the magnetic field configuration in the inner magnetosphere. As we saw in Section 7.2.5, the magnetic field produced by this current system is generally directed northward and thus makes the magnetic field configuration 'dipolar' in the equatorial plane. Further, the 'landing points' of the field lines for a given equatorial crossing distance shift toward higher latitudes and thus the plasma flow caused by the deflation of the plasma sheet is also deflected toward higher latitudes as the current grows stronger along the circuit. It is expected that it is this process which is responsible for the poleward expansion of the auroral bulge.

On the other hand, the first deflation stage is rather uneventful in the distant magnetotail ($X > -15\, R_E$), except that the plasma sheet simply thins. This is despite the fact that it has been widely speculated and believed by many that the magnetosphere achieves the conversion of the magnetic energy by forming a magnetic neutral (X-) line in the near-Earth plasma sheet, at a geocentric distance of about 10 to 15 R_E, where reconnection of the open field lines (namely, the geomagnetic field lines which have merged with the interplanetary magnetic field lines) takes place. In Chapter 6, it was shown that there is no indication of the speculated large-scale change of the magnetic field structure in the magnetotail during magnetospheric substorms.

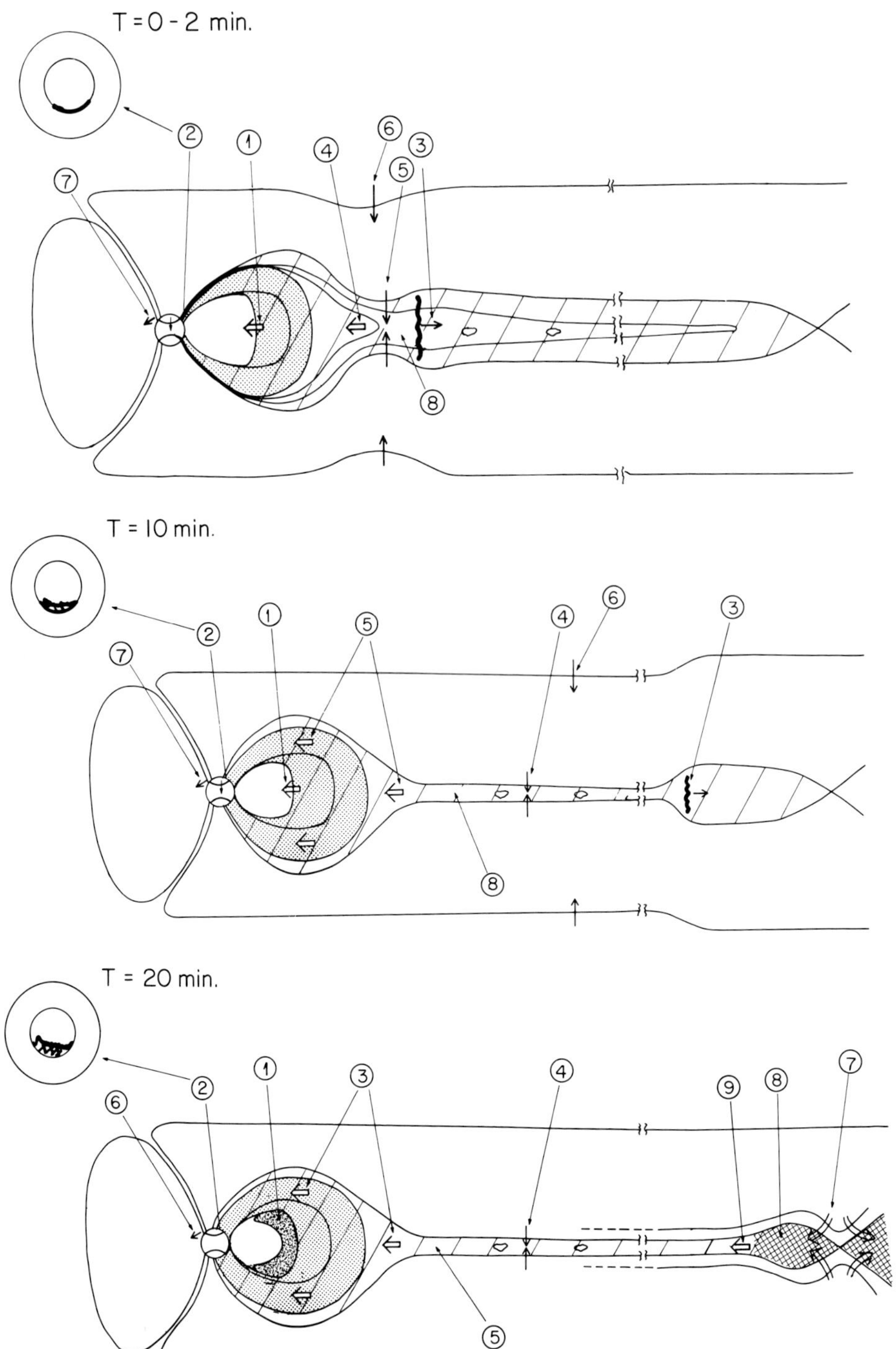

Fig. 9.13(a), (b) and (c). Schematic illustrations, showing major substorm features at (a) $T = 0$–2 min, (b) $T = 10$ min and (c) $T = 20$ min, respectively.

Figure 9.13(a), (b) and (c) show schematically the major features of the magnetospheric substorm at $T = 0$–2 min, 10 min and 20 min. The major features are also listed below:

$T = 0$–2 min.

1. Sudden earthward displacement of plasma in the near-Earth plasma sheet (injection).
2. Sudden brightening of an auroral arc and the electrojet formation.
3. Generation and propagation of a fast rarefaction wave.
4. Earthward plasma flow.
5. Plasma sheet thinning.
6. Magnetopause motion.
7. Equatorward motion of the cusp.
8. Disruption and diversion of the cross-tail current.

$T = 10$ min.

1. Earthward displacement of plasma.
2. Expanding auroral bulge.
3. Propagation of the fast rarefaction wave.
4. Plasma sheet thinning.
5. Earthward plasma flow.
6. Magnetopause motion.
7. Equatorward motion of the cusp.
8. Disruption and diversion of the cross-tail current.

$T = 20$ min.

1. Ring current formation.
2. Expanding auroral bulge.
3. Plasma flow.
4. Plasma sheet thinning.
5. Disruption and diversion of the cross-tail current.
6. Equatorward shift of the cusp.
7. Enhanced reconnection.
8. Production of a hot plasma.
9. A high speed plasma flow.

In the following, we shall consider in some detail the propagation of rarefaction waves in the plasma sheet in order to illustrate how the magnetotail as a whole responds to the waves (Chao *et al.*, 1976). First of all, we must consider the pressure balance in the magnetotail. It is assumed that the boundary of the plasma sheet is a tangential discontinuity. Therefore, the total pressure (including both the thermal and magnetic pressure) has to be balanced across the boundary; see Figure 9.14. In the high latitude lobe, plasma pressure is small compared with magnetic pressure. Therefore, the lobe pressure P_L can be approximated by the magnetic pressure $(B_T^2/8\pi)$. On the other hand, inside the plasma sheet, the ratio $\beta = P_0/(B_0^2/8\pi)$ is appreciably greater than unity.

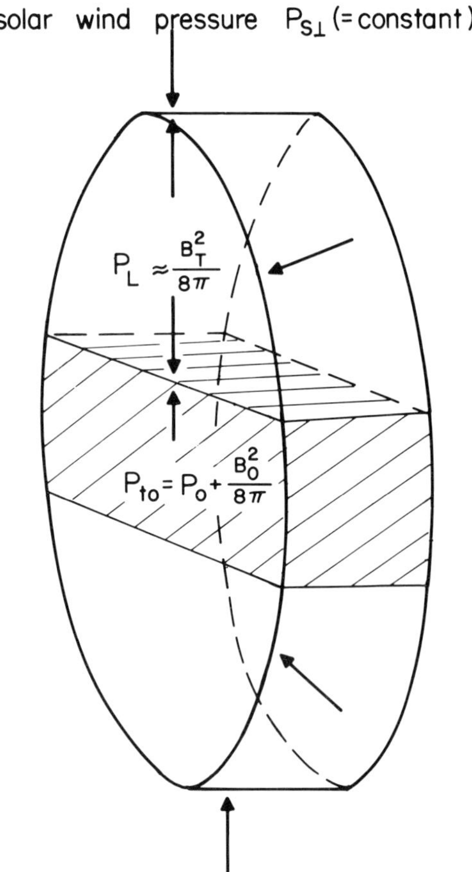

Fig. 9.14. Pressure balance in the magnetotail.

At the onset of a substorm, the inner edge of the plasma sheet moves rapidly earthward, resulting in a rarefaction wave which propagates in the anti-solar direction. The rarefaction wave reduces the pressure in the plasma sheet, causing the lobe magnetic field to expand toward the midplane of the plasma sheet. As a result, the plasma sheet thins. The above thinning process can be modeled in three steps as shown below.

The plasma sheet will be assumed one-dimensional. The magnetic field is parallel to the midplane along the X-axis as shown in Figure 9.15. One end of the plasma sheet is bounded by the inner edge of the plasma sheet, while the other end is bounded by the X-type neutral line. Before the substorm onset, the plasma sheet is in static equilibrium,

$$P_{t0} = P_0 + \frac{B_0^2}{8\pi}$$

where we assume the plasma pressure to be isotropic. The subscript 't' denotes the total pressure and the subscript '0' indicates the initial state before any

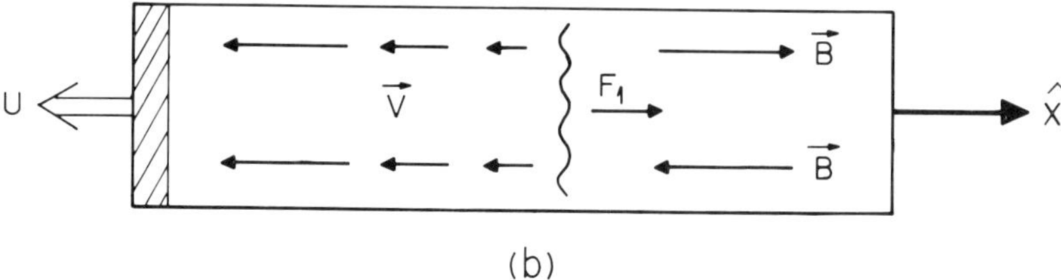

Fig. 9.15. The plasma sheet modeled as a one-dimensional slab. A rarefaction wave generated in the plasma sheet due to earthward motion of the inner edge at a speed U. F_1 is the front of the rarefaction wave propagating along the tailward direction.

disturbance is generated in the plasma sheet. The total pressure P_{t0} must be equal to the lobe pressure P_L which must remain constant during thinning because P_L is balanced by the lateral solar wind pressure $P_{s\perp}$ which is assumed to be constant, namely $P_{s\perp} = P_L = P_{t0}$.

At $T = 0$, the inner edge of the plasma sheet moves earthward at a speed U in the negative X-direction as shown in Figure 9.15. A rarefaction wave is generated and is propagated along the plasma sheet. The front F_1 of this rarefaction wave propagates at the sound speed a_0 in the X-direction. Since β $(= 8\pi P_0/B_0^2)$ is greater than unity in the plasma sheet, the rarefaction wave along $\pm\boldsymbol{B}$ must be the fast mode of MHD waves. There is a plasma flow V in the negative X direction induced by the rarefaction wave. The pressure P_{t1} and the density ρ_1 behind the rarefaction wave are lower than the ambient pressure and density respectively; the subscript '1' indicates the state after rarefaction. The magnetic field will not be changed after rarefaction because the wave propagates along $\pm\boldsymbol{B}$.

The solution of the MHD rarefaction wave propagating along the magnetic field is identical to that of the gas dynamics. The results are given by Landau and Lifshitz (1959) as follows.

If the inner edge of the plasma sheet moves at a speed U where $U < 2a_0/(\gamma - 1)$, then the solution for the speed of the plasma in the rarefaction wave is shown in the upper part of Figure 9.16, where γ is the ratio of the specific heat of the plasma. If the inner edge moves faster than $2a_0/(\gamma - 1)$, then the solution for the speed of the plasma flow is shown in the lower part of Figure 9.16. There occurs a vacuum region between the trailing edge of the wave and the inner edge if the inner edge moves faster than $2a_0/(\gamma - 1)$. The other parameters inside the wave are given by

$$\rho_1 = \rho_0[1 - \tfrac{1}{2}(\gamma - 1)|V|/a_0]^{2/(\gamma-1)}$$
$$P_1 = P_0[1 - \tfrac{1}{2}(\gamma - 1)|V|/a_0]^{2\gamma/(\gamma-1)}$$

where $\gamma = 5/3$ for the rarefaction wave propagated in an isotropic plasma along the magnetic field.

Although the plasma in the unperturbed plasma sheet is collisionless, the plasma pressure has been assumed to be isotropic in our model. Behind the rarefaction wave, the pressure along the X-direction will be reduced much more

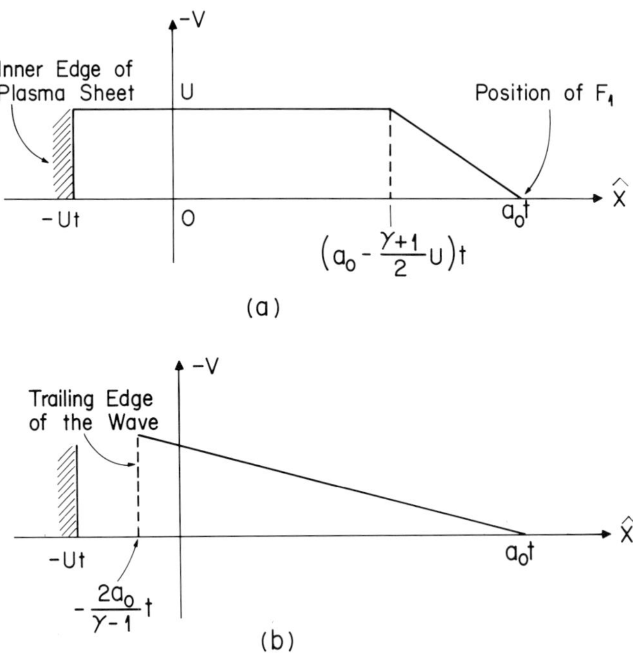

Fig. 9.16. Upper diagram: The velocity of the plasma behind the rarefaction wave for a speed of the inner edge $U \leqslant 2a_0/(\gamma - 1)$. The flow is in the earthward direction. The lower diagram: The velocity of the plasma is the same as the above for $U \geqslant 2a_0/(\gamma - 1)$.

than that in the perpendicular direction. This pressure anisotropy is unstable for the mirror instability and will be reduced. Therefore, it is assumed that the plasma is isotropic behind the rarefaction wave.

Since the total pressure behind the rarefaction wave is lower than that across the plasma sheet boundary, the boundary is forced to move toward the midplane. Thus, the plasma sheet is compressed by the high latitude lobe pressure in the region behind the front of the rarefaction wave. Since $P_L = P_{s\perp}$(= constant), the cross-section of the magnetotail should be reduced, and indeed the reduction has been observed (Section 6.4.2). Such a pressure balance process makes it difficult to determine changes of magnetic energy in the magnetotail in terms of changes of B_T, although Caan *et al.* (1975) claimed that the magnetic energy decreases significantly during substorms. The pressure balance process would mask the decrease; it is most likely that the observed decrease of B_T was simply due to the diamagnetic effect of the expanding plasma sheet during the recovery phase.

In this situation, Chao *et al.* (1976) showed that one can solve for B_2, P_2 and ρ_2. The results are given by

$$\frac{B_2^2}{8\pi} = \frac{P_{s\perp}}{(1+\beta_\perp)}$$

$$P_{\perp 2} = \frac{P_{s\perp}}{(1+\beta_\perp)}\beta_\perp$$

$$\rho_2 = \left(\frac{P_{s\perp}}{1+\beta_\perp}\right)^{1/2} \frac{(8\pi)^{1/2}}{B_0}\rho_1$$

$$P_{\parallel 2} = P_1\left(\frac{P_{s\perp}}{1+\beta_\perp}\right)^{1/2} \frac{(8\pi)^{1/2}}{B_0}$$

$$\frac{P_{\perp 2}}{P_{\parallel 2}} = \left(\frac{P_{s\perp}}{1+\beta_\perp}\right)^{1/2} \left(\frac{B_0^2}{8\pi}\right)^{-1/2}$$

where $\beta_\perp = 8\pi P_1/B_0^2$.

Figure 9.17 is a plot of the magnetic pressure $B_2^2/8\pi$, the plasma pressures $P_{\perp 2}$, $P_{\parallel 2}$, the density ρ_2 and the pressure anisotropy $P_{\parallel 2}/P_{\perp 2}$ in units of the unperturbed quantities $B_0^2/8\pi$, P_0, ρ_0, respectively, as a function of β for $U = 0.2\ a_0$ and $0.8\ a_0$. The rarefaction and compression change the unperturbed state by only 10 to 20% for $U = 0.2\ a_0$. However, for $U = 0.8\ a_0$, the change over the unperturbed state is 50% or more. From these plots, one can determine the degree of rarefaction and compression for a given U and β.

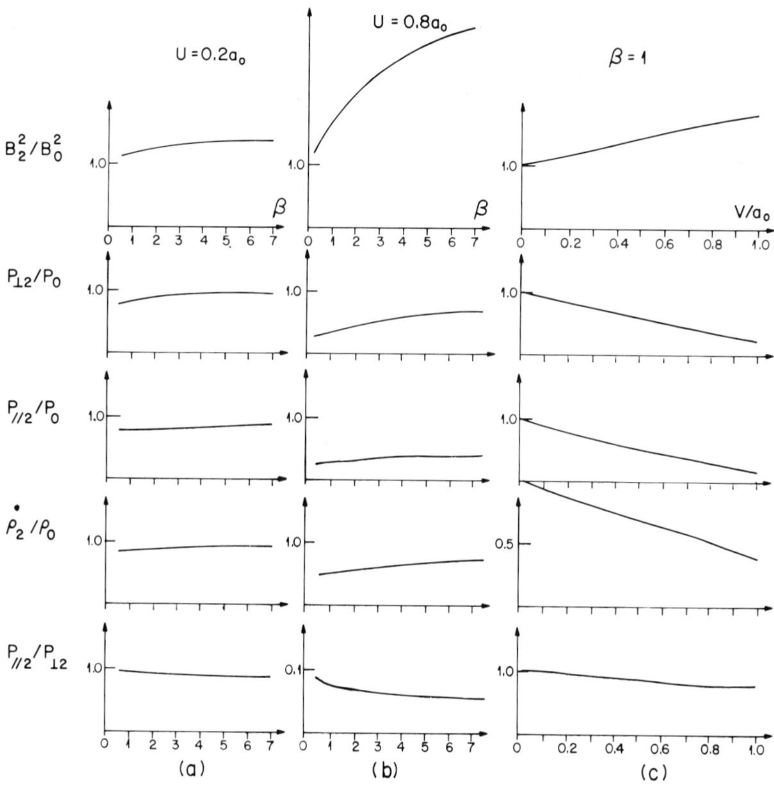

Fig. 9.17. (a) Plots of the ratio of the magnetic field pressure, perpendicular and parallel pressure, density and pressure anisotropy against $\beta(=8\pi P_0/B_0^2)$ for $U = 0.2a_0$. (b) Plots of the ratio are the same as (a) for $U = 0.8a_0$. (c) Plots of the ratio of magnetic field pressure, perpendicular and parallel pressure, density and pressure anisotropy against v/a_0 for $\beta = 1$.

Figure 9.17 shows also the ratios of all the quantities against the ratio V/a_0, where V is the speed of the plasma flow behind the rarefaction wave. For large V/a_0, the ratios of all quantities deviate more from the unperturbed quantities.

On the basis of these plots, one can find all the quantities behind a rarefaction wave when the unperturbed quantities are known. Suppose the magnetic field inside the plasma sheet varies according to $B_0 = B_\infty \tanh(Z/3)$. The half thickness of the plasma sheet is assumed to be 3 R_E. Figure 9.18 is a plot of $B_0(Z)$ as a function of the distance Z from the neutral sheet. As can be seen, a discontinuity in B_0 at 3 R_E is assumed which is taken to be the plasma sheet boundary. In the high latitude lobe, the magnetic field B_T is the major quantity which balances the pressure in the plasma sheet. Assuming $U = 0.8\ a_0$, the final profile of the magnetic field B_2 in the thinned plasma sheet is given by the dashed curve. The half thickness of the thinned plasma sheet is reduced from 3 to 1.5 R_E. In estimating the thickness of the thinned plasma sheet, we have assumed that the change in thickness is proportional to the change in the magnetic field (*i.e.*, $\delta Z \sim \delta B$) because of the conservation of magnetic flux in a one-dimensional plasma sheet model. The changes in all the other quantities can be calculated as shown in Table 9.2 where Z is the coordinate for the unperturbed plasma sheet and Z_2 is the

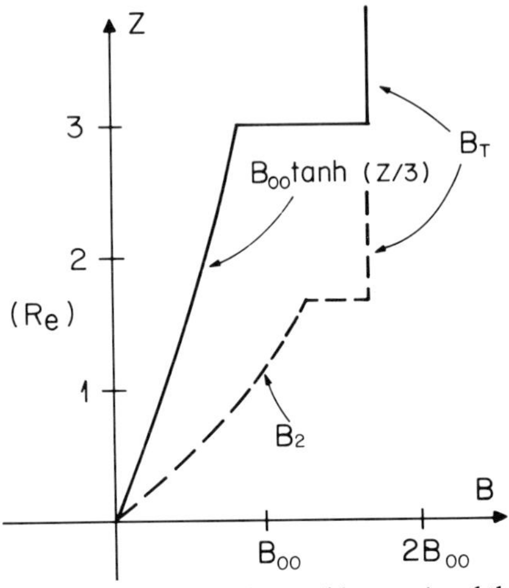

Fig. 9.18. The magnetic field profiles before thinning (solid curves) and the same profile after thinning (dashed curves). Note the half thickness reduces from 3 to 1.6 R_E in this case where $U = 0.8a_0$

TABLE 9.2
Changes of plasma parameters in the plasma sheet after the passage of a rarefaction wave

$Z(R_E)$	$Z_2(R_E)$	B_2^2/B_0^2	$P_{\perp 2}/P_0$	$P_{\parallel 2}/P_0$	ρ_2/ρ_0	$P_{\parallel 2}/P_{\perp 2}$[a]
0.6	0.28	4.4	0.93	0.44	0.83	0.48
1.2	0.58	3.7	0.79	0.41	0.76	0.52
1.8	0.92	3.1	0.65	0.37	0.69	0.57
2.4	1.28	2.6	0.55	0.34	0.63	0.62
3.0	1.67	2.3	0.48	0.32	0.59	0.66

[a]For $Z_2(R_E)$ only.

coordinate for the thinned plasma sheet. The changes in profile of the observable quantities can be obtained from this table. We can see that $P_{\perp 2}/P_0$, $P_{\parallel 2}/P_0$ and ρ_2/ρ_0 are less than one, while the magnetic field ratio B_2/B_0 is greater than one.

(b) *Magnetic Energy Conversion*

In the present model, it is proposed that the second stage of the magnetospheric substorm begins when the rarefaction wave finally reaches the magnetic neutral line which marks the outermost boundary of the plasma sheet ($T = 20$ min). By this time the poleward boundary of the auroral bulge and the auroral electrojet reach almost the highest latitude. The plasma sheet in the distant magnetotail is at its thinnest. In the model presented here, the earthward plasma flow generated by the rarefaction wave supplies the energy for the expansive phase of the substorm. Thus, when the rarefaction wave reaches the outer boundary of the plasma sheet and the plasma sheet is completely deflated, the first phase ends. It is as late as this epoch of the magnetospheric substorm when the magnetic energy conversion process begins in the magnetotail. However, the actual magnetic energy conversion process has so far not been definitely identified.

Thus, we shall first examine numerically the important plasma and magnetic parameters in the recovering plasma sheet (cf. Section 1.4.2).

(i) *Inflow region.*

$$V_A = \text{The Alfvén wave speed in the high latitude lobe.}$$
$$= B/\sqrt{4\pi nm}$$
$$\simeq 2\times 10^8 \text{ cm s}^{-1}$$

where $B = 10\gamma = 10^{-4}$ G

$n = 10^{-2}$ cm^{-3}

$$v = (E\times B) \text{ drift speed of plasma in the high latitude lobe.}$$
$$= \frac{E(\text{mV/m})}{B(\gamma)}\, 10^3 \text{ km s}^{-1}$$
$$\simeq 30 \text{ km s}^{-1}$$

where $E = \dfrac{\Phi}{2\,R_T} = \dfrac{50\text{ kV}}{30\,R_E} = \dfrac{5\times 10^7\text{ mV}}{2\times 10^8\text{ m}}$

$B = 10\ \gamma$

$$M = \text{the merging rate}$$
$$= \frac{v}{V_A} = \frac{30\text{ km s}^{-1}}{200\text{ km s}^{-1}} = 1.5\times 10^{-2}$$

(ii) *Outflow region.* The observed plasma and magnetic field parameters in the recovering plasma sheet are

$$V_R = 500\ \mathrm{km\ s^{-1}}$$
$$B_z = 5 \times 10^{-5}\ \mathrm{G}$$
$$n = 10^{-1}\ \mathrm{cm^{-3}}$$
$$kT = 10\ \mathrm{keV}$$

In Section 5.4.5, it was noted that Sonnerup's theory is applicable only for a high β plasma, so that the released energy will not be sufficient in supplying the needed energy for a substorm, in spite of the fact that his theory gives the largest value of the merging rate $M \simeq (1+\sqrt{2})$ among reconnection theories. Petschek's theory gives a value of the merging rate M which agrees with what can be estimated from a reasonable set of magnetotail parameters. However, the plasma flow along the field lines is directed toward the X-line in the region between the two wave fronts (OL, OT), and in the main outflow region it is an $(E \times B)$ flow. The presently available observations of plasma flow do not seem to support such a flow pattern; the field-aligned component of the flow is directed toward the Earth. Parker's theory (1963) appears to be most consistent with the presently available observations.

As enhanced reconnection begins along the X-line, a considerable heating of plasma takes place. The pressure differential associated with the heating process forces a hot plasma to flow away from the reconnection region. As a result, the plasma sheet expansion proceeds rapidly toward the Earth as the plasma flows with a speed of 500–1000 km s^{-1}, ($T = 40$ min). This phenomenon can be identified as the recovery (or the expansion) of the plasma sheet. During the recovery phase, the entire plasma sheet is inflated again, while the auroral bulge begins to contract equatorward in the polar ionosphere, since the auroral electrojet and the disruption of the cross-tail current begin also to subside during this stage.

The energy flux F_R of the plasma flow in the recovering plasma sheet is thus given by

$$\begin{aligned} F_R &= V_R(\tfrac{1}{2}\rho V_R^2 + \tfrac{5}{2}n\kappa T) \\ &\simeq 2.7 \times 10^{-1}\ \mathrm{erg\ cm^{-2}\ s^{-1}} \end{aligned}$$

Assuming the half thickness and the width of the plasma sheet to be 1 R_E and 30 R_E, respectively, the total energy flux in the northern half of the magnetotail becomes of order 3.2×10^{18} erg s^{-1}. Note that the magnetic energy density in the recovering plasma sheet is much less than the kinetic energy density of plasma.

Assuming that the reconnection process takes place along a distance of order 100 R_E, the $(E \times B)$ convection speed v which is required to generate the energy flux of 3.2×10^{18} erg cm^{-2} s^{-1} may be estimated by

$$\begin{aligned} v &= (3.2 \times 10^{18}\ \mathrm{erg\ cm^{-2}\ s^{-1}}) \Big/ \left(\frac{B_T^2}{8\pi}\right) \times (100\ R_\mathrm{E}) \times (2\ R_T) \\ &\simeq 60\ \mathrm{km\ s^{-1}} \end{aligned}$$

This estimate may be compared with the value estimated earlier ($\simeq 30$ km s^{-1}).

It is important to emphasize that a high speed flow of hot plasma is observed at $X = -10 \sim -45\ R_E$ at about the maximum epoch and after the substorm, not during the expansive phase. In Section 6.6, it was repeatedly stressed that the recovery of the plasma sheet begins at about the time when the auroral bulge and the auroral electrojet attain the highest latitude. This fact alone is an important indication that the conversion of magnetic energy into plasma energy does not take place during the expansive phase at $X \simeq -10 \sim -15\ R_E$. It should be stressed here also that there is no way to explain characteristics of plasma in the thinning plasma sheet in terms of the reconnection process and thus of the conversion of magnetic energy into plasma energy by either the large-scale deformation of the magnetic field structure or the bubble formation suggested by Schindler (1974). The plasma temperature in the plasma sheet may even decrease during thinning. Figure 9.19 shows a model of the magnetospheric substorm, proposed by Hones (1967), in which it is suggested that the X-line is formed in the near-Earth plasma sheet (marked by a star). In such a situation, we would expect a high speed flow of *hot* plasma at $X \simeq -18\ R_E$ during the expansive phase. Such observations are very rare. Further, in his model, the recovery of the plasma sheet is caused by a sudden outward shift of the X-line. The reasons for the initial stationary state and this sudden shift are not clear.

Schindler (1974) suggested that the ion-tearing mode instability plays an important role in converting the magnetic energy and in forming the suggested large-scale change of the magnetic field structure. However, it is difficult to predict the proposed growth on the basis of a linear theory of the instability. The growth of the instability results in the north-south component of the magnetic field across the midplane, which tends to suppress the growth. Thus, the instability may be able to produce only small-scale 'bubbles' in the magnetic field structure, but not the needed large-scale structure.

Schindler (1974) suggested also that when the turbulence grows, the magnetic energy inside the separatrix (namely, the bubble structure hatched in Figure 9.20) will be dissipated, and the corresponding plasma particles will undergo heating and pitch-angle diffusion, leading to enhanced precipitation. However, it is not difficult to see that the magnetic energy available in the bubbles embedded in the plasma sheet (as illustrated in Figure 9.20), is rather small ($B \sim 5\ \gamma$), compared with the magnetic energy available in the high latitude lobe. Thus, the extent of the bubble would have to be very large ($> 100\ R_E$) to supply the required energy flux. Certainly, it is difficult to contain such a mechanism within $X \simeq -15\ R_E$.

Thus, in the model presented here, the magnetospheric substorm is not a simple relaxation process of the magnetosphere toward a dipolar configuration. The plasma energy which is contained in a newly 'excited state' magnetosphere by the IMF or is converted from the magnetic energy during one substorm is dissipated as substorm energy during the expansive phase of a new or the next substorm; the energy conversion and the resulting inflation of the plasma sheet occur again during the recovery phase of the third substorm, and so on, until the whole excess energy in the magnetotail is dissipated. That is to say, it is through such a repeated process of deflation and inflation that the magnetosphere achieves finally the 'relaxation', returning to the 'ground state'. We suggest that the magnetic energy conversion

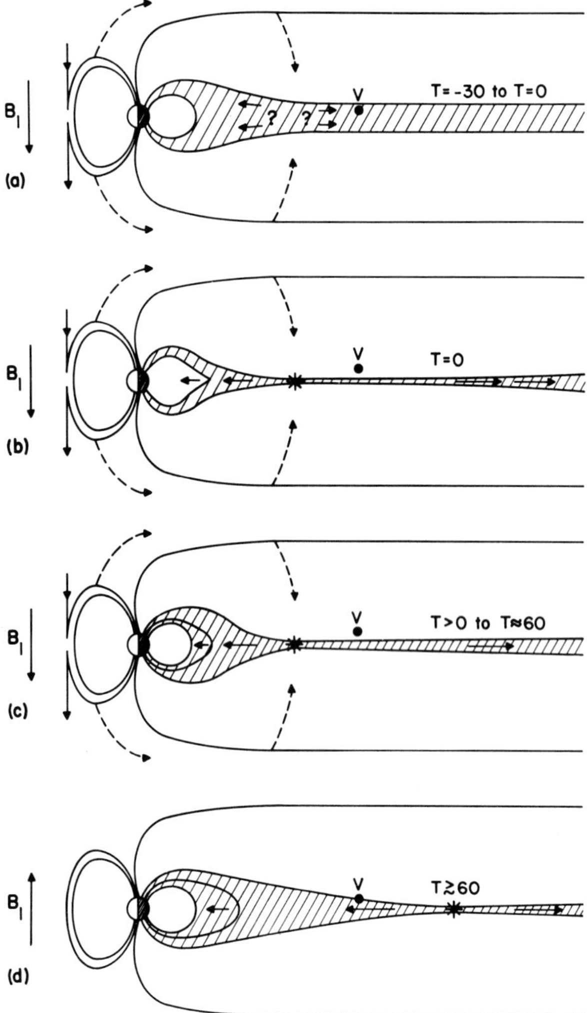

Fig. 9.19. Model of the magnetospheric substorm proposed by Hones. The dot labeled by *V* represents a Vela satellite in the plasma sheet which is the shaded region. B_I represents the NS component of the IMF. The star mark indicates the location of the proposed *X*-line. (Hones, E. W. Jr.: *Radio Sci.* **8**, 979, 1973.)

takes the form of the reconnection, described in the theory proposed by Parker (1963), which occurs beyond the distance of the moon for most substorms. Figure 9.21(a) and (b) show schematically the major substorm features at $T = 40$ min and 2 h, respectively. They are also listed below:

Magnetospheric changes about 40 min after the onset of the magnetospheric substorm:

1. Enhanced reconnection.

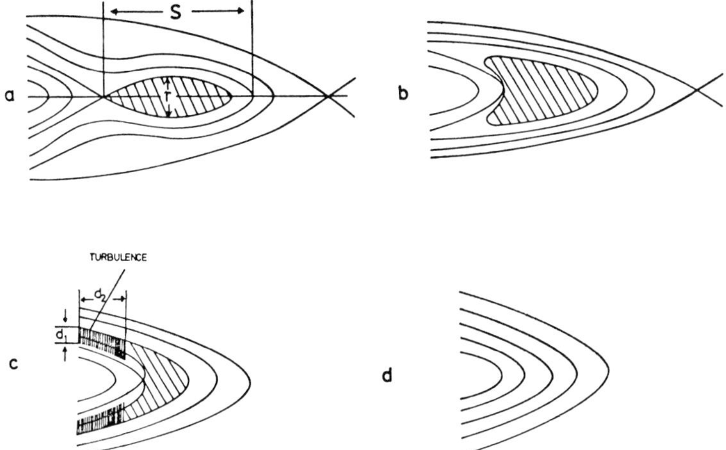

Fig. 9.20. Schematic illustration of the proposed processes occurring after the ion tearing mode instability has led to regions of closed field lines. (Schindler, K.: *J. Geophys. Res.* **79**, 2803, 1974.)

2. High speed plasma flow.
3. Recovery of the plasma sheet.
4. Increase of the B_z component.
5. Magnetopause motion.
6. Expanding bulge reaching the highest latitude.
7. Formation of the ring current.

Magnetospheric changes about 2 h after the onset of the magnetospheric substorm:

1. Recovery of the plasma sheet.
2. Contraction of the expanding bulge.

9.2.3. CRITICAL TESTS AND UNSOLVED PROBLEMS

As mentioned in the previous subsection, the proposed model is meant to be simply an exercise in synthesizing a variety of magnetospheric and ground-based observations in terms of the concept of magnetospheric substorms. We are still far from constructing even a first approximation model which has a sound physical ground. Thus, as a part of the exercise, we shall examine several aspects of the model.

(i) *Responses of the Magnetosphere to the Southward Turning of the IMF*

It was shown in Sections 5.2 and 9.2 that a series of magnetospheric substorms results when the magnetosphere is 'pulsed' once by the southward turning of the IMF vector. At present, the time constants, τ_1, τ_3, and τ_4 (Section 5.2; Figure 5.5) are not quantitatively established. If there is a definite value for τ_3, the magnetos-

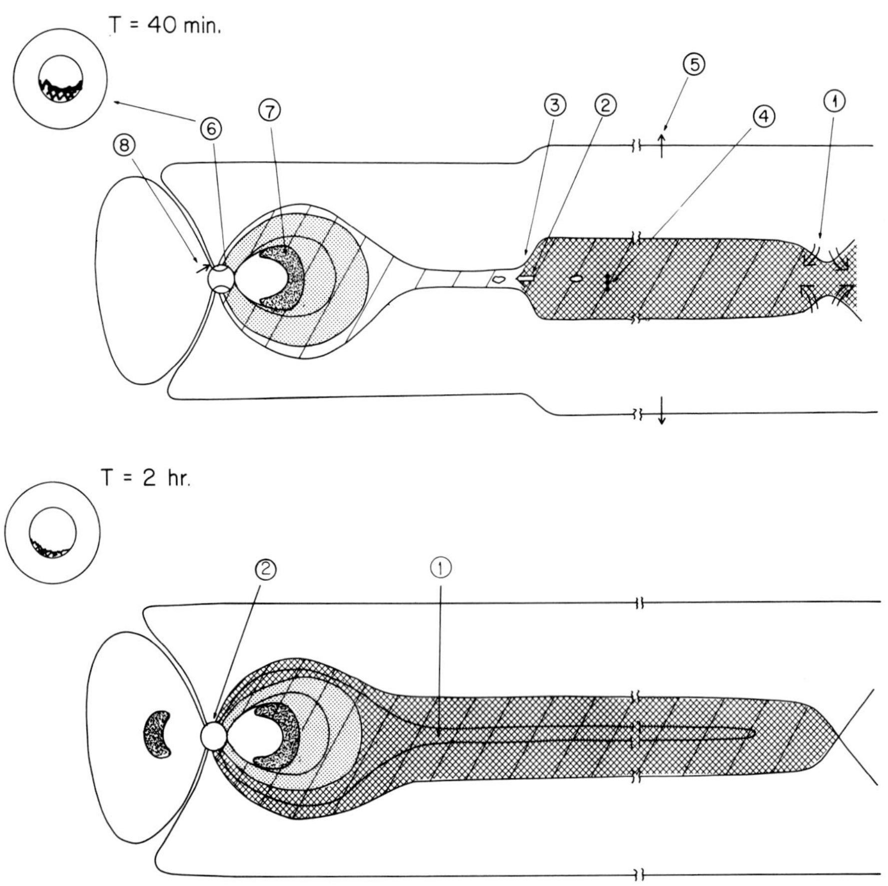

Fig. 9.21. Schematic illustrations, showing major substorm features at (a) $T = 40$ min and (b) $T = 2$ h, respectively.

phere is considered to have some internal structure (an internal oscillator) which tends to trigger substorms *quasi-periodically*. If there is no definite value for τ_3, one must conclude that substorms have characteristics of *random fluctuations*.

(ii) *Triggering Process*

Most of the triggering processes proposed so far assume that microscopic plasma processes (such as ion-cyclotron waves or ion-acoustic waves) play a major role in disrupting the coupling between the ionosphere and the magnetosphere and thus in allowing the proposed instabilities or electric fields to grow rapidly.

In the mechanism suggested by Coroniti and Kennel (1972), which is adopted in the proposed model, the ionosphere is the source region of the discharge current which decouples the magnetosphere from the ionosphere when the discharge current intensity reaches a threshold value. Swift (1967) and Liu (1970) examined the interchange or flute type instabilities on the outer surface of the ring current belt or the inner surface of the plasma sheet. However, the ionosphere tends to discharge the growing space charges associated with the instabilities. Thus, they proposed that when the discharge current along the field lines becomes intense enough, ion-acoustic waves are generated and the field-aligned current circuit becomes resistive, decoupling the magnetosphere from the ionosphere and allowing the instabilities to grow rapidly.

It is thus important to examine the growth of plasma waves along the field lines and in the plasma sheet. On the other hand, the observations alone may not be able to distinguish whether the plasma waves are a cause or an effect of the triggering process.

(iii) *Generation of the Rarefaction Wave*

Coroniti and Kennel (1972) considered only the equatorward directed electric field in the ionosphere. However, in order for the rarefaction wave to propagate in the anti-solar direction, the westward component of the electric field is needed. Indeed, a large westward electric field of order 20–40 mV m^{-1} has been observed during substorms (Section 7.7.2). It is therefore of great importance to examine how the observed westward electric field can develop in the ionosphere. It is also important to observe directly the inward speed of the motion of plasma at about the synchronous distance to see whether it is substantially greater than what is expected from the convection speed. Note in this connection that auroral arcs drift rapidly equatorward *within* the auroral bulge, in spite of the fact that the bulge expands poleward (Section 6.5.1).

The observed growth of the westward electric field of order 20–40 mV m^{-1} in an early epoch of substorm may be an important indication that the ionosphere plays an active role in generating the rarefaction wave. Unfortunately, the total potential drop associated with the westward electric field is not known. On the other hand, even if the electric field is present along the oval in the $0 \sim 6$ MLT sector, the total potential drop will be of order 150 kV which is appreciably greater than the expected potential drop across the magnetotail.

(iv) *Thinning of the Plasma Sheet*

In the previous section, it was possible to predict changes of various plasma and magnetic field parameters associated with the propagation of the rarefaction wave. It was noted there that the problem is complicated by the fact that the magnetotail tends to adjust itself in maintaining the pressure balance, so that the identification of the rarefaction wave becomes a complicated problem. Nevertheless, the prediction was the first quantitative study of thinning of the plasma sheet. It remains to be seen whether the predicted changes of various plasma parameters will be found in future observations of plasma sheet thinnings.

In this respect, it is also important to determine accurately the flow pattern of plasma in the thinning plasma sheet. In the model presented in the previous section, the flow should be directed toward the Earth, and the energy needed for the formation of the ring current belt and the auroral substorm during the expansive phase must be supplied by the plasma flow generated by the rarefaction wave.

(v) *Conversion of the Plasma Flow Energy into Auroral Particle Energy*

In any models which rely on plasma flow as the source of energy for auroral activity, it is necessary to have a mechanism which converts the flow energy of plasma into the energy of precipitating auroral electrons. It should be noted in this connection that an auroral arc is associated with a local current *system* which is much more intense than the current system observed by the TRIAD satellite (Sections 1.3.2 and 3.9.3). Therefore, the flow energy of plasma must be fed into the current system; that is to say, there must be a local dynamo associated with an auroral arc (Section 3.9.3).

(vi) *Enhanced S_q^p Current System*

Several important substorm features can be explained in terms of the observed enhancement of the S_q^p current. The model presented in the previous section cannot provide the mechanism by which such an enhancement is induced. Atkinson (1971) suggested that it results from the disruption and the subsequent diversion of the cross-tail current to the polar upper atmosphere by thinning of the plasma sheet. At present, however, it is not known whether the thickness of the plasma sheet is related to the current density (per unit length of the magnetotail). Some workers consider that current-driven instabilities will disrupt the cross-tail current and subsequently divert it to the polar upper atmosphere.

(vii) *Delayed Reconnection Enhancement*

All the reconnection theories proposed so far consider a two-dimensional situation in which the electric field becomes *a priori* constant. It is still not proven whether such steady state, two-dimensional reconnection theories are applicable in magnetospheric problems, namely the interaction between the magnetosphere and the IMF and the magnetic energy conversion in the magnetotail.

In the proposed model, it was noted that the plasma characteristics in the recovering plasma sheet are consistent with what are expected from the reconnection theory proposed by Parker (1963). This fact is by no means a conclusive proof that the reconnection process takes place in the magnetosphere. Other processes which can lead to the magnetic field energy dissipation should also be examined quantitatively. The current disruption theory of solar flares, proposed by Alfvén and Carlqvist (1967), can offer another way that the magnetic field energy can be dissipated. It remains to be seen if the reconnection theory will eventually be proven by future satellite observations of plasma parameters in the recovering plasma sheet. In fact, regardless of models in study, if one assumes an enhanced reconnection during either the expansive phase or the recovery phase, one must make a serious effort to examine whether the observed plasma parameters would agree with what are expected from a reconnection theory.

One of the problems in Parker's theory is that so long as the electric conductivity estimated on the basis of a simple binary collision theory is used, the predicted merging rate M is several orders of magnitude less than what is expected from magnetotail data.

Parker (1973) proposed that the intercharge instability can speed up the escape of plasma from the region between opposite fields and leads to rapid reconnection. Kan and Chao (1976) extended Parker's idea into a model and showed that rapid reconnection in the magnetotail, driven by the interchange instability, can occur only when the ionosphere and the plasma sheet are decoupled.

In the model previously presented, it was proposed that the observed delay of the appearance of the hot plasma flow is due to the fact that the enhanced reconnection does not begin until the rarefaction wave reaches the X-line which marks the anti-solar end of the plasma sheet boundary. Thus, the hot plasma should appear first at $X \simeq -100\,R_E$ and then later at $X \simeq -15 \sim -45\,R_E$. As mentioned in Section 6.6, there are some reports that the recovering plasma sheet appears first nearest to the Earth and later at greater distances. Thus, the proposed model contradicts those reports. However, this particular problem of timing of the recovery of the plasma sheet as a function of geocentric distance is a difficult one, because the recovery time is also a function of Z. Thus, it is not at all a settled question, and a more detailed study is needed.

One of the ways to examine whether an enhanced reconnection and subsequent transfer of the newly formed closed field lines take place during the expansive phase is to follow the location of the dayside cusp as a function of substorm time.

The newly reconnected field lines are supposed to be convected toward the dayside magnetopause during the expansive phase, so that one would expect that the latitude of the last closed field lines shifts poleward. However, there have been a number of observations which suggest that the latitude of the 'foot' of the last closed field line actually decreases during substorms. The location of the 'foot' of the last closed field line has been identified or inferred from various phenomena:

(i) The location of the cusp where the magnetosheath-like plasma is found by polar orbiting satellites.

(ii) The equatorward boundary of the oval area in the polar cap where solar electrons impinge.

(iii) The location of the auroral oval in the midday sector, that is, of midday auroras.

(iv) The location of the F2 irregularity zone (FLIZ).

(v) The location where the daytime (magnetic) agitation is intense.

Figure 9.22 shows, as a function of Kp: the latitudes of the poleward and equatorward boundaries of the 'soft precipitation zone' (observed by OGO-4) near local noon (Hoffman, 1971); the cusp location identified by locating the region where magnetosheath-like protons are found (Winningham, 1972); the latitudes of the equatorward boundary of the area of solar electron and proton bombardment (McDiarmid *et al.*, 1972); and the location of the midday part of the auroral oval (Feldstein and Starkov, 1967; Feldstein, 1972; Vorobjev *et al.*, 1975). The equatorward shift of midday auroras was also noted from a polar orbiting satellite which carried an ultra-violet photometer (Chubb and Hicks, 1970). The figure also includes the location of the F2 irregularity zone (FLIZ) as a function of Kp (Pike, 1972) and the latitudes of the most intense daytime (magnetic) agitations as a function of Kp. All these phenomena show clearly that the cusp region shifts equatorward during magnetospheric substorms.

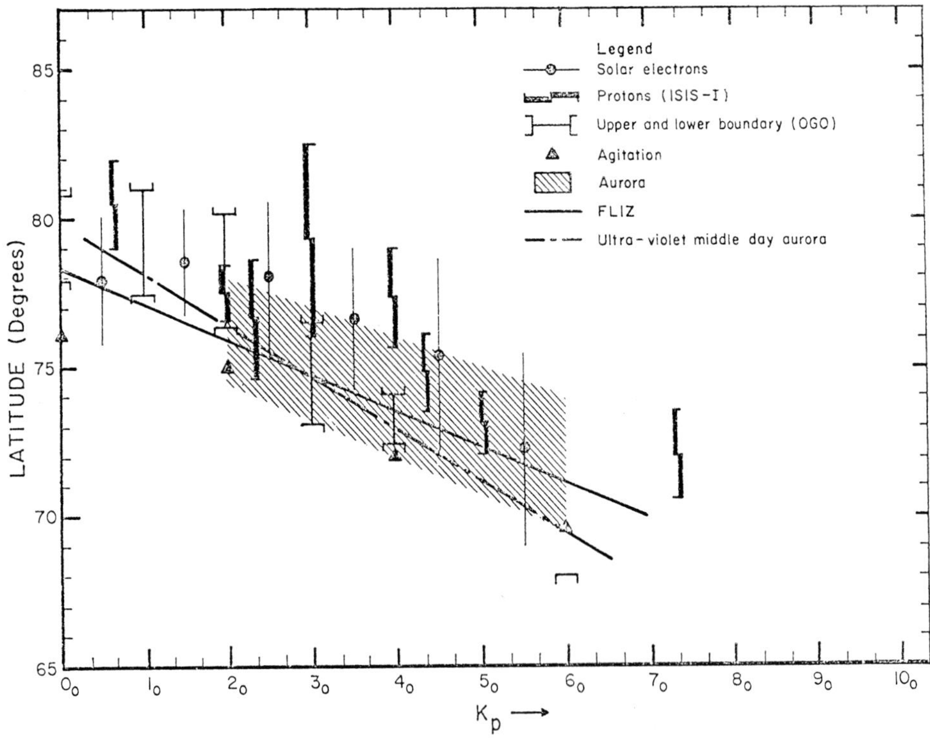

Fig. 9.22. Latitude of the cusp as a function of the Kp index. The cusp locations are determined by a variety of methods, as indicated in the legend.

As mentioned in Section 4.4.3(b), the IMF B_z component also controls the north-south shift of the cusp location. Therefore, it is important to examine whether or not the above Kp dependence is due partially to the IMF effect. This problem was examined recently by Kamide *et al.* (1976) who concluded that the cusp location is certainly lowered by substorm effects. It is not possible to explain such a delay in terms of the delay of flux return toward the dayside magnetosphere, since it is known that the westward electrojet develops and extends into the morning sector without such a long delay. Therefore, these studies suggest strongly that the reconnection process takes place in a late epoch of substorms.

9.3. Concluding Remarks

The main purpose of *Polar and Magnetospheric Substorms*, the predecessor to this book, was to present a synthesizing and unifying study of various ground-based data of polar upper atmospheric phenomena such as auroral, magnetic activities and ionospheric disturbances, and to show that these complicated phenomena can be understood as manifestations of a single phenomenon which takes place intermittently in the magnetosphere – the magnetospheric substorm. When that book was being written, the first satellite observations of particles and magnetic fields during substorms were just becoming available.

The main purpose of the present book is to attempt to synthesize both satellite and ground-based data of magnetospheric phenomena by extending the concept of a magnetospheric substorm which was established in the previous book. This task has been more difficult than the preceding one. On the other hand, a number of workers have joined the author in this venture. It is gratifying to find that many complicated magnetospheric phenomena can indeed be synthesized and understood as manifestations of a single phenomenon, the magnetospheric substorm.

It appears that the magnetospheric substorm is not a unique phenomenon to the Earth, but is, perhaps, a quite universal phenomenon in nature. Siscoe, Ness and Yeates (1975) presented interesting evidence that some of the features observed in Mercury's nightside magnetosphere bear striking resemblances to substorm phenomena in the Earth's magnetosphere. It is likely that a similar phenomenon will also be found in the Jovian magnetosphere. As suggested in the previous book, there is a close phenomenological similarity between magnetospheric substorms (auroral substorms) and solar storms (solar flares). This suggestion has since been considered seriously by a number of workers (cf. De Feiter, 1975).

Therefore, there is no doubt that what we learn about the magnetosphere and the magnetospheric substorm of the Earth will be of great value in understanding a variety of cosmic electrodynamic phenomena.

References

Akasofu, S.-I.: 1968, *Polar and Magnetospheric Substorms*, D. Reidel Publ. Co., Dordrecht-Holland.

Alfvén, H. and Carlqvist, P.: 1967, 'Currents in the Solar Atmosphere and a Theory of Solar Flares', *Solar Phys.* **1**, 200.

Arnoldy, R. L.: 1971, 'Signature in the Interplanetary Medium for Substorms', *J. Geophys. Res.* **76**, 5189.

Atkinson, G.: 1971, 'Magnetospheric Flows and Substorms', Review paper presented at the Advanced Study Inst. on Magnetosphere-Ionosphere Interactions, Dalseter, Norway, April 14–23.

Bartels, J.: 1932, 'Terrestrial Magnetic Activity and Its Relations to Solar Phenomena', *Terr. Magn.* **37**, 1.

Belcher, J. W. and Davis, L. Jr.: 1971, 'Large-Amplitude Alfvén Waves in the Interplanetary Medium, 2', *J. Geophys. Res.* **76**, 3534.

Belcher, J. W., Davis, L. Jr., and Smith, E. J.: 1969, 'Large-Amplitude Alfvén Waves in the Interplanetary Medium: Mariner 5', *J. Geophys. Res.* **74**, 2302.

Burlaga, L. F.: 1974, 'Interplanetary Stream Interfaces, *J. Geophys. Res.* **79**, 3717.

Burlaga, L. F.: 1975, 'Interplanetary Streams and Their Interactions with the Earth', *Space Sci. Rev.* **17**, 327.

Burlaga, L. F. and Ogilvie, K. W.: 1970, 'Heating of the Solar Wind', *Astrophys. J.* **159**, 659.

Caan, M. N., McPherron, R. L. and Russell, C. T.: 1975, 'Substorm and Interplanetary Magnetic Field Effects on the Geomagnetic Tail Lobes', *J. Geophys. Res.* **80** 191.

Chao, J., Kan, J., Lui, A. T. Y. and Akasofu, S.-I.: 1976, 'A Theory of Plasma Sheet Thinning', *Planet. Space Sci.* (submitted, 1976).

Chubb, T. A. and Hicks, G. T.: 1970, 'Observations of the Aurora in the Far-Ultraviolet from OGO-4', *J. Geophys. Res.* **75**, 1290.

Coleman, P. J. Jr., Davis L. Jr., Smith, E. J. and Jones, D. E.: 1966, 'Variations in the Polarity Distribution of the Interplanetary Magnetic Field', *J. Geophys. Res.* **71**, 2831.

Coroniti, F. V. and Kennel, C. F.: 1972, 'Magnetospheric Substorms', *Cosmic Plasma Physics*, K. Schindler (ed.), p. 15, Plenum Press, New York.

De Feiter, L. D.: 1975, 'Chromospheric Flares or Chromospheric Aurorae?', *Space Sci. Rev.* **17**, 181.

Dryer, M.: 1975, 'Interplanetary Shock Waves Recent Development', *Space Sci. Rev.* **17**, 277.
Durney, B. R. and Pneuman, G. W.: 1975, 'Solar-Interplanetary Modeling; 3-D Solar Wind Solutions in Prescribed Non-Radial Magnetic Field Geometries', *Solar Phys.* **40**, 461.
Feldstein, Y.-I. and Starkov, G. V.: 1967, 'Dynamics of Auroral Belt and Polar Geomagnetic Disturbances', *Planet. Space Sci.* **15**, 209.
Feldstein, Y.-I.: 1972, 'Auroras and Associated Phenomena', *Solar-Terrestrial Physics*, 1970, Part III, E. R. Dyer (ed.), p. 152, D. Reidel Publ. Co., Dordrecht-Holland.
Gosling, J. T., Hildner, E., MacQueen, R. M., Munro, R. H., Poland, A. I. and Ross, C. L.: 1974, 'Mass Ejections from the Sun; a View from Skylab', *J. Geophys. Res.* **79**, 4581.
Gosling, J. T., Hildner, E., MacQueen, R. M., Munro, R. H., Poland, A. I. and Ross, C. L.: 1975, 'Direct Observations of a Flare Related Coronal and Solar Wind Disturbance', *Solar Phys.* **40**, 439.
Gulbrandsen, A.: 1973, 'On the Possibility of Inferring the Solar and Interplanetary Sector Structure from Statistics of Geomagnetic Storms and Solar Activity', *Planet. Space Sci.* **21**, 2003.
Gulbrandsen, A.: 1974, 'Coronal λ5303 Intensity, Geomagnetic Activity and Solar Sources of High-Speed Plasma Streams', *Planet. Space Sci.* **22**, 841.
Gulbrandsen, A.: 1975, 'The Solar M-Region Problem – An Old Problem Now Facing Its Solution?', *Planet. Space Sci.* **23**, 143.
Hansen, R. T., Hansen, S. F. and Sawyer, C.: 1976, 'Long-Lived Coronal Structures and Recurrent Geomagnetic Patterns in 1974', *Planet. Space Sci.* (in press).
Hirshberg, J., Bame, S. J. and Robbins, D. E.: 1972, 'Solar Flares and Solar Wind Helium Enrichments: July 1965–July 1967', *Solar Phys.* **23**, 467.
Hirshberg, J., Nakagawa, Y. and Wellck, R. E.: 1974, 'Propagation of Sudden Disturbances Through a Nonhomogeneous Solar Wind', *J. Geophys. Res.* **79**, 3726.
Hoffman, R. A.: 1971, 'Properties of Low Energy Particle Impacts in the Polar Domain in the Dawn and Dayside Hours', Goddard Space Flight Center', *Rep. X-646-199.*
Hollweg, J. V.: 1975, 'Waves and Instabilities in the Solar Wind', *Rev. Geophys, and Space Phys.* **13**, 263.
Hones, E. W. Jr.: 1973, 'Plasma Flow in the Plasma Sheet and Its Relation to Substorms', *Radio Sci.* **8**, 979.
Horwitz, J. L. and Akasofu, S.-I.: 1976, *Planet Space Sci.* (submitted 1976).
Hundhausen, A. J.: 1972, *Colonal Expansion and Solar Wind*, Springer-Verlag, New York.
Kamide, Y., Burch, J., Winningham, J. D. and Akasofu, S.-I.: 1976, 'Dependence of the Latitude of the Left on the Interplanetary Magnetic Field and Substorm Activity', *J. Geophys. Res.* **81**, 698.
Kan, J. R. and Chao, J. K.: 1976, 'Rapid Reconnection Driven by the Interchange Instability During Substorms', *J. Geophys. Res.* (submitted).
Krieger, A. S., Timothy, A. F. and Roelof, E. C.: 1973, 'A Coronal Hole and Its Identification as the Source of a High Velocity Solar Wind Stream', *Solar Phys.* **29**, 505.
Landau, L. D. and Lifschitz, E. M.: 1959, *Fluid Mechanics*, Pergamon Press, Addison-Wesley Publishing Co., Inc.
Liu, C. S.: 1970, 'Low Frequency Drift Instabilities of the Ring Current Belt', *J. Geophys. Res.* **75**, 3789.
McDiarmid, I. B., Burrows, J. R. and Wilson, M. D.: 1972, 'Solar Particles and the Dayside Limit of Closed Field Lines', *J. Geophys. Res.* **77**, 1103.
McPherron, R. L., Russell, C. T. and Aubry, M. P.: 1973, 'Satellite Studies of Magnetospheric Substorms on August 15, 1968, 9, Phenomenological Model for Substorms', *J. Geophys. Res.* **78**, 3131.
Mustel, E. R.: 1964, 'Quasi-Stationary Emission of Gases from the Sun', *Space Sci. Rev.* **3**, 137.
Ness, N. F. and Wilcox, J. M.: 1967, 'Interplanetary Sector Structure, 1962–1966', *Solar Phys.* **2**, 351.
Neupert, W. M. and Pizzo, V.: 1974, 'Solar Coronal Holes as Sources of Recurrent Geomagnetic Disturbances', *J. Geophys. Res.* **79**, 3701.
Newkirk, Jr., G.: 1975, 'Recent Perspectives in Solar Physics: Elemental Composition Coronal Structure and Magnetic Fields, Solar Activity'. Presented at International Cosmic Ray Conference, Munich, Germany, August 22, 1975.
Olbert, S.: 1968, 'Summary of Experimental Results from M.I.T. Detector on IMP-1, *Physics of the Magnetosphere*, R. L. Carovillano, J. F. McClay and H. R. Radoski (eds.), D. Reidel Pub. Co., Dordrecht-Holland.
Parker, E. N.: 1963, 'The Solar-Flare Phenomenon and the Theory of Reconnection and Annihilation of Magnetic Fields', *Ap. J. Suppl.* **8**, 177.
Parker, E. N.: 1973, 'The Reconnection Rate of Magnetic Fields', *Astrophys. J.* **180**, 247.
Pecker, J.-C. and Roberts, W. O.: 1955, 'Solar Corpuscles Responsible for Geomagnetic Disturbances', *J. Geophys. Res.* **60**, 33.

Pike, C. P.: 1972, 'Equatorward Shift of the Polar F Layer Irregularity Zone as a Function of the Kp Index', *J. Geophys. Res.* **77**, 6911.
Pneuman, G. W.: 1973, 'The Solar Wind and the Temperature-Density Structure of the Solar Corona', *Solar Phys.* **28**, 247, 1973.
Roelof, E. C., Cuperman, S. and Sternlich, A.: 1975, 'On the Correlation of Coronal Green-Line Intensity and Solar Wind Velocity', *Solar Phys.* **41**, 349.
Rossi, B. and Olbert, S.: 1970, *Introduction to the Physics of Space*, McGraw-Hill Book Co., New York.
Sakurai, K.: 1973, 'Solar Flare Emissions and Geophysical Disturbances', Goddard Space Flight Center Publication, *X-693-73-277.*
Sakurai, K. and Chao, J. K.: 1973, 'Expansion Pattern of Helium-Enriched Shell Associated with Solar Flares', *Nature* **246**, 72.
Schindler, K.: 1974, 'A Theory of the Substorm Mechanism', *J. Geophys. Rev.* **79**, 2803.
Severny, A., Wilcox, J. M. and Scherrer, P. H. and Colburn, D. S.: 1970, 'Comparison of the Mean Photospheric Magnetic Field and the Interplanetary Magnetic Field', *Solar Phys.* **15**, 3.
Sheeley, N. R. Jr., Harvey, J. W. and Feldman, W. C.: 1976, 'Coronal Holes, Solar Wind Streams and Recurrent Geomagnetic Disturbances: 1973–1976', *Solar Phys.* (in press).
Siscoe, G. L., Ness, N. F. and Yeates, C. M.: 1975, 'Substorms on Mercury?', *J. Geophys. Res.* **80** 4359.
Snyder, C. W., Neugebauer, M. and Rao, U. R.: 1963, 'The Solar Wind Velocity and Its Correlation with Cosmic Ray Variations and with Solar and Geomagnetic Activity', *J. Geophys. Res.* **68**, 6361.
Svalgaard, L., Wilcox, J. M. and Duvall, T. L.: 1974, 'A Model Combining the Polar and the Sector Structured Solar Magnetic Fields', *Solar Phys.* **37**, 157.
Svalgaard, L., Wilcox, J. M. and Scherrer, P. H. and Howard, R.: 1975, 'The Sun's Magnetic Sector Structure', *Solar Phys.* **45**, 83.
Swift, D. W.: 1967, 'The Possible Relationship Between the Auroral Breakup and the Interchange Instability of the Ring Current', *Planet, Space Sci.* **15**, 1225.
Thompson, W. B.: 1962, *An Introduction to Plasma Physics*, Pergamon Press, Addison-Wesley Publishing Co. Inc.
Timothy, A. F., Krieger, A. S. and Vaiana, G. S.: 1976, 'The Structure and Evaluation of Coronal Holes', *Solar Phys.* (in press).
Uchida, Y., Altschuler, M. D. and Newkirk, G. Jr.: 1973, 'Flare-Produced Coronal MHS-Fast-Mode Wavefronts and Morton's Wave Phenomenon', *Solar Phys.* **28**, 495.
Volk, H.: 1975, 'Microstructure of the Solar Wind', *Space Sci. Rev.* **17**, 255.
Vorobjev, V. G., Gustafsson, G., Starkov, G. V., Feldstein, Y. I. and Shevnina, N. F.: 1975, 'Dynamics of Day and Night Aurora During Substorms', *Planet. Space Sci.* **23**, 269.
Wilcox, J. M. and Svalgaard, L.: 1974, 'Coronal Magnetic Structure at a Solar Sector Boundary', *Solar Phys.* **34**, 461.
Winningham, J. D.: 1972, 'Characteristics of Magnetosheath Plasma Observed at Low Altitudes in the Dayside Magnetospheric Cusps', *Earth's Magnetospheric Processes*, B. M. McCormac (ed.), p. 68, D. Reidel Pub. Co., Dordrecht-Holland.

INDEX OF NAMES

(Page numbers for first and co-authors are in roman type, while those for all others [whose names are not usually mentioned in the text] are in italics.)

Aarons, J. 124
Aarsnes, K. 21, *116*, 512
Ackerson, K. L. 93, 99, 100, 168, *283*, 336, 358, 370
Aikin, A. C. 460
Akasofu, S.-I. 7, 10, *17*, 20, *22*, 24, *27*, *29*, *38*, 48, 49, 54, 62, 71, 72, 75, 76, 77, *83*, 88, *93*, *101*, 109, *126*, 137, *165*, 172, *192*, *193*, *201*, 217, *221*, 232, 233, 239, *242*, 245, 247, *248*, 250, 255, 256, 267, 272, 274, 278, 288, 294, 296, 300, *315*, 316, 318, 320, 321, 326, *333*, 336, 347, *348*, 362, 363, *367*, 370, 371, 382, *383*, 385, *392*, *394*, 396, 405, 410, 411, 413, 414, 421, 428, *447*, 454, 455, 476, 478, 486, 512, 562, *569*, *584*
Alfvén, H. 1, 56, 155, 156, 158, 178, 288, 582
Alksne, A. *294*
Allen, J. H. 249
Altschuler, M. D. *556*
Amayenc, P. *460*, *479*
Amundsen, R. 21, 116, *512*
Anderson, A. D. 458
Anderson, C. W. III *424*
Anderson, H. R. 26, 111, 117
Anderson, K. A. 20, *38*, 137, 153, 210, 315, 347
Andrews, M. K. 109, 120, 123
Anger, C. D. 17, 76, 88, *93*, *104*, 116, *398*, *526*
Arendt, P. R. 458
Arens, J. F. *522*, *537*
Armstrong, J. C. 1, 21, 22, 24, *51*, 396, 398, 446
Armstrong, T. P. *369*
Arnoldy, R. L. 26, 104, 111, *165*, 168, 228, *237*, 247, 248, *414*, 518, *526*, 551
Asbridge, J. R. *137*, *147*, *191*, *316*, *348*, *367*
Ashour-Abdalla, M. 524
Atkinson, G. 44, 175, 230, 288, 289, 290, 396, 490, 582
Aubry, M. P. *208*, 217, 218, 224, 225, 230, 235, 255, *288*, *300*, 316, *340*, 347, 392, 396, *498*, *559*
Axford, W. I. 5, 37, 56, 117, 119, 154, 263, 281, 340, *524*
Axisa, F. *21*

Badhwar, G. D. 117
Bahnsen, A. *146*, *169*, *447*
Balsley, B. B. 244, *425*
Bame, S. J. 137, *147*, 153, *191*, *232*, *294*, *316*, 347, *348*, 367, *556*
Banks, P. M. 37, 112, 124, *414*, 434, 441, 447, 455, 456, 460, 536
Barcus, J. R. 345, 526, 530
Barfield, J. N. *390*, *473*, 478, 501, 537
Baron, M. J. *37*, 441, *447*
Bartels, J. 7, 549
Bates, H. F. 121, 456, 459
Bauer, S. J. 458
Beard, D. B. 13, 160
Belcher, J. W. 6, 551, 552
Belon, A. E. 81, 116, *121*
Benedict, P. C. 116
Ben'kova, N. P. 459
Bennett, G. *104*
Berkey, F. T. 524, 525
Berko, F. W. 21, 91, 108, 111, 512
Bering, E. A. 169, 415
Bernstein, W. *116*, 117, 512
Berthelier, A. 205
Berthelier, J. J. *205*
Bewersdorff, A. *526*
Bewick, A. *21*
Bewtra, N. K. *160*
Bhargava, B. N. 205
Biermann, L. 1
Bird, M. K. 160
Birkeland, K. 7, 382, 405
Bjordal, H. 526
Bjordal, J. *526*
Blake, J. B. *19*, *20*
Blamont, J. E. 458
Blanc, M. *460*, *479*
Block, L. P. 179
Bogott, F. *213*, 255, 447, 501, 506, *527*
Bondareva, T. V. 518
Bonnevier, B. 396, 416
Borg, H. 116, 512
Bornatici, M. 158
Bosqued, J. M. 104, 169
Bostrom, C. O. *359*
Boström, R. A. 45, 46, 47, 48, 169, 288, 289, 396
Bowers, E. C. 283, 284
Bowling, S. B. 154, 284, 315
Bowman, G. G. 121
Boyd, J. S. *253*
Brace, L. H. *76*, 537
Bradbury, J. N. *370*, *414*
Bravo, S. 21

Brekke, A. 278, 414, 435, 437, 440, 455, 456
Brice, N. 513, 524
Brinton, H. C. *139*, *533*, *534*
Brommundt, G. *37*
Brønstad, K. *526*, *527*
Brown, N. B. 83
Brown, R. R. 192, *526*
Brown, W. L. 522
Bruce, R. W. *458*
Bryant, D. A. 104, 169, 170
Buchau, J. 17, *72*, *76*, 77, 83, *115*, 120
Buck, R. M. 326, 348
Budzinski, E. E. *92*
Bühler, F. *119*
Burch, J. *37*, 92, 115, 192, 194, 220, 221, 247, *478*, *501*, 518, *584*
Burke, W. J. 151, 302, 359, 367
Burlaga, L. F. 191, 551, 552
Burnell, S. J. 121
Burrows, J. R. 19, 21, *76*, *92*, *93*, *100*, *146*, 336, *583*
Burton, R. K. *21*, 294, 297

Caan, M. N. 209, 227, 228, 333, 572
Cahill, L. J. Jr. 161, *165*, 192, 414, 415, 453, 511, *512*, *522*
Callen, J. D. *524*
Camacho, L. *21*
Camfield, P. A. *424*
Camidge, F. P. 301
Campbell, W. H. 200
Cantarano, S. *15*, *145*
Cardona, G. *104*
Carlqvist, P. 169, 288, 582
Carovillano, R. L. 513
Carpenter, D. L. *31*, 37, 41, 139, *450*, 459, 478, 533, 537
Carpenter, L. A. 41, 478
Carter, D. A. *427*
Casserly, R. T. Jr. 111
Cauffman, D. P. 31
Caverly, R. S. 100
Chan, K. W. 518
Chandra, S. 458
Chao, J. K. 556, 569, 572, 582
Chapman, J. H. 123
Chapman, S. 1, 7, 13, 294, 296, 390, 405, *413*
Chappell, C. R. 104, *112*, 139, *146*, 164, *177*, 533, 534, 535
Chase, L. M. *38*, 104
Chen, A. J. *41*, 164, 396, 508, 534, 535
Cherevko, T. N. *512*
Chimonas, G. 456, 459
Ching, B. K. 458
Chivers, H. J. A. *119*, *524*
Chmyrev, V. M. 177
Choy, L. W. 104, 165, 168, *414*
Christensen, A. B. 526
Christophersen, P. *116*, *169*, *447*, *512*
Chubb, T. A. 116, 583
Church, S. R. *533*
Clark, M. A. 116
Clark, T. A. *526*
Cloutier, P. A. 111
Coffey, T. 177, *425*
Colburn, D. S. *200*, *218*, 222, 227, 232, *235*, 294, *347*, *367*, *392*, *548*
Cole, K. D. 455, 456, 513
Coleman, P. J. Jr. *208*, 330, *390*, 392, 512, 551
Collard, H. R. *15*, *145*
Cooper, W. A. 21
Coppi, B. 283, 291
Cornwall, J. M. 512, 513, 514
Coroniti, F. V. 44, 289, 290, 293, *512*, 513, 524, 565, 566, 580, 581
Courtier, G. M. *104*, *169*
Cowley, S. W. H. 56, 155, 158, 195, 196, 197, 524
Craven, J. D. 91, 522
Croley, D. R. Jr. *104*
Crooker, N. U. 390, 401, 512
Crystal, T. L. *37*
Cummings, W. D. 54, 330, 390, 392, 401, 512
Cuperman, S. *550*

D'Angelo, N. 146, *453*
D'Arcy, R. G. Jr. *326*
Davies, K. 458
Davis, L. Jr. 6, 13, *551*, 552
Davis, L. R. *522*
Davis, M. J. 456
Davis, T. N. *37*, 38, 41, 72, *76*, *79*, *81*, 87, 88, 169, 249, *253*, 335, *405*, 406, 414, *453*, 511
Debabov, A. S. *404*
Deehr, C. S. 101, 113, 336
DeForest, S. E. 139, *256*, 331, 476, 488, 497, 501, 509
Deney, C. L. *117*
Denholm, J. V. 87
Derblom, H. 84
Deshmukh, A. R. *534*
Dessler, A. J. *45*, 56
DeVries, L. L. 458
DeWitt, R. N. 72, 405, 455
Doering, J. P. *104*, 113
Dolginov, Sh. Sh. 512
Domingo, V. 21
Donahue, T. M. 455
Doupnik, J. R. 37, *414*, 434, 441, 447
Driatskiy, V. M. 192, 524
Dryer, M. 552, 556
Dungey, J. W. 1, 13, 14, 155, 263
Durney, A. C. 21
Durney, B. R. 548
Duvall, T. L. 548
Dyson, P. L. 124

Eastwood, J. W. 155
Eberhardt, P. *119*
Eather, R. H. 77, 83, 88, 109, 114, 115, *370*, *414*, 513
Ecklund, W. L. *425*, 427

Egeland, A. *117*, *169*, *447*, *526*
Engelmann, J. 21
Evans, D. S. 104, 111, *117*, 167
Evans, J. E. 116
Evans, L. C. 19

Fahleson, U. V. 31, 169, *450*, 453
Fairfield, D. H. 15, 145, 192, 217, *248*, 316, 330, 337, 343, 382, 392
Fälthammar, C.-G. 158, 179, 288, 289
Fanselow, J. L. 19
Farley, T. A. *340*, *498*
Farthing, W. H. 111
Fedder, J. A. 455, 456
Fejer, J. A. 45
Feldman, P. D. 104, 113
Feldman, W. C. *548*
Feldstein, Y. I. 2, 15, 17, 21, 48, 91, 200, *205*, 250, *336*, *512*, *536*, 583
Fennell, J. F. 19, *21*, 369
Fenner, M. A. *360*
Ferraro, V. C. A. 1, 13
Feygin, V. M. *104*
Fiske, K. J. *533*
Flindt, H. R. 19
Fogle, B. T. 462
Föppl, H. 37
Forbes, T. G. 217
Formisano, V. *191*
Foster, J. C. 248, *459*
Francis, S. H. 456
Frank, L. A. 93, 99, 100, 109, 137, 139, 146, 162, 165, 168, 172, *283*, 336, 358, 370
Frank-Kamenetskiy, A. I. *524*
Fredricks, R. W. 146, 177, *221*, 293, *513*, 524
Freeman, J. W. Jr. 37, *41*, 44, *45*, *143*, 501, 506
Friis-Christensen, E. 200, 201
Fremouw, E. J. 427
Fritz, T. A. *473*, *501*, *507*, *512*, *514*
Fukao, S. 59, 60, 280
Fukunishi, H. 116, 370, 371, 514
Fukushima, N. 31, 51, 382, 390, *394*, 396, 401, 404
Furman, D. R. 455
Furth, H. P. 283

Gall, R. 21
Galperin, Yu. I. 37, 38
Gardner, E. H. *239*
Garrett, H. B. 360
Gassmann, G. J. *77*
Geiss, J. *119*
Gendrin, R. 293, 513, 514, 524
Glass, N. W. *81*
Glazhevska, A. 512
Goldstein, B. E. *367*
Gonzalez, W. D. 56, 61, *200*, 213, 277
Gosling, J. T. 191, 555
Gough, D. I. *424*
Grafe, A. *246*, 294, 512
Grebowsky, J. M. 164, 534, 535
Green, D. W. *37*, *104*
Green, I. M. *146*
Greenwald, R. A. 425
Gringauz, K. I. 137
Grünwaldt, H. *137*
Guerin, C. 205
Gueth, K. 37
Gulbrandsen, A. 550
Gurevich, A. V. 518
Gurnett, D. A. 31, 33, 169, 172, 175, 239
Gustafsson, G. *336*, 524, *583*

Haerendel, G. *31*, 37, 38, 140, *169*, 383, *450*, 452
Hajeb-Hosseinieh, H. *459*
Hall, D. S. 170
Hallinan, T. J. 169
Hansen, R. T. 550
Hansen, S. F. *550*
Hanson, W. *37*, 124
Harang, L. 193, 405, 413, 414
Hardy, D. A. 143, 148, 153, 154
Hargreaves, J. K. 456, 524, 526
Harper, R. M. 456
Harris, K. K. *139*, *164*, 533, *534*
Harvey, J. W. *548*
Hasegawa, M. 48
Haser, L. *37*
Haskell, G. P. 21
Hauge, R. 514
Häusler, B. *21*, 345, 522
Hays, P. B. 104, 455, 458, 460
Heacock, R. R. 514, 515
Heaps, M. G. 456
Henderson, D. J. *514*
Hedgecock, P. C. *38*, 162, *383*
Hedin, A. E. 458
Heelis, R. 37, 38
Heikkila, W. J. 53, 56, *76*, 83, 93, 108, 109, 146, *165*, *336*
Henriksen, K. 113, *524*
Heppner, J. P. 31, *37*, *191*, 199, 201, 215, 401, 404, 405, 414, 434, 452, *455*
Herman, J. R. 458
Hess, W. N. 162
Heuring, F. T. *21*, *51*
Hicks, G. T. 116, 583
Higbie, P. R. *38*, *369*
Hildner, E. *555*
Hill, T. W. 146, 217, *360*
Hills, H. K. *143*
Hilton, H. H. *121*
Hines, C. O. 5, 37, 154, 456, 459
Hirshberg, J. 248, 294, 297, 556
Hoerner, V. *37*
Hoffman, J. *76*, *537*
Hoffman, R. A. *21*, 91, 108, 111, *160*, 336, 473, 507, 508, 509, 511, 514, 518, 583
Hollweg, J. V. 552
Holzer, R. E. *21*
Holzer, T. E. 124, 175, 176, 218, 219, 248
Holzworth, R. H. 242, 277

Hones, E. W. Jr. *37*, 38, *137*, 147, 148, 150, 151, 165, *192*, *232*, 300, 301, 315, 316, 317, 318, 327, 332, *333*, 347, 348, 352, 357, 362, *363*, 366, *367*, 369, *392*, 393, *453*, 507, 526, 577
Hook, J. F. 455
Horng, J.-T. *501*
Horning, B. L. 416
Hovestadt, D. *21*, *345*
Howard, R. *548*
Hron, M. 404
Hruska, A. 336, 343
Huang, Y. H. 160
Hughes, G. F. *81*
Hultqvist, B. 91, 93, 116, 512, 513, *524*
Hume, W. D. *248*
Hundhausen, A. J. *191*, *294*, 552, 556
Hunsucker, R. D. *121*, 456, 459
Hurley, J. 13
Hynds, R. J. *21*

Iglesias, G. E. 111, 117
Iijima, T. 194, 223, 224, 239, 246, 332, 389
Imhof, W. L. 532
Innanen, W. G. 21
Intriligator, D. S. 15, 145
Isaacson, P. O. *111*
Isaev, S. I. *246*
Ivanov, K. G. 191, 248
Iwasaki, N. 404

Jacchia, L. G. 458
Jackson, D. D. *416*
Jacobs, J. A. 514
Jaggi, R. K. 41, 44, 232, 473, 480, 482, 486, 508
Jani, K. G. 458
Jeffries, R. A. 37, 38, 453
Jelly, D. H. 524
Jentsch, V. 526
Jimenez, J. *21*
Johnson, R. E. 456
Johnson, R. G. *100*, 119, *234*, *477*, *532*
Johnstone, A. D. 109, 169, 253
Joki, E. G. 116
Jones, A. V. 116
Jones, D. E. *551*
Jones, K. L. 459
Jones, R. A. 112, *455*
Jørgensen, T. S. 200, *453*

Kamide, Y. *22*, 24, *27*, *29*, 31, 221, 256, 274, 278, *315*, *383*, 390, 394, 395, 396, 401, 404, 428, 440, 476, 584
Kamiyama, H. 112
Kan, J. R. 62, 155, 172, 176, 213, 215, 217, *569*, 582
Kangas, J. *526*, *527*
Kaplon, M. F. *117*
Kappler, H. *31*, *169*, *450*
Karas, R. H. *362*, 526
Kasymov, U. 518
Katz, L. *522*
Kaufmann, R. L. 192, 498, 501
Kavanagh, L. D. Jr. 41
Kawasaki, K. 48, 49, 193, 201, 205, *233*, 247, *248*, *316*, *392*, 394, 512
Keath, E. P. *359*
Kelley, M. C. 31, 34, 169, *213*, 243, 425, 450, 451, 453, 533
Kennel, C. F. 11, 44, 109, 176, 289, 290, 293, 524, 565, 566, 580, 581
Keys, J. G. 427
Khorosheva, O. V. 246
Killeen, J. *283*
Kimball, D. S. *405*
Kimura, I. 524
Kindel, J. M. 176
Kintner, P. *165*, *414*
Kirchoff, V. W. J. H. 41, 478
Kirsch, E. 522
Kisabeth, J. L. *256*, 385, 396, 398, 414, 416, *419*, 421, 443, *497*
Kivelson, M. *177*, *208*, *217*, *218*, 221, 340, *347*, *390*, *392*, 473, 498, 518
Kivinen, M. *514*
Klobuchar, J. A. *459*
Knudsen, W. C. 126
Koch, L. *21*
Kodama, M. *526*
Kokubun, S. 48, 239, 245, 246, 348
Konradi, A. 473, *501*, 507, 510
Kotadia, K. M. 458
Kremser, G. 526, 527
Krieger, A. S. 548
Krimijis, S. M. *369*
Krishnamurthy, B. V. 458
Krochl, H. W. 249
Kropotkin, A. P. 56, 289, 291
Krylov, A. L. 44
Kuznetsov, B. *246*
Kvifte, G. J. *526*

Laird, A. R. *536*
Lakshmi, D. R. 458
Lam, H.-L. *248*
Landau, L. D. 571
Lange-Hesse, G. 424
Langel, R. A. 51, 201, 512
Lanzerotti, L. J. *20*, 162, *362*, *459*, 522, *526*
Lassen, K. 86, 165, *200*
Laval, G. *283*
Lazarus, A. J. *15*, 145, *191*, 316
Leadabrand, R. L. *427*
Lebeau, A. F. 77
Ledley, B. G. 111, *191*
Lee, Y. C. 161
Legrand, J. P. *526*, *527*
Leinbach, H. 192
Leont'yev, S. V. *31*, 199, 245, 404
Lepping, R. P. *283*
Lewis, P. B. *111*
Lezniak, T. W. 237, 498, 501, 518, *526*
Lichtenstein, P. R. 524

Liemohn, H. 514
Lifshitz, E. M. 571
Lin, C.-A. 193
Lin, C.-S. *239*
Lin, R. P. 20, *38*
Lindalen, H. R. *116*, *117*, 512, 514
Liperovskiy, V. A. 177
Lipovetskiy, V. A. *104*
Liu, C. S. 289, 290, 580
Lucas, C. 513
Lucht, P. 31, 34, 450
Luckey, D. 114
Lui, A. T. Y. 17, 76, 93, 116, *137*, 175, 192, 230, 232, 316, 320, 321, 322, 333, 348, 349, 354, *569*
Lundin, R. *169*, *447*
Lüst, R. 37, 452
Luton, J. M. 458
Lyatskiy, V. B. 31, 195, 199, 245, 404
Lyon, E. F. 137, 151, 316
Lyons, L. R. 513, 522, 524

MacQueen, R. M. *555*
Maeda, H. 404
Maeda, K. 112, 460
Maekawa, K. 404
Maezawa, K. 14, 217, 333
Maggs, J. E. *81*
Maguire, J. J. *501*
Maier, E. J. *76*, *537*
Malingre, M. 456
Mal'tsev, Yu. P. *31*, 44, 195, 404
Manka, R. H. 31, 449, 450
Mansurov, S. M. 200
Maral, G. 526, 527
Mariani, F. 15
Martin, J. H. *21*
Maseide, K. *117*
Mason, R. H. *145*
Mather, K. B. *81*
Matsushita, S. 200, 244, 458, 459
Matuura, N. 458
Mauk, B. H. 506
Maynard, N. C. 34, 109, 169, 434, 447
Mayr, H. G. 458, 459
McClure, J. P. 124
McCoy, J. E. 38, 41
McCune, J. E. *524*
McDiarmid, I. B. 19, 21, *37*, 92, *100*, 104, 111, 115, 117, 146, 336, *453*, 583
McEwen, D. J. 104, 106
McGuire, R. E. *38*
McIlwain, C. E. 38, 40, 139, 175, *232*, 331, 473, 476, 488, 495, 497, 501, 504, 506, 509, 510, 518, 522
McKibbin, D. D. *15*, *145*
McNamara, A. G. 424
McPherron, R. L. 208, 209, *218*, 224, 225, 227, *228*, 230, 235, 237, 253, 256, 288, *294*, 300, 316, 317, *333*, 345, *347*, 390, *392*, *416*, 559, *572*
Mead, G. D. 13
Meek, J. H. 405
Megill, L. R. 456
Melzner, F. *37*
Mende, S. B. 83, 109, 114, 115
Mendillo, M. 44, 459
Meng, C.-I. 104, 109, 137, 153, *193*, 210, 222, 227, 232, 242, 248, *250*, 277, 315, 316, 318, 347, 369, 383, 385, 392, *394*, 396, *405*, *413*, 536
Meriwether, J. W. 455
Metzger, P. H. 116
Meyer, B. *37*
Midgley, J. E. 13
Mihalov, J. D. 153
Mikerina, N. V. 191, 248
Mikkelsen, I. S. 453
Miller, J. R. *104*, 117
Mishin, V. M. 86
Mizera, P. F. 104, 116, *121*, 512, 513
Montbriand, L. E. 116, 370
Montgomery, M. D. *137*, *146*, 153, *232*, *235*, *316*, *348*, *367*, 369
Moorcroft, D. R. 425
Moore, J. H. *104*
Morfill, G. 21
Morse, F. A. *104*, 121
Mozer, F. S. 31, 34, 61, 169, 213, 239, *243*, 277, 415, 425, 447, 449, 450, *451*, 453, 501, 506, 527, *533*
Münch, J. W. *522*
Munro, R. H. *555*
Mustel, E. R. 549

Nagata, T. 48, *116*, 223, 224, 239, 246
Nagayama, N. 235, 301, 302, 332
Nagy, A. F. *112*, *458*
Nakagawa, Y. *556*
Nakano, G. H. *532*
Nambu, M. 524
Narcisi, R. S. 113
Nelms, G. L. 123
Ness, N. F. 1, 15, *21*, 145, 284, 315, 316, 343, 392, 522, 548, *586*
Neugebauer, M. *146*, *177*, *221*, *551*
Neupert, W. M. 548
Neuss, H. *37*
Newell, R. E. *239*
Newkirk, G. Jr. 555, *556*
Nielsen, E. 20
Nishida, A. 14, 44, 137, 151, 233, 235, 243, 244, 246, 247, 256, 265, 301, 302, 316, 327, 332, 394, 404

Obayashi, T. 458
Oelbermann, E. J. Jr. *20*
Ogawa, T. 175
Ogilvie, K. W. 104, 191, *248*, 552
Oguti, T. 116, 348, 370
Olbert, S. 551
Oldenburg, D. W. *424*
Olesen, J. K. 453
Olson, W. P. 229
Omholt, A. 116, *526*

Ondoh, T. 191, 192
Orens, J. *425*
Orozco, A. *21*
Osipova, I. L. *404*
Ossakow, S. L. 425
Oya, H. 38

Page, D. E. 21
Palmer, I. D. 38, 40, *369*
Papadopoulos, K. 177, *425*
Papagiannis, M. D. 44, 459
Parady, B. 511
Park, C. G. 163, 460, 535, 536, *537*
Park, R. J. 111
Parker, E. N. 1, 46, 56, 555, 576, 578, 582
Parkinson, T. D. *455*
Parks, G. K. *235*, *237*, 239, 518, 526
Paschmann, G. 100, 111, *137*, 140
Patel, V. L. 191, 192
Paulikas, G. A. 19
Pazich, P. M. 111
Pecker, J.-C. 549
Peek, H. M. 116
Pegov, L. A. *246*
Pellat, R. 239, *283*
Peltier, W. R. 456
Perkins, F. W. 178
Perreault, P. *76*, *250*, 294, *347*, *414*
Peterson, R. W. *79*, *348*
Peterson, V. L. *536*
Petriceks, J. *37*, *447*
Petschek, H. E. 11, 56, 58, 283, 524, 576
Pettersen, H. *526*
Petviashvily, V. I. *536*
Pfitzer, K. A. 229, 237, 518
Pharo, M. W. II *139*, *533*
Piddington, J. H. 1, 20, 61, 283
Pierson, J. D. *526*
Pike, C. P. 76, *77*, 83, 115, 121, 123, 242, 583
Pilkington, G. R. 526
Pizzo, V. 548
Pneuman, G. W. 548
Poland, A. I. *555*
Pomerantz, M. A. 20
Pongratz, M. B. *348*
Ponomarev, V. N. *37*
Porath, H. *424*
Poros, D. J. 162
Potter, W. E. 31, *165*, 414, 453
Prasad, S. S. 455
Presnell, R. I. *427*
Primdahl, F. 453
Prölss, G. W. 514
Pudovkin, M. I. 31, 155, 246, 404
Pushkova, G. N. 536

Quenby, J. J. 21

Rabben, H. H. *37*
Rahman, N. K. *139*
Raitt, W. J. 458
Rangarajan, G. K. 205
Rao, N. N. 456
Rao, U. R. *551*
Rassbach, M. E. 217
Reagan, J. B. *532*
Rearwin, S. 104
Reasoner, D. L. 104, 151, 302, 359, *367*
Reber, C. A. 458
Reddy, B. M. 458
Reddy, C. A. *456*, 458
Rees, M. H. 77, 109, 112, 114, 116, 454, 455, 474, 514
Reid, G. C. 77, 218, 219
Reid, J. S. *514*
Rème, H. *104*, 169
Rich, F. J. 151, 154, *367*
Richmond, A. D. 458, 459
Riedler, W. 91, 116, *336*, 512, *526*, *527*
Rieger, E. *37*, *447*, 455
Rino, C. L. *37*, 447
Roach, W. H. *37*, *453*
Robbins, D. E. *556*
Roberts, C. S. *522*
Roberts, W. O. 549
Roble, R. G. *458*, 462, 474, 514
Roederer, J. G. 38, 507
Roelof, E. C. 359, *548*, 550
Roldugin, V. K. 192
Romick, G. J. 83, 116
Romishevskii, E. A. 13
Rosen, L. H. 524
Rosenbauer, H. 137, 140, *146*
Rosenberg, T. J. *248*, *459*, 526
Rosenbluth, M. N. *283*
Ross, C. L. *555*
Rossberg, L. 336
Rostoker, G. *24*, 45, 47, 248, 301, 385, 396, *398*, 404, 414, 416, 419, 421, 424, 443, 497
Rothwell, P. 347, 522
Royrvik, O. 83, 524
Rugge, H. R. 458
Russell, C. T. 146, *177*, *208*, *209*, *217*, *218*, *221*, *228*, 230, *288*, *294*, *300*, *333*, *347*, *390*, *392*, *559*, *572*
Rüster, R. 536
Rycroft, M. J. 121

Saeger, K. H. *526*
Saifudinova, T. I. *86*
Saito, T. 514
Sakurai, K. 556
Sandel, B. R. *111*
Sandford, B. P. 77
Sarris, E. T. 369
Sato, T. 175, 176
Sawchuk, W. *17*, *88*
Sawyer, C. *550*
Scarf, F. L. 146, 177, 221, 283, 284, 524
Scearce, C. S. *15*, *145*
Scherrer, P. H. *548*
Schield, M. A. 45, 162, 165

Schieldge, J. P. 193
Schindler, K. 155, 158, 279, 280, 281, 283, 284, 291, 300, 301, 315, 577
Schmidt, R. J. *192*
Scholer, M. 21, *345*, *522*
Schulz, M. 162
Schusterman, L. *458*
Schutz, S. *213*
Sckopke, N. *137*, 522
Sears, R. D. *441*
Semar, C. L. 473
Serlin, R. 31, 34, 449, 450
Severny, A. 548
Shabanskiy, V. P. 512, 518
Sharova, V. A. *512*
Sharp, G. W. *139*, *164*, *533*, *534*
Sharp, R. D. *100*, 116, *119*, *234*, *419*, *477*, *497*
Sharp, W. E. 104
Shcherbakov, V. P. 44
Shchuka, T. I. *524*
Sheeley, C. W. 548
Shefov, N. N. 458
Shelley, E. G. *100*, 119, 234, *419*, 477, *497*
Shepherd, G. G. 76, 84, *88*, 117, *537*
Sherman, C. *113*
Shevnin, A. D. 512
Shevnina, N. F. *205*, *336*, *583*
Shumilov, O. I. 524
Silva, R. W. *145*
Singer, S. *147*, *232*, *316*, *347*, *348*, *363*, *526*
Siren, J. C. *31*, *37*, *450*
Siscoe, G. L. 53, 54, 191, 193, *316*, 512, 585
Sivjee, G. G. 104, 106
Skillman, T. L. *191*
Skovli, G. *336*
Sletten, A. 526
Slowey, J. 458
Smith, E. J. 532, 533, 551
Smith, M. J. *169*
Smith, P. H. 160, 161, 473, 507, 508, 509, 514
Snyder, A. L. Jr. 72, 76, 77, *126*, *250*, 255, 335, 405, *414*
Snyder, C. W. 551
Sonnerup, B. U. Ö. 56, 58, 60, 155, 212, 213, 214, 215, 280, 283, 576
Soop, M. 155, 279
Søraas, F. *116*, 117, *512*, 514
Sørensen, J. 526
Southwood, D. J. 473, 518
Sozou, C. 512
Spangslev, F. *453*
Specht, H. *526*, *527*
Speiser, T. W. 56, 155, 217, 283, 284, 288, 315
Spiger, R. J. 111
Spurling, P. H. 459
Stadsnes, J. *526*
Starkov, G. V. 205, 250, *336*, 583
Starr, J. A. *243*, *451*
Stauning, P. *336*
Sten, T. A. *117*
Stenbaek-Nielsen, H. C. *37*, 79, 81, *453*
Stern, D. P. 14, 198
Sternlich, A. *550*
Stöcker, J. *37*
Stoffregen, W. *37*
Stolarik, J. D. *37*, *401*, *452*, *455*
Stone, E. C. *19*
Stone, K. *37*, 478, 533
Storey, L. R. O. 139
Stringer, W. J. 116
Strömman, J. R. 109
Strong, I. B. *137*, *191*, *316*
Strunnikova, L. V. *512*
Stubbe, P. 458
Su, S. Y. 213, 215, 501
Sugiura, M. 23, 162, 175, 191, 249, 383
Sumaruk, P. V. 200
Svalgaard, L. 200, 201, 265, 548
Sweeney, R. E. 512
Sweet, P. A. 55
Swider, W. 113
Swift, D. W. 41, 45, 172, 176, 179, 289, 290, 456, 480, 482, 484, 486, 580
Syrovatskii, S. I. 283, 284

Tanskanen, P. *526*, *527*
Taylor, H. A. Jr. 139, 163, 164, 533, 534
Testud, J. 456, 459, 460, 479
Theander, A. *524*
Thirkettle, F. W. 84
Thomas, B. T. 162
Thomas, G. R. *336*
Thomas, J. O. 120, 121, 123
Thomas, L. 458
Thorne, R. M. *512*
Timothy, A. F. 548
Tohmatsu, T. 116, 370
Toichi, T. 155
Trefall, H. *526*, *527*
Trerskaya, L. V. 518
Troitskaya, V. A. 514
Troshichev, O. A. 246
Truttse, Yu. L. 458
Tsedilina, Ye. Ye. 518
Tsuda, T. 59, 60, 280
Tsunoda, R. T. 425, 427
Tsurutani, B. *248*, 255, *447*, *527*, 532, *533*
Tsyganenko, N. A. 155
Tulinov, G. F. *104*
Tulinov, V. F. 104
Tulanay, Y. K. *164*, *534*
Turtle, J. P. 20
Tverskaya, L. V. 246
Tveten, L. H. 456

Uchida, Y. 556
Ullaland, S. *526*
Ulwick, J. C. *113*
Unti, T. 230
Unwin, R. S. 424, 427

Vaiana, G. S. *548*
Vampola, A. L. 19, *20*, *104*, 522
Van Allen, J. A. 20, 21, 38
Van'yan, L. L. 404
Van Zandt, T. E. 536
Vasseur, G. 456
Vasyliunas, V. M. 41, 45, 56, 58, *154*, 331, 480, 482, 484, 486, 497
Venkatarangan, P. 100
Venkatesan, D. *17*, *93*, *104*, *232*, *316*
Verniani, F. *458*
Verschell, H. J. *37*, *453*
Vestine, E. H. 390
Vij, K. K. 104
Villante, U. *15*, 145
Volk, H. 552
Volland, H. 199, 458, 459
Volosevich, A. V. 177
Vondrak, R. K. 26, 111, 117
Vorobjev, V. G. 192, 336, 583

Wagner, R. A. *76*, *77*, 126
Walker, D. N. 192
Walker, R. C. 15, 145, 146
Walker, R. J. 518
Wallington, V. 347
Wallis, D. D. 398, 414
Walsh, W. J. 163, *164*
Wang, T. N. C. 425
Warren, C. S. *501*
Watkins, B. 120, 126
Wax, R. L. 117
Webb, V. H. *522*
Wehrenberg, P. J. *239*
Welcott, J. H. 348
Wellck, R. E. *556*
Wescott, E. M. 37, *79*, *401*, 452, *453*, *455*
West, H. I. Jr. *326*
Westerlund, L. H. 104
Whalen, B. A. 37, 38, 104, 109, 111, 117, 453
Whalen, J. A. *17*, 76, 77, 83, 85, *115*, *242*
White, R. *513*
Whitteker, J. H. *76*, *537*
Wickwar, V. B. 441, *447*, 455
Wiens, R. G. 421, 424
Wiens, R. H. 116
Wilcox, J. M. *200*, 548
Wilhelm, G. *526*
Wilhjelm, J. *200*, 201
Williams, D. J. *359*, 473, *478*, *501*, *507*, 509, 513, 514, 518, 522, *537*
Wilson, C. R. *250*, *414*, 456
Wilson, M. D. *21*, *146*, *583*
Winckler, J. R. 237, 498, 501, 518, 524, 526
Windle, D. W. 512
Winningham, J. D. *37*, *83*, *93*, *101*, 108, 109, 146, 165, *221*, 336, *453*, 583, *584*
Wlodyka, L. E. *113*
Wolf, R. A. 41, 44, 154, *193*, 232, 473, 480, 482, 486, 508, 534
Wolfe, J. H. *15*, 145

Yasuhara, F. 22, 27, 29, 88, *101*, *109*, *165*, *193*, *201*, 232, *233*, *250*, 315, *336*, 383, *394*, 398, 446, 480
Yeates, C. M. *585*
Yeh, T. 56, 281
Yliniemi, J. *524*
Young, D. T. *193*
Young, T. S. T. 524
Yudovich, L. A. *536*
Yurchenko, O. T. 458

Zaitzev, A. W. 48
Zaytseva, S. A. *246*, 512
Zevakina, R. A. 459
Zhigalov, L. N. *512*
Zhigulev, V. N. 13
Zhuchenko, Yu. M. 104
Zhulin, I. A. *86*
Zipf, E. C. *455*
Zmuda, A. J. 1, 21, 22, 51, 396, 398, 446
Zosimova, A. G. *37*

INDEX OF SUBJECTS

AE index (definition) §4.4.7(b)
Alfvén layer 43–45
Alfvén wave
  Interplanetary 6, 7, 552, 558
  Magnetotail 57, 575
Annular auroral belt 76, 128 §3.8.2
Anomalous resistivity
  Cross-tail current 289
  Field-aligned current 175, 290, 562
Atmospheric emission §2.6
Auroral bulge 8, 71, §6.5, 453, 567, 569, 577, 581
Auroral electrojet
  Auroral substorm §7.4
  Cross-section §7.5
  Cross-tail current disruption 288
  Field-aligned current §7.2, §7.3
  Interplanetary magnetic field §7.6
  Radar aurora §7.5.3
  Substorm intensity 276
  X-ray substorm 527
Auroral electron
  Acceleration process §3.9
  Atmospheric emission §2.6
  Auroral bulge §6.5.2
  Drift motion §8.5.2
  Field-aligned current §2.5
  Plasma injection §8.3
  Spectra §2.4
  Statistical precipitation pattern §2.3
Auroral kilometric radiation 175
Auroral oval 15
  Barium cloud release §7.7.4
  Continuity §1.2
  Electric field §1.3.4
  Field-aligned current §1.3.2
  Interplanetary magnetic field 241
  Magnetospheric plasma §3.8.1
  Minimum size 7, 263, 266, 270, 279
  Model 28
  Morphology §2.1
  Neutral line 2
  Polar cap 13
  Polar ionosphere §2.9
  S_q^p variation §1.3.7
Auroral proton §2.7, 116, 370, 517
Auroral substorm 10
  Auroral bulge §6.5
  Auroral electrojet §7.4
  Definition 7
  Electric field 447–449
  Interplanetary magnetic field §5.2.2
  Model §9.2
  Morphology 7–9, §2.1
  Plasma injection §8.3
  Plasma sheet thinning §6.3.2
  Substorm function §5.2.2
  Substorm intensity §5.3
Auroral type Es 126–127
Away sector 199

Barium cloud release 37, 38, 382, §7.7.4
Boström–Rostoker (BR) model 45
Boundary layer 137, 139
BPS precipitation (region) 98, 108, 127, 137, 336
Buneman two-stream instability
  Auroral electrojet 425
  Field-aligned current 176

Chapman–Ferraro current 190, 215
Chapman–Ferraro theory 1, 13, 14
Chorus 532
Conjugacy of northern and southern auroras 81
Continuous aurora 77
Convection of magnetospheric plasma
  Axford–Hines' model 5
  Equatorial plane and the polar cap projection §1.3.5
  Interplanetary magnetic field §4.4.5, §8.2
  Meridian plane projection §3.5.1
  New model 5, 263
  Plasma injection 293, §8.2
  Polar ionosphere 126
  Substorm process 562
  Theoretical model (quiet time) §1.3.5
  Theoretical model (substorm) §8.2.2
  Westward electric field 447, 452, 478
Coronal green line 550
Coronal hole 548, 551
CPS precipitation (region) 98, 99, 108, 127, 137, 165, 336
Current continuity equation 27, 41
Current disruption theory of solar flares 582
Current-driven instability 176, 562–564
  Buneman two-stream instability 176, 425
  Ion-acoustic instability 176, 425
  Ion-cyclotron instability 176

Current system
  Boström–Rostoker (BR) model 45
  Fejer–Swift–Vasyliunas–Wolf (FSVW) Model 44
  Substorm current system 381, §7.1, 564
Cusp 13, 108, 137, 165, 220, §3.3, 394, 583
Cyclogenesis 263

Dawn-dusk asymmetry of energetic particle distribution §6.8.4
Deflation model of magnetospheric substorms §5.5.1, §9.2.2
Deflation of the plasma sheet 292, 373, 565
Diffuse aurora 24, 26, 127, 165, §2.2.1, 342, 402, 406, 486
Discrete aurora 4, 17, 24, 25, 94, 127, 165, §2.2.1, §3.9.2, 342, 406
Disruption of the cross-tail current 285, 288, 292, 300, 334, 381, 393, 395, 567, 569, 576, 582
Double layer 175, 178
DP-2 current (variation) 233, 239, 243, 255, 265; (See also S_q^p current)
Drift flute mode instability 289
Drift motion
  Electron §8.5.2
  Proton §8.5.1

Electric field
  Interplanetary magnetic field EW component §4.3
  Interplanetary magnetic field NS component §4.4
  Quiet time §1.3.4
  Substorm §5.5.3, §7.7, §8.2.1, §8.2.2
Electrical circuit analogy 218, 404
Energy production rate §5.3, 561
Exosphere 139
Expansive phase 7, 561; (See also Reconnection model and Deflation model)

F2 layer irregularity zone (FLIZ) 121, 583
Fault line §8.4
Fejer-Swift-Vasyliunas-Wolf (FSVW) Model 44
Field-aligned current
  Auroral arc §1.3.2, 127, 128, 168, 172, 175, 177
  Auroral electrojet §7.3
  Auroral electron §2.5
  Current system §1.3.6
  Electrical circuit analogy 289
  Ionospheric current §1.3.3
  Ionospheric electric field §1.3.5
  Magnetic field §7.2, 446
  Plasma convection §1.3.5
  Solar wind-magnetosphere dynamo 4
  S_q^p current 49
FLIZ; (See F2 layer irregularity zone)
Flute instability 289, 562
Forward shock 551

Geomagnetic storm 120, §5.6, §6.8.3, 522
Geomagnetic storm particle §6.8.3
Gradient drift instability 425
Gravity wave 382, 459
Growth phase 208, 224, 225, 232, 235, 237, 239, 254, 255, 333, 345, 360, 559

Hall conductivity and current 27, 402, §7.7.2, 484
Harang discontinuity 34, 401, 404, 414, 434
Helium ion §2.8.1
Helium-rich shell 556
Helmet streamer 548, 551
High speed solar wind stream 548, 550, §9.1.2
Hydrogen aurora; (See Proton aurora)

Incoherent scatter radar 37, 38, 382, §7.7, 456, 458
Inflation of the inner magnetosphere 567
Infrasonic wave 293, 456
Injection; (See Plasma injection)
Injection boundary 506, 510
Interchange instability 289, 562
Interplanetary discontinuity 190, 551
Interplanetary sector structure 199, 548
Interplanetary shock wave 190, 264, 551
Inverted V precipitation 99, 100, 107, 127, 169, 172, 370, 371
Ion drag effect 382, 459
Ionospheric current §1.3.3, §7.7; (See also Auroral electrojet)
Ionospheric substorm 10, §7.8, §8.6
Ion-acoustic wave (instability)
  Auroral electrojet 425
  Cross-tail current 284
  Field-aligned current 176
Ion-cyclotron wave instability 176, 513
Ion-tearing instability 280, 281, 283, 301, 577
IPDP pulsation 514
Irreversible disturbance 7, 264, 265

Jovian magnetosphere 585

Light ion trough 163
Lobe plasma 139
Ly-α emission 116

M region 549, 550
Magnetic bubble 555
Magnetic field line arcade 548, 551
Magnetic Reynolds number 57, 60, 281
Magnetopause apex distance 62, 192, 217, 230, 560, 569
Magnetosphere
  Closed model 1
  Open model 1, 14
Magnetospheric storm 264
Magnetotail (definition) 15
Main phase of a geomagnetic storm 294, 522
Mantle aurora 77
Mercury 585

Merging
  Dayside merging = merging of the interplanetary and geomagnetic field lines
  Nightside merging = reconnection of the open field lines
Merging model; (See also Reconnection)
  Alfvén's model 158
  Petschek's model 58, 283, 576
  Schindler's model; (See Ion-tearing mode instability)
  Steady state model §1.4.2
  Sweet–Parker model 56, 283, 576
  Time dependent model 279
Merging rate
  Conventional definition (M) 57, 280, 575, 582
  Rate of production of closed field lines (Φ_N) §4.4.1, §5.2, 292
  Rate of production of open field lines (Φ_D) §4.4.1, 263, §5.2
Micropulsation substorm 10, 420, 514
Midday aurora §2.2.2; (See also Cusp)
Mid-latitude red arc 165, 474, 513
Mid-latitude trough 163
Minimum size oval 7, 266, 270, 279

Neutral point 13, 196
Neutral wind 293, §7.8
Noctilucent cloud 462
Nose structure 507

Open field line (definition) 1
Oxygen ion §2.8.2
Ozone (mesospheric) 460, 461

Pedersen conductivity and current 27, 402, §7.7.2, 484
Piston gas 556
Pitch-angle diffusion 513, 522
Plasma flow in the magnetotail §6.7, 372, 566, 576
Plasma injection 139, 460, 473, §8.2, 560, 563, 565, 566, 569
Plasma mantle §3.2
Plasma sheet 15, 137, §3.4
  Auroral particles 165
  Convection 43, §3.5.1
  Cross-tail current §3.5.2
  Equilibrium §3.5.2
  Lunar distance 151
  Origin §3.5
  Quiet time §3.4.1
  Substorm; (See Plasma sheet thinning and Plasma sheet expansion)
Plasma sheet expansion §6.6, 372, 396, 575–579
Plasma sheet recovery; (See Plasma sheet expansion)
Plasma sheet thinning 165, 230, §6.3, 372, 560, 564, 565, 581
Plasma tail 165, §8.6
Plasmasphere 137, 139, §3.7
  Electric field 41, 478
  Ring current 473
  Substorm 293, 460, §8.6
Polar cap
  Aurora §2.2.3
  Auroral bulge 342
  Barium cloud release 452–453
  Conductivity model 28
  Convection 35, 452–453
  Definition 13
  Electric field 33–35, 239, 452–453
  Ionosphere §2.9
  Particle precipitation §2.4.4
Polar cap aurora §2.2.3, §2.4.4
Polar cusp; (See Cusp)
Polar elementary storm 7
Polar ionosphere 293, §2.9
  E region §2.9.2
  F region §2.9.1
Polar magnetic substorm 10, 382
Polar rain 109
Polar shower 109
Polar squall 109
Polar substorm 10
Polar wind 124
Positive bay (low latitude) 256, §7.2.4, 394
Proton aurora 77, 116, 370, 517
Proton aurora substorm 10, 517
Proton belt; (See Ring current)
Proton-cyclotron wave instability 513

Quasi-reversible disturbance 6, 264

Radar aurora 381, §7.5.3
Rarefaction wave 290, 564–566, 569, 575, 581
Reconnection model of magnetospheric substorms §5.5.1, 300, 334, 340, 356, 373, 569, 577
Recovery phase 7, 14; (See also Reconnection model and Deflation model)
Recurrent (27-day) disturbance 548, 549
Retarded type Es 126, 127
Reversible disturbance 6, 264
Ring current
  Current distribution 161
  Drift motion of protons §8.5.1
  Energy 274
  IPDP pulsation 514
  Interplanetary magnetic field 235
  Magnetic field 511
  Mid-latitude red arc 165, 475
  Plasma 139
  Plasma injection §4.4.5, §8.2
  Plasmasphere 165, 474
  Quiet time §3.6.1
  Substorm model §9.2.2
  Substorm theory 289

S §4.4, 263, 266, 267
Sector structure 548, 551
Solar electrons 19, 583
Solar flare 279, 555, 582, 585

Solar proton 19, 345
Solar storm 554
Solar wind-magnetosphere dynamo 1, 15, 23, 51, 52, §1.4, 263, 265
Sporadic E layer 126
S_q^p current (variation) §1.3.7, 200, 233, 245, 256, 265, 292, 381, 389, 474, 560, 582
SPS spectrogram 96, 103, 123, 342, 343
ssc; (See Storm sudden commencement)
Storm sudden commencement (ssc) 191, 194, 294
Stream interface 551
Sudden impulse (si) 191

Tearing instability 283, 291
Thermospheric disturbance §7.8
Toward sector 199
Traveling wave disturbance 293, 456

Van Allen belt; (See also Ring current)
  Aurora §3.8.2
  Electron belt §3.6.2, 293, §8.5.2
  Proton belt §3.6.1, 293
Vela sphere 142, 151, 152
VLF emission substorm 10

Wave-particle interaction 4, 11, 175, 176, 293, 473, 513, 532
Whistler 37, 41, 478, 535

X-ray substorm 10, §8.5.2

ASTROPHYSICS AND SPACE SCIENCE LIBRARY

Edited by

J.E. Blamont, R.L.F. Boyd, L. Goldberg, C. de Jager, Z. Kopal, G.H. Ludwig, R. Lüst, B.M. McCormac, H.E. Newell, L.I. Sedov, Z. Švestka, and W. de Graaff

1. C. de Jager (ed.), *The Solar Spectrum. Proceedings of the Symposium held at the University of Utrecht, 26–31 August, 1963.* 1965, XIV + 417 pp.
2. J. Ortner and H. Maseland (eds.), *Introduction to Solar Terrestrial Relations, Proceedings of the Summer School in Space Physics held in Alpbach, Austria, July 15–August 10, 1963 and Organized by the European Preparatory Commission for Space Research.* 1965, IX + 506 pp.
3. C.C. Chang and S.S. Huang (eds.), *Proceedings of the Plasma Space Science Symposium, held at the Catholic University of America, Washington, D.C., June 11–14, 1963.* 1965, IX + 377 pp.
4. Zdeněk Kopal, *An Introduction to the Study of the Moon.* 1966, XII + 464 pp.
5. B.M. McCormac (ed.), *Radiation Trapped in the Earth's Magnetic Field. Proceedings of the Advanced Study Institute, held at the Chr. Michelsen Institute, Bergen, Norway, August 16–September 3, 1965.* 1966, XII + 901 pp.
6. A. B. Underhill, *The Early Type Stars.* 1966, XII + 282 pp.
7. Jean Kovalevsky, *Introduction to Celestial Mechanics.* 1967, VIII + 427 pp.
8. Zdeněk Kopal and Constantine L. Goudas (eds.), *Measure of the Moon. Proceedings of the 2nd International Conference on Selenodesy and Lunar Topography, held in the University of Manchester, England, May 30–June 4, 1966.* 1967, XVIII + 479 pp.
9. J.G. Emming (ed.), *Electromagnetic Radiation in Space. Proceedings of the 3rd ESRO Summer School in Space Physics, held in Alpbach, Austria, from 19 July to 13 August, 1965.* 1968, VIII + 307 pp.
10. R.L. Carovillano, John, F. McClay, and Henry R. Radoski (eds.), *Physics of the Magnetosphere, Based upon the Proceedings of the Conference held at Boston College, June 19–28, 1967.* 1968, X + 686 pp.
11. Syun-Ichi Akasofu, *Polar and Magnetospheric Substorms.* 1968, XVIII + 280 pp.
12. Peter M. Millman (ed.), *Meteorite Research. Proceedings of a Symposium on Meteorite Research, held in Vienna, Austria, 7–13 August, 1968.* 1969, XV + 941 pp.
13. Margherita Hack (ed.), *Mass Loss from Stars. Proceedings of the 2nd Trieste Colloquium on Astrophysics, 12–17 September, 1968.* 1969, XII + 345 pp.
14. N. D'Angelo (ed.), *Low-Frequency Waves and Irregularities in the Ionosphere. Proceedings of the 2nd ESRIN-ESLAB Symposium, held in Frascati, Italy, 23–27 September, 1968.* 1969, VII + 218 pp.
15. G.A. Partel (ed.), *Space Engineering. Proceedings of the 2nd International Conference on Space Engineering, held at the Fondazione Giorgio Cini, Isola di San Giorgio, Venice, Italy, May 7–10, 1969.* 1970, XI + 728 pp.
16. S. Fred Singer (ed.), *Manned Laboratories in Space. Second International Orbital Laboratory Symposium.* 1969, XIII + 133 pp.
17. B.M. McCormac (ed.), *Particles and Fields in the Magnetosphere. Symposium Organized by the Summer Advanced Study Institute, held at the University of California, Santa Barbara, Calif., August 4–15, 1969.* 1970, XI + 450 pp.
18. Jean-Claude Pecker, *Experimental Astronomy.* 1970, X + 105 pp.
19. V. Manno and D.E. Page (eds.), *Intercorrelated Satellite Observations related to Solar Events. Proceedings of the 3rd ESLAB/ESRIN Symposium held in Noordwijk, The Netherlands, September 16–19, 1969.* 1970, XVI + 627 pp.
20. L. Mansinha, D.E. Smylie, and A.E. Beck, *Earthquake Displacement Fields and the Rotation of the Earth. A NATO Advanced Study Institute Conference Organized by the Department of Geophysics, University of Western Ontario, London, Canada, June 22–28, 1969.* 1970, XI + 308 pp.
21. Jean-Claude Pecker, *Space Observatories.* 1970, XI + 120 pp.
22. L.N. Mavridis (ed.), *Structure and Evolution of the Galaxy. Proceedings of the NATO Advanced Study Institute, held in Athens, September 8–19, 1969.* 1971, VII + 312 pp.
23. A. Muller (ed.), *The Magellanic Clouds. A European Southern Observatory Presentation: Principal Prospects, Current Observational and Theoretical Approaches, and Prospects for Future Research, Based on the Symposium on the Magellanic Clouds, held in Santiago de Chile, March 1969, on the Occasion of the Dedication of the European Southern Observatory.* 1971, XII + 189 pp.

24. B.M. McCormac (ed.), *The Radiating Atmosphere. Proceedings of a Symposium Organized by the Summer Advanced Study Institute, held at Queen's University, Kingston, Ontario, August 3–14, 1970.* 1971, XI + 455 pp.
25. G. Fiocco (ed.), *Mesospheric Models and Related Experiments. Proceedings of the 4th ESRIN-ESLAB Symposium, held at Frascati, Italy, July 6–10, 1970.* 1971, VIII + 298 pp.
26. I. Atanasijević, *Selected Exercises in Galactic Astronomy.* 1971, XII + 144 pp.
27. C. J. Macris (ed.), *Physics of the Solar Corona. Proceedings of the NATO Advanced Study Institute on Physics of the Solar Corona, held at Cavouri-Vouliagmeni, Athens, Greece, 6–17 September 1970.* 1971, XII + 345 pp.
28. F. Delobeau, *The Environment of the Earth.* 1971, IX + 113 pp.
29. E. R. Dyer (general ed.), *Solar-Terrestrial Physics/1970. Proceedings of the International Symposium on Solar-Terrestrial Physics, held in Leningrad, U.S.S.R., 12–19 May 1970.* 1972, VIII + 938 pp.
30. V. Manno and J. Ring (eds.), *Infrared Detection Techniques for Space Research. Proceedings of the 5th ESLAB-ESRIN Symposium, held in Noordwijk, The Netherlands, June 8–11, 1971.* 1972, XII + 344 pp.
31. M. Lecar (ed.), *Gravitational N-Body Problem. Proceedings of IAU Colloquium No. 10, held in Cambridge, England, August 12–15, 1970.* 1972, XI + 441 pp.
32. B.M. McCormac (ed.), *Earth's Magnetospheric Processes. Proceedings of a Symposium Organized by the Summer Advanced Study Institue and Ninth ESRO Summer School, held in Cortina, Italy, August 30–September 10, 1971.* 1972, VIII + 417 pp.
33. Antonin Rükl, *Maps of Lunar Hemispheres.* 1972, V + 24 pp.
34. V. Kourganoff, *Introduction to the Physics of Stellar Interiors.* 1973, XI + 115 pp.
35. B. M. McCormac (ed.), *Physics and Chemistry of Upper Atmospheres. Proceedings of a Symposium Organized by the Summer Advanced Study Institute, held at the University of Orléans, France, July 31–August 11, 1972.* 1973, VIII + 389 pp.
36. J. D. Fernie (ed.), *Variable Stars in Globular Clusters and in Related Systems. Proceedings of the IAU Colloquium No. 21, held at the University of Toronto, Toronto, Canada, August 29–31, 1972.* 1973, IX + 234 pp.
37. R. J. L. Grard (ed.), *Photon and Particle Interaction with Surfaces in Space. Proceedings of the 6th ESLAB Symposium, held at Noordwijk, The Netherlands, 26–29 September, 1972.* 1973, XV + 577 pp.
38. Werner Israel (ed.), *Relativity, Astrophysics and Cosmology. Proceedings of the Summer School, held 14–26 August, 1972, at the BANFF Centre, BANFF, Alberta, Canada.* 1973, IX + 323 pp.
39. B. D. Tapley and V. Szebehely (eds.), *Recent Advances in Dynamical Astronomy. Proceedings of the NATO Advanced Study Institute in Dynamical Astronomy, held in Cortina d'Ampezzo, Italy, August 9–12, 1972.* 1973, XIII + 468 pp.
40. A. G. W. Cameron (ed.), *Cosmochemistry. Proceedings of the Symposium on Cosmochemistry, held at the Smithsonian Astrophysical Observatory, Cambridge, Mass., August 14–16, 1972.* 1973, X + 173 pp.
41. M. Golay, *Introduction to Astronomical Photometry.* 1974, IX + 364 pp.
42. D. E. Page (ed.), *Correlated Interplanetary and Magnetospheric Observations. Proceedings of the 7th ESLAB Symposium, held at Saulgau, W. Germany, 22–25 May, 1973.* 1974, XIV + 662 pp.
43. Riccardo Giacconi and Herbert Gursky (eds.), *X-Ray Astronomy.* 1974, X + 450 pp.
44. B. M. McCormac (ed.), *Magnetospheric Physics. Proceedings of the Advanced Summer Institute, held in Sheffield, U.K., August 1973.* 1974, VII + 399 pp.
45. C. B. Cosmovici (ed.), *Supernovae and Supernova Remnants. Proceedings of the International Conference on Supernovae, held in Lecce, Italy, May 7–11, 1973.* 1974, XVII + 387 pp.
46. A. P. Mitra, *Ionospheric Effects of Solar Flares.* 1974, XI + 294 pp.
48. H. Gursky and R. Ruffini (eds.), *Neutron Stars, Black Holes and Binary X-Ray Sources.* 1975, XII + 441 pp.
49. Z. Švestka and P. Simon (eds.), *Catalog of Solar Particle Events 1955–1969. Prepared under the Auspices of Working Group 2 of the Inter-Union Commission on Solar-Terrestrial Physics.* 1975, IX + 428 pp.
50. Zdeněk Kopal and Robert W. Carder, *Mapping of the Moon.* 1974, VIII + 237 pp.
51. B.M. McCormac (ed.), *Atmospheres of Earth and the Planets. Proceedings of the Summer Advanced Study Institute, held at the University of Liège, Belgium, July 29–August 8, 1974.* 1975, VII + 454 pp.
52. V. Formisano (ed.), *The Magnetospheres of the Earth and Jupiter. Proceedings of the Neil Brice Memorial Symposium, held in Frascati, May 28–June 1, 1974.* 1975, XI + 485 pp.
53. R. Grant Athay, *The Solar Chromosphere and Corona: Quiet Sun.* 1976, XI + 504 pp.

54. C. de Jager and H. Nieuwenhuijzen (eds.), *Image Processing Techniques in Astronomy. Proceedings of a Conference, held in Utrecht on March 25–27, 1975.* 1975, XI + 418 pp.

55. N.C. Wickramasinghe and D.J. Morgan (eds.), *Solid State Astrophysics. Proceedings of a Symposium, held at the University College, Cardiff, Wales, 9–12 July 1974.* 1976, XII + 314 pp.

57. K. Knott and B. Battrick (eds.), *The Scientific Satellite Programme during the International Magnetospheric Study. Proceedings of the 10th ESLAB Symposium, held at Vienna, Austria, 10–13 June 1975.* 1976, XV + 464 pp.